APPLIED SYSTEMS AND CYBERNETICS

Pergamon Titles of Related Interest

Eykhoff *TRENDS AND PROGRESS IN SYSTEM IDENTIFICATION*
Iserman *IDENTIFICATION AND SYSTEM PARAMETER ESTIMATION*
Janssen *DYNAMIC MODELLING AND CONTROL OF NATIONAL ECONOMIES*
Titli/Singh *LARGE SCALE SYSTEMS: Theory and Applications*
Yager *FUZZY SET AND POSSIBILITY THEORY: Recent Developments*

Related Journals*

BULLETIN OF SCIENCE, TECHNOLOGY AND SOCIETY
CONSERVATION AND RECYCLING
SOCIO-ECONOMIC PLANNING SCIENCES
TECHNOLOGY IN SOCIETY

*Free specimen copies available upon request.

APPLIED SYSTEMS AND CYBERNETICS

Proceedings of the International Congress on Applied Systems Research and Cybernetics

Edited by G.E. LASKER

VOLUME V
Systems Approaches in Computer Science and Mathematics

Pergamon Press
New York Oxford Toronto Sydney Paris Frankfurt

Pergamon Press Offices:

U.S.A.	Pergamon Press Inc., Maxwell House, Fairview Park, Elmsford, New York 10523, U.S.A.
U.K.	Pergamon Press Ltd., Headington Hill Hall, Oxford OX3 OBW, England
CANADA	Pergamon Press Canada Ltd., Suite 104, 150 Consumers Road, Willowdale, Ontario M2J 1P9, Canada
AUSTRALIA	Pergamon Press (Aust.) Pty. Ltd., P.O. Box 544, Potts Point, NSW 2011, Australia
FRANCE	Pergamon Press SARL, 24 rue des Ecoles, 75240 Paris, Cedex 05, France
FEDERAL REPUBLIC OF GERMANY	Pergamon Press GmbH, Hammerweg 6 6242 Kronberg/Taunus, Federal Republic of Germany

Library of Congress Cataloging in Publication Data

International Congress on Applied Systems Research
and Cybernetics (1980 : Acapulco, Mexico)
Applied systems and cybernetics.

1. System theory--Congresses. 2. Cybernetics--
Congresses.. I. Lasker, George, 1935- II. Title.
Q295.I577 1980 003 81-13765
ISBN 0-08-027196-0 (set) AACR2
ISBN 0-08-027202-9 (Vol.5)

Printed in the United States of America

APPLIED SYSTEMS AND CYBERNETICS

Proceedings of the International Congress on Applied Systems Research and Cybernetics
edited by Dr. George Lasker

VOLUME I:	THE QUALITY OF LIFE: SYSTEMS APPROACHES
VOLUME II:	SYSTEMS CONCEPTS, MODELS and METHODOLOGY
VOLUME III:	HUMAN SYSTEMS, SOCIOCYBERNETICS, MANAGEMENT and ORGANIZATIONS
VOLUME IV:	SYSTEMS RESEARCH IN HEALTH CARE, BIOCYBERNETICS and ECOLOGY
VOLUME V:	SYSTEMS APPROACHES IN COMPUTER SCIENCE and MATHEMATICS
VOLUME VI:	FUZZY SETS and FUZZY SYSTEMS, POSSIBILITY THEORY and SPECIAL TOPICS IN SYSTEMS RESEARCH

All papers submitted for the presentation at the Congress and for the subsequent publication in the Proceedings have been reviewed, first on the basis of extended abstracts and then on the basis of final manuscripts. The authors submitted camera-ready copies and their papers appear here essentially as we received them. Inclusion of the paper in the Proceedings in no way constitutes the endorsement by the Congress Council of the authors' views and opinions.

THE INTERNATIONAL CONGRESS

ON

APPLIED SYSTEMS RESEARCH AND CYBERNETICS

held

December 12-16, 1980

Acapulco, Mexico

sponsored by

The School of Computer Science, University of Windsor

Society for General Systems Research

American Society for Cybernetics

Society for Applied Systems Research

Asociacion Mexicana de Sistemas y Cibernetica

Computer Science Association

Association Internationale De Cybernetique

Canadian Information Processing Society

Austrian Society for Cybernetic Studies

Societe Internationale de Biologie Mathematique

Information Systems Division of International Communication Association

Societe d'Informatique Fondamentale

and

supported by

The Mexican Ministry of Education - S.E.P.,

IBM Corporation,

BATA, Ltd.,

&

Dome Petroleum

Congress President & Organizer	Honorary Congress President
Prof. Dr. G.E. Lasker	Prof. Dr. G.J. Klir

Organizing Committee and Congress Council

Prof. S. Braten	- University of Oslo, Norway
Prof. G. Klir	- SUNY at Binghamton, U.S.A.
Prof. G. Lasker	- University of Windsor, Canada
Prof. G. Pask	- Systems Research Ltd., England

Program Committee

Dr. J.P. Antun (Mexico)	Prof. A. Locker (Austria)
Dr. B. Banathy (U.S.A.)	Prof. C. Lucena (Brazil)
Prof. R. Billinghurst (Canada)	Prof. A. McMahon (Australia)
Prof. G. Broekstra (Netherlands)	Dr. T. Moulin (France)
Prof. C. Carlson (Finland)	Dr. M. Nowakowska (Poland)
Prof. V.B. Cervin (Canada)	Prof. K. O'Mara (Canada)
Dr. J.L. Elohim (Mexico)	Dr. J. Prawda (Mexico)
Prof. R. Ericson (U.S.A.)	Dr. C.L. Sheng (Canada)
Prof. H. Haken (Germany)	Dr. B. Teiling (France)
Dr. L. Kohout (U.K.)	Prof. G. Boulanger (Belgium)
Dr. J. Krispin (Canada)	Prof. S.A. Umpleby (U.S.A.)
Dr. I. Lerou (Chile)	Prof. R.R. Yager (U.S.A.)
Prof. J. Echlin (Canada)	Prof. M. Zeleny (U.S.A.)

Liaison Committee & Advisory Board

Dr. J.P. Antun (Mexico)	Dr. F. Lara-Rosano (Mexico)
Prof. R. Billinghurst (Canada)	Prof. M. Marasinghe (Canada)
Dr. J.L. Elohim (Mexico)	Dr. K. McCaulley (U.S.A.)
Prof. J.P. van Gigch (U.S.A.)	Dr. J. Prawda (Mexico)
Prof. W.C. Hoffman (U.S.A.)	Prof. V. Slamecka (U.S.A.)
Prof. J. Kratochvil (W.Germany)	Prof. S. Umpleby (U.S.A.)

TABLE OF CONTENTS

Preface.......................................xix

Acknowledgements..............................xxvii

Alphabetical Listing of Authors...............xxx

Index of Major Subject Areas..................xxxiii

I. NATIONAL INFORMATION RESOURCES AND SYSTEMS

Information as a National Economic Resource
by M. Therese Flaherty............................2064

Information-Based Measures of Society: An Index of the Information Utilization Potential
by M. J. Menou....................................2071

Measuring the National Information Economy
by Michael R. Rubin...............................2080

Information Services of Egypt: Outline of Project ISE
by M. A. K. Madkour and V. Slamecka...............2084

National Scientific and Technical Information Systems in Saudi Arabia
by L. G. Burchinal and A. Al-Mazi.................2092

II. DATABASE SYSTEMS

The Impact of Auxiliary Information and Update Operations on Database Computer Architecture
by M. J. Menon and David K. Hsiao.................2104

An Introduction to Sabre, A Multi Processor Data Base Computer
by Georges Gardarin...............................2112

On the Long-Term Implications of Database Machine Research
by David Elliot Shaw....2122

Analysis of Backend Database Systems Architectures
by Fred Maryanski and Madhav Marathe....2127

Methodological Problems for Data Modelling Concepts in RM/T's Data Base Design
by R. R. Wagner and A. M. Tjoa....2138

On the Semantics of Data Dependencies in Relational Data Bases
by S. K. Arora, S. R. Dumpala and K. C. Smith....2144

Recent Research Results on the Infoplex Intelligent Memory System (IMS) Project
by Stuart E. Madnick....2151

Design of an Integrated Medical Database
by P. De and W. D. Haseman....2157

A Network Structure Study of the Errors in Databases
by A. J. Payne....2162

Data Base Alerters and Message Filters
by Shi-Kuo Chang....2169

III. COMPUTERS, MICROCOMPUTERS AND MICROPROCESSORS

Computers: Technological Miracle and Societal Failure?
by Edward Alexander Tomeski and Konrad E. Sadek..2174

Informatics as a System-Science
by Ernst P. Billeter....2181

Videotex, Cybernetics, and HUC
by Howard J. Hilton....2189

An Option for a Microcomputer Course in the Physics Department at the Facultad De Ciencias, Unam
by Héctor Guerrero G. and Fernando Favela V. 2197

Prospects and Problems of the Impact of Microcomputers on Human Life
by S. Raman and U. Deo....2200

Microprocessor Applications to Assist in Repertory Grid Elicitation and Termination
by Terence R. Keen....2207

IV. COMPUTER SOFTWARE TECHNOLOGY

The Future of Computer Software Technology and its Effects on Computer Program Development
by F. Gunther, F. Siems and G. Goldman..........2212

Interactive Programming as a CAD Task; Some Remarks on System Structures, Prerequisites, and Implications
by Hans-Jürgen Hoffman..........2217

TOPCAG: Top-Down Computer-Aided Graphics
by N. Magnenat-Thalmann and D. Thalmann..........2223

Some Systems for Modeling Computer Security Requirements
by David C. Rine..........2228

V. DISTRIBUTION SYSTEMS AND PROCESSES

Transportability of Data and Related Software in Electrical Distribution System Modeling
by John C. Thompson and Turan Gonen..........2246

A Network Editor and Shell Program for Electrical Distribution System Planning
by John C. Thompson and Turan Gonen..........2251

Rockwell AIM 65/AIM 680 Applications in Distributed Process Control
by Mohamed Rafiquzzaman..........2256

VI. NATURAL AND COMPUTER LANGUAGES

Identifying Aspects of Human Language Processing to Aid in Computer Processing of Human Language
by Lawrence J. Mazlack..........2261

Speakeasy -- An Evolutionary System, Designed by its Users
by Stanley Cohen..........2266

On the Problem of the Mathematical Theorization of Human Systems Dynamics
by Guy Jumarie..........2272

Bridging the Gap Between Natural and Computer Language
by Douglas R. Skuce..........2277

On the Development of a Recursive Model of Word Structure in the English Language
by K. S. O'Mara, W. M. Jaworski and S. Klasa.....2284

VI. NATURAL AND COMPUTER LANGUAGES (Cont'd)

Design Issues in Natural Language Processing Systems
by F. Oppacher....2291

VII. MAN-MACHINE INTERACTION AND COMMUNICATION

Man-Computer Communications by Voice
by C. Y. Suen....2298

Pragmatics of Information and Man-Machine Cooperation
by Ingbert Kupka....2303

On the User Interface of Computing Software
by Robert F. Ling....2310

Building User Interfaces with an Extensible Language System
by Ronald E. Mills....2315

Semiotics in the Real World: Making Interactive Computer Systems Accessible to Ideographic-Language Users
by Ian H. Witten....2320

Towards a Programmer's Assistant for Novice Users
by M. A. Bauer and D. M. Goldstein....2325

Messages from Systems to Users: A Proposal for Standardization
by William F. Eddy....2331

VIII. ARTIFICIAL INTELLIGENCE AND AUTOMATA THEORY

A Microcomputer Based Artificial Intelligence Laboratory
by Michael J. Magee....2336

Design and Evaluation of an Image Generator with Information Exchange
by Akira Ishikawa and Hiroshi Mieno....2341

Coordination of Two Adjacent Service Networks
by Niv Ahituv and Oded Berman....2346

Knowledge Structures Based on General Automata Compared with Other Types
by Eldo C. Koenig....2350

Formal Neurons to Synthesize Arbitrary Probabilistic Automata
by R. Moreno-Díaz, F. Hernández-Guarch and J. M. Limiñana....2360

IX. INFORMATION RETRIEVAL

How Personal Computers Will Affect the Future of Libraries and Information Centers
by Eugene Garfield.....2365

Accessibility of Local Information Retrieval Systems
by N. Howden and K. Lee.....2370

SPRINT - A System for Information Retrieval in Law
by R. Pavlov, V. Kiskinov and G. Angelova.....2374

Retrieval Aspects of the Universal Relation for Attribute-Oriented Query Languages
by A. M. Tjoa and R. R. Wagner.....2378

Need of a Reliable Message in Biblioinformatics Retrieval
by A. Perales de Mercado.....2384

Application of Systems Approach to Automatic Interpretation of Free Text Documents
by W. Delaney, L. Guerriero, C. Mero, E. Cannavò and E. Vaccari.....2387

Some Aspects in the Design of an Automated System for Terminology Services "Termservice"
by B. Nikolova and R. Pavlov.....2392

X. WORD PROCESSING

A Systems Approach to Word Processing
by P. A. Land.....2395

XI. PATTERN RECOGNITION AND SCENE ANALYSIS

Automatic Interpretation of Symbolic Data on Maps and Engineering Drawings
by J. Kittler.....2400

Visual Object Isolation and Recognition by a Camera-Computer System
by O. Santana, I. de la Nuez and J. Fortes.....2405

A Microcomputer-Based Letter Recognizer: Discussion of Some Problems and Solutions and Results Obtained With an 8085
by M. P. Beddoes and Wolfram Lunscher.....2411

Use of a Rangefinder for 3-Dimensional Scene Analysis
by Z. U. Islam and C. N. Shen.....2416

XI. PATTERN RECOGNITION AND SCENE ANALYSIS (Cont'd)

A Recursive Algorithm for Smoothing by Spline Functions
by C. S. Kim and C. N. Shen....................2421

XII. SIMULATION METHODOLOGY

Experimental Optimization by Simulation Technique
by Alain Checroun and Motaz Khorshid............2428

SIMBAS
by Armando Guerra T. 2447

On the Estimation of a Discrete Parameter in the M/M/m/n Queueing System
by Jacob E. Samaan and Derrick S. Tracy..........2450

Simulating New Materials Processes Required for the Realization of an Alternative Energy Source
by Alexander Kran....................2457

A Simulation Model for the Yucatan Peninsula
by Manfred Kudlek and Ingo Schwarz..............2462

XIII. HUMAN FACTORS

Computer Angst
by Mark G. Graff and Roger B. Johnson............2467

User Satisfaction and Teleprocessing
by Kornel Terplan....................2472

User- and Administrator- Friendly Architecture for Interactive Software Development
by H. Kitagawa, T. L. Kunii, M. Azuma and Y. Mizuno....................2477

Computerese and the Information Barrier
by Sandra Eveland....................2484

Videodisc: A New Technology for Human Factors
by Nancy H. McDonald....................2489

Human Factors Issues in Office Information Systems
by F. H. Lochovsky....................2497

Human Production and Detection of Errors in Data Systems
by I. B. Turksen....................2502

XIV. CAI AND EDUCATION

The Making of an Instructional Program Writer
by Ernest F. Garcia 2509

CAI Systems and Teaching Material Portability
by M. Ferraris, V. Midoro and G. Olimpo 2515

Forests of Oriented Trees as Metastructures for a CAI System
by Ian R. Greenshields 2521

Teaching Statistics and Computers
by Walter W. Zwirner 2526

XV. COMPUTERS AND THE HUMANITIES

The Estate of Art: Asking More of Computers in the Humanities in the Eighties
by Jeffrey F. Huntsman 2530

French Symbolist Literary Theory and Computer-Aided Analysis of Novels
by Paul A. Fortier 2535

Verbs: An Affordable Approach to Computer-Aided Studies of Narrative Techniques
by B. B. Elias 2541

The Idea Behind Computer Stylistics
by L. T. Milic 2547

Statistical Software Tools for Computer-Assisted Analysis of Music
by Luigi Logrippo and Bernard Stepien 2553

Computerized Content Analysis of Narrative in Annual Reports: A Computer in the Humanities Looks at the Corporate Word
by Robert L. Oakman 2564

A Typology of Melodic Contours Implemented by Computer
by Ramon Pelinski 2570

Statistics and the Computer in Formula Analysis of Serbo-Croatian Heroic Verse
by Rudy S. Spraycar 2576

The Role of Language in the New Communication Technology
by Henry Kučera 2581

Trends in Soviet Cybernetics
by Ladislav Matějka 2587

XVI. MATHEMATICAL PROGRAMMING

Nonlinear Programming: A Survey of its Use in Improving the Quality of Life
by Louis G. Birta....2592

A Mixed-Integer Nonlinear Programming Model Well Suited for Use in the Analysis and Control of Energy Systems
by Dudley V. Goetschel....2603

Solving a Capital/Technology Constraint Problem by Implicit Enumeration
by H. T. Lau....2608

Identification of Stochastic Parameter Inequalities
by Dobrin Burev....2613

XVII. MODELING AND CONTROL OF LARGE SCALE SYSTEMS

Large Scale Systems: Structure, Stability, Reliability
by D. D. Šiljak....2619

Stability of Distributed Parameter Systems With Finite Dimensional Compensators Via Singular Perturbation Methods
by Mark J. Balas....2627

A System Engineering Methodology for Evaluating the Behaviour of Large Scale Power Systems Under Small Perturbations
by M. Eremia, H. Crişciu and A. Gheorghe....2634

Stability Analysis of Ecological Systems
by Helen A. Ryaciotaki-Boussalis and Steven D. Martinez....2639

Computational Methods for Distributed Parameter Systems: Identification and Control
by H. T. Banks....2644

A Compensation Method for Slowly Time-Varying Discrete Dynamic Systems
by M. de la Sen and S. Dormido....2646

Describing Functions Methods for High-Order Highly Nonlinear Systems
by James H. Taylor....2651

Some Recent Qualitative Results on Models Based on Difference Equations
by E. F. Infante....2658

XVII. MODELING AND CONTROL OF LARGE SCALE SYSTEMS (Cont'd)

Guaranteed Ultimate Boundedness Control for Uncertain Dynamic Systems
by G. Leitmann....2660

On the Numerical Analysis of Dynamical Systems
by Eusebius Doedel....2662

Real-Time Solution of Stiff Linear Differential Systems
by Karl H. Agar and Wiley E. Thompson....2667

Fundamental Structures of Algorithmic Systems in Industrial Organization
by H. G. Lauenroth....2672

XVIII. STOCHASTIC SYSTEMS AND PROCESSES

Asymptotic Stability of a Class of Stochastic Interconnected Systems
by Dhemetrios Boussalis....2678

Stochastic Stability of a Class of Decentralized Systems
by Dhemetrios Boussalis and Wiley E. Thompson....2683

On the Choice of the Number of Integrations in Arima Models
by N. K. Sinha and O. Abul-Haggag Ibrahim....2688

Stochastic Model Relating High School Performance to that in University Education in Engineering
by N. K. Sinha and O. Abul-Haggag Ibrahim....2694

Nearly-Completely Decomposable Stochastic Processes
by G. Louchard and G. Latouche....2701

XIX. SPECIAL MATHEMATICAL STRUCTURES

Classifications Endowed by Transfer of Structures
by J. P. Auray, G. Duru, A. Dussauchoy and M. Mougeot....2705

A New Lattice Theoretical Method in the System Research
by K. Seitz....2710

The Application of the Concept of Complex Space/Time to a Model of Electric Charge
by David W. Brisson....2713

PREFACE

The Proceedings contain a selection of papers presented at the International Congress on Applied Systems and Cybernetics, which was held December 12-16, 1980 at the Acapulco Convention Center in Acapulco, Mexico. The papers cover a large variety of subject areas and describe current trends and advances in the application of Systems Science and Cybernetics to many different fields.

The interdisciplinary oriented Congress provided a forum for presenting and discussing scientific works in the areas of applied systems research, cybernetics and several other fields. More than 1200 scientists, engineers and professionals from 42 countries presented in parallel sessions over a five-day period nearly 1000 scientific papers.

Since the main theme of the Congress was: THE QUALITY OF LIFE AND HOW TO IMPROVE IT, a large number of papers that have been included in the Proceedings deal with Quality of Life related issues. The papers examine how systems research and cybernetics should be used to help us solve various technical, socio-political, ecological, and other problems that we are facing at present and prepare us for the challenges of the future. Many papers also examine how modern sciences can more effectively utilize systems methodology in order to improve the quality of human life in our society, to promote individual freedom, social justice, civilized discourse, peace, and the dignity of man. Topical coverage of these Proceedings clearly reflect and confirm a newly emerging trend in the evolution of Systems Science and Cybernetics, characterized by a shift from the value-free sciences in the past towards value-oriented sciences at present and in the future. In my opinion, it is through this evolution that Systems Science and Cybernetics will acquire a new important human and spiritual dimension, which will tremendously expand the domain and the scope of research activities of these two sciences in the years to come.

The Congress Proceedings are divided into six volumes. Each volume covers one or more subject areas that are relevant to the major themes of the Congress.

Volume I is devoted to the Quality of Life issues and also includes the inaugural addresses that were presented at the Opening Session of the Congress. The volume contains papers that present diverse points of view and concepts of quality of life, and show how quality of life can be measured and analyzed in various social settings through a variety of methods and approaches. These papers also examine the impact of modern technology and automation on the human environment and on the quality of life. Many of these papers study the effect of the development of personality and the development of knowledge on the quality of life. They also examine how changes in man and the human society affect the quality of life. Some papers explore the quality of life in terms of the quality of social interaction and participation, and in terms of quality of education. Quality of life is also explored here in the context of social values, beliefs, ethics, lifestyles and basic needs.

Several papers in this volume present cross-national and cross-cultural analysis of the quality of life in the U.S.A., Canada, France and elsewhere. These papers examine the effect that income, energy consumption, economic growth, transportation and other factors have on the perception of well-being.

One set of papers studies the quality of life from the political perspective. Another set of papers examines the quality of life in the context of law from a human rights perspective.

A special section of the volume is devoted to the quality of work life from the systems point of view. The volume also examines the contribution of sports to the quality of life and analyzes the quality of life in the context of human kinetics. The concluding section of the volume studies the impact of emerging technologies and communication systems on the quality of human life in the future.

Volume II of the Proceedings focuses on the conceptual and methodological foundation of General Systems Theory and Applied Systems Research. The current thinking in Systems Science and Cybernetics is reflected in the content of the first few introductory sections of the volume. These introductory sections primarily deal with new systems concepts, models and theories of general systems. Some of these concepts are further explored in the section on the epistemological issues in Systems Science and Cybernetics.

Several papers in the subsequent two sections are devoted to the issues of human behavior modeling and to human learning systems. The papers contained in the following sections are devoted to Decision Systems and Decision Analysis, and to Management of Information Systems. These sections cover a broad spectrum of topics ranging from the theoretical decision systems design to the applied decision systems research.

A special section of the volume is devoted to the Systems Research Methodology. It contains many important contributions to the methodology of systems modeling and systems analysis. Several papers in this section also deal with some theoretical issues of general problem solving. Volume II concludes with a section that deals with methodological issues of systems reliability and maintenance analysis.

Volume III of the Proceedings is devoted to studies of Human Systems, Sociocybernetics and Psychocybernetics. The first section of this volume deals with Philosophic and Policy Issues in Institutional Articulation, and with modeling and analyzing various human systems and institutions. The second section on Management and Organization in Human Systems includes a large variety of papers that examine different approaches to human systems management and study the problem of managerial efficiency. Some of these papers also examine problems related to the activities and functioning of organizations.

The third section of Volume III is devoted to Sociocybernetics. It contains papers that deal primarily with theoretical and philosophical issues of sociocybernetics. Some of these papers also examine sociocybernetics in relationship to the quality of life. The next section, devoted to Psychocybernetics and Behavioral Science, presents several important papers that introduce new systems models of personality. Some papers in this section also examine the developmental changes in the evolution of the individual person.

The following section on Social Interaction and Human Development is closely related to its predecessor. This section contains some excellent papers on the analysis of human evolution and examines in great detail various behavior contingencies in social interaction. This section also presents new developments in conversation theory and addresses the issues of time and dualities in self-reflective dialogical systems.

The subsequent section on Human Service-Delivery Systems contains papers that deal with the planning, organization and functioning of various human service-delivery systems and examines the effect of various information systems on the quality of social work. Several papers of this section also study self-service trends in our society and explore the information needs and infrastructure design in social welfare.

The following section on School Systems and Education presents papers on applied systems research that deal with the application of systems approaches in Educational Environments. The next section contains papers on human values and beliefs. These topics are primarily investigated from a systems perspective and address several important philosophical, ethical, and social issues. Another section on Cross-Cultural Perceptions presents a variety of papers dealing with ways of thinking and seeing among peoples of different cultures of the world.

The subsequent section on Social Development and Urbanization examines the processes of migration, urbanization, industralization and social development from a systems point of view. This section also examines the factors that determine the success of the community development.

The next section deals with the Behavioral and Systems Approach to Historical Analysis and presents a systems assessment of specific events from ancient and modern history. The final section of Volume III is devoted to Studies of the Future. The papers presented here examine the future scenarios of human evolution and introduce methodologies for forecasting future developments.

Volume IV is devoted to Systems Research in Health Care, Biocybernetics and Ecology. The first section of this volume contains papers that present various concepts of health care and reflect current thinking in this area. The second section introduces various health care systems and models and addresses some issues on managing the long-term health care facility. It also describes a special information system for public health.

The third section on Systems and Cybernetic Models of Nursing contains several papers on systems models of nursing care and on the application of General Systems Theory to nursing. The papers in this section also discuss various concepts of nursing and examine the quality of health care.

The next section on Health Care Systems Analysis contains papers that deal with the application of systems analysis to the design, development, and operation of certain types of health care systems. The following section contains the result of research work in the area of Medical Diagnostics and Therapy. The papers of this section also show how certain statistical methodology and computer-based information systems can aid the physician in his daily diagnostic and therapeutic work.

The subsequent section is devoted to Stress and to the Behavioral Aspects of Health. The papers presented here propose various strategies for stress management and examine the effect of common stressors on daily living. The section also contains an important paper on cybernetic foundations of preventive behavioral health science.

In the next section on Computer-Aided Analysis and Modeling of Biosystems, the authors describe recent advances in modeling and simulation of spatial tumor growth, the computer-assisted chromosome karyotyping and a new approach to constructing agonistic-antagonistic models of biosystems. The following section on Systems Research in Biocybernetics contains papers that investigate special biological systems and processes. Several papers study cybernetic models of neural systems, retinal processes and vertebrate skeletal muscles. This section also examines the control mechanism of fermentation processes from the systems point of view.

A special part of Volume IV is devoted to Systems Research in Ecology. This part consists of the last four sections of the volume. The first section on ecology deals with Modeling and Simulation of Ecological Systems. The second section deals with Systems Approach to Environmental Quality and contains several important papers on environmental systems analysis. The third section on the Management of Ecological Systems deals primarily with decision-making in ecosystems management. Finally, the last section of the volume is devoted to the Assessment of the Environmental Impact. The papers presented in this section examine the effects of competition and predation on diversity of communities, the role of ecosystems analysis in risk assessment of environmental chemicals, and the large scale environmental impact from oil shale development.

Volume V of the Proceedings examines Systems Approaches in Computer Science and Mathematics and concentrates on several major areas of current systems research in computer science and theoretical and applied mathematics. The first section of the volume is devoted to the studies of National Information Resources and Systems. The papers presented in this section examine various information-based measures of society and study the information as a national economic resource. The presented papers also describe the national information services of certain countries.

The second section on Database Systems contains papers that deal with the design, modeling, analysis and methodologies of integrated database systems. The papers also study the impact of auxiliary information and update operations on database computer architecture and examine the long-term implications of database machine research. The third section on Computers, Microcomputers, and Microprocessors contains papers that deal with various problems of applications of computer- and microprocessors-based systems. The following section on Computer Software technology presents papers that deal with interactive programming, computer-aided graphics and computer security requirements. This section also explores the future of computer software technology and its effects on computer program development.

The next two sections are devoted to Distribution Systems and to Natural and Computer Languages. In the language section, several papers are presented that explore the gap between natural and computer language and that deal with design issues in natural language processing systems. The following section is devoted to Man-Machine Interaction and Communication. It contains papers on man-computer communication by voice, on user interface of computing software, and on other related topics. The subsequent section on Artificial Intelligence and Automata Theory describes a microcomputer based artificial intelligence laboratory, knowledge structures defined through the concepts of general automata and a design of an image generator with an information exchange.

Another section on Information Retrieval contains papers that discuss biblioinformatics retrieval, accessibility of local information retrieval systems, retrieval aspects of the universal relation for attribute-oriented query languages and other related topics. The subsequent section deals with a System Approach to Word Processing.

The following section of Volume V is devoted to Pattern Recognition and Scene Analysis. This section also deals with automatic interpretation of symbolic data on maps and engineering drawings, visual isolation and recognition of objects by a camera-computer system and with use of a range-finder for three-dimensional scene analysis. The next section on Simulation Methodology incorporates papers that introduce several new simulation systems models, methods and approaches. This section also illustrates how certain simulation techniques can be used for experimental optimization.

The following two sections are devoted to Human Factors, to Computer-Aided Instructional Systems and to Computer Science Education. The papers in these sections deal with human factor issues in office information systems, human production and detection of errors in data systems, user satisfaction and with other related topics. The last section on computer science is devoted to the Application of Computers in the Humanities. This section contains papers on a large variety of topics ranging from computer-aided analysis of novels, through computer-assisted analysis of music to the computerized content analysis of narrative in annual reports.

Four sections of Volume V are devoted to Mathematics. The first mathematics section of the volume deals with Concepts, Models and Techniques of Mathematical Programming. The second section deals with the Methodology of Mathematical Modeling and Control of Large Scale and Distribution Parameter Systems. The third section is devoted to the Theory and Applications of Stochastic Systems and Processes, and the last section of this volume contains papers on Special Mathematical Methods and Models.

Volume VI of the Proceedings is devoted to Fuzzy Sets and Fuzzy Systems, to Possibility Theory and it also covers Special Topics in Systems Research. The first section of the volume deals with New Concepts of non-quantifiable and ill-defined (fuzzy) data and systems. The papers presented here address many important problems in the conceptual foundations of fuzzy set theory. They deal with a mathematical theory of uncertainty, fuzzy experiments and with precision and relevancy of fuzzy sets. They also examine how to cope with ill-defined (fuzzy) problems.

The second section of Volume VI reports on Recent Developments in Fuzzy Set Theory. The papers contained in this section describe recent developments in fuzzy arithmetic and fuzzy topology, fuzzy-lip functions and some other related topics. They also deal with the analysis and synthesis of fuzzy mappings and explore problems of information transmission by a system of fuzzy events. The following two sections of this volume deal with Statistical Analysis of Fuzzy Data and examine different notions of Fuzzy Measures.

The subsequent two sections are devoted to Fuzzy Logic and Fuzzy Relations. The papers in these sections examine fuzzy logic knowledge bases and automated fuzzy reasoning, linguistic approach in fuzzy logic, a fuzzy propositional model of concept structure and various types of fuzzy relations on fuzzy sets. The next section on Fuzzy Systems deals with fuzzy linear models, with optimal control of fuzzy systems and also addresses some problems of modeling of uncertain systems. The following section of Volume VI is devoted to Possibility Theory and its applications. It deals with model semantics in fuzzy set theory, with possibilistic approach to the analysis of evidence and with applications of fuzzy sets and possibility theory to systems management. The subsequent section is devoted to Decision Making. It deals primarily with various aspects of fuzzy decision theory and its applications.

The next section contains papers that are focused on fuzzy set approach to repertory grid analysis. These papers study fuzzy entailment analysis and examine how fuzzy semantics improve precision in the repertory grid. The final section on fuzzy sets presents a variety of papers describing Applications of Fuzzy Set Theory to different fields such as earthquake engineering, ecosystem analysis, computer software design, systems management, etc.

The following part of Volume VI is devoted to Special Topics in Systems Research. The first section of this part presents Descriptive Models of various political systems and processes. The second section deals with Systems Approaches to the Analysis of Socio-economic Development. It presents some important papers on actor-oriented systems theory and on the role of institutional actors in social development. It also addresses several problems of regional economic development. The following section in special topics is devoted to the Modeling of Economic Systems. Several papers presented here introduce various microeconomic and macro-economic models for economic policy analysis.

The subsequent two sections are devoted to the Systems Analysis of Agricultural Development and to the Systems Modeling for Land Use and Resources Planning. The papers in these two sections examine various aspects of agricultural development and present several methods and models for land-use evaluation and land-use planning.

The following section on Modeling of Life Support Systems presents papers that deal with food distribution in relation to government policy. These papers also present systems models for food and nutrition policies, and study the inter-national determinants of nutrition levels. The final section of the volume deals with global socioeconomic problems. In particular, it studies developing countries' capital needs, debt problems and growth prospects. It also examines the role of financial institutions for meeting basic shelter needs and develops the design characteristics of a supporting institution that can effectively deliver affordable housing to low income target groups.

The wide spectrum of topics covered in these Proceedings indicates that systems research methodology and principles of cybernetics are being applied to an increasingly larger number of disciplines where they are helping us to successfully solve many intricate real-life problems.

However, the most significant contribution of these Proceedings, especially of the papers included in Volume I, is that they help us create greater awareness of the quality of life issues among scientists, engineers and professionals from diverse fields. Hopefully, this will provide a higher motivation and greater inspiration for each of us to get more deeply involved with these issues in our everyday work so that we can make an even greater contribution towards the happiness and well-being of all.

ACKNOWLEDGEMENTS

I would like to take this opportunity to express my thanks to all the participants of the Congress for making this memorable event such a great success and for submitting for publication papers of such an excellent quality. I would also like to express my thanks to all members of the Congress Council, Organizing Committee, Program Committee and Liaison Committee for their kind assistance with the organization of the Congress.

My special thanks belongs to Prof. Stein Bråten from the University of Oslo in Norway for his encouragement, inspirational ideas, and helpful guidance which aided me greatly in the planning and preparation of the Congress, and in topical categorization of the Congress Proceedings. Many thanks belongs also to Prof. George Klir and Prof. Gordon Pask whose helpful counsel and organizational experience helped me significantly throughout the preparation of the meeting.

My special thanks and deep gratitude go also to all my colleagues and friends who organized and chaired a great number of individual Congress Sessions, Panels and Symposia and who reviewed papers that were submitted for presentation at the Congress and for subsequent publication in the Proceedings. These colleagues and friends did a marvelous job and contributed significantly to the success of the Congress. They are: Prof. Ronald Yager who organized the largest Symposium on Fuzzy Sets and Fuzzy Systems; Prof. Jean Echlin who organized a Symposium on Cybernetic Models of Nursing; Prof. Vladimir Slamečka who organized a Symposium on National Information Resources and Systems; Prof. A. K. S. Jardine who organized a Session on Systems Reliability and Maintenance Analysis; Prof. F. A. Shull who organized a Symposium on Managing Human Service Organizations; Prof. R. Perret who organized a Symposium on Modeling and Control of Microbiological Systems; Drs. John A. Busch and Gladys M. Busch who organized a large Symposium on Sociocybernetics; Dr. Stuart Umpleby who organized a Panel on Current Trends in Cybernetics; Prof. James W. Haefner who organized a Symposium on Ecosystems Analysis; Prof. Nancy McDonald who organized a Session on Human Factors in Computer Science; Prof. Jan Kryspin who organized a Symposium on Health Care Systems; Prof. Kan Chen who organized a Session on Decision and Modeling; Prof. Verna Willis who organized a Symposium on Management of Variety in Human Systems; Dr. M. L. Howell who organized a Symposium on Human Kinetics; Prof. Kozma Balkus and Prof. Bruce Grindal who organized a Session on Peace and Quality of Life Through Cultural Development; Prof. Luigi LoGrippo, Prof. Ladislav Matejka and Prof. Richard W. Bailey who organized several Sessions on Computers in the Humanities; Prof. Thomas Casstevens who organized a Symposium on Descriptive Models in Political Science; Dr. Bela H. Banathy who organized a Symposium on Enriching the Quality of Life: A Challenge to Education; Prof. Robert F. Ling who organized a Session on the User Interface of Computing Software; Dr. Barry Smith who organized a Session on Systems Modeling for Land Use and Resources Planning; Prof. M. L. Marasinghe who organized a Symposium on Law and Quality of Life; Dr. G. A. Daneke who organized a Session on Life-Quality Accounting and Alternative Assessment Strategies; Prof. S. Light who organized a Session on Strategic Planning and Adaptive Management; Dr. Kevin O'Mara who organized a Symposium on Systems Research in Health Care; Prof. David K. Hsiao who organized a Session on Database Computers; Dr. J. A. Wojciechowski who organized a Symposium on Development of Knowledge and the Quality of Life; Prof. Stebelsky who organized a Session on Agricultural Resource Development; Dr. J. M. Scandura who organized a Symposium on Instructional Theory; Prof. James Gripton who organized a Symposium on Quality of Life and Social Welfare; Dr. J. J. Pia who organized a Session on Systems Thinking in Language Sciences; Prof. G. Romsa who organized a Session on Urban Futures, Social Justice and Technology; Dr. Robert V. Varadarajan and Prof. R. L. Tummala who organized a Session on Systems Approach to Environmental Quality; Prof. Sakari T. Jutila who organized a Symposium on Modeling of Social Development; Prof. John A. Beckett who organized a Panel on Ways of Thinking and Seeing Among Peoples of Different Cultures of the World; Prof. R. E. Sykes and Prof. E. Brent who organized a Session on Application of Systems Theory to the Study of Social Interaction; Prof. Mark L. Balas who organized a Symposium on Modeling and Control of Large Scale Systems; and Prof. W. C. Hoffman who organized a Session on Environmental Psychology.

In addition, I would also like to express my thanks to many of my colleagues from Canada, U.S.A., Europe and elsewhere, who helped us review many papers that were submitted to the Congress. They did a fine job ensuring that only papers of the highest quality were accepted for publication in the Proceedings.

I am very grateful for the support and help received from the University of Windsor. In particular, I wish to thank Dr. Paul V. Cassano, Vice President-Academic; Prof. John W. Whiteside, Secretary of the Board of Governors; Dr. C. L. Sheng, past Director of the School of Computer Science; and my colleague from the School, Dr. K. Fung, who acted as an Assistant Director of the Congress. My special thanks belongs also to my fourth-year students from the School of Computer Science whose occasional assistance helped me keep a good part of the Congress work well ahead of schedule and thus offset the disruptive effects of the Canadian postal strike.

My gratitude belongs also to all sponsoring societies for their support and for promoting the Congress amongst their members.

I am particularly grateful to the Mexican Ministry of Public Education (SEP) - especially to its Viceminister for Educational Planning, Dr. Emilio Rosenblueth, whose moral and financial support greatly contributed to the success of our Congress.

On behalf of the Congress Council, I would also like to gratefully acknowledge the moral and financial support given to the Congress by the IBM Corporation, BATA, Ltd., and Dome Petroleum.

With a deep gratitude, I would like to express my thanks to the Social Sciences and Humanities Research Council in Ottawa whose support enabled many of my colleagues from Canada to actively participate in the Congress.

I am also very grateful to Pergamon Press, especially to its Vice President, Basil Larthe, for his very helpful and friendly advice, encouraging words, and extensive assistance with the publication of the Proceedings.

My gratitude also belongs to all those on the secretarial staff of the Congress who volunteered to help me with many administrative matters and with typing. They are: Hanny Lang from West Germany; Martha Vanderlind, Chief of the Congress Secretariat from the U.S.A.; Magali Riverón from Mexico; and Gisele Lacroix from Canada.

My deepest gratitude belongs to Mary Wakeman from the U.S.A., an Executive Assistant to the Congress President, who helped me most extensively with preparing the Proceedings for publication. Mary typed most of the preliminary materials and proofread it. She worked on these Proceedings with me every day after her regular job, devoting all her free time including weekends to the completion of this publication. She did an outstanding job with great professional competence, and it is largely to her credit that the Proceedings could be sent to the publisher on schedule.

George-Eric Lasker
Congress President and Editor
University of Windsor
Windsor, Canada
May 1981

ALPHABETICAL LISTING OF AUTHORS

Name	Page
Agar, K. H.	2667
Ahituv, N.	2346
Al-Mazi, A.	2092
Angelova, G.	2374
Arora, S. K.	2144
Auray, J. P.	2705
Azuma, M.	2477
Balas, M. J.	2627
Banks, H. T.	2644
Bauer, M. A.	2325
Beddoes, M. P.	2411
Berman, O.	2346
Billeter, E. P.	2181
Birta, L. G.	2592
Boussalis, D.	2678, 2683
Brisson, D. W.	2713
Burchinal, L. G.	2092
Burev, D.	2613
Cannavò, E.	2387
Chang, Shi-Kuo	2169
Checroun, A.	2428
Cohen, S.	2266
Crişciu, H.	2634
De, P.	2157
Delaney, W.	2387
de la Nuez, I.	2405
de la Sen, M.	2646
de Mercado, A. Perales	2384
Deo, U.	2200
Doedel, E.	2662
Dormido, S.	2646
Dumpala, S. R.	2144
Duru, G.	2705
Dussauchoy, A.	2705
Eddy, W. F.	2331
Elias, B. B.	2541
Eremia, M.	2634
Eveland, S.	2484
Favela, V., F.	2197
Ferraris, M.	2515
Flaherty, M. T.	2064
Fortes, J.	2405
Fortier, P. A.	2535
Garcia, E. F.	2509
Gardarin, G.	2112
Garfield, E.	2365
Gheorghe, A.	2634
Goetschel, D. V.	2603
Goldman, G.	2212
Goldstein, D. M.	2325
Gonen, T.	2246, 2251
Graff, M. G.	2467
Greenshields, I. R.	2521
Guerra, T., A.	2447
Guerrero G., H.	2197
Guerriero, L.	2387
Gunther, F.	2212
Haseman, W. D.	2157
Hernández-Guarch, F.	2360
Hilton, H. J.	2189
Hoffman, H. J.	2217

Howden, N. 2370
Hsiao, D. K. 2104
Huntsman, J. F. 2530

Ibrahim, O. Haggag..2688, 2694
Infante, E. F. 2658
Ishikawa, A. 2341
Islam, Z. U. 2416

Jaworski, W. M. 2284
Johnson, R. B. 2467
Jumarie, G. 2272

Keen, T. R. 2207
Khorshid, M. 2428
Kim, C. S. 2421
Kiskinov, V. 2374
Kitagawa, H. 2477
Kittler, J. 2400
Klasa, S. 2284
Koenig, E. C. 2350
Kran, A. 2457
Kučera, H. 2581
Kudlek, M. 2462
Kunii, T. L. 2477
Kupka, I. 2303

Land, P. A. 2395
Latouche, G. 2701
Lau, H. T. 2608
Lauenroth, H. G. 2672
Lee, K. 2370
Leitmann, G. 2660
Limiñana, J. M. 2360
Ling, R. F. 2310
Lochovsky, F. H. 2497
Logrippo, L. 2553

Louchard, G. 2701
Lunscher, W. 2411

Madkour, M.A.K. 2084
Madnick, S. E. 2151
Magee, M. J. 2336
Magnenat-Thalmann, N. 2223
Marathe, M. 2127
Martinez, S. D. 2639
Maryanski, F. 2127
Matějka, L. 2587
Mazlack, L. J. 2261
McDonald, N. H. 2489
Menon, M. J. 2104
Menou, M. J. 2071
Mero, C. 2387
Midoro, V. 2515
Mieno, H. 2341
Milic, L. T. 2547
Mills, R. E. 2315
Mizuno, Y. 2477
Moreno-Díaz, R. 2360
Mougeot, M. 2705

Nikolova, B. 2392

Oakman, R. L. 2564
Olimpo, G. 2515
O'Mara, K. S. 2284
Oppacher, F. 2291

Pavlov, R. 2374, 2392
Payne, A. J. 2162
Pelinski, R. 2570

Rafiquzzaman, M. 2256
Raman, S. 2200

Rine, D. C. 2228
Rubin, M. R. 2080
Ryaciotaki-Boussalis,H.A..2639

Sadek, K. E. 2174
Samaan, J. E. 2450
Santana, O. 2405
Schwarz, I. 2462
Seitz, K. 2710
Shaw, D. E. 2122
Shen, C. N. 2416, 2421
Siems, F. 2212
Šiljak, D. D. 2619
Sinha, N. K. 2688, 2694
Skuce, D. R. 2277
Slamecka, V. 2084
Smith, K. C. 2144
Spraycar, R. S. 2576
Stepien, B. 2553
Suen, C. Y. 2298

Taylor, J. H. 2651
Terplan, K. 2472
Thalmann, D. 2223
Thompson, J. C. 2246, 2251
Thompson, W. E. 2667, 2683
Tjoa, A. M. 2138, 2378
Tomeski, E. A. 2174
Tracy, D. S. 2450
Turksen, I. B. 2502

Vaccari, E. 2387

Wagner, R. R. 2138, 2378
Witten, I. H. 2320

Zwirner, W. W. 2526

INDEX OF MAJOR SUBJECT AREAS

I.	NATIONAL INFORMATION RESOURCES AND SYSTEMS	2064
II.	DATABASE SYSTEMS	2104
III.	COMPUTERS, MICROCOMPUTERS AND MICROPROCESSORS	2174
IV.	COMPUTER SOFTWARE TECHNOLOGY	2212
V.	DISTRIBUTION SYSTEMS AND PROCESSES	2246
VI.	NATURAL AND COMPUTER LANGUAGES	2261
VII.	MAN-MACHINE INTERACTION AND COMMUNICATION	2298
VIII.	ARTIFICIAL INTELLIGENCE AND AUTOMATA THEORY	2336
IX.	INFORMATION RETRIEVAL	2365
X.	WORD PROCESSING	2395
XI.	PATTERN RECOGNITION AND SCENE ANALYSIS	2400
XII.	SIMULATION METHODOLOGY	2428
XIII.	HUMAN FACTORS	2467
XIV.	CAI AND EDUCATION	2509
XV.	COMPUTERS AND THE HUMANITIES	2530
XVI.	MATHEMATICAL PROGRAMMING	2592
XVII.	MODELING AND CONTROL OF LARGE SCALE SYSTEMS	2619
XVIII.	STOCHASTIC SYSTEMS AND PROCESSES	2678
XIX.	SPECIAL MATHEMATICAL STRUCTURES	2705

I. NATIONAL INFORMATION RESOURCES AND SYSTEMS

INFORMATION AS A NATIONAL ECONOMIC RESOURCE

M. Therese Flaherty*

*Dept. of Economics, Stanford University, USA, and Graduate School of Business Administration, Harvard University, USA

ABSTRACT

This paper suggests that scientific and technical information (STI) naturally divides into two sets which must be treated differently for national policy purposes: fast-diffusing, which is a pure public good, and slow-diffusing, which is not, but which is the basis of competition in key high technology industries. It is argued that STI policy must be as mindful of the imbalances and bottlenecks which prevent STI from being transformed into national wealth as it is of the levels of STI produced. Skilled labor supply, capital availability, tariffs and patents as well as direct government subsidy for the production and dissemination of STI are discussed.

KEYWORDS

Scientific and technical information; technology diffusion; know-how; public good; information policy; balanced growth.

INTRODUCTION

In the twentieth century national wealth seems to be based on a nation's scientific and technical capabilities as much as on its endowment of conventional natural resources like bauxite and crude oil. Since World War II the United States has earned its imports primarily by exporting from its R&D intensive industries (Gruber, Mehta, Vernon, 1967.) However, despite the indispensability of scientific and technical information (STI), most nations are only just beginning to consider how this resource should be managed and exploited. It should be noted that this paper treats the problem mainly from the point of view of an industrialized country like the U.S. However the considerations discussed here should be relevant for STI policy formulation anywhere.

The policy problem,of course, appears at least as difficult for information resources as for mineral natural resources. Like minerals, pure information only produces value when it is used and loses value through depreciation and obsolescence. Unlike minerals, pure information (1) can be inferred and exploited by anyone after it has been implemented; (2) can be combined with conventional inputs in production of any scale by many agents without diminishing its use value. This public good nature of pure information along with the uncertainty associated with its returns makes it difficult for firms to recover their costs of developing it, makes it possible for non-inventors to capture much of the returns from it, and, of course, for governments to manage their national resources.

The next section of this paper suggests that scientific and technical information naturally divides into two sets which must be treated differently for policy purposes: fast-diffusing, with both pure public good properties, and slow-diffusing, without the first pure public good property since rivals understand it only after considerable time and investment. Basic scientific results and many breakthrough innovations, like the planar process and the triac in semiconductors, fall into the first category. They diffuse quickly, rarely induce revenue sufficient to cover their costs of production, and frequently generate significant positive externalities in fast-diffusing scientific and slow-diffusing discoveries which build on them. National policy which subsidizes both the production and diffusion of this type of information can be justified. This is already reflected in the programs of most industrialized countries.

Slow-diffusing scientific and technical information, on the other hand, is not easily inferred and exploited by rivals after its first use, although it can be applied without extra cost by its developer in production of any scale. This sort of information includes technological know-how and can be a revenue-producing resource to both its developer and the nation where it is developed. The more valued the product in which slow-diffusing STI is embodied, the more widely the firm can sell it, and the fewer the rival firms exploiting the STI, the greater the revenues. Firms fund the production of this sort of information and it forms the basis of much of their commercial activities nationally and internationally. Governments have had essentially no policy with respect to this sort of information which directly generates so much of their wealth. (Patent policies, which might fall in this category, are notoriously ineffective.)

Later sections discuss fast and slow-diffusing STI and a number of general policy issues relating to their production, dissemination and implementation. The need for balanced development of STI resources and conventional resources if a country is to attain its potential national income is emphasized. While it is unlikely that most nations can identify and attain the jointly optimal levels of fast and slow-diffusing STI and conventional inputs, they can spot bottlenecks and competitive imbalances in their process of technological development and application. Awareness that such bottlenecks can prevent their economies from deriving benefit from expensive investments in science and technology should motivate policy-makers to pay as much attention to the dissemination and implementation of STI as they might to its production. Skilled labor supply, tariffs, capital availability, and patents as well as direct government subsidy for the production and dissemination of STI are discussed.

THE DUAL NATURE OF STI

It has become commonplace for business and policy experts to point out that information differs considerably from conventional inputs to productive processes and that, as a consequence of these differences, the market may neither adequately allocate resources to the production of STI nor adequately disseminate and implement STI. The economic model of pure information emphasizes that (1) it can be inferred and exploited by anyone after it has been implemented (i.e., noone can be excluded from using it once it has been produced), and (2) the scale of its use does not impair its use value. These pure public good properties of STI entail several classic allocation problems. First, many people other than the one who incurred the cost of production will benefit from the information (a positive externality in economics jargon); since these prople do not pay for the information and they would have been willing to pay something, society would have been willing to invest more in producing more or better STI. Second, the value of the information is usually appropriated not by direct consumption - as is the case with a television program - but by implementing the information so as to make profits in a business venture. This makes the resource allocation problem maybe even worse. If all the firms in an industry implement the new information immediately and if the industry is perfectly competitive and composed of identical firms (in such an industry all the firms make zero economic profits at equilibrium because they compete to bid prices down to their average, equal to marginal costs) then the inventor firm will not be able to make enough profit as a result of the invention to cover the costs of producing the invention and, realizing this ahead of time, the potential inventor will not invent. In this case the private net benefits of inventing to any firm in the industry are zero or negative while the social net benefits of lower prices or new products for consumers are significantly positive. If the market is less than perfectly competitive with non-identical firms, then the inventor may be able to recover some, all, or even more than his costs of information production. How the potential cost recovery compares with the cost of production in general is an unanswered question although the presumption is usually toward underinvestment and lower returns.

A second generic force which might foster underinvestment by risk averse firms relative to the socially optimal level in the production of STI is the considerable uncertainty about which STI will payoff and when. The uncertainty and thus the size of the mistake increases as one moves conceptually from development of existing technologies to basic science.

Both of these factors lead to areas of potential benefit from government intervention. However, further understanding of how STI contributes to national wealth is a prerequisite to policy specification. A central suggestion of this paper is that on these considerations STI divides naturally into two exclusive sets for which different national policies are appropriate: fast and slow-diffusing. Of these, only fast-diffusing STI has the pure public good properties which impair the market's ability to allocate resources to its production.

FAST-DIFFUSING STI

Basic scientific information is fast-diffusing and has the two pure public good properties and the accompanying management problems. Many breakthrough innovations like the planar process and the triac design in semiconductors also belong in the fast-diffusing group. Scientists' rewards are in large part their reputations which depend on general and rapid recognition of their accomplishments; rapid dissemination and acquisition of new science information is an integral part of the scientists' efforts. Breakthrough innovations have the same property because they are very similar in form to scientific propositions and because rival businesses are quite interested in acquiring them.

All the quick-diffusing information has the public good problem that it rarely generates enough income to pay for its production. Some - but by no means all - science generates considerable positive externalities, so that society as a whole (or the firms in an industry) would in total be willing to pay for more of it than would be produced by if only the natural market mechanism is there to "pay" for it. As a result, science was for centuries subsidized by rich patrons and by scientists' accepting lower salaries than they would earn in alternate employment. More recently governments subsidize the production of most scientific information. Since the benefits of science accrue to the nation and the world over a period and in a way that prevents the scientist from covering his salary and expenses, this seems appropriate.

It is also socially optimal for governments to charge interested people less than or equal to the marginal cost of distribution for fast-diffusing scientific information whose production it funds. This will promote wide dissemination of the information to those who can implement it in more science or in technology. One might argue that scientific information should not be disseminated to foreign nationals on the grounds that it might then be incorporated in products which compete with those of the home country. However, it seems that the value of the positive externalities generated by foreign nationals to increase the world stock of science will by far outweigh the few new products generated abroad - especially if the home country resources in slow-diffusing technology and conventional resources are balanced with those of fast-diffusing STI.

Private firms have, of course, funded a great many fast-diffusing technical breakthroughs whose property rights have not been adequately protected by patents. However, the poor protection patents afford many technological breakthroughs is not reason for great concern. Firms fund the breakthroughs because they are frequently joint products with slow-diffusing STI on which returns are appropriable. There is no a priori reason to expect that the production of technical breakthroughs is suboptimal.

These comments not withstanding, there does seem to be a consensus that the stock of fast-diffusing STI in many industrialized countries must grow faster in order to support desired levels of economic growth. However, two contrasting examples belie that idea: the U.K. which has gathered so many Nobel prizes but has experienced very slow economic growth and Japan which has gathered so few Nobel prizes but has managed to grow very rapidly indeed. Rather than indicating the adequacy of fast-diffusing STI provision, the examples focus attention on the need for balance among scientific and technological breakthroughs, technology development and conventional inputs for sustained and successful economic growth. (Abernathy and Townsend, 1975, emphasize this same rule for a firm's management of a particular production process.) The need for balanced development of resources which must be combined to yield value is, of course, an old theme in the literature of development economics, but it is also an important one for developed countries contemplating the emerging world marketplace in so many industries. Balance entails disseminating and implementing STI for the national benefit as well as producing it. It also entails coordinated management of STI and conventional resources, like skilled and unskilled labor and energy.

Balance also leads to consideration of slow-diffusing STI as a conduit of fast-diffusing STI to implementation. Slow-diffusing STI does not have the first pure public good property since there is a period after implementation during which rivals are unable to use the STI and the inventor can exploit it in profitable business.

SLOW-DIFFUSING STI

The distinguishing feature of slow-diffusing information is that it is complex and its implications are not easily appreciated by professionals other than its producer. This may arise in a large part because this "know-how" is frequently closely related to the production details of particular business. Even firms competing in the same high technology business do not use the same process. For example, each manufacturer of novalac epoxies for integrated circuit encapsulation uses a different formulation and production process. (Coughlan and Flaherty, 1980). So there may be several kinds of barriers to quick adoption of an improvement made in one process other than secrecy about the technical description of the improvement. First, the jargon associated with the processes differ. Second, managers may not be able to quickly distinguish improvements which apply to several processes from improvements which apply only to one. Third, the improvement may need to be adapted before application to other processes. Fourth, resources devoted to investigating and adopting a rival's improvement of uncertain value must be diverted from on-going, independent efforts to improve the process at hand. Overcoming all of these obstacles - including learning a technical description of the improvement - requires considerable time, effort, and expense. This is especially true if the improvement is in the largest category which we generally call "know-how." Technical improvements usually are numerous, very small, and very specific to the process while the intelligence effort may be largely a fixed cost per unit of information (Rosenberg, 1976). Reinforcing the point, Teece (1979) presents some persuasive empirical evidence that the time and expense required for know-how transfer between plants of the same petrochemical firm are considerable.

For firms and for countries the central problem in production of managing slow-diffusing STI is not the public good problem of providing incentives to produce the optimal level of STI. Rather, it stems from the facts that in the presence of considerable uncertainty (1) several high-technology industries - like electronics and semiconductors - are technologically essential in the development of most other industries (especially the service sector) in industrialized countries and (2) these basic high-technologies are progressing technically so very quickly that missing a technical step can render a national industry at a permanent disadvantage. The disadvantage is not only technical, but is based on the ability of indigenous firms in these and related industries to compete effectively in world markets. A technical disadvantage translates into relatively low scale production, high costs, inability to support indigenous firms in related industries to develop state of the art products. This is a dynamic version of the balanced development problem. This is very serious since if a bottleneck (e.g., a technical step is missed) occurs in a basic industry then many of the nation's industries will suffer in international competition for a long time.

Of course, government intervention with large subsidies and so on may be sufficient to achieve balance and a level of competence sufficient for viable international competition in the key industries. This is precisely the kind of effort now being made by governments in Japan, France, and the U.K. for their semiconductor industries. The policies of Japan are very well known now, but the policies of France in semiconductors are just as striking and illustrative. The French government has spent $63 million (U.S.) since early 1978 to stimulate the French electronics industry. The 1977 "Plan Composants", allocated $133 million (U.S.) over a 5 year period for a program to develop French IC capacity. The first $44 million was split between RTC (La Radiotechnique Compiloc, the Philips/France branch) and Thomson-CSF (a French firm). It enables Thomson to develop bipolar technology and MSI linear products. It enables RTC to develop LSI TTL and ECL devices. In 1978, the plan has given $44 million to a Motorola (U.S.) - Thomson-EFCIS (joint venture between Thomson and the French Atomic Energy Commission) combination within which Motorola will give MOS technology to the French. The final $44 million was given to National Semiconductor (U.S.) and Saint-Gobain-Pont-a Mousson for an NMOS and C-MOS plant.

How much national investment in slow-diffusing STI for these industries is "optimal" is too large and too complex a question to be settled here. However, balancing and easing obvious bottlenecks in the development of key high-technology industries can be advocated with the certainty that these are essential to effectively exploiting existing STI in a nation's interest. Several points where bottlenecks or competitive imbalances are likely to occur can be discussed.

First, STI may be produced but not implemented and exploited by the firms within a nation. For example, numerical control was well-understood in the U.S. but not implemented by machine tool builders for quite a few years. The military and NASA in the U.S. are also concerned that the STI they fund is not implemented by industry. Government-level intervention to promote wider adoption of the technologies embodied in the STI are surely warranted, although the particular efforts undertaken thus far have not been distinctly successful. Surely a better understanding of the decision to implement slow-diffusing STI is required before such direct intervention can be expected to succeed.

Second, a skilled labor force in balance with technical abilities of firms is essential if a nation is to exploit its STI internationally. If there is a shortage of labor and if capital cannot in the short run be further substituted for labor, then the domestic industry - even if it has superior technology - cannot expand production even if there is large, unsatisfied demand. If rival firms in other nations can acquire the domestic technology or an acceptable substitution for it along with sufficient conventional inputs, then the rival might acquire loyal customers and the international market. This would, of course, make later international expansion by the domestic firm more difficult than the earlier expansion in contest with a weaker rival would have been. Both profits to the domestic firm and economic growth for the domestic economy would be permanently lowered as a result of labor shortage, not a technology shortfall.

Shortages of labor skilled in particular specialties are likely in Western economies where students chose their training with little guidance from firms or government and in fields where training is time-consuming so that shortages and the high salaries they induce are not apparent when they should be to induce field selection. These problems might be alleviated by collective or government action to forecast industry needs for professionals and make that known to students as they select their fields of study. Excess demand for electrical and computer engineers in the U.S. is now developing that appears to require not only more students in the areas but also significantly larger facilities for training them. Since expanding university staff and facilities requires very long term financial commitments by universities (which already have tight resources), they are reluctant to take the first step. Perhaps government intervention is warranted. In contrast, Japan now graduates twice as many engineers each year as the U.S. while it has only half the population of the U.S.

Third, capital formation may differ significantly among nations. If some nations restrict the use of their capital supply to domestic firms, then the cost of capital may differ significantly among nations. This would, of course, induce firms of the lower capital cost nation to invest more in STI, plant, and equipment and to undertake more risky projects in the international market. Such low capital cost industry could well drive other nation's industries from the world market. Governments should realize that this competitive imbalance reduces the value of their nation's STI and consider lowering domestic capital costs as well as working to ensure equal access to international capital markets and protecting domestic markets.

Fourth, tariff and non-tariff barriers to trade can create competitive imbalances similar to those created by differential costs of capital. Governments should consider both working to ensure equal access to important foreign markets as well as protective measures if the efforts fail.

There is a final question of whether a nation should invest to speed dissemination of foreign STI to its industry and inhibit dissemination of domestic STI to foreign industry. It seems clear that speeding the dissemination of foreign STI to domestic industry can only improve the competitive position of the industry and the wealth of the nation. Since the cost of acquiring foreign STI need be born only once if it is disseminated by a national group, there is reason for the government to provide that service if firms for some reason cannot or do not join together for that purpose. Many nations - e.g., Sweden - already provide their industries with surveys of foreign STI.

Should a nation try to inhibit dissemination of domestic STI to foreign industry? Flaherty (1980) shows that firms in the U.S. semiconductor equipment, materials and components industries have disseminated their technology internationally simultaneously with promoting profit-making businesses overseas. There has been very little licensing of slow-diffusing technologies in that industry, and it is difficult to make the case that technology has been given away. While there have been acknowledged management mistakes, it seems clear that the overwhelming effort by U.S. firms has been to exploit technology and earn repatriatable profits. The effort has nevertheless built up strong customers and, increasingly, strong competitors. It is clear that doing no business internationally leaves the world market completely open to foreign competitors. This suggests that effective national policies toward the key technologically progressive industries would strengthen the abilities of indigenous firms to implement their slow-diffusing STI and appropriate more of the returns due their STI internationally.

ACKNOWLEDGEMENT

The work for this paper was prepared in connection with the International Scientific and Technical Information Policy Group of the U.S. Commerce Department Technical Advisory Board during the 1979-1980 academic year.

REFERENCES

Abernathy, W.J., and P.L. Townsend (1975). Technological Forecasting and Social Change, 7, 379-396.
Coughlan, A.T., and M.T. Flaherty (1980). "A Short Business and Technology History of the Plastic Encapsulation of Integrated Circuits", Working Paper, Stanford University.
Flaherty, M.T. (1980). "Regularities in the International Transfer of U.S. Semiconductor Technology", Working Paper, Stanford University.
Gruber, W., D. Mehta, and R. Vernon (1967). Jour. of Pol. Econ., 75, 20-35.
Rosenberg, N. (1976). Perspectives on Technology. Cambridge University Press, New York.
Teece, D.J. (1977). Econ. Journ., 87, 242-261.

INFORMATION-BASED MEASURES OF SOCIETY; AN INDEX OF THE INFORMATION UTILIZATION POTENTIAL

M. J. Menou

Consultant to UNESCO/PGI

ABSTRACT

The formulation of national information policies, planning of information systems, comparative studies, in particular those aiming at the transfer of information technology or the appraisal of achievements in information activities at the national or international level, call for an objective, accurate and comprehensive characterization of the situation prevailing in individual sectors, countries or regions. The paper discusses the possible derivation of an index of the information utilization potential (IUP).

KEYWORDS

Socio-economic indicators; information indicators; comparative studies in information; national information systems; national information policies; evaluation of national information systems.

NEED FOR AN INFORMATION-SPECIFIC MEASURE OF THE COUNTRIES

During the last thirty years the information sector has experienced a dramatic development in most countries. Information is now considered as a vital resource both for national development and in international relations. Consequently, organized efforts in order to plan and monitor all activities related to the provision and distribution of information and establish national information policies are deserving increased attention at the national and international levels (UNCSTD, 1979; UNISIST II, 1979).

At the same time, using national resources as well as bi- and multi-lateral aid resources, the less developed countries are attempting to prevent the information sector from becoming another major gap between them and the industrialized countries by building up or strengthening their information infrastructures. More often than not, this takes the form of a transfer of information technology from the more advanced countries.

Such efforts require that a meaningful assessment could be made of the situation prevailing in a particular country or region, or within a given economic sector, as to the production, dissemination and use of information, as well as of its evolution. Furthermore the evaluation of previous endeavours, as well as the appraisal of the relevance of experiments undertaken in other countries, the eventual replication of which is considered in a different environment, require a comprehensive and meaningful characterization of the informational situation of particular countries.

For the time being, only macro-economic indexes, indicators or groupings are available in order to characterize countries, such as the Gross National Product, Gross National Product per capita or income groups. Not only their internal significance has been questioned but also their ability to provide a measure for more specific concerns, as for instance by Morris (1979). As a result the construction of particular indices or indicators has been attempted in a variety of sectors such as development, science and technology, human resources, quality of life, etc. (Baster, 1972; Liu, 1973).

In the field of information, the need for a systematic approach in the formulation of policies and plans, the need for evaluation, and, consequently, the need for relevant data and indicators has been recognized for a long time, as exemplified in Keren's (1975) guidelines. However, this is still to be concretized. At the 3rd UNISIST meeting on the planning and implementing of national information activities, a specific recommendation was made towards the establishment of an informatio-specific index (Slamecka, 1979). More recently, the chairman of the high level conference on information, computer and communication policies for the 1980's convened by OECD, noted in his concluding statement the "need for more timely and more relevant data in order to improve our understanding of the changes taking place and as an essential input to the policy making process." (OECD, 1980).

PREVIOUS WORKS ON INDICATORS OF SCIENTIFIC AND TECHNICAL INFORMATION

D. King and co-workers (1976, 1977) made a substantial contribution in their NSF sponsored study of indicators of scientific and technical communication in the United States and some other countries: USSR, Japan, West Germany, France, and United Kingdom (Roderer and Schell, 1977). They considered a communication cycle in eleven steps and analyzed a large number of variables related to these various steps. A similar attempt was initiated in Brazil under the guidance of Lancaster (1978).

In the framework of studies of indicators of scientific and technological development, indicators of scientific communication or information were also considered, such as number of libraries, collections, patents, and more particularly citations. Solla Price (1969; 1975) correlated citations with economic indicators like GNP per capita and consumption of electricity, which appear to be close at the two extremes of the development scale but much less significant for the countries in an intermediate position.

Some work was also carried out on specific aspects of information activities like publication in foreign and national journals by Sandoval (1976), relationship between the number of libraries, librarians, books and loans by McCarthy (1980). The Latin-American center for economic and social documentation--CLADES--(1979) conducted an intensive study of the information infrastructures in the region related to socio-economic development. A number of indicators for the appraisal of the information infrastructure was developed.

Although their interest goes without question, these studies share, in our point of view, a number of limitations:

- they keep with the traditional definition of the scope of information activities as scientific and technological information, while it is now recognized that most countries and users constituencies are interested in a much wider and more comprehensive definition (UNISIST II, 1979);

- they pay more or exclusive attention to the information sector itself, neglectin geographical, social, cultural and other environmental factors, which are not less important;

- they do not integrate the various scores into a single measure.

STATISTICS ON THE INFORMATION SECTOR

For several years UNESCO (1970) and ISO (1974) have promoted the collection of statistics on libraries and book production according to internationally agreed standards. These data can be found in the UNESCO Statistical Yearbook, at least for those countries which are providing them. Vilentchuk (1975) has developed for UNESCO guidelines for the inventory of scientific and technological information and documentation facilities, which have been up to now very seldom used. Murphy (1979) was contracted by UNESCO to study the problem of statistics on scientific and technical information and documentation. On the basis of his report, pilot projects are currently being conducted in several countries.

Such data are obviously a basic ingredient, and a badly needed one, for any set of indicators or index of the information situation in any country. However, they limit the perception of information activities to the information institutions, which cover only a part, and often, particularly in developing countries, the smallest part of the phenomenon. On the other hand, they are of little significance per se, unless they are related to the other basic characteristic features of the countries such as the size, population, production, etc. For instance, a well developed school library system is not so significant for a country which gives priority to industrialization.

PROBLEMS OF CALCULATING INFORMATION INDICATORS

Assuming that information is fundamentally a social phenomenon, information indicators are social indicators, as opposed to economic ones. Fisher (1967) indicates that the accuracy of an indicator depends upon four characteristics: the choice of variables, the number of variables, the original data, and the calculation formula.

In the absence of a recognized model of the society or the particular concern under study, the choice of the variables is a difficult one, as noted by Andrews (1973), who defines social indicators as a limited yet comprehensive set of coherent and significant statistics. Liu (1973) indicates that the selected variables should be relevant for the majority of people to whom the index applies, be adaptable to changing conditions, be easily understood, and be flexible enough to encompass any life style. Morris (1979) also stresses the need for universality, simplicity of construction, intelligibility and objectivity.

Ideally the number of variables should be as inclusive as possible of the component features of the phenomenon studied. According to Fisher (1967), 50 is a minimum significant number, above which the improvement is gradual, although the extra burden of handling a number of variables in excess of 200 does not necessarily provide a worthy return in increased accuracy.

Reliability of data is of course a natural requirement, but the influence of inaccurate data on an index is diminishing with the number of variables considered (Fisher, 1967). Since for a large scale study direct data collection is unthinkable, one has to rely upon secondary sources in which, more often than not, data are not available for the year of reference or not available at all. On the other hand, making use of different sources proves rather cumbersome because of the many inconsistencies in definitions and coverage. One has also to bear in mind that, particlarly in developing countries, natural and social pheonmena generate rapid changes which have to be taken into account in the structure of an index.

Because the variables considered in a social index are being measured by many different yardsticks, all scores have to be standardized. Liu (1973; 1975) used the Z score method for standardizing the values and the standardized additive method, the adjusted standardized method and factor analysis for the construction of the index, and found that the indexes did not differ so much that the more complicated method would be worth the effort. In a further study (1976) he came to the conclusion that the calculation made on groups of factors, each represent-int a distinct parameter or indicator, was more meaningful and that intermediate scores should be presented together with the resulting sum. Morris (1979) scaled the countries performances on a 0 to 100 scale, in which minimum and maximum values were based on historical and predictible scores, this is only feasible for a small number of variables of the type he was using (literacy, life expectancy, mortality). Another concern is the relative importance of the different variables in the making of the total social situation and the need to give them a proper weight, but, in the absence of a single and stable representation of such complex relationships, equal weighting is the most common solution.

PROPOSED CALCULATION METHOD FOR THE IUP

It seems appropriate to standardize values for the different variables using the Z score method and then use the standardized additive method to calculate the index. The adjusted standardized method would not allow an account of the slight variations and seems therefore inappropriate. In order to eliminate the Z scores on a given variable, a constant value of one shall be added to all scores. Missing data shall be replaced by the minimum value found for the given variable. Variables which correspond to negative factors, eg. number of native languages, shall be represented by negative scores and subtracted in the later calculation of the corresponding parameter. Whenever acceptable, the latest year for which data are available shall be used when data could not be found for the year of reference commonly used on a given variable. Variables shall receive an equal weighting in the absence of any authoritative interpretation of their relationship. Variables shall be first summed up by group, each group representing a distinctive facet, or parameter, of the information related situation, what is more appropriate for further interpretation.

CHARACTERISTICS OF THE IUP

The variables to be taken into account in the calculation of the IUP should:

- correspond in a fair proportion to all steps of the communication cycle, as expressed by King (1976), but also consider information needs as they result from actual and priority areas of socio-economic activities, as well as the national capacity for control and development of the national information system;
- include all significant aspects of information work, but also all relevant environmental factors such as geographical, demographic, social, cultural, etc.; the need for and possibility of taking into account cultural factors in particular has been demonstrated by Cochrane (1979) and in the information field by Kotei (1977);
- correspond to all forms of communications in different societies;
- give an account not only of the situation as it stands but also reflect the potentialities.

In order to help in the interpretation of the IUP scores as well as to provide a checklist for analysis and data collection in national planning and policy formulation, not only the IUP should be composed of the sum of the intermediate scores found for the various groups of variables considered but also country profiles be available with figures and scores for the individual variables.

To the extent data are available at the level of socio-economic sectors or areas of interest, and variables could be identified which have a more direct implication on sectoral information activities or specific information functions, we feel that particular secondary indices could be calculated in the same way and country profiles should present the necessary data for doing so.

The literature on the planning or evaluation of national information systems and policies is far from being consistent as to the parameters to be considered particularly outside information infrastructures. The same applies to the monographic literature on the information situation in various countries, as noted by Keren (1980). Consequently, at this stage, the choice of variables could not avoid some bias and relied partly upon the author's own experience.

At present, some 200 variables are considered, of which examples are given below together with the corresponding group:

- physical environment, eg. free access to sea, continuous territory, temperature and humidity;
- demography, eg. active population, population growth rate, rural population;
- education, eg. literacy rate, enrollment ratios;
- socio-cultural, eg. number of native languages, number of ethnic groups, number of years since the first university was established;
- economics, eg. percentages of Gross Domestic Product coming from the major sectors, imports and exports of manufactured products, size of agricultural exploitations, energy consumption, external debt, imports of paper;
- public administration, eg. number of civil servants, number of local authorities, federal structure;
- communications, eg. density of roads, passengers traffic, mail traffic;
- telecommunications, eg. number of radio receivers, of hours of radio broadcasts, of telephones;
- information generation, eg. percentage of Gross Domestic Product allocated to research and development, number of scientists and engineers, number of enterprises and agricultural exploitations;
- production of primary information, eg. number of books, periodicals, newspapers, patents deposited;
- information infrastructure, eg. number of information institutions, collections, number of information specialists;
- information use, eg. circulation of periodicals, items loaned by information institutions;
- information processing facilities, eg. value of computing or micrographic equipment installed, number of computer specialists.

PRELIMINARY RESULTS

Attempts have been made to calculate an IUP like index, mainly in order to test the various calculation methods. For this purpose a set of 41 pertinent but not comprehensive variables was selected in consideration of the availability of data. These variables are distributed as follows: demography 2, education 4, economics 10, communications 5, telecommunications 5, information generation 4, production of primary information 5, information infrastructures 3, information use 2. Countries were selected so that they represent all possible conditions, be included in major statistical sources and do not have too many missing data. 93 countries met, to some extent, these conditions.

Calculations were made summing up the standardized scores of each variable and replacing the missing data by 0, by the minimum and by the mean value, in order to assess the influence of the chosen substitution on the ranking. It was found that when missing data are replaced by the mean value, the ranking of the countries with missing data was notably raised, this benefiting more those in the lower part of the scale. Substitution by 0 or minimum value yielded rather similar and less distorted results, though when minimum value is used, those countries with the larger number of missing data are slightly advantaged. Table I shows some examples of the variation in ranking according to the substittuion used.

TABLE I Rank Variation According to Substitution of Missing Data

Country code	No of missing data	Rank with missing data replaced by 0	Minimum	Mean
9	0	46	46	61
16	1	3	3	3
54	2	40	40	50
59	2	86	84	92
12	5	7	7	7
61	5	33	33	40
35	5	73	74	35
91	11	17	19	21
67	17	29	28	25
5	17	49	52	27
90	18	76	72	42

Test calculations were also made summing up all the variables together and summing up the sums of groups of variables, the results of which were consistants. The changes in ranking were not significant when the prototype IUP was based on the sum of group scores. Although the coverage of the parameters affecting the information situation was by far not comprehensive with the preliminary list of variables, comparisons were attempted between GNP per capita and prototype IUP rankings in order to assess whether the IUP would be able to discriminate between countries with close economic status and whose information situation would intuitively be considered as similar. Table 2 shows some examples of this analysis.

ACKNOWLEDGEMENTS

The study reported here is conducted jointly with Professor H. Borko of the Graduate School of Library and Information Science, University of California at Los Angeles. It has been initiated with the support of a research grant from the University and is now developed under contract from UNESCO, General information programme.

REFERENCES

Andrews, F. M. (1973). Social indicators and socio-economic development. Journal of developing areas, 8, 3-12.

Baster, N. (1972). Measuring development; the role and adequacy of development indicators. F. Cass, London.

CLADES (1979). Informacion y documentacion para el desarollo en Costa Rica. CEPAL CLADES, Santiago de Chile.

Cochrane, G. (1979). The cultural appraisal of development projects. Praeger, New York.

Fisher, I. (1967). The making of index numbers. Kelley, New York.

ISO (1974). International library statistics. ISO 2789.

Keren, C. (1975). Guidelines on the planning of national scientific and technological information systems. UNESCO, Paris.

Keren, C. (1980). Information services issues in less developed countries. In Williams, M. E., ed. Annual review of information science and technology, vol. 15. Knowledge Industry Publications, White Plains, NY.

Kino, D. W.; Lancaster, F. W.; McDonald, D. D.; Roderer, N. K.; Wood, B. L. (1976). Statistical indicators of scientific and technical communications (1960-1980); 3 vol. King Research Inc., Rockville, Md.

King, D. W.; McDonald, D. D.; Roderer, N. K.; Schell, C. G.; Scheuller, C. G.; Wood, B. L. (1977). Statistical indicators of scientific and technical communication (1960-1980); 1977 Edition. King Research Inc., Rockville, Md.

Kotei , S. I. A. (1977). Some variables of comparison between developed and developing library systems. International library review, 9 (3), 249-267.

Lancaster, F. W. (1978). Plan for the collection of data to form a set of statistical indicators of scientific and technical communication in Brazil. IBICT, Roi de Janeiro.

Liu, B. C. (1973). Quality of life indicators: a preliminary investigation. Social indicators research, I, 187-208.

Liu, B. C. (1975). Quality of life: concepts, measure and results. American journal of economics and sociology, 34 (I), 1-13.

Liu, B. C. (1976). Quality of life indicators in US metropolitan areas. Praeger, New York.

TABLE 2 Comparison of GNP per capita and IUP Rankings

COUNTRIES	GNP per Capita	IUP
Low income countries		
Togo	69	65
Sierra Leone	73	75
Pakistan	72	71
Sri Lanka	75	67
Middle income countries		
Jordan	46	50
Syria	48	54
Dominican Republic	49	76
Guatemala	50	45
Zambia	62	31
Cameroon	63	50
Philipines	59	49
Thailand	60	70
Industralized countries		
USA	4	1
Fed. Rep. of Germany	5	18
Norway	6	11
Canada	7	6
Capital surplus oil exporting countries		
Saudi Arabia	12	57
Libya	15	32
Countries with centrally planned economy		
Poland	23	20
Hungary	25	16

CONCLUSIONS

Although our work is still at a preliminary stage, it appears that one will be able to calculate a meaningful IUP. As a matter of fact, such an index would not be an exact measure of each parameter nor of the whole information situation of the countries, would it be only because of the inaccuracy or unavailability of too many desirable data. Like similar index or indicators, it may however noticeably reduce the uncertainty in analysis, initiate a process of data collection and analysis at the international level, and introduce a systematic quantitative approach in the planning and assessment of national information systems. It is anticipated that the IUP will be calculated for a sample of countries during a pilot project in 1981 and submitted for practical evaluation to a small group of specialists.

McCarthy, C. (1980). Bibliotecas brasileiras: exemplos de dados e indicadores com base em dados do IBGE. Revista de biblioteconomia da UFMG, 8 (I), 52-59.

Morris, D. M. (1979). Measuring the condition of the world's poor. Pergamon, New York.

Murphy, D. (1979). Statistics on scientific and technical information and documentation. UNESCO, Paris.

OECD (1980). High level conference on information, computer and communication policies for the 1980's. Paris, 6-8 October 1980. Concluding statement by the chairman. OECD, Paris.

Price, D. de S. (1969). Measuring the size of science. Yale University Press, New Haven, Conn.

Price, D. de S. (1975). Some statistical results for the number of authors in the States of the United States and the Nations of the World. In Who's publishing in science. Institute of scientific information, Philadelphia.

Roderer, N. K.; Schell, C. G. (1977). Statistical indicators of scientific and technical communication worldwide. King Research Inc., Rockville, Md.

Sandoval, A. (1976). A current Latin-American bibliography compiled from non Latin-American journals; a bibliometric study. FID Congress, Mexico City, 1976.

Slamecka, V. (1979). Implications of modern information technology for national information policy and planning; main issues discussed. 3rd UNISIST meeting on the planning and implementing of national information activities in science and technology, Warsaw, Poland, 23-26 January 1979. UNESCO, Paris.

UNCSTD (1979). United Nations Conference on Science and Technology for Development. Final report. A/CONF. 81/16. United Nations, New York.

UNESCO (1970). Recommendation concerning the international standardization of library statistics. 16th session of the General Conference. UNESCO, Paris.

UNISIST II (1979). Intergovernmental conference on scientific and technological information for development. Final report. UNESCO, Paris.

Vilentchuk, L. (1975). Guidelines on the conduct of a national inventory of scientific and technological information and documentation facilities. UNESCO, Paris.

MEASURING THE NATIONAL INFORMATION ECONOMY

Michael R. Rubin

U.S. Department of Commerce, Office of Science and Technology, Washington, D.C. 20230

Any opinions, or findings expressed here are those of the author and do not necessarily reflect the views of the Department of Commerce.

ABSTRACT

According to an Input-Output table based upon 1972 data, the "information sector" of the United States economy grew at a pace roughly equal to the other sectors of the economy in the period from 1967 to 1972. While the primary information sector accounted for 25.1 percent of the Gross National Product (GNP), in 1967, it had declined slightly to 24.8 percent by 1972. In international trade, the sector accounted for roughly 12 percent of the nation's exports. Some 98 percent of U.S. requirements for information goods and services were met domestically.

KEYWORDS

Information economics; the primary information sector; input-output economics; employment trends; international trade statistics.

INTRODUCTION AND DEFINITIONS

This paper summarizes the findings gathered from the construction of an Input-Output table showing the information sector within the context of the national economy for the year 1972. (Rubin, 1980a).

The industries of the 1972 table correspond to those of a similar table for the year 1967, which was constructed for an earlier study. (Porat and Rubin, 1977.)

The definition used here for the information sector follows the definitions established by Porat (1977b). Eight major categories of industries are included in the information sector:

- Knowledge Production and Inventive--which include R&D.
- Information Distribution and Communications--such as education and telecommunications.
- Risk Management--primarily insurance and finance industries.
- Search and Coordination--advertising and the like.
- Information Processing and Transmission--computers and so on.
- Information Goods--pencils, semiconductors, and other similar goods.
- Selected Government Activities--public education, Postal Service and others.

- Support Facilities--including buildings, furniture and maintenance.

This broad typology has been used here to identify the particular elements of the Input-Output table which comprise the information sector; in addition, the typology has been applied to individual occupations to define the "information" elements of the U.S. work force.

THE INPUT-OUTPUT STUDY

Between 1967 and 1972, the information sector's share of GNP declined from 25.1 to 24.8 percent of the total. Table 1 compares the 1972 results obtained here with the earlier findings of Porat and Rubin. (Porat, 1977a). While the sector's share of Government, finance and insurance, and real estate remained essentially static, the sector's share of manufacturing and services declined.

TABLE 1 Gross Product Originating by Industry (in millions of dollars)

Industry	Total Value Added 1972	Information Value Added 1972	Information Percent of Total 1972	Information Percent of Total 1967
All Industries, Total	1,182,766	293,796	24.8	25.1
Agriculture, Forestry and Fisheries	32,163	0	0	0
Mining	18,881	0	0	0
Construction	76,107	6,613	8.6	23.6
Manufacturing	293,400	39,508	13.5	14.6
NonDurable goods	122,870	14,571	11.9	13.0
Durable goods	170,530	24,937	14.6	15.7
Transportation	44,875	0	0	0
Communications	28,357	28,357	100.0	100.0
Electric, Gas and Sanitary Services	25,281	0	0	0
Wholesale & Retail Trade	166,103	16,386	9.9	12.4
Finance and Insurance	43,970	42,659	97.0	99.5
Real Estate and Rental	141,084	27,955	19.8	18.6
Services	161,267	66,414	41.2	49.4
Government	146,609	65,904	45.0	45.4
Rest of World	6,918	0	0	0
Household Industry	5,349	0	0	0
Inventory Valuation Adjustment	-7,591	0	0	0

The Input-Output table and supporting materials have been published in the Journal of Information Processing and Management. (Rubin, 1980a)

INTERNATIONAL TRADE IN INFORMATION GOODS AND SERVICES

Some four dozen products and services are identified as constituting the international trade of the information sector. In a recently constructed time series of United States exports and imports (Rubin, 1980b), the nation's international trade was shown to have an appreciably smaller percentage of information goods and services than does the domestic economy. Over the fifteen years documented by the study, the sector accounted for roughly 12 percent of the nation's exports, and 8 percent of its imports. In 1978, for example, the percentages of exports are as shown in Table 2.

TABLE 2 Sectoral Shares of U.S. Exports

Sector	Percentage
Manufactured Products	43
Military Transfers/Investment	23
Agriculture	10
Noninformation Services	9
Information Goods	8
Information Services	4
Minerals	2
Other	1
Total Exports	100

A similar pattern holds for imports.

TABLE 3 Sectoral Shares of U.S. Imports

Sector	Percentage
Manufactured Products	45
Minerals	17
Investments	13
Noninformation Services	10
Information Goods	7
Agriculture	4
Information Services	1
Other	3
Total Imports	100

The preponderance of U.S. exports and imports are in noninformation related commodities. However, the corollary to a small market share overseas, for the information sector, is a large market for its domestic sales. Approximately 97 percent of the goods and services produced by the information sector are consumed in this country. Conversely, roughly 98 percent of purchases, by the U.S., of information goods and services are provided by the domestic economy.

The composition of the sector's imports and exports are also telling. The bulk of the sector's international trade is concentrated in the purchase and sale of a very few goods and services.

TABLE 4 Leading Export and Import Information Commodities

Product	Percentage
Exports:	
Electronic Computing Equipment	16
Miscellaneous Information Services	14
Mechanical Measuring Instruments	8
Photographic Equipment	7
Radio and TV Equipment	6
All other information sector exports	49

Imports:	
Radio and TV Receivers	25
Semiconductors	9
Photographic Equipment	8
Paper	8
Radio and TV Equipment	7
All other information sector imports	43

EMPLOYMENT TRENDS

There has been a steady increase in the proportion of workers in the U.S. labor force employed in information occupations, rising from 8 to 41 percent of the total, over the course of the century.

TABLE 5 Percentage of Information workers, 1870 to 1970

Year	Percentage of Total Labor Force
1870	8.0
1890	10.9
1910	18.2
1930	24.9
1950	30.5
1970	41.0

This vast change in occupational patterns appears more attributable to societal changes than to technological changes. Two percent of the national work force was employed in 1970 in occupations which were created by innovative technologies such as radio, television, telephones and computers. In contrast, 31 percent of the 1970 work force was employed in administrative, educational and clerical occupations which had existed in 1870.

REFERENCES

Machlup, F. (1962). The Production and Distribution of Knowledge in the United States. Princeton University Press, New Jersey.

Porat, M.U. (1977a). The Information Economy: Definition and Measurement. Government Printing Office, Washington, D.C.

Porat, M.U. (1977b). The Information Economy: Sources and Methods for Measuring the Primary Information Sector. Government Printing Office, Washington, D.C.

Porat, M.U., and M. R. Rubin (1977). The Information Economy: The Interindustry Transactions Matrices. Government Printing Office, Washington, D.C.

Rubin, M.R. (1980). The Information Sector and GNP: An Input - Output Study. Inf. Process. & Manage. Accepted for publication.

Rubin, M.R. (1980b). The Roles of Information Goods and Services in the U.S. Economy. Inf. Process. & Manage. Accepted for publication.

INFORMATION SERVICES OF EGYPT: OUTLINE OF PROJECT ISE

M.A.K. Madkour* and V. Slamecka**

*Academy for Scientific Research and Technology, Cairo, Arab Republic of Egypt
**School of Information and Computer Science, Georgia Institute of Technology, Atlanta, Georgia, USA

INTRODUCTION

In January 1980, the Egyptian Academy of Scientific Research and Technology (ASRT) embarked, under a bilateral agreement with the United States, on an ambitious effort to establish a nationwide system of information services, aimed at ensuring availability and utilization of accumulated knowledge in the socioeconomic development of the Arab Republic of Egypt. This paper outlines the content of and approach to the first phase of Project ISE ("Information Services of Egypt"), devoted to the design of such a system.

For the past three decades, the question of accumulation and effective use of knowledge has been of increasing interest to those concerned with socioeconomic development. Sir Arthur Lewis (1954) offered, in 1954, an ingenious discussion of the general role of knowledge in socioeconomic development, and suggested that the accumulation of knowledge is one of three indispensable elements for the development process to occur, along with the accumulation of capital and the "effort to economize." More recently, interest in the exploitation of available knowledge has permeated into the forefront of social and economic planning, as attested to by the 1979 United Nations Conference on Science and Technology for Development.

Over the years attempts at inducing and accomplishing information transfer have increasingly assumed formal strategies and character. At present the preferred strategy in developing countries for harnessing available knowledge for purposes of socioeconomic development is through the design and implementation of information systems and services that are of nationwide scope. The development of these systems and services proceeds on a coordinated basis, partly justifying their designation as "national information systems." These systems are viewed as the principal mechanism for the management and utilization of knowledge in the service of national development.

It is important to note that the phenomenon "national information system" is a new one. Whereas in industrialized countries the information sector (comprising the generators, vendors and users of knowledge) has evolved gradually over the last three centuries, primarily in response to the growing demand for information, national information systems in developing countries are artificial creations being planned and designed *de novo*. As such, these systems are innovative constructs of modern social planning, pioneered largely by the developing countries themselves and implemented in environments that differ at many levels — cultural, linguistic, economic, political and technological — from the environment in industrialized countries.

In modern Egypt, the importance of knowledge in social and economic development was evident in the early 1960's when the country established a focal point for scientific and technical documentation. This unit was transformed, in 1968, into the National Information and Documentation Centre (NIDOC) and attached to the Academy for Scientific Research and Technology (ASRT). The origins of Project ISE date from 1974, when the US-Egyptian Agreement on Technology, Research and Development identified information services as a priority project within the collaborative program of Applied Science and Technology Research. In April 1976, a symposium and workshop of Egyptian and US specialists developed a set of recommendations that defined the scope of Project ISE, and led to the formation of a US-Egyptian program committee on scientific and technical information. A year later, ASRT signed a project agreement with the National Science Foundation of the United States which had been asked by the US Agency for International Development to assume responsibility for the management of Project ISE grant activities. During the subsequent 18 months, NSF and ASRT jointly selected the Georgia Institute of Technology as the US contractor on Project ISE.

NSF contract INT-7924187 with the Georgia Institute of Technology was signed in November 1979. The contract provides for an initial two-year effort, the primary objective of which is to establish the basis for the implementation of a nationwide network of scientific and technological information services that can effectively support the socioeconomic development of the Arab Republic of Egypt. The contract is referred to as Phase I of the overall collaborative effort between the two countries, the total duration of which is estimated at five years. Phase II will be concerned with the implementation of the services.

Project ISE is organized around two advisory groups. On the Egyptian side is a Steering Committee which consists of senior executives from the public and private

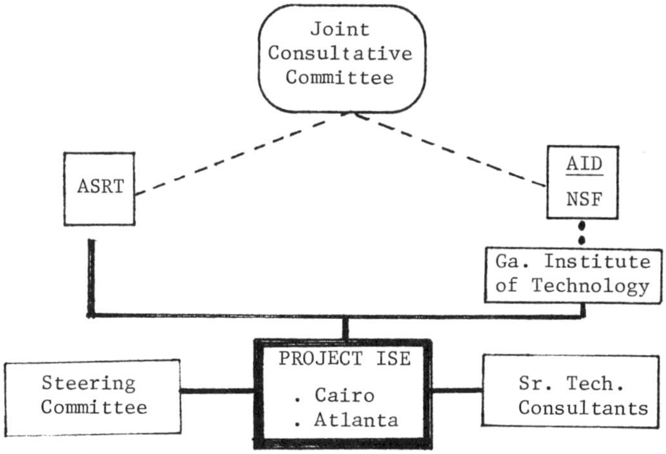

Fig. 1. Project ISE Organization.

sectors of the ARE; they represent the more important agencies and constituencies of the country affected by the prospective information activities. On the US side there is a small group of senior technical consultants. The Cairo project staff consists of senior technical specialists on loan from various agencies and organizations, notably the ASRT and its National Information and Documentation Centre (NIDOC), the National Research Centre (NRC), Cairo University, as well as selected organizations from the private sector. Understandably, the staffing level of Project ISE fluctuates in accordance with the personnel requirements of the various technical tasks.

METHODOLOGY OF PHASE I

The methodology for national information systems design has made reasonable progress during the past 25 years, since the first attempts at planning national documentation and information services were made in countries with planned economies. During recent years this methodology has been intensely scrutinized and formulated in the form of a logical process of general steps and activities.

The major steps in this process, representing activities of Phase I of project ISE, are depicted below. They consist of two clusters of activities designated as "systems analysis" and "systems design." The approximate duration of each of these activities is nine months; the systems analysis was initiated in January 1980, and is scheduled for completion in fall of that year; the systems design activity, which is to culminate with a detailed plan of implementation, schedule and costs, is slated for completion in October 1981.

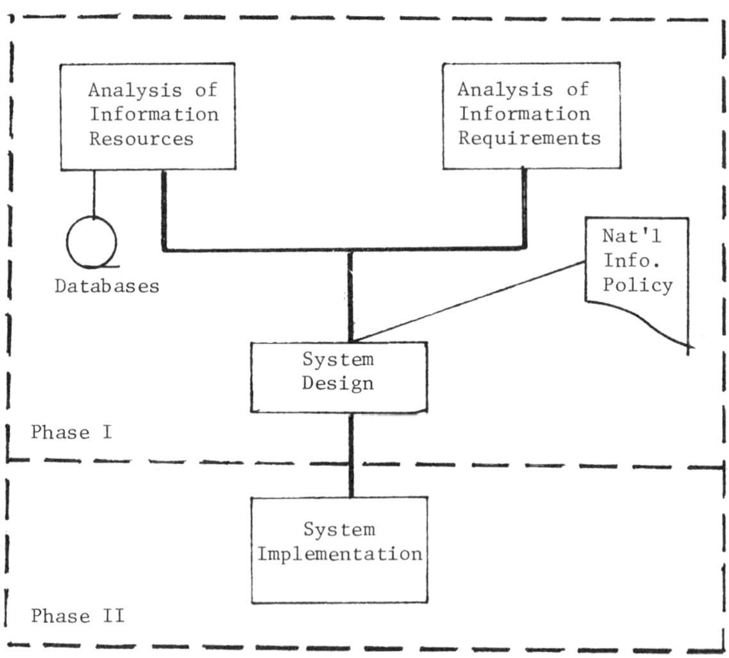

Fig. 2. National Information System: Phases I and II.

In order to accelerate the actual development of information services in Egypt, it is foreseen that the second phase of Project ISE (implementation phase) will overlap with Phase I so as to permit rapid initiation of activities that are relatively independent of the final system design (for instance, manpower development). Thus despite its contractual division into two phases, Project ISE is conceived and operated as a homogeneous enterprise.

The remainder of this paper outlines the major tasks of the analysis and design efforts of Phase I.

SYSTEMS ANALYSIS

As depicted in Fig. 2, the analysis consists of two major clusters of tasks: assessment of the current situation pertaining to information activities in Egypt; and assessment of the national requirements for information services. These tasks are referred to, respectively, as the analysis of resources and the analysis of requirements, and are described in more detail below.

Analysis of Resources

The assessment of the current information situation in Egypt, underway since January 1980, concerns a variety of resources and proceeds through a number of parallel approaches. The objective is to obtain as accurate a picture as possible of the existing information sector of the country. Accordingly, the analysis covers the following.

Information sector organization. The purpose of this activity is to describe the structure of the information sector in Egypt that generates and produces information, manages indigenous and foreign information, and provides information services. Activities in both the public and the private sector are considered.

Inventory of recorded information. The purpose of this task is to assess the available data/information resources in Egypt, both in terms of types of information and specific holdings. As regards the latter, Project ISE subcontractor, King Research, Inc., has compiled core lists of serial titles, databases, and reference materials for the primary areas relevant to national socioeconomic development, and is comparing against these lists the inventories of all major Egyptian libraries and information agencies.

Manpower. The assessment of Egyptian information manpower seeks to obtain data on professional and paraprofessional staffs, their education, activities, etc. in information organizations of both the public and the private sectors.

Facilities. The object of this task is to assess the level of existing technological support for information work, and the availability of physical space for the entire range of information functions.

Education. This effort surveys the status of education and training programs in Egypt relating to the broad range of activities concerned with information production, management and use. University programs, in-service training, continuing education, and private sector training programs are covered.

Legislation. An attempt is made to identify the laws, decrees, and regulations that either support or hinder information activities such as publishing, communications, and distribution of information.

Social/administrative constraints. Project ISE seeks to identify the major constraints that affect the design of information services in Egypt. An example of social constraints is the system of education, in which the training for problem solving largely ignores the existence of information resources. As an example of administrative constraints, one can cite some practices of library management which effectively impede resource sharing.

Finances. To complete the assessment of the existing situation Project ISE is attempting to obtain an overall picture of funding of information activities in Egypt, covering both Egyptian and foreign assistance funds.

The primary method of data collection is the survey. A major survey of the above resources is being conducted by an Egyptian subcontractor organization on behalf of Project ISE in approximately 400 agencies and organizations in the public and private sectors, using multipart survey instruments and employing sixty professional surveyors especially trained for this survey by Project ISE consultants. The collected and verified data is processed for aggregate analysis by Cairo University Computer Center.

Survey of Requirements

The identification and analysis of information needs is the critical component of the systems analysis effort of Project ISE. The difficulty of this undertaking is due not only to the well-known problems encountered while interpreting user requirements; it is accentuated by the fact that the primary objective of Project ISE is to design a system of information services supportive of national socioeconomic development, and that there does not exist a well-established methodology for clearly determining the information requirements for such an objective. The large majority of information systems design efforts at the national level have had as their objective the support of science and technology, and have concerned themselves with the management of recorded scientific information. In contrast, Project ISE surmises that other types and categories of information — policy-making, economic, industrial, marketing, business, operational, and legal — have the highest utility and priority in socioeconomic decision-making and problem solving; and, as is well known (Saracevic, 1979), much of this resource is experiential and not easy to locate in traditional data and information resources.

Another major factor affecting national requirements analyses are differences in the predisposition of different peoples for using information. Whereas in industrialized countries, information is the very lifeblood of social and economic existence and their societies are prolific information consumers as well as producers, in developing countries, the proclivity to use information is weaker. Their societies often lack strong motivation to use information for improvement and in interpreting and adapting information not presented in easily digestible forms; and infrequently share indigenous experiences (Slamecka, 1979). These factors compound the difficulties of the systems analyst's attempt to determine national information requirements.

To reduce the effects and uncertainties due to these factors, Project ISE is approaching the analysis of national information requirements for the socioeconomic development of Egypt from several parallel directions. These are as follow.

Analysis of socioeconomic plans. The general strategy for socioeconomic development of Egypt, formulated in the "October Paper" of 1974, is the basis for a series of five-year plans starting with the 1979-1983 plan. Project ISE is analyzing this plan from the viewpoint of information requirements. The national plans in agriculture, industry, commerce, transportation, health, other service industries, as well as the science and education sectors, are covered by the analysis, using the general approach due to Gray (1978).

Top management interviews. Lengthy, indepth interviews have been conducted with individuals having executive responsibility for 30 large-scale projects accorded priority in Egypt's Five-Year Plan. The interviews sought to assess their attitude toward and awareness of the data and information required by their projects, and to determine project information management styles.

User survey. In what is perhaps the more traditional approach, Project ISE has interviewed three categories of problem-solvers in some 400 agencies and organizations: the administrative decision maker, the research staff member, and the operations staff member. The interviews were guided by a structured questionnaire.

Agency function analysis. Project ISE has gathered the objectives and assignments of major Egyptian institutions, such as ministries and R&D institutions, and has analyzed their information-related implications.

R&D projects analysis. A database is being generated describing Egypt's R&D projects in a format compatible with the Smithsonian Science Information Exchange (SSIE) format. The projects fall into three categories: R&D projects funded by Egypt, projects funded jointly by Egyptian and foreign monies, and academic research projects.

Citation analysis. The half-life of citations of major Egyptian journals has been determined and compared with the half-life of citations in leading international journals in the same fields. The analysis (McCarn, 1980) is indicative of a relative lack of awareness of current information resources by Egyptian scientists; parenthetically, it provides a strong economic rationale for the development of science-oriented information services.

Cumulatively, the above interviews, surveys and analyses should provide the data and the basis on which informed judgments will be made concerning the information requirements of the country, and on which priorities among these requirements will be determined. This comprehensive approach not only attempts to assess information needs for social and economic development but it also seeks to assert the roles of such functional categories of information as current manufacturing and process/production data; equipment data; operational experience; economic financial and demographic data; data for planning, policy-making, and management; and training materials; as well as information emanating from research projects.

The data generated by the resource and the requirements analyses is coded from multiple viewpoints and computer processed for aggregate analysis. It is expected that this approach will yield a comprehensive picture of the current information sector of Egypt.

SYSTEM DESIGN

For purposes of clarity, we may distinguish in the design effort between two sets of functions: intrinsic and extrinsic.

Intrinsic Functions

These functions are the technical activities proper to the system of information services. They are as follows.

Information carrier production. Once generated, information enters the communication flow by means of a variety of carriers. The designer's task is to ensure the availability and viability in Egypt of those information carriers that have been shown to effectively communicate useful information; in Egypt there is a notable absence of the technical report medium, for example. Another task is to foster efficiency in the production process; for instance, the publication of scientific journals in Egypt is particularly slow.

Information provision. Given a comprehensive statement of national information requirements in Egypt, Project ISE expects to have to augment the information resources available in the country at the present time. Lists of serial and reference titles and of databases recommended for procurement by Egypt will be compiled. Procurement of these resources will provide an incentive toward the development of resource sharing practices in the country, currently almost non-existent, and of initiating cooperative acquisition practices.

Information organization. Project ISE foresees the creation of inventories of national information resources in Egypt. Cooperative acquisition of materials provided is a step toward the establishment and maintenance of a union list of current materials. Union lists of existing holdings of non-Arabic materials can be created with the assistance of the MARC and REMARC files, the by-product being machine-based catalogs of such holdings in participating libraries. Bibliographic control of indigenous (Egypt-generated) serial literature will be pursued through the development of a database of abstracts of Egyptian scientific and technical documents.

Information services. The crux of Project ISE is the development of nationwide information services; clearly, these services must support the information requirements determined in the systems analysis phase. The range of information services to be considered in the design effort will include: information availability announcements; products generated from the databases of Egyptian information (such as an Egyptian abstracting journal); referral services; database searching and document delivery from within Egypt and abroad; and advanced information services such as those provided by information liaison officers and information analysis centers.

Extrinsic Functions

The purpose of these functions is to support a healthy environment in which the technical functions of the national information system can operate and flourish. In project ISE, the intrinsic functions to be considered in the design phase are the following:

The information environment. It is difficult to anticipate an intensive application of available, accumulated knowledge in any developing country without a concomitant, extensive effort to raise the proclivity of the country's people to generate, seek out, and apply such knowledge. Project ISE recognizes the critical importance of this factor, and intends to propose a range of carefully formulated strategies in this regard, ranging from activities in the educational system, methods of promoting information use in everyday activities, to intense marketing of information services. Understandably, the effect of many of these strategies is long term; nevertheless, it is a crucial area, insofar as the effectiveness of information services is positively related to the information propensity of the users.

Manpower development. A more formal approach to the development of manpower in the information sector is required in Egypt. Such an approach will encompass the development of degree programs at Egyptian universities; continuing education training for practicing information workers; and inhouse training of paraprofessional and clerical staffs. In parallel, project ISE will support the organization of the information profession in Egypt, and the educational and training activities of its society or societies. A new post-graduate course in information and computer systems is being inaugurated in Fall 1980 at Cairo University; and a non-degree part-time course has been designed for practicing professionals. A massive manpower development in Egypt is a sine qua non.

National-level organization and management. The organizational structure of a national information system has many ramifications; its ability to provide adequate budgets, its effectiveness in raising the social and economic status of the manpower of the sector, its ability to orchestrate coordination among diverse elements of the system, and others. At issue is what is the optimal organizational pattern at the national level that is conducive to the greatest cost-effectiveness of the information industry of the country. Undoubtedly, the traditional arguments between a centralized and a decentralized approach to the organization structure of national information systems are too simplistic. Whereas centralization is typically conducive to efficiency at the expense of effectiveness, and decentralization favors effectiveness at the cost of efficiency, the designer's task is to optimize both parameters, and to do so within such constraints as the relationship between the public and private sectors, traditional sentiments toward control, coordination and collaboration, the flexibility of the legislative apparatus, etc.

The product of the design effort of Project ISE is intended to be a series of management and technical propositions cumulatively describing the organization and operations of the future system of nationwide information services of Egypt. These documents, together with schedules and cost data, will provide Egypt with a rational basis for the decision to implement an information support facility for the country's long-term social and economic development.

REFERENCES

Gray, J. (1978). Information Policy and Planning for Economic and Social Development. Paper prepared for first UNISIST Meeting on Regional Information Policy and Planning in West Africa, Accra, Ghana.

Lewis, A.W. (1954). Theory of Economic Growth. Allen and Unwin, London.

McCarn, D. (1980). Citation Patterns of Selected Egyptian Journals. Georgia Institute of Technology, Atlanta, and Academy of Scientific Research and Technology, Cairo, U.A.R.

Saracevic, T. (1979). Perception of the needs for scientific and technical information in less developed countries. In V. Slamecka (Ed.), Scientific and Technical Information Services for Socioeconomic Development. International Science and Technology Institute, Inc., Washington, DC. pp. I-1 - I-68.

Slamecka, V. (1978). Information Transfer between the US and Developing Countries: An Assessment and Suggestions. International Science and Technology Institute, Washington, DC.

NATIONAL SCIENTIFIC AND TECHNICAL INFORMATION SYSTEMS IN SAUDI ARABIA

L. G. Burchinal* and A. Al-Mazi**

* Div. Int. Prog., National Science Foundation
Washington, D.C., USA
** Div. Inf. Syst., Saudi Arabian National Center
for Science and Technology, Riyadh, Saudi Arabia

ABSTRACT

A number of criteria are used to assess the cost-effectiveness, from a national perspective, of three major models for organizing national data bases and information systems in developing countries. The national-centralized model, judged superior to either the laissez-faire or international-decentralized models, was selected as the basis for science and technology information systems development in Saudi Arabia. Developments since 1978 are described. Guidelines for other developing countries are presented.

KEYWORDS

National information systems; S&T data bases; international S&T information systems; developing countries; user services; national networks; international networks.

ISSUE

In establishing national scientific and technical data bases, developing countries face many issues. Of these, technical questions related to data base design and management and hardware and software considerations receive the most attention. Another, far more critical, set of issues is often neglected. These concern the conceptualization, scope, and organizational placement of data base activity and how these activities will be integrated with the information activities of various public and private bodies in the country and with broader international efforts.

Considerable information exists about technical issues. Literature abounds. Salesmen and technical specialists are additional sources of information. In contrast, relatively little literature relates to conceptualization and broader political-organizational issues related to national S&T data base development. What there is, is exhortive rather than analytical.

Consequently, many developing countries find themselves implicitly committed to a given approach to data base development, without having explicitly looked at the options and making a conscious choice.

Saudi Arabia was at this juncture in 1978. A deliberate choice was made.

This choice process and actions since then are described with the hope that they may be useful to other developing countries as well.

OPTIONS

There are three main options for organizing national S&T data bases:

1) Laissez-faire -- market oriented
2) Centralized -- national oriented
3) Multiple -- international oriented

Laissez-Faire Oriented

This option is best seen in the United States, although the European Economic Community also is following this approach. Under this approach, organizations, public or private, may and, in many countries, are encouraged to offer data base services. Generally, market forces are relied on to determine viability of services, although some data bases may be sheltered, through subsidies, for various reasons (U.S. Department of State, 1979).

The most obvious result of this approach is a profusion of data bases, many of which overlap one another.

Centralized-National Oriented

Under this approach, countries establish one organizational locus and seek to maintain a comprehensive S&T data base, perhaps incorporating more limited specialized files, to cover all S&T and related literature in the country.

This approach has long been championed by UNESCO and is favored by most developing countries. Beginning in the 1950s, UNESCO aided the establishment of national documentation centers, such as NIDOC in Egypt and INSDOC in India. In the 1970s, UNESCO teamed up with the International Council of Scientific Unions to establish UNISIST--a program to stimulate development of a worldwide S&T information system, based on strong national systems (UNESCO, 1975). For a while, UNISIST and a library-oriented program in UNESCO (NATIS), sparred for international leadership. This issue was publicly defused by merging the two efforts, and dedicating the combined, though still limited resources, to creating national focal points and national systems embracing S&T information services, national archives, and library services.

UNESCO's programs follow naturally from the dominance and increasing assertiveness of developing countries in its forums. This approach also fits well with the strategy behind the New Information and Communication Order espoused by the Third World as a special case of its New Economic Order.

There are obvious consequences of adopting this model. One is the establishment of a single or at least dominant data base covering all S&T areas in a country. The other is the presence of a single major point for national-international interfaces.

Multiple-International Oriented

Though initially promoted by the early group of nuclear powers to spread the peaceful uses of atomic energy, more recently this option has become the organizing principle for various international-decentralized data bases. Under this approach, each participating country forwards its input in a structured form to a central international center. This center edits and merges the input, maintains the international data base, and provides the integrated output back to participating centers in tape or printed forms.

Specialized agencies of the UN family, and the Canadian Development Research Center, provide the impetus for this model (International Development Research Center, 1980). Today there are a dozen or more international-decentralized systems in operation or at various stages of promotion and development, including:

INIS--the International Nuclear Information System, International Atomic Energy Agency, the granddaddy of this kind of system;

AGRIS--the International Information System for Agricultural Sciences and Technology, Food and Agriculture Organization (FAO);

TCDC/INRES -- Information Referral System for Technical Cooperation Among Developing Countries, the United Nations Development Program;

DEVSIS -- Social and Economic Development Information, International Development Research Center, Canada;

POPIN -- International Information Network for the Population Field, Department of International Economic and Social Affairs, United Nations;

INPADOC -- International Patent Information, International Patent Documentation Center, Vienna;

INTIB -- Industrial and Technological Information Bank, United Nations Industrial Development Organization (UNIDO), Vienna;

SPINES -- Scientific and Technological Policy, UNESCO.

An obvious result of this approach is that countries can find themselves participating in various international cooperative efforts, through a number of different national centers, with little coordination of activities within the borders.

Assessment of Options

The laissez-faire approach requires a number of prerequisites, including the following:

- A structure of scientific and technical organizations, established information centers, clearinghouses and publishing organizations that can undertake data base services;
- Output of S&T materials in sufficient quantity to sustain generation of a number of specialized information services (bibliographies, indexes, directories, data bases, etc.) across a variety of disciplines and problem areas;
- A large, skilled, and mobile population of information managers, professionals, and technicians;

- o A set of market forces that place monetary value on information products and services.

Few of these conditions exist in most developing countries, including Saudi Arabia. Therefore, this laissez-faire approach was not considered.

The other two options were compared, using criteria listed in Table 1.

Clearly, the centralized model was the most effective choice for Saudi Arabia. This comparative analysis also suggests this model is probably the most cost-effective one for most developing countries.

In reality, in most countries, developing or industrialized, senior S&T managers or political decision-makers are not presented with analyses of choices for organizing information systems. Instead, committments to participate in one or another international-decentralized system are made at lower levels among specialized S&T bodies. Before long, a country can find itself committed to participation in several international systems, that within a country, result in redundancy of coverage and reduced cost-effectiveness in terms of the total national effort. Once committed, pressures exist at the national and international levels to continue.

In contrast, dedication to establishment of a single national S&T data base and information system reduces the fragmentation tendency and assures the most cost-effective use of resources as well as optimal services to users.

With a single national system, comprehensive, non-redundant and quality-controlled output can be maintained. Specialized products can be generated easily and at far less cost to meet the needs of a specific ministry.

Moreover, by initially adopting the centralized-national option, countries do not preclude choice among future options for national and international systems development. At the national level, countries have the choice of maintaining a single, centralized, comprehensive S&T data base (or set of them) or opting for a centrally managed, but decentralized system, with specialized centers operating as part of a distributed data base system. A top-down approach, in this case, has many advantages over a bottoms-up approach. Standards, data base management procedures, protocols, etc., can be more easily established, perhaps, though, at the loss of greater innovation that often occurs in less controlled environments. At the international level, countries can exercise a wide variety of options, depending upon their policies and interests. They can enter into commercial arrangements with online services for use of their tapes, establish barter arrangements with other national or commercial systems, join international-decentralized systems, or, in various ways, use all these arrangements.

In the "information-era" that we are believed to be entering, having these options will be important to developing countries. If information is the valuable commodity of the post industrial world, as we are told, industrialized countries will develop systems appropriate for various domestic and international markets. International-decentralized systems are not likely to fare well in this competition. Countries with information of value internationally, however, will be in a position to bargain for cooperative arrangements with already established international systems.

TABLE 1 Comparison of Centralized-National and International-Decentralized Options

Criteria	Options: Centralized National (CN)	Options: International Decentralized (ID)	Comments
Management of National S&T Information			
Comprehensive multidisciplinary coverage	Superior		CN clearly superior. ID gives spotty coverage with serious gap and substantial overlap
Consistent use of standards	Superior		CN provides greater assurance of consistent use of national/international standards
Quality control	Probably Superior		With one operation, greater quality control of CN more likely: actuality, depends upon staffing and management
Cost/effectiveness of total national information processing and building data bases	Superior		CN more likely to be most cost/effective because of no redundancy, single management, single data base design; economies of scale; continuity better assured
Specialized coverage (national and foreign)		Superior	Being tied to a specified discipline/field, ID has clear advantage
Derivation of specialized products, especially involving multi-disciplines	Superior		With its more comprehensive base, CN offers the most cost/effective generation of specialized products
Response to changing national needs and priorities	Superior		CN, with less committment to a given discipline or field, can move more quickly in adaptive fashion to meet new requirements
Representation in international forums	Superior		Representatives from CN, especially if from national S&T authority, carry greater status, can address broader range of issues and represent country more effectively
Human Resource Use			
Use of available trained personnel	Superior		Scarce talent can apply their expertise across broader fields in CN; benefit for country
Advance training and development of personnel	Superior		CN more likely to have national responsibility for training as well and more able to promote and arrange training

Criteria	(CN)	(ID)	Comments
Technological Factors			
Use of existing hardware/software	Probably Superior		CN offers increased cost/effectiveness in use of hardware/software; prorated over larger data base; greater user community; economies of scale
Development/purchase of new hardware/software	Probably Superior		For same reasons as above. Also, greater "critical mass" at CN aid in development or selection of hardware/software
Transfer of information technology into the country	Superior		CN offers a visible entry point and learning center re use of technology; more likely diffuse through S&T and management circles more quickly
National network development	Superior		With one data base and associated system, CN is more likely to foster effective national networking through use of its standards, software, codes, etc.
Specialized international network development		Superior	ID definitely superior, unless CN also serves each of the major decentralized ID networks
Generalized international network development	Superior		With its comprehensiveness, CN definitely superior, has greater potential value to more foreign information suppliers
Movement to numeric, textual data bases	Probably Superior		With one data base, concentration of talent, and national authority, CN more likely in position to integrate bibliographic, numeric, and textual data bases
User Considerations			
Access to all relevant national information	Superior		CN definitely superior; especially when so many problems require multi-disciplinary results
Ease of use, especially online	Superior		With CN, users have to learn only one system; under ID users may be faced with conflicting vocabularly control, commands, etc.
Creation of outputs across multiple fields	Superior		With its comprehensive coverage, CN also better able to generate wider subject range of general and specialized products
Connections to specialists and interest groups		Superior	ID being located usually in a specialized S&T organization has better ties to corresponding special interest group nationally and internationally

Criteria	(CN)	(ID)	Comments
Ease in becoming established		Superior	ID can be strated informally at lower levels in specialized S&T organization; CN, being more visible at national level, often requires specific authorization from national political levels
Innovation in processing and services to users	--	--	Arguments can be made both ways: hard to draw judgement on this criteria

SAUDI ARABIAN DEVELOPMENTS

S&T information systems development in Saudi Arabia is the responsibility of the Saudi Arabian National Center for Science and Technology (SANCST). Within several months after its establishment in November 1977, SANCST committed itself to building a national S&T information system, based on access to foreign data bases and development of Saudi Arabian S&T data bases, based on the centralized approach. Since then, SANCST has established a unique capability that, in certain critical aspects, surpasses S&T information systems available in any other country.

Some of the main features of SANCST's information activities are:

1) Committment to information systems development as the highest initial priority of the Center. (What other newly-established scientific institute has made such a committment?)

2) Integration of all computer- and information-related activities, such as described next, under one management.

3) Establishment of the most comprehensive, integrated, single set of national S&T data bases.

4) Establishment of a computer center at SANCST.

5) Establishment of user-conscious services.

6) Development of the SANCST library, using automated systems.

7) Development of a national network of library and information services to increase access to information available within the Kingdom.

8) Extension of the national network through international networking.

9) Development of specialized S&T information manpower.

10) Development of an R&D program for development of Arabic computer, information and translation services.

11) Development of a program to disseminate reports generated by SANCST and its grantees.

12) Creation of public and user awareness of the significance and value of computer, telecommunications, and information technology and services.

Brief comments for each activity follow.

Committment

The size and range of the activities underway or planned attest to SANCST's committment to information services. Without this committment, little would have been accomplished. Lack of similar committment by top management and decision-makers more than anything else, remains a deterent to orderly development of information services in many developing countries.

Management

SANCST's decision to place management of all computer and information-related activities in one organizational unit and under one manager is also a key factor in its initial success. This arrangement permits the Director, SANCST's Information Systems, to integrate short- and long-term planning, allocate resources effectively, monitor, and, as necessary, modify activities, with maximum effectiveness. The usual bureaucratic impediments, common when activities are scattered across various managers, are removed.

Comprehensive S&T Data Bases

When usually used, the phrase "S&T information" system or data base refers, in reality, to only a bibliographic system, based on a set of citations, abstracts, and index terms about documents stored in machine-retrieval form.

SANCST has gone far beyond this limited conception. Its S&T data base is comprised of eight separate, but related data bases. Six are publicly available and provide information about critical aspects on the S&T enterprise in the Kingdom, such as:

1) Research in progress -- with summaries of research sponsored by SANCST and other organizations throughout the Kingdom as well as foreign projects of interest to Saudi Arabia;

2) S&T publications -- with abstracts and descriptive information of publications generated by Saudis, Saudi Arabian organizations, published in Saudi Arabia or prepared for Saudi Arabian organizations by foreign organizations as well as selected foreign publications in S&T areas of particular significance for Saudi Arabia;

3) Libraries and information centers -- with descriptions of their holdings and services and similar information for foreign libraries and information centers in S&T areas of interest to the Kingdom;

4) S&T manpower -- with descriptions of educational background and work experiences of persons involved in S&T activities in or on behalf of Saudi Arabia;

5) Laboratory, research and data processing facilities -- with descriptive information about their goals, activities and accomplishments; and

6) Major equipment -- of the preceding organizations with descriptive information, including availability to researchers from other organizations.

These data bases were designed to accomplish several purposes. First, unlike so many other systems, they are intended to provide data for national S&T planning and management. Decision-makers and planners for instance, are able to pool information across data bases to create a comprehensive picture of current national efforts and capabilities with regard to a given problem. Second, the data bases were designed for use by individual researchers and users to identify research and researchers in closely related areas, to select reviewers of projects, to identify relevant literature, to locate facilities and equipment needed to carry out research, and so on. Information on libraries and information centers is included for several reasons: 1) to help librarians and information specialists aid users in locating needed information; and 2) to aid in developing a national network for such services.

Two other data bases are for internal SANCST use only. These cover:

1) SANCST Research in Progress -- with management information about receipt and review of preliminary proposals, similar data for formal proposals, and data and ticklers for monitoring funded projects; and

2) The SANCST Records Management System -- with information about all records (correspondence, telexes, memos, etc.) sent, received or generated for use within SANCST.

These eight data bases share certain features that enhance their use and effectiveness:

1) All were designed at one time; hence they share common data elements and definitions, mnemonics, codes and coding procedures for each comparable field.

2) Where appropriate, international coding sltandards were used (country codes, etc.); otherwise, common conventions were used for various fields (personnel names, corporate authors, place names).

3) A powerful, flexible software is used for data base maintenance and retrieval (Informatics' RECON IV/STIMS IV) for all data bases.

4) The Informatics software also permits aggregation of data bases for various purposes. For example, in the case of routine searching, data bases for research in progress, bibliographic, and information centers and libraries can be combined into one macrodata base. So can data bases for research facilities and equipment. Various combinations can also be made for internal management purposes: for example, to show all fields with missing data.

5) Common subject indexing is used for appropriate fields across all the data bases.

6) There are numerous extra fields in most data bases, beyond those routinely used by researchers, that can be used by decision-makers and managers.

SANCST Computer Center

The data bases just described are mounted at the Saudi Arabian National Computer Center, which uses an IBM 370/158. In addition, SANCST has begun building its own computer center, and has acquired an IBM 4331 with various peripherals.

User-Conscious Services

SANCST offers two types of online searching: 1) from the SANCST data bases at the NCC; and 2) from data bases accessible through foreign online services, such as Systems Development Corporation (ORBIT) and Lockheed Information Services (DIALOG) in the United States. A leased line, with connections to the

U.S. telecommunication network, is used to access U.S. online services.1/

Arrangements are being made to address European hosts as well.

To help users use these services, SANCST has prepared a variety of user awareness and search-assistance materials, including a Search Request Form, Search Analysts Guide, User Evaluation Form, a folder and notebook for returning search results to users, and pamphlets, announcements, and posters describing the services. In addition, demonstrations of searchers are conducted regularly.

Also, as part of its service committment to users, SANCST arranges for fast delivery of documents identified from searches, but not available in users' local libraries. To the extent possible, loans or copies are obtained from other libraries in the Kingdom; however, considerable material is obtained through sources in the United States and Europe. The leased line is also used to place orders for materials with U.S. suppliers from terminals at SANCST, through use of an electronic mailbox service. This same service, incidently, is used for a variety of management-related communications between SANCST and the NSF/SANCST Program Office. Also, early 1981, FAX services will be installed between SANCST and its major information supplier, Warner-Eddison Associates in Cambridge, Massachussetts, and the NSF/SANCST Program Office. Then, much of the information transfer between SANCST and U.S. offices, in the form of mail message, and copies of materials, will be transmitted electronically.

SANCST Library

SANCST is now in the midst of developing its Master Plan for construction of its facilities. Among the first buildings to be erected will be the SANCST Library and Information Center. Meanwhile, SANCST information services are housed in temporary quarters.

National Network Development

Several projects, now in advanced planning, are directed toward building a national network. These include: 1) development of a National Union Catalog of holdings of major libraries; 2) an automated interlibrary loan program; 3) stimulation of standards for network development to make both preceding tools and others (SANCST authority files for place names, corporate authors, etc.) available at major libraries and information centers throughout the Kingdom and for accessing the SANCST and foreign data bases.

1/ This line is maintained as a common untility for all activities carried out under the U.S.-Saudi Arabian Joint Commission on Economic Cooperation. The joint project between SANCST and the NSF and Department of the Treasury is only one of 19 cooperative arrangements between Saudi Arabian ministries and corresponding U.S. agencies. On the U.S. side, the Department of the Treasury, as the lead agency, also is responsible for the operation of the leased line and associated processes such as multiplexing, digitizing, telemail, and FAX.

International Networking

SANCST already is connected to international networks for searching foreign data bases. In addition, SANCST looks forward to cooperative arrangements with other countries in the Middle East and to creating access to its data bases through various national and international networks and services.

Manpower Development

Recruitment and development of skilled manpower remains a critical challenge. Long-term solutions rest on the development of appropriate undergraduate and graduate programs in computer, information and library sciences at Saudi Arabian universities. In the shortrun, SANCST is trying to meet its manpower requirements in all conventional ways, through recruitment of trained and experienced specialists, short courses, attendance at conferences, on-the-job training, and by visits of staff to other centers.

R&D Program

SANCST's R&D program in this area is focused upon development of Arabic computer, information, library, and translation capabilities. In terms of hardware and software, the objective is to develop fully Arabic computer systems based on international standards, including all programming in Arabic. With reference to library and information activities, the objective is to create standardized, automated bibliographic and information exchange in Arabic. Translation research includes: development of an English-to-Arabic S&T terminological data bank; testing and enhancement, as necessary, of English-to-Arabic computer-assisted translation programs; conversion of the English language data bases cited previously to Arabic; and dual maintenance and operation of SANCST data bases in Arabic and English.

SANCST Dissemination Program

SANCST's objective is to foster research that will contribute to achieving the Kingdom's development objectives. As part of this effort, SANCST provides grants to organizations for support of research. SANCST is also planning the establishment of several national research institutes. SANCST's Information Systems Division is also responsible for publication and dissemination of reports from grantees and inhouse researchers. Word and image processing will be used to prepare reports with dissemination of reports from grantees and inhouse researchers. Word and image processing will be used to prepare reports with dissemination occurring in printed, micrographic, and digital forms.

Public Awareness

So far, discussion has focused on technical considerations. SANCST also realizes that the citizens of the Kingdom and potential users of SANCST's services must be acquainted with the significance and economic value of computer and information services if the Kingdom is to achieve its long-term goals. Therefore, SANCST is developing a public and user awareness program, using printed, videotape, and other educational materials.

DISCUSSION

The preceding description of SANCST's information activities was intended to suggest a conceptual approach that may be useful to other developing countries as well. This approach is based on the following key concepts:

1) Commitment to a centralized, national information system, based on a comprehensive, integrated set of data bases embracing all scientific and technical disciplines and areas;

2) Assignment of responsibility for this effort to one institution at the national level;

3) Within that institution, integration of all computer and related information functions, ranging across paper, micrographic, and electronic media, in one organizational unit and under the supervision of one manager;

4) Commitment to using state-of-the-art information processing technologies and process;

5) Integration of major libraries and information centers within the country in an evolving national network, with connections to international networks and major foreign data bases; and

6) Recognition of the importance of creating informed public awareness of the significance and economic value of computer and information technologies and their uses.

While Saudi Arabia obviously can apply considerable resources in carrying out its program, the concepts and practices described are suggested as guides for other developing countries, even if the scale and pace of activities may be different.

REFERENCES

International Development Research Centre (1980). _International Cooperative Information Systems: Proceedings of a Seminar Held in Austria, Vienna, 9-13 July 1979_. IDRC, Ottawa.

UNESCO (1971). _UNISIST: A Study Report on the Feasibility of a World Science Information System_. UNESCO, Paris.

UNESCO (1975). _UNISIST Guidelines on the Planning of National Scientific and Technical Information Systems_. UNESCO, Paris.

U.S. Department of State (1979). _The Information Resources and Services of the United States: An Introduction for Developing Countries_. U.S. Department of State, Washington, D.C.

II. DATABASE SYSTEMS

THE IMPACT OF AUXILIARY INFORMATION AND UPDATE OPERATIONS ON DATABASE COMPUTER ARCHITECTURE*

M.J. Menon, David K. Hsiao

Dept. of Computer and Information Science
The Ohio State University
Columbus, Ohio

ABSTRACT

In this paper, we classify various database machine organizations into eight different catagories on the basis of whether they support a hardware directory to reduce the search space and whether they are SISD, MISD, SIMD or MIMD machines. It is felt that database machines that fall into the four directory-less categories will prove to be either cost-prohibitive or to have I/O bottlenecks. We also show that among machines with directories, MIMD machines may prove to be no better than SIMD machines owing to a problem known as update blocking which is unique to database machines with directories. The conclusion is that SIMD database machines with directories may provide the most promise for the future.

KEYWORDS

Database computers; hardware directories; updates; update blocking; SISD, MISD, SIMD and MIMD architectures.

INTRODUCTION

Advances in technology and database research have prompted considerable attention to the design and implementation of database machines (Baum, 1976; Hsiao, 1977, 1979; Lowenthal, 1977). Designs of database machines have been proposed and prototype constructions of database machines are already underway (Banerjee, 1979b; Copeland, 1973; Lin, 1976; Ozkarahan, 1977). In comparison with software database management systems, these machines will perform the basic database management functions with improved reliability and performance. However, there is still considerable disagreement among the various researchers as to which one of the many suggested designs is the most cost-and-storage-effective and provides the greatest throughput improvement over the software systems.

In this paper, we make a case for DBC-like (Banerjee, 1979b) machines with single-

*The work reported herein is supported by the Office of Naval Research under contract N00014-75-C-0573.

instruction-stream-and-multiple-data-stream (SIMD) architecture. We begin by classifying the existing designs of database machines. We then describe the DBC architecture and capabilities, and indicate why it is more viable than alternative designs. Finally, we show that the attempt to change the DBC architecture to a multiple-instruction-stream-and-multiple-data-stream (MIMD) architecture may not be worthwhile. In this regard, we identify the problem associated with the MIMD architecture as the <u>update blocking problem</u>.

CLASSIFICATION OF DATABASE MACHINE ORGANIZATIONS

The first classification we shall adopt for database machines is according to whether or not they use hardware directories for search space reduction. DBC (Banerjee, 1979b) is a machine with a directory. On the other hand, RAP (Ozkarahan, 1977), CASSM (Copeland, 1973) and DIRECT (DeWitt, 1978) are directory-less machines. Within this broad classification, database machines may be further classified by whether they are of SISD, MISD, SIMD or MIMD organizations.

CASSM and RAP both use the cellular logic approach consisting of an array of identical cells and a controller. Each cell consists of a circular memory element (e.g., a disk track of a fixed-head disk or its electronic replacement such as CCD modules) and a processing element. Data in the memory elements are searched simultaneously by the logic of the cells. No auxiliary information about the data is used for the search. Thus, both of them are directory-less SIMD machines.

In DBC, the database store can content-address an entire cylinder. This is accomplished by accessing all the tracks of a cylinder of the moving-head disk in one disk revolution, once the access arm of the disk has been positioned on the proper cylinder. Furthermore, enough processing elements are available so that the incoming data streams from all track surfaces can be processed in the same revolution against the same query. To minimize unnecessary cylinder accesses, the search space must be reduced to as few cylinders as necessary to satisfy the query. The auxiliary information, i.e., indices, used to reduce the search space is stored in a directory called the structure memory. Thus, DBC is a SIMD machine with a directory.

In the following space, we show our classification of CASSM, RAP, DBC and other

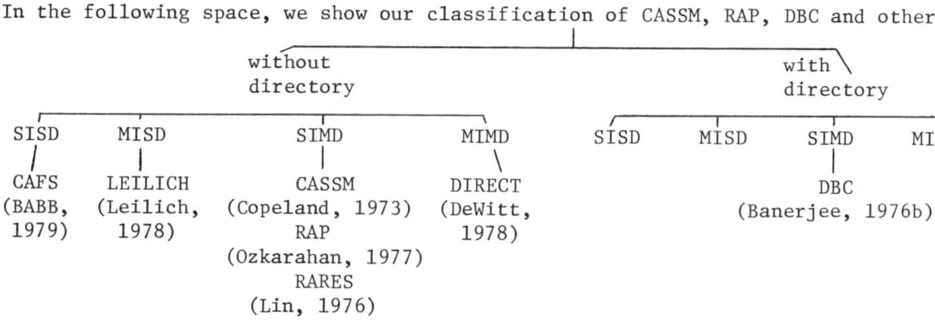

database machines. It is seen that DIRECT is a directory-less MIMD machine and that the database machine prototyped at Braunschweig is a directory-less MISD machine.

PROBLEMS WITH DIRECTORY-LESS DATABASE MACHINES

In the absence of indices, the directory-less database machines must employ one of the two possible approaches to content-addressing of the database. The first approach requires that the entire database be content-addressed. For a truly large database, these machines do not seem to be cost-effective. For example, consider the cost of a billion-byte CASSM-like machine. Assuming each CASSM cell has a capacity of 24K bytes, the machine would require 40,000 such cells. The cost per bit of head-per-track devices (or its electronic equivalent) is about 0.05 cents. Thus, the storage cost is about 5 million dollars. In addition, let us assume that the cost of the logic associated with each cell is about $180 (Kannan, 1978). The logic cost will then be 7.2 million dollars. Therefore, the total cost of a billion-byte system will be 12.2 million dollars. A 10 billion byte CASSM-like machine will cost 122 million dollars.

The second approach involves the use of conventional secondary storage such as disks and mass storage. This can be found in DIRECT. Thus, these machines serve as staging devices which load the database from conventional mass memories for database management and processing. However, staging requires high-speed buffer memories. Furthermore, high-bandwidth I/O buses must be available for moving large chunks of data from the conventional moving-head disks to the stager for processing; otherwise, the stager would be idle most of the time. Unless high-volume data are quickly moved to the stager, no useful work can be accomplished by the database machine.

Thus, we see that both approaches employed in the case of directory-less machines are not satisfactory. The former technique is cost prohibitive and the latter has loading/unloading, i.e., I/O, problems.

This section has indicated to us that directory-less machines may not be viable. Thus, we feel that approaches to machine design must concentrate on machines with directories rather than on directory-less machines. In the remaining sections, we will show that due to a problem inherent in directory machines, the throughput achievable by using a MIMD machine may be no better than that achievable by using a SIMD machine. Since we are most familiar with DBC, we will try to indicate the problem inherent in machines with directories in the context of DBC.

A SIMD MACHINE ARCHITECTURE - DBC

The smallest unit of data in DBC is a <u>keyword</u> which is an attribute-value pair. Information is stored in and retrieved from DBC in terms of records; a <u>record</u> is made up of a collection of keywords and a record body. An example of a record is (<Relation, EMP>,<Job, MGR>,<Dept, TOY>,<Salary,3000>). DBC recognizes several kinds of keywords: simple, security and clustering. Simple keywords are intended for search and retrieval purposes. Security keywords (Banerjee, 1979a) are used for access control. Clustering keywords are utilized for placing records, with a high probability of being retrieved together, in close proximity in the database store. A <u>keyword predicate</u> is of the form (attribute, relational operator, value). A relational operator can be one of [=, ≠, >, ≥, <, ≤]. A <u>query conjunction</u> is a conjunction of predicates. An example of a query conjunction is (Salary>5000) ∧ (Dept≠TOY) ∧ (Name = JAI). A <u>query</u> is a Boolean expression of predicates in the disjunctive normal form. Thus, a query is a disjunction of query conjunctions. Each entry in the structure memory of DBC consists of a keyword followed by a set of indices. An <u>index</u> is a pair of the form (cylinder number, security atom number) (Banerjee, 1979a). The cylinder number indicates where in the database store records containing the keyword may be found and the security atom number indicates the security privileges accorded to these records.

THE NATURE OF THE UPDATE OPERATIONS

In DBC, the format of an update command consists of three parts as shown below.

UPDATE QUERY MODIFIER

The first part specifies the command name. The second part is a query. All records that satisfy this query must be updated. Optionally, the second part may contain 'ALL' to indicate that all records in DBC are to updated. The modifier can be of one of the following three types.

Type 0: <attribute, newval=constant>
Type I: <attribute, newval=f(oldval)>
Type II: <attribute1, newval=f(attribute2, oldval2)>

As a pair, the modifier specifies an attribute, which is present in the keywords of the records that are to be modified, and a value, that this attribute will take. The new value is either a constant, a function of the old value of the attribute being modified, or a function of the old value of some other attribute in the record.

THE NOTION OF UPDATE BLOCKING

An Illustrative Example

Let us imagine a particular database with many records. One of the attributes that appears in some of these records is, say, Salary. Consider, now, that the following two requests are issued to DBC, one after the other.

UPDATE ALL <Salary, newval=oldval+25>

RETRIEVE (Salary=75)

After execution of the first command and after the records in the database store are properly updated, the structure memory must also be updated. A view of the structure memory content is shown below.

SM Entries before update

<Salary, 75>	(2,2) , (3,2)
<Salary, 50>	(1,1) , (4,1)
⋮	⋮

SM Entries after update

<Salary, 100>	(2,2) , (3,2)
<Salary, 75>	(1,1) , (4,1)
⋮	⋮

Consider what happens if the retrieve request is executed before the structure memory is finished with the update. Since the update process is not yet complete, the structure memory may still indicate that relevant records are present in cylinders 2 and 3. However, due to the earlier update command executed by the database store, only records that contain the keyword <Salary, 100> are present in cylinders 2 and 3. Thus, the retrieve request will fail even though records which contain the keyword <Salary, 75> are, nevertheless, present in cylinders 1 and 4. Thus, whenever there is an update request that modifies a directory keyword, we must block the execution of all subsequent requests that may be affected by the update request until the structure memory has been modified to reflect the changes in the database store caused by the update request. We shall define what we mean by requests that may be affected by an update request in a later section.

It is not possible to update the index terms in the structure memory before the actual execution of the command in the database store except in certain special situations as follows.

(1) The attribute to be modified is not part of a directory keyword.
(2) The attribute to be modified is part of a directory keyword, but it is not a security or a clustering attribute and the query part of the update request contains 'ALL' and the modifier of the update request is of Type 0 or Type I.
(3) The attribute A to be modified is part of a directory attribute, but it is not a clustering or a security attribute and the query part of the update request consists of a single predicate using the attribute A and the modifier is of Type 0 or Type I.

Clashing and Blocking

We shall refer to two queries such that the set of records that satisfy one query overlaps with the set of records that satisfy the other query as <u>clashing queries</u>. If two queries are clashing, all the attributes in either query are called <u>clashing attributes</u>. We observe that two queries are nonclashing if at least one of the attributes in either query is nonclashing. If two queries cannot be determined to be nonclashing, we say that the queries <u>may be clashing</u>. We are now in a position to give a definition of what we mean by 'may be affected by update'. A request issued subsequent to the update, i.e., [UPDATE QUERY1 <ATTRIBUTE1, NEWVAL=CONSTANT>], [UPDATE QUERY1 <ATTRIBUTE1, NEWVAL1=f(OLDVAL1)>] or [UPDATE QUERY1 <ATTRIBUTE1, NEWVAL1=f(<ATTRIBUTE2, OLDVAL2>)>], where ATTRIBUTE1 is not a security or clustering attribute, <u>may be affected</u> by the update if it is a retrieve, delete, or update request and the query used in the request contains ATTRIBUTE1 and may be clashing with QUERY1 on all attributes in the two queries other than ATTRIBUTE1. The case where ATTRIBUTE1 is a security or clustering attribute is not considered here (see (Hsiao, 1980)).

Requests Being Perpetually Blocked by Updates

Consider that the following three requests are issued one after the other.

(1) UPDATE ((Salary>50) ∧ (Name=MENON)) <Salary, newval=oldval+25>
(2) RETRIEVE ((Salary>75) ∧ (Dept=20)
(3) UPDATE (Name=MENON) <Dept, newval=oldval+20>

The retrieve request is blocked by the earlier update request and must wait until the update completes. However, the second update request is not blocked by the first update request and, hence, it is immediately scheduled for execution. Therefore, by the time the first update request completes, the second update request is at some stage of execution. It is seen that the second update request also blocks the retrieve request. Thus, the retrieve request continues to be blocked until the second update request is also completed. The above argument can be extended to show that the retrieve request may be perpetually blocked if updates are allowed to proceed simultaneously.

It is now clear that not only are MIMD machines with directories unable to utilize the multiple-query-execution capability fully that they are designed for, but also may cause the perpetual blocking of some requests with the use of such capability.

A CASE AGAINST MIMD MACHINES WITH DIRECTORIES

Consider what happens when we try to modify the DBC architecture to make it MIMD.

Lengths of Order Queues

First of all, more storage is needed to maintain the queues of orders. In the MIMD design, one queue of orders will have to be maintained for every track (as opposed to every cylinder for SIMD) in the database store for which one or more orders are awaiting execution. Let

C = total number of cylinders in the database store;
TPC = number of tracks per cylinder;
T = number of tracks that will have to be accessed by a typical order;
CO = number of cylinders that will have to be accessed by a typical order;
QSIMD = total size of order queues in SIMD case;
QMIMD = total size of order queues in MIMD case.

If the total number of orders awaiting execution is the same in the two cases, QMIMD = T/CO * QSIMD. The ratio QMIMD/QSIMD has been tabulated in Table 1 for various typical values of T and CO. It is noticed that the order queues are larger in the MIMD case by a factor of between 1.67 and 20.

Table 1. Order Queue Ratio QMIMD/QSIMD

T \ CO	1	2	3
5	5	2.5	1.67
10	10	5	3.33
15	15	7.5	5
20	20	10	6.67
25	*	12.5	8.33
30	*	15	10
35	*	17.5	11.67
40	*	20	13.33
45	*	*	15
50	*	*	16.67
55	*	*	18.33
60	*	*	20

Sizes of Directories

The second change in requirements in the case of an MIMD architecture is in the size of the directory. A MIMD machine will need a directory with entries containing track numbers rather than cylinder numbers. Let

k = number of bits needed to represent a keyword;
M = maximum number of cylinders that will be occupied in any file;
S = maximum number of security atoms (Menon, 1980) that will be needed for any file;
P = average number of index terms per keyword in the SIMD case;
q = average number of index terms per keyword in the MIMD case;
TPC = number of tracks per cylinder;
DSIMD = total directory size in the SIMD case;
DMIMD = total directory size in the MIMD case.

In the SIMD case, log M + log S bits will suffice to represent an index term

(Hsiao, 1976a). In the MIMD case, each index term requires log (M * TPC) + log S bits. Thus, assuming that we have the same number of keywords in the directories in the two cases,

$$DMIMD/DSIMD = (k + q*(\log(M*TPC) + \log S))/(k + p*(\log M + \log S))$$

Using K=40, M=256, TPC=16, S=1024,

$$DMIMD/DSIMD = (40 + 22q)/(40 + 18p)$$

The value of the ratio DMIMD/DSIMD is tabulated, in Table 2, for various typical values of p and q. It is seen from there that the MIMD machine has much larger directory requirements than an SIMD machine.

Table 2. Directory Size Ratio DMIMD/DSIMD

q \ p	5	10	15	20
4p	3.692	4.182	4.387	4.5
8p	7.077	8.182	8.645	8.9
12p	10.462	12.182	12.903	13.3
16p	13.846	16.182	17.161	17.7

q is represented as multiples of p

Processing Complexity

Let us now consider the processing requirements in the two cases. In the present design of DBC, a controller is present in order to issue the orders to the different processing elements. When the processing elements finish executing one order, the controller will broadcast the next order to be executed to all the processing elements. Thus, relatively simple communication is needed between the controller and the processing elements. In the MIMD case, however, the controller must keep track of the progress of orders on each of the processing elements separately. Also, since the controller must issue instructions to each of the processing elements individually, it must issue many more instructions than in the SIMD case. All this requires a more sophisticated controller than that required for the SIMD case.

Along with the above disadvantages there is the further problem of blocking, thereby disallowing the full use of the multiple-query-execution capability inherent in an MIMD machine and causing some requests to be perpetually blocked. Thus, it is our analysis that SIMD machines with directories are the best possible solution towards future database computer architecture.

CONCLUSIONS

In this paper, we have considered various database machine organizations and have classified them into eight different categories; first on the basis of whether they support a hardware directory to reduce search space and then on the basis of whether they are SISD, MISD, SIMD, or MIMD machines. It is felt that database machines that fall into the four directory-less categories will prove either to be cost-prohibitive or to have I/O bottlenecks. Additional analysis shows that MIMD machines with directories may prove to be no better than SIMD machines with directories owing to a problem known as update blocking which is unique to database machines with directories. The conclusion is, therefore, that SIMD machines with directories provide the most promising architecture for future database computers.

REFERENCES

Babb, E. (1979). Implementing a relational database by means of specialized hardware. ACM Trans. on Database Systems, 4, 1, 1-29.

Banerjee, J., D.K. Hsiao, and M.J. Menon (1979a). The clustering and security mechanisms of a database computer (DBC). OSU Technical Report, 79-2.

Banerjee, J., D.K. Hsiao, and K. Kannan (1979b). DBC - A database computer for very large databases. IEEE Trans. on Computers, 28, 6, 414-429.

Baum, R.I., and D.K. Hsiao (1976). Database computers - A step towards data utilities. IEEE Trans. on Computers, 25, 12, 1254-1259.

Copeland, G.P., G.J. Lipovsky, and S.Y.W. Su (1973). The architecture of CASSM - A cellular system for non-numeric processing. Proceedings of the First Annual Symposium on Computer Architecture, 212-228.

DeWitt, D.J. (1978). DIRECT - A multiprocessor organization for supporting relational data base management systems. Proceedings of the Fifth Annual Symposium on Computer Architecture, California, 182-189.

Hsiao, D.K., and S.E. Madnick (1977). Database machine architecture in the context of information technology evolution. Proceedings of the Third International Conference on Very Large Databases, New York, 63-84.

Hsiao, D.K. (1979). Database computers are coming! Database computers are coming. Computer, 12, 3, 7-11.

Hsiao, D.K., and M.J. Menon (1980). Design and analysis of update mechanisms of a database computer (DBC). OSU Technical Report, 80-3.

Kannan, K. (1978). The design of a mass memory for a database computer. Proceedings of the Fifth Annual Symposium on Computer Architecture, California, 44-51.

Leilich, H.O., G. Steige, and H. Ch. Zeidler (1978). A search processor for database management systems. Proceedings of the Fourth International Conference on Very Large Databases, 280-287.

Lin, C.S., D.C.P. Smith, and J. Smith (1976). The design of a rotating associative array memory for a relational database management application. ACM Trans. on Database Systems, 1, 1, 53-65.

Lowenthal, E.I. (1977). A survey - The application of database management computers in distributed systems. Proceedings of the Third International Conference on Very Large Databases, New York, 85-92.

Menon, M.J., and D.K. Hsiao (1980). The access control mechanism of a database computer (DBC). Fifth Workshop on Computer Architecture for Non-Numeric Processing, California, 17-28.

Ozkarahan, E.A., S.A. Schuster and K.C. Sevcik, (1977). Performance evaluation of a relational associative processor. ACM Trans. on Database Systems, 2, 2, 175-195.

AN INTRODUCTION TO SABRE, A MULTI PROCESSOR DATA BASE COMPUTER

Georges Gardarin
INRIA-SIRIUS and INSTITUT DE PROGRAMMATION
BP 105, 78150 LE CHESNAY
FRANCE

ABSTRACT

The objectives and basic concepts of the SABRE multi-processor data base computer are presented in this paper. SABRE is a functionally complete data base machine whose main objective is efficiency for managing relational data bases. This objective is achieved by using parallelism, data filtering and multi-dimensional clustering. This paper also contains a description of the SABRE organization in terms of processors and interconnection network.

KEYWORDS

Database machine; multi-processor; filtering; clustering; relational database; join; restriction; projection; query decomposition.

INTRODUCTION

The development of (distributed) data base systems on general purpose computers is limited by several factors : disk I/O bottleneck, low degree of real parallelism, system overhead, difficulties to achieve data integrity, lack of data security. Moreover, it appears to be almost impossible to integrate existing heterogeneous systems into general purpose distributed systems. The SABRE project intends to develop a multi-processor data base machine which manipulates relational data bases by the way of a high level assertional data manipulation protocol. In addition, SABRE data base machines and processing computers can be interconnected through a network to constitute a distributed data base system.

First, this paper focus on the objectives of SABRE which are the following : facilitate distributed data base management, supply to the host systems a high level data manipulation language, utilize parallelism for improving response time, minimize I/O overhead, accelerate search evaluation by the way of special access paths, maintain data integrity and security.

Then, the basic concepts sustaining the project are introduced : associative partition, filtering on the fly, multi-dimensional clustering, parallel sort and join, query decomposition, defered update, data manipulation protocol.

Finally, the architecture of SABRE is portrayed. The SABRE data base machine is in fact characterized by two types of architectures. The functional architecture describes the functions performed and their imbrications in terms of levels, layers, interfaces and protocols. The operational architecture defines the same machine in terms of processors and interconnection network. The functional architecture is mapped into the operational one by assigning processors to each function and by selecting a type of interconnection network. In SABRE, a certain degree of independence between the functional and operational architecture is achieved by a distributed execution kernel which controls the cooperation of processors and the correct execution of functions.

Throughout the paper, some of the problems and original solutions proposed in SABRE will be briefly suggested : automaton based filtering techniques, parallel join algorithms, parallel query decomposition strategies, multi-dimensional clustering. Some more thorough descriptions can be found in (GARD80a),(FAUD80),(KARL80),(VALD80),(GARD80b).

SABRE OBJECTIVES

Distributed Data Base Management

The SABRE machine is first a tool for facilitating distributed data base management. SABRE is thought to be included in a distributed data base management system as SIRIUS DELTA (LEBI80). Such a system can be built with two types of sites : distributed data base computers and transaction processing computers (GARD79a). SABRE can be seen as a distributed data base computer where the distributed functions are optional. The specifications of these functions of course depend on the host distributed system. However, in general they should at least include replicated data maintenance (CHU 80) and distributed query processing (BERN80) facilities.

Assertional Relational Data Base Manipulation

To assure a high level of data independence and a good optimization of data accesses, SABRE offers to the users commands with which they define sets of data to be selected without specifying the method to select them. Therefore, the external interface of SABRE is derived from an assertional data base manipulation language for querying and updating relational data bases (STON76). A data base is seen by the host computer network as a set of permanent relations. By combining domains of these permanent relations, it is possible to define temporary relations. They are created by a transaction and destroyed at transaction end or latter if requested.

A transaction can then select data from permanent relations into temporary relations. A selection expression is in its simplest form a logical combination of predicates of the form (V1.D1 o V2.D2) where V1 and V2 are variables representing relations, D1 and D2 are domains and o is an operator chosen among $<$, $\leq$, $=$, $\neq$, $>$, $\geq$. A transaction can also read, insert and update tuples from/into a temporary relation. At transaction end (or latter) updated temporary relations are reintegrated into the permanent relations if the updating transaction has been correctly executed. This allows isolation of transaction errors from the permanent data base.

Finally, to simplify the data base machine interface, SABRE does not supply a specific data description interface. Data description is performed by inserting and modifying tuples into a meta data base which contains the data base schema. Also, the interface is extended to allow text processing oriented commands. A first draft of this interface will be found in (GARD80b).

Parallelism

SABRE is a data base machine composed of a set of micro-processors. The functions of the data base machine are distributed among the various processors. Each processor possesses a private memory and realizes a specific function (HSIA78). However a same function can be performed by several processors which then share the load due to that function and which can replace each other in case of failure of one of them. This multi-processor aspect allows two types of parallelism : intra-request parallelism which introduces parallelism to perform a unique data base manipulation and inter-request parallelism which consists in carrying out simultaneously several requests from different users.

I/O Time Reduction

The first bottleneck of classical data base management systems is the I/O time. To reduce I/O overhead, SABRE includes two facilities :
(1) A large memory (some millions of bytes) is utilized as a cache where intermediate and final results of data manipulation can be kept. This memory is shared by different processors and can be extended on disks by a special memory management processor.
(2) Data are filtered in parallel to disk transfers, that is selections are performed on the fly by special processors associated with each disk units or more generally disk loops (BANC80,FAUD80). Therefore, only data satisfying at least a part of the user query are transfered into the cache memory. Also, selections are performed simultaneously to disk I/O, thus reducing the overhead.

Search Acceleration

Also in order to limit the number of I/O necessary to scan a large file, we suggest the development of a search accelerator. The search accelerator manages access paths to tracks. Three types of access paths are proposed :
(1) Multi-dimensional search trees which permit multi-attribute clustering.
(2) Classical hashing function and index tables.
(3) Link tables between different relations which memorizes pre-evaluated joins.
The access path memorization data (tables,trees) are separated from the data base and stored in access path files. That simplifies filtering and renders access path optional.

Data Base Integrity

A data base satisfies a set of assertions called integrity constraints. Integrity constraints are threatened by three factors (GARD80c) :
(1)Erroneous user updates which can introduce inconsistent data in the data base.
(2) Concurrency conflicts which can induce improper sequences of operations leading to inconsistencies.
(3) Hardware and software failures which can prevent the completion of certain transactions or worst destroy a part of a data base.

SABRE intends to supply devices to control these three factors. First, incorrect updates should be trapped when introduced or at transaction commit. This should be done by verifying a predefined set of integrity constraints associated to each data base (GARD79b). Second, to avoid inconsistencies induced by concurrency conflicts, the simultaneous execution of transactions is restricted and only equivalent to serial schedules are permitted (GARD77). Two concurrency control methods are offered and can be selected by the data base administrator : time stamp ordering (BERN80) and locking (ESWA76). Finally, failures are controlled by two ways :

replication of harware and transaction atomicity. On one hand, at least two processors are able to perform a given function. Thus, when one fails, the distributed execution kernel asks the other to carry out the crashed task. On the other hand, transaction atomicity allows only commitment of correctly executed units of updates thus guaranteeing data base consistency.

Data Security

SABRE protects data against unauthorized or malicious accesses. Two devices classically realize such protections : a view definition facility and an operation controller. Views are subset definitions of a data base which can be restructured as usual (STON76). Authorizations are given to authentificated transactions and can be transmitted from a transaction to another (GRIF76).

SABRE BASIC CONCEPTS

Associative Partition

Data are stored in associative partitions. An associative partition is a content addressable memory where data are sequentially stored and associatively retrieved by specifying a multi-attribute criteria. The current technology of SABRE entails that a partition is assimilated to a disk track. Inside the partition, the notion of address is inexistant. Thus, as noticed in (HSIA78), inside a partition data can be moved about without any need to modify index tables, trees and links.

The unit of partition content addressable by associativity is the attribute. An attribute can be seen as a variable length data item. Attributes are grouped into tuples. The unit of insertion in a partition is a tuple. Tuples are of variable length. When a partition is full and that a new tuple has to be inserted inside, the partition is split in two partitions according to the values of a given tuple attribute.

Filtering

Filtering is the process which realizes associativity. More precisely, when a track is read on a disk, the data flow is filtered by a processor associated to the disk unit (or loop if several track can be read simultaneously on the same disk unit). That means that only attributes satisfying the given search criteria are selected. Finally, filtering can be characterized as the basic method which makes a disk track looks like an associative partition.

Clustering

To reduce the period of time necessary to elaborate frequent query answers, we utilize a multi-attribute clustering. Clustering consists in grouping tuples having similar characteristics in tracks which can be accessed together in a minimum delay. In SABRE, for any clustered relation R, an ordered set of attributes (A1,A2,...An) is defined by the data base administrator. Also, to each attribute Ai the data base administrator associates a maximum split number Si. When tuples are inserted in R, first a partition is allocated to R. When this partition is full, it is split in two partitions according to the median value of A1 in the original partition. When a new partition is full, it is also split in two partitions. However, when a partition has identical values for the clustering attribute or when the number of splits performed in the range of an attribute Ai reaches the maximum number given by the data base administrator Si, the next clustering attribute Ai+1 is utilized to split the partition. Finally, splitting partitions at different levels according to different attributes leads to the construction of a multi-dimensional search tree which is very helpfull for accelerating data retrieval.

Parallel Sorting

Sorting is a very important operation in data base management systems. Sorting algorithms (KNUT73) perform external sorting on disks and internal sorting in main memory. In SABRE, several disk channels and several specialized processors make possible to execute in parallel both external and internal sorting. For external sorting, we propose to utilize a distributive sorting algorithm which successively divides classes in sub-classes according to the value of the sorting attribute(s). The equilibrium between classes is kept by a special function included in the filtering process (FAUD80).

Parallel Join

Join is a very important operation in relational data bases. If R and S are two relations, the equi-join of R and S on attributes A of R and B of S noted R(A=B)S is the relation obtained by combining each tuple of R with each tuple of S and by keeping only tuples satisfying R.A = S.B . Three algorithms are proposed in SABRE to perform parallel join (VALD80) : filtering in parallel the cartesian product of R and S, sorting in parallel R on A and S on B and then merging the two results, hashing R on A and accessing the hashed R for each tuple of S in parallel. The last algorithm is optimized with bit arrays (VALD80). Each algorithm has a domain of application which is determined by the size of R and S and the size of the join.

Query Decomposition

Whenever SABRE received a query which is expressed in a QUEL like language (STON76) a process generates a tree of relational algebra operations which has to be executed to answer the query. This process is called query decomposition. Query decomposition optimizes the query cost, that is mainly the I/O execution time. For that, it tries to maximize the utilization of the cache memory and also parallelism.

Defered Update

SABRE does not enable direct updating of a data base. Updating requires the definition of a temporary relation which can be updated. Updated temporary relation can be integrated to the data base at transaction commitment, thus allowing data base modifications. Consequently, updates are defered to transaction ends where they are performed under the direction of a two step commitment protocol (BAER80).

Data Manipulation Protocol

Finally, SABRE presents an external interface which can be considered as an application layer protocol of the ISO distributed system architecture (ISO 79). This protocol specifies all the interactions between SABRE and the host computers and is called a data manipulation protocol. A first draft of this protocol can be found in (GARD80b).

SABRE ARCHITECTURE

Functional Architecture

The functional architecture describes SABRE in terms of layers and interfaces between layers. This architecture is composed of five layers which are represented figure 1. Each layer manipulates data objects simpler and simpler. The five layers are the following :
(1) Distributed Data Bases. This layer is optional. It performs the integration of SABRE in the distributed data base system.
(2) Local Views. This layer is also optional. It performs the definition and manipulation of external schemas and the security controls. It translates an external request manipulating virtual relations defined in an external schema into one or several internal requests manipulating relations really implemented on disks. It also allocates and controls transaction access rights.
(3) Local Data Bases. This basic layer supplies the necessary tools to manage a data base local schema and performs query decomposition.
(4) Relations. This layer manages the access paths and transforms a relation name in a set of partition addresses. That is where the multi-attribute clustering is performed and where the multi-dimensional search tree is constructed and utilized.
(5) Partitions. This layer performs filtering in parallel on several partitions and also includes the basic tools for parallel joins and sortings of partitions. We also include here the management of the cache memory.

A layer is characterized by its external interface and the external interface of the inferior layer. The external interface of layer (1) is the data manipulation protocol. The external interfaces of layer (2) and (3) are a data manipulation object language directly deduced from the data manipulation protocol and which is not far from a compiled version of QUEL (STON76). The external interface of layer (4) is the relational algebra. Finally, interface of layer (5) is also the relational algebra, but at the level of a set of partitions.

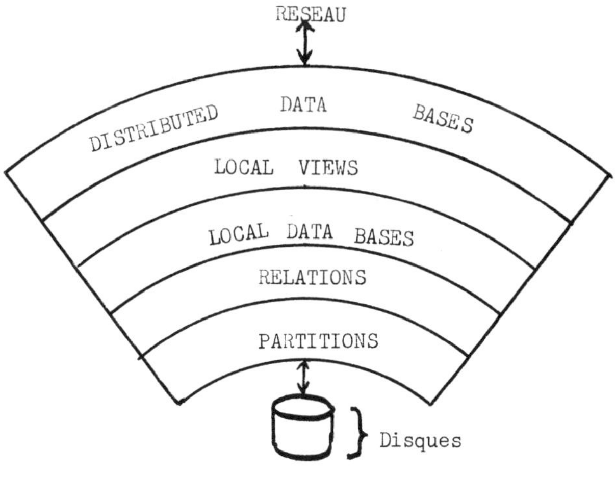

Fig. 1. The functional architecture

Operational Architecture

The operational architecture is composed of processors interconnected by an interconnection network. In the following, we first survey each of the proposed processor types which are included in SABRE.

Filter. A filter processor (FIP) is associated to each disk loop, that is to each flow of data which comes from a disk. The role of a filter is to perform the selection of relevant attributes of tuples on the fly during the reading of a disk track. We propose a new class of filters derived from the general pattern of finite state automata and permitting the evaluation of general expressions (FAUD80). Filters represents an implementation of a part of the partition layer.

Joiner. Joiners (JSP - Join and Sort Processor) perform joins and sortings. Several joiners generally works in cooperation to carry out efficiently joins and sortings. The algorithms introduced above (VALD80) are implemented in each joiner. Joiners represents an implementation of another part of the partition layer.

Memorizer. The memorizer manages the cache memory. It allocates pages to store the filtered data and the results of joins. It also extends the memory on disk when saturated. The memorizer represents an implementation of another part of the partition layer. The abbreviation which designates the memorizer is MMP (Memory Manager Processor).

Logger. The logger is a special processor which is added to improve data reliability. It records partition updates into logs, receives checkpoints and performs restarts and recovers. It is designated by the abbreviations RLP (Reliable Log Processor).

Accelerator. An accelerator (RAP - Relation Access Processor) manages access paths and is a complete implementation of the functions associated to the relation layer. Therefore, an accelerator performs the multi-attribute clustering and manages the multi-dimensional search tree (KARL80).

Decomposer. A decomposer (QDP - Query Decomposition Processor) performs the query decomposition. It is a complete implementation of the local data base layer.

Protector. A protector (SVP - Security and View Processor) manages security and views. It corresponds to an implementation of the local view layer.

Transport Station. A transport station (TSP - Transport Station Processor) performs the basic function of the distributed data base layer, that is emission and reception of messages. For the time being, that is the only processor specified at this layer.

Interconnection Network

We have now defined the different processors composing the operational architecture. This architecture also includes an interconnection network. Different kinds of interconnection network can be considered: bus, star, multi-loops, cross-bar (HSIA78,DEWI79). In SABRE, for the time being, we assume the simplest interconnection network which is a bus. Therefore, the operational architecture of SABRE is finally represented figure 2. The common bus could appear as a bottleneck : that is only a question of speed. Moreover, the current development of SABRE are independent of this assumption as we are going to see in the next paragraph.

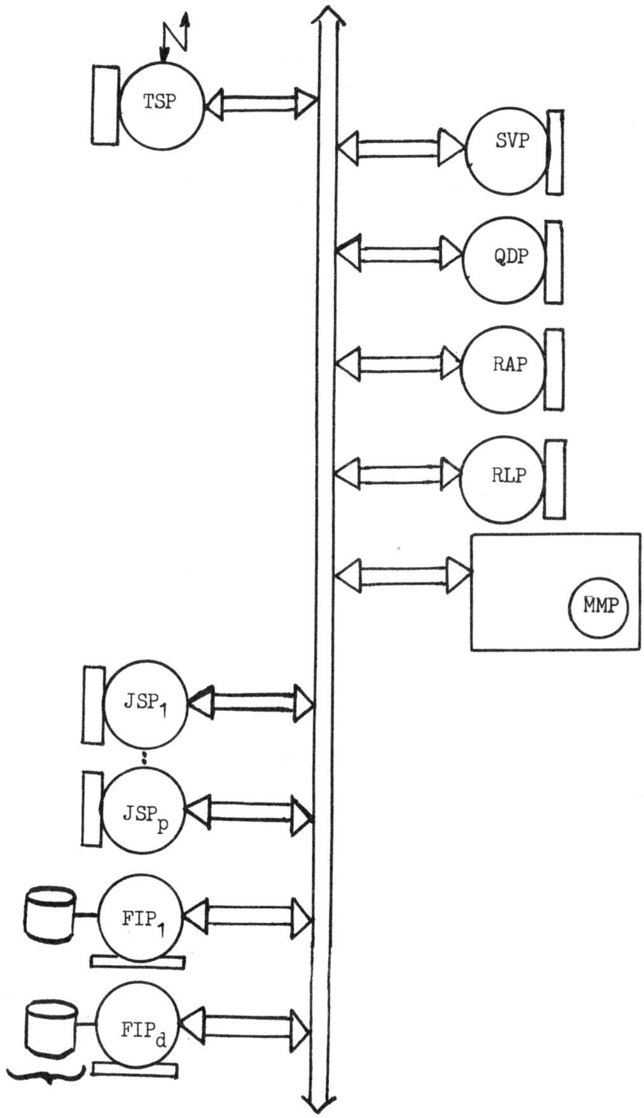

Fig. 2. The operational architecture

Execution Kernel

The last part of the operational architecture is the execution kernel. It is a distributed operating system which controls the distributed execution of a request. When a request is submitted to SABRE, it is first given to the decomposer which produces an execution plan composed of a sequence of actions and synchronization expressions. The execution kernel then evaluates the synchronization expressions and initiates actions whose synchronization expressions are true. Each initiated action is executed by one (or several if parallelism is necessary) specialized processors representing the associated function. When the action is executed, the specialized processor which represents the function returns to the execution kernel the result parameters and a set of synchronization variables to set. New actions can then be activated. Thus, a request is finally processed.

CONCLUSION

In this paper, we have presented the objectives, basic concepts and the architecture of SABRE, a multi micro-processor data base computer. The SABRE design allows for a number of specialized micro-processors to cooperate in order to perform efficiently data base management functions. This project explores several new concepts as parallelism, filtering, multi-dimensional clustering in a relational data base context. We are also currently emphasizing the fault tolerance aspects of the project.

On the implementation point of view, we are currently developing an emulation of the SABRE data base computer. Each processor is programmed in PASCAL. The execution kernel is programmed in extended PASCAL in order to enable parallelism. The emulation should work at the beginning of 1982. Then, we will envision to bring SABRE on a real multi micro-processor system.

ACKNOWLEDGMENTS

Ten persons are currently working in the SABRE project : Z. BELLASHENE, P. BERNADAT, P. FAUDEMAY, G. GARDARIN, K. KARLSSON, G. KUBRYCK, N. TEMMERMAN, M. THONNET, P. VALDURIEZ, Y. VIEMONT. R. GOMEZ and S. SEDILLOT are also consultants for the project.

REFERENCES

(BAER80) J.L. BAER, G. GARDARIN, C. GIRAULT, G. ROUCAIROL
The Two-step Commitment Protocol: Modeling, Specification and Proof Methodology. 5th International Conference on Software Engineering, San Diego, 1981

(BANC80) F. BANCILHON, M. SCHOLL
Design of a Backend Processor for a Database Machine. ACM-SIGMOD International Conference on Management of Data, Los Angeles, 1980

(BERN80) P.A. BERNSTEIN, N. GOODMAN
Timestamp-based Algorithm for Concurrency Control in Distributed Database Systems. 6th International Conference on Very Large Data Bases, Montreal, 1980, Proc. IEEE, 285-300

(CHU 80) G. GARDARIN, W. CHU
A distributed Control Algorithm for Reliably and Consistently Updating Replicated Databases. To appear in IEEE Transactions on Computers, Dec. 1980

(DEWI79) D. J. DEWITT
A Multiprocessor Organization for Supporting Relational Database Management Systems. IEEE Transactions on Computers, Vol 28/N6, June 1979, 395-406

(ESWA76) K. P. ESWARAN, J. N. GRAY, R. A. LORIE, I. L. TRAIGER
The Notions of Consistency and Predicate Locks in a Database System. Commun. ACM, Vol 19/N11, Nov. 1976, 624-633

(FAUD80) P. FAUDEMAY
Une Nouvelle Classe de Filtres Multi-expressions et Multi-fonctions. Journees Machines Bases de Donnees, Sophia-Antipolis, France 1980

(GARD77) G. GARDARIN, P. LEBEUX
Scheduling Algorithms for Avoiding Inconsistency in Large Databases. 3rd International Conference on Very Large Data Bases, Tokyo, 1977, Proc. IEEE, 501-506

(GARD79a) G. GARDARIN
A Unified Architecture for Data and Message Management. National Computer Conference, New-York, 1979, AFIPS Vol 48, 681-688

(GARD79b) G. GARDARIN, M. MELKANOFF
Proving Consistency of Database Transactions. 5th International Conference on Very Large Data Bases, Rio, 1979, Proc. IEEE, 291-298

(GARD80a) G. GARDARIN et al.
Objectifs,Principes et Architecture d'une Machine Bases de Donnees Reparties. Journees Machines Bases de Donnees, Sophia-Antipolis, France 1980

(GARD80b) G. GARDARIN, J. BEGUE
Un Protocole de Manipulation de Donnees Relationnel. Congres AFCET 1980, NANCY, France

(GARD80c) G. GARDARIN
Integrity,Consistency,Concurrency,Reliability in Distributed Database Management Systems. Distributed Data Bases,North-Holland 1980, 335-351

(GRIF76) P.P. GRIFFITHS, B. W. WADE
An Authorization Mechanism for a Relational Database System. ACM Trans. on Database Systems, Vol 1/N 3, 242-255, Sept. 1976

(HSIA78) J. BANERJEE, D. K. HSIAO, R. I. BAUM
Concepts and Capabilities of a Database Computer. ACM Trans. on Database Systems, Vol 3/n 4, 347-384, Dec. 1978

(ISO 79) ISO/TC97/SC16 N227
Reference Model of Open System Interconnection. June 1979

(KARL80) K. KARLSSON
Le Modele d'Acces de la Machine Bases de Donnees SABRE
Journees Machines Bases de Donnees, Sophia-Antipolis, France 1980

(KNUT73) D. KNUTH
Sorting and Searching- The Art of Computer Programming
Vol 3, 1973

(LEBI80) J. LE BIHAN et al.
SIRIUS : A French Nationwide Project on Distributed Data Bases.
6th International Conference on Very Large Data Bases, Montreal, 1980, Proc. IEEE

(STON76) M. STONEBRAKER et al.
The Design and Implementation of INGRES.
ACM Trans. on Database Systems, Vol 1/N 3, Sept. 1976, 189-222

(VALD80) G. GARDARIN, P. VALDURIEZ
Algorithmes Multiprocesseurs de Jointures de Relations. Journees Machines Bases de Donnees, Sophia-Antipolis, France 1980.

ON THE LONG-TERM IMPLICATIONS OF DATABASE MACHINE RESEARCH

David Elliot Shaw
Department of Computer Science
Columbia University
New York, New York 10027

ABSTRACT

It is argued that the long-term import of current research on specialized hardware for database management may extend far beyond what is now considered the immediate province of database management systems. In particular, the potential support which "database machines" may provide for very high level mechanisms for the description of information processing tasks is considered and exemplified with an operational system for knowledge-based information retrieval.

KEYWORDS

Database machines; mathematical logic; parallel architectures; VLSI systems; program description; artificial intelligence.

INTRODUCTION

During the past few years, research involving the design of specialized hardware adapted to the requirements of large-scale database management has emerged as one of the most rapidly growing research areas in the field of computer science. It seems clear that the very considerable attention which this research has recently begun to attract within the academic, industrial and government communities stems largely from the enormous and rapidly increasing reliance of our society on the systematic management and manipulation of large collections of data. Indeed, the fact that such machines may, in the near or intermediate future, lead to much higher cost-to-performance ratios in conventional database management applications is of great interest in itself. It is our suspicion, however, that the long-term implications of certain recent advances in this area may extend far beyond the scope of contemporary database management systems.

This paper will examine and exemplify the potential long-term significance of certain investigations regarding database machines in the context of some very fundamental problems involving the description and execution of information processing tasks. Our conjectures regarding the potential implications of recent findings in this area derive largely from the observation of a very interesting relationship between the general and powerful descriptive mechanisms of predicate logic and the primitive operations of a relational algebra. In particular, we will examine the "high-level" import of nonstandard machine architectures which support the highly efficient evaluation of relational algebraic operators.

DESCRIBING INFORMATION PROCESSING TASKS

One of the most consistent trends in the history of computer science can be characterized as a gradual progression from procedural to declarative mechanisms for the description of information processing tasks. Feigenbaum (1974), for example, has noted that the human user is interested in specifying what the computer is to do, without regard for the details of how the desired task is to be performed. The chronological sequence beginning with machine language programming, extending through the use of assembly languages and the subsequent development of high-level language compilers, and leading to contemporary research on artificial intelligence and automatic programming, he argues, can be viewed as a systematic journey from the "how" to the "what" end of this spectrum.

Allowed to fantasize about the kind of computing tools we would some day like to have at our disposal, we might ask for the ability to communicate our information processing needs to the computer system in much the same way as a data processing manager now communicates with a human programmer, making reference to domain-specific entities and relationships in the "real world" -- such constructs as customers, inventories and accounts, for example -- in a natural, "human-like" manner. As the cost of computer hardware becomes increasingly dominated by the time and expense involved in translating "high-level" task specifications into working computer programs, the immense potential value of even modest advances in this direction becomes increasingly obvious.

As software tools begin to claim a larger share of our attention, and as the cost-to-performance ratio for digital hardware continues to drop dramatically, it is natural to ask whether there is any way that our unprecedented capabilities in the area of computer hardware might be mobilized to help meet such unprecedented challenges in the area of computer software. In this regard, it is interesting to observe that although the cost of computer hardware has decreased by a factor of many thousands during the past few decades, the fundamental architecture of the computer itself has remained essentially the same, based closely on what is generally referred to as the von Neumann model of computation. Our own work during the past few years has been based in large part on the suspicion that characteristics not intrinsic to the process of computation, but associated instead with the particular constraints of the von Neumann machine, may have made it difficult for computer scientists to conceive of tools which might well support major advances in program description. It is our feeling that the most important long-range implications of current research on non-von Neumann "database machine" architectures may involve not improvements in the efficiency of existing information processing applications, but support for very high level program description tools which might otherwise be impractical to implement at all.

A KNOWLEDGE-BASED RETRIEVAL LANGUAGE

A surprisingly large share of the kinds of information processing activities with which both human and automated data processors are charged may be viewed as involving various kinds of knowledge-based matching or retrieval problems. Although meaning-based matching may, in a COBOL-based inventory control system for airplane parts, be deeply embedded within program loops and sort routines, and hence difficult to recognize as such, the fact remains that a large proportion of the CPU cycles which will be expended in any particular run might be thought of as identifying appropriate records for manipulation on the basis of domain-specific criteria -- in our example, criteria involving, say, airplanes, motors and aircraft-specific part-whole relationships. Our recent research has focused on a concrete task of manageable scope, but which nonetheless captures many of the most important characteristics of this important aspect of program description: the knowledge-based document retrieval application.

The user of a knowledge-based document retrieval system formulates a high-level pattern description to be matched against all descriptions contained in a large target collection (the set of books in a computer science library, for example). In contrast to a conventional retrieval system, in which mechanical, "syntactic" criteria are adequate to identify all matching documents, the success of a match in a knowledge-based document retrieval system depends in general on domain-specific entities and relationships. In the case of the computer science library, for example, the system might be required to "know about" such entities as computers, algorithms, programmers and storage devices. Certain characteristic attributes of these entities (the "storage medium" attribute, whose values differ for different kinds of storage devices, for example) might also be included in this domain-specific knowledge. Among the typical kinds of relationships which might be embodied in the knowledge base of such a system is the fact that a tape drive is a particular kind of storage device whose storage medium is always magnetic tape; one simple deductive inference involving this relationship might establish the fact that a pattern description in which the subject of the document is described as involving a storage device with magnetic tape as its medium would be satisfied by a target description in which a tape drive appeared in the corresponding position.

The description language which we used in our experimental implementation of a knowledge-based retrieval system (Shaw, 1980a) is based closely on the language KRL (for Knowledge Representation Language) (Bobrow and Winograd, 1977). While space will not allow a detailed definition of this knowledge-based description language, the following sample document description, presented without explanation, may suggest some of its most important features:

```
a Document with
  authors =
    set-with-all-of Thompson, Walters
    set-of an Engineer
  countries-of-publication =
    set-with-any-of USA, Great-Britain
  subject involves
    an Invention with purpose =
      or Power-generation, Power-transmission
  copyright-dates =
    set-with-exactly 1935, 1962
a Survey-article
```

MATHEMATICAL LOGIC: THE ESSENTIAL INTERMEDIARY

In the interest of general applicability to problems other than document retrieval, the rules defining the semantics of matching within our knowledge-based description language were not embedded inextricably within the code of the retrieval system, but were instead formulated as an independent, separable set of axioms expressed in a restricted first-order predicate calculus. Twenty-two such axioms were required to describe the matching rules for our knowledge-based description language. One of these axioms (which embodies the basis on which a target description of the form "a Storage-device with medium = Magnetic-tape" would successfully match a less restrictive description of the form "a Storage-device", for example) is presented below in an attempt to convey the general flavor of the matching axioms:

```
Perspec-imp-perspec (tar-perspec, pat-perspec) ≡
  ∃ proto .
     (*Perspec-proto* (pat-perspec, proto)
      ∧ *Perspec-proto* (tar-perspec, proto))
    ∧ ∀ pair-name .
        (∃ pat-fill .
           *Obj-slot-fill (pat-perspec, pair-name  pat-fill)
         ∧ ∃ pat-fill, tar-fill .
             *Obj-slot-fill* (pat-perspec, pair-name, pat-fill)
             ∧ *Obj-slot-fill* (tar-perspec, pair-name, tar-fill)
             ∧ Dtion-imp-dtion (tar-fill, pat-fill))
```

The set of matching axioms serves not only as a modular, perspicuous, and easily specified description of the semantics of the knowledge-based description language, but as an executable "program" for actually carrying out the matching process which it describes. More precisely, the matching process is carried out by a procedure called LSEC (for Logical Satisfaction by Extensional Constraint), which interprets the set of axioms to identify all target descriptions which match the user-specified pattern description according to the matching rules for the description language, making reference to the domain-specific knowledge base. (Details of the LSEC algorithm are described in Shaw (1980a).)

THE DATABASE MACHINE AS A LOGIC ENGINE

Unfortunately, the LSEC algorithm relies very heavily on the execution of several operations which, on a von Neumann machine, are quite expensive when their operands comprise a large amount of data (as is typically the case in the sort of application with which we are concerned). Specifically, the algorithm repeatedly evaluates the most expensive operations of a relational algebra -- in particularly, the equi-join operation. Thus, while predicate calculus serves a vital role as an "intermediate form" in our approach -- acting as a very general and powerful descriptive tool, on the one hand, and (through its close mathematical relationship to the relational algebra) as a computationally effective mechanism for evaluation, on the other -- we have been forced to consider various alternatives to the von Neumann architecture in order to justify the promise of our approach in realistic applications.

The particular machine architecture we have developed (Shaw, 1979, 1980) which is based on a hierarchy of associative storage devices, evaluates the "difficult" relational algebraic operators in a highly efficient manner, particularly when the argument relations are large. Specifically, this architecture permits an asymptotic improvement of O(log n), with very favorable constant factors, over the best known evaluation methods for these operators on a conventional computer system, without the use of redundant storage, and using currently available and potentially competitive technology. Although it was the efficient implementation of large-scale knowledge-based operations which originally motivated our design, these results have recently begun to attract attention within the database management community by virtue of the important role played by the more complex relational algebraic primitives within database management systems based on the relational model of data (Codd, 1970). It is this connection with the concerns of relational database systems, together with the close relationship between our work and earlier research on specialized hardware for database management (for example, the RAP (Ozkarahan, Schuster and Smith, 1974), CASSM (Su, Copeland and Lipovski, 1975), DBC (Hsiao, 1977) and DIRECT (DeWitt, 1979) machines) which has motivated our use of the phrase "relational <u>database</u> machine".

It is our conjecture, however, that this area of research may in the long run have an impact extending far outside the immediate province of data base management.

ACKNOWLEDGEMENTS

The author is indebted to Bob Floyd, Don Knuth, Gio Wiederhold and Terry Winograd, of Stanford University, for substantial contributions to the work reported in this paper. This research was supported in part by the Defense Advanced Research Projects Agency under contract MDA903-77-C-0322 to the Knowledge Base Management Systems Group at Stanford.

REFERENCES

Bobrow, D. G., and Terry Winograd (1977). An overview of KRL-0, a knowledge representation language, <u>Cognitive Science</u>, <u>1</u>, no. 1.

Codd, E. F. (1970). A relational model of data for large shared data banks, <u>Commun. ACM</u>, <u>13</u>, no. 6.

DeWitt, David J. (1979). DIRECT--A multiprocessor organization for supporting relational database management systems, <u>IEEE Trans. Comput.</u>, <u>C-28</u>, no. 6.

Feigenbaum, Edward A. (1974). Unpublished lecture.

Hsiao, David K. (1977). The architecture of a database computer--a summary, <u>Proc. Third Workshop on Non-Numeric Processing</u>.

Ozkarahan, Esen A., Stewart A. Schuster, and Kenneth C. Smith (1974). A data base processor, Technical Report CSRG-43, Computer Systems Research Group, University of Toronto.

Shaw, David Elliot (1979). A hierarchical associative architecture for the parallel evaluation of relational algebraic database primitives, Stanford Computer Science Department Report STAN-CS-79-778.

Shaw, David Elliot (1980). A relational database machine architecture, <u>Proc. 1980 Workshop on Computer Architecture for Non-Numeric Processing</u>.

Shaw, David Elliot (1980a). <u>Knowledge-Based Retrieval on a Relational Database Machine</u>, Ph.D. Dissertation and Stanford Computer Science Department Report (in press).

Su, Stanley Y. U., George P. Copeland, and G. J. Lipovski (1975). Retrieval operations and data representations in a content-addressed disc system, <u>Proc. Int. Conf. on Very Large Data Bases</u>.

ANALYSIS OF BACKEND DATABASE SYSTEMS ARCHITECTURES

Fred Maryanski and Madhav Marathe
Digital Equipment Corporation
Maynard, Mass., USA

ABSTRACT

The concept of offloading the data management function to a dedicated processor has been suggested many times as a quick and economical method for improving the performance and resource utilization of a data processing installation with a heavy database workload. A backend database machine becomes more cost effective when it is constructed using a number of cheap mini or microprocessors. In this paper we present two multiprocessor backend database architectures and investigate their performance using queueing network analysis methods. The first architecture uses multiple processors in a pipelined fashion to construct an intelligent disk controller. The second employs multiple loosely coupled backend processors each with its own disk. We specifically do not consider any special purpose hardware devices because it is interesting to see how much performance gain can be achieved by using just the existing components. The models are parameterized based upon information available from previous studies and descriptions of state-of-the art hardware. An on-line transaction processing workload is assumed. The analysis indicates the potential superiority of the multiple backends architecture over the pipelined controller architecture. The methodology presented here can be extended for other workloads and configurations. In addition to the performance projections, the paper provides analysis of the impact of the various backend configurations upon some of the basic objectives of data management systems such as robustness, security and reduction of application complexity.

KEYWORDS

Backend computer, intelligent controller, database performance

INTRODUCTION

There has been a general industry trend towards large databases and large, complex systems for managing these databases. The emergence of the small business systems will extend the trend in the direction of medium size databases and easy to use, fast systems for managing them. It has been suggested that the best approach to solving the database performance problem is to dedicate significant processing and memory resources to the database management function. The developments in disk technology have permitted the on-line storage of large amounts of data. However, this has compounded rather than alleviated the database performance problem. Thus recent research has focused upon dedicated data management processors of either the hardware [HSIA79], [LANG79] or software [CANA74], [MARY80], [PASS80] variety.

If high performance backend database systems can be constructed using slow but cheap mini or microprocessors rather than having to develop special purpose hardware, the resulting systems can be very cost effective. In general, if multiprocessor systems fail to give a linear performance improvement, the reasons are excessive communication between processors and excessive process synchronization overhead. The architectures we study are designed to avoid these problems.

EVALUATION OF THE ARCHITECTURES

The major portion of our evaluation effort was dedicated to the development of queueing network models that characterized the performance of the backend architectures. Thus, we will separate the performance issues from the other evaluation criteria and discuss these in more detail. Queueing network analysis is ideally suited for this kind of analysis since a quick and rough analysis is sufficient but the systems to be compared are quite different from one another. An effective comparison of the architectures requires selection of a representative workload, system level performance metrics and relevant detail in the system model. The following sections discuss these three in the context of our study.

The Workload Model

A representative workload is essential for comparing the architectures since each architecture can work better than the others for some specific workload. The workload we consider is based on a transaction processing type application. We decided to model three types of transactions- enquiry, update and queries. To simplify the model, all users generating the transactions are considered identical. The amount of load placed on the system is varied by varying the number of users. The following are some workload parameters and assumptions common to all the three transaction types.

1. The average think time between the end of one transaction and the beginning of the next is 10 seconds.

2. In each system, the four disks are accessed with the same frequency on the average.

3. The effects of concurrency control algorithms in multiprocessor systems can be ignored.

4. The effects of swapping and operating system overheads can be ignored.

5. Communication with the backend requires 10000 instructions each in the host and the backend processor.

6. 10000 instructions of application code are executed in the host for each transaction in addition to the database code which will be executed in the host or in the backend processors.

Table 1 below shows the workload parameters specific to the three request types. Note that this workload assumes that the number of database instructions executed is directly proportional to the number of disk accesses. This assumption is expected to be true for any reasonable implementation which avoids making unnecessary disk accesses. The actual number of disk accesses for a request is drawn at random from a geometric distribution. The actual number of instructions to be executed for each disk access is drawn at random from an exponential distribution with the mean as specified in table 1. This model of the transaction processing workload represents our estimate of the actual workload rather than any measurements. Refining the model of the workload is certainly one of the avenues of further study.

Table 1. Transaction Parameters

Transaction type	Percent of total transactions	Memory required K bytes	Average number of disk accesses	Average number of instructions per access
Enquiry	64	1	5	10000
Update	35	2	10	10000
Query	1	10	100	10000

The Performance Metrics

Since a transaction processing system is being modelled, the important system level performance metric is the average response time for each type of transaction. There are other minor metrics such as the percentage of the time the main processor is idle (i.e. available to do other work); but we will not consider them in detail.

The System Models

We first list the assumptions common to all the architectures under study and then discuss specific architectures in detail. The common assumptions are:

1. The host processor executes instructions at the rate of 1 million instructions per second.

2. The processors in both the backend systems execute instructions at 0.333 million instructions per second.

3. The communication lines between the host and the backend do not cause excessive delays.

4. All systems have four disks and each disk has an average service time of 30 milliseconds. The actual service time is exponentially distributed with this mean. The disk service time can be reduced by an intelligent controller as discussed below.

In the queueing network terminology, a terminal users thinks for a 'think time' and makes a request at time T1. The request propogates through the network of queues, consumes the specified amounts of the resources and returns to the terminal user at time T2. The response time (T2 - T1) is recorded at this point.

It can be seen that the intelligent controller can be used with any of the three data models - Codasyl, Hierarchical or Relational. We assume that all the three models can be implemented such that a request results in more or less the same number of instructions or disk accesses.

Single Processor System

This is the baseline architecture in the study. It is also representative of the current state-of-the-industry. Figure 1 depicts the internal structure of a single processor DBMS system. In subsequent architectures the modules of this system will be distributed among multiple processors. It is the implications of the various distribution strategies that we will analyze.

In a single processor system, the resources for which queueing may take place are the main memory, the host processor and any of the four disks. For simplicity, we ignore the effects of swapping. However, if these are deemed important, they can be modelled by increasing the resource requirements of the requests, by adding extra jobs or by slowing down the servers at higher loads.

The Intelligent Controller

An intelligent controller takes over the database functions from the host software. The host is still used to run the applications software. The database processing can be tightly coupled to the physical data access methods by an intelligent controller. The basic structure of the configuration is presented in fig. 2. The intelligent controller appears to the host processor as an intelligent device rather than as a cooperating processor.

Inside the controller, the processors function in a pipelined fashion. The communications processor handles the protocols for the incoming requests and the outgoing responses. It creates the necessary data structures in the shared memory within the controller and invokes the next processor in the pipeline. The exact decomposition of the database functions among the remaining processor will vary with the implementation. However, we study a typical pipeline consisting of a pre/post processor, a physical access processor and a search

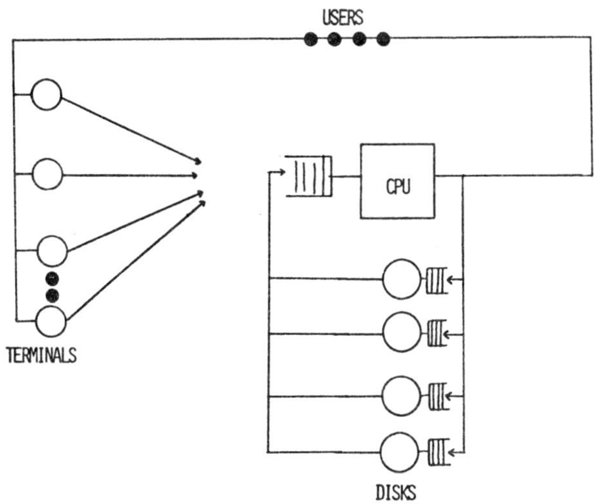

Figure 1. Single Processor System

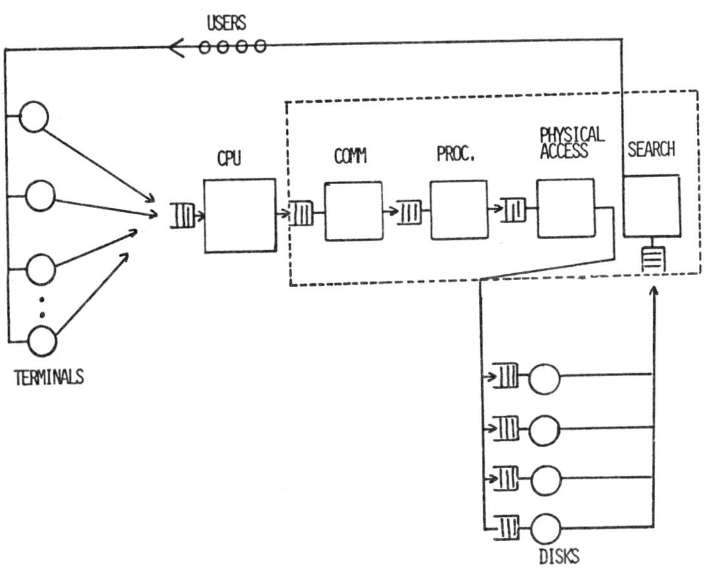

Figure 2. Pipelined Intelligent Controller

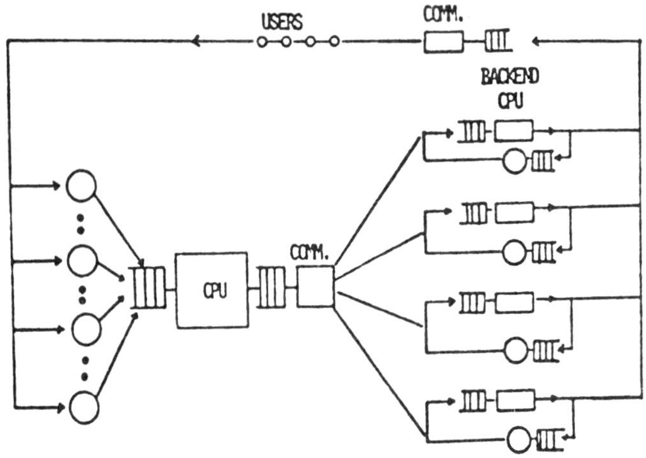

Figure 3. Multiple Backend Processors

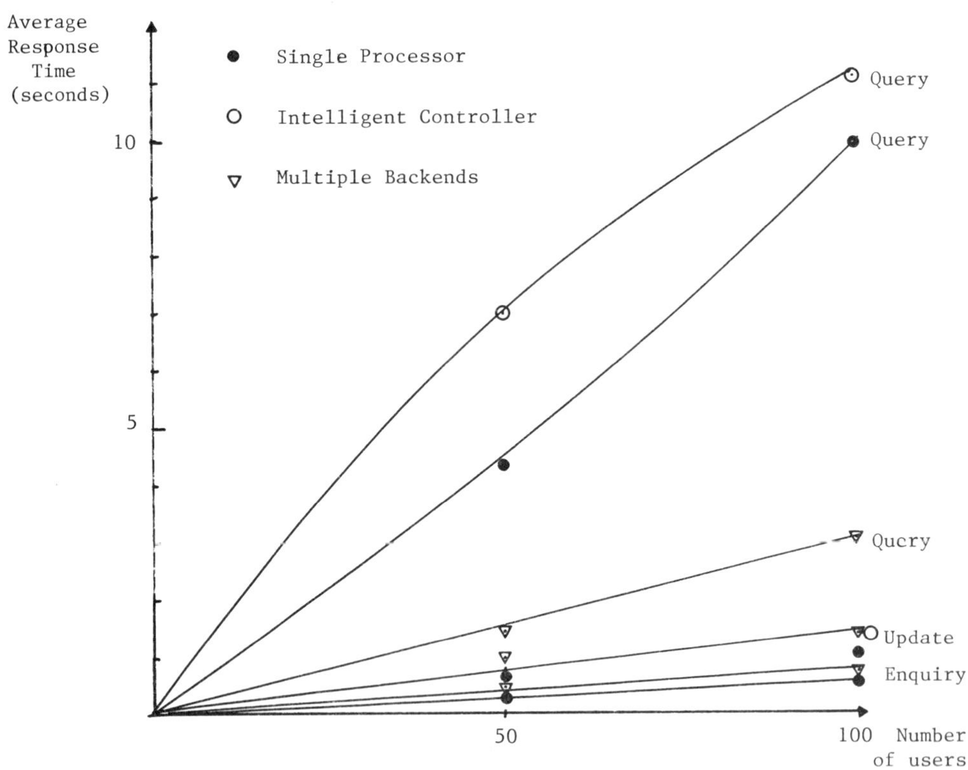

Figure 4. Average Response Times in the Three Configurations

processor. Roughly speaking, the pre/post processor will handle the file system processing and the logical to physical address translation. The physical access processor will handle the the seek and the rotational latency optimizations. It wil also initiate the actual disk transfers. The search processor will perform a search within a page to obtain the required index or record.

We assume that an efficient functional decomposition has been made so that the database instructions are evenly distributed among the pre/post processor, the physical access processor and the search processor. We further make some realistic assumptions regarding the processors and memory that can be provided in an intelligent controller. We assume that these processors execute instructions at the rate of 0.333 million instructions per second. This is certainly in line with the types of microprocessors available today. It is difficult to construct a pipeline involving more than four processors due to the contention for the shared memory and excessive interprocessor communication. With more processors in the pipeline, each processor has to do only a part of the whole job thus requiring more communication of the state and special case information among processors. We therefore consider a pipeline of only four processors. We assume 100 Kbytes of data memory inside the controller, which is a reasonable amount with today's technology. In the future, an intelligent controller can be constructed using faster processors and more data memory.

The effects of seek and rotational latency optimizations are modelled by reducing the average disk service time by 20% when there are three requests pending for a disk and by 40% when four or more requests are queued up. These assumptions need to be validated with actual experiments.

The Multiple Backends

A natural extension of the basic two processor backend DBMS is to incorporate additional database processors to increase the concurrency and hence the processing power of the system. Figure 3 depicts the multiple processor backend DBMS congiuration. The primary difference in terms of information flow between this architecture and its two processor counterpart is that the host processor must contain a mechanism for parcelling out the database requests to the backend processors. Several strategies for this allocation of requests and their associated architectural structures are described in Reference [MARY80].

In order to effect a fair comparison with the intelligent controller system, we study a system with four backend processors. Again we assume the backend processors executing instructions at the rate of 0.333 million instructions per second. Keeping the total amount of memory the same, we provide 25 Kbytes of data memory in each of the four backend systems. It should be noted that since the database software will have to be replicated in each of the backends, the total amount of memory for this configuration will be more than that for the intelligent controller.

We assume that enquiry and update types of transactions are directed at the proper backend processor by the host. This eliminates the need for explicit synchronization among the backend processor while processing these types of transactions. We further assume that the database is carefully distributed over the four backends so that all query type transactions can be simultaneously executed by all the backend processors. This kind of parallelism cannot be achieved by the pipelined processors in an intelligent controller. We therefore expect the multiple backends system to perform better for the query transactions.

Results of the Queueing Model

Figure 4 displays the estimated average response times for the three configurations and for the three types of transactions. It can be seen that it is indeed possible to use multiple mini or micro processors to effectively create a backend database system with performance comparable to the more expensive dedicated single processor system. In the case of the pipelined intelligent controller, the response times are slightly larger than those for the dedicated system. This is due to the difficulty in keeping all the processors in the pipeline busy when uneven resource demands are being made by the transactions. This effect is more prominent for the query type transactions since they traverse the pipeline many more times.

Table 2 displays the average length of the queue for a disk. Since the average length is small, we can conclude that the seek and rotational latency optimization algorithms are not going to benefit this workload. However, the effects of these algorithms can be observed in the smaller disk utilizations for the intelligent controller configuration (see table 2).

In the case of the multiple backends configuration, again the response times for the enquiry and update type transactions are similar to those for the dedicated single processor case. The small increase over the single processor case can be attributed to the additional communications overhead and the time spent waiting for the data memory to become available. The respone time for query type transactions is however much better. This is due to the fact that queries can be processed in parallel by all the backends. The performance improvement obtained in this case is encouraging since it has been obtained in spite of the uneven resource requirements being made on the backend processors. Since the backend processors work on independent parts of the database while processing a query, their execution has to be synchronized only at the end of their respective processing. The synchronization overhead is therefore small compared to some other parallel algorithms.

Table 2 points out that because the 100 Kbytes of the data memory has to be partitioned into four systems, queuing for data memory takes place. This is due to the fact that buffering requirements are typically not reduced when the database is partitioned. This suggests that more data memory should be provided for the multiple backends configuration.

ADDITIONAL EVALUATION

Here we consider the implications of the architectures in terms of the evaluation parameters not explicitly treated in the preceding section. In all cases we will consider the single processor system as the baseline and deal with the other architectures in terms of their potential for enhancing the particular capability under discussion relative to the single processor configuration.

Technological Feasibility

Since we have explicitly not specified any requirements for special-purpose hardware, all of the architectures in our taxonomy are realizable from off the shelf components. The feasibility of utilizing special-purpose hardware for database processing has been under investigation for several years [HSIA79], [LANG79]. The economic viability of these systems has yet to be verified although the concept appears to have merit.

Reduction in Software Complexity

With the exception of the operating system, database management systems and their associated application programs are typically the most complex pieces of software in a data processing environment. Complexity implies costly development and maintenance. Unfortunately, in this respect, the differences in the backend database architectures will be transparent to the application program. This indicates that the problems of database application program complexity will not be alleviated by any of the backend systems. Instead the solution lies with the development of data models that correspond more closely to real world environments and the integration of the data model with the application programming language. The languages PASCAL-R [SCHM77], PLAIN [WASS78] and THESUS [SHOP79] represent steps in the later direction.

Preservation of Integrity and Robustness

Integrity of the database is threatened by hardware or software failure or a poorly designed database manager. The database system architecture can only address the integrity problems associated with failure. Features such as range checking can be implemented in virtually the same manner on any of the configurations. Generally, multiprocessors are more prone to failure since more components and interconnections are involved. However, multiprocessors can be designed with an ability to detect failures and operate in a degraded state using the surviving processors. The two configurations utilizing multiple database processors have the greatest potential for robustness if a reasonable recovery subsystem is developed. The intelligent controller configuration can take advantage of the tight coupling with the disk access and error correction mechanisms to provide slightly greater robustness than the multiple backends configuration.

Multi-User Update Capability

One of the most complex problems faced by database management system implementors is the maintenance of the integrity of shared data. The classical solution is to maintain a centeralized lock monitor which controls access to data that may be referenced by multiple users. When multiple database processors are introduced the problem of lock management in a distributed environment arises. This is currently one the major research issues in database management. Numerous schemes have been offered as theoretical solutions [BERN79] but none have been demonstrated to be feasible in terms of performance. The lack of detailed information on the concurrency control mechanism is a weak point of the models that we have developed in this study. However, this is an issue that could be addressed in future work.

Security

A key feature of the backend architectures in terms of security is that all processors are closely-coupled. Thus the problems of security of public communication lines does not arise. The primary security benefit of the backend concept lies in the isolation of the database function at the backend processors which do not execute any application tasks. Thus all access to the database must originate on the host and appear to the database manager as a legimate request. Attempts at penetration are therefore limited to disguising illicit requests and monitoring the host communication buffers and system security mechanisms can focus upon these areas of possible penetration. The use of multiple backend processors does not radically change the security posture of the architectures.

CONCLUSION

We have analyzed the problem of improving database system performance by using a multiprocessor backend system. Our analysis shows that such systems can indeed perform as well as traditional dedicated single processor systems and at a lesser cost. In fact, a multiple backends configuration can outperform the dedicated host by exploiting the inherent parallelism in a query type transaction.

We have demonstrated the applicability of the queueing network analysis to the problem of database systems performance. The queueing network methodology employed in this study is easy to use and it can be refined as more experience is gained with specific backend implementations.

Several avenues of further work can be persued. Refining the workload model and validating the assumptions made in the study is an obvious one. The special purpose hardware database machines can also be studied in this framework after suitable models of these systems are developed. Finally, the analysis presented here can be extended to include the effects of concurrency control algorithms and communications protocols.

REFERENCES

[BERN79] "Approaches to Concurrency Control in Distributed Data Base Systems" P.A. Bernstein and N. Goodman. Proc. NCC, June 1979.

[CANA74] "A Backend Computer for Data Base Management" R. Canaday, et al. Communications of the ACM Oct. 1974.

[HSIA79] Special Issue on Database Machines D. Hsiao (ed.) IEEE

Computer Mar. 1979.
[LANG79] Special Issue on Database Machines G. Langdon (ed.) IEEE Transactions on Computers June 1979.
[MARY80] "Backend Database Systems" F. Maryanski. ACM Computing Surveys Mar. 1980.
[PASS80] "Providing Network Data Services Using a Backend Data-Base Machine" J.A. Passafiume and J.A. Ravan. IEEE Compcon March 1980.
[SCHM77] "Some High Level Language Constructs for Data of Type Relation" J. Schmidt. ACM Transactions on Database Systems Sept. 1977.
[SHOP79] "Theseus - A Programming Language for Relational Databases" J.E. Shopiro. ACM Transactions on Database Systems Dec. 1979.
[WASS78] "A Software Engineering View of Data Base Management" A.I. Wasserman. Proc. Conf. on Very Large Data Bases, Sept. 1978.

Table 2. Results of the Queueing Network Analysis

Quantity	Number of users	Single Processor	Intelligent Controller	Multiple Backends
Host CPU util.	50	43%	4.5%	3.8%
	100	79%	8.5%	8.7%
Pipeline util.	50	-	36%	-
	100	-	63%	-
Backend util.	50	-	-	22%
	100	-	-	59%
Avrg disk queue	50	0.1	0.1	0.1
	100	0.6	0.35	0.5
Avrg proc queue	50	0.3	0.2	0.1
	100	2.4	1	0.7
Avrg mem queue	50	0	0	0
	100	0	0	0.1

METHODOLOGICAL PROBLEMS FOR DATA MODELLING CONCEPTS IN RM/T's DATA BASE DESIGN

R. R. Wagner* A M. Tjoa**

* Institut für Informatik, Universität Linz, Linz, Austria

** Institut für Statistik und Informatik, Universität Vienna
Vienna, Austria

ABSTRACT

The extended relational model (RM/T) was proposed by Codd to capture more meaning of the data. It will be shown in this paper that besides the semantical relevance of RM/T, the extensions seem also to be very useful for the low-level-interface design of a relational data base system like the implementation concept GAMMA-0. Furthermore in this paper some design problems (e.g. surrogates) used for the description of RM/T will be discussed.

KEYWORDS

Extended relational model RM/T; relational model; low-level-interface; gamma-0; surrogates; data modelling concepts; data base design.

INTRODUCTION

In (Codd,1979) the RM/T was proposed as an extension of the relational model (Codd 1970,1971,1974) to capture more meaning of the data. In this paper it will be shown, that the extensions proposed by Codd (1979) are not only of semantical relevance but are also very important with respect to implementation considerations. Some of these extensions seem to be very useful for the low-level-interface design of a relational database system. Of course many of these extensions represent a significant improvement to other implementation concepts like GAMMA-0 (Bjørner and co-workers, 1973) on the one hand. On the other hand implementation concepts in RM/T can be described by characteristics of GAMMA-0.

It is the opinion of the authors that besides the semantical and implementation relevance of the extended relational model, RM/T could also be important for the implementation of user-oriented query languages like Codd's (1974) RENDEVOUZ since e.g. the different graphs defined in RM/T seems to be very useful for supporting the search of access paths.

SELF DESCRIBING ASPECT IN RM/T AND GAMMA-0

One of the main characteristics which may be required from a data model is that it is capable of describing itself (Abrial, 1974). This requirement is important since the model itself can be considered as a part of an "object" to be modeled.
In GAMMA-0 (Bjørner and co-workers, 1973) this property is fulfilled by the introduction of a so called masterrelation, which is a relation describing relations by its tuples. In this concept the masterrelation itself is described by a tuple of it. Of course at the initial state the masterrelation consists only of this tuple, whereas the other relations are empty. A simplified GAMMA-0 concept for the initial state of the masterrelation, and the description of the masterrelation and a regular relation is demonstrated in Fig. 1 and Fig. 2.

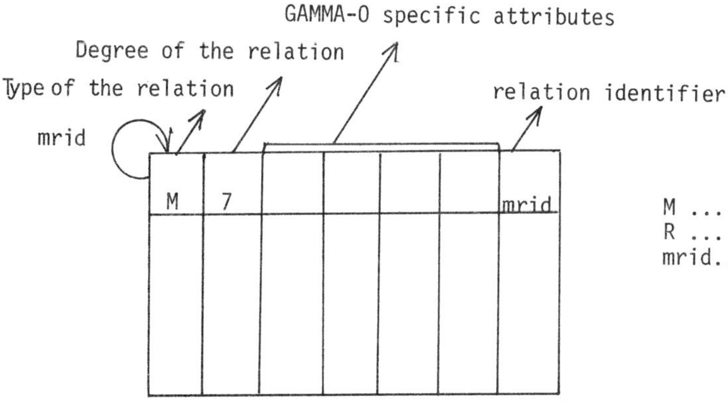

M ... Masterrelation
R ... Regular relation
mrid..Masterrelation identifier

Fig. 1. Intitial state of the masterrelation in GAMMA-0.

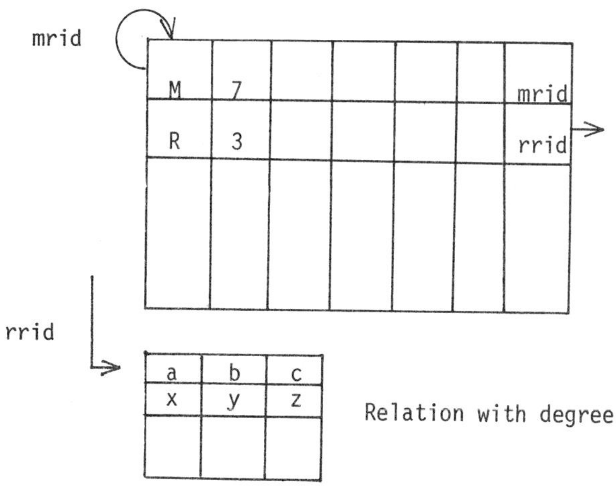

rrid...is the identifier of this regular relation described by this tuple.

Relation with degree 3

Fig. 2. Description of the masterrelation and a regular relation.

In the following the description of RM/T by itself is given. The catalogue-structure for RM/T is shown below.

CATR(R¢ RELNAME RELTYPE)

CATRA(RA¢ R¢ A¢)
CATA(A ¢ ATTNAME USERKEY)
CATAD(AD¢ A¢ D¢)
CATD(D¢ DOMNAME VTYPE ORDERING)
CATC(C¢ PERNAME)
CATRC(RC¢ R¢ C¢)

with:
RELNAME ... relanme of relation
ATTNAME ... attname of attribute
DOMNAME ... domname of domain
PERNAME ... categorie label
RELTYPE ... type of object represented by relation
USERKEY ... indicates whether attribute participates in a user defined key for corresponding relation
VTYPE ... syntactic type of value
ORDERING... indicates whether > is applicable between values in corresponding domain.

R¢,A¢,D¢, C¢ are attributes defined on the E-domain and contain surrogates for entities of type:relation,attribute,domain,category label.
RA¢, AD¢, RC¢ are attributes defined on the E-domain and contain surrogates for associative entities of type:relation-attribute,attribute-domain,relation-category-label.

In the following the initial state of the catalogue-relations is described.
Note: All elements from E-domain are writte in italics.

CATR	R¢	RELNAME	RELTYPE
	CATR	CATR	CAT
	CATRA	CATRA	CAT
	CATA	CATA	CAT
	CATAD	CATAD	CAT
	CATD	CATD	CAT
	CATC	CATC	CAT
	CATRC	CATRC	CAT

CATRA	RA ¢	R ¢	A ¢
	CATR_R ¢	*CATR*	*R ¢*
	CATR_RELANME	*CATR*	*RELNAME*
	CATR_RETYPE	*CATR*	*RELTYPE*
	CATRA_RA ¢	*CATRA*	*RA ¢*
	CATRA_R ¢	*CATRA*	*R ¢*
	CATRA_A ¢	*CATRA*	*A ¢*
	CATA_A ¢	*CATA*	*A ¢*
	CATA_ATTNAME	*CATA*	*ATTNAME*
	CATA_USERKEY	*CATA*	*USERKEY*
	CATAD_AD ¢	*CATAD*	*AD ¢*
	CATAD_A ¢	*CATAD*	*A ¢*
	CATAD_D ¢	*CATAD*	*D ¢*
	CATD_D ¢	*CATD*	*D ¢*
	CATD_DOMNAME	*CATD*	*DOMNAME*
	CATD_ORDERING	*CATD*	*ORDERING*
	CATC_C ¢	*CATC*	*C ¢*
	CATC_PERNAME	*CATC*	*PERNAME*
	CATRC_RC ¢	*CATRC*	*RC ¢*
	CATRC_R ¢	*CATRC*	*RC ¢*
	CATRC_C ¢	*CATRC*	*C ¢*

CATA	A ¢	ATTNAME	USERKEY
	R ¢	R ¢	true
	RELANME	RELNAME	false
	RELTYPE	RELTYPE	false
	RA ¢	RA ¢	true
	R ¢	R ¢	false
	A ¢	A ¢	false
	A ¢	A ¢	true
	ATTNAME	ATTNAME	false
	USERKEY	USERKEY	false
	AD ¢	AD ¢	true
	A ¢	A ¢	false
	D ¢	D ¢	false
	D ¢	D ¢	true
	DOMNAME	DOMNAME	false
	VTYPE	VTYPE	false
	ORDERING	ORDERING	false
	C ¢	C ¢	true
	PERNAME	PERNAME	false
	RC ¢	RC ¢	true
	R ¢	R ¢	false
	C ¢	C ¢	false

CATAD	AD ¢	A ¢	D ¢
	R ¢_ED	*R ¢*	*ED*
	RA ¢_ED	*RA ¢*	*ED*
	RC ¢_ED	*RC ¢*	*ED*
	RELNAME_NAME	*RELNAME*	*NAME*
	RELTYPE_T	*RELTYPE*	*T*
	ATTNAME_NAME	*ATTNAME*	*NAME*
	USERKEY_BOOLEAN	*USERKEY*	*BOOLEAN*
	VTYPE_NAME	*VTYPE*	*NAME*
	ORDERING_BOOLEAN	*ORDERING*	*BOOLEAN*
	PERNAME_NAME	*PERNAME*	*NAME*

CATD	D ¢	DONAME	VTYPE	ORDERING
	ED	E_DOMAIN	E_TYPE	no
	NAME	NAME	STRING	no
	BOOLEAN	BOOLEAN	BOOLEAN	yes*
	T	RM/T_TYPES	TYPE**	no

* false <true

** Symbol {CAT,A,C,E,G,I,K,L,N,P,T}

CATC	C ¢	PERNAME
	Empty	

CATRC	RC ¢	R ¢	C ¢
	Empty		

The relation type CAT is defined for these relations. Of course theoretically these relations can also be classified as P-(property-relation), A-(associative entity type relation), C-(characteristic-relation) etc. (e.g. CATD is a property relation). The CAT-type-relations have their own properties (e.g. cannot be deleted) and some additionally constraints:

- Every relation in the database must be identified by a tuple CATR.
- Every attribute of a relation must be defined by a tuple in CATRA.
- Every attribute used in CATRA must be defined in CATA.
- Every attribute has a domain which must be defined in CATA.
- Every domain used in CATAD must be defined in CATD.

Note that all catalogue relations are base relations and of course the two integrity rules (rule 1: entity integrity and rule 2:referential integrity) of RM/T are also valid for the catalogue relations.

CLASS RELATIONS AND P-RELATIONS

The two integrity rules of RM/T mentioned in the chapter before are also very useful for an implementation concept like GAMMA-0.

Rule 1 (entity integrity):No primary key value of a base relation is allowed to be null or to have a null component.

Rule 2 (referential integrity):Suppose an attribute A of a relation R is defined on a primary domain D. Then, at all times, for each non-null value v of A there must exist a base relation (say S) with a simple primary key (say B) such that v occurs as a value of B in S.

Suppose that there exist the two RM/T-relations R_1(Supplier ¢ City ¢) and R_2(City ¢ name of city). The concept which is illustrated in Fig. 3 corresponds to the class-relation-concept of GAMMA-0, in which character string attribute values are expressed indirectly to conserve space.

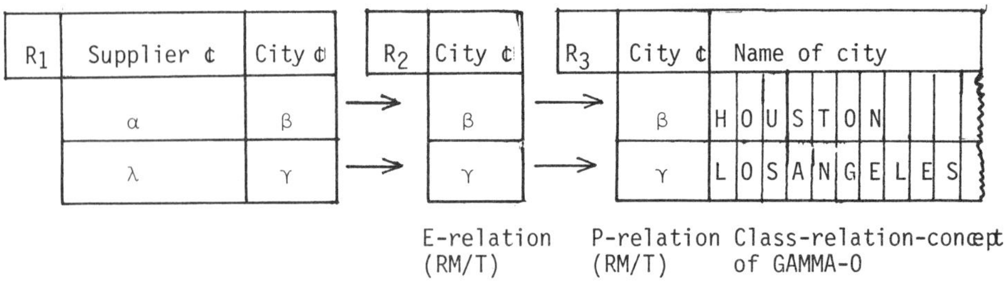

Note that supplier ¢ is defined on a primary domain.

Fig. 3. For the class-relation-concept in GAMMA-0 the referential rule is valid.

Since every surrogtae of city ¢ defined in the relation R_1 must be defined in R_2 (E-relation) and (usually) also in a P-relation (in this example R_3) the referential integrity rule of RM/T is valid for the class-relation-concept in GAMMA-0.

Furthermore in GAMMA-0 each distinct tuple of any relation has associated with it a unique tuple-identifier (as we have seen it in Fig. 1 and Fig. 2.: e.g. mrid, rrid), whereas in RM/T the concept of surrogates, which identify entities unique in the entire database, is introduced. Therefore surrogates can be interpreted in some way as tuple-identifier of GAMMA-0. While there is a resemblance between tuple identifiers and surrogates, they are not the same. An RM/T surrogate generally serves as key value in at least two tuples (one in an E-relation and one in a P-relation), and in more if the "immediate properties" of an entity are not grouped into one relation. The use of surrogates on a user level has interesting implications for implementation.

ASSOCIATIONS

An other problem arises in the application of the association rule (rule 6 of Codd's RM/T).

Rule 6 (association integrity): Unless there is an explicit integrity constraint to the contrary, an associative entity can exist in the database (i.e. there is a corresponding surrogate in the appropriate E-relation), even though one or more entities participating in that association are unknown. In such a case the surrogate E-null is used to indicate that a participating entity is unknown.

In contrary to rule 6 it is the opinion of the authors that the existence of an entity cannot be allowed, if one of the participating entity is unknown. In this case the identification of the circumstances in which nulls are used must be made.

Example:

Assign

Assign ¢
λ
γ

Assign_Employee_Project

Assign ¢	Employee ¢	Project ¢
λ	ω	μ
γ	ω	ω

Assign_Date

Assign ¢	Start Date
λ	80-1-1
γ	80-2-1

For the associative entity assign with surrogate λ it cannot be determined which employee has started the project μ at the start_date 80-1-1. For the entity with the surrogate γ there is neither the possibility to determine the employee not the project of the assignement with the start_date 80-2-1.

CONCLUSION

Although Codd's RM/T was proposed to capture more meaning of the data, this extended relational model is also very usefule for the low-level-interface design of a relational database system. It was shown in this paper by means of the RM/T-catalogue-relations that RM/T is a self describing model like the master-relation concept of the GAMMA-0. In the same section it was also shown that implementation concepts in RM/T can be described by characteristics of GAMMA-0 on the one hand. On the other hand many extensions of RM/T represent a significant improvement of GAMMA-0. In a further discussion it was shown that the surrogates, which are very similar to the tuple-identifier-concept of GAMMA-0, are very useful for the implementation, but encounters problems in the reality.

REFERENCES

Abrial,J.R. (1974). Data semantics. In J.W.Klimbie, and K.L. Koffemann (Eds.), Data Base Management,North-Holland,Amsterdam.

Bjørner,D., Codd E.F., Deckert K.L., and Traiger I.L. (1973). The GAMMA-0 n-ary relational data base interface specification of objects and operations. IBM-Research-Report RJ 1200, San Jose.

Codd, E.F. (1970). A relational model of data for large shared data banks. Commun.ACM , 13, 6,pp.23-88.

Codd, E.F. (1971). Further normalization of the database relational model. In Rustin (Ed.),Database Systems, Prentice Hall, Englewood Cliffs N.J. .

Codd, E.F. (1974). Recent investigations in relational database systems. Proceedings of Information Processing, North-Holland, Amsterdam.

Codd, E.F. (1974). Seven steps to RENDEZVOUS with the casual user. In J.W.Klimbi, and K.L.Koffmann (Eds.),Data Base Management, North-Holland, Amsterdam.

Codd,E.F. (1979). Extending the database relational model to capture more meaning. ACM Trans.Database Systems, 4,4, pp.397-434.

ON THE SEMANTICS OF DATA DEPENDENCIES IN RELATIONAL DATA BASES

S.K. Arora*, S.R. Dumpala* and K.C. Smith**

*Department of Electrical and Computer Engineering
McMaster University, Hamilton, Canada
**Department of Electrical Engineering
University of Toronto, Toronto, Canada

ABSTRACT

The design of "good" data bases is the aim of dependency and normalization theory. However in the past three years several dependencies and normal forms have appeared in the literature which are of purely theoretical interest. We believe a data base designer need consider only a subset of dependencies, root dependencies, and normal forms based on them in his design process. Semantic overtones of root dependencies and their normal forms are presented.

INTRODUCTION

Dependencies are a systematic way of dealing with semantically useful situations that arise in data. People think in terms of semantic connections among data rather than in terms of relational algebra which is highly mathematical. For example, in a company each employee may have a distinct employee number. The data base designer is faced with the task of modelling this semantic connection between an employee and his employee number. He must further ensure that no two employees ever get the same employee number during operations on the data base. For this situation, we have the Functional Dependency (FD), see for example [Armstrong 1974, Bernstein 1975, Codd 1971]. The data base designer uses the FD between employee number and employee in his design process and also specifies it as an integrity constraint which must not be violated during operations on the data base.

For many years, FD was the only known dependency. However, the need for other dependencies was felt when it became apparent that FD is not adequate to model many semantically useful situations in data. (See for example [Schmid, Swneson 1975].) In the last three years, several new dependencies have appeared in the literature. The various dependencies in the literature are

(i) Functional Dependency (FD) [Codd, 1971]
(ii) Mutlivalued Dependency (MVD) [Fagin, 1977a]
(iii) Mutual Dependency (MVD) [Nicolas, 1978]
(iv) Hierarchical Dependency (HD) [Delobel, 1978]
(v) Mixed Dependency [Arora, Smith, 1979]
(vi) Co-Dependency (COD) [Arora, Smith, 1979]
(vii) Template Dependency (TD) [Sadri, Ullman, 1980]
(viii) Transitive Dependency (TrD) [Paredaens, Gallaire, 1980]
(ix) General Dependency (GD) [Janssens, Paredaens, 1979]
(x) Subset Dependency (SD) [Sagiv, Walecka, 1979]
(xi) Boolean Dependency (BD) [Fagin, 1977b]
(xii) Generalized Mutual Dependency (GMD) [Mendelzon, Mairer, 1979]
(xiii) Implication Dependency (ID) [Fagin, 1980]

(xiv) Null Multivalued Dependnecy (NMVD) [Lien 1979]
(xv) Operator Multivalued Dependency (OVMD) [Arora, Smith 1980b]
(xvi) Extended Functional Dependency (EFD) [Vassihiou 1980]
(xvii) Algebraic Dependency (AD) [Yannakakis, Papadimitrious 1980]

SEMANTICS OF DEPENDENCIES

During the explosion of research in the last three years, people have tended to study dependencies purely for theoretical interest. Hence not all the dependencies in the literature are useful to a data base designer for controlling and modelling the semantics of data. In the previous section we have pointed out that FD is very useful semantically. Schmid and Swenson, (1975), point out that FD is not adequate to model some semantic situations that arise in data. They give the following example. Suppose an employee works in several departments, then EMPLOYEE $\nrightarrow$ DEPT. (EMPLOYEE does not functionally determine DEPT.), and suppose that <u>each</u> department has only one contract type, DEPT $\rightarrow$ CTYPE. Clearly there is <u>knowledge</u> about employees and contract types on which they are working. But it cannot be expressed as an FD. We can look at this example in two different ways. If we wish to look at <u>each</u> employee and his relation to DEPT and CTYPE then it can be expressed by a mutual dependency (MUD) i.e., EMPLOYEE DEPT|CTYPE. The data base designer can now specify this MUD as an integrity constraint. However, if we wish to look at <u>each</u> department and its relation to EMPLOYEE and CTYPE then it can be expressed <u>by a</u> multivalued dependency (MVD), i.e., DEPT $\twoheadrightarrow$ EMPLOYEE|CTYPE. We extend Schmid and Swenson's (1975) example as follows. Suppose each contract type is divided into several portfolios and each employee of the department has access to all the portfolios in that department. We can model the knowledge about departments, employees, contract types and portfolios by a hierarchical dependency (HD), i.e., DEPT:EMPLOYEE|CTYPE|PFOLIO. Here we are looking at <u>each</u> department and its relation to EMPLOYEE, CTYPE and PFOLIO.

We again extend Schmid and Swenson's (1975) example as follows. Suppose an employee works on several projects and each project is assigned to one department. Clearly there is knowledge about PROJ, DEPT, CTYPE and EMPLOYEE which cannot be modelled by HD. This knowledge can be modelled for <u>each</u> employee by a COD, i.e., EMPLOYEE $\equiv$ PROJ|DEPT|CTYPE. The MD is a special case of COD. If we modify this example and say that an employee can work only on one contract type at a time then we can model this knowledge for each employee by an MD, i.e., EMPLOYEE $\simeq$ PROJ|DEPT/CTYPE.

It can now be pointed out that an FD is a dependency with two sets of attributes (one root-attribute set on the left hand side and antoher attribute set ont he right hand side which we choose to call a branch-attribute set). An MVD has three attribute sets, one root and two branch. An HD is an extension of MVD to "n" attribute sets, one root and "n-1" branch. An MUD has three attribute sets, one root and two branch. A COD is an extension of MUD to "n" attribute sets, one root and "n-1" branch. An MD is also defined on "n" attribute sets, one root and "n-1" branch. However, it exhibits features of both MVD and MUD. FD, MVD, MUD, HD, MD and COD form a class of dependencies which we choose to call <u>root-dependencies</u>.

We give formal definitions for each root dependency. A functional dependency, $X \rightarrow Y_1$, exists in a relation, $R[X, Y_1]$, if in every instance of the relation some function $f: X \rightarrow Y_1$ exists.

In a relation, $R[X, Y_1, Y_2]$, the MVD, $X \twoheadrightarrow Y_1|Y_2$ holds if and only if, $R[X, Y_1, Y_2]$ is the natural join of its projections, $R[X, Y_1]$ and $R[X, Y_2]$. Here X, Y_1, and Y_2 are disjoint sets of attributes.

In a relation, $R[X, Y_1, Y_2]$, the MUD, X $Y_1|Y_2$, holds if and only if $R[X, Y_1, Y_2]$ is the natural join of its three projections - $R[X, Y_1]$, $R[X, Y_2]$ and $R[Y_1, Y_2]$.

In a relation, $R[X, Y_1, Y_2, \ldots, Y_n]$, the HD, $X\colon Y_1|Y_2|\ldots Y_n$, holds if $R[X, Y_1, Y_2, \ldots, Y_n] = R[X, Y_1] * R[X, Y_2] * \ldots * R[X, Y_n]$ where $*$ is the natural join. Here X, Y_1, Y_2, ..., Y_n are mutally disjoint sets of attributes and $R[X, Y_i]$ are projections of $R[X, Y_1, Y_2, \ldots, Y_n]$.

The COD $\quad X \equiv Y_1|Y_2|\ldots Y_n$
holds in a relation $R[X, Y_1, Y_2, \ldots, Y_n]$ if
$R = R[X, Y_1] * R[X, Y_2] * \ldots * R[X, Y_n] * R[Y_1, Y_2] * R[Y_2, Y_3] * \ldots * R[Y_{n-1}, Y_n]$.
Here X, Y_1, Y_2, ..., Y_n are mutally disjoint sets of attributes and each relation on the right hand side is a projection of $R[X, Y_1, Y_2 \ldots Y_n]$.
The MD $\quad X \simeq Y_1 \int Y_2|Y_3 \int Y_4|Y_5| \ldots \int Y_n$
holds in a relation $R[X, Y_1, Y_2, \ldots Y_n]$ if
$R = R[X, Y_1] * R[X, Y_2] * \ldots * R[X, Y_n] * R[Y_1, Y_2] * R[Y_3, Y_4] * \ldots * R[Y_{n-1}, Y_n]$
Here X, Y_1, Y_2, ..., Y_n are mutaly disjoint sets of attributes and each relation on the right hand side is a projection of $R[X, Y_1, Y_2, \ldots Y_n]$.
We give examples of semantic modelling for each root dependency.

(i) Functional Dependency

$$R[X, Y_1] = \begin{array}{ll} a_1 & b_1 \\ a_2 & b_1 \\ a_3 & b_2 \end{array}$$

$X \rightarrow Y_1$; $R[X, Y_1] \equiv$ REPRESENT[AGENT, PRODUCT]
"An agent represents only one product at a time".

(ii) Multivalued Dependency

$$R[X, Y_1, Y_2] = \begin{array}{lll} a_1 & b_1 & c_1 \\ a_1 & b_2 & c_2 \\ a_1 & b_1 & c_2 \\ a_1 & b_2 & c_1 \\ a_2 & b_3 & c_3 \\ a_2 & b_3 & c_4 \end{array}$$

$X \twoheadrightarrow Y_1|Y_2$; $R[X, Y_1, Y_2]$ = REPRESENT [AGENT, PRODUCT, COMPANY]
"When an agent works for several companies and is familiar with several products then each company, he works for, must be producing all the products, he is familiar with".

iii) Mutual Dependency.

$$R[X,Y_1,Y_2] = \begin{array}{lll} a_1 & b_1 & c_1 \\ a_1 & b_2 & c_1 \\ a_1 & b_3 & c_2 \\ a_2 & b_1 & c_1 \\ a_2 & b_4 & c_3 \\ a_3 & b_4 & c_4 \end{array}$$

$X \quad Y_1|Y_2$; $R[X,Y_1,Y_2]$ = REPRESENT [AGENT, PRODUCT, COMPANY].
"When an agent works for several companies, the products he represents for one of them must not be produced by the others".

iv) Hierarchical Dependency.

$$R[X,Y_1,Y_2,Y_3] = \begin{array}{llll} a_1 & b_1 & c_1 & d_1 \\ a_1 & b_1 & c_1 & d_2 \\ a_1 & b_2 & c_2 & d_2 \\ a_1 & b_2 & c_2 & d_1 \\ a_1 & b_1 & c_2 & d_1 \\ a_1 & b_1 & c_2 & d_2 \\ a_1 & b_2 & c_1 & d_1 \\ a_1 & b_2 & c_1 & d_2 \\ a_2 & b_3 & c_3 & d_3 \end{array}$$

$X\colon Y_1|Y_2|Y_3$; $R[X,Y_1,Y_2,Y_3]$ = REPRESENT [AGENT,PRODUCT, COMPANY,STATE].
"When an agent works for several companies and is familiar with several products and works in several states then each company he works for, must produce all the

products he is familiar with and have offices in all the states in which he works".

v) Mixed Dependency

$$R[X,Y_1,Y_2,Y_3] = \begin{array}{cccc} x_1 & y_{13} & y_{22} & y_{32} \\ x_1 & y_{13} & y_{21} & y_{32} \\ x_1 & y_{12} & y_{21} & y_{32} \\ x_1 & y_{13} & y_{21} & y_{31} \\ x_1 & y_{13} & y_{22} & y_{31} \\ x_1 & y_{12} & y_{21} & y_{31} \\ x_2 & y_{12} & y_{21} & y_{32} \\ x_2 & y_{11} & y_{22} & y_{32} \end{array}$$

$X \approx Y_1 \int Y_2 | Y_3$; $R[X,Y_1,Y_2,Y_3]$ = MARKETS[Agent, Company, Product, Area]. "If a company makes any product with which one of its agents is familiar then he must be one of the representatives for that product in every area where he operates and the product must be distributed in all areas where such an agent is available".

vi) Co-Dependency.

$$R[X,Y_1,Y_2,Y_3] = \begin{array}{cccc} x_1 & y_{13} & y_{22} & y_{32} \\ x_1 & y_{12} & y_{21} & y_{31} \\ x_1 & y_{12} & y_{21} & y_{32} \\ x_1 & y_{13} & y_{21} & y_{32} \\ x_1 & y_{13} & y_{21} & y_{31} \\ x_2 & y_{11} & y_{22} & y_{32} \\ x_2 & y_{11} & y_{22} & y_{32} \end{array}$$

$X \equiv Y_1 | Y_2 | Y_3$; $R[X,Y_1,Y_2,Y_3]$ = MARKETS[Agent, Company, Product, Area] "If a company makes any product with which one of its agents is familiar then he must be one of the representatives for that product in every area where he operates and in which the product is distributed".

Finally we mention in passing that FD's and MVD's have been defined with null values - EFD's and NMVD's. Also OMVD's are defined using Theta Join and Projection operations. These are the beginnings of potentially new areas of research and we do not consider them in this paper.

NORMALIZATION

In the earlier section, it was pointed out that root dependencies are all semantically useful. It is our belief that a data base designer need consider only the root dependencies in the design process. Other dependencies in the literature may prove to be of theoretical interest only in the long run. The data base designer must control the pattern of root dependencies in a relation to avoid semantically undesirable effects - anomalous behavior of the data base. Codd (1971) proposed normalization as a process to eliminate anomalous behavior of the data base, namely "Insertion Anomaly", "Deletion Anomaly" and "Consistency on Update" (Codd 1971).

The various normal forms in the literature are 1NF, 2NF, 3NF, BCNF, 4NF, 5NF, PJ/NF, (3,3)NF, PSJU/NF, DK/NF, TP/NF (Codd 1970, Codd 1971, Fagin 1977, Fagin 1979, Fagin 1979a, Smith 1978, Armstrong 1980a, Armstrong 1980, Fagin 1980). The 1NF controls the format of the relation. The normal forms based on root dependencies are 2NF, 3NF, BCNF, 4NF and 5NF. It is our belief that a data base design need consider only these normal forms in his design process. We show the semantic usefulness of each of these normal forms based on root dependencies. Before we do that, we mention in passing that (3,3)NF is also based on FD but applied to generalization and aggregation hierarchies (Smith 1978) of relations. We do not consider this normal form here. It however represents a potentially new area of research in the theory of data bases.

i) First Normal Form (1NF):

"A relation is in 1NF if every attribute in the relation is based on simple domains".

Ex. CLASS [TEACHER, STUDENT, CODE]
CODE [SECTION, SUBJECT]

Here CODE is a non simple attribute in relation CLASS. The 1NF will be,

CLASS (TEACHER, STUDENT, SECTION, SUBJECT)

The representation makes it easier to define simple operations on the relations.

ii) Second Normal Form (2NF):

"A relation is in 2NF, if it is in 1NF and each non-prime attribute is fully dependent upon every key".

Ex. CLASS [STUDENT NO., YEAR, PROGRAM, AGE GROUP]

Here, (STUDENT NO., YEAR) → PROGRAM
YEAR → AGE GROUP.

The key is (STUDENT NO., YEAR) and a non-prime attribute AGE GROUP is partially dependent on the key. So CLASS is not in 2NF. It will have insertion anomaly as a side effect - when the first STUDENT is entered in the data base the relationship between YEAR and AGE GROUP is also entered. Also when the last STUDENT in a YEAR is deleted, the reltionship between that YEAR and AGE GROUP disappears giving rise to deletion anomaly. These effects disappear when CLASS is decomposed as follows;

CLASS I [STUDENT NO., YEAR, PROGRAM]
AGE [YEAR, AGE GROUP]
These two relations are in 2NF.

iii) Third Normal Form (3NF):

"A relation is in 3NF if it is in 2NF and no non-prime attribute is transitvely dependent on any key".

Ex. CLASS [STUDENT NO., YEAR, AGE GROUP]
Let,
STUDENT NO. → YEAR → AGE GROUP

Here AGE GROUP is transitively dependent on STUDENT NO. and AGE GROUP is non-prime. So CLASS is in 2NF but not in 3NF. Here the first student in a YEAR created the relationship between YEAR and AGE GROUP as a side effect. Similarly we have deletion anomaly. Further if a student's YEAR is changed, his AGE GROUP must also be changed for otherwise the data base will show an inconsistency. So CLASS should be decomposed as,

CLASS I [STUDENT NO., YEAR]
AGE [YEAR, AGE GROUP]
These relations are in 3NF.

iv) Boyce-Codd Normal Form (BCNF):

"A relation is in BCNF if it is in 1NF and, for every set of attrributes C, if any attribute, not in C, is functionally dependent on C then each and every attribute is functionally dependent on C".

Ex. CLASS[STUDENT NO., YEAR, PROGRAM, TEACHER]

Let

(STUDENT NO., YEAR) → PROGRAM
(STUDENT NO., YEAR) → TEACHER
TEACHER → YEAR.

This relation is in 3NF but the prime attribute YEAR is transitively dependent on (STUDENT NO., YEAR) as follows;

STUDENT NO., YEAR) → TEACHER → YEAR

This poses consistency problems on update. Suppose YEAR of a TEACHER is changed along with the STUDENT he teaches, a user might look for STUDENT NO., YEAR combinations for that TEACHER and change the YEAR in each one of them. This of course entails the side effect that PROGRAM for this YEAR is now incorrect. Also the data base may have two tuples with the same STUDENT NO. and YEAR combination thus destroying the key. Hence CLASS should be decomposed as,

CLASS I [STUDENT NO., YEAR, PROGRAM]

TEACHER I [TEACHER, YEAR]

These relations are in BCNF.

v) Fourth Normal Form (4NF):

"A relation is in 4NF if, whenever a non-trivial MVD, X →→ Y, holds then so does the FD, X → S, for every attribute A".

This means that a relation in 4NF cannot have any non-trivial MVD that is not the result of a key.

Ex. CLASS [STUDENT NAME, TEACHER, PROGRAM]

There are no FDs in CLASS and the relation is all key. So it is in BCNF. But, let,

TEACHER →→ STUDENT NAME

and TEACHER →→ PROGRAM

This means that all students of a TEACHER participate in all the PROGRAMS with which he is associated. No CLASS has the undesirable property that a TEACHER and all his PROGRAMS are repeated for each of his STUDENTS. If one PROGRAM of a TEACHER is changed then it affects several tuples. Making the update in only one tuple leads to inconsistencies in the data base. So CLASS should be decomposed as follows;

CLASS I [STUDENT NAME, TEACHER]

PROGRAM [TEACHER, PROGRAM]

These relations are in 4NF.

vi) Fifth Normal Form (5NF):

"A relation $R[Y_1, Y_2, \ldots, Y_n]$ is in 5NF when every non-trivial COD - i.e., a COD with at least two branch-segments - in R contains a key of R in its root-segment".

Ex.

CLASS [STUDENT NAME, TEACHER, PROGRAM, LEVEL]

There are no FDs or MVDs in CLASS. So it is 4NF. But let,

TEACHER ≡ STUDENT NAME|PROGRAM|LEVEL

Now CLASS has the undesirable property that a STUDENT NAME, PROGRAM and LEVEL combination may be repeated for several TEACHERS who are associated. A change in one tuple affects several tuples causing update problems. So CLASS should be decomposed as follows:

CLASS I [STUDENT NAME, TEACHER]

CLASS II [TEACHER, PROGRAM]

CLASS III [TEACHER, LEVEL].

These relations are in 5NF.

CONCLUDING REMARKS

We have presented an overview of the dependency and normalization theory area in this paper. A large class of dependencies, namely, root dependencies are very useful for semantic modelling in relational data bases. The pattern of root dependencies must be controlled in a relation to avoid semantically undesirable effects - insertion anomaly, deletion anomaly and consistency on update. The normal forms for this purpose are 1NF, 2NF, 3NF, BCNF, 4NF and 5NF. other dependencies and normal forms are by and large of theoretical interest only.

ACKNOWLEDGEMENT

This work was supported in part by the Science and Engineering Research Board Grant Number 214-7248.

REFERENCES

Armstrong, W.W., (1974). Dependency Structrues of Data Base Relationships, Proc. of IFIP '74, North Holland, 580-583.

Arora, S.K., and K.C. Smith, Graphical Normal Forms Based on Root Dependencies in Relational Data Bases, Submitted for publication.

Arora, S.K. and K.C. Smith, (1980). A Normal Form Based on Theta-Join and Projection for Relational Data Bases, IEEE COMPCON FALL, Washington, D.C., 295-300.

Arora, S.K. and K.C. Smith, (1979). Well Connected Relations in Dependency Structures of Data Bases, Proc. of Conference 1979 CIPS/DPMA/FIG, Quebec City, 95-101.

Bernstein, P.A., (1975). Normalization and Functional Dependencies in the Relational Data Base Model, Ph.D. Thesis, Dept. of Computer Science, University of Toronto.

Codd, E.F., (1971). Further Normalization of the Data Base Rational Model, Courant Computer Science Symposium 6, Data Base Systems, (Ed. R. Rustin), Prentice Hall, May 1971, pp. 33-64.

Delobel, C., (1978). Normalization and Hierarchical Dependencies in the Relational Data Model, ACM TODS, Vol. 3, No. 3, 201-222.

Fagin, R., (1977). Multivalued Dependencies and a new Normal Form for Relational Databases, ACM TODS, Vol. 2, No. 3, 262-278.

Fagin, R., (1979). Normal Forms and Relational Operators, ACM SIGMOD, BOSTON, 153-160.

Fagin, R., (1979). A Normal Form for Relational Data Bases that is Based on Domains and Keys, IBM Res. Rep., RJ 2520(32950).

Fagin, R., (1980). Horn Clauses and Database Dependencies, Proc. 1980 ACM SIGACT Symp. on Theory of Computing, IBM Res. Rep. RJ 2741.

Fagin, R., (1977). Functional Dependencies in a Relational Database and Propositional Logic, IBM J. Res. Develop., 534-544.

Janssens, D. and J. Paredaens, (1979). General Dependencies, University of Antwerp Res. Rep. 79-35.

Lien, Y., (1979). Multivalued Dependencies with Null Values in Relational Databases, Proc. of Fifth International Conference on Very Large Data Bases, Rio De Janeiro, Brazil, 61-66.

Mendlezon, A.O., and D. Mairer, (1979). Generalized Mutual Dependencies and the Decomposition of Database Relations, Proc. Fifth Int. Conf. on Very Large Data Bases, Brazil, 75-82.

Nicolas, J.M., (1978). Mutual Dependencies and Some Results on Undecomposable Relations, VLDB 78, Berlin, 13-15.

Paredaens, J. and M. Gallaire, (1980). Transistive Dependencies in a Database Scheme, R.A.I.R.O. Informatique/Computer Science, Vol. 14, No. 2, 149-163.

Smith, J.M., (1978). A Normal Form for Abstract Syntax, Proc. VLDB, West Berlin, 156-162.

Sadri, F. and J.D. Ullman, (1980). A Complete Axiomatization for a Large Class of Dependencies in Relational Databases, Proc. Twelfth Annual ACM Symp. on Theory of Computing, Los Angeles, 28-30.

Sagiv, Y. and S. Walecka, (1979). Subset Dependencies as an Alternative to Embedded Multivalued Dependencies, Dept. of Computer Science, University of Illinois at Urbana-Champaign, UIUCDCS-R-79-980, UILU-ENG 79 1732.

Schmid, M.A. and J.R. Swenson, (1975). On the Semantics of the Relational Data Model, ACM SIGMOD International Conference on Management of Data, San Jose, California, 211-223.

Yannakakis, M. and C.M. Papadimitriou, (1980). Algebraic Dependencies, M.I.T. Research Report.

Vassiliou, Y., (1980). Functional Dependencies and Incomplete Information, Proc. VLDB, Montreal, 260-269.

RECENT RESEARCH RESULTS ON THE INFOPLEX INTELLIGENT MEMORY SYSTEM (IMS) PROJECT

Stuart E. Madnick

Center for Information Systems Research
Alfred P. Sloan School of Management
Massachusetts Institute of Technology
Cambridge, Massachusetts 02139

ABSTRACT

This paper presents the results on some recent performance models of the Data Storage Hierarchy (DSH) subsystem of the INFOPLEX Intelligent Memory System (IMS).

In the INFOPLEX IMS, the theory of hierarchical decomposition is applied to produce a highly parallel and pipelined database computer architecture which can be implemented by means of complexes of multiple-microprocessors. In particular, a data storage hierarchy is used to handle the storage and retrieval of a large volume of data efficiently. The information management functions are decomposed into a functional hierarchy implemented by a hierarchy of microprocessors. Decentralized control mechanisms are used to coordinate the activities of both the storage hierarchy and the functional hierarchy.

KEYWORDS

Database computer; databases; storage hierarchies; performance evaluation; performance models; simulation.

INTRODUCTION

Confronted by an ever more complex and information intensive society, it has become increasingly difficult to make decisions without more effective information management tools. Due to their enormous storage capacity and processing speed, computers have become a powerful tool for storing, processing, and retrieving information.

In our view of the future, we can foresee the evolution of an information utility (Hsiao and Madnick, 1977; Madnick, 1977) where personal computers, as well as large-scale conventional computers, can be connected to a network of information nodes. The INFOPLEX Intelligent Memory System (IMS) (Lam and Madnick, 1978; Madnick, 1977;

Work reported herein was supported, in part, by the National Science Foundation through grant MCS77-20829 and, in part, by the Naval Electronic Systems Command through contract N00036-78-G-0160.

Madnick, 1979) is aimed at meeting the above requirement for a high performance information node with a throughput capability of 10-100 times that of conventional architectures.

THE INFOPLEX IMS ARCHITECTURE

The key concepts of the INFOPLEX architecture are hierarchical decomposition and distributed control. The conceptual organization of the INFOPLEX is depicted in Fig. 1.

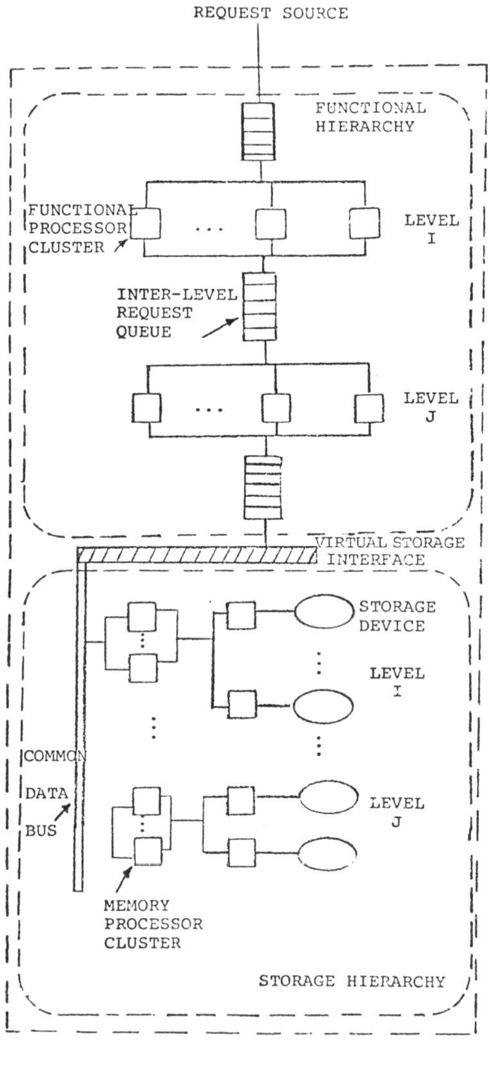

Fig. 1. INFOPLEX IMS Conceptual Organization

Techniques of hierarchical decomposition are applied to organize the information management functions to obtain a highly modular functional hierarchy. Each level of the functional hierarchy is implemented using multiple microprocessors.

The techniques of hierarchical decomposition are also applied to organize the storage subsystem to obtain a modular storage hierarchy capable of supporting the storage requirements of the functional hierarchy. Microprocessors are used at each level of the hierarchy to implement the storage management algorithms so the hierarchy appears as a very large, highly reliable, high performance virtual storage space.

Due to the high modularity of these organizations, both the functional hierarchy and the storage hierarchy can take advantage of distributed control mechanisms. Each level in a hierarchy only communicates with its adjacent levels and each module within a level only communicates with its adjacent modules. Thus, no central control mechanism is necessary.

This paper will address recent performance studies of the storage hierarchy subsystem. For more information on the functional hierarchy, the reader should refer to Madnick (1979).

THE INFOPLEX IMS STORAGE HIERARCHY

To provide a high performance, highly reliable, and large capacity storage system, INFOPLEX makes use of an automatically managed storage hierarchy.

Rationale for a Storage Hierarchy

The requirement of high performance storage at low cost is best satisfied by a mixture of technologies combining expensive high-performance devices with inexpensive lower-performance devices.

There are many ways that such an ensemble of storage devices may be structured, but in our research (Madnick, 1973; Madnick, 1975a), we have found the technique of hierarchical decomposition to be very effective. Using this technique, the ensemble of heterogenous storage devices is organized as a hierarchy. Information is moved between storage levels automatically depending upon actual usage such that the information most likely to be referenced in the future is kept at the highest (most easily accessed) levels.

General Structure of the INFOPLEX Storage Hierarchy

The lowest level of the hierarchy always contains all the information of the system. A high level always contains a subset of the information in the next lower level. To satisfy a request, the information in the highest (most easily accessed) level is used. The addressed information will be moved up the hierarchy if it is found in other than the highest level.

Strategies for Information Movement

Various storage management and movement techniques, such as page splitting, read through, and store behind, can be distributed within the hierarchy of levels. This facilitates parallel and asynchronous operation in the hierarchy. Furthermore, these approaches can lead to greatly increased reliability of operation. For example, under the read-through strategy (Fig. 2), when data currently stored at level i (and all lower performance levels) is referenced, it is automatically and simultaneously copied and stored into all higher performance levels. The data itself is moved between levels in standard transfer units, also called pages, whose size N (i-1, i) depends upon the storage level from which it is being moved.

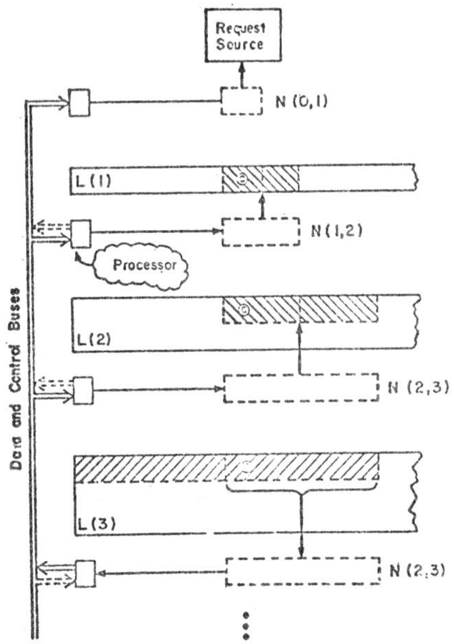

Fig. 2. Read Through Data Movement Strategy

For example, suppose that the datum "a", at level 3, is referenced (see Fig. 2). The block of size N(2,3) containing "a" is extracted and moved up the data bus. Level 2 extracts this block of data and stores it in its memory modules. At the same time, level 1 extracts a sub-block of size N(1,2) containing "a" and level 0 extracts the sub-block of size N(0,1) containing "a" from the data bus. Hence, under the read through strategy, all upper storage levels receive this information simultaneously.

In a storage behind strategy all information to be changed is first stored in L(1), the highest performance storage level. This information is marked "changed" and is copied into L(2) as soon as possible, usually during a time when there is little or no activity between L(1) and L(2). At a later time, the information is copied from L(2) to L(3), etc. A variation on this strategy is used in the MULTICS Multilevel Paging Hierarchy (Greenberg and Webber, 1975). This strategy facilitates more even usage of the bus between levels during idle bus cycles. Furthermore, the time required for a write is only limited by the speed of the highest level memory.

More details on the INFOPLEX IMS storage hierarchy algorithm can be found in Madnick (1975a, 1975b, 1979).

INFOPLEX STORAGE HIERARCHY PERFORMANCE MODEL

A series of simulation studies of the INFOPLEX IMS Data Storage Hierarchy (DSH) system have been recently completed (Lam and Madnick, 1979). A key objective of these simulation studies was to assess the feasibility of supporting very large transaction rates requiring millions of storage subsystem reads and writes per second with good response time using the current design of the INFOPLEX IMS storage hierarchy and the read-through and store-behind algorithms.

An initial GPSS/360 simulation model was developed for a configuration with three storage levels. The results obtained from this model proved interesting. It was found that, at very high locality levels, when most of the references are satisfied by the highest performance storage level, the store-behind algorithm interacts with the DSH-11 buffer management algorithms to create a system deadlock. This was not anticipated in the design of the storage hierarchy protocols and led to a redesign of the buffer management scheme to be used.

Another GPSS/360 simulation model was developed for a configuration with four storage levels. This model makes use of deadlock-free buffer management algorithms. Results from this model revealed further interesting properties of the store-behind algorithm. It was found that at high locality levels, the store-behind requests form a pipeline. Thus the rate of write operations that can be serviced is limited by the slowest stage in the pipeline, i.e., the slowest storage device. It was also found that a bottleneck may be developed at the lowest level when the block size of that level is too large.

A better balanced system was obtained by increasing the degree of parallelism in the lower storage levels and by decreasing the block sizes used by these storage levels. This system was then used as a basis to compare the performance of the INFOPLEX IMS storage hierarchy architecture under different technology assumptions. It was found that using 1979 technologies, a throughput of .7 million operations per second with mean response time of 60 microseconds was obtainable for a mix of storage references consisting of 70 percent read requests and 30 percent write requests. Using 1985 technology assumptions, the same storage reference mix produced a throughput of 4 million operations per second with a mean response time of 10 microseconds. These results are depicted in Fig. 3.

CONCLUSION

The need for highly reliable information management systems that can support transaction volume and storage capacity several orders of magnitude higher than current information systems calls for new architectures, both in information management functions and the hardware that supports these functions.

The INFOPLEX Intelligent Memory System architecture is a major effort in this direction. The performance studies reported in this paper indicate that its automatic storage hierarchy should be able to support the high performance, high capacity storage requirements of INFOPLEX IMS.

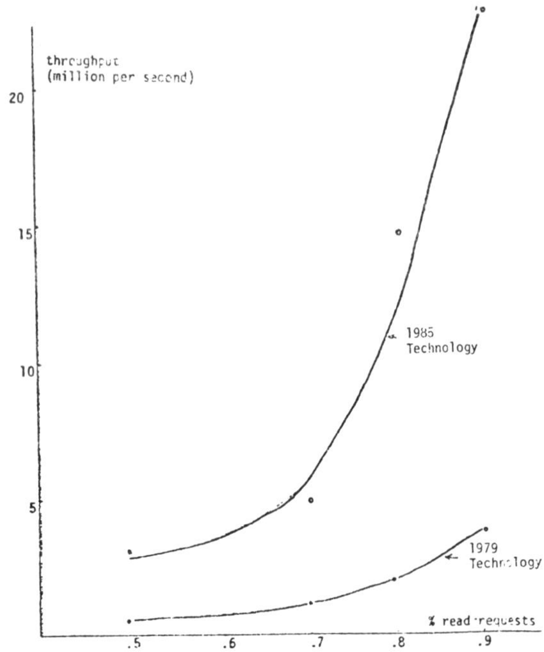

Fig. 3. Throughout Projections for Data Storage Hierarchy

REFERENCES

Greenberg, B.S., and Webber, S.H. (1975). Multics Multilevel Paging Hierarchy. IEEE INTERCON.

Hsiao, D.K., and Madnick, S.E. (October 6-8, 1977). Data Base Machine Architecture in the Context of Information Technology Evolution. Proceedings of the Third International Conference on Very Large Data Bases, 63-84.

Lam, C.Y., and Madnick, S.E. (August, 1978). INFOPLEX Data Base Computer Architecture - Concepts and Directions. M.I.T. Sloan School of Management, C.I.S.R. Internal Report No. N010-7808-01.

Lam, C.Y. and Madnick, S.E. (September, 1979). Simulation Studies of the DSH-11 Data Storage Hierarchy System. C.I.S.R. Internal Report No. N010-7909-04.

Madnick, S.E. (1973). Storage Hierarchy Systems. M.I.T. Project MAC, Report No. TR-105.

Madnick, S.E. (1975). Design of a General Hierarchical Storage System. IEEE INTERCON, 1-7.

Madnick, S.E. (May 1975). INFOPLEX Hierarchical Decomposition of a Large Information Management System Using a Microprocessor Complex. National Computer Conference, 44, 581-587.

Madnick, S.E. (1977). Trends in Computers and Computing: The Information Utility. Science, 195, 5283, 1191-1199.

DESIGN OF AN INTEGRATED MEDICAL DATABASE

P. De* and W. D. Haseman**

*College of Administrative Science, Ohio State University, Columbus, OH 43210

**School of Business Administration, University of Wisconsin-Milwaukee, Milwaukee, WI 53201

ABSTRACT

This paper presents an integrated approach to Database Design. It describes an efficient procedure to generate the logical structure of a database. A model for the physical implementation is also developed. The methodology has been used in a real world medical application. A description of that study is included.

KEYWORDS

Database; logical design; physical design; medical system.

INTRODUCTION

Of late the subject of database design has attracted a lot of attention in the research community. However, most of the reports published to date concentrate either solely on the logical aspects of the design or solely on its physical aspects, but rarely on both. While such research has its own merit, more emphasis needs to be placed on the overall design. In this paper we describe an attempt in that direction.

We first present a procedure for generating an efficient logical structure of a database. The structure so formed is shown to possess a number of interesting and useful properties. Once the logical design is complete, a model for its physical implementation is also developed. The objective of the model is to minimize the total cost, subject to the operating constraints that may be present in the system under consideration. The entire process has been applied in the development of a database for a moderately large medical system, using real life data.

LOGICAL STRUCTURE

Our methodology takes advantage of the properties of both the relational model and the network model (Date, 1977; Haseman, 1977) of database design. Judged individually, the relational model appears to be superior in terms of the logical description of a database, while the network model looks better and more practical from the point of view of the actual design. We use the first one for the initial description of the database and the second one for its final structure, and thereby capture the best of both worlds.

Our procedure consists of two algorithms. A detailed description of these algorithms can be found in (De, 1978). Here we will just summarize what they perform.

The input to the first algorithm is the set of all relations in the database and the set of attributes (Date, 1977; Haseman, 1977) that form these relations. The output is a collection of attribute groups such that (i) every attribute in each group occurs in more than one relation, (ii) the attributes of the same group always occur together, i.e., if one of them occurs in some relation, then the others also occur in that relation; if one does not, the others do not either.

The input to the second algorithm is the output of the first one, as well as the relations themselves and the information about how their attributes depend on one another. The output is a network structure which is equivalent, in information content, to the original relations. The algorithm uses two special operations that we have developed, namely SPLIT and MERGE. The first one takes, from each relation, the attribute groups mentioned above, and establishes the proper links between these groups and the remaining portion of the relation. The effect is to convert the relations into a network. The second operation combines all occurrences (Date, 1977; Haseman, 1977) of the same group, without keeping any value more than once. Duplications of the same value are thereby avoided.

The network generated has some interesting properties (De, 1979). They can be stated as follows: (i) The number of data items is minimized in the network. In fact, each data item now appears only once, even though the original relational description might contain repetitions; (ii) Given that the number of data items is minimized, the number of records (Date, 1977; Haseman, 1977) as well as the number of their occurrences are also minimized in the network; (iii) If we start with Third Normal Form (3NF) relations (Codd, 1972), then the network generated will maintain the advantages of third normality.

From properties (i) and (ii), we can conclude that the network contains no redundant information. Property (iii) is particularly important. The relational literature shows that if a relation is not in 3NF, there are some potential problems in the insertion, deletion or updating of information; however, the problems do not arise if third normality is maintained. Property (iii) says that our network structure is also free from these problems, once we make sure that all input relations are converted to 3NF. Such conversion is straightforward (Codd, 1977).

PHYSICAL IMPLEMENTATION

Once the logical structure of the database is established, the next obvious question is how we can implement it in practice. There are several techniques available for implementing each <u>set</u>[1] of the network. Some of these techniques are preferable in terms of the storage requirement, while some others are preferable in terms of the access time. The objective is to determine which particular technique to use for each individual <u>set</u> so that the sum of storage and access costs is minimized for the entire network. In the process of doing so, we must not exceed the maximum allowed levels on the total storage space and the expected access time, if such levels exist.

We have formulated the problem as an integer programming model (De, 1978). A brief description of the model appears in Appendix 1. The following is a description of a solution procedure. Suppose g_i (j,k) = minimum cost for implementing i <u>sets</u>

---

[1] Here "<u>set</u>" refers to a set-structure as defined in the network model (Date, 1977; Haseman, 1977), not a usual mathematical set. To avoid confusion, we will underline network sets in our discussion.

(1 through i) with no more than j amount of storage space and no more than k amount of expected access time. Then the model can be solved by the following dynamic program:

$$g_o(j,k) = 0 \text{ for all } j \text{ and } k$$

$$g_i(j,k) = \min_{m=1,\ldots,NUM_i} \{C_{mi} + g_{i-1}(j-SP_{mi}, k-w_i AT_{mi})\}$$

where $C_{mi} = C_1 SP_{mi} + F_i C_2 AT_{mi}$.

It is shown (De, 1978) that this solution procedure is computationally efficient. Actually, the efficiency can be regulated by varying the units of storage space and access time. Larger units make the process faster, at the expense of the accuracy of the result.

MEDICAL APPLICATION

We have applied the above methodology to design a database to be used by the Veteran's Administration Hospital in Pittsburgh for its infectious disease control program.[2] Among other capabilities, this database system will provide the microbiology department with an automated system for storing, recording and reporting bacteriology culture results. It will reduce the amount of clerical, fiscal and administrative work currently performed by medical technologists, thereby expediting the delivery of laboratory test results to patient-charts. The database will also enable researchers in related fields to readily access patient related information. The following is an outline of the design.

To start with, we derived a list of all data elements necessary to describe the system. This was done with the help of the Infection Control Officer and the Nurse Epidemiologist.

The next task was to find out how these various data elements depended on one another, or in other words, to identify the functional dependencies (Date, 1977; Haseman, 1977) present in the system. Once we did that, it was easy to represent the system in terms of a relational model. First, the basic relations were formed, and then they were converted to their 3NF (Codd, 1972) counterparts following the decomposition approach (Codd, 1972). Twenty-four 3NF relations resulted. We used the above set of relations as input to our network generating algorithms discussed earlier. The resulting network contains seventy-six <u>sets</u>. As mentioned before, this network possesses some interesting "optimal" properties.

The next phase of the design process involved the use of our implementation model. For that, we had to estimate the values for the following parameters: (i) storage utilization for each <u>set</u> as a function of <u>set</u> implementation; (ii) cost per unit of storage per unit time; (iii) storage capacity of the system; (iv) each <u>set's</u> contribution to access time as a function of set implementation; (v) cost per unit of access time; (vi) maximum allowable expected access time, weighted over all queries.

Of these parameters the ones that required usage information, were derived from the data for a typical year's activity at the infectious disease control center. Cost figures were based on the typical rates charged by the computation Center of Carnegie-Mellon University, where the computer runs took place. Four different kinds of implementation techniques were used. These techniques differed from one another only in the number of pointers (Haseman, 1977) used. Implementation 1 used only the "next" pointer, 2 used "next" and "prior" pointers, 3 used "next" and "owner" pointers, and 4 used all three pointers. Limits on storage space and access time were varied over a number of plausible values in order to study the behavior of the model. The output of one of those trials is shown in Appendix 2.

---

[2] A detailed narrative of the text environment can be found in (Fedorowicz, 1980).

The output indicates that the bounds on storage utilization and access time have been adjusted upwards. This is a reflection of the difficulty imposed by the use of a finite step size in the dynamic program calculations. During each iteration of the DP, we utilize partial solutions (for a number of storage space-access time "pairs") that were derived in the previous iteration. Because a finite step size is used, it is necessary to round down the desired "pair" of bounds to the nearest "pair" for which we have a solution. Thus we "lose" some storage space and access time (averaging half of the step size) during each iteration. The unfortunate result is that the originally intended bounds on storage utilization and access time are in effect diminished. We have attempted to alleviate this problem by adjusting the original bounds upward, with the intent that the effective bounds will then approximate those desired.

ACKNOWLEDGEMENT

This work was supported in part by grants from the Information Science and Technology Division of the National Science Foundation (IST-7918472 and IST-8019259). The authors would like to thank Jane Fedorowicz and Mike Golibersuch.

REFERENCES

Codd, E. F. (1972). Further normalization of the database relational model. In R. Rustin (Ed.), Database Systems, Courant Computer Science Symposia Series, #6, Prentice-Hall, pp. 33-64.

Date, C. J. (1977). An Introduction to Database Systems, 2nd ed. Addison-Wesley, Reading, MA.

De, P., W. D. Haseman, and C. H. Kriebel (1978). Toward an optimal design of a network database from relational descriptions. Oper. Res., 805-823.

De, P. (1979). Essays on Scheduling and Database Systems. Ph.D. Thesis, Carnegie-Mellon University.

Fedorowicz, J., and W. D. Haseman (1980). Database design for infectious disease control. J. of Med. Systems, 4, 2.

Haseman, W. D., and A. B. Whinston (1977). Introduction to Data Management. Irwin, Homewood, IL.

APPENDIX 1

Notations

N = Total number of sets in the network.

NUM_1 = Number of possible implementation alternatives for set i, i=1,2,...,N.

SP_{mi} = Storage space required for set i if alternative m is chosen for i, m=1,2,...,NUM_i.

C_1 = Cost per unit storage space per unit time.

M_i = Number of possible types of queries that access set i.

AT_{mij} = Access time for set i corresponding to the jth type of query if alternative m is chosen for i, j=1,2,...,M_i.

f_{ij} = Frequency of the jth type of query for set i.

$$F_i = \sum_{j=1}^{M_i} f_{ij}$$

= Total frequency of the queries that access set i.

$P_{ij} = f_{ij}/F_i$

$AT_{mi} = \sum_{j=1}^{M_i} P_{ij} AT_{mij}$

= Average access time for set i if m is chosen for i.

C_2 = Cost per unit of access time.

w_i = F_i/n, where n is the cumulative frequency of all queries together.

MAX_1 = Upper bound on the available storage space for the entire network.

MAX_2 = Upper bound on the expected access time required to process a query.

X_{mi} = Implementation indicator = $\begin{cases} 1 & \text{if alternative m is chosen for set i} \\ 0 & \text{otherwise.} \end{cases}$

Model

$$\text{Minimize} \sum_{i=1}^{N} \sum_{m=1}^{NUM_i} X_{mi}(C_1 SP_{mi} + F_i C_2 AT_{mi})$$

$$\text{Subject to} \sum_{i=1}^{N} \sum_{m=1}^{NUM_i} X_{mi} SP_{mi} \leq MAX_1$$

$$\sum_{i=1}^{N} w_i \sum_{m=1}^{NUM_i} X_{mi} AT_{mi} \leq MAX_2$$

$$\sum_{m=1}^{NUM_i} X_{mi} = 1 \text{ for all i}$$

$$X_{mi} = 0 \text{ or } 1$$

APPENDIX 2

Actual storage capacity is:	560000.00
Adjusted storage capacity is:	649702.80
Minimum storage space required is:	441970.00
Steps size for storage is:	2360.60
Actual maximum average access time is:	11000.00
Adjusted maximum average access time is:	13546.88
Minimum average access time required is:	7648.84
Steps size for access time is:	67.02

Per unit storage cost is: 0.0003564450 in units of \$ per word year
Per unit access cost is: 0.0000027700 in units of \$ per millisecond

Implementation #1: 1,2,4,5,6,8,9,10,12,13,14,15,17,20,21,23,25,26,27,28,29,32,33
35,36,39,40,41,43,45,46,47,48,49,50,51,52,53,54,55,56,57,58,59
62,63,64,65,66,67,68,69,70,71,72,73,74,75,76

Implementation #2: 11,16

Implementation #3: 3,7,19,22,24,30,31,34,37,38,42,44,60,61

Implementation #4: 18

Total storage used is:	549920.00 words	Total access cost is:	\$104.04
Average access time is:	7811.95 milliseconds	Total cost is:	\$300.06
Total storage cost is:	\$196.02		

A NETWORK STRUCTURE STUDY OF THE ERRORS IN DATABASES

A. J. Payne,

University of Waikato, New Zealand

ABSTRACT

We have derived a model of a database based on graph theory which will evaluate the usage of particular elements. The model has been extended to derive the error probability in the database and a technique has been presented to determine values from a run time system which will give the precise values for a particular database to indicate where it needs to be improved.

INTRODUCTION

This paper analyses the structure of a database as a network and shows how a study of error properties can be derived by the inclusion of a sink to cover the set of error cases that can arise in a database definition. The first part of the paper gives a short survey of database definition techniques. The second section enumerates some results that are derived from graphs and networks. In part three the results from the first two sections are combined to give a set of useful results to give a measure of success for a database system. The fourth portion summarises the results that were derived in the preceding sections. Finally a list of references is given to help with library searches.

DATABASE BACKGROUND

Databases and files have long been a matter of research. They are known to have units, properties and measures. They range in structure from sequential files through indexed files, scatter storage techniques [Sperry Rand Research Centre 1971] to network [Metaxides 1969] and relational structures [Codd 1970]. The actions include update, replacement, insertion, deletion, ordering, duplication and recovery and are applied to all the structures. Van der Pool [1972, 1973a, 1973b] and Schneiderman [1973, 1974] have studied the optimum organisation of a database.

RESULTS FROM GRAPHS AND NETWORKS

Graph theory [Beztiss 1975] has been studied from the early times of modern mathematics on the basis of Lattice Theory and the ordering of elements. They have been processed in terms of path matrices to determine loops and the flow between nodes. If i and j are nodes of a graph then $a_{ij} = 1$ where there is a connection between nodes i and j, and 0 otherwise. The matrix A is known as the path matrix.

The raising of the matrix to a power gives the lengths of paths between nodes and when a diagonal element has a value there is a cycle in the graph.

Shortest path problems have been derived from the path matrices when multiple paths join nodes. These tend to have little application to the theory of database. The last technique found with graph theory is that of flow quantities through the graph. If we define a matrix to indicate the proportion of in flow from one node to another then we can define a matrix A corresponding to the path matrix. From the input flow vector B we can calculate the outflow and flow through nodes of the graph by the equation $C = (I-A)^{-1}B$.

COMBINED RESULTS

The definition of a database can be represented as a set of syntactic and semantic rules. Each of the rules can be redefined as a set of conjunctive elements or disjunctive elements in separate sets of rules.

eg. $A = B|C|D$
$B = D.E.F$

Let us try to map these relations onto a graph structure - we obtain two structures shown Fig. 1 and 2 respectively. (Notice the analogy of data access with the tree structure). The head of a subtree is the owner of the tree. The multiple exits shown in Fig. 1 demonstrate the division of flow along the different paths selected from an 'or' operation. The single path given in Fig. 2 shows how each element is used with equal frequency under the 'and' operation.

We can exemplify the use of chain rules by the rule

$A = B.A$

The network for this operation is shown in Fig. 3. We see how the definition is the loop within the network with the definition being held at the head of the tree.

The simple sets of rules combine together very easily in the database context but not so easily when we try to combine them in terms of networks. When we describe the flow through the database we can define an enumeration of nodes and a set of linear equations which show rates of use of certain paths. As examples of this equation let us consider the set of rules

$A = B|C|D$
$B = D.E.F$
$D = B.A$

We make the enumeration

A ⟶ 1
B ⟶ 2
C ⟶ 3
D ⟶ 4
E ⟶ 5
F ⟶ 6

If we use b_i to represent the flow/usage of node i and a_{ij} to give the portion of the flow at node i to node j, thus:

$$b_1 = b_4$$

$$b_2 = a_{12}\ b_1 + b_4$$

$$b_3 = a_{13}\ b_1$$

$$b_4 = a_{14}\ b_1 + b_2$$

$$b_5 = b_2$$

$$b_6 = b_2$$

An additional constraint on the a matrix is

$$a_{12} + a_{13} + a_{14} = 1$$

We note that the above set of equations is inconsistent and this situation is alleviated by specifying either the usage quantity of the head nodes or the usage quantity of the bottom nodes. For example we can modify the usage of the head node to become $b_1 = b_4 + 1$ which is a consistent equation. Alternatively we may change the usage of the bottom nodes to give

$$b_3 = 1$$

$$b_5 = 1$$

$$b_6 = 1$$

which leads to inconsistent equations.

We can use the technique to evaluate the usage of elements of the database. The frequency of usage will help us determine where the accent should be placed on designing the database. If we know the access time to elements, we have a measure of the runtime of the database operations.

Let us modify the basic elements of the database graph to include an 'error' sink. We can show the effect of errors occurring in the database by defining the error element in the alternative rules. For example in our elementary database we would find that the rules would become

A = B|C|D|G
B = D.E.F
D = B.A

where G is the error element.

The enumeration of elements would be (A,B,C,D,E,F,G) ——>(1,2,3,4,5,6,7) and the equations of usage assuming that the head was specified would be

$$b_1 = b_4 + 1$$

$$b_2 = a_{12}\ b_1 + b_6$$

$$b_3 = a_{13}\ b_1$$

$$b_4 = a_{14}\ b_1 + b_2$$

$$b_5 = b_2$$

$$b_6 = b_2$$

$$b_7 = a_{17}\, b_1$$

$$a_{12} + a_{13} + a_{14} + a_{17} = 1$$

The value of the usage of the error element G is the value of b_7 so we have a way of assessing how much a database is subject to errors. If we wish to include the effect of errors into the basic elements then we can add the extra rules

$$E = H \mid G$$
$$F = I \mid G$$

where H and I are the true values of E and F respectively in the original database. H and I map onto nodes 8 and 9 giving the equations for b_7 b_8 and b_9

$$b_7 = a_{17}\, b_1 + a_{57}\, b_5 + a_{67}\, b_6$$

$$b_8 = a_{58}\, b_5$$

$$b_9 = a_{69}\, b_6$$

where $a_{57} + a_{58} = 1$

and $a_{67} + a_{69} = 1$

Summarising the results so far

(a) we can express a database as a graph structure

(b) we can express the flow from each element relative to others as a matrix A where A_{ij} is the use of element i relative to element j

(c) we can calculate the absolute use of each database element as $B = (I-A)^{-1}C$ where B_i is the usage of element i and C_i is the usage of basic elements of the database

(d) the average execution time of a data access can be defined by DB where D_i is the process time of each element in the base element set

(e) we can modify the database definition to include a base element known as 'error' which can be included in all alternative rules and introduced into all positions using basic elements

(f) we can evaluate relative chances of errors, and absolute chances of errors by evaluating the usage of the error element.

From the practical viewpoint we can use a monitor of the database to determine the values of the A matrix and the C vector for both the use of data elements and the occurence of errors. We can calculate the values of the D vector from the database definition so that we can evaluate the expected process time of the database and the expected error rate.

Let us consider how we can apply the model defined above to draw some conclusions about a database using empirically simple values (see Fig. 4,5,6,7,8 and 9). We note that every alternative branch decreases the usage of the elements. Each recursion rule gives an increase of usage. When it comes to error evaluation the conjugate rules decrease the probability of correctness in a data access to a much greater degree than alternative rules. Recursion influences correctness to the extent in which recursion causes conjugation of elements. The processing time of the database is increased by the conjugate laws and the number of times conjugate laws are activated by a recursion rule. The disjunctive rules have little effect on the process time of a database as such.

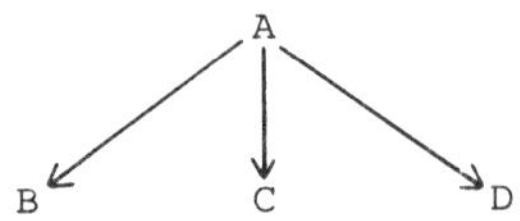

Fig. 1 Disjunctive tree A = B|C|D

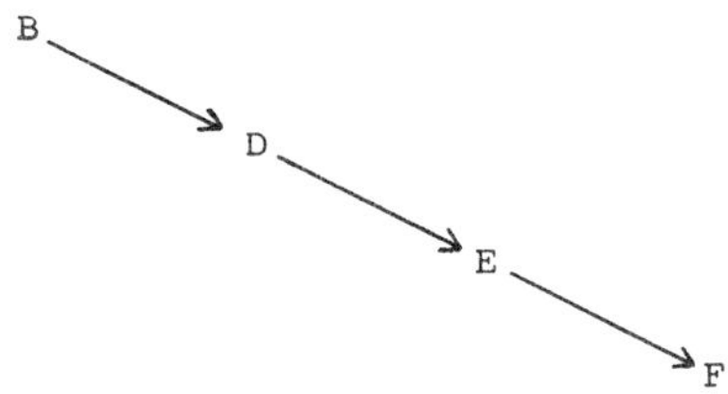

Fig. 2 Conjunctive tree B = D.E.F

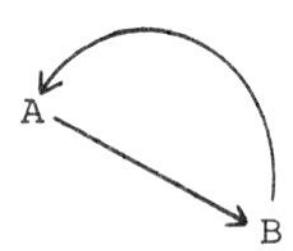

Fig. 3 Recursion network A = B.A

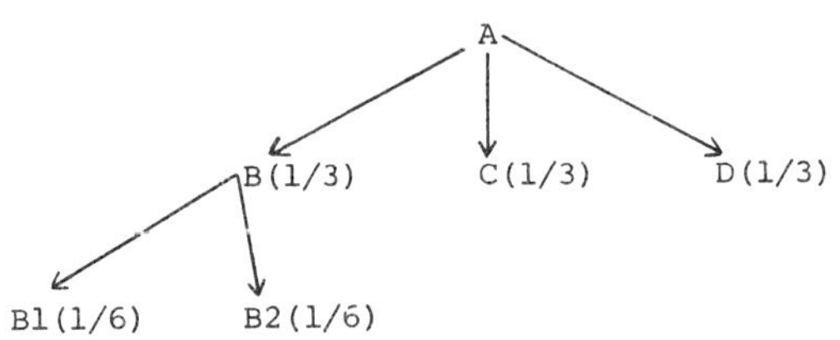

Fig. 4 Flow - disjunctive rules A = B|C|D y
B = B1|B2

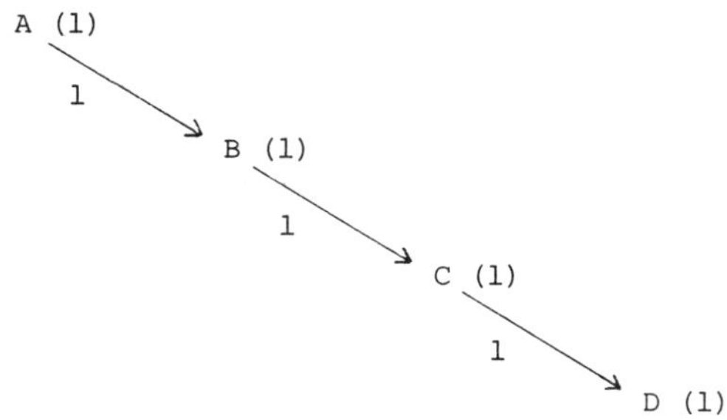

Fig. 5   Flow - conjunctive rules A = B.C.D

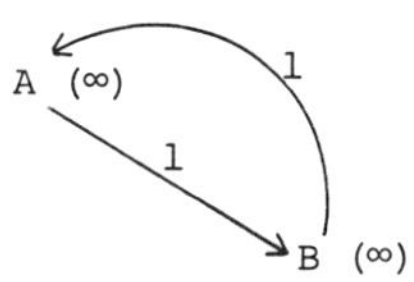

Fig. 6   Flow - recursive rules A = B.A

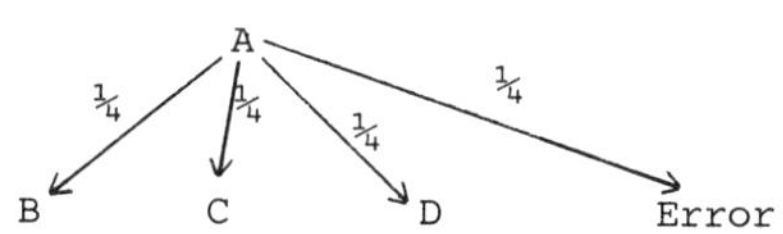

Fig. 7   Correctness - disjunctive rules A = B|C|D|Error

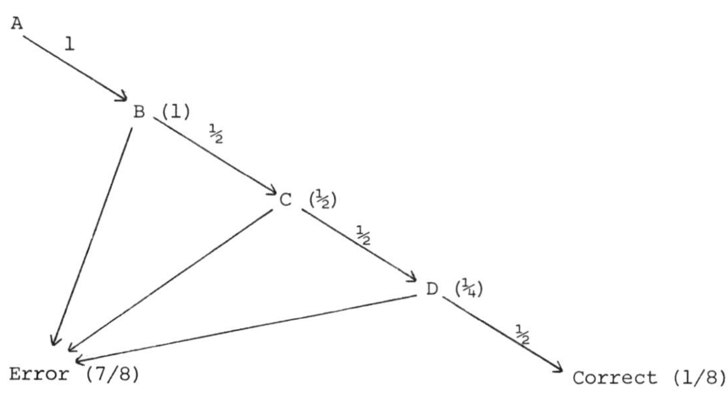

Fig. 8   Correctness - conjunctive rules A = B.C.D

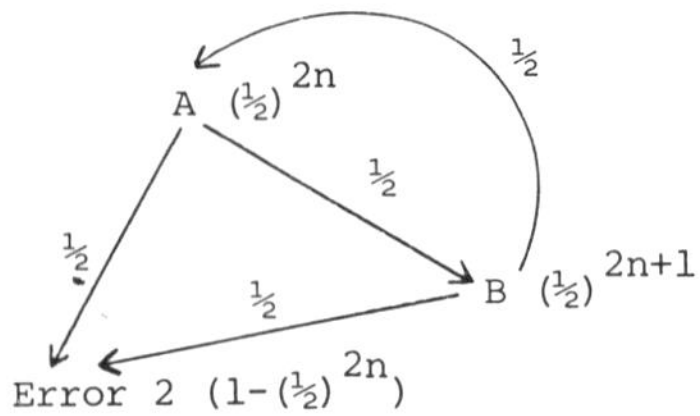

Fig. 9 Correctness - recursive rules.

CONCLUSIONS

We have derived a model of a database based on graph theory which will evaluate the usage of particular elements. The model has been extended to derive the error probability in the database and a technique has been presented to determine values from a run time system which will give the precise values for a particular database to indicate where it needs to be improved.

REFERENCES

Beztiss, A. T. (1975). Data Structures - Theory and Practice. Academic Press, New York.

Codd, E. F. (1970). A Relational Model of Data for Large Shared Data Banks. CACM 13, 377-587.

Metaxides, A. (1969). 1969 Report of the Codasyl Database Task Group.

Schneiderman, B. (1973). Optimal Storage Allocation for a File with Open Addressing. IBM J. Res. Develop 17, 106-114.

Schneiderman, B. (1974). A Model for Optimising Indexed File Structures. International J. Comp. Inf. Sciences 3, 93-103.

Sperry Rand Research Centre Working Meeting. (1971) Sudbury, Mass., U.S.A. October 1971.

Van der Pool, J. A. (1972). Optimal Storage Allocation for Initial Loading of a File. IBM J Res. Develop 16, 579-586.

Van der Pool, J. A. (1973). Optimal Storage Allocation for a File in Steady State. IBM J. Res. Develop 17, 27-38.

Van der Pool, J. A. (1973). Optimal Storage Allocation for a File with Open Addressing. IBM J. Res. Develop 17, 106-114.

DATA BASE ALERTERS AND MESSAGE FILTERS

Shi-Kuo Chang
Knowledge Systems Laboratory
University of Illinois at Chicago Circle

ABSTRACT

This paper presents the basic concepts of data base alerters and message filters. Research problems on the specification and decomposition of user models are discussed. The relevancy of fuzzy sets theory to the design of fuzzy message filters is then discussed.

KEYWORDS

Data base alerters, message filters, fuzzy sets theory, linguistic message filters, information exchange theory

INTRODUCTION

A theory of information exchange has been presented in [CHANG79]. In this paper, we present some preliminary concepts and design issues on designing data base alerters and message filters, using the framework of information exchange theory. The relevancy of fuzzy sets theory to the design of fuzzy message filters is also discussed.

<u>Data base alerters</u> are programs (or processes) associated with a data base management system, to dynamically monitor occurrences of interest in a data base [ANDER76, BUNEM77, BUNEM79,, HAMME76, MCDON75]. Data base alerters attempt to inform the user about changes of state in the data base, or the occurrences of certain events.

Data base alerters can be regarded as message filters. Incoming messages to the alerter are data base update messages, which may include "before" and "after" values of data base objects being updated. Outgoing messages from the alerter are messages to the various agents (or user programs, user processes) about data base state changes. With no loss of generality, we can restrict our attention to an alerter which serves only a single user agent. The alerter filters out irrevalent data base update messages, and sends (transformed) data base update messages to the user agent. This is accomplished by referring to "alerting conditions" set by the user agent. From the perspective of information exchange theory, the alerter uses an alien-model to filter messages and identifies messages relevant to the alien agent. The alien-model is constructed by the alerter based upon knowledge about the interests of the alien agent. In practice, this is accomplished by having the alien user agent (user program or user process) sending "set-alert" messages to the alerter [BUNEM77].

The designers of computer information systems are also interested in constructing message filters, so that an information node will not be swamped by irrevalent network messages. For example, the alerters may send out too many alerting messages, which should be masked out when necessary. Such filters are similar to alerters described above, except that instead of an alien-model, a self-model is employed. In other words, an agent knows its own interests, and uses a self-model (which reflects his interests) to filter out irrelevant messages.

Therefore, both data base alerters and message filters utilize (alien or self) user models to filter out irrelevant incoming messages, and send out (transformed) relevant outgoing messages. Data base update messages are sent to the alerter, and the messages are filtered according to alien-models of the users, i.e. S = (Sr,-,Sa), where "-" denotes "don't-care". On the other hand, incoming messages are sent to the filter, and the messages are filtered according to self-model of the agent, i.e. S = (Sr,Ss,-).

SPECIFICATION OF USER MODELS

A block diagram of the linguistic filter is illustrated in Figure 1. The decoder accepts incoming message xi, and transforms xi to decoded message x'i. The incoming message xi is first sent to a feature extractor, which generates a feature vector yi. yi is then classified by a classifier, which generates a message prototype class ci. xi, yi, and ci are then processed by a linguistic filter, which generates the decoded message x'i.

As an example, the feature extractor may extract the following information: (a) message freshness (system clock reading minus message time stamp), (b) message prioirty class (predefined in the message), (c) sender group-id (predefined), (d) relevancy score, (e) message clarity, (f) message ambiguity, (g) message strength, and (h) message specificity [YAGER78, 79]. If a message is fresh, has high priority, belongs to certain sender group of interest, is relevant, clear, unambiguous, strong and specific, then it is classified to be in message prototype class c1 (acceptance). If a message is not fresh (obsolete) and irrelevant, it is classified to be in message class c2 (rejection). Otherwise it is in class c3 (conditional acceptance). Messages in class c1 will pass through the linguistic filter unchanged. Messages in class c2 will be supressed by the linguistic filter. Messages in class c3 will be modified (similar to signal conditioning) by the filter. Therefore, the linguistic filter is functionally similar to a classical band-pass filter in systems theory.

In classical band-pass filters, signals are filtered according to a filter's frequency spectrum characteristics. In the linguistic filter, we can regard the filter as a function f, from X x S x Y x C to X, where X is the message space, S is the state space, Y is the feature space, and C is the prototype message space.

For example, the filter may be as follows: f(xi,s,y,c1) = xi, f(xi,s,y,c2) = x0, f(xi,s,y,c3) = x'i, where c1, c2, and c3 are the acceptance class, rejection class, and conditional acceptance class, respectively, s is any state, y is any feature vector, x0 is a null message, and x'i is the transformed message.

From the above discussions, it can be seen that the decoder's actions are guided by the user model. The central problem then is how to specify a user model. There are several approaches: using logical predicates to define user's interests [BUNEM77, HAMME76]; using mathematical equations to define user's regions (or clusters) of interest (assuming messages can be transformed into vector representation); using a rule-based approach [ANDER76, CHANG76, MCDON75]; or using formal grammars and linguistic specifications.

We will present a specification scheme related to formal grammars, to express the relevancy score of a message. We can then determine the relevancy of messages using relevancy classes, which should not be confused with prototype message classes. A message may be relevant according to relevancy classes, and yet not useful according to other features (for example, it may be obsolete). These other criteria are considered to be ad-hoc criteria, and not reflected in the user model based upon relevancy class specifications.

A relevancy class Rj is defined to be a set of messages having the same relevancy score. We can describe a relevancy class Rj using a formal grammer Gj, so that a message xi is in relevancy class Rj, if it is in L(Gj). The relevancy classes need not be disjoint, i.e. L(Gj) and L(Gk) may have nonempty intersection. In fact, this may even be desirable, so that there is a continuum in relevancy scoring. Each relevancy class Rj has associated with it a relevancy score rj. We define the relevancy score of a message xi, as the maximum of all rj, such that xi is in Rj.

A _specification_ of relevancy classes, or a user model, is a formal system R = (G1,r1; G2,r2; ... , Gn,rn), where Gj are the formal grammers specifying the set of messages in Rj, and rj are their relevancy scores. For example, we can use regular expressions to specify the relevancy classes: if R1 represents all messages containing a substring w, its specification is (Z)*w(Z)*, where Z is the alphabet set for the message space X.

The justification for this approach is based upon the following observations: (1) a data base consists of (possibly hierarchically organized) object-value pairs; and (2) a message is a linear string of keywords, object names, object-value pairs, etc. Therefore to specify relevancy classes, we can specify certain keywords, object names, object-value pairs that are of interest. The grammer rules then enable us to specify the structure of messages of interest. In the simplest case, we can specify a set of keywords and their relevancy scores. For simple data base alerters, the specification could be strings consisting of command-keywords (such as update, insert, delete, etc.), object names (such as database-name, file-name, etc.), and object-value pairs (such as (OLD temp, '20'), (NEW temp, {..,0,..,19}), etc).

Given a relevancy specification R, we often want to determine the approximate relevancy score of a message xi, by looking at only the initial portion of the message -- human beings can often determine the relevancy of a message (say, a letter) by looking at a portion of it. We propose the following definition: a message xi is k-relevant for Rj, if INIT(k;xi)w is in Rj, where INIT(k;xi) is the k-prefix string of xi, and w is some string in (Z)*. The k-relevancy score of xi is then defined to be the maximum of all rj's, such that xi is k-relevant for Rj.

A relevancy specification R is _k-t-adequate_ for W, where k is an integer, t is a threshold value, and W is a subset of X (the message space), if for every message xi in W, the absolute difference of its relevancy score and k-relevancy score is no greater than t. R is k-t-adequate, if it is k-t-adequate for X. Finally, R is k-adequate, if it is k-0-adequate.

If a specification R is k-t-adequate, this means that we need only look at k-prefix of messages in determining their relevancy scores, and the errors can never be greater than t. We can investigate the following: (a) Given W, what properties must R have, so that R is k-t-adequate for W, for some k and t? What rules should one impose on the specification of Gj, or what subclasses of formal grammers should one use, so that R can become k-t-adequate? (b) Given R, for what W will R be k-t-adequate? Can there be a constructive procedure to find the set of messages whose relevancy can be (easily) determined using R?

The first investigation will shed light on the design of relevancy specifications for user models. The second investigation will shed light on the design of messages (or message format specification) whose relevancy can be (easily) determined.

The linguistic filter f is defined to be a function from X x S x Y x C to X. If the relevancy specification R is k-t-adequate, then we could use f(xi,s,y,c3) = INIT(k;xi), where c3 denotes the conditional acceptance class. In other words, only the initial portion of the (less relevant) messages is retained. By definition of k-t-adequacy, this initial portion INIT(k;xi) already determines the message's approximate relevancy score. The receiver, if really interested, can inspect INIT(k;xi) and decides whether to ask the sender to retransmit the full message (or the full message may be stored somewhere, so that the receiver can retrieve it, if needed). We can further investigate the design of such restricted linguistic filters: If Gi are regular (context-free, context-sensitive, etc.) grammers, can the linguistic filter be designed in a procedural manner?

We can also investigate alternate definitions for k-relevancy: instead of looking at k-prefix of xi, one can also look at (for example) every m-th symbol in xi and use at most k symbols to determine k-relevancy; or one could adaptively look at k symbols, with a constraint on total length of prefix looked at (say, km for some constant m). These alternate definitions can be evaluated for their k-t-adequacy conditions.

DECOMPOSITION OF USER MODELS

In the previous section, we have discussed how to specify a user model using a specification of relevancy classes. Linguistic filters can then be designed, based upon user models. In this section, we discuss research problems in user model decomposition.

The motivation for user model decomposition is as follows. We have seen that both alerters and filters use user models in messages filtering. For an agent in an information network, we always have a pair (Ra, Rs). The alerter uses Ra (alien-model) to initially select relevant messages, which are transmitted over the network, to the user agent. The user agent's message filter then uses Rs (self-model) to further select relevant messages, which are finally presented to the agent.

Suppose the true model Rt is given (in terms of a relevancy class specification). In the ideal case, Ra = Rs = Rt, i.e. both alien-model and self-model are identical to the true model. In which case, either the alerter or the filter is redundant. In practice, however, Ra and Rs could be different, and their combination is Rt. We write Rt = Ra + Rs, to denote that Rt can be decomposed to Ra and Rs.

One extreme is that Ra = Rt, and Rs is a null specification. The advantage is that there will be reduced message traffic over the network. The disadvantage is that the alerter becomes quite complex. Since alerters are usually implemented within a data base management system, this may represent excessive updating overhead. Moreover, it may be difficult to keep Ra up to date. Since user's interests change in time, Rt chages. If Ra = Rt, then any change in Rt must be reflected in Ra, resulting in network traffic overhead and Ra updating overhead.

The other extreme is that Rs = Rt, and Ra is a null specification. The advantage is that no alerter is needed for this agent, and Rs can be updated locally, to reflect changes in user's interests. The disadvantage is that there might be excessive message traffic, since all update messages must be sent to the agent, before being screened by the filter.

Clearly, there are many intermediate cases, whereby part of the screening power is in the alerter, and part in the filter. We can further investigate this problem of user model decomposition.

Let R = Ra + Rs be a decomposition. Let X denote the message space, and ft(X), fa(X), and fs(X) denote the message sets passed through a filter using models R Ra, and Rs, respectively. A decomposition is <u>lossless</u> if ft(X) = fs(fa(X)). We can investigate techniques to find lossless decompositions. If ft(X) contains fa(X), i.e. the alerter prescreens messages but does not supress any relevant messages, then one can design a filter to screen remaining irrelevant messages. We can first investigate such design approaches, namely, to find Ra that guarantees losslessness, and then find Rs.

If ft(X) does not contain fa(X), i.e., the alerter may supress relevant messages, then the decomposition cannot be lossless. This problem is of interest, because if an user agent shifts his interests, then the original alien-model may not guarantee losslessness. We can investigate this problem of testing whether ft(X) contains fa(X), and means of doing this test efficiently. We can also investigate the use of message protocols, so that discrepency between Ra and Rs can be detected and corrected.

We can then investigate techniques to find the optimal lossless decomposition, which is the decomposition with minimal C1(Ra) + C2(Rs) + C3(fa(X)), where C1(Ra) is the cost of maintaining Ra, C2(Rs) is the cost of maintaining Rs, and C3(fa(X)) is the network message transmission cost.

Another topic for research is concerned with semantics of relevancy specifications. As mentioned above, data base objects might be hierarchically structured -- we have data bases, files, records, attributes, etc. If "attribute" is mentioned in Ri, then "records", "files" in same Ri should take lower structural precedence. On the other hand, data base objects may have predefined semantic precedence relations. For example, we may have a file to store time information (so that timed-based alerters can be installed), which has attributes: year, month, week, day, hour. A natural precedence among these objects are defined. If "week" is mentioned in Ri, then "month", "year" in same Ri should take lower semantic precedence. Thus, one can "focus" attention to important substrings in a message. We can also investigate the problem of incorporating such structural/semantic information in a user model, and how to utilize such information for relevancy scoring.

FUZZY MESSAGE FILTERS

The above definition of data base alerters and message filters can be enhanced, if we consider the alerters and message filters to be fuzzy alerters and fuzzy message filters. In the case of alerters, the alerting condition will be a fuzzy logical predicate. In the case of message filters, the grammers Gj can be replaced by fuzzy grammers G'j. Since real-life events are often imprecisely defined, such an extension will be quite advantageous, and further research is needed to explore the specification and design issues related to such fuzzy alerters and fuzzy message filters.

REFERENCES

[ANDER76] R. H. Anderson and J. J. Gillogly, "Rand Intelligent Terminal Agent (RITA): Design Philosophy", Rand Report R-1809-APA, February 1976.

[BUNEM77] O. P. Buneman and H. L. Morgan, "Implementing Alerting Techniques in Database Systems", Proc. of IEEE COMPSAC Conference, November 8-11, 1977, 463-469.

[BUNEM79] O. P. Buneman and E. K. Clemons, "Efficiently Monitoring Relational Databases", ACM Trans. on Database Systems, Vol. 4, No. 3, September 1979, 368-382.

[CHANG76] S. K. Chang, "Intelligent Coupling of the User to Medical Information Systems", Proceedings of Third Illinois Conference on Medical Information Systems, Chicago, Illinois, November 4-5, 1976, 29-46.

[CHANG79] S. K. Chang, "Toward a Theory of Information Exchange", Proceedings of First International Symposium on Policy Analysis and Information Systems, June 28-30, 1979, 252-264. [GROSZ77] B. J. Grosz, "The Representation and Use of Focus in a System for Understanding Dialogs", Proc. of Fifth IJCAI, Cambridge, Massachusetts, 1977, Vol. 1, 67-76.

[HAMME76] M. Hammer, "Error Detection in Data Base Systems", Technical Report, M.I.T. Laboratory for Computer Science, 1976.

[MCDER76] J. McDermott, A. Newell and J. Moore, "The efficiency Of Certain Production System Implementations", September 1976.

[MCDON75] C. McDonald, B. Bhargava, D. Jeris, "A Clinical Information System for Ambulatory Care", Proc. of National Computer Conference, May 1975, Anaheim, California.

[SIKLO79] L. Siklossy, "Passive vs. Active Question-Answering", Technical Report KSL-30, Knowledge Systems Laboratory, University of Illinois at Chicago Circle, 1979.

[YAGER78] R. R. Yager, "Toward a Theory of Signal Response based On Fuzzy Sets", Proc. Int. Conf. on Cybernetics and Society, Tokyo, 921-925, 1978.

[YAGER79] R. R. Yager, "On Modeling Interpersonal Communications", in Fuzzy Sets: Theory and Applications to Policy Analysis and Information Systems, S. K. Chang and P. P. Wang (eds.), Plenum Publishing Co., 1979.

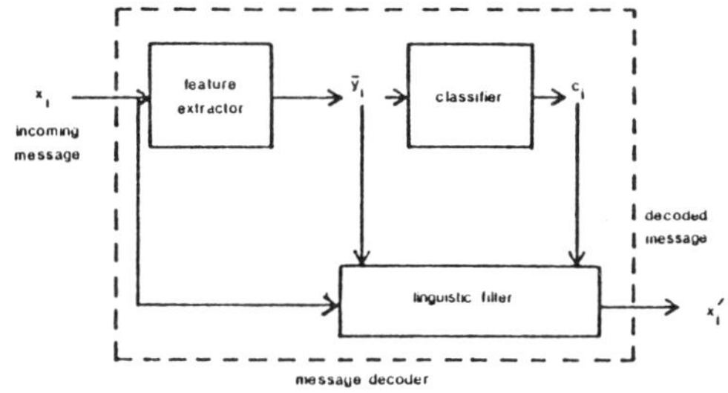

Figure 1 The Linguistic Filter

COMPUTERS: TECHNOLOGICAL MIRACLE AND SOCIETAL FAILURE?

Edward Alexander Tomeski* and Konrad E. Sadek**

* Professor of Management, Barry College, Miami, Florida, U.S.A.

**Information Consultant, World Vision International, Monrovia, California, U.S.A.

ABSTRACT

This article probes areas of social responsibility which have been neglected in the computer field. The computer industry, one of the largest and fastest growing, has great impact on our society since the computer pervades practically every organization unit and every occupation. The authors point out some of the detrimental aspects of computers and isolate areas needing constructive improvement by the computer industry and computer users. Also, they identify some of the possible steps that should be taken to overcome the lag in social responsibility in the computer field.

KEYWORDS

Computer; social responsibility; unemployment; privacy; crime; paperwork; productivity; automation; software; computer technicians.

INTRODUCTION

We live in a technological age that creates modern miracles to solve problems. Yet we have, as a people, become increasingly aware that technological developments (e.g., the automobile, nuclear energy, petro-chemicals, etc.) are unfortunately usually accompanied by negative impact on society (e.g., pollution, threat of extinction of the human society, illness and genetic disturbances, etc.).

The modern electronic computer has been touted by some esteemed thought leaders (e.g., Herbert Simon, Ashley Montague, Herman Kahn) as one of the greatest technological developments in the history of mankind. The computational and data processing capabilities of the computer have greatly extended the intellectual reach of people. For instance, many scientific and engineering breakthroughs (e.g., the space program, atomic energy) would not have been feasible without the accompanying analysis and control made possible by high speed, vast storage computer devices.

The automotive industry has been encouraged to become more concerned about air pollution and passenger safety. Companies that are using our nation's natural resources find pressure to be more conservation conscious and to be more ecology-minded. Pressure mounts to assure that food processors and drug and cosmetic manufacturers market products that are not harmful to the user's health and well being. The time may be coming ripe for the computer industry to also be exposed to a searching study of that industry's negative impact on society and how any deficiencies may be corrected. This is becoming particularly urgent because of the rapidly growing pervasiveness of computers in literally millions of organizations and homes. Only token attention has thus far been directed to these issues. One of the giants of cybernetics, Norbert Wiener, anticipated many of these impending problems with computers in his classic treatise: The Human Use of Human Beings.

WORK AND COMPUTERS

Automation and computerization have been changing the composition and distribution of work, with the computer and automation equipment absorbing both some of the menial tasks and also taking on higher order activities. Although blue collar and white collar workers are the most obvious individuals to be directly affected by automation and computerization, increasingly managers and professional people are experiencing the impact of the new technology.

To some degree the essentiality of humans is diminished when powerful automation and computer systems absorb vast segments of work processing, decision-making, and even control itself. While such systems may seem, on the surface, to improve efficiency -- there may be an overall reduction in organizational effectiveness. This is true since research studies often report that such systems are accompanied by some deterioration in employees' loyalty, moral, and motivation as they become cogs in the automation and computer system.

This dilemma is not new. It has been reported that workers tried to destroy textile machines installed in factories during the Industrial Revolution in England. We are instructed that we cannot stand in the way of technological progress, and that people must adapt to the technological marvels. Unfortunately, too many computer technicians view this as a license to automate or computerize everything possible consistent with obtaining cost advantage and benefits to justify the investment. It is rare when human factors are carefully considered in such systems; little work has been devoted to the best balance of work between people and equipment. Computer manufacturers, when they design equipment and software, also appear to be out-of-touch with human needs. For example, programming languages are still awkward to use and pose a learning problem for many persons. We are reminded of Charlie Chaplin's pathetic experience with automation in "Modern Times" and Hal's domination of the humans in the space ship in "2001: A Space Odyssey." In both cases, man became a victim of the machine systems man created!

It is interesting to note that although the U.S.A. is the most heavily computerized nation its productivity has been sagging and its economic strength has weakened in recent years. Behaviorists are likely to suggest that we have devoted toomuch attention to the machine component of the man-machine system, and not enough constructive attention to the motivation of employees. Nations with higher productivity increases and more buoyant economies appear to have devoted more attention to building better relations of a continuing type with their employees. Such nations include: Switzerland, West Germany, and Japan. This suggests that perhaps the business and government leaders, in the U.S.A., should take a broad look at the role of the worker in organizations, for it is suspected that our current systems are not conducive to developing and maintaining employee loyalty to organization and high levels of job performance. It is somewhat of a paradox that in a nation that prides itself on being the most democratic nation in the world -- little democracy is practiced in most of the business enterprises in the U.S.A.!

The eminent psychologist Erich Fromm has said: "Computers should become a functional part of a life-oriented computer system and not a cancer which begins to play havoc and eventually kills the system. Machines or computers must become means for ends which are determined by man's reason and will. The values which determine the selection of facts and which influence the programming of the computer must be gained on the basis of the knowledge of human nature, its various possible manifestations, its optimal forms of development, and the real needs conducive to this development. That is to say, man, not technique, must become the ultimate source of values; optimal human development and not maximal production the criterion for all planning."

PRIVACY AND COMPUTERS

The continued technological improvement in computers and decline in the cost of computers have reached a point where large-scale data banks have become feasible. Data which previously were scattered, unintegrated, and uneconomic to collect, store, retrieve, and analyze, can now become centralized, integrated, and relatively inexpensive to obtain. This development has raised debates over the balance between the need for improved information by government and business, and the protection of the individual against invasion of privacy as well as other potential misuse of such concentrated information. Despotic leaders of a Machiavelian ilk have long recognized the power of the control of information. And Lord Action said: "Power corrupts, and absolute power corrupts absolutely." This was dramatized by Orwell in his classic _1984_, which was prophetic in many ways as we enter the decade of the 1980's.

The computer joins a series of technical developments that potentially attack the principle of the individual's right to privacy: polygraphying, personality testing, truth drugs, wire tapping devices, and even manipulating an individual's personality by drugs and brain surgery.

It is true that, in the U.S.A., the federal government has at least temporarily shelved a previous ambitious plan for a national data bank which could have been used for tracking every citizen's activities. Such an instrument in the hands of an "imperial" president's administration (such as Nixon's) could result in harrassment and persecution of minorities who differed from the administration's views. Congress and some state legislators have passed a number of laws which provide some limited protection to the privacy rights of citizens. Yet, knowing the inadequacy of enforcing laws when the administrator is power-hungry, the threat remains that government and business can use their computer networks (perhaps on a sub rosa basis) to monitor the activities of citizens and employees. Recall how in the frightening satire "Dr. Strangelove" a despotic leader was able to computer control an attack force against the Soviet Union? We periodically learn of alarming events taking place in the federal bureaucracy which are illegal or harmful to our national interest; only recently an audit committee discovered that a top secret computer facility was being used, by employees, for such purposes as computer games, designing Christmas cards, selling unauthorized time to outside users, etc.

Particularly disturbing is the condition that inaccurate records maintained by computers are too often part of a closed-loop system, or such computer-generated information is accepted as "gospel" by a user who may be largely computer-illiterate. In Florida, a police information data bank reported a speeding car as one that was stolen. Through an unfortunate circumstance the police officer stopping the car shot and killed the driver who he assumed was a perpetrator who had stolen the vehicle. Sadly, the data bank had not been updated for the fact that the stolen car had been found some days before and had been returned to its rightful owner. The owner was driving the car, rushing to the airport to catch a plane to visit with his dying father, when the police officer stopped him. Such inaccurate data banks may be the cause of citizens being denied credit and bank loans, losing out on job opportunities, etc. Or the result may be a lost life!

CRIME AND COMPUTERS

The computer, in a number of ways, breaks down conventional control methods. This opens avenues for illegal acts to occur by employing the computer as an electronic accomplice in crime.

Notwithstanding distributed data processing, computerization has fostered the centralizing of data processing records. Paperwork that was once performed in a variety of offices (e.g., accounting, purchasing, payroll, personnel) is now frequently processed to a large extent in a single physical location -- the computer department -- and under a single administrator. This has weakened the check and balance protection inherent when different departments and individuals perform different phases of an overall job. The centralization of control of information under one authority can be an inducement for taking unilateral steps to manipulate records.

When work is placed on the computer, many of the conventional visual records are modified or, indeed, no longer exist. A considerable amount of data, in a computer system, may be stored and processed in electronic form; such records are not visible to a human's sight. Traditional procedures for catching problem areas, often involving visual audits, are no longer as effective in computer systems. Moreover, computer technicians design computer applications and programs that are frequently mysterious to managers and auditors who lack the knowledge and skills of the computer technician. To an extent the organization is at the "mercy" of the computer technocrats.

Studies by Stanford Research Institute indicate that computer-assisted bank frauds are much more lucrative than those of the old fashioned manual frauds. The average computer-assisted fraud nets the culprit about $450,000 versus only $19,000 for the manual kind. A Miami bank that claims "you are worth more than you think" was reported to have accused one of its vice presidents of embezzling $350,000; he used knowledge of computers to divert funds from internal bank accounts to his own account. The famous (or infamous) Equity Funding Insurance case involved top management use of its computers to generate some 64,000 fictitious life insurance policies -- which bloated the apparent assets of the corporation and thus allowed management to conduct a number of multi-million dollar rip-off schemes in the financial community. In another case, a computer technician "broke" the code of a telephone company's telephone/communication system and was able to direct the computer to have delivered to his home about a million dollars in telephone equipment. Paradoxically, when this computer technician was eventually sent to jail for his crime -- he found a new and profitable career. Like most white collar criminals, his jail term was very modest (of course, many never go to jail) and when released he set up a prosperous consulting firm -- advising organizations how to avoid being ripped off by computer technicians!

INFORMATION POLLUTION AND COMPUTERS

Information overload has been one of the results from high speed office equipment such as computers, word processing machines, and copying devices. Citizens are bombarded by junk-type mail generated by computer processed mailing lists that appear to be used indiscriminately by marketers. Managers and employees are saturated with seemingly endless reports (some of which lack clarity to the non-computer person) which add to the paperwork pollution that abounds in most offices. Useless data and reports seem to clutter files, desks, windowsills, and other locations -- and this paperwork avalanche undoubtedly interferes with more productive activities.

It is reported that something like $40 billion a year of taxpayer's money is used to pay the bills for the U.S.A. federal paper-shuffling. About half of this cost is absorbed by business that must fill out the forms, and the other half is the cost of the government's printing, processing, and filing requirements. If one included the paperwork of state and local governments, in addition to the federal government, the total paperwork cost of all government might well be about $200 billion a year! And the problem keeps on growing. In 1965, there were 3,803 different federal forms contributing to the paperwork explosion. By 1971, there were 5,298 forms and by 1975 it was estimated that the figure was over 6,000. These forms are generated by some 3,500 government bureaus and agencies that maintain some 8,000 separate record systems. Just to print a list of these requires some 3,000 pages of close-set type in the Federal Register.

Business enterprises have their own labyrinth information overload. Some organizations are attempting, often in vain, to get a handle on this information avalanche. Some try a charge-out procedure so unit managers carefully assess the cost of information before requesting it. Beyond the sheer volume of paperwork pollution is the serious problem of computer errors which plague management, employees, and citizens. The computer can indeed usually process data faster than human beings -- but it also can create and compound errors at a rate beyond the capacity of humans. Some of these errors can be almost amusing to the casual observer, but usually they can cause hardship or psychological suffering to those affected and they also can be costly. A few examples follow:

- A young Arizona pneumonia patient received a bill from a hospital computer charging him not only for his chest ailment but also for delivery room nursery costs -- alleging that he had given birth to a baby girl.

- One man gave twenty magazine subscriptions to friends as gifts, but the donor was sent all the magazines -- while his friends got all the bills.

- One of the largest stockholders of a billion dollar industrial organization had the name of Alfred A. Smit. One of his dividend checks was misprinted by the computer; his last name was printed in such a way as to form a four-letter obscene word.

- A retired man, being treated for cancer of the bladder, had been receiving his treatment under the medicare program. All of a sudden he received a computer-generated notice that payments would cease because he was dead.

- Government investigators attempted to see whether bogus payroll checks were being processed by a government computer. They submitted a paycheck request to be issued to Mickey Mouse, Disney-World, Florida. They were appalled when the computer system generated the check for issuance.

- A customer of a national department store was dunned for many months to pay a supposedly past due amount of under $100. The customer repeatedly wrote and called the organization advising them that their records were incorrect. He continued to receive harrassing bills and warnings. Finally, he sued the organization and the court awarded the customer about $50,000.

CONCLUDING COMMENTS

The computer industry and computer users may be facing an impending crisis in the area of social responsibility, as the computer's success in invading government offices, businesses, and homes proceeds unabated. We expect that this technological marvel is simultaneously creating an enormous societal problem that is only being addressed in an inadequate and piecemeal fashion. The headlong rush to exploit the rising demand for this new technology has caused manufacturers and users to neglect these largely human and social problems. Perhaps by 1984 we shall see events occur which will impel a more balanced approach to computers, or pressure for government regulation?

REFERENCES

Bell, D. (1973). Coming of Post-Industrial Society. Basic Books, New York.

Brown, H. (1954). Challenge of Man's Future. Viking Press, New York.

Chase, S. (1968). The Most Probable World. Harper and Row, New York.

Clarke, A. C. (1973). Profiles of the Future. Harper and Row, New York.

Heilbroner, R. (1974). An Inquiry Into the Human Prospect. W. W. Norton, New York.

Kahn, H. (1967). The Year 2000. Macmillan, New York.

Meadows, D. L., et. al. (1973). Dynamics of Growth in A Finite World. Wright-Allen, Cambridge, Massachusetts.

Orwell, G. (1949). 1984. Signet, New York.

Sadek, K.E. and E. A. Tomeski. (1980). User/Systems Knowledge/ Skill Considerations in Information Systems Design. Presented at the Joint National Meeting of ORSA/TIMS, Colorado Springs, Colorado.

Sadek, K.E. and E. A. Tomeski. (1979). Challenges for MIS. Journal for Systems Management. Cleveland, Ohio.

Sadek, K.E. and E. A. Tomeski. (1980). Macro Level Systems and Their Influence Upon Society. Presented at the Southeastern Regional Meeting of the Society for General Systems Research, Birmingham, Alabama.

Tomeski, E. A. (1975). The Computer Industry's Social Responsibility. Computers & People. Newtonville, Massachusetts.

Tomeski, E. A. (1975). Building Human Factors Into Computer Applications. Management Datamatics. Netherlands.

Tomeski, E. A. and H. Lazarus. (1975). People-Oriented Computer Systems. Van Nostrand Reinhold, New York.

Tomeski, E. A. (1979). Fundamentals of Computers in Business: A Systems Approach. Holden-Day, San Francisco, California.

Toffler, A. (1970). Future Shock. Random House, New York.

Wiener, N. (1950). The Human Use of Human Beings. Houghton Mifflin, Boston, Massachusetts.

INFORMATICS AS A SYSTEM-SCIENCE

Paper presented at the International Congress on Applied System Research and Cybernetics, Acapulco, December 12-15, 1980

by Ernst P. Billeter
University of Fribourg/Switzerland

ABSTRACT

Informatics is usually considered as the science of computer applications in very different domains. The complexity of hardware and software is such that people working in informatics have to be carefully trained. Now, this training is mainly a mathematical training. But informatics has much closer connections to systems theory and cybernetics than to mathematics. It is shown in this paper that informatics is first of all a system science. This conclusion has some influence on the informatics curricula.

KEYWORDS

Informatics; System theory; Cybernetics.

Informatics is no longer a domain addressing itself to big companies utilizing big computers. This domain is extending more and more to the lower end of computer size. In this domain we distinguish nowadays mini-computers, micro-computers, pocket-size computers. The prices went down so that a private person could easily buy such a computer and use it. This is the origin of personal computers, a slice of the computer market which cannot be neglected anymore.

Informatics which deals with such a growing domain can be defined as the science dealing with optimal design (hard- and software) and use of computers for practical problems (design of information systems). It embraces also the means of making the computers work, i.e. apart from the hardware side, the programming or software side of this domain. Informatics deals with a great variety of subjects which can be categorized into two main groups, namely computer-based subjects and application-based subjects.

Computer-based subjects range from the construction of suitable computers on the hardware side to the development of appropriate languages on the software side. Application-based subjects are concerned with the special nature of practical applications. Here we can distinguish between medical-oriented informatics, business-oriented informatics, technical-oriented informatics etc.

As the first computers appeared, mathematical problems were mainly solved on them (applied mathematics, engineering science etc.). But the computer developed more and more from a purely calculating machine to a data-processing machine able to treat numbers and letters alike. This was a breakthrough into the domain of business applications, a field of application in which the computer was and still is very successful. Now, about 90% of all the computers installed worldwide are used for business applications in a wide sense.

The computer applications became so predominant that it was necessary to introduce special lectures in conventional curricula at universities or to establish special computer-oriented curricula. Given the fact that the first computer applications were concerned with mathematical problems, it is understandable that the first computer oriented curricula were mainly based on mathematics. But for the emerging business and other non-mathematical applications, these curricula were too abstract and theoretical. What practitioners were looking for was a practical curriculum in informatics for special fields of applications. Unfortunately, these curricula have difficulties in developing. As in the early days of computers, informatics as a mathematical computer science is still predominant. Of course, this kind of informatics is still needed for the further development of computer science. But a purely theoretical development of informatics without a comparable development on the application side will sooner or later destroy itself.

This very slow development on the practical side of computer applications, compared to the mathematical oriented use of computers, is understandable. In general, a mathematical problem can directly be translated into a suitable form for computers, given the fact that mathematical problems, expressed by mathematical formula, lend themselves straightforwardly to computer treatment. On the other hand, business problems (or other non-mathematical problems) need a more or less intensive preliminary study before reaching the maturity of computer adaptability. In general, non-mathematical problems have to undergo a preliminary treatment which is known as systems analysis. It follows from what we have said that non-mathematical problems need a special philosophy of approach.

Whereas mathematical problems can, in general, be programmed directly without preliminary adaptation, this is not true for non-mathematical problems. The knowledge of a programming language (e.g. Fortran) is, in general, sufficient to solve mathematical problems on a computer. But the knowledge of a programming language has not the same importance for solving non-mathematical problems. This fact has not been sufficiently recognized in the field of non-mathematical applications on computers.

During the almost 30 years of computer development the hardware development ran away, whereas the software development was very slow. On the hardware-side the computer speed became faster and faster, computer architecture has changed and will continue to change, smaller and smaller circuitery was introduced which is known als Large Scale Integration (LSI), Very Large Scale Integration (VLSI) and Josephson-Technology. The computer will be of a very small size, for it is now possible to build a computer in a cube having sides of a few inches.

On the software-side we had a big jump in the early years of the computer era. The first computers had to be programmed by using the machine code. This very uncomfortable situation from the point of view of the user changed radically through the introduction of problem-oriented languages (Fortran and others). This change happened already in about 1954. The first meeting on automatic programming, as it was called at that time, was held in 1954 and was sponsored by the Office of Naval Research. In June 1956 a further meeting on "Advanced Programming Methods for Digital Computers" was held under the sponsorship of the Navy Mathematical Computing Advisory Panel and the Office of Naval Research. A year later, in 1957, a symposium was held at the Franklin Institute in Philadelphia on "Automatic Coding". Fortran was first introduced in November 1954 (Preliminary Report, Specification for the IBM Mathematical Formula Translating System, Fortran). In April/May 1959 there were meetings at the University of Pennsylvania Computing Center to develop a common business language, later called Cobol. This language was approved in November/December 1959.

But since that time no further radical change has occured. The development of software was limited to some grammatical and syntactical adaptations, the fundamental idea of the programming languages however remained unchanged. Without exaggeration we can say that the software development has not followed the breakneck speed of hardware development. For this reason we are now realizing a development gap between hardware and software.

But this is not the only gap we are facing. Another, perhaps more important gap is coming up now. It is the difference between the conception of the application field of informatics and the software. We are looking at hardware, software and the application fields as selfcontained entities with very loose connections.Hardware is developed with little regard to software; software is developed with little regard to the organization of a firm if we refer to a business application. In introducing a computer into a firm, the people charged with the organization work have to take a given software and a given hardware. Is Cobol and PL/1 really the appropriate language to solve the computation problems of a firm or do we need another generation of languages (e.g. non-procedural languages)? How many programmers are suffering from the problem they experience when they have to program a specified business application for a specific computer. If this programmer is looking for an optimal program, he should go down to assembler or even to machine language. These few remarks will have shown that there is almost no accordance between hardware, software and organization from the user's point of view.

The reason for these difficulties lies, in my opinion, in the fact that these three domains are not viewed from a systems or cybernetical point of view. We have only three domains which, taken together, form a system, but every domain is itself a system or subsystem.

Whereas nowadays we are aware that the computer is a system and the firm with its organizational and hierarchical structure is also a system, we are not yet aware that software too is a system. Software is an integral part of the problem to be solved and of the computer on which the problem has to be solved. Software is not just a way to get a solution, but this way has to be adapted to a given situation. This has been felt by some researchers who have introduced the so-called structured programming and modular programming. By these techniques a program is taken as an entity which is built up by several parts and modules. But until now structured and modular programming have remained on the level of a programming technique. In my opinion we have to search for what lies behind these techniques.

Therefore we can experience an emerging tendency to systematize programming. Systematic programs have also the advantage of making the debugging of programs much easier. Furthermore, the adaptation of program packages to real situations creates no major difficulties if these packages are developed on a systematic basis.

Fundamentally, these and other similar programming techniques show that a program has to be viewed as a system. Programming is not merely a technique to get results from a computer, but it is a system which has to be in accordance with the problem and with the computer. Taking a business oriented program, it has to be in accordance with the firm in which it should be used and with the computer used in this firm. The firm, the computer and the software are a big system. The first question we are confronted with concerns the problem of whether a program or a set of programs can be taken as systems. To analyze this question, we have to go back to the fundamental meaning of a program. Essentially, a program is, from the viewpoint of cybernetics, a transformation. The program has to transform certain input-information A into certain output-information B. If T stands for transformation or program, we can draw the following diagram:

$$T \downarrow \begin{array}{l} A = \{ a_1, a_2, \ldots a_n \} \\ B = \{ b_1, b_2, \ldots b_m \} \end{array}$$

A is the set containing the special information (or groups of information) a_i ($i = 1, 2, \ldots n$) and B is the set of all resulting information (or groups of information) b_j ($j = 1, 2, \ldots m$).

This is a very rudimentary picture of a program T. Usually we get, by using a specific program, certain results B, which sometimes are again transformed by other programs into other results. In general, we have several input-sets, several programs and several output-sets. According to this generalization we can draw a more general diagram:

	a_1 a_2 .... a_n
	i_1, i_2, ... i_n
T_1	b'_1, b'_2, ... b'_m
T_2	b''_1, b''_2, ... b''_p
..	....
T_r	$b_1^{(r)}$, $b_2^{(r)}$ .... $b_s^{(r)}$

The meaning of this diagram is the following. We have an input vector $A = \{a_1, a_2, \ldots a_n\}$. This vector takes on the place of the general vector $I = \{i_1, i_2, \ldots i_n\}$. The original input vector undergoes a transformation given by the program T_1, and we get the intermediary results $B' = \{b'_1, b'_2, \ldots b'_m\}$. This output vector takes the place of the general input vector I and is treated by a further program T_2, which gives us the output vector $B'' = \{b''_1; b''_2; \ldots b''_p\}$ and so on (T_2 can be the same program as T_1). The given diagram is the cybernetical model of a general data processing activity which can be defined e.g. as containing a billing program, a stock-control program, a stock-replacement program etc.

If the transformation T or the program is a system, it should conform to the definition of a system. In systems theory, we usually define a system as the following quintuple:

- $A = \{a_1, a_2, \ldots a_n\}$
- $B = \{b_1, b_2, \ldots b_m\}$
- $Z = \{z_1, z_2, \ldots z_q\}$
- $P: B \times Z \rightarrow Z$
- $G: Z \times A \rightarrow B$

This definition contains three sets, the input set A, the output set B and the set of the states of the system Z, and two functions P and G. The function P tells us, that the following state of the system is given by the output and the state given before. Mathematically, the output and the state given before are connected by the cartesian product. If $B = \{b_1, b_2, \ldots b_m\}$ and $Z = \{z_1, z_2, \ldots z_q\}$ we have the new state vector defined as $Z = \{z'_1, z'_2, \ldots z'_{qm}\}$ where $z'_1 = (b_1 z_1)$, $z'_2 = (b_1 z_2)$ etc. The second function G explains that the combination of the states and the input give us a new output, i.e. the new output is dependant on the state of the system and the input information.

The given cybernetic model of a data processing activity is based on several programs $T_1, T_2, \ldots T_n$. All or some of these programs taken together are necessary to perform a given data processing activity and can be taken as a system. They stand for the different states Z of the system. It follows that the program or states of this system are defining the general data processing activity within a business enterprise.

This situation can be visualized by the following graphs (for simplicity's sake we take a system of three programs):

	A
	I
T_1	B'
T_2	B"
T_3	B'" = $B^{(n)}$

or $A \rightarrow T_1 \rightarrow B' = B_1^{(n)}$

$A \rightarrow T_1 \rightarrow B' \rightarrow T_2 \rightarrow B'' = B_2^{(n)}$

$A \rightarrow T_1 \rightarrow B' \rightarrow T_2 \rightarrow B'' \rightarrow T_3 \rightarrow B''' = B_3^{(n)}$

$A \rightarrow T_1 \rightarrow B' \rightarrow T_3 \rightarrow B''' = B_4^{(n)}$

or

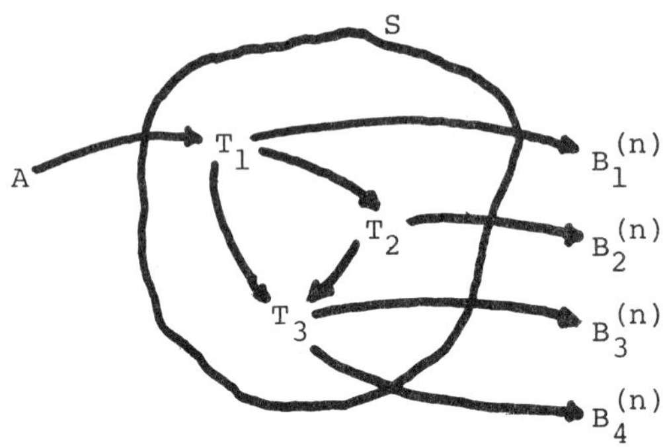

In the last graph we see easily that together the three programs of our example can be taken as a system S, which is necessary to get from the input information A the output information (final results) $B_i^{(n)}$. In our example there are several possible states of the system, according to the fact that all or just some of the programs are activated for the data processing work. So the state of the system can be defined as having just one program activated (T_1), or as having two programs activated (T_1 and T_2 or T_1 and T_3), or as having all three programs activated. According to these different states of the system, we get different results ($B_1^{(n)}$, $B_2^{(n)}$, $B_3^{(n)}$, $B_4^{(n)}$).

The P-function states, in this case, that the combination of the output ($B_i^{(n)}$) and the given state of the system (specific sequence of programs) gives us a new sequence of programs to be utilized during the next computation step. The G-function on the other hand tells us that a given combination of programs (state of the system) and a given input are creating a specific output ($B_i^{(n)}$).

But the individual programs can also be defined as systems (sub-systems of the system S). Here the input is given by all the orders which can be realized on a given computer (list of orders). The output is given by the combination of orders to perform a specific data processing activity (program). The states Z can be defined as the specifications of the problems within a business enterprise to be solved by a given program.

The P-function of the definition of a system says that the combination of orders (program) taken together with one or more practical problem specifications induces the problem specifications of the next data processing activity, because the results of such an activity give an indication of the next problem to be treated. The meaning of the G-function is the following. We have given problem specifications and a given list of orders; the combination of these sets gives us a new program.

Under this aspect we can draw the following graphs:

	A
	I
Q_1	T_1
Q_2	T_2
....	...

Q = problem specification

or

$$A \rightarrow Q_1 \rightarrow T_1 = T_F^{(1)}$$
$$A \rightarrow Q_1 \rightarrow T_1 \rightarrow Q_2 \rightarrow T_2 = T_F^{(2)}$$
$$A \rightarrow Q_1 \rightarrow T_1 \rightarrow Q_3 \rightarrow T_3 = T_F^{(3)}$$
$$A \rightarrow Q_1 \rightarrow T_1 \rightarrow Q_2 \rightarrow T_2 \rightarrow Q_3 \rightarrow T_3 = T_F^{(4)}$$

or

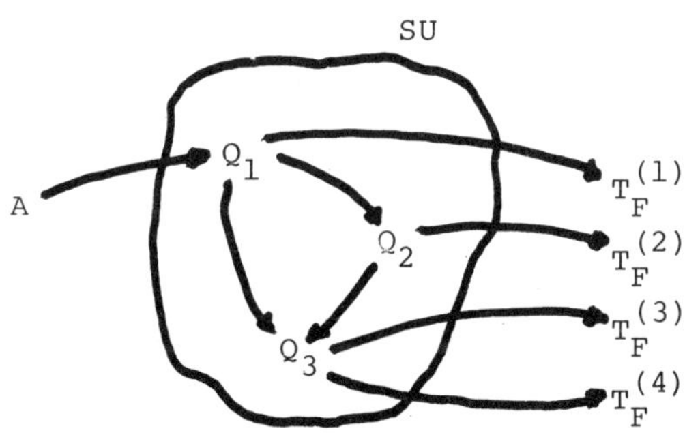

SU = Sub-System

The connection of the different programs T_i within the system S depends, as we have seen, on the goal we are aiming at in performing a data processing activity within a firm. But this goal is given by the business enterprise doing this data processing activity.By this we see the connection between software and software-packages on the one and the aims of a firm on the other hand.

Furthermore, the subsystems we were talking of are themselves dependent on the list of orders as input, i.e. on the hardware utilized. So we see that the business enterprise, the software and the hardware are strictly connected and have to be analyzed as a unity, i.e. from a cybernetical point of view.

These ideas can be concentrated in the following table:

	System	Sub-system
Element	Information	order
System definition	all programs	one single program
Input A	Information to be treated	list of orders
Output B	Results obtained	Combination of orders (programs)
State Z	Programs	Problem specifications

The ideas developed in this paper make it clear that the domains of hardware, software and the firm have to be taken as one entity and one system. Practical informatics is, therefore, strongly dependent on the cybernetical interpretation of a data processing system. For these reasons the actual curricula in informatics have some deficiencies which should be eliminated when establishing such curricula. They should first of all give to the student a fundamental knowledge in systems theory and cybernetics. On this basis it will be possible to build up an interactive system, the components of which are the problem specifications given by the firm (application side), the hardware (list of orders) and the software (programs).

VIDEOTEX, CYBERNETICS, AND HUC

Howard J. Hilton

ABSTRACT

Videotex is a development that will involve millions of people in the daily process of selecting and retrieving, with the aid of computers, information in their homes. Cybernetics, being the science of control and communication, will have a multitude of new applications and data for study. HUC, an information concept using a universal code for the unique identification of all recorded knowledge and information, can provide an interface and feedback mechanism between the videotex user and the computer.

Videotex and cybernetics are terms that are subject to various definitions. This paper presents definitions in the context of the information process, reviews the present state of development, and suggests some of the implications for the future. It also reviews the experiment with "HUC Access to Knowledge" on the British viewdata, later Prestel, which was the beginning of videotex.

Other videotex systems have been developed in France, Canada, and Japan. Various experiments are in progress in the United States. A comparison of various major systems is provided.

Some applications of cybernetic principles to these developments are indicated together with a description of the HUC hardware and software.

KEYWORDS

Videotex; cybernetics; HUC; information system; cybernetic center; access to knowledge; software; hardware; Prestel; viewdata; Antiope; Telidon; universal code; unique identification; encrytion; cable systems; interactive broadcast; satellite; geographic index; proper name index; subect term index; type of material; level of material.

INTRODUCTION

At first glance there would appear to be little in common among videotex, cybernetics, and HUC other than the fact that they are new terms for concepts which are not to be found in the current edition of the Oxford English Dictionary. In fact, only cybernetics is defined in the current editions of such dictionaries as Webster or Random House. Even in these cases the definition is not sufficiently precise for use in the various applications to be considered in this paper.

It is perhaps interesting to note that Ashby (1964) in his book An Introduction to Cybernetics does not formally define cybernetics. He states "Whenever a word is formally defined it is printed in bold-faced type." Cybernetics is not so indicated; only terms such as operand, operator, transform, transaction, transformation, element, etc. are formally defined.

DEFINITIONS

Videotex is a method of delivering alpha, numeric and graphic information to television devices using digital instead of analog decoders. The first major videotex development took place in England. Before the term videotex was coined the terms teletext and viewdata were used to describe two different aspects of this delivery method---broadcast and interactive. The broadcast or teletext delivery was known as Ceefax and Oracle, two systems used in England to broadcast digital signals using the vertical blanking lines of the TV signal. Viewdata, later known as Prestel, was an interactive system using a modified television set and the telephone with a modem.

Originally, videotex was used to describe both broadcast and interactive systems. At one stage there was agreement that it would refer only to interactive systems. However, the European Economic Community uses it without modification to include both delivery methods. As the discussion on applications below will show, broadcast videotex and interactive videotex are both necessary in an effective public information system. Videotex as defined above would include both delivery techniques.

Cybernetics, although not formally defined by Ashby (1956 and 1964, p.4 might, according to him, "...in fact, be defined as the study of systems that are open to energy but closed to information and control---systems that are 'information-tight'." This, it might be noted, does not resemble any of the common dictionary definitions, one of which defines cybernetics as "a science dealing with the comparative study of the operations of complex electronic computers and the human nervous system." Weiner (1948 and 1961) who is generally given credit for the first use of the term defined it as "the entire field of control and communication theory, whether in the machine or in the animal" He was apparently unaware (Dechert, 1966) that Ampere in 1838 had applied the French translation of the Greek to the science of government.

Control which implies feedback and decision, and communication, which involves the transmission of information, are the central elements of cybernetics. But more than being involved "in the machine or in the animal", they are at the heart of the social institutions and coming ever to the fore in what is known under various descriptive phrases as the information revolution, the electronic revolution, the technical revolution or the post-industrial society. Thus, the artificial intelligence of the machine, the human use of machines, and social interaction are becoming important areas for the application of cybernetics to current problems in addition to those in the fields of psychology and biology.

The definition of cybernetics to be used in the discussion of videotex and HUC expands the original expression of Wiener by adding the two elements implicit in his elucidation of the concept, as follows:

Cybernetics is the science of self-regulating systems, which (1) exist in living matter and its relationship to its environment, (2) as the interaction among living things, (3) in machines, and (4) in the interaction between living things and the machine.

Self-regulation (Maturana and Varela, 1975) in the context of cybernetics includes processes that maintain organisms or organizations as viable entities and that enable machines to perform selection and control operations. The specification of the types of self-regulating systems confirms that cybernetics is a science that embodies the theories and techniques of the social, natural, and physical sciences as they relate to the study of systems.

HUC, which stands for the Hilton Universal Code for the unique identification of all recorded knowledge and information, is the designation of an information system designed ultimately to provide global access to all public knowledge and information. A system of this magnitude which is the product of interaction between man and machines has to embody elements of self-regulation and artificial intelligence.

Thus, videotex is a means for communicating information and HUC is the concept for creating the social institutions and for establishing the basis for the necessary interaction between man and machine. The combination of the two provide an information system which will enable individuals to have access to knowledge and information on a scale far beyond anything that exists today.

The social system involved in the production, organization, distribution, and use of knowledge and information constitutes a self-regulating system that encompasses a portion of the entire field of control and communication theory and that is closed to information and control but open to energy. This is an area that offers a multitude of opportunities to apply the science of cybernetics.

APPLICATIONS

Videotex--description

Videotex is going to herald a new era in the production, organization, distribution and use of knowledge and information. The previous eras were those of the scribes, the printing press, and the online databases. If the thesis of this paper is confirmed the new era will be that of HUC-videotex. The distinction between videotex and the CRT's (cathode ray tubes) or VDU's (video display units) of the online database era is the capability of videotex units to receive and display television signals as well as the digital signals of the videotex delivery.

There are several types of videotex systems. The characteristics, as of the present, which distinguish them are the method of forming graphic images and delivering the signals. There are two methods presently utilized by the various systems for forming graphic images. These are alpha mosaic and alpha geometric. Alpha mosaic creates the images and colors from the bit patterns of character blocks, whereas the alpha geometric generates spatial areas and colors with PDI's (position description indicators). There will likely be other systems which will have greater flexibility than either of these systems. Likewise, there are two types of delivery systems, line and packet.

Change in this field is very rapid. For example, as of this writing, November 1980, changes are being made in two major systems Telidon and Antiope, to achieve compatibility. Thus, Telidon, the Canadian alpha geometric system, may adopt the control set of Antiope and Antiope the French alpha mosaic system, may also provide for the PDI's of the Canadian system.

The major reason why the online database system of today will be replaced by the HUC-videotex era lies in the complexity, speed, and cost of distribution of knowledge and information. The least expensive method for mass distribution of information is the broadcast of a signal which is available to anyone. The cost-benefit aspects of the many variations of this technique require analysis beyond the scope of this paper. Suffice it to state that for a new cybernetic city, as discussed under that heading, a broadcast signal delivered over a cable system to a videotex unit in every home would provide the least expensive per unit, per person, production and distribution cost.

HUC-description

HUC is an important element in any videotex system that is designed to provide access to public knowledge and information. It is the interface between the human and the machine, between the user and the system. As such, it is structured to facilitate the operation of the system by the user and to provide the foundation for a more efficient database organization and system architecture. It does this by using a unique, mnemonic, and informative code for the unique identification of all recorded knowledge and information. This code, in a structured format has a maximum of 61 alpha numeric characters divided into the following 15 fields (Hilton, 1980A):

1. Type of material -- 2 numeric, or for official national or international publications or documents, or for statistical series, 2 alpha characters;
2. Source -- 6 alpha and 5 numeric characters;
3. Check character -- 1 alpha character;
4. Status of material -- 1 numeric character;
5. Time by year, month, day, hour, minute, and second, or edition or number of a document or work within any of these periods from 1700 to 2700 -- 7 alpha characters;
6. Primary storage identification -- 2 numeric characters;
7. Check character -- 1 alpha character;
8. Special media designation -- a numeric character;
9. Special media identification -- 3 alpha characters;
10. If microform, type or other media designation -- 1 numeric character;
11. Publisher, producer, copyright royalty account, or author in special situations -- 6 alpha and 5 numeric characters;

12. Language fields -- 2 alpha characters;
13. Special field designation -- 2 numeric characters;
14. Special field classification, document numbers, mass density storage addresses etc. -- 2 alpha and 13 numeric characters; and
15. Check character -- 1 alpha character.

The 61 alphanumeric characters of the code utilize the HUC character set which replaces some of the punctuation marks and symbols of the standard character set with special characters which extend the alpha characters to 63 and the numeric characters to 20 with 45 neutral and control characters.

All material identified by this code can be located by the use of the five category index of the HUC system. These categories are (1) subject terms and phrases, (2) proper names, (3) geographic names and places, areas, or segments, (4) type of material, and (5) level of material. The last two categories contain only one entry for each document or piece of material whereas the first three can contain as many entries as necessary.

The software required for the full implementation of the HUC information system concept includes the following major elements: (1) organization, (2) identification, (3) user guidance and assistance, (4) selection, (5) access, (6) manipulation, (7) accounting and payment, (8) entry, (9) communication, and (10) control and security.

As a test of the capability of a videotex system to provide access to all recorded knowledge and information, "HUC--Access to Knowledge" was established as a database on Prestel, the British interactive videotex system. This database consisted, in part, of a portion of the U.S. public database ERIC. This was identified and indexed on the basis of HUC by Langton Information System and was financed by a contribution from Harte-Hanks Communications, Inc. The conclusion was that with software modifications designed to accommodate more powerful terminals, Prestel could be utilized to provide access for which HUC is designed. The Antiope system subsequently incorporated nearly all of the elements deemed necessary for the effective operation of HUC. These include broadband, high speed delivery, multi-screen access, multi-digit selection, and multi-address packet delivery capability. (Le Moign, 1980)

Cybernetic City

The United Nations estimates that 10,000 new towns will have to be built before the end of this century. Each of these could be a cybernetic city. At the heart of a cybernetic city is a cybernetic center which in the early stages of development could serve as the planning and later as the operations center. Since the science of cybernetics focuses on the information and communication involved in the process of feedback and decision, a cybernetic city is one in which the latest information technology is utilized to maximize the benefit to the inhabitants in the functions of work, leisure, and exercise of social responsibility.

HUC-videotex, as an information system concept, is designed to provide people in a cybernetic city with all knowledge and information they need and desire, if available, for work, leisure, and exercise of social responsibility. More specifically this would involve private and protected information for work, copyrighted information, provided with payment for copyright royalties, to meet entertainment, educational, and general information demands, and public domain information provided to facilitate participation in local government and social activities, contribution to the attainment of social objectives such as conservation, protection from fire, natural disasters, and criminal activity, alleviation of poverty and misery, and protection and enhancement of the environment. HUC also provides a communication network which relates the cybernetic city to the international flow of information and communication among individuals, organizations, and governmental institutions.

The study of man-machine interaction in the production, distribution, and use of knowledge and information forges the principal link between HUC and the science of cybernetics. Cities have been built through the ages, some with planned development and some without. Information systems have been developed to utilize the power of computer technology. What distinguishes the cybernetic city from those of the past is the focus on the information component and the interaction between the inhabitants and the developing technologies. What distinguishes HUC from other information systems is the focus on the requirements of the user and the integration of the various aspects of electronic technology to meet the requirements of individuals and society.

The individual, far from being surfeited by knowledge and information, is unable to act intelligently in many areas because information which exists is not readily available. With the pace of current developments, the speed with which certain types of information become available needs to be increased. HUC and present technology make this possible today and in the process will make the cybernetic city a more efficient and enjoyable place in which to live and work. Although there are many architectural conceptions for such cities, there is none, other than HUC, that describes in detail how individuals would obtain access ultimately to all public knowledge and information and the private information which they need on the job.

In order for individuals to obtain this access a cybernetic city should have certain capabilities. The basic communication link would be a fiber optic cable into each home and office. The front end of this cable would contain retransmission capability for television signals, links for interactive satellite communication, a computerized HUC database, and switching capability for interactive videotex. The connections at the terminal end would provide for television, broadband broadcast, videotex reception, and high speed interactive videotex, digital and analog high fidelity reception, security and other monitoring signals, encrypted transmission and reception, and for buffered storage of received signals. All private and protected information exchanged among terminals would be in encrypted form, if desired. Copyrighted information, likewise, would be in encrypted form if desired by the copyright owner. Public domain information, advertising material, company reports and brochures, schedules, tourist information, letters (which people now write to an editor) etc. would be broadcast in clear text and retrieved, if missed on broadcast, on an interactive basis.

Three different cases provide examples of the operation of the system: (1) current information on a subscription from a selective disseminator of information (SDI service), (2) working environment using private, local, national, and international databases, broadcast encrypted and clear material, and specialized retrieval from copyrighted and non-paid public databases, and (3) browsing for entertainment, education, vacation planning, shopping, and for income.

SDI (Selective Dissemination of Information)

The example used here of an SDI system is that of a newspaper providing such service to subscribers. The subscribers would provide information profiles covering their areas of interest. The initial parameters might be based on the number of words either by percentage of subjects or by subject coverage. The subscribers would receive via broadcast videotex during the night only the volume requested. If the specification were by percentage such as national, international, and local news, sports, editorials, features, and business, the daily material would be similar to the newspaper of today. If the specification covered a variety of other subjects, such as legal cases, government regulations, journal articles, reports, etc. the daily material might be headlines and abstracts with HUC identifiers for the full text which could be accessed using the interactive videotex.

The delivery of the material by broadcast videotex would be addressed to the subscribing terminals and encrypted in a general code for all current subscribers. Thus, there would be a common core of headline news that would probably go to all subscribers and then finally the special references and items of interest for individual subscribers. The SDI broadcast delivery would utilize the full bandwidth of one channel and would be dumped into terminal memory for transfer to other storage depending upon the volume of material and the terminal configuration.

Work Environment

The terminal in a work environment might be a part of a XIBUS ring which in turn would be a node on a cable system. This node might be receiving and transmitting data via satellite, microwave links, telephone and a local cable system. The ring controller could have a large optical, digital, disc storage and a capabity to sort data from the various inputs to determine items of interest to the terminals on the ring.

If, for example, a geographic area were of interest, the HUC geographic index would automatically trigger the capture of any document with a document profile containing the desired geographic segment. The processor could be programmed to accept data for any set of geographic parameters or for any of the other four index categories. This information could then be used for SDI or stored to augment an existing database. Since the terminals would have access to various databases, the local storage could be devoted largely to the HUC index. All that would be required for retrieval from the databases would be the HUC identifiers. Payment for copyrighted material and for database access could be made by use on the basis of the HUC identifiers.

Browsing

Many people fear that mass density storage and rapid retrieval will put an end to the pleasure of browsing, but it should extend it to the pleasure of all who like to explore the ramifications of a subject. With modern technology and citations using HUC identifiers it should be possible for an individual to obtain immediate access to all of the citations in a state-of-the-art report. Now obtaining all of the citations to such a report is a monumental task.

Also, browsing can be turned into a profitable one person enterprise. As one explores a body of literature, gaps are noticed and new reports and analyses are indicated. Since all information is readily at hand, it is possible for an individual to prepare material and with a touch of a key put it into the system for inclusion in the HUC index and database. With a global market, any intellectual effort which meets a need would provide a substantial return. Browsing also provides a field for advertisements, which can be organized to meet the needs of the users for shopping. Likewise the information for vacation planning such as lists of hotels, schedules, etc. would be readily available.

CONCLUSION

According to Ashby (1956 and 1964), "Cybernetics offers the hope of providing effective methods for the study, and control, of systems that are intrinsically extremely complex." A videotex cable system with over 100 channels and communication with other systems is in itself a complex system. Coupling that with a global interactive network and the HUC concept for providing database support for a global videotex system produces a large and complex system.

It is appropriate to look to cybernetics for aid in solving some of the problems which can be foreseen in the achievement of the objective of providing access to all public knowledge and information. These would include the architecture of the system, the structure of the database, the ergonomics of the terminal and use in the home and office, the legal and ethical problems, and the software for developing the system intelligence to anticipate and to meet the needs of the user.

REFERENCES

Ashby, W. Ross (1956 and 1964). An Introduction to Cybernetics, First published in 1956 by Chapman and Hall Ltd. and then in 1964 by Methuen and Co. Ltd.

Dechert, Charles R. (1966). The Social Impact of Cybernetics, Simon and Schuster

Hilton, Howard J. (1980) HUC - A User Designed System for All Recorded Knowledge and Information

Le Moign, D. (1980). Presentation de STARTEL: Un Serveur Videotex Raccorde aux Reseaux par Paquets

Maturana, Humberto and Varela, F. (1975). Autopoietic Systems Biological Computer Laboratories, Univ. of Illinois, Urbana

Wiener, Norbert, (1948 and 1961). Cybernetics or Control and Communication in the Animal and the Machine, The M. I. T. Press

AN OPTION FOR A MICROCOMPUTER COURSE IN THE PHYSICS DEPARTMENT AT THE FACULTAD DE CIENCIAS, UNAM.

Héctor Guerrero G. and Fernando Favela V.

Laboratorio de Cibernética, Facultad de Ciencias
UNAM

ABSTRACT

Due to the evident necessity for learning the use of microcomputers in everyday life and scientific work, we have tried to laid the basis to permit the physical science student develop a sound criteria for the solution of specific problems related with computing. Basis such that will permit the student use, within his environment, such powerful tool as the microcomputer. Didactical material specially prepared for an easy understanding and use of microcomputers was employed. That permits an emphasis on application, particularly on physics.

KEYWORDS

Microcomputer applications; software; hardware; microcomputer design; education.

PREFACE

This course is a new point of view for the course called "digital machines" which is offered in the Department of Physics at the Facultad de Ciencias, UNAM. This lecture has been traditionally directed to a general learning of digital machines. At the present time computers - and particularly microcomputers - have become a fundamental part of the course. We think that this rise from the evident versatility and power - greater everyday - that microcomputers present in the solving of scientific and daily life problems.

APPROACH

Since this course is a part of the physical sciences curricula, we give special importance to the "How to use a microcomputer" and "How to solve problems with a microcomputer" parts. That is, the student is not going to become a professional on computing science through this course, but he will acquire the necessary skill to manipulate efficiently the microcomputer and to present solutions to a problem with it, all that pointing to the observation, analysis and control of physical phenomena.

PROGRAM'S STRUCTURE

I.- Introduction to digital machines.
II.- Computing machines general structure.
III.- Computing machines programming.

IV.- Peripherical devices.
V.- Special devices.
VI.- Aplications.

Theme I.- The student reviews digital circuits and includes concepts such as funcional modules (ALU, registers, etc.) as an introduction to the next theme.

Theme II.- The basic architecture of a CPU and other items like TSL, memory, I/O devices, etc. are presented to give the idea of the basic structure of a microcomputer. The architecture of the most popular microprocessors (6800, Z80, 8080) is given. This makes the student capable of choose among the available microcomputers, founding the election on the advantages of the different CPUs.

Theme III.- It is very important to point out that the computer is an invaluable help in modeling. But this capacity requires a series of techniques and procedures to make it effective. More than that, it is necessary that the student acquire a way of thinking according to that capacity, in order to get the best results. Therefore, we have dedicated an important part of the course to give the student a sistematic way for solving problems. So in this part the student is iniciated in the actual handling of the microcomputer by programming. We call this an initiation because the student will acquire more experience through out the entire course.

Themes IV and V.- This themes are the most important ones related to hardware. Here the student learns that microcomputers have to be coupled to peripherical devices like TTY, CRT display, Hard and Floppy disks, tape-cassette, A/D, D/A, transducers, etc., without wich we cannot interpret the data from the microcomputer and the microcomputer is unable to understand the data from the analog world.

Theme VI.- Finally, we reach the point where all the concepts, elements and methods showed on the preceding themes are used all together by the student. In this theme the student is asked to develop a proyect on microcomputers, always under the teacher's supervision.

AUXILIARIES

We planed an open lab where the student performs activities related to the course and in which the student familiarizes himself with the machines. It has been designed and constructed a processor with didactic purpose, that will help in the understanding of the actual functioning of real machines. This processor was made with standard integrated circuits. It has a simple instruction set which allow the student quickly understanding of the computer handling.

Alongside the hardware, it was designed an assembler-simulator for this processor, which makes easier the actual observation of the program's execution, because it gives the option of running the program step by step. It also indicates for each step the changes that took place as a result of the instruction just executed.

We want to point out that the software's commands and instruction set are made in spanish.

This program is made relocatable so everyone who has a system based on a 6800 microcomputer can use it. One of the most important features of the program is that allows the student to work with a simple assembler program in spanish so he can understand its functioning faster. The simpler the faster the student can understand it. The program can display all the instructions that the processor can perform with a small explanation on what the instruction does.

The processor is a very simple computer that can perform the most usual instructions of real machines. The student can make a program in Mnemonics and assemble it; then he has two options: to run the program on the simulator, or to list the assembled program, load it on the processor and run it, that allows the student to work with software or hardware depending on his preferences and abilities.

BIBLIOGRAPHY

Bishop, R. (1979). Basic Microprocessors and the 6800. Hayden Book Company, Inc. New Jersey.

Davis, G.B. (1980). Introducción a los Computadores Electrónicos. CECSA, México.

Taub, H., and Schilling, D. (1977) Digital Integrated Electronics. McGraw-Hill, New York.

Zaks, R., and Lesea, A. (1979). Microprocessors Interfacing Techniques. Third Edition. Sibex, Berkeley.

PROSPECTS AND PROBLEMS OF THE IMPACT OF MICROCOMPUTERS ON HUMAN LIFE

S. Raman, Ph.D. and U. Deo, M.Eng.

Department of Epidemiology and Community Medicine
University of Ottawa
1461 Heron Road
Ottawa, Ontario
K1V 9V8

ABSTRACT

Microcomputer technology is the fastest growing area of scientific research and probably one which has the highest potential for changing human way of life and behaviour in this century. It is predicted that every individual in the western world will use directly or indirectly a microcomputer based device within the next five years and the usage is expected to expand by a factor of ten within a decade. The current application of microcomputers span such conventional areas as communication, education including scientific and business computing, recreation and health care which includes a host of microcomputer based medical devices. Technological advances currently underway will touch practically all facets of everyday life with a promise in the very near future of such exciting possibilities as electronic mail, artificial human organ replacement parts, 'intelligent' device controls for household appliances and automobile and the robotization of industrial and routine housekeeping chores which could cover as much as 80% of the traditionally human activities. The implication of these changes towards the control and modification of human behaviour would be prophetically far-reaching. The impact of the shrinking work week and the enormous enhancement of the entertainment potential of this technology which could drastically alter social values and ethics ought to be carefully assessed.

KEYWORDS

Microprocessors; impact on human life; applications of microcomputers.

INTRODUCTION

Computer models of the world and the sub-system relating to population, agriculture, industry, pollution and energy are uniformly pessimistic with variations in predictions only in reference to the degree and the duration to the ultimate doom [Forrester, 1971; Meadows, 1972]. In respect of the demographic transition it is inevitable that the population will age and consequently the dependency rate will increase to a large proportion. Automation probably will be the only solution to the problem. Tendency currently underway towards this transition may be noted in the microcomputer revolution.

COMPUTER BASICS

In any computer, there are five functional blocks interconnected in the following manner.

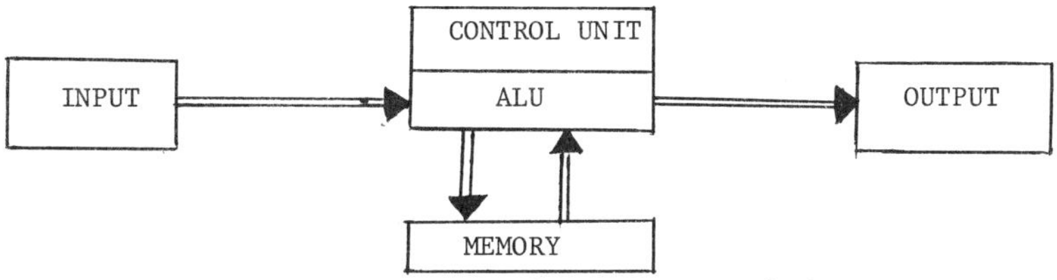

The Control Unit sequences the operation of the entire system. Arithmetic-Logic Unit (ALU): As the name suggests this unit performs arithmetic and logic operations on data flowing through it.
Control Unit together with ALU is called CPU (Central Processing Unit). It will generate and manage all control signals necessary to synchronize all types of operations.

Memory: This in any computer is used to store information. Two main categories are, RAM (random access or read-write memory) and ROM (read-only memory). Contents of RAM are lost on power failure or this is a volatile memory. Therefore, it is appropriate to store intermediate data results in this type of memory. The second type of memory, ROM is non-volatile. Also, its contents can not be altered, thus programs are stored in this.

Input-Output: These serve as an interconnection between the computer and the outside world. Some typical input devices are: keyboards, sensors (temperature sensor, presence detector, pressure sensor). A few output devices are: LED's, LCD's (these are often used in pocket calculators), printers, lamps, any control mechanism like relay of a motor.

A microprocessor is basically a CPU on a single chip. A microcomputer is a microprocessor together with Memory. There are currently available excellent references which provide an insight in basic principles of microprocessors [Zaks, 1977].

BRIEF HISTORY OF MICROPROCESSORS

The emergence of microprocessors has been accidental. At the end of 1971 the first microprocessor (Intel 4004) was introduced in the market. Its design resulted from the contract between Intel Corporation and a Japanese desk calculator manufacturer. The next significant event was introduction of 8008, by Intel, 8-bit general purpose microprocessor, in 1972. Around 1976, Intel 8080, successor of 8008 was available in the market, at the same time Motorola introduced 6800.

We now have a third generation, or the successors of 8080 and 6800, which are Z80 from Zilog, F8 from Fairchild and Mostek, PPS4 from Rockwell, 9940 from Texas Instrument etc. Presently more than 30 different types of microprocessors are on the market.

The new generation of 16-bit machines was first generated by Texas Instrument. This helped to extend memory capacity to 64K. Now there are known 16-bit products by Intel (Intel 8086), Zilog (Z8000), Motorola (MC6800). The manufacturer's list is growing fast. Further design and production of 32-bit based microprocessors which will be highly efficient are currently underway. A lot of computer terminals and peripheral devices like CRT, floppy disk controllers, paper-tape readers, keyboards, printers, keypunches make notable use of microcomputers. They function as slaves under the control of the CPU which enables one to off-load the CPU and free it for more important tasks. A considerable advancement has been made in graphic terminals with the advent of microprocessors. Some important advantages of microprocessors are:

- Very small number of external components required.
- Reduced volume and miniaturization of the system.
- Less power consumption and dissipation.
- Increased reliability due to less inter-connections.
- Programmability.
- Lower cost. A general purpose chip sells for $15 to $30 a piece today. It costs $10 for 100 chips and $1 or $2 for quantities in bulks of 10,000.

Advantages of programming: The three advantages are (1) Flexibility, (2) Speed, (3) Cost-effectiveness. We shall briefly state what each one means.

(1) Flexibility: A program can be easily changed according to requirements. Debugging a program is much faster than finding errors in an equivalent hardware system. A well-written program is divided into several subprograms or procedures. This enables someone to add an extra function or remove another one.

(2) Speed: This does not refer to execution speed but, this is the human speed required to solve a problem by developing and debugging a program. This speed is high where a High-Level language (e.g. FORTRAN or BASIC) is used.

(3) Cost-effectiveness: Because the programs can run on the same hardware module one does not need to make any hardware changes to the system. The same module can be used for variety of applications. The mass-production of such standardized modules makes it possible to achieve very low costs for the hardware portion of the system.

CURRENT APPLICATIONS

Current applications of microprocessors span the whole spectrum of human activity. The most common applications are in:

- Communications.
- Military.
- Industry.
- Business and offices.
- Education.
- Medicine and health care.
- Home and entertainment.

It will take pages to expand on all these applications, so we will state some examples in each of the above major areas and expand a few of them in some detail.

Communication: The microcomputer in conjunction with the home television and the telephone network has currently revolutionized the communication technology. Telephone networks with microprocessor based elements will provide new services such as tele-photocopying, electronic mail, home surveillance against break-ins, fire and enhance conventional services through automatic dialing and routing the calls to designated locations. In addition, the system will have high reliability, low-cost and flexibility.

Two way cable communication system for the home television that have been developed such as, Telidon, Vista, Viewdata and Oracle place at the finger tips of the consumer huge computer data banks which with adequate satellite links could be truly international.

Military: The army is the biggest consumer of microcomputer products. These components find application in sophisticated weapon systems which make them highly accurate and more deadly. While the applications to instrumentation control are generally known, many others are classified and shrouded in secrecy. It is claimed that the future battles will be waged on the computer front aimed at the destruction of the enemy's arteries of commerce. Microcomputers will play no mean role in such a battle.

Industry: Many aspects of the industrial activity from design and manufacture of the products to the packaging and distribution are amenable to efficient automation through the use of microcomputers. The fully 'automated factory' is a current technological possibility and the tendency is already evident in most of the western countries including Japan.

The automation of the production processes involves:

- Process control and control systems.
- Instrumentation controls and robots.
- Quality control and statistical checks.

Process control and control systems: Microcomputers by themselves or as a part of minicomputers can control input and output of instructions controlling devices in the manufacturing process which may control system parameters such as, temperature, pressure, pH, speed and flow rate of fluids, gases and chemicals. The ability to incorporate low-cost process controllers using microcomputers is of significant advantage to the industry. Further the software flexibility of the microprocessors enables the use of the same system to control multiple tasks without any additional hardware components.

Instrumentation controls and robots: Microprocessor based instruments currently render measurements accuracy and control of processes, which were hitherto impossible using strictly human controls. Further the development of robots controlled by microcomputers have mechanized repetative activities in industries thus eliminating danger to the work force and human errors and improving the efficiency and quality of production enormously. Many complex robots have been designed for specific industrial jobs such as, welding, spray painting and metal castings which with microprocessor capabilities display flexibility and adaptability [Engineers Digest, 1979]. There are significant developments in the field of sensors for instrumental control systems [Sneller, 1979].

Quality controls and statistical checks: Microprocessor based devices can perform routine quality checks on finished products and even provide early warnings during the production process if the tolerance limits show a tendency to go out of control. In addition the enormous computing power of the micro-computer permits complex statistical calculations to predict the future performance of the product and hence optimise the production system in its entirety.

Finally microcomputer availability has helped industry in automation of production monitoring and recording, automatic warehousing, distribution and efficient accounting and sales management.

Business Applications: To justify the use of an equipment for business, it must be reliable, simple to operate, and provide a satisfactory return on the investment. In the past, several discouraging factors prevented the extensive use of computers in business and office environment, viz, expense, bulkiness, need for back-up facilities, unacceptable time delays in accessing the computer and the need for a trained professional to program and operate the computers. All these have changed with the advent of microcomputers. Small desk-top systems with adequate memory and speed are currently available for a cost of less than $2000 which can meet most of the needs of a small to medium business enterprise. Some of the typical business applications are, payroll and inventory control, maintenance of accounts, word and text processing, automated addressing and mailing, sales projections, and financial modelling. Popular commercial microcomputer systems are marketed based on TRS-80 by Radio Shack, APPLE II, and PET by the Commodore groups of companies.

Education: Educational institutes increasingly use microprocessors. These devices range from relatively simple programmable calculators to special purpose sophisticated educational systems. Computer Aided Instruction (CAI) is currently on an explosive phase which includes as an introduction the use of computer itself. Microprocessors are used in both learning and evaluation and can also be part of a large system where an instructor is enabled to relate to a large audience widely distributed through the aid of telephone and television.

Medicine and health care: These can be broadly divided into two groups. 1) Implanted on human carrier. 2) Medical instruments or external devices. Let us consider examples of some of them.

Artificial vision: A system for artificial vision for blind is under development in the University of Utah Institute for Biomedical Engineering [Electronics, 1974]. A miniature semiconductor image array is mounted in the eyesocket. A microprocessor scans the array elements in a serial manner and formats the data for transmission to circuitry implanted in the skull.

Microtutor for the mentally retarded: Rehabilitation for mentally retarded individuals often includes training to perform simple tasks to allow them to work constructively in the society. This training involves frequent one-to-one contact with a supervisor who must evaluate trainees in terms of quality of work. Due to the number of required contact hours, a supervisor can effectively train only a limited number of clients. Since any technological solution to this problem must be cost effective, microcomputer is preferred over expensive minicomputer aids.

Speech communication aid: An Intel 8080 microprocessor is used in a communication aid for aphonic and motor deficient patients. The system has two opposite facing screens. So user and respondent can maintain eye contact. Screen displays up to 20 characters and a thermal printer provides hard copy.

Independent living for elderly: As the age advances, loss of short-term memory frequently accompanies. Due to this, there is a significant threat for elderly people who require dietary restrictions or self-administered medications. A microcomputer information retrieval aid using 6502 has been developed [Houge, 1979].

It works in a simple manner. The individual presses a start button. On the TV screen is received a list and times for the day's schedule.

for example,

1) breakfast menu.
2) four capsules of medicine X.
3) take and record your blood pressure at 9 a.m., 3 p.m., and 9 p.m.
4) dentist appointment at 5 p.m.

If a person forgets to perform an item by scheduled time or to cancel it, an alarm rings until it is cancelled. A caretaker periodically enters new information into the system. Instead of TV, a notebook storage can also be used. Battery operated 8085 microcomputer will display notebook page numbers on a LCD.

Patient monitoring: physiological monitoring in intensive care units is done by computers. A microcomputer at the bedside can do data acquisition, display, reduction and transmission to a central computer at the intensive care unit.

Physician's office: A stand alone microcomputer system equipped with a CRT, mini-floppy disk and printer is developed by Cameron (1978). The programs are written in Basic and are interactive. This helps the doctor to obtain all information regarding that patient on the screen, and keep track of medication prescribed and the patient's problems.

Home and entertainment: Manufacturers are particularly tempted to use microcomputers in electronic games and domestic appliances. Some of the growing range of domestic appliances are dishwashers, sewing machines, knitting machines, microwave ovens, telephone answering systems and so on. The micro-based games are either manual games or electronic or video games. Besides games there are other devices in housing a microprocessor such as, cassettes, hi-fi, recorders, televisions, etc. Microprocessors are becoming standard components in automobiles which improve energy efficiency and safe driving.

DISCUSSION

The impact of microcomputer revolution on society will be far-reaching. It has been claimed that the invention of the computers will revolutionize human brain power as much as the invention of the steam engine has transformed the use of human muscle power. It is legitimate to claim hence that the microcomputer will spur a revolution in individual life as much as the use of the personal automobile. While the personal automobile is still a 'status symbol' in many cultures, in the west as a means of transportation it has changed our leisure life and our sexual mores and practices; spawned suburban living; changed our shopping habits; introduced highway construction at an unprecedented rate; enhanced pollution generated health hazards and even introduced new holiday and recreational modes with accompanying increases in death toll [Bauer, 1966]. None of these events at the beginning of the century were anticipated nor all of the consequences desired. It is possible to draw a similar parallel for the impact of microcomputers on social life.

An immediate consequence of the micro-revolution is the enormous leisure an individual will have which presents the problem of utilization. We have already reached the limits in the use of conventional recreational activities as may be seen from the overcrowding of resorts and the participation rates in games. It is conceivable that the microcomputer itself will be instrumental in devising new means of entertainment. Already the micro-based entertainment industry is expanding at a phenomenal rate but inevitably the human interests are likely to decline and the distinct possibility of increase in crimes as a second order consequence of the micro-revolution looms large.

Already the computers have given rise to a new breed of criminals. The new criminal known as 'computer thief' is very different from the conventional ones in that he is highly intelligent, probably an introvert and who brings to crime the zeal of a problem solver. A clever computer thief can swindle millions of dollars without the financial systems ever being aware of it and could even create chaos in the international monetary system by selectively sabotaging by computer installations. The microcomputer in league with the telephone endows every clever human being with the potential for such damage.

On a positive note, it is possible to visualize in the near future of implantable microcomputer devices to amplify human intelligence. While the human brain is the product of a few million years of evolution, the hardware hybridization of the human with the computer - a 'cyborg' - could accelerate intellectual evolution that may well auger the arrival of the 'superman'.

ACKNOWLEDGEMENTS

The authors would like to express their gratitude to the National Sciences and Engineering Research Council of Canada for supporting this research and to the Health Care Research Unit of the Department of Epidemiology and Community Medicine, University of Ottawa for travel assistance (U. Deo).

REFERENCES

Bauer, R.A. (1966). In Social Indication. M.I.T. Press.

Electronics. (1974). Data processing, LSI will help to bring sight to blind. 42(2): pp 81-86.

Engineers Digest. (July, 1979). p 21.

Forrester, J.W. (1971). World Dynamics. Wright Allen Press, Cambridge, Mass.

Houge, J.C. (1979). A microprocessor-based prompting and information system for support of the elderly in independent living situations. Instrumentation Systems Centre, University of Wisconsin.

Meadows, D.H. (1972). The Limits to Growth. Universe Books, New York.

Sneller, J. (1979). New sensors are starting a second revolution in extrusion control, Modern Plastics International, August 1979, pp 56-59.

Zaks. R. (1977). Microprocessors from Chips to Systems, Sybex.

MICROPROCESSOR APPLICATIONS TO ASSIST IN REPERTORY GRID ELICITATION AND TERMINATION

Dr. Terence R. Keen

Vice Principal, Garnett College, Roehampton Lane,
London, SW15 4HR, England, U.K.

ABSTRACT

In this paper the author considers the problems facing a repertory grid administrator with respect of determining when to stop eliciting further constructs from the client. An intraclass correlation coefficient (ICC) is computed using a small microprocessor (RML 380Z with 56K core) every time a new construct is added. It is hypothesised that elicitation may cease when three successive ICCs remain constant. Evidence is provided to support this hypothesis.

KEYWORDS

Repertory Grids; Intraclass Correlation; Elicitation termination

It cannot be denied that the 1970s saw a steady increase in the body of academics who were using methodologies based in Personal Construct Psychology. (Fransella 1976.) As a member of that body one became increasingly aware of the inefficiencies associated with the technique. Two in particular are worthy of note.

In eliciting a repertory grid (Kelly 1955), the researcher will find himslef in 'conversation' with his client for many hours. In one project with which the author was concerned, (Keen 1978), some two hundred grids were elicited in a total of over 600 hours. Much of this time was wasted because the clients had really exhausted their repertoire of constructs relatively early in the process, subsequent 'forced' elicitation of new constructs merely compels the client to use new verbal labels for constructs already used. The alert administrator may identify some of these during elicitation but no formal methods exist to help him decide when to stop eliciting new constructs from the client.

The second area of inefficiency relates to the elicitation process itself. The whole process is a complex conversation which needs to be carefully monitored by a competent administrator. Many users of repertory grid based methodologies have expressed concern over the way in which potential users might acquire these 'administrative skills'. Whilst computer assisted elicitation has been possible for as long as grid methodology, it has generally not been available at the time and place required. The work of Shaw and Thomas (1976) resulted in the development of PEGASUS which was the first elicitation programme which would run on a small microprocessor and overcome many of the difficulties of personal administration. Keen and Bell (1979) developed a further programme, DYAD, which also addressed itself to this problem. The second problem, then, namely that of elicitation, had been solved relatively easily by means of easy access to small computers but in generating this solution the former problem was exacerbated. The computer could never 'intuitively' know when to stop and asking the client has proved to be a less than adequate way of stopping.

This paper then considers a method by which a repgrid administrator might know when to terminate an elicitation by an analysis of the expanding grid during elicitation by means of a small computer to hand. The principle is described by Bell and Keen (1979) in a paper entitled, "A statistical aid for the grid administrator." At the time of writing, large volumes of data had not been explored, but the author is now able to report some interesting further evidence to support the use of the same statistic (Element Intraclass Correlation as a measure of cognitive complexity) as a means of knowing when to stop expanding a grid either as a human administrator or when using computer aided elicitation programmes such as DYAD or PEGASUS.

A grid consists of a series of responses by the client, one for each element with respect to each construct. The aim of grid analysis is to show the relationships among constructs and elements which have resulted in the responses made, and there have been two major approaches to this analysis. One has been the use of clustering algorithms, such as the approach of Shaw and Thomas (1976); and the other, a singular value decomposition or "principal components" approach epitomised in the work of Slater (1977).

The intraclass correlation among elements can be easily computed (by a small computer) as each construct is elicited, and may be said to provide a measure of cognitive complexity at that stage of elicitation.

There has been substantial interest in this idea for some time and measures of Intensity of construct relationship have been developed, e.g. Bannister (1960, 1962) and used in a number of studies, e.g. Warren (1966), and Mair (1964). Bannister considered that a high Intensity score might indicate a high degree of organization in the area of the subject's component space being investigated, and thereby represent what Kelly termed 'tight' construing. Low Intensity, he hypothesised, might indicate a relative lack of clear-cut conceptual structure. Bonarius (1965), Adams-Webber (1969, 1970) and Landfield (1971) have all considered the concept of cognitive complexity, but a review of this literature provides a less than clear picture of what could be called a definition of the term, indeed measures purporting to be of 'cognitive complexity' show amazing differences in nature. In an attempt to clarify the position, Vannoy (1965) produced evidence suggesting that the concept of cognitive complexity is in itself multi-dimensional, and that all the different indices measured different aspects of it. The author claims that this measure can be a useful summary index that can be used in decision making during the elicitation of a grid. For example in grid elicitation to know when to stop eliciting constructs.

Cognitive complexity, like other grid measures, depends on the constructs elicited. Not only the quality (range of convenience, etc.) but also the quantity. The cognitive complexity of a 2 construct elicitation is likely to be less than the complexity of a 12 construct elicitation.

Having looked at the results with respect to the original analysis, it seems that the final conclusions drawn from a grid analysis will not be significantly different if construct elicitation is ceased when three successive values of the element intraclass correlation do not change. Thus a microprocessor to hand during an elicitation, and into which the growing grid is fed, enables a practitioner to know what is happening with respect to the cognitive complexity of the respondent, and to assist him (or her) in managing the elicitation as well as giving advance warning of the time when further construct elicitation is unlikely to be of additional value.

The Bell and Keen paper (1980) tested the idea of terminating grid elicitation after three successive values of intraclass correlation do not change significantly. However, this testing was on a limited sample and did not include the re-analysis of the truncated grid. Since then the author has taken a sample of 30 grids elicited several years ago all containing more than 12 constructs. In each case the ICP was run (Intraclass Correlation Programme) and a line drawn across the grid under the construct which represented the third additional construct for which intraclass correlation remained relatively static. The grid "above the line" was then analysed using GRIDDLE as was the original whole grid. In each case the element correlations were compared between the full grid and the "shortened" grid.

Over the total sample the agreement between element relationships always exceeded the 0.1 level of significance. So, one may conclude that relationships between elements change less than 10% when grid elicitation is terminated at the point determined by the "no change" intraclass correlation method described. Table 1 gives the data for one typical respondent.

TABLE 1 Interelement Correlations of (a) Full Grid and (b) Truncated Grid

E	1	2	3	4	5	6	7	8
1		0.573	1.307	1.452	1.067	1.126	0.701	0.722
2			1.282	1.394	0.994	1.056	0.854	0.713
3				0.942	0.986	0.834	1.105	1.493
4					0.692	0.696	1.222	1.347
5						0.523	0.933	0.996
6							0.890	1.199
7								0.845
8								

(a) Full Grid

E	1	2	3	4	5	6	7	8
1		1.93	.69	.31	1.16	1.18	1.84	1.71
2			.48	.15	1.00	.85	1.79	1.89
3				1.07	1.00	1.61	1.00	0.17
4					.86	1.00	.21	.40
5						1.00	.71	1.00
6							1.32	0.5
7								1.47
8								

(b)

agreement between (a) and (b): r = +0.603 ^{2} test exceeds 0.1 level of significance

CONCLUSIONS

Adding further constructs after the element intraclass correlation coefficient has remained stable for three successive new constructs (recalculated as each new construct is added) makes little noticeable difference to the way relationships between elements emerge from the grid analysis, these element relationships may change by less than 10%.

A listing in BASIC of the programme which readers of this paper are welcome to use free of charge providing the usual acknowledgements are made to the author is available from the author who would welcome comments from any users of the system.

Adams-Webber, J. (1969, 1970). Cognitive complexity and sociality. British J. of Social and Clinical Psychology, Vol. 8, 211-216. An analysis of the discriminate validity of several repertory grid indices. British J. of Psychology, Vol. 61, 83-90.

Bannister, D. (1960, 1962). "Conceptual Structure" Thought disordered schizophrenics. J. Mental Science, Vol. 106, 1230-1249. Personal construct theory: a summary and experimental paradigm, Acta Psychologica, Vol. 20, 104-120.

Bell, R. and T.R. Keen (1980). A statistical aid for the grid administrator. International J. of Man-Machine Studies.

Bonarius, J.C.J. (1965). Research in the personal construct theory of George A. Kelly: role construct repertory test and basic theory. In B.A. Maher (Ed.), Progress in Experimental Personality Research, Vol. 2, 2-46. Academic Press, New York and London.

Dyad (See Keen and Bell 1980).

Fransella, F. and D. Bannister (1977). A manual for repertory grid technique. Academic Press, London.

GRIDDLE. A principal component based grid analysis package written in BASIC on a 56K microprocessor developed by the author together with R. Bell

Hopwood, W. and T. Keen (1978) TARGET: A new approach to the Appraisal of teaching PLET, Vol. 15 No. 3.

Keen, T.R. (1979). Pedagogic styles in physics education: an attitude scaling and repertory grid study. Ph.D. thesis, Open University.

Keen, T.R. and R. Bell (1980). One thing leads to another: A new approach to elicitation in the repertory grid technique. International J. of Man-Machine Studies. Special edition June 1980.

Kelly, G.A. (1955). The Psychology of Personal Constructs, Vols. I and II. Norton, New York.

Landfield, A.W. (1971). Personal Construct system in Psychotherapy. Rand McNally, Chicago.

Mair, M. (1964). The deviation, reliability and validity of grid measures, some problems and suggestions. British Psychological Society Bulletin, 17, 55.

Shaw, M and L. Thomas (1976). FOCUS on education - an interactive computer system for the development and analysis of repertory grids. International J. of Man-Machine Studies, Vol. 10, 138-173.

Slater, P. (1977). The measurement of Intrapersonal space by grid technique, Vol.2. Dimensions of Intrpersonal space. Wiley, London.

Vannoy, J.S. (1965). Generality of cognitive complexity - simplicity as a personal construct. J. of Personality and Social Psychology, Vol. 2, 385-396.

Warren, N. (1966). Social Class and Construct Systems: Examination of the cognitive structure of two social class groups. British J. of Social and Clinical Psychology, Vol. 4, 254-263.

IV. COMPUTER SOFTWARE TECHNOLOGY

THE FUTURE OF COMPUTER SOFTWARE TECHNOLOGY AND ITS EFFECTS ON COMPUTER PROGRAM DEVELOPMENT

Dr. F. Gunther, F. Siems, G. Goldman

Booz·Allen & Hamilton Inc.
4330 East West Highway
Bethesda, Maryland, U.S.A.

ABSTRACT

As computers progressively play a more important role in the lives of every individual, the demand for efficient computer program development is becoming vital. This paper investigates state-of-the-art software technology and its implementation through the 1980's. A mathematical methodology to evaluate the effectiveness of various software tools is presented as well as a matrix showing productivity factors associated with a representative sample of the tools. The most efficient tools are then applied to a normalized workload from which proposed architectures for a software support facility are derived.

KEYWORDS

Software engineering process; software development tools; software support facility; software development; software maintenance; computer program; software.

INTRODUCTION

This paper presents the results of an investigation delving into the establishment of a center to efficiently develop and maintain computer programs. A crucial need for such a software support facility (SSF) has evolved from the proliferation and high cost of such development. As computer processing permeates our lives, software development costs are rapidly becoming a dominant system cost factor.

Following the lead of hardware, software technology has undergone a cyclic evolution. Machine code was first replaced with assembly language in the 1950's, halving the required number of instructions for a given application. During the 1960's and early 1970's, high order languages (HOLs) further reduced the number of instructions for a given programming application up to a factor of ten. Assembly languages later enjoyed a renewed popularity in achieving required higher processing speeds. However, increased processing speed and performance resulting from advances in semiconductor technology have helped foster the predominant use of HOLs and sometimes even specialized HOLs, known as very high order languages (VHOLs).

TOOLS OF THE SOFTWARE ENGINEERING PROCESS

The software engineering process (SEP) can be defined as the engineering methodologies and tools used to support software development, maintenance, and management. Further refinement of this rudimentary definition yields six stages of the process: (1) requirements definition, (2) design, (3) coding and implementation, (4) testing and integration, (5) maintenance, and (6) project management. Software tools are being developed to improve the quality and efficiency of the outputs of each of these stages. The tools, presented by stage, are described in the following subparagraphs.

Requirements Definition

This first phase entails conversion of a system concept into a detailed description of user and system resource requirements. Decisions made during this most difficult stage significantly impact the quality of the final software product. Software tools used typically raise the level of effort during this stage, resulting in up to a one hundred fold effort reduction during the maintenance phase. Three types of tools have been developed: (1) requirements analysis tools, (2) requirements verification tools, and (3) requirements statement languages.

Requirements analysis tools are usually non-automated methodologies which aid in the initial generation of system requirements. Requirements verification tools check system requirements for consistency or check a given system specification against all requirements. Requirements statement languages provide a means for formal, unambiguous expression of system requirements in a machine-processable language. Efforts are underway to extend this automated concept to directly translate a requirements statement language into executable code.

Design

The design of a software system consists of those efforts which translate user requirements into explicit program specifications. Currently a manual, creative process, design depends heavily on the available system resources. Tools developed to ease the design burden include: (1) structured design methodologies, (2) program design languages (PDLs), (3) design teams and walkthroughs, and (4) program design and documentation tools.

Structured design methodologies, long realized to be essential for a quality software product but usually deferred by schedule and funding pressures, provide an orderly top-down approach to the design process. PDLs are a family of languages capable of expressing design ideas in a structured form for easy translation into a programming language. When design teams and walkthroughs are used, a peer group team is responsible for the design, review, and approval of each software module. Finally, program design and documentation tools are automated packages aiding in the design, design clarification, and preparation of reports and graphic output. Use of these four types of tools will raise the level of effort required for design, but will significantly reduce the subsequent SEP phase efforts.

Coding and Implementation

The coding and implementation phase is the well-known process of actual programming. As such, more tools have been developed for this stage of the SEP than for any other phase. These tools have been categorized into: (1) structured languages and preprocessors, (2) code teams and walkthroughs, and (3) code audit and restructuring tools.

Structured languages and code teams/walkthroughs extend these design concepts to the coding phase. Preprocessors convert unstructured code to structured code. Code audit and restructuring tools analyze code inputs and enforce coding standards. The use of these tools can produce immediate savings.

Testing and Integration

The quality and quantity of software testing have historically been nonuniform. Comprehensive system verification and validation would require a prohibitively large number of tests; compromise approaches usually test every logical path with representative data. Three types of testing and integration tools are: (1) automatic test data generators, (2) path analysis and module testing tools, and (3) test verification tools.

Automatic test data generation tools define type and range specifications of program input parameters for comprehensive testing. Path analysis tools scan program files for inconsistencies or errors in module construction, flagging unreachable modules and poor construction. Some path analysis tools analyze the execution characteristics of programs through dynamic testing. Test verification tools examine and measure the effectiveness of software test cases. Overall, using these tools can reduce the efforts of testing and integration.

Maintenance

The maintenance phase, spanning as much as a twenty year cycle after initial system installation, includes debugging and correction of errors uncovered during operational use as well as system enhancements introduced as a result of changing user requirements. The high relative cost of changes made during this phase, as indicated under requirements above, has enlightened software developers to the value of structured programming. However, the need for maintenance will always exist, and two categories of tools have been identified: (1) code examination and upgrade tools, and (2) program portability tools.

Code examination and upgrade tools examine program interface structures by tracing their use of data elements. These tools may also perform structure changes to source code for improved readability. Program portability tools greatly reduce the effort of transporting software among dissimilar systems through the use of language standardization processors or tools, machine emulators, and machine simulators.

Project Management

Program management functions are performed throughout the system life cycle and provide the cost and schedule visibility necessary to determine project status. Automated tools developed to aid in this process serve to eliminate the manual preparation of management documentation. These have been grouped into: (1) cost and manpower estimation tools, (2) configuration control tools, (3) scheduling and milestone management tools, and (4) SEP support tools.

Cost and manpower estimation tools aid in the generation and dissemination of price, cost, and work breakdown data. Configuration control tools regiment the organization, generation, revision, and audit of data files to provide for consistency and traceable histories. Automated scheduling and milestone tools provide review capabilities, some with graphic displays and plots. The SEP support tools consist of methods, as well as specific tools, to provide project management functions throughout the software development process.

THE EFFECT OF TOOLS ON THE SEP

The effects which the types of tools presented in this paper have on the SEP have been estimated by both vendors and other independent agencies. These data are not presented here as they are incomplete and unreliable due to loosely controlled experiments. For the purpose of this paper, productivity factors were estimated for each category of tools presented in the previous section, based on field interviews and a literature search. A factor greater than one reflects increased productivity, while a factor less than one reflects decreased productivity. For illustrative purposes, the design productivity factors are presented in Table 1. In the table, phase geometric factors (p_k explained below) are attributed to each phase of the SEP with an overall savings in the right-hand column. This overall savings on the SEP, effected by solely implementing a tool category, was calculated by applying the productivity factors in proportion to the industry observed distribution of relative effort for the six SEP stages: (1) Requirements definition (R) - 12 percent, (2) Design (D) - 12 percent, (3) Coding/implementation (C) - 9 percent, (4) Testing/integration (T) - 15 percent, (5) Maintenance (M) - 32 percent, and (6) Project Management (P) - 20 percent.

TABLE 1 Design Phase Productivity Factors

Tool Category	Productivity Factor By Phase						Total
	R	D	C	T	M	P	Savings (%)
Structured Design	1.0	0.5	1.2	1.7	1.7	1.0	9
PDLs	1.0	1.0	1.2	1.1	1.1	1.0	6
Design Teams/Walkthroughs	1.0	0.7	1.1	1.2	1.2	1.0	4
Design/Documentation Tools	1.0	1.2	1.1	1.0	1.1	1.1	8

To evaluate the effects of employing multiple tools over a period of time, the level of effort (E) required for software development for a "tool-less" environment is first expressed mathematically for the six stages as a function of time "t":

$$E(t) = R(t) + D(t) + C(t) + T(t) + M(t) + P(t) \tag{1}$$

The adjusted workload for each phase is calculated by applying the productivity factor corresponding to the tools to be implemented during that phase. The adjusted design phase efforts, $\bar{D}(t)$, are then calculated as:

$$\bar{D}(t) = \frac{\left[\sum_{j=1}^{n} \frac{a_j}{p_j}\right]}{\left[\prod_{k=1}^{m} p_k\right]} D(t) \tag{2}$$

where $\sum_{j=1}^{n} a_j = 1$

a_j = portion of phase affected by p_j

p_j = phase arithmetic factor, attributed to a tool which affects only one portion of a particular phase of the SEP

p_k = phase geometric factor (PGF), attributed to a tool which has a multiplicative effect on the whole of a particular phase

The adjusted level of effort for each of the remaining five stages is similarly derived. The total adjusted effort is calculated for each year by summing each adjusted phase effort and applying the development geometric factors p_i, attributed to those tools having a multiplicative effect on the whole development cycle (excluding maintenance and project management), as follows:

$$\bar{E}(t) = \left[\prod_{i=1}^{r} \frac{1}{p_i}\right]\left[\bar{R}(t) + \bar{D}(t) + \bar{C}(t) + \bar{T}(t)\right] + \bar{M}(t) + \bar{P}(t) \quad (3)$$

Difficulty still arises when attempting to quantify the effects of employing multiple tools. As was previously indicated in discussing productivity estimates made by others, terms are missing in equations (2) and (3). Many, if not all, of these tools are interactive, and complex dependent terms should be included. These terms are difficult to theoretically derive; a need exists for further tightly controlled experimentation to arrive at reliable productivity factors and quantitative measures of the interrelationships among different tools.

However, conservative estimates were made to approximate the adjusted workload of an SSF using tools in nearly all the categories presented. (Tools within categories not used are in themselves effective, but when employed with the rest of the tools proposed, their effect is minimal or even negative on productivity.) Using these tools, it is estimated that for any given year, the level of effort expended for requirements definition will be increased by a PGF of 0.56, design efforts will not change, and coding/implementation, testing/integration, maintenance, and project management efforts will be decreased by PGFs of 2.00, 2.00, 2.00, and 1.25, respectively. This results in an overall effect of 77.6 percent, or nearly a 25 percent savings every year.

CONCLUSION: CREATING AND OPERATING AN SSF

A software support facility can, of course, be implemented on any scale; however, the initial capital investments (both hardware and software) cause an SSF to only appeal to large organizations. Three possible architectures for an SSF are: (1) centralized mainframe, (2) centralized, multiple host, and (3) distributed processing. As a start-up facility or one for a relatively smaller organization, the centralized mainframe approach is appropriate. For a somewhat larger organization, the concept is extended to interconnecting multiple processors (either homogeneous or heterogeneous) under the control of a centralized management information system (MIS). However, the most desirable architecture for the large organization is based on a distributed processing approach. In this way, software is developed at different locations on remote host computers. All sites and the SSF are interconnected through a communications network. Through its MIS, the SSF would make available to all users a complete repository of development tools. Using the network, tools and any other information of effort-saving interest may be exchanged with any other user.

Implementing an SSF may not result in immediate savings; learning to function within this new environment may produce short periods of reduced productivity. Furthermore, sudden imposition of new software tools and programming standards could be catastrophic on overall programmer productivity; tools and methodologies must be incrementally implemented. The SSF personnel should also keep abreast of the state-of-the-art to take advantage of emerging tools and methodologies. In addition, every programming environment is unique; any single tool or methodology may affect each environment differently. To evaluate the effects of such differences, data should be collected internally to measure the effectiveness of the selected tools and methodologies. This data collection may increase spending in the short term, but will establish an effective iterative process in choosing the exact tools and methodologies to be used by the SSF in developing and maintaining more economical, efficient computer programs.

INTERACTIVE PROGRAMMING AS A CAD TASK; SOME REMARKS ON SYSTEM STRUCTURES, PREREQUISITES, AND IMPLICATIONS

Hans-Jürgen Hoffmann

Unit Programming Languages and Compilers
Department of Computer Science, Techn. University at Darmstadt,
Steubenplatz 12, D - 61oo Darmstadt

ABSTRACT

Design of interactive programs covers more than only program and data design. There are further concepts involved: system structure design, dialogue structure design, display structure design. Computer aided, interactive program design has to support design according these lines, too.

The paper discusses such concepts and some approaches to cover them in future. Results are reported from recent experimental work in the area done in the unit Programming Languages and Compilers of Technical University at Darmstadt.

KEYWORDS

Interactive programming; system structure; dialogue Structure; display structure; refinement in dialogue structure design.

INTRODUCTION

The main theme of the congress is:

The quality of life and how to improve it.

These are extraordinary words; for an engineer or computer scientist -- whatever one prefers -- even more difficult to define in strong terms he is used to consider in his thinking than for an researcher in humanities or natural sciences. What is the measure for quality of life and how is the scaling which indicates improvement set by the community of human beings?

Computers are an important matter which has been related to the main theme of the congress. Certainly not free from influence of pbulic discussions everywhere in the world the organizers have considered computers to be instruments the usage of which is heavily interwoven with the grade of quality of life achieved so far by mankind. There may be positive aspects, there may be negative aspects, there may be both when asking for improvement. As an engineer and computer scientist I am strongly convinced that the technical and systematic work we do in design of computers and in installment of applications on computers is one of the prerequisites of improved quality of life; is one of the prerequisites -- if I am allowed to give some examples -- to support research in medicine, to make production processes more safe, to make the best of natural resources, to reduce the burden of routine work, to allow fair decision-making in politics and economics, to develop human capabilities otherwise restricted by limited physical rigourness of the human body and/or by limited processing and memorizing power of the human brain, even to entertain people in their spare time, in a hopefully, intellectually challenging manner. As a citizen who is looking around what is going on, obviously, I have to notice that there are also applications which are far from being well installed, even far from being good in the ethical scaling which I personally can accept. However, again as an engineer and computer scientist, I think I am certainly able to see some more reasons which led to a bad installation or which do not prevent abuse than a computer layman; and I think that explaining my work -- even if somebody may consider it as narrow-minded, technical, specialized -- and discussing it in an interdisciplinary audience like the one assembled here increases the probability that in the future the positive aspects of computer usage will definitely outweigh negative aspects.

My work is in reasoning about programming (other people call it programming methodology) and in means to support the programming process. Today, I will concentrate on structural aspects relevant during design of programmed systems, especially during design of ways of interaction with such systems.

Computers presently do not show and, in my strong opinion based on my experience in computer science and in sharp contrast to prophecies of writers in some science fiction literature, they will never show the ultimate, self-embedded objective to act upon their environments by themselves as it is shown, from men's origin, in the development of mankind by human actions. *Computers have to be programmed.* If we observe negative aspects in computer usage it is the programmer's fault or mistake or it is the fault of men which direct a programmer dependent in his work into a wrong direction. I cannot banish men with bad objectives from directing programmers (apart from what being bad objectives is a matter of an old, possibly never closed discussion in mankind). What I can do, what I can try to do, is preventing programmers from their mistakes caused by thoughtlessness, by stress, by bad circumstances. Understanding structural aspects relevant during system design which I above have announced to concentrate upon is a means to achieve this goal.

Due to space limitations it is not possible to include the full text of the manuscript and figures. The full text is available from the author (Hoffmann, 1980d).

ON PROGRAM DESIGN

Computers have to be programmed. You have already heard of this <u>thesis</u>.

We may think of *throw-away programs*. They are not worth to be considered deeply in the scope of this paper as their influence on the environment the computer is used in will be low and transient. And we may think of, so to say, *systems programs,* application packages, "software" or what you like to call them. Programming a systems program is a hard task which has to be performed with responsibility for all consequences of the usage of the program; you have to consider all situations the computer possibly will act with upon its environment when it is controlled by your program; and the environment may, in most cases will involve not only you alone. Addressing computer science people one may emphasize problems of correctness of the program, its performance, its maintainability etc. However, the behavioural aspects of computer usage chargeable to the programmer are as important as those. A mistake with respect to behavioural aspects is not a "cavalier's delict" not worth mentioning.

<u>Thesis:</u> Systems programs have to be consciously designed, they may not come into existence incidentally.

The thesis is not only true for computer science people. They are aware of the problems, no doubt, since the end of the sixties when the catchword *software crisis* appeared, and when they started much research in problems of program design. A new discipline, *software engineering,* is well established in the meantime. Also the casual programmer, the researcher in a different field, has to be addressed. *He must learn to design programs.* His programs are worth more if they are systems programs, not only throw-away programs.

As one result of the research in programming methodology, under the discipline of software engineering, we learned to use our computers not only to run our programs but also to get support from them in designing the programs. *Systems programming became a computer aided design task,* or CAPD task for short ("P" stands for "program").

There is a special, although increasingly very important kind of systems programs: programs with which *a user interacts* permanently from a dedicated terminal. The user may be a scientist who delegates some routine calculations to the computer; an engineer who controls some apparatus via the computer; a draftsman who materializes his ideas in a "description" of an object, in a kind of blue-print if you prefer this engineering term; a clerk who collects and interrogates data; a programmer who designs programs supporting such activities, who -- towards this end -- employs a combination of activities as listed, e. g., some editing work, controlling his files, materializing steps in an algorithm as a statement sequence, collecting and interrogating a cross-reference list of names of program variables. Thus, most CAPD systems of today are interactive; and many, if not most interactive application systems have at least a touch of a CAPD system.

Design of a program which is intended for user interaction involves another task not yet mentioned: design of the "dialogue" the user has to perform for interaction, i. e., the conditions for and the format, contents, succession, and effects of messages to be exchanged amongst himself and the computer. The *man-machine interface* is a prime factor in considering aspects which relate computer and computer applications with quality of life.

Thesis: Designing an interactive program implies designing the dialogue which a user has to perform for interaction.

It is not as obvious as with program and data design how to get support for dialogue design from the computer. *Computer aided dialogue design,* or CADD for short, is a new, and in my opinion a very essential field for research and integrated CAPD/CADD system development. We start by investigating structuring concepts for interactive programs.

STRUCTURE OF AN INTERACTIVE PROGRAM

Computers do not act upon their environments by themselves. They react upon an action of their environment. The statement is valid in case of user interaction, too. Progress in computation achieved by a computer controlled by an interactive program is in reaction to a message received from the user.

Thesis: We may structure an interactive program by
- a *common driver component* which analyses messages received; which autonomously handles "misunderstandings" (to speak of "error" is not the right term in this context); which helps the user on demand or in an emergency, respectively; and which checks the applicability of the command derived from a message. By
- a set of *command handler routines* each of which involves just the "semantics" of the command a routine is associated to. And by
- an aggregation of data accessible to the command handler which we call the *state* component of the program.

In the first glance the common driver component looks like being application independent. This is not true. It is here where application dependent dialogue design begins. There is a general, application independent *schema for action* the driver component has to perform in every *dialogue spot,* after receiving a message and before transmitting the reply; for details cf. (Hoffmann, 1980c, 1980d).

SUPPORT FOR DIALOGUE DESIGN

In recent methodological work on dialogue design (Oberquelle, 1980; Hoffmann/Kupka, 1978) as well as in descriptions of some implementations (e. g., Budde/Sylla, 1980; Barron, 1980) it is common practice to employ *Petri-nets* (cf. Peterson, 1977) or modifications thereof to specify the dialogue aspects of an interactive program. Useful for our purposes are Petri-nets in the form of agency/channel-nets; they specify system structure.

I will refine agency/channel-nets in two directions.

Firstly, according to the work of Oberquelle a marked Petri-net is derived. *Agencies are refined to transitions and channels to places.* Some places model dialogue spots (as introduced above). This direction of refinement allows, to give some significant examples, to study problems of
- reachability of all dialogue spots,
- reachability of the terminating place from all spots reachable from the initializing place,
- possible collaterability in firing of transitions (implying duplex mode of channels, access control to the state component, and process synchronization).

We presently investigate techniques to make these studies. I consider this to be one approach to support dialogue design. It is important to know of such properties during the dialogue design phase to avoid dialogue collapse when running the program.

There is another direction for refinement (Hoffmann, 1980c). It is a *refinement of the dialogue spots*. The refinement serves to exhibit there the details of action as contained in the schema for action (introduced above).

In his diploma thesis (Knorz, 1978) one of my students investigated and experimentally implemented some aspects of structural analysis, extraction and symbolic resolution of operands, acceptability and applicability checks, and reply compilation to be performed by a common driver component. The common driver component accepts and handles messages in a nonformalized *professional terminology of catchwords* (in natural language), covering a suitably restricted mini world (Knorz, 1979b). The reason for referencing the work in the scope of this paper -- setting aside the notational interface of catchwords which per se may be very attractive for applications you are interested in -- is the fact that he realized an interpretative strategy for controlling the aspects listed above and that he realized a generative strategy for preparing the interpreter control table (Knorz, 1979a).

There is another aspect which should be supported during design. It is the *layout of messages* to be displayed *on the video screen*. Although it is only of notational significance it shows a high rate of correlation with user acceptance of an interactive program. Decisions about layout heavily depend on hardware characteristics of the video device. To overcome maintenance and portability problems it is important to control and isolate all decisions which, of necessity, are hardware dependent from those which are related to the dialogue structure and the semantically relevant tokens of messages. A specific agency, the *screen handler*, has been identified. Its purpose is to map messages originated in one of the application agencies or otherwise passing through the discussion leader (e. g., keyboard input echoed upon the screen) into a display.

I am aware of two basic approaches to understand the effect of screen handling. The first one -- I call it *static screen handling* -- is a one-to-one mapping of message formats to displays putting semantically relevant tokens, e. g., numbers, text items, into reserved fields of the display; standard texts, graphical signs as lines, frames etc. are automatically added to the display. A variant of static screen handling involves *subdisplays* of a size just overlaying display fields set aside for that purpose. It becomes a runtime decision what subdisplay overlays a given field depending on the specific structure of the message to be handled. One of my students has recently developped a supporting system in his preparatory thesis (Straub, 1979) which allows to define displays and subdisplays (of up to two layers) in an interactive manner on the screen itself, which will later be used in the application.

The other approach, *dynamic screen handling,* considers the screen to be a resource for displaying messages (Hoffmann, 1980a, 1980b). Our work on this approach is just starting.

LITERATURE REFERENCES

The reports PU1... are reports of the unit Programming Languages and Compilers. They are available upon request from the author.

Budde, R. and K. H. Sylla (1980). KAW - ein Kommunikationsabwickler im BS 2ooo. GMD-Spiegel, 10, 17 - 30.

Barron, J. L. (1980). Dialogue organization and structure for interactive information systems. Techn. Report, Computer Systems Research Group, Univ. Toronto, CSRG-108.

Hoffmann, H.-J. (1980). Project application to the German Research Foundation, unpublished.

Hoffmann, H.-J. (1980). Über den Einsatz des Kommunikations-Betriebsmittels Bildschirm oder Grundregeln zur Gestaltung strukturierter Bildschirmdialoge, an einem schlechten Beispiel gezeigt. Research Report PU1R2/80.

Hoffmann, H.-J. (1980). Unstructured dialogues considered harmful or how to design dialogues, some lessons from programming methodology. Forthcoming Research Report PU1R7/80.

Hoffmann, H.-J. (1980). Interactive programming as a CAD task; some remarks on system structures, prerequisites, and implications. Research Report PU1R5/80.

Hoffmann, H.-J. and I. Kupka (1978). Arbeiten mit interaktiven Systemen, erläutert am Vorgang des Programmierens - Theoretische Konzepte und Gesichtspunkte der Realisierung - (How to work with interactive systems, explained by the process of programming - theoretical conceptions and implementation aspects). Notizen zum Interaktiven Programmieren, 1, 5 - 30.

Knorz, G. (1978). Entwicklung und Implementierung einer Dialogkonzeption für eine anwendungsbezogene Programmdateneingabe (Design and implementation of a conception of dialogue for application-oriented data input). Diploma thesis, PU1-S066.

Knorz, G. (1979). Metasystem für ein System zur Dialogführung, das anwendungsbezogene Eingaben in einer stichwortartigen Fachsprache akzeptiert (Metasystem for a system of dialogue control which accepts application-oriented messages in a professional terminology of catchword). Notizen zum Interaktiven Programmieren, 2, 5 - 15.

Knorz, G. (1979). Ein Dialogsystem für anwendungsbezogene Eingabe in einer stichwortartigen Fachsprache (A dialogue system for application-oriented input in a professional terminology of catchwords). Angewandte Informatik, 21, 11, 495 - 503.

Oberquelle, H. (1978). 'Grobe' Beschreibung von Systemen durch Netze ('rough' description of systems by nets). GI-Workshop über Petri-Netze, May 1978, Erlangen. (extended version as Research Note of Dept. Comp. Science, University in Hamburg, Germany, IFI-HH-M-62/78,1978).

Peterson, J. L. (1977). Petri nets. Compt. Surv., 9, 223 - 252.

Straub, W. (1979). Ein Programmsystem zur leichteren Erstellung von Dialogprogrammen mit PL 360 (A programmed system to facilitate the implementation of dialogue programs in PL 360). Preparatory thesis (Studienarbeit), PU1-S070.

TOPCAG: TOP-DOWN COMPUTER-AIDED GRAPHICS

N. Magnenat-Thalmann*, D. Thalmann**

*Département des Méthodes Quantitatives, Ecole des Hautes Etudes Commerciales, Montréal, Canada
**Département d'Informatique et Recherche Opérationnelle, Université de Montréal, Montréal, Canada

ABSTRACT

A Top-down Computer-aided Graphics (TOPCAG) system is an interactive graphics system based on top-down methodology. A TOPCAG system allows the user to create graphics, then to choose a region of the drawing in order to create a new picture of this region with more details. In that way, a general picture is drawn at a first level, then parts of it can be detailed at the second level and the process can be repeated more times. In this paper, a user-oriented TOPCAG system is presented. The major features of this system are explained and the inclusion process is emphasized by the means of a complete example.

KEYWORDS

Top-down methodology; graphics edition; structured design; user-oriented; abstract graphical data type.

WHAT IS TOP-DOWN COMPUTER-AIDED GRAPHICS?

Producing graphics by computer for many types of applications, especially those involving massive amounts of data or repetition, is much more efficient than by traditional manual methods. People need to use graphics to represent concrete situations or diagrams to illustrate text in reports. For that purpose, a graphics editor is necessary to create and modify figures, in the same way as a text editor allows the user to create and modify text. The authors have designed and implemented such a graphics editor, which has been especially designed for office automation - (Magnenat, Thalmann, 1980a). However, - this editor permits to create a picture at only one level.

In this paper, we present a system which is based on top-down methodology: Top-Down Computer-Aided Graphics (TOPCAG). Top-down design (Yourdon, Constantine, 1979) is based on the following characteristics: we make decisions at the highest level first and then iteratively make decisions at the lower levels. In Computer Graphics, such an approach is very fundamental, because it allows the user to create graphics at a first level without details, then to decide which regions (called static windows) have to be created at a second level with more details. Subsequent levels may be created in the same way. At each level, graphics may be modified by using graphics editing commands.

Applications of a TOPCAG system are very numerous: flowcharts, system flowcharts, engineering graphics, business applications.

PRESENTATION OF A TOPCAG SYSTEM

The First-Level Graphics

This TOPCAG system has been designed with the fundamental criterion: it has to be very easy to use for a non-specialist in computer science. The user has only a few number of commands to learn. Firstly, he has to declare what is the purpose of the session: CREATE a graphics, LOOK at a graphics or MODIFY a graphics.
In the phase of creation, the user builds the picture with the graphics editor. He chooses primitives in a catalog and the location where they have to be drawn. At this stage, he can still change the location, the angle or the size of the figures. He can also type a text to be included in his figures or link different objects by pointing them out. In fact, 20 different commands are available; they are very user-oriented because of their simplicity.
When this first-level graphics is finished, the user leaves the graphics editor mode by entering the command END. The next step is a structuration phase.

Creation of Lower Levels

To create a new picture of a region with more details, the user has to specify this region. This is performed by a rectangle which can be moved: the dynamic window. This rectangle in dashed lines appears when the user types the command WINDOW. The standard size of this window may be modified by another command SIZE. When the user has moved the dynamic window to the right place, he decides that this region has to be detailed by creating a copy of the dynamic window: a static window. The command to perform this action is STRUCTURE (identifier), the identifier is the name of the static window.
The screen is cleared and the window (that means the region selected before) is represented on the screen. The user may modify it as he wants with the graphics editor. He may also use again the command WINDOW and the command STRUCTURE to create graphics at a third level. If he then wishes to come back to the more general picture, he can use the command UP and the preceeding picture of the upper level appears on the screen again. At this stage, the user may then select a new region and again use the graphics command STRUCTURE.

THE INCLUSION PROCESS

When the user creates a static window, he would often like to include in this window visible diagrams of the upper level. A diagram is considered as visible if at least one of its edge (or a part) is included in the window. The clipping is done by using the Cohen-Sutherland algorithm. The decision of including diagrams in the window is performed when the user types the command STRUCTURE. The static window increases at the maximum screen size, then the system asks: UPPER LEVEL INCLUSION? Y = YES, N = NO, A = ALL.
"N" means that any diagrams has to be included (or it has been already done in a previous session). "A" means that all visible diagrams have to be included. When the user answers "Y", for each visible diagram the system asks.
INCLUSION OF THIS DIAGRAM? Y = YES N = NO.
The user may select which diagrams are useful at the lower level.

An Example

The following 4-level example shows the inclusion process. At each level, a static window is created and all visible diagrams are included in the lower level.

At the main level, five persons have been drawn using the editor facilities.

Fig. 1. Five persons with the dynamic window in dashed lines.

The user has created a static window at this position and has included all the visible diagrams (a person only in our case). With the editor, the user adds the following details: mouth, nose, eyes, ears, hair and belt; the picture is shown in Fig. 2.

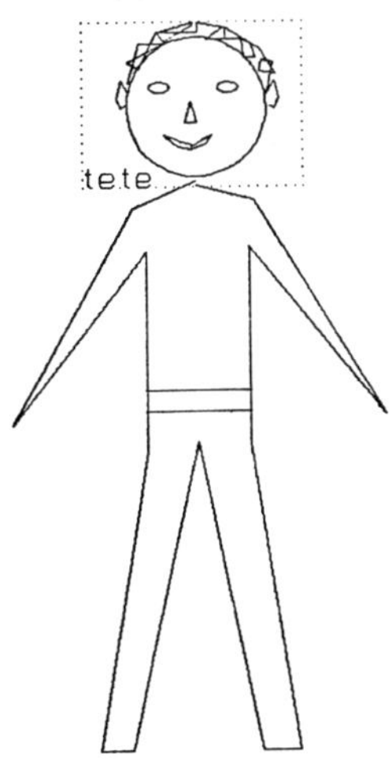

Fig. 2. A person with the dynamic window at the head.

A new static window is created with inclusion and the user modifies the eyes, the mouth and the ears (Fig. 3) then he creates a new static window on the right eye.

Fig. 3. A head with the dynamic window at the eye.

Fig. 4 shows this eye with new details added by the user.

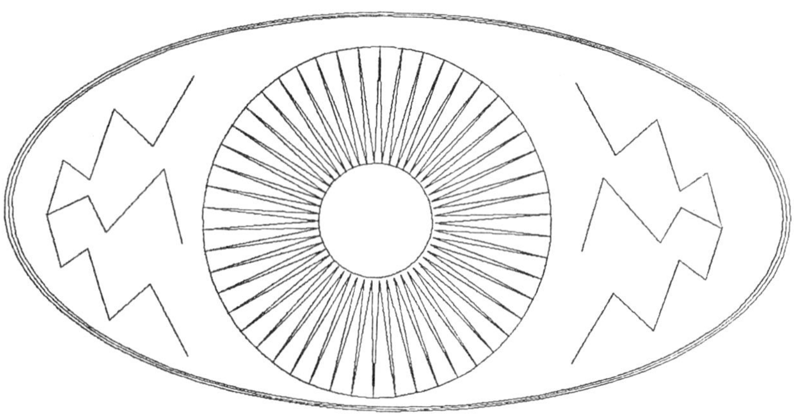

Fig. 4. An eye.

THE PROCESSING OF AN EXISTING GRAPHICS

It is also possible to process an existing graphics by using the command MODIFY. In this case, the user may enter the command DETAILS. Static windows, that have been selected at the creation stage, appears in dashed lines on the screen. The user points out one of these windows and with the command DOWN, the corresponding detailed picture appears on the screen.

By using the editor, the user may create new diagrams, modify other ones, change sizes or orientation and so on. He may also include the new diagrams in lower static windows by using the inclusion process. A command DELETE permits to suppress a window and all the lower level windows which are dependant on this one.

IMPLEMENTATION

Our system has been implemented on a CDC Cyber with a graphical terminal HP2648A. It is an interactive program which consists of 3000 source lines of MIRA-2D (Magnenat, Thalmann, 1980b), a graphical PASCAL (Jensen, Wirth, 1974) extension based on abstract graphical data types (Thalmann, Magnenat, 1979). Interactive commands are immediately analyzed and clear error messages are provided. The top-down graphics structure is supported by a hierarchy of files. The system has the following limitations: 10 levels and 50 static windows. A program for producing hard copies of all static windows on a VERSATEK printer has been also developed.

CONCLUSION

The system that has been presented is a system which may be used in a great number of applications, because the top-down approach is fundamental in all areas. For example, it allows to build complex circuits in Electrical Engineering, house maps in Architecture or logical data flow diagrams (Gane, Sarson, 1979) in systems analysis. The TOPCAG method which is the basis of this system is essential and systematic to develop complex drawings like the stepwise refinement method is surely the best to build a large program (Wirth, 1971) and our system was developed by using this method.

ACKNOWLEDGEMENT

The authors would like to express their gratitude to André Larouche who has worked for this project. This research has been supported by Natural Sciences and Engineering Research Council Canada and l'Ecole des Hautes Etudes Commerciales de Montréal.

REFERENCES

Gane, C., and Sarson, T. (1979). Structured Systems Analysis. Prentice-Hall.

Jensen, K., and Wirth, N. (1974). PASCAL User Manual and Report. Springer-Verlag.

Magnenat-Thalmann, and Thalmann, D. (1980a). Graphics Editing in Office Automation. Proc. Canadian Information Processing Society Conference, CIPS Session '80, Victoria.

Magnenat-Thalmann, and Thalmann, D. (1980b). A Graphical PASCAL extension Based on Graphical Types. Software-Practice and Experience, Vol. 10

Thalmann, D., and Magnenat-Thalmann, N. (1979). Design and Implementation of Abstract Graphical Data Types. Proc. COMPSAC '79, Chicago, IEEE Press, 519-524.

Wirth, N. (1971). Program Development by Stepwise Refinement. Commun. ACM, Vol. 14,4, 221-227.

Yourdon, E., and Constantine, L.L. (1979). Structured Design. Prentice-Hall.

SOME SYSTEMS FOR MODELING COMPUTER SECURITY REQUIREMENTS

David C. Rine

Western Illinois University
Macomb, Illinois 61455 U.S.A.

ABSTRACT

This paper presents four models of use in analyzing a system's security requirements in the abstract. A discussion of relations between security and information bindings is included, and a number of features for study are outlined.

INTRODUCTION AND REVIEW OF SOME PREVIOUS MODELS

Early Protection Mechanisms

In the late 1940's and early 1950's, early computers were machine-language programmed by single users with programs being processed one at a time. Users could protect confidential programs and data against unauthorized access by simply taking the programs and data home (in the form of notes, punched cards, or tapes) after each use.

By the late 1950's and early 1960's, with the introduction of multiuser programing and operating systems, programs and data began to fall into two classes, either application programs which utilize user libraries, or privileged programs which could only be processed by the executive operating system. Hence, computer system protection schemes were of two levels and used devices such as switches and keys to isolate application programs/data from privileged programs.

In addition to privileged programs/data, it was soon discovered that some application programs and data demanded a greater degree of protection. Computer system protection schemes using devices such as protection rings were employed by users to give these more sensitive application programs/data a greater degree of security.

By the late 1960's and early 1970's, with more levels of software separating the user-programmer from the operating systems and with more diverse and complex applications and query programming security requirements, it was discovered that the protection mechanisms afforded by operating systems did not often match well, or enforce properly, the security requirements of the user community.

This problem still exists. It is, therefore, of primary importance to study the relation between computer security policies (rules, requirements) and protection mechanisms (features, techniques) in order to exhibit and analyze those design features and programming techniques of a computer system influencing various security policies (Zadeh, 1977). This is what we discuss herein.

Thus, several means are introduced for studying relations between computer security specifications and protection techniques in order to find ways by which features of a computer system relating to security specifications may be designed, checked or evaluated. The study herein is concerned with both theoretical studies and modelling of security policies and protection mechanisms. One formal basis of the research includes a model introduced in the Basic Model section of this paper. Research objectives include design of algorithms and system programs which translate user program source code into corresponding protection mechanism code; using lattice techniques to show how to iteratively construct consistent, sound protection mechanisms which are more nearly "complete" then given ones.

The research also outlines linkage-edit, load, and call time mechanisms to certify programs, in comparison to previous work which has been limited to run-time and/or compile-time mechanisms. For each of these certification mechanisms the relationships between bindings to security classes and information bindings between names and objects are outlined and described in the Relations Between Security and Information Bindings section of this paper.

Suppose that as in (Rine, 1978) and (Jones, 1975) M(Z,I) denotes a protection mechanism system program for a given application program Q and a given accompanying security policy I. Hence, M(Q,I) enforces I. Then, an example of the input and output features of a system program which would translate user program source code into corresponding protection mechanism code and which could be implemented by using instances of functions F and S given earlier by (Conway, 1972) for ASAP follows (see Fig. 1): a system program could be built which would take the assignment statement "$V \leftarrow E(W_1,\ldots,Wp)$" of application program Q and generate the following code for protection mechanism M(Q,I): $V \leftarrow E(W_1,\ldots,Wp)$

```
proc;
  .
  .
  .
call S_t (u,V); if Ṽ then do; Ṽ ← null; V ← 1; end;)
else do; call F_t (u,E,W_1,...,Wp; call S_t(u,v);
   Ṽ ← 0; end;
...  F_t: proc (u,d);
if (read-ck ^ di) then do;/* at translation
time */ call gen-fetch-code return (fetch-code);
end;
else if (¬ read-ck ^ di) then call threat-monitor;
else if (read-ck ^ dd) then /* at run time */
call F_r(u,d,read-ck); return;
end F_t;
  .
  .
  .
```

```
. . . S_t: proc (u,d);
      if (write-ck ^ di) then do; /* at translation time */
      call gen-store-code; return (store-code); end;
      else if (¬ write-ck ^ di) then do; call threat-monitor;
      Ṽ ← 1; end; else if (write-ck ^ dd) then /* at run time
      */ call S_r(u,d, write-ck); return;
      end S_t;
. . . F_r: proc (u,d, check);
      if check then call fetch (d);
         else return (null);
      end F_r;
. . . S_r:  proc (u,d, check); if check then call store (d);
           else return (null);
           end S_r;
end;
```

Fig. 1. A Translation System Program.

Language Concepts

More recent database management systems have, for the most part, chosen to implement access control and protection mechanisms at the translation step, using for example S_t's and F_t's, because it is easier to do so. For example, (Held, 1975) has implemented access control in its query language QUEL in the following way: the "RESTRICT" QUEL command is appended to a user's qualification by the AND operator. An IBM query language similar to QUEL called SEQUEL, a query language for System-R, does translation time access control checking. From (Conway, 1972) one may conclude that some efficiency is gained if the appending is done at query translation time and if one assumes complete data independence. However, there is still the matter of the Trojan Horse Problem of application program access wherein object code inside the translation step leads to illegal access (the Trojan Horse Problem does not normally apply to query language access) and the Observability Postulate Assumption.

It may be helpful to differentiate the problem of controlling access for reasons of security from locking data to control simultaneous updating (a problem of deadlocking and integrity as opposed to security). This problem of locking for reasons of security is covered in (McCauley, 1975-76). The important ideas are the following:

> data dependency "A process should be able to lock a subset of the database or a 'critical region' without having to know (and to address) each element in this subset."
>
> lock data to control simultaneous updating (versus security control) protected sets specified by different processes need not be disjoint.
>
> simultaneous access by two processes to a record satisfying two "elementary Boolean conditions" is not possible.

Programming language names are symbols used to represent user information entities, i.e. the things that the users of programming languages talk about.

Information binding (Elson, 1973) refers to a set of programs which maps each user program name to a corresponding object such as a location value pair.

There are various types of bindings and there are various times at which these bindings can be performed. Program creation binding is analyzed during compile-time, and execution reference binding is analyzed during run-time.

A certification and authorization of a particular information binding implies that a user query or user application program has legal access to information values derived from a particular set of locations in storage(s).

There is usually a security policy which attempts to describe which security class a name and its associated object belongs to. Recent work (Denning, 1976; 1976; 1977) presents a compile-time mechanism that certifies no flows in violation of the flow policy. This work also supposes the "binding" of each "object" to a security class is statis (as opposed to dynamic) and can be determined from the declaration contained in a program. Their "objects" are actually what we refer to as names.

In the study described herein we will also outline linkage-edit, load, and call time mechanisms to certify programs, in comparison to previous work which has been strictly limited to run-time mechanisms (Graham, 1968; Rine, 1976-77) and compile-time mechanisms (Denning, 1976; 1976; 1977). For each of these cases we will investigate the relationships between bindings to security classes and information bindings between names and objects, and their relations to the corresponding certification mechanisms.

Current Research

Basic Model. In this section a basic model is described which associates with a user or program both a security policy and an associated protection mechanism. This model is similar to one proposed by (Jones and Lipton, 1975) but includes some additional components worthy of further investigation. The model is introduced as consisting of the following components:

Basic Model

OB) Objects to be protected, logical and physical.

OP) Operations* on objects, logical and physical.

EN) Environment:** subset of Cartesian product OP X OB, logical and physical.

MT) Enforcement: protection mechanisms - software and hardware, as a function of time.

N) Environment Modification: Logical and physical data-base migration.

G) Migration: $EN_1 \rightarrow EN_2$ or EN_1/restriction $\rightarrow$ EN_2/ restriction

*Of which subjects' processes are composed; e.g. authorized accesses are operations.

**Associated with each user is an environment. Dynamic alteration or modification of the database should preserve security.

<u>Program</u> (user or subject-program) $Q: D_1 x \ldots x D_k \to E_1 x \ldots x E_n$

D_i: range of <u>ith</u> input

E_j: range of <u>jth</u> output

$X_i, \ldots, X_k \varepsilon$ OB

$Y_i, \ldots, Y_n$ output variables

$$Q(X_1, \ldots, X_k) = (Y_1, \ldots, Y_n) = (Q_1(X_1, \ldots, X_k), \ldots, Q_n(X_1, \ldots, X_k)).$$

Let n be a positive integer. Security policy I_j, 1 j n, for Y_j is a collection of subsets $\{Xj_1, \ldots, Xj_m\}$ such that $\forall \, Xj_s{}^j \exists Xp \varepsilon \{X_1, \ldots, X_k\}^j$ such that $Xj_s \subseteq X_p$. Security policies $\{I_1, \ldots, I_n\}$ is the <u>security policy</u> I of Q. Given Q and I introduce <u>protection mechanism</u> $M(Q,I)$: $D_1 x \ldots x\, D_k \to E'_1 x \ldots E'_n$ where F(M) <u>violation notices</u>, $E'_i = E_i \cup F(M)$, $E_i \cap F(M) = \emptyset$, and $M_i(Q,I)(X_1, \ldots, X_k) = Q_i(X_1, \ldots, X_k) = Y_i$ or $M_i\,(Q,I)(X_1, \ldots, X_k) \varepsilon F(M)$. From the motivation by (Conway, 1972) and (Lampson, 1973) one introduces "soundness". Consider Q, I, M(Q,I). M(Q,I) <u>sound</u> if $\forall j, Y_j$, $I_j = \{Xj_1, \ldots, Xj_m\} \exists$ restricted mapping of M(Q,I) denoted $M'(Q,I)$: $Dj_1 x \ldots x Dj_m \to E_1' x \ldots x E'_n$ such that $M_j(Q,I) = M'j(Q,I) = Y_j$. i.e. $M'_j(Q,I)$ <u>only</u> gives information derived from I_j, authorized information; $M_j(Q,I)$ <u>could</u> give information derived from either I_j or not I_j if there is a <u>leak</u> as illustrated by Fig. 2 below.

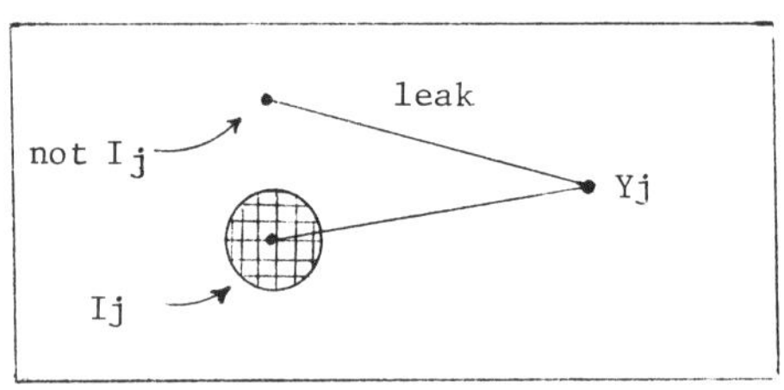

Fig. 2. Sound Protection Mechanisms: unsound mechanism, indicating a possible leak to output variable Y_j.

While it is true that a sound protection mechanism enforces a security policy for a user program it is also true that it may over-enforce the policy thereby generating unnecessary violation notices. Consider the following graphic remark.

<u>Remark</u>: Suppose $\exists$ M(Q,I) such that $\forall i$ $M_i(Q,I)(X_1, \ldots, X_k) = C \varepsilon$ Constant F(M) fixed. Then M(Q,I) is useless and gives too many violation notices. Hence, there is a dichotomy between over-enforcement and soundness.

To address this dichotomy, the concept of "Completeness" is introduced in order to use lattice operations to build up, by the meet and join operations from sound protection mechanisms, a protection mechanism which neither over-enforces nor allows leaks.

Informally, completeness in the general sense implies that all of these and only those requests Q that are security sensitive must pass through the internal protection mechanisms M(Q,I) before being honored.

As a compromise one may only be able to get a new protection mechanism which is sound and which does not over-enforce, i.e. generate as many violation notices, as much as the old protection mechanism. One introduces "completeness", similar to (Jones and Lipton, 1975) as follows:

Completeness (lattice concepts)

Let $M^1(Q,I), M^2(Q,I)$. Then, $M^1(Q,I) \geq M^2(Q,I)$ whenever $M^1(Q,I)(a) \neq Q(a)$ implies $M^2(Q,I)(a) \neq Q(a)$.

Let $M^1(Q,I)$, $M^2(A,I)$ sound. Define $M = M^1 \cup M^2$ join such $\forall$ a

$$M(a) = \begin{cases} Q(a), & \text{If } \exists\ i \varepsilon 1, 2 \text{ such that } M^i(a) = Q(a), \\ M^i(a), & \text{otherwise.} \end{cases}$$

Conclusion: M is sound and $\forall i\ M \geq M^i$.

Pf. Induction.

A total access control mechanism M(I) is said to be complete when every attempt Q to access information during the progress of a computation is validated. This is, generally, impossible since it is not known in advance what all Q would be. However, completeness does compare the extent to which the security requirements I, if compiled with, provide assurance that the information system is not vulnerable to penetration (i.e. compares which mechanism is better). An example covering this model, which was just described, appears below.

Example. It is helpful to be able to view the model just described in the Basic Model section of this paper in terms of a real commercial database system (DBS) and associated protection mechanism. Because of its popularity, let us consider an example like IBM's IMS DBS (IBM).

Security policies may be implemented via IMS protection mechanisms in the following ways:

1. segment sensitivity (SENSEG),
2. processing options (PROCOPT=),
3. data encryption before storage, and
4. terminal and password security.

Figure 3 illustrates the basic architecture of an IMS System. At the physical storage level data encryption provides data access protection. At the communications level data encryption and terminal protection provides data access protection. At the applications program level password protection provides data access protection. At the DL/I calls level from the host language record segment sensitivity and authorized processing options provide data access protection.

In terms of our model Q is a DL/I call, an Xj_s is a program communication block (PCB), an S_p is a database description (DBD), and, indeed, $\forall s \exists p$ such that $Xj_s \leq Xp$, i.e. the segments of each PCB reside in some DBD. Moreover, the set of PCB's assigned legally to a given applications group, $\{Xj_1, \ldots, Xj_m\}$, describes the complete restricted view of the database (data submodel), $\{X_1, \ldots, X_k\}$, for which they have authorized access; and, therefore, it is indeed the realization of their security policy Ij. Hence, for all programming groups who may wish to issue a given type of DL/I call, Q: CALL PLITDLI(FØUR,*,ØP,I/O Area, qualifier), security policies $\{I_1, \ldots, I_n\}$ is, indeed, the security policy I of Q.

Y_j is precisely the I/O Area to which users programs' authorized PCB's $\{Xj_1, \ldots, X_{im}\}$ may return data from a DL/I call without raising an error flag, i.e. violation notice. $Q(Xj_1, \ldots, Xj_m) = Yj$.

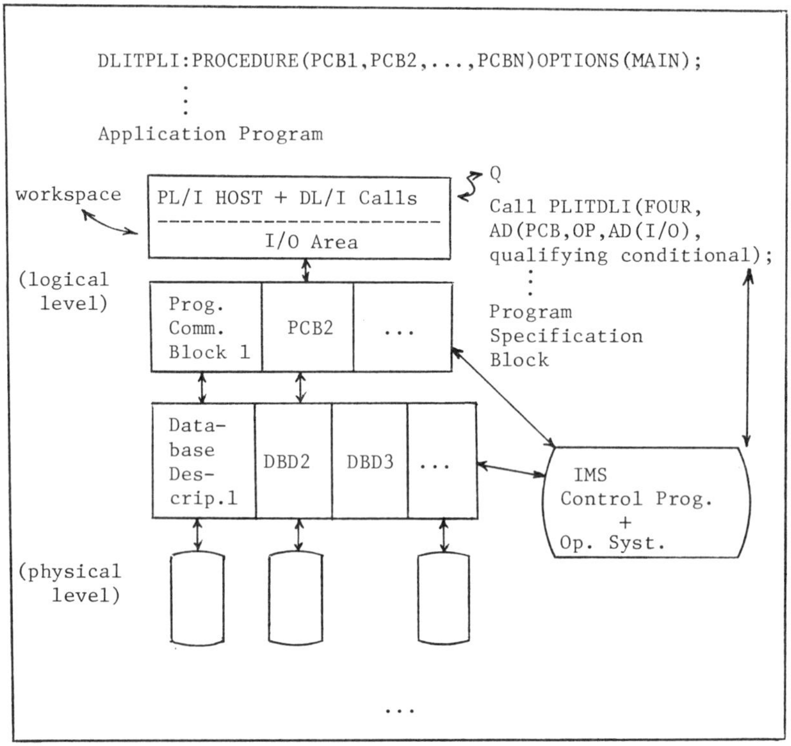

Fig. 3. IMS Data Bases Architecture of IMS System

We raise the question as to whether or not IMS under SENSEG and PROCOPT is sound, as a protection mechanism, given a DL/I call type, Q, and a designated security policy, I, over access to all PCB's via DBD's. That is, does IMS really enforce a policy formalized in SENSEG/PROCOPT notation? Consider the employee database described in Fig. 4.

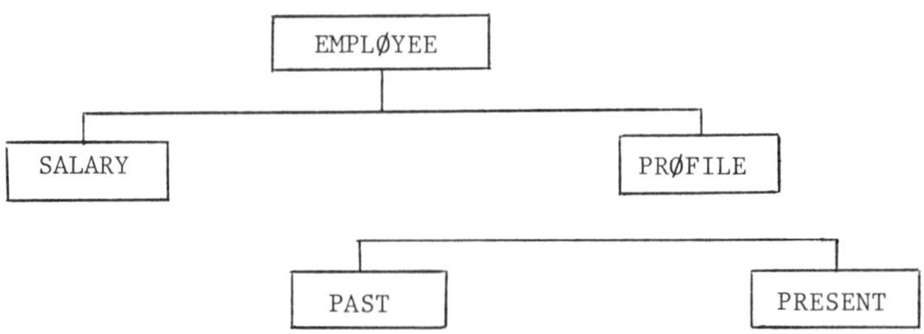

Fig. 4. Employee IMS DB, Physical Database PDB1.

Let EMPLØYEE, PRØFILE, and PRESENT be sensitive segments authorized to PCB1; and, let SALARY, PAST be non-sensitive segments unauthorized to PCB1; and protected in theory by PROCOPT=K. Therefore, the restricted view of the database is as in Fig. 5.

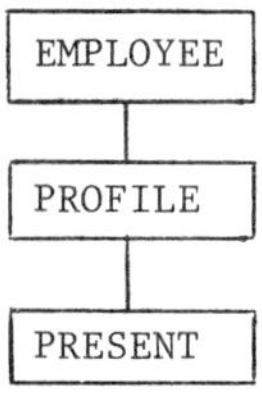

Fig. 5. Employee IMS Logical Database PCB1.

Suppose that a query DL/I call, Q, requests data about PRESENT information regarding an EMPLOYEE.

Then, internally, IMS must follow from the root (EMPLOYEE), segment occurrences according to the hierarchical sequence (tree preorder traversal) from EMPLOYEE occurrences, through SALARY, PROFILE, PAST, and PRESENT occurrences, until the desired PRESENT segment occurrence is obtained, as in Fig. 6.

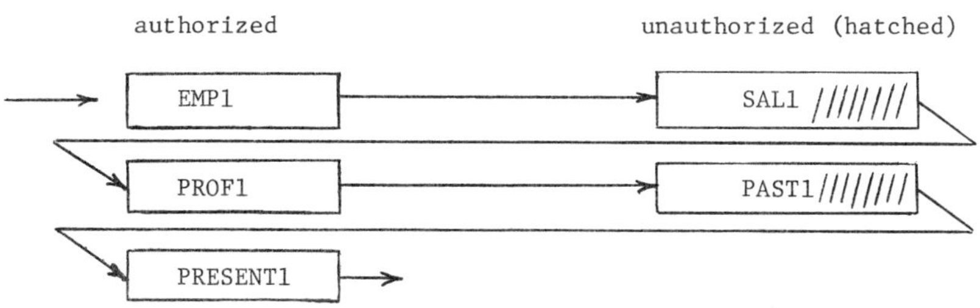

Fig. 6. Internal Data Structure Traversal.

Thus, the accessing program must <u>pass through</u> physical records which are unauthorized to the user.

Should a system program mechanism be developed which could, while the access program is processing, leak physical records which are being passed through, then the protection mechanism would not be sound, as depicted in Fig. 7. Thus, there may not exist M'(Q,I) such that Mj(Q,I) = M'j(Q,I) = Yj. A better protection mechanism would involve a data structural arrangement <u>prohibiting pass throughs</u>.

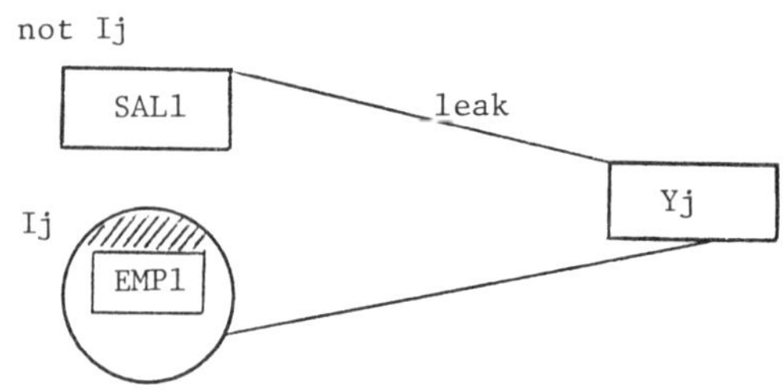

Fig. 7. A Possible Leak to I/O Area Yj.

Further Study.

Relations Between Policies and Mechanisms. In (Jones and Lipton, 1975) the high water mark protection mechanism $M_h(Q,I:G, > G_o)$ was introduced as a model for protection features. Recall that if Q is a flowchart representation of a user program and I is a security policy, then (G, >) is a linearly ordered set with grading > and G_o: $\{X_1, \ldots, X_k, Y, \ldots\} \to G$ is a function from the variables of Q to G. An assumption is that there exists $p \in G$ such that $\forall v \in I$, $w \notin I$, then $G_o(w) > p \geq G_o(v)$. The following are example interpretations:

1. G is a set of classifications (military system of clearances) such as "top secret", "secret", "confidential", "public", ..., which are ordered such that top secret > secret > confidential > public ...;

2. initially all the variables in a security policy I are given some classification equal to or lower than p, and all other variables are given some classification higher than p;

3. G = {0,1} = {deny, permit}.

Recall the following connection from (Rine, 1978) between security policies and security matrices (Conway, 1972):

$$\lambda(Q,I_u)(d) = \begin{cases} 0, & \text{d denied to u} \\ 1, & \text{d permitted to u.} \end{cases}$$

This connection could be extended to multiple-values (Rine, 1975) as follows:

$$\lambda(Q,I_u)(d) = \begin{cases} 0, & \text{d denied to u} \\ 1, & \text{d usually denied to u} \\ 2, & \text{d sometimes permitted to u} \\ 3, & \text{d usually permitted to u} \\ 4, & \text{d permitted to u.} \end{cases}$$

0 corresponds to the highest cost (penalty) to break the protection mechanism, 1 to the next highest cost (penalty), and 4 to the lowest cost (no penalty). For the protection mechanism the values 0, ..., 4 could also be interpreted as protection ring numbers (Held, 1975).

Military classification schemes seem to be completely binary; either access is permitted or it is denied. However, in formulating security policies, even in the military, human use of policy statements is subject to vague interpretations; and managers use such terms as "sometimes permitted" in trying to describe security policies.

Another way of viewing what takes place in passing from the security policy in the first example (1.) above to its protection mechanism is to observe that "confidential" is a fuzzy set having an initailly vague meaning. Likewise, are top secret, secret, public, denied, usually denied, sometimes permitted, usually permitted, and permitted fuzzy sets.

Continuing the notion of fuzzy sets, suppose that U = {0,1,2,3,4,5,6,7,8,9,10} is a universe of discourse consisting of all security classes or protection ring numbers (Graham, 1968). Let N be a variable taking values in U and u the generic element of U. Then N = 2 signifies that N is assigned the ring number 2. Suppose that the "grade" of membership in fuzzy set "confidential" is approximately .8, $.8 \varepsilon (G,>)$. Next, from (Zadeh, 1977) let "Poss" denote the possibility distribution.

Continuing the concept of possibility distributions, consider the "degree of possibility that 'dataset is 2' given the proposition 'dataset is confidential'". This can be denoted as follows:

Poss (dataset is 2 | dataset is confidential) = .8, where 'confidential' is a fuzzy subset of U.

In the area of management, administrators often must express security policies by using imprecise terminology. The following are examples:

dataset A is confidential.
dataset B is denied access.
dataset C is restricted but dataset D is permitted access.

Possibility distribution statements such as Pass (A is 2 | A is confidential) = .8 and Poss (B is 1 | B is denied) = .2 illustrate that Poss is a connection between policies and enforcements, where for example .2 and .8 are meanings (results of an evaluation) of enforcement.

A general possibility distribution statement is Poss (X is r | X is F) = g, "X is F" is a policy, F fuzzy subset, "X is r" is a desired enforcement, and $g \varepsilon (G,>)$ is an evaluation of enforcement.

Consider the following four types of possibility distribution statements. For each type below (1-4) there are two extreme cases (deny, permit), and one of the cases is a problem. In type (1) if r is small and g is small, then leaking of information may occur; in type (2) if r is large and g is large, then leaking of information may occur; in type (3) if r is small and g is large, then over enforcement may occur; and if r is large and g is small, then over enforcement may occur.

Thus, from the previous section, Basic Model, of the paper, types (1) and (2) illustrate a lack of soundness for the protection mechanism; whereas types (3) and (4) illustrate a lack of completeness, i.e. the protection mechanisms for types (3) and (4) are less than optimal with respect to the lattice ordering defined in the last section for sound protection mechanisms in particular.

Graphical interpretations of these relations were introduced in (Rine, 1978).

1. Poss (X is 0 |└──────┘) = 0.
deny (leak) — permit (OK)

2. Poss (X is 10 |└──────────┘) = 1
deny (leak) — permit (OK)

3. Poss (X is 0 |└──────┘) = 0.
deny (OK) — permit (over enforcement)

4. Poss (X is 10 |└──────────┘) = 1
deny (OK) — permit (over enforcement)

The ability to answer questions relating to the extracting of meanings of information under security policies that is stored in a database is being discussed. A theory of possibility may help lead to a model for illustrating how this can be done with a complete and sound protection mechanism.

In conclusion we have indicated how important it is to study the correlation(s) between security policies (rules, requirements, specifications) and protection mechanisms (features, techniques) in order to define and analyze specific architectural features of a computer system corresponding to specific security policies. We have seen that it may be advantageous to measure strengths or weaknesses of such correlations through possibilities. Important tools may include the usage of possibility curves, such as those previously illustrated (Rine, 1978).

Relations Between Security and Information Bindings. The systems programmer is the most difficult individual to control, and the position must be supervised. Data files should not be open to processing by anyone, for example the systems programmer, except at specified times and under supervision. More than one individual should be involved in writing programs, and all logs should be available for review. Modifications to programs should not be made on a demand basis, but instead should be accumulated and made at one time when adequate review and testing can be provided. Controls should be presented as providing protection to the employees so they cannot be blamed if something happens. An effective internal audit activity is mandatory, and users of the computer applications should be the only ones allowed to submit transactions, not systems programmers. It is especially important, therefore, to specify security policies and design protection mechanisms for systems programs such as schedulers, loaders, linkage editors, and compilers.

Programming language names are symbols used to represent user information entities, that is the things that programming languages talk about. We elaborate on this in terms of information binding as follows.

Information binding refers to an, often complicated, algorithm which maps each program name to corresponding objects such as a location-value pair.

There are various types of bindings (Elson, 1973) such as name-declaration, declaration-declaration, declaration-object, declaration-constraints, declaration-description, location-value; and there are various times at which these bindings can be performed such as program creation, linkage-edit, load, call, execution reference. Program creation binding is analyzed during compile-time, and execution-reference binding is analyzed during run-time.

A certification and authorization of a particular information binding implies that a user query or user application program has legal access to information values derived from a particular set of locations in storage. Such certification of a program and information bindings can, therefore, be performed at various times such as program creation, linkage-edit, load, call, and execution-reference.

There is usually a security policy which attempts to describe which security class a name and its associated object belongs to. Recent work (Denning, 1976; 1976; 1977) presents a compile-time mechanism (checking being done during the analysis phase of a compiler) that certifies a program only if it specifies no flows in violation of the flow policy. This work also supposes the "binding" of each "object" to a security class is statis (as opposed to dynamic) and can be determined from the declarations contained in a program. Their "objects" are actually what we refer to as names. If n_1, n_2 are names and s_1, s_2 are security classes, then this work refers to the certification of flows $s_1 \rightarrow s_2$ where n_1 is bound to s_1 and n_2 is bound to s_2. Certifying flows $n_1 \Longrightarrow n_2$ at compile-time (symbolic analysis) implies that information binding between n_1 and object o_1, between n_2 and o_2, binding between o_1 and s_1, between o_2 and s_2, and flow $o_1 \Longrightarrow o_2$ have all been authorized (see Fig. 8).

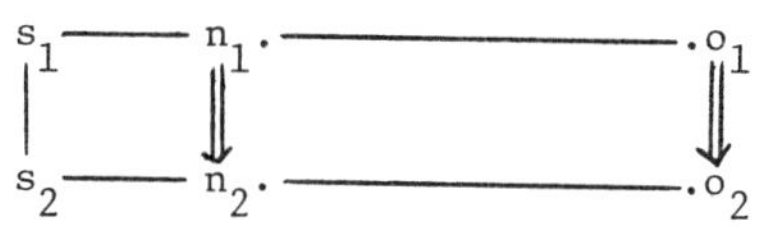

Fig. 8. Certification Analysis by Binding Models: Binding and Information Binding.

There is a very strong assumption, and there are many other alternatives to investigate.

Let us consider seven relevant examples based on the general set of relations illustrated in Fig. 9.

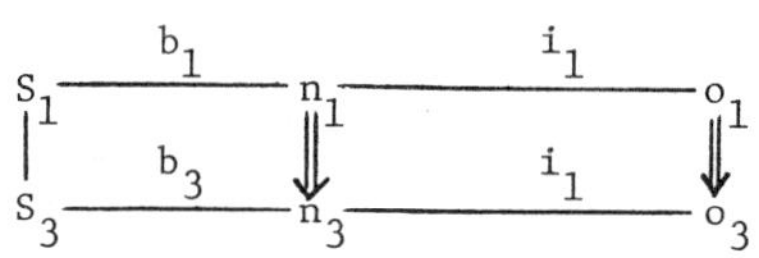

S_k: security class

o_k: object (locations, values) pair

n_k: symbolic name

b_k: binding

i_k: information binding

t_k: information binding time

Fig. 9.

Example 1.

i: name - declaration

t: linkage edit

Let p, q be the machine code object modules of two procedures of a compiled program. Let name n appear in p, and let the declaration of n, dcl(n), appear only in q as illustrated in Fig. 10. The linkage editor is responsible for linking p and q together during formation of the load mudule. Suppose that the linkage editor chooses q and dcl(n) where dcl(n) contains the security class s information associated with n. Then, the user could specify one class s at program creation time, and the linkage editor system program to another class s' at linkage edit time, for example corresponding to another procedure as pictured in Fig. 10.

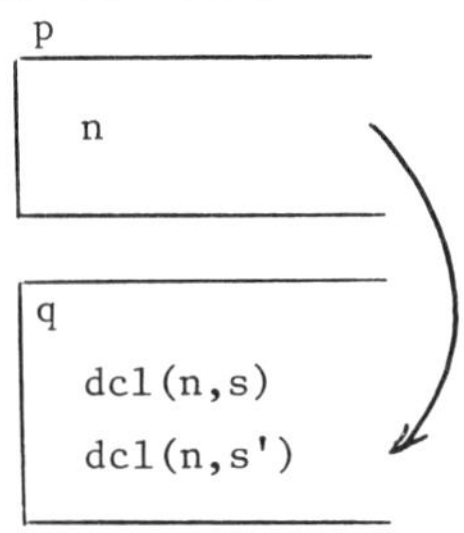

Fig. 10

Example 2.

i: declaration - declaration

t: linkage edit

Let customer, service 1, service 2 be the machine code object modules of three procedures of a compiled program such that customer calls service 1, service 2, as service routines from the system program library. Let n_1, n_3 be names appearing with their declarations in customer, and let $n_1 \Rightarrow n_3$ denote an information flow from n_1 to n_3. For the call from customer to service 1, let n_1 be a "COMMON" variable, and for the call from customer to service 2, let n_3 be a "COMMON" variable. Let $n_1 \Rightarrow n_2$ denote an information flow to an intermediate name n_2 declared within service 1, and let $n_2 \Rightarrow n_3$ denote an information flow from this intermediate name, known also in service 2, within service 2. Now the syntax describing the intermediate information flows $n_1 \Rightarrow n_2$, $n_2 \Rightarrow n_3$ are unknown at program creation and compile times, since they are realized within service routines stored in the system library as machine code. Due to possible leaking of information and a break in security the user may not want this flow authorized. Figure 11 describes the situation.

Example 3.

i: declaration - object

t: load time

Let name n appear in a program, and let T denote the symbol table (see Fig. 12) of the program generated by the compiler. A typical record entry in T would contain the symbolic name n, the declaration dcl(n), a field for the location (address(es)) loc(n) of n to be filled in and computed at load time, and a field for the value(s)

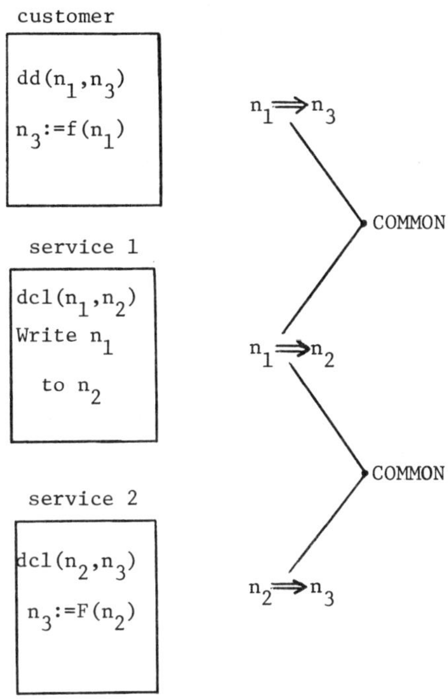

Fig. 11. Certification Analysis by Binding Models: Information Flow and Declaration - Declaration Binding

value (n) of n to be filled in before or during execution-reference time. Since loc(n) is computed at load time, the loader system program could be made to contain code which illegally computes an address, contrary to user-designated security, which could be accessed by another unauthorized program.

T

n	dcl(n)	loc(n) - value (n)

Fig. 12.

<u>Example 4</u>.

i: declaration - constraint

t: program creation vs. linkage edit

Let name n appear in a program, and let T denote the symbol table of the program generated by the compiler. A typical record entry in T would contain the symbolic name n, the declaration dcl(n) of n, and a collection of fields constituting a list of constraints of n (see Fig. 13). Examples of constraints include, but aren't limited to, length of storage to be allocated, possible data types, possible allowable operations involving n, security class(es), and time duration of n in the system. At linkage edit time the linkage editor system constraints, for example legal operations involving n, before loading and executing the program.

T

n	dcl(n)	list of constraints: length, types, operations, security, time duration

Fig. 13.

Example 5.

i: declaration - description

t: call

Let name n appear in a program, and let T denote the symbol table of the program generated by the compiler. A typical record entry in T would contain the symbolic name n, the declaration dcl(n) of n, a collection of fields constituting a list of constraints of n, and from the list of constraints the current description of n. When a particular procedure block involving n is referenced by a call as illustrated in Fig. 14, the current description of n for that block could be changed to an illegal one so that the block could perform an unathorized operation with n, for example writing to rather than reading from some file.

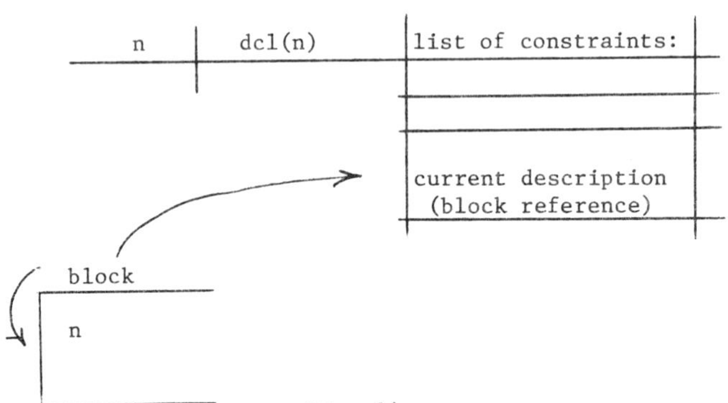

Fig. 14.

Example 6.

A constant's symbolic name is such that its location must contain a fixed value, that is not variable. This value is fixed at program creation time. Suppose that a program constant is associated with the computation of a special set of record keys. The value of this constant could be changed by the loader system program at load time so as to affect values of keys computed during execution, thus implying accesses which the user had not described.

Example 7.

i: argument - parameter

t: call

Unknown to the user and prior to call execution the relation between arguments and parameters could be changed from call-by-value to call-by-address so as to illegally leak information through a program variable.

The following Table 1 summarizes the relations of these examples to one another.

information binding / times & system programs	name-declaration	declaration-declaration	declaration-object	declaration-constraint
program creation (compiler)				
linkage edit (linkage editor)				
load (loader)				
call (scheduler)				
execution reference-(hardware)				

⋮

(continuing)	declaration-description	location-value	argument-parameter
program creation (compiler)			
linkage edit (linkage editor)			
load (loader		example 6	
call (scheduler)	example 5		example 7
execution reference (hardware)			

Table 1. Certification Analysis of Binding Models: Relations of information bindings and binding times.

security class binding / times and system programs	L Low	A	B	C	D	H High
linkage edit (linkage editor)						
load (loader)						
call (scheduler)						
execution-reference (hardware)						

Table 2. Certification Analysis by Binding Models: Relations Between Security Class Bindings and Binding Times.

Table 2 describes a set of relations between times, system programs, and security class bindings. Table 3 outlines specific cases to investigate for a given security design problem.

accessing properties / times & system programs	high (name) activity (file) and low security	high (name) volatility (file) and use of system library	high (name) response (high) time and high security
program creation (compiler)	▨		
linkage edit (linkage editor)		▨	
load (loader)			
call (scheduler)			
execution-reference (hardware)			▨

Table 3. Certification Analysis by Binding Models: Specific Relations Between Accessing Properties and Binding Times.

The linkage editor could be designed to check on certification and authorization of system programs. The checking, at first glance, seems to be a difficult procedure, especially if it involves examination of machine code.

REFERENCES

Conway, R. W., et al, On the implementation of security measures in information systems, CACM, April, 1972.

Denning, D. E., On the derivation of lattice structured information flow, CSD TR 180, Computer Sciences Department, Purdue University, W. Lafayette, March, 1976.

Denning, D. E., A lattice model of secure information flow, CACM 19, 5 (May, 1976), 236-243.

Denning, D. E. and P. J. Denning, Certification of programs for secure information flow, CACM 20, 7 (July, 1977), 504-513.

Elson, M., Concepts of programming languages, Science Research Associates, Chicago, 1973.

Graham, R. M., Protection in an information processing utility, CACM, May, 1968.

Gries, D., Compiler construction for digital computers, John Wiley and Sons, New York, 1971.

Held, G. D., INGRES - a relational data base system, AFIPS - Conf. Proc., vol. 44 and NCC, 1975.

Information Management System (IMS) Application Programming Reference Manual and Utilities Reference Manual, IBM Corporation, Data Processing Division, White Plains, New York.

Jones, A. K. and R. J. Lipton, The enforcement of security policies for computation, Proc. Fifth Symp. on Operating Systems Principles, Operating Syst. Rev. (ACM SIGOPS Newsletter), Vol. 9, 5 (November, 1975), 197-206.

Lampson, B. W., A note on the confinement problem, CACM, October, 1973.

McCauley, E., A model for data secure systems, 1975-76, Ph.D. Thesis, The Ohio State University, Columbus.

Rine, D. C., Possibility theory: as a means for modeling computer security and protection, Proc. Eighth International Symposium on Multiple-valued logic, IEEE Computer Society and ACM, Chicago, 1978.

Rine, D. C., Multiple-valued logic and flowcharting, Australian Computer Science Journal, vol. 7, no. 2, July, 1975.

Rine, D. C., Simple models for security policies and associated protection mechanisms in data management, technical report, Computer Science, University of Texas, San Antonio, 1976-77.

Smith, L., Architectures for secure computing systems, MTR-2722, MITRE Corporation, June, 1974.

Zadeh, L. A., Fuzzy sets as a basis for a theory of possibility, memorandum no. UCB/ERL M77/12, Electronics Research Lab., University of California, Berkeley, February, 1977.

V. DISTRIBUTION SYSTEMS AND PROCESSES

TRANSPORTABILITY OF DATA AND RELATED SOFTWARE IN ELECTRICAL DISTRIBUTION SYSTEM MODELING

John C. Thompson Turan Gonen

School of Electrical Engineering
and Computer Science
University of Oklahoma
Norman, Oklahoma

ABSTRACT

This paper discusses some results of a research project whose goal was to develop better methodology for distribution system planning (Gonen and Thompson, 1979). One of the design objectives of the implemented system was to make its operation as independent as possible of the host hardware/software. In order to achieve this portability, given the present technology, it was necessary to simulate many of the functions of a computer's operation system with the distribution system planning model (DSPM) itself (Thompson and Gonen, 1980). These functions are located in the shell program which provides an interactive interface between the distribution system planner and the functions and programs of the DSPM.

KEYWORDS

Portability; database management system; operating system; optimizer; syntax analysis.

INTRODUCTION

Portability is a requirement having two major aspects. The first is the portability of the data itself; this is the easiest condition to satisfy. Data in character form may be ported easily from one computing environment to another. The second is the portability of the software which processes the data and this is complicated by the nature of the application to which the software is to be devoted. A database management system, such as the one supporting the DSPM, generally requires the use of the full repertoire of machine instructions for a given computer. Current software technology suggests that only high-level languages be used to develop a database management system but the most widespread high-level languages were not seen as satisfactory for such an effort.

As a solution to the implementation language problem, the language C (Kernighan and Ritchie, 1978), was chosen. Once the software was fully implemented, it was then translated (by hand) into Ratfor (Kernighan and Plauger, 1977), which in turn, was then translated into Fortran IV by the Ratfor translator. Finally, the resulting Fortran source was optimized using an approach suggested by Knuth (1974).

This effort has resulted in the implementation of four software products: a canonifier program which allows the database to be software-compatible with all of the application programs; a database management system supporting the suggested database; an optimizer program which transforms the Fortran output of the Ratfor translator into efficient code; and lastly, a breadboard version of the shell program (Thompson and Gonen, 1979).

THE CANONIFIER PROGRAM

A systems view of the canonifier is shown in Fig. 1. The program runs under interactive control via a terminal but the usual source of control data is from the <u>input specification file</u>. This file contains statements which may be thought of as very general format statements. The program functions in two modes. In one mode, it reformats data which has been extracted from the database using queries. After being reformatted, the data may serve as input to an analysis program. Thus this mode is called <u>input mode</u>. The other mode is the converse of input mode. It accepts output from an analysis program and reformats it so that it may be incorporated into the database.

The error file (Fig. 1) is used to note any specification errors encountered either in the terminal input or in the data found in the input specification file. If any errors are detected, no output file is generated. Instead, the user uses the text editor to correct the input specification file or reenters data from the terminal (or both) until no further errors are found.

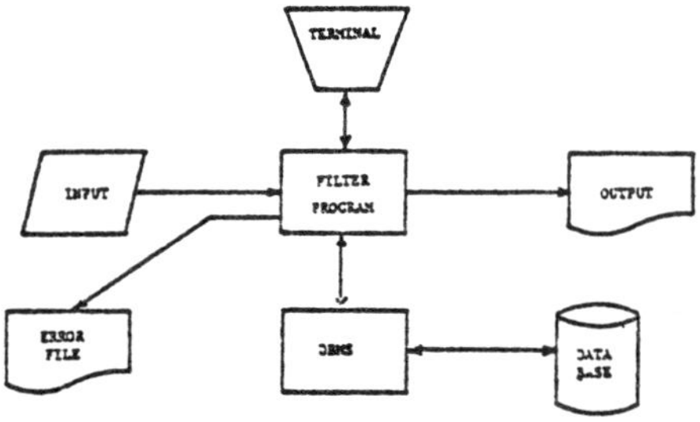

Fig. 1. A Systems View of the Canonifier

THE DATABASE MANAGEMENT SYSTEM

The database management system (DBMS) was implemented using the PATRICIA data structure as a directory. This structure has a number of theoretical advantages over some of the popular schemes currently being used (Thompson, 1980). In particular, it can be completely characterized syntactically, which means that correctness issues of implementation can be treated by methods of Formal Language Theory. As part of the DBMS, a user language was implemented based on the relational algebra. This language allows one to write queries in prefix form so that nesting is straightforward.

FORTRAN SOURCE OPTIMIZER

Portability requirements place a heavy burden on the system designer. At the present time, the only suitable language which is truly portable is Fortran IV. However, Fortran's rather primitive control structures make the programming of an application such as a DBMS extremely difficult. Such programming tends to be difficult even in a language such as Algol which has much more powerful control structures. To obtain more powerful control structures within a Fortran context, the implementers of this project turned to Ratfor, a dialect of Fortran which has powerful control structures. Portability for Ratfor is obtained by translating those Ratfor statements which are not in themselves Fortran statements into Fortran statements. The major drawback to this translation process is that the resulting Fortran program is not nearly as efficient as an equivalent program coded by hand would be. Since the database software is extensive (it is comprised of more than forty subprograms), it is not feasible to do the translation by hand. Thus, it is clear that what is needed is a tool to assist the programmer and that tool is the Fortran Source Optimizer.

Major Features

The information flow is illustrated in Fig. 2

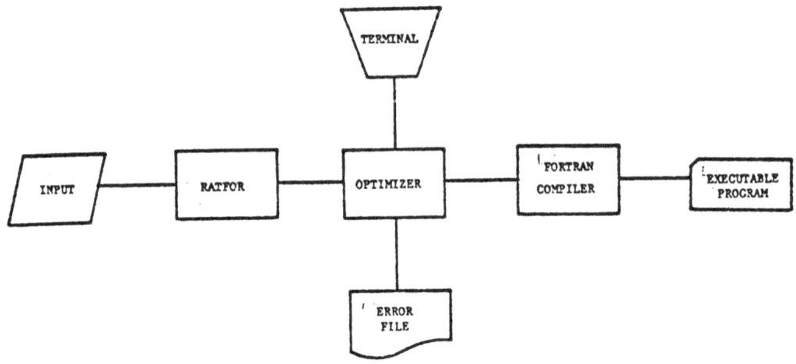

Fig. 2. System Information Flow

The optimizer accepts Fortran statements generated by the Ratfor processor and performs the following actions.

(1) Detects any syntax errors. The Ratfor processor does little or no syntax checking. Any statement which it does not recognize is assumed to be a Fortran statement (not a Ratfor statement) and is passed as it was encountered to the output file, which in turn is passed to the optimizer.

(2) Finds common subexpressions. Many expressions can be "factored" so that certain arithmetic expressions need to be evaluated only once.

(3) Detects "dead" code. Some portions of a program may not be reachable during normal execution because the programmer has made errors in his algorithm design or because the algorithm was not faithfully translated into a correct Ratfor program.

(4) Eliminates redundant control transfers. The Ratfor control statements which are all single entrance-single exit constructs are implemented using Fortran goto statements. The implementation is done in a simple-minded way with the result that no testing is done for situations such as a transfer to statement number 10 which is itself a goto statement jumping to number 50. This kind of redundancy is eliminated.

(5) Allows interactive program restructuring. The programmer executes the optimizer interactively. After initial processing of the Fortran source, the optimizer is prepared to accept commands from the programmer and rearrangement of the source can be done. This rearrangement may result in more sophisticated code than the optimizer could produce unassisted. At the same time, the routine processing referred to above has already been done and is thus not a concern of the programmer.

Underlying Principles

Many of the optimizing steps depend on the generation of the flow graph which is a digraph whose edges represent transfers of control within the source program and whose vertices represent groups of source statements in which there are no transfers. The creation of such graphs is well-understood and standard methods are available (Aho and Ullman, 1977). These graphs permit the detection of loops, and unreachable (dead) code. Another important graph which the optimizer constructs is the Directed Acyclic Graph (DAG). This graph is used to analyze the vertices of the flow graph. Such analysis shows how the values computed in one basic block (vertex) are used in subsequent blocks. This leads to the detection of common subexpressions and loop optimization.

The syntax analysis is done using simple precedence methods for expression validation and recursive descent parsing of statements. After the programmer has modified the source program, he may again run both the syntax checker and the analyzer. This allows him to check for errors which might have inadvertantly crept in when the modifications were made.

SHELL IMPLEMENTATION

The Shell program implemented by the work reported here did not meet the portability goals originally set for it. The diversity and peculiarities of several operating systems under which the total system might be expected to function demanded of the staff far greater resources of time and effort than were available. This was recognized early in the life of the project and that recognition prevented any large expenditures of manpower which ultimately would have been wasted. Instead of tilting with such a windmill, the staff decided to implement the shell using the inherent features of Unix on an available PDP-11/70. It is possible for a programmer to write his own command processor (shell) to drive the underlying functions of Unix and this was what was done. It was felt that it was important to experiment with the features proposed in the original functional description and these were the ones implemented.

SUMMARY AND CONCLUSIONS

The Shell program itself is the only system component which has not been implemented in a portable manner. The canonifier program has proven to be the key to making all of the applications/analysis programs compatible without requiring any substantial modification of them. The canonifier program works well as a front end to the DBMS which was implemented using the PATRICIA data structure. As part of the DBMS, a user language was implemented based on the relational algebra. Lastly, as part of the approach taken to achieve portability, a Fortran source optimizer was implemented which allows the system programmer to optimize the portable but inefficient code produced by the Ratfor translator. Using Ratfor allows one to program in a more comfortable language than standard Fortram.

REFERENCES

Aho, A. V., and J. D. Ullman (1977). Principles of Compiler Design. Addison-Wesley, Reading, MA.

Gonen, T., J. C. Thompson and B. L. Foote (1979). Research Final Report for Electrical Energy Distribution Systems. Submitted to the U. S. Department of Energy, University of Oklahoma, Norman, OK.

Kernighan, B. W. and K. Plauger (1977). Software Tools. Prentice-Hall, Englewood Cliffs, NJ.

Kernighan, B. W. and D. M. Ritchie (1978). The C Programming Language. Prentice-Hall, Englewood Cliffs, NJ.

Knuth, D. E. (1973). The Art of Computer Programming, Vol. 3, 490-500. Addison-Wesley, Reading, MA.

Knuth, E. E. (1974). Structured programming with goto statements. Computer Surveys, 6 261-301.

Thompson, J. C., and T. Gonen (1979). An interactive distribution system planning model. Proceedings of the Modeling and Simulation Conference, University of Pittsburgh, PA.

Thompson, J. C. and T. Gonen (1980). A database design of a distribution system planning model. Proceedings of the Modeling and Simulation Conference, University of Pittsburgh, Pittsburgh, PA.

Thompson, J. C., (1980). A Syntactical Specification of the Data Structure PATRICIA. Technical Report, School of Electrical Engineering and Computer Science, University of Oklahoma, Norman, OK.

A NETWORK EDITOR AND SHELL PROGRAM FOR ELECTRICAL DISTRIBUTION SYSTEM PLANNING

John C. Thompson Turan Gonen

School of Electrical Engineering and Computer Science
University of Oklahoma
Norman, Oklahoma

ABSTRACT

Recent research has led to the development of new tools for the electric energy distribution system planner. Two of these new tools are discussed in the paper. The first is a Network Editor which allows the planner to manipulate network objects much as one manipulates text using a text editor. The second is the overseer program for the planning system which provides the planner many of the features of an operating system.

KEYWORDS

Network; distribution system; operating system; electric energy; filter.

INTRODUCTION

This paper discusses some innovative results of a research project whose goal was to develop better methodology for electrical energy distribution system planning. A model for the planning process was developed and a computerized interactive system was implemented to provide the planner with a set of sophisticated tools. (Gonen and Thompson, 1980), (Thompson and Gonen, 1979)

THE NETWORK EDITOR

The Network Editor is one of the important conceptual tools in the methodology espoused by some recent research (Gonen, Thompson and Foote, 1979). The Distribution System Planning Model (DSPM) as the planning tool was called, consists of these major components: the applications/analysis programs, the database management system, the Network Editor and the shell program. In this section of the paper, a fuller description of the Network Editor will be given with more emphasis on its conceptual bases rather than on its operational capabilities.

The Need for Such a System

One of the chief concepts of the methodology developed in the DSPM is that of a network. The intention is that since the engineer/planner is used to thinking in terms of network concepts, the dialogues with the planning system should be in network terms. For this reason, two ideas were amalgamated: a vehicle for manipulating network objects combined with concepts from the single most successful interactive computer tool available, the text editor. Therefore, the intent behind the Network Editor is to create a tool which allows him to create or alter networks as easily as text editors allow their users to manipulate text.

Another important reason for wanting to include the Network Editor in the methodology derives from the sheer diversity of network objects and the complexity of their interconnections. The designer's representational abilities must span the range from simple components such as individual transformers or customer meters to entire substations which consist of subnets. For this task, one would like to have a conceptually elegant and simple to use tool. To understand the Network Editor as it is proposed here, it is first necessary to examine the concepts upon which it rests.

Guiding Concepts

There is an important distinction between primitive network objects and compound network objects. Descriptions of the former require no subnet information. Thus the individual transformer or capacitor or even line element is not comprised of a network but is to be found in a network i.e., is atomic from the viewpoint of the network being described. On the other hand, a substation, consisting as it does, of a collection of transformer, busses, meters, etc., can only be described by the inclusion of network information. For this reason, it must be considered a compound object.

A network is modeled by means of a directed graph whose nodes or vertices are physical components and whose edges are descriptors which define how the physical components are interconnected. Edges connect one node to another via ports. Ports may or may not have an internal structure of their own. If a feeder carries three phases, the source and load ports will be described by triples, each element of the triple being referred to as a "pin." In some cases, a single multiport may join two or more ports with a smaller number of pins. For example, in a urban residential neighborhood, a three-phase feeder may divide into three single phase feeders, each of which serves say, twenty-five families. Thus an edge is defined not only by the source and target functions described above but also by the type of port found at each end of the edge. In this way some integrity constraints may be built into the system, preventing the user of the Network Editor from joining network objects with incorrectly specified ports.

In a graph theory sense, each edge has a color as do most nodes. Some colors represent single phases. Let us say that red, green and blue each represent a distinct phase while white arbitrarily represents a three-phase component. In addition, define the "sum" of the colors red, green and blue to be the color white and the sum of any two of the colors to be a color which is not one of any of the four. Junction rules for the way in which ports can connect may be expressed very simply (for most components): an edge with a given color must connect two network objects with that same color. In cases where there is a separation of phase (as in some residential areas) then the "color sum" must be constant on either side of the junction. Thus a white feeder can meet three subfeeders so long as one is red, one is green and one is blue. In this way the Network Editor prevents any phase confusion.

Some network components have their non-topological parameters completely defined by constants. For example, a capacitor may have its rating fixed at 20 kVA. Typically however, many network components have one or more variable parameters; the tap settings on a transformer may be variable; the level for a voltage regulator may be adjustable. This means that specifying the class to which such a network object belongs does not specify the values for its variable parameters. One way these values may be set is for them to be associated with state variables. Such state variables may assume integer, floating point or Boolean values. More than one object may have its parameters specified by the same variable. Thus the designer may alter the setting of several devices by changing the value of a single state variable.

State variables are clearly useful for defining the settings of switches. A generalization of this notion leads to an even more powerful idea, that of design switches. Suppose it is required to consider the effects on performance of making some relatively minor modifications to a network. For example, suppose there are several alternative locations in the network for a combination of voltage regulators and capacitor banks. The question to be answered is which combination at which locations, gives the best performance. The straightforward way to determine the answer is to make a list of all the possibilities to be tested and them implement the network corresponding to each in turn. A more elegant solution makes use of the design switch. At the time an object is created (becomes the instantiation of an element from an object class), a state variable can be assigned as a design switch associated with the object. When this switch is "on," the object appears in the network just as any other object. When the design switch is "off," the object is virtual and for all applications purposes, disappears. Thus by turning design switches off and on, the configuration of the network can be greatly varied. In the above example of capacitors and voltage regulators, these objects may be placed in the network all at one time to be controlled by design switches. By successively turning switches off and on, the various alternatives may be created and analyzed by the applications programs.

Another important concept incorporated into the Network Editor is that of the cursor. A cursor or pointer is used as a variable in a program to allow the planner to refer to various elements in the network. When, for example, an object is created, a cursor variable is associated with the object. The operations to be described all use cursors to refer to the objects upon which they operate.

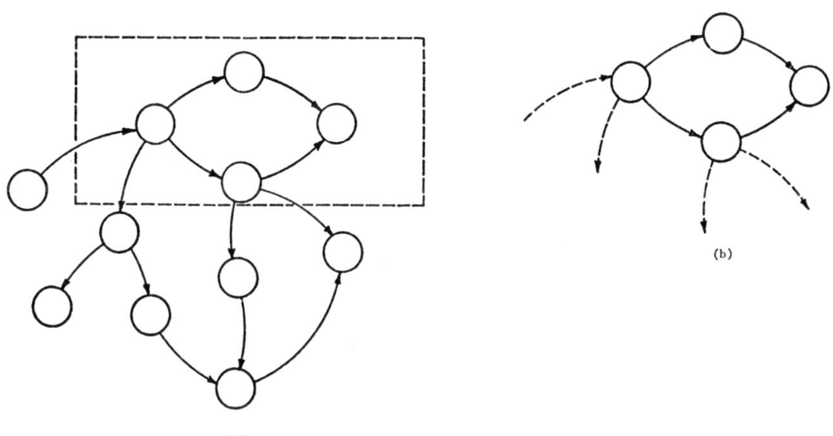

Fig. 1 A network component to be identified by a context search.

Underlying the notion of an editor for a network is the fact that networks can be described in terms of text strings. These strings, which are known as K-formulas, permit one to visualize some network operations as operations on a text string. In particular, one may specify some collection of network components and invoke the Network Editor to find all places in the network where this configuration is present. Figure 1 illustrates the general situation. The subnet shown in the insert (b) is sought in the larger network (a). When it is found, it may be replaced by another network configuration, just as one text string may be substituted for another by a text editor. For example, suppose one has identified a particular subnetwork as something to be represented by a compound object. Using context searching, the Network Editor can locate all locations where the subnet appears then replace the collection of network objects comprising it with compound objects. This might be done to study the variation of several parameters in the compound object and could be done more easily this way than by direct variation of all occurrences of the subnet in the original network.

THE SHELL PROGRAM

Every system has its executor or monitor. For the DSPM, that overseer is called the shell. Its general purpose is to serve as a command line interpreter to which the user directs his commands and inquiries. In effect, however, it is a miniature operating system, implementing some of the more basic functions traditionally associated with operating systems. Some of its features have been borrowed from the Unix operating system developed at the Bell Laboratories (Ritchie and Thompson, 1974). In what follows, the basic notions incorporated into the shell will be discussed.

Purpose of the Shell

In concept, the Planning System is to be highly portable. However, standing in an almost dichotomous position to this requirement are the necessities to create files, execute programs in a fixed sequence, monitor exceptional conditions and support several interactive users simultaneously. These latter functions conventionally have been assumed by operating system, which logically speaking exist at a lower level than so-called user or applications programs.

When one considers raising what have generally been operating system functions to the applications program level, as the methodology embodied in the DSPM suggests, the first question to be examined is whether that is even possible. The answer which one can make to this question is weakly dependent on time--our views concerning operating systems ten years ago were much different than they are today--but to be objective any answer given must be in the present tense. In consequence, the answer today is that it depends on the native operating system on top of which, the Planning System is to run. All of the prototype system implemented to date has been implemented on a Unix-based system. As a result, few actual difficulties have been encountered. Implementing the system on an operating system native to IBM 370 machines would be considerably more difficult. Thus while portability remains a laudable ideal, it appears at present to be an unreachable one.

Conversely, it is expected that in the next five years, as general understanding of what constitutes a hospitable user environment becomes more widespread, that implementation difficulties which today might exist for an arbitrarily chosen operating system will be greatly overcome and this time scale is sufficiently short that considerations expressed here might have a timely effect on emerging technology. Seen from this view, the shell represents a collection of functions which the distribution planner needs to have available and which form an umbrella covering the other system components.

Shell Functions

Each user is known to the shell via a userid. This is an alphanumeric string which serves to associate all files, relations networks and databases with the user. The shell allows the user to create files and remove files which are associated with his userid. This file processing is done in such a way that the user is isolated from the normal file handling features of the native operating system. Thus, the user is not burdened with learning the syntax which accompanies file processing on the native system. He has only to learn the simple syntax accepted by the shell.

One of the most significant problems solved through use of the shell mechanism is that of software compatibility with respect to the applications/analysis programs. Since these programs were written by different individuals at different companies and institutions, there is no uniformity of format for input and output data. In some cases, the same data is required by different programs but in different units! Two approaches suggested themselves. In one, the internals of each program would be modified so that data requests were programmed as calls to the database management system (DBMS). This clearly would be a difficult task, requiring the intimate understanding of the internals of each program. The other approach required the use of a "filter" which would be placed between the program and the DBMS. The filter would function as a very sophisticated Format statement specifying queries to the database and conversion of data from the form stored in the database to a form compatible with the requirements of the program (Thompson and Gonen, 1980). A request to the shell by the user for the execution of a program results in the identification by the shell of the proper filters to be used. The filter program queries the database and creates a file which matches the demands for the input data to the applications program. The shell then causes this program to execute using the input file created by the user. The output from the applications program is collected on a file which is run through a second filter program that reformats the data into a form acceptable to the database. The only data necessary for this filtering to take place are the specifications on the formats for input (output) to (from) the applications program. These specifications are part of the database.

SUMMARY AND CONCLUSIONS

The network Editor is an interactive program which allows the planner to deal with networks in a manner similar to the way in which one deals with word processing using a text editor. One may define, modify or delete network components using simple commands. Of particular power are the features which permit context searching in the network, modification of an object's status using state variables and the forming of compound objects from simpler network components.

The shell is the executor of the DSPM; it makes available to the planner many of the features of a real-time operating system. In principle, these features would be independent of the machine on which the Planning System was running and would provide a fully portable basis for executing planning programs.

REFERENCES

Gonen, T., J. C. Thompson and B. L. Foote (1979). Research Final Report for Electrical Energy Distribution Systems, submitted to the U. S. Department of Energy, University of Oklahoma, Norman, OK.

Gonen, T., and J. C. Thompson (1980). Analysis of distribution system planning data requirements. Proc. of the Modeling and Simulation Conference, Univ. of Pittsburgh, Pittsburgh, PA.

Ritchie, D. M., and K. Thompson (1974). The UNIX time-sharing system. Comm. ACM 17 365-375.

Thompson, J. C. and T. Gonen (1979). An interactive distribution system planning model. Proc. of the Modeling and Simulation Conf., Univ. of Pittsburgh, Pittsburgh, PA.

ROCKWELL AIM 65/AIM 680 APPLICATIONS IN DISTRIBUTED PROCESS CONTROL

Professor Mohamed Rafiquzzaman, Ph.D.
Electrical and Computer Engineering Dept.
California State Polytechnic University
Pomona, CA 91768 USA

ABSTRACT

The purpose of this paper is to provide a representative approach to the implementation of Rockwell AIM 65 and AIM 680 as a tool for various types of applications in distributed process control. A number of applications have been identified by Rafiquzzaman [1] for applying AIM 65 in process control under the following configurations: (1) AIM 65 with one card add-on module; (2) AIM 65 with a four slot card cage. The applications presented in this paper involves systems in which a number of AIM 65 and AIM 680's are linked together through a data hiway to form a distributed control system. This system can then be supervised by another AIM 680 as a host for optimization etc.

The main advantages of using AIM 65/AIM 680 configuration over the currently popular system such as the Honeywell TDC 2000 is that both continuous and batch processes can be handled using the distributed concept by the AIM 65/AIM 680 system. The other advantages are cost effectiveness due to availability of a number of inexpensive single card modules and also ease of implementation.

INTRODUCTION

The introduction of the Single Board Computers (SBC's) as a tool for the system designer has opened the way for many varied application areas for benefit from the advantages of computer utilization.

The AIM 65 was originally designed as a university kit for use in the microprocessor laboratories. AIM 65 is a Rockwell 6502 based microcomputer system operating at 1 MHz. It has 65K addressing capability, 13 addressing modes and 56 instructions. It features on-board thermal printer, alphanumeric display, a full size alphanumeric keyboard. Two installed 4K ROM's hold an advanced interactive monitor program for program debugging, generating error messages etc. A 44-pin application connector provides for interfacing a teletype and one or two audio cassette recorders. A 44-pin expansion connector has full system bus capability. AIM 65 can be interfaced to a card cage via an adapter/buffer module through the expansion connector. A detailed description of the capabilities of AIM 65 is given in [2].

The AIM 680 is a Motorola 68000-based (16-bit microprocessor) Single Board Computer (SBC) module. It consists of an SBC module with a 40 character display, 40 column printer and a composite video display. It includes the following packages on the basic system:

- Single user operating system

- Symbolic Debug

The software option packages consist of:

- Multiuser operating system
- 68000 Assembler
- 6500 Cross Assembler
- 'C' Language Compiler
- PASCAL Interpreter and/or Compiler
- Fortran Compiler
- Cobol Compiler
- Basic Interpreter and/or Compiler
- FORTH Interpreter

The AIM 680 also provides:

- Eight 28-pin ROM sockets for 4 to 32K words of ROM
- Thirty-two 16-pin sockets for 64K words of dynamic RAM
- Bus Interface: Motorola VERSABUS
- Two R6522 VIA's: one for the printer and display, one for the user
- Two Asynchronous Communications Interface Adapters (ACIA): R6551 - One for a system terminal and the other for the user
- One Peripheral Interface Adapter (PIA): One R6520 for the keyboard

The AIM 680 is not provided with any single step hardware since the 68000 has a built-in feature to trace after each instruction. Also there is no hardware breakpoints since this feature can be implemented in software by replacing code with a trap vector and replacing it back after going through the supervisory program.

TYPICAL AIM 65/AIM 680 PROCESS CONTROL APPLICATIONS

The basic characteristics of the distributed process control were considered by Watson [3] which led to the architecture of the totally distributed control system (TDC 2000) developed by Honeywell. A typical TDC 2000 configuration is shown in Figure 1.

Each basic controller containing a microprocessor and data base memory provides control of 8 analog loops. These microprocessors can be programmed with a number of control algorithms (PID, cascade, ratio, etc.).

The operator's station consists of CRT's displaying process variables, alarms, set points, valve outputs, etc., and thus provides the man-machine interface of the overall control system.

The process interface unit contains multiplexers, A/D and D/A converters. etc.

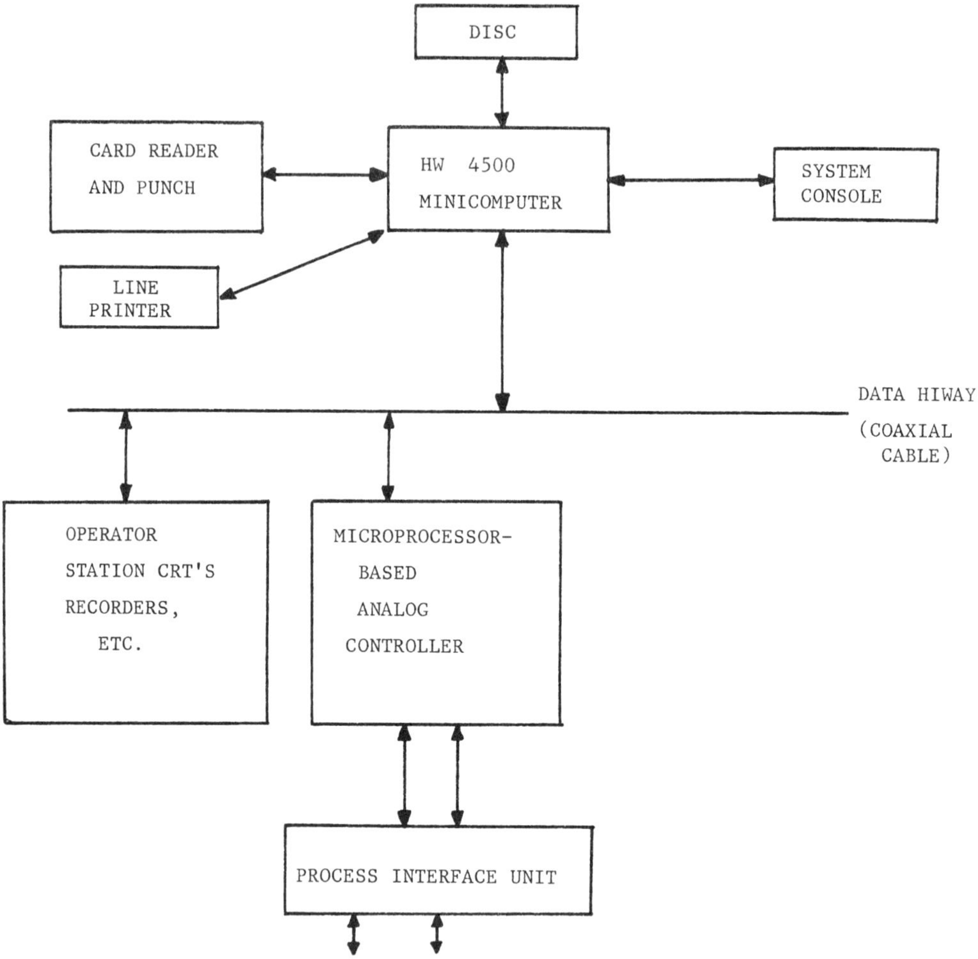

Fig. 1. Basic Honeywell TDC-2000 configuration.

The data hiway is a system of serially multiplexed communication medium between the system elements named above. It is a single coaxial cable along with digital messages (data) are sent by one unit and received by the appropriate unit to which they are addressed. Units use the data hiway one at a time and in accordance with a protocol and priority system built into the units and under control of the Hiway Traffic Director (not shown in the figure). Provision is made for two redundant coaxial cables for a redundant system.

The Honeywell 4500 minicomputer is used as a supervisory computer for the overall system. It also provides the software for optimization and batch process control.

Honeywell TDC-2000 is only capable of controlling and monitoring continuous processes (temperatures, flows, etc.) over the data hiway. Control of the batch or sequential processes (on-off devices) must be handled by the 4500. These discrete signals are brought into the 4500 through its digital Input/Output. Therefore, the batch processes cannot be controlled in a distributed manner. This problem is overcome by using the AIM 65/AIM 680 configuration shown in Figure 2.

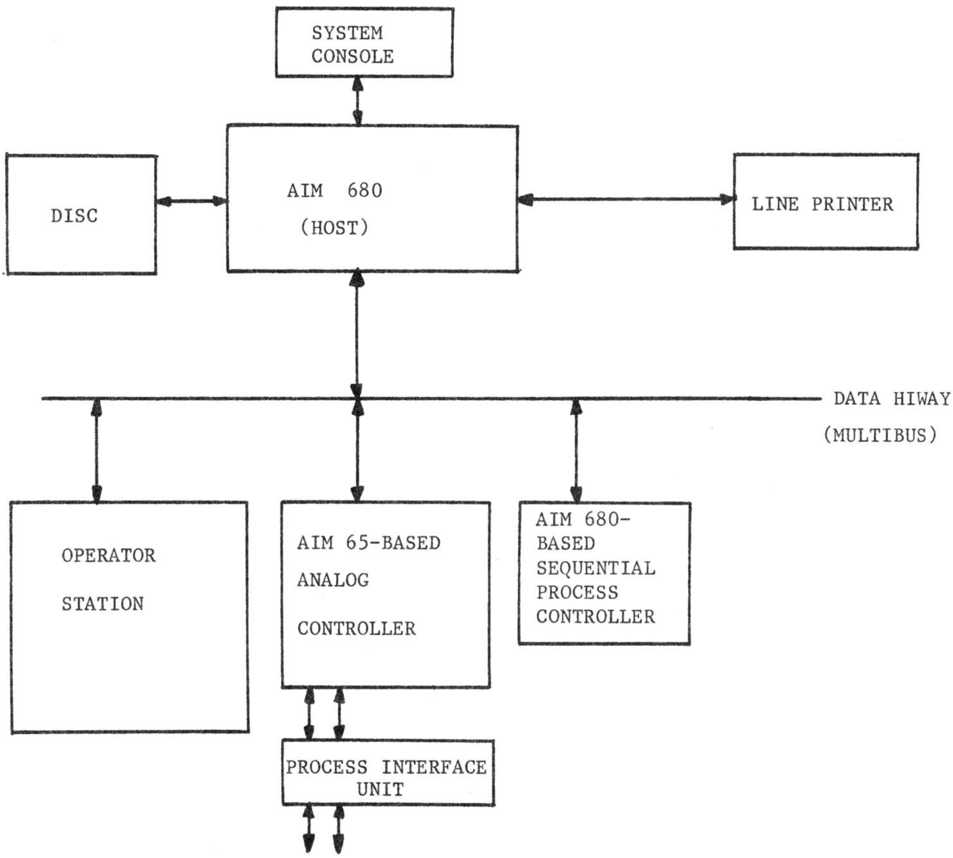

Fig. 2. AIM 65/AIM 680-based configuration for controlling both analog and sequential process control in a distributed manner.

The availability of high level language such as PASCAL with the AIM 680 provides the AIM 65/AIM 680 configuration with the capabilities of distributed analog and batch process control over the data hiway. In Figure 2, the operator station consists of CRT's, line printer, etc. The AIM 65 based analog controller is programmed for controlling temperatures, flows, etc., using a number of algorithms. Each AIM-65 is capable of controlling 25 analog loops. The host AIM 680 is used for supporting a number of peripherals and also for optimizing the overall process. The host AIM 680 can be used to compute the optimum values of some of the process variables such as temperatures, flows, etc., for maximizing the production. Any sequential process consisting of on-off devices such as pumps, motor-operated valves (MOV'S) can be controlled using another AIM 680. A number of AIM 680's can thus be distributed over the whole plant under the supervision of the host AIM 680.

CONCLUSIONS

The concepts described in this paper show the capabilities of AIM 65/AIM 680 configuration in the distributed process control areas. The main advantage of AIM 65/AIM 680 configuration over the existing systems such as the Honeywell TDC 2000 is that the AIM 65/AIM 680 can be used to control both continuous and batch processes over the same hiway using distributed concept. Other advantages are cost effectiveness, ease of implementation, etc. Finally, work is under progress to apply AIM 65/AIM 680 configuration in a refinery distribution system.

REFERENCES

1. Rafiquzzaman, M. "Rockwell AIM 65 Applications in Process Control." University of Pittsburgh Modeling and Simulation Conference, USA, May, 1980.
2. "Rockwell International's Microelectronic Devices Data Catalog," Rockwell International, Anaheim, California, USA.
3. Watson, J.M., Chem. Eng. (London), March, 1976.

VI. NATURAL AND COMPUTER LANGUAGES

IDENTIFYING ASPECTS OF HUMAN LANGUAGE PROCESSING TO AID IN COMPUTER PROCESSING OF HUMAN LANGUAGE

Lawrence J. Mazlack
Information Systems
University of Cincinnati
Cincinnati, Ohio 45221

ABSTRACT

The ability to use language is one generally specified attribute of an intelligent entity. So far, machines have only been fully successful in manipulating restricted languages.

Some of the restricted machine languages that men have developed have been designed to provide access to information stored in computer databases. As more information has become stored in these databases, interest has grown in providing a nontechnical trained user to access database stored information through a human or "natural language."

This paper examines some of the major viewpoints as to how humans acquire and structure language. This information is then extended and combined with other language processing knowledge into an identification of the major elements comprising a mechanism to interface between natural languages and the formal language used to form database queries.

HUMAN LANGUAGE STRUCTURE AND ACQUISITION

As a specifically human activity is under consideration, it would seem reasonable to seek help from studies of the human processor and how it comes to acquire language and develop representational structures. Unfortunately, studies of human language have not precisely identified the mechanisms involved. Indeed, there is a wide divergence of opinion. It may be that one of these structure/acquisition theories is correct; it may not be. Machine NL processing may find it useful to draw from these speculations. But, as there is a considerable disagreement between practitioner of human NL processing, machine NL practioners have a great deal of freedom to use whatever appears to be useful whether or not it corresponds to one of these theories. It also should be noted that all of the different approaches can find a parallel in some machine intelligence project.

The following are the three principal viewpoints as to the nature of human NL processing. The viewpoint accepted with regard to acquisition has a lot to do with how an investigator proposes that a human NL be represented and utilized.

STRICTLY LEARNED

Behaviorial theorists have adopted the position that language can be accounted for by traditional learning principles (Heatherington, Parke, 1979). They view language learning as an empirical experience. Skinner (1957) stressed reinforcement. Bijou (1976) gave more weight to imitation. According to the reinforcement (operant) view, the parent selectively reinforces those parts of a child's babbling sounds which sound like adult speech, thus increasing vocalization of these sounds. Those proposing that a child learns more through imitation hold that the child picks up words, phrases, and sentences through imitation and then through reinforcement and generalization learns when it is appropriate to use and combine responses.

The behaviorial view is often considered to be inadequate as a complete basis for language acquisition because traditional learning theories cannot adequately explain how each individual obtains the grammatical structure capability of language (Brown, 1973). (Brown pointed out that children are just as likely to be rewarded by their parents for truthful, but grammatically incorrect statements as they are for grammatically correct utterances. In this case, it is difficult to see how reinforcement can easily explain the acquisition of correct grammar.) Also, language learning theory does not explain the regular sequence in which language develops, irrespective of language learned. Children in human cultures seem to learn the same types of grammatical rules in the same order (Heatherington, Parke, 1979).

BIOLOGICALLY DEFINED AND STRUCTURED, GENETICALLY PREPROGRAMMED

Chomsky (1968) proposed that the human structure includes an innate concept of human languages. He (1966) suggested that a language user must possess innate devices or mechanisms predisposed to acquire the rules and skills that constitute human language, and that these cannot be acquired solely through learning. Human language is said to have a finite structure with an infinite potential to generate new linguistic utterances. Language "nativists" have not spelled out exactly what they consider to be innate. They do agree that normal humans are biologically "predisposed" to learning a language. Lennberg (1967) stated a precise case for biological determinants. He argued that language is based on highly specialized mechanisms which not only predispose but actually shape language development. Nativists learned by traditional learning principles.

Chomsky's views are widely admired by portions of the linguistic community. His and other like work forms a significant point of view. His views are not universally admired by those who are interested in machine NL processing. Lakoff (1978) discusses some of the more general disagreements between some linguists and some AI workers. Charniak (1976) discusses Chomsky's transformational grammar in a tutorial and provides a detailed discussion of some possible deficiencies.

STRUCTIONAL-FUNCTIONAL

Piaget is the leading proponent of this viewpoint. He uses the term "schemata" to refer to the cognitive structures underlying and supporting organized patterns of behaviour. These cognitive structures provide the processess which organize and respond to experience. Even in infants, the child's behaviour is said to reflect some type of organized pattern of relating to the environment.

The early schemata are rapidly modified as the infant interacts with the surrounding environment. There is a gradual shift with age and mental activities to symbolically represented schemata. In older children schemata are more internalized and more mental. Piaget calls these mental equivalents of behaviorial schemata operations.

Certain unlearned principles of processing and responding to experience result in continuous modification of the schemata. One principle (organization) integrates and coordinates physical or psychological data into more complex systems. In the assimalation process, the child relates and modifies (i.e., assimilates) new experiences in relation to an existing schemata. (In language, this action corresponds to mediation.) Accommodation is the process complementary to assimilation. It modies the schemata to cope with environmental demands.

With reference to language, Piaget feels that, while language involves distinct mechanisms and exhibits a greater degree of biological specialization than some other complex human processes, there exist important continuities between language and certain non-language capabilities. He accepts Chomsky's suggestion that language necessitates the user possessing a language acquisition device that has a strong inner structure, but disagrees with the assumption that the structure is genetically preprogrammed, and with the lack of any attempt to link language behavior to the preceeding sensorimotor forms (semiotic) of behavior (Inhelder, 1971).

Piaget accepts Chomsky's reasoning that language requires complex and stable mental structures, but questions the assumption that because such structures are complex and stable, they need to be governed by innate mechanisms. Piaget claims that Chomsky incorrectly sees only two alternatives, either an innate schema, or acquisition from the outside world through simple learning. Piaget indicates a third possibility: a process of "internal equilibration." In this, self-regulatory processes involved in sensorimotor schemas progressively interact with, or use, environmental events to form the kinds of inner structures that Chomsky shows to be necessary for the human use of grammatical (rule governed) language.

Piaget's view of human knowledge representation is much like that of the machine intelligence mainline view of how to represent knowledge in frames, scripts, etc. Minsky (1975, page 229) thinks "that there is a similarity between Piaget's idea of a concrete operation and the effects of applying a transformation between frames of a system." An understanding of Piaget would seem to be useful to those developing knowledge based systems. To the extent that a NL system uses mediating information, a NL system needs a knowledge base mechanism. Frame equivalent systems are currently the most popular strategy among machine intelligence workers for the representation of mediated information.

DEFINING AN ACTIVITY AREA: MAPPING

All this is fine, but where does it lead to in defining an activity area within human/database communication?

A. SIMPLIFICATION - WRITTEN COMMUNICATION

As a first simplification step, it is clear that it is easier to deal with written communication as opposed to spoken language. Written language has greater cohesion and rigor. It has less need of redundancy and various spoken null devices such as "huh." The use of written language also removes problems of phonology from primary consideration. This is useful as besides the language processing problems found with written language, spoken language requires the capability to separate words from words, and sentences from sentences. This separation capability has not proven to be simple as it requires a considerable knowledge of the world just to do the word from word separations, let alone handling parsing problems. For example, it often turns out the pause in the middle of a word is longer than the pause separating two words. A most useful simplication is to consider only written communication.

B. SIMPLIFICATION - USING EXISTING DBMS FACILITIES

There are a host of existing databases (DB) and attendant DB query languages. More are currently in production. Curreny utilized DB query languages provide a mechanism to access data stored in a DB structure. The DB structures are generally correspondent to the structure problem that they are being used for.

Considerable sophistication has been employed to develop the structures of these DB. The DB access languages that have been developed have been designed to access these structures. A secondary goal of the languages has been the ease of interface with human users.

In accessing an existing DB from a NL, there are two possible approaches. One, the NL mechansim can directly access the data base, perhaps using some of the DBMS facilities. Or, two, the NL mechanism can _interface_ with the DBMS access language(s). The interface solution would appear to be the simplest because it is making full use of existing DBMS facilities.

The interface approach is not the strategy followed by the best known NL DB project, PLANES (Waltz, 1976). In PLANES, Waltz developed both a NL access language and a special purpose relational database. ROBOT (Harris, 1978) does utilize existing DBMS structures, but requires more connection to the DBMS than an interface to a query language.

C. SIMPLIFICATION - QUERIES ONLY

To further narrow the area of interest, only data base queries or questions will be considered. This eliminates DB data storage or modification and structural questions. A good portion of NL DB investigations, in fact, implicitly operate under this restriction. But, it is always nice to state your assumptions. Also, the specification of the pragmatics does formally exclude large bodies of ambiguities requiring deep semantic data. For example, the NL interface does not have a great need to recognize that the following are equivalent, if the NL interface is only used for resolving queries and not for modifying stored data.

(1) GOLD WAS THE COLOUR OF JANE'S HAIR.
(2) JANE'S HAIR WAS GOLD.
(3) JANE'S HAIR WAS THE COLOUR OF GOLD.
(4) JANE'S HAIR WAS GOLDEN.
(5) JANE HAD GOLD HAIR.
(6) JANE HAD HAIR OF GOLD.
(7) JANE HAD GOLDEN HAIR.

Of course, the difficulty with the proper understanding of (1) - (7) is the semantic knowledge that living human beings may have hair the colour of gold, but that the hair of living human beings is not made out of the metal gold.

Obviously, a NL interface which cannot modify stored data is not as useful as one which can. However, if the goal is to be able to actually do something, simplifying the problem in this manner would seem to be moving toward a more doable task. If it actually can be done in another question.

F. SIMPLIFICATION - INTERNAL VOCABULARY REDUCTION

A NL interface which accepts a rich natural input and reduces it to a constrained semi-formal output, must obviously reduce the variability of the words in the language. First, this is so as the DBMS query language only has a limited number of possible commands in its vocabulary. For example,

(1) TELL ME MARY SMITH'S AGE.
(2) WHAT IS MARY SMITH'S AGE?
(3) REPORT THE AGE OF MARY SMITH.
would probably be reduced to one semi-formal statement such as
(4) PRINT AGE ('MARY SMITH').
where PRINT is the command into which TELL ME, WHAT IS, and REPORT are mapped.

How statement (4) is derived from (1), (2), or (3) is more than a simple table look up with an attendant vocabulary reduction. However, recognizing such mapping or reduction relationships are the responsibility of an NL interface.

Additionally, an early reduction in the vocabulary will help in the recognition of the syntactic elements of the sentence. For example, a pattern matching procedure would be more efficient if only the pattern
(1) PRINT ATTRIBUTE NAME
had to be considered instead of the set
(2) PRINT ATTRIBUTE NAME
(3) LIST ATTRIBUTE NAME
(4) PROVIDE ATTRIBUTE NAME
This can be done by reducing the input words to a smaller vocabulary.

Perhaps, a reduction down to a very minimal list of words similar to Wilks (1975) primatives is not required. Wilks was looking for a minimal list that could be used to define the senses of other words. Of more interest here, is a vocabulary which is not obviously redundant with two words covering the same or nearly the same subject area. Whether the reduced vocabulary should be oriented toward the subject at hand or be general in nature is unclear.

Besides word role and pattern recognition, the reduced vocabulary version of queries could be useful when verifying the machine's understanding of queries.

REFERENCES

Bijou, S., Child Development: The Basic Stage of Early Childhood, Prentice-Hall, Englewood Cliffs, 1976.

Brown, R., "Development of the first Language in the Human Species," Am. Psychol., 28, 1973, pp. 97-106.

Charniak, E., "Syntax In Linguistics," in E. Charniak, Y. Wilks (eds.) Computational Semantics, North-Holland, Amsterdam, 1976.

Chomsky, N., Topics in the Theory of Generative Grammar, Mouton, The Hague, 1966.

Chomsky, N., Language and the Mind, Harcout, Brace & World, New York, 1968.

Harris, L. R., "A System for Primitive Natural Language Acquisition," Int. J. Man-Machine Studies, 1977, V. 9, pp. 153-206.

Inhelder, B., "The Sensory-Motor Origins of Knowledge," in D. N. Walcher and D. L. Peters (eds.), Early Childhood: The Development of Self-Regulatory Mechanisms, Academic Press, New York, 1971.

Lakoff, G., "Some Remarks on AI and Linguistics," Cognitive Science, V. 2, N. E, July-September, 1978, pp. 267-275.

Minsky, M., "A Framework Representing Knowledge," in The Psychology of Computer Vision, P. H. Winston (ed.), McGraw-Hill, 1975.

Piaget, J., "Piaget's Theory" in P. H. Mussen (ed.), Carmichael's Manual of Child Psychology, 3rd ed., V. 1, Wiley, New York, 1970.

Skinner, F. F., Verbal Behavior, Appelton-Century-Crofts, 1957.

Wilks, Y., Seven Theses on Artificial Intelligence and Natural Language, ISSCO Memo no. 17, 1975.

SPEAKEASY -- AN EVOLUTIONARY SYSTEM, DESIGNED BY ITS USERS

Stanley Cohen

Speakeasy Computing Corporation
222 West Adams Street, Chicago, Illinois 60606

ABSTRACT

The utility of any computer language is best determined by how easily the language is adapted to the actual need of its users, rather than the needs as perceived by the language designers. Speakeasy, from its onset, was a system designed by its users, with growth potential of the system being its most powerful asset. The paper explores the historical growth of the system, the problems which have been and are being faced and their solutions.

KEYWORDS

Computer language, user-oriented system, library-oriented system, system design.

INTRODUCTION

The real value of any computer system must ultimately be judged by a single, simple criterion: "How well does it suit the needs of its users?" Other merits in some sense are irrelevant. Specific or unique features contained in any system may be of academic interest, but in any long view contribute to development only in the sense that they relate to the ultimate objective of providing tools for users.

One of the major problems in the design of a computer program for use by others is one of psychological perspective. Can the designers really play the role of the user in the construction of the system? While it may seem that such a question can surely be answered in the affirmative, in practice this has proven less than obvious.

While it is true in many cases that the designer comes from the user environment and, therefore, initially is able to provide a user-oriented viewpoint, maintaining such a perspective in a large and long-term development while at the same time keeping pace with the constant and rapid evolution of both computer technology and the user's discipline is a difficult and often impossible task. A great temptation is to join the bulk of systems that have become standard and invariant and that expect users to understand the system as it exists. A frozen design no longer responsive to the changing needs of the users results.

This paper describes a system which attempted to face such problems from the onset.

That it has succeeded in surviving and evolving for a very long period with most of its goals intact is less important as a testimonial to the current Speakeasy System than it is to the original design specifications. Given that perspective, the original design should be viewed as a possible prototype for new systems so that other designers might benefit from lessons learned.

HISTORY

The Speakeasy System was created in the early 1960's. For proper perspective, it is best to understand that this was the period when monitor schemes were in their infancy and the whole concept of an operating system to manage a job stream was just beginning. Time-sharing and the use of terminals as communication devices were not even discussed seriously.

Providing a language that was oriented toward a user unfamiliar with a computer was not viewed seriously since it was not addressing the "real" problem concerning the use of the machine. Neither was this particular development viewed as a serious undertaking since the resources available were meager. It was more of an intellectual exercise, viewed by all as more of a hobby than any real project. No funding, manpower or long-term commitment for support was available. Given such an environment, necessity led to an unconventional development, tuned to abnormal needs. That it led to as powerful a system as Speakeasy is less a matter of chance than a consequence of the original design specifications. Speakeasy, from its onset, was designed for growth by accumulation of features. That it has grown so large and still remained coherent is no accident, but rather a direct result of an environment designed to enlist the aid of users and, at the same time, satisfy their needs.

THE LANGUAGE DESIGN

Speakeasy is a system designed with the goal of making a computer usable by people knowledgeable in their own disciplines, aware of the capabilities of a computer and interested in carrying out calculations on a machine. In designing such a system, it was important from the onset to provide a natural means of using the computer so as to attract the end-user community. To do otherwise would have provided a new computer language but destroyed the potential source of contributors. For this reason, the communication format was simple. The Speakeasy language syntax was modeled on conventional, unaltered mathematical notation as much as possible. Commonly known words were used to describe operations in the language. These may appear as obvious statements for any language, but it is well to realize that most computer languages are designed for computers and require that the user become aware of hardware limitations. To give just a few examples, a statement such as SIN(1) is illegal in Fortran because the argument to SIN must be a floating point number and A = 3/2 + 5.3 yields a result of 6.3 because of integer truncation. Similarly, 25 + 1/3 will result in 5.333 in PL/1. Certainly a format statement such as (3F4.3,7X,3I3) is not an obvious syntactical form to a person untutored in computer languages. The reasons for the propagation of such burdensome language forms is that they have become natural to the designers of languages because of their own educational histories. It requires a great deal of extra effort on the part of the designer to begin anew and question even the basic parts of a language syntax. Completely untutored users are more likely to note the obvious unnatural features of any language. If they are able to communicate these ideas to the designers, all subsequent users benefit.

For Speakeasy, the language specifications were simple. If a statement was clear to a user and if it was not ambiguous, then the language processor should have been able to interpret and carry out the calculation. Needless to say, this definition was not always met, for if it were, we would have designed the ultimate language. The philosophy implied in this approach is what is important. Complaints by users about the lack of understanding on the part of the system were viewed by the designers as failures in the Speakeasy System. Attempts were then made to enhance the system so that such misunderstandings would not reoccur. This interaction with the users of the system rather than the presentation of a complete system was of prime importance. This design philosophy has been maintained not only for the system as a whole, but for additions to the system as well.

Speakeasy is a word-oriented language. Commonly understood words are used to invoke operations in the language. Carefully chosen words convey almost all that is necessary to enable a novice to begin using Speakeasy. Speakeasy is built on a very, very large vocabulary of words that are simple, direct and easily understood. Words such as SUM, AVERAGE, TABULATE, GRAPH and QUIT carry meanings that are obvious even to the untrained. That these words do what the user expects is one virtue of Speakeasy; it means that even the first-time user can carry out calculations in an unintimidating atmosphere and identify with the friendly, responsive mode of operations of this system. The use of natural words coupled with standard mathematical notation also means that the user has important visual feedback in calculations. The user, therefore, is able to concentrate on the problem that is to be solved, rather than the formation of the problem in a foreign computer language.

The choice of words, such as those illustrated above, is relatively straightforward, but for more specialized operations, the decisions relating to syntax are often best decided by actual users after the operation has been produced in preliminary form. From experience, we have found that even the most obvious additions are prone to designer limitations, and that a preliminary trial introduction of a new word will lead to proposed alternatives, additions and modifications. It is this flexible approach which has permitted the system to maintain an overall integrity that includes a great deal of human-oriented decisions. To further aid in the evolution of the system, the approach has been to specify to users the means for extending the language, permitting them to add new words to satisfy their own needs.

THE SYSTEM DESIGN

To facilitate the inclusion of user-designed operations, the Speakeasy System was designed around the concept of modular components, with separate components maintained in libraries. In most systems, the benefit of each addition must be weighed against the consequence that it will have on the overall system performance. It is therefore unlikely that a feature of great benefit to only a few users will ever be adopted if it degrades the system's use for others. The modular library-oriented approach of this system minimizes the overhead of additions and, therefore, encourages additions of marginal interest.

The structure of the Speakeasy System is illustrated in Fig. 1. The system interprets the requests of the user and breaks them down into basic syntactical components. A scratch-pad storage facility available to the processor enables it to maintain transient information for the duration of the particular application. The processor and this scratch pad provide the mechanism for a user to define and address structured objects that represent transient information in the tasks being assigned to the system.

Libraries of pre-compiled routines are attached to the processor to provide the system with its working vocabulary. When a specific operator is requested of the system, all of the libraries attached to the system are searched. If the requested operator is found, it is dynamically linked to the system so that it may perform its function. The use of attached libraries makes it possible to provide an ever in-

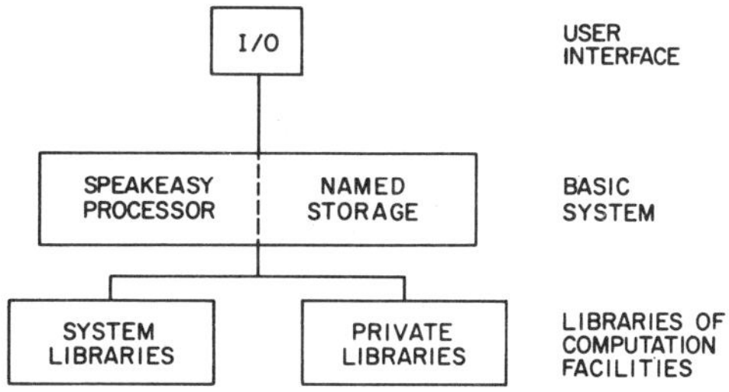

Fig. 1. A schematic view of the Speakeasy System. The bulk of the computational facilities are contained in the modules attached to the system.

creasing set of such operators without creating a system of immense size and complexity. In most cases, such operations involve operands previously defined and saved in the transient scratch-pad area. New results are also saved there and therefore are available for later use by other operations. This operator-operand formulation with a wide variety of operators on structured objects is the heart of the Speakeasy System.

DOCUMENTATION

Any computer system must be properly documented in order for it to be usable. Documentation for a major system is not an easy task. If approached solely by conventional means, it would be even more difficult in a system like Speakeasy that is designed for growth. Any printed documentation would be outdated before it was published. Fortunately, the system itself is capable of providing facilities that not only supply the documentation, but do so in a way that is guaranteed to be self-sustained.

A library of documents referred to as the Speakeasy Help documents is attached to the system. Each word, concept or facility available in the system is described in a member of the Help document library. The library is designed as a tree structure so that a user probing the library in an interactive session is led quickly to the specific concept of interest.

The Help documents are of particular importance in an evolving system such as this, since each operation introduced is first presented to users by this new document. The person adding the word is forced to concisely define the operation and it use. It is at this introductory stage that users testing new contributions become critical of both the operation and the documentation. Therefore, it provides the developer with quick feedback, which is necessary for the operation to be effectively modified to satisfy the users.

EXAMPLES OF SPEAKEASY

Speakeasy is designed to be unintimidating. The example shown in Fig. 2 illustrates Speakeasy in the desk calculator mode of operation. In this mode, the system is responsive to the request of the user. Each time a line is entered, Speakeasy responds by answering the request and then printing a :_ to indicate that it is waiting for the next request. The use of Speakeasy in a desk calculator mode with a rich vocabulary is often all that is needed by some users. Programming as such is not

```
:_2+2
2+2 =  4
:_3*SQRT(3)
3*SQRT(3) =   5.1962
:_3**3
3**3 =  27
:_ANSWER+20
ANSWER+20 =   47
:_X=9 ; Y=18
:_X*Y-5
X*Y-5 =   157
:_4*X+|3*5!-7!|
4*X+|3*5!-7!| =   4716
:_(3*LOG(4))*(SQRT(3.5*8/6.555)+17.6)
(3*LOG(4))*(SQRT(3.5*8/6.555)+17.6) =   81.792
:_ACOS(-1)
ACOS(-1) =   3.1416
:_SIGNIFICANCE 15
:_ANSWER
ANSWER =   3.14159265358979
```

```
:_SIGNIFICANCE 5
:_ANGLES IN DEGREES
:_RATIONALIZE
:_2/3+1/7*SIN(30)
2/3+1/7*SIN(30) =   31/42
:_(-32)**(-3/5)
(-32)**(-3/5) = -1/8
:_(-2)**(3/2)
IN LINE " (-2)**(3/2)   "  ENTERED COMPLEX DOMAIN.
:_DOMAIN COMPLEX
:_RETRY
(-2)**(3/2)      = -2.82841
:_(2-3I)**2
(2-3I)**2 = -5-12I
:_SQRT(3-4I)
SQRT(3-4I) =   2-1I
:_ANSWER**2
ANSWER**2 =   3-4I
```

Fig. 2. Speakeasy in the desk calculator mode of operation.

an initial requirement for learning to use the system. When the need arises though, a sophisticated stored program mode with an editor is able to answer the extended need.

Users are able to add new features. A complete graphics package was developed by John Reynolds of COMSAT Laboratories. It has provided users with an interactive facility that is made very powerful by the presence of other operations in the system. These are shown in Fig. 3. The ability of the Speakeasy System to grow by the inclusion of user-developed operations means that the development team is supplemented by a large number of helpers. The overall system environment, designed for growth, thus answers the needs of users by harnessing their contributions.

```
:_graphics tek4012
:_x=grid(0,10)
:_y=sin(x)*exp(-x/5)
:_graph(y:x)
```

produces the first curve and the axis labeling

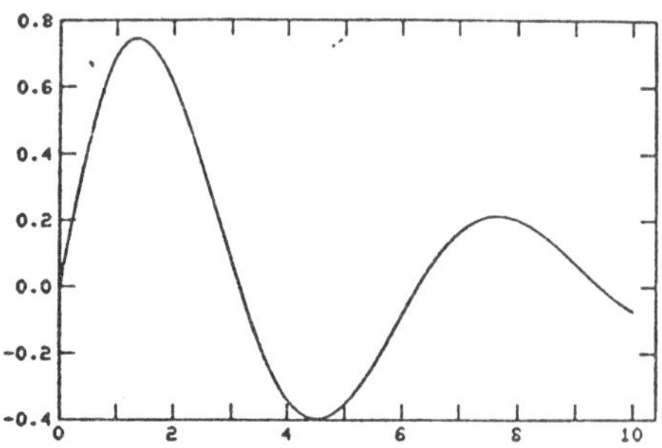


```
:_addgraph(deriv(y:x):x)
```

adds the second curve to the original display.

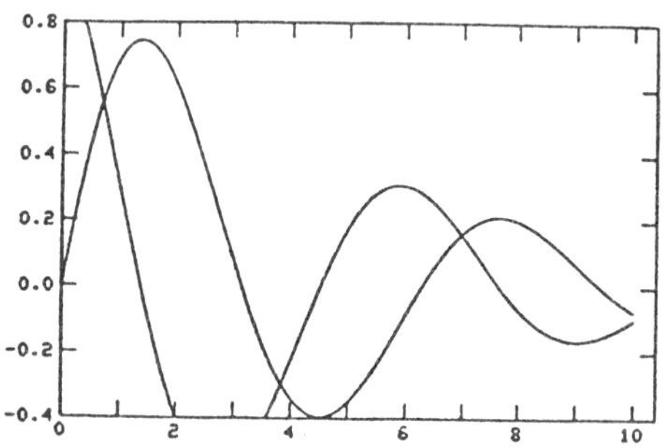

Fig. 3. The powerful graphics package in Speakeasy.

Speakeasy is able to operate on structured objects and thus provides a concise yet powerful command language. Figure 4 shows the definitions and use of arrays and matrices. The operations are so obvious that any explanation would be redundant. Any person who has ever programmed equivalent operations in conventional compiler languages will be aware of the simplicity of the Speakeasy approach. The clear, unambiguous notation is much of the attraction of Speakeasy. That it combines power with such simplicity is why its use is becoming so widespread. For example, the single word EIGENVALUE provides even a novice user with access to EISPACK, the most advanced matrix eigensystem available today.

IT CONTINUES

While the availability of statistical operations, libraries of contributed operations designed for their needs and internal matrix algebra has attracted economists to Speakeasy, it is only now that the language itself is being extended to include data types specifically oriented toward such users. During this period, the additions to the system are being tested by users as they are introduced. Following the procedure outlined in this paper, new operations are being evaluated by actual users and modifications are being made in response to their requests. As a consequence of learning by experience, the additions have been revised several times already and are once again under evaluation to make sure that this special audience will find within the new system a tool which really satisfies their needs.

```
:_X=ARRAY(3,3:1,2,3,5,2,7,3,1,6)
:_X

  X (A 3 BY 3 ARRAY)
  1  2  3
  5  2  7
  3  1  6

:_1/X

1/X (A 3 BY 3 ARRAY)
  1       .5      .33333
  .2      .5      .14286
  .33333  1       .16667
```

```
:_X=MATRIX(3,3:1,2,3,5,2,7,3,1,6)
:_X

  X (A 3 BY 3 MATRIX)
  1  2  3
  5  2  7
  3  1  6

:_1/X

1/X (A 3 BY 3 MATRIX)
 -5/16  9/16 -1/2
  9/16  3/16 -1/2
  1/16 -5/16  1/2

:_NORATIONALIZE
:_1/X

1/X (A 3 BY 3 MATRIX)
 -.3125  .5625 -.5
  .5625  .1875 -.5
  .0625 -.3125  .5
```

```
:_ANGLES IN DEGREES
:_X=GRID(0,360,15)
:_SINE=SIN(X)
:_COSINE=COS(X)
:_TABULATE X SINE COSINE

  X     SINE      COSINE        X     SINE      COSINE

  0     0         1             195   -.25882   -.96593
  15    .25882    .96593        210   -1/2      -.86603
  30    1/2       .86603        225   -.70711   -.70711
  45    .70711    .70711        240   -.86603   -1/2
  60    .86603    1/2           255   -.96593   -.25882
  75    .96593    .25882        270   -1        0
  90    1         -0            285   -.96593   .25882
  105   .96593    -.25882       300   -.86603   1/2
  120   .86603    -1/2          315   -.70711   .70711
  135   .70711    -.70711       330   -1/2      .86603
  150   1/2       -.86603       345   -.25882   .96593
  165   .25882    -.96593       360   0         1
  180   -0        -1
```

```
:_X=7,3,4,SQRT(3),1,8
:_AVERAGE(X)
AVERAGE(X) =   4.122
:_CUMSUM(X)

CUMSUM(X)  (A 6 COMPONENT ARRAY)
  7        10       14       15.732   16.732   24.732

:_ORDERED(X)

ORDERED(X)  (A 6 COMPONENT ARRAY)
  1        1.7321   3        4        7        8
```

Fig. 4. There are a large number of functions such as CUMSUM and AVERAGE. The explicit looping so common in other languages is rarely used in Speakeasy.

ON THE PROBLEM OF THE MATHEMATICAL THEORIZATION OF HUMAN SYSTEMS DYNAMICS*

Guy Jumarie

Department of Mathematics
Université du Québec à Montréal
P.O. Box 8888
Montreal, QUE., H3C 3P8
CANADA

ABSTRACT

It is shown how, by using very simple arguments in linguistics, one arrives at the geodesic of Minkowski to describe communication in human systems.
KEYWORDS: Information, Subjectivity, Natural Languages

OVERVIEW

In this paper, we examine the problem of modelling human systems dynamics. One of the specific features of such systems appears to be the absence of absolute reference frames which is caused by the subjectivity involved by human behaviour. If we further take for granted that communication is essential in the evolutions of human media, we are led to consider the modelling of both evolution and information in a subjectivistic framework. We herein describe the main elements of an approach which directly arrives at the special and the general relativity.

NATURAL LANGUAGES AND COMMUNICATION

We think we can assert, with no risk of making a grave mistake, that information is conveyed between human beings mainly via *natural languages* (NL in the following). They are not the only channels; pictures and motion too convey information in human systems; but they are the main ones. In which way, how to describe NL in order that they provide a suitable approach to the dynamics of human systems ?

Natural Languages

Every NL can be considered as defined by a set $\Omega = \{\beta\}$ of symbols generically denoted by β, call them *words* to fix the thought, and a set $\Omega' = \{b\}$ of meanings whose the generic element is denoted by b. The structure of Ω, or else the *internal structure of the* NL is organized by a set of rules, the grammar rules, which are referred to as the *syntax* of the NL. In the same way Ω' is ruled by the so-called *semantics* which somewhat define all the admissible combinations of its elements. Each word in Ω is associated with one or several meanings in Ω', ant it is customary in linguistics to refer to this correspondence as a *lexem*: shortly a lexem is a word considered in a given sense.

*Research supported by the National Research Council of Canada

Formally, a NL is an application $L : \Omega \to \Omega'_p$ which maps Ω onto the set Ω'_p of the subset of Ω'.

Let us consider a second NL, $l : \Sigma \to \Sigma'_p$. If there exist two one-to-one applications $h : \Omega \to \Sigma$ and $h' : \Omega'_p \to \Sigma'_p$ such that the commutative diagram in Fig. 1

$$\begin{array}{ccc} \Omega & \xrightarrow{\quad L \quad} & \Omega'_p \\ {\scriptstyle h}\downarrow & & \downarrow{\scriptstyle h'} \\ \Sigma & \xrightarrow{\quad l \quad} & \Sigma'_p \end{array}$$

Fig. 1. Equivalent languages

holds, we shall say that L and l are *equivalent*. Remark that this diagram is exactly the basic diagram which defines the equivalence of applications in differential geometry (catastrophe theory) !

Natural Languages and Communication

A modelling of communication in human systems can be related to equivalent languages as above defined, in the following way. The speaker S says something in a language (L,Ω,Ω') to a listener R who himself has his own language (l,Σ,Σ'). We shall say that there is communication between S and R provided that R can define two one-to-one applications h and h' such that the diagram of Fig. 1 holds.

It is clear that with such a definition, there may be communication without mutual understandability, but in the present framework, communication with understandability appears as being a subclass of communication processes in the general sense of this term, so that we can restrict ourselves to this broad acceptation in a first step.

Subjectivity in Communication

Subjectivity, what is it ? We shall refer to subjectivity as to this ability that an observer, because of a certain prior internal model which is built in his own, has to interpret and to recover the content of a message which is receiving, in such a manner that the understood meaning is not the actual meaning of the message as received, but is rather what the observer thinks it should be. In this way, a human receiver can rectify mistakes in an erroneous transmission, but it may also give a wrong significance to a message which is nevertheless correctly transmitted.

One of the striking feature of subjectivity is the following. When a message is sent by means of a certain natural language NL to the observer who is receiving this message, this latter is defined locally in syntax and semantics, and it is precisely the role of subjectivity to define which part of the syntax and of the semantics in the NL is involved by the message in question. The message contains a certain amount of information which can be resolved into the information related to its syntax and the information related to its semantics. In the same way, there is a certain uncertainty on the meaning of the message: even when the message is suitably technically transmitted, the human receiver may have a certain doubt on its content. *Subjectivity works in such a way that when information increases then uncertainty decreases and conversely.*

This is a modelling of this feature which is given in the following subsection.

identification between wyntax and semantics so that the message is completely determined by its syntax only. When $H(b) > H(\beta)$, we have not a one-to-one correspondence between Ω and Ω'; but in the average, a given word in Ω has $(H(b)-H(\beta))$ lexems, therefore an uncertainty which is specific to the structure of the NL.

Subjectivity, information, Uncertainty

We are now in a position to introduce subjectivity in communication processes via natural languages.

In the absence of subjectivity, it seems right to assume that both $H(b) + H(\beta)$ and $H(b) - H(\beta)$ are constant. This assumption no longer holds in the presence of subjectivity. Indeed, in such a case, the structures of Ω and Ω' are not fixed as before, but they are rather changing with the subjectivity of the observer R. All takes place as if this latter were defining his own field of probability on Ω and Ω' respectively, so that the various entropy functions above depend explicitely upon R via these probabilities, say $H(b/R)$ and $H(\beta/R)$. It is then difficult to assume again that $H(b/R) + H(\beta/R)$ and $H(b/R) - H(\beta/R)$ are constant. But since it is a current matter that uncertainty decreases when information increases, and conversely; we shall assume that it is the product

$$(H(b/R) + H(\beta/R))(H(b/R) - H(\beta/R)) = H^2(b/R) - H^2(\beta/R) \tag{4.2}$$

which remains constant.

As a word of caution about the physical dimensions of $H(b/R)$ and $H(\beta/R)$, we shall introduce a constant c, depending upon the measurement unit, and such that the quantity

$$c^2H^2(b/R) - H^2(\beta/R)$$

is not affected by the subjectivity. Locally in syntax and semantics, the differential geodesic

$$s^2(L/R) = c^2H^2(b/R) - H^2(\beta/R) \tag{4.3}$$

is not affected by the presence of subjectivity.

So, according to the well known lorentz equations, the condition

$$c^2H^2(b/R) - H^2(\beta/R) = c^2H^2(b) - H^2(\beta) \tag{4.4}$$

is satisfied when one has

$$H(\beta/R) = \rho[H(\beta) + u\,H(b)] \tag{4.5}$$

$$H(b/R) = \rho[H(b) + (u/c^2)H(\beta)] \tag{4.6}$$

with

$$\rho := (1-(u^2/c^2))^{-1/2} \tag{4.7}$$

and this is exactly the relativistic fuzzy set model that we have suggested previousl [1929] [1980]. □

So, by using very simple and direct assumptions regarding natural languages, we arrive at the use of the geodesic

$$d\sigma^2 = c^2\,dH^2(b) - dH^2(\beta) \tag{4.8}$$

DERIVATION OF THE MINKOWSKI GEODESIC

The Framework

Despite it is by no way a strict necessity, we shall herein use the concept of *entropy* as defined by Shannon, and se mainly to fix the thought. We bear in mind: given a random experiment β which can yield the results $\{B_1, B_2, \ldots, B_n\}$ with the respective probabilities $p(B_1), p(B_2), \ldots, p(B_n)$; the entropy $H(\beta)$ of β is defined as

$$H(\beta) := -\sum_{i=1}^{n} p(B_i) \log p(B_i)$$

(the symbol := means that the quantity on the left-hand side is defined by the expression on the right, while =: means that the right-hand side is defined by the expression on the left) where the logarithm function may have any basis.

$H(\beta)$ measures the amount of uncertainty that an observer has about the result of the experiment β; and also, it defines the amount of information which is contained in β.

Information and Uncertainty in Natural Languages

A NL is pictured on the diagram of Fig. 2.

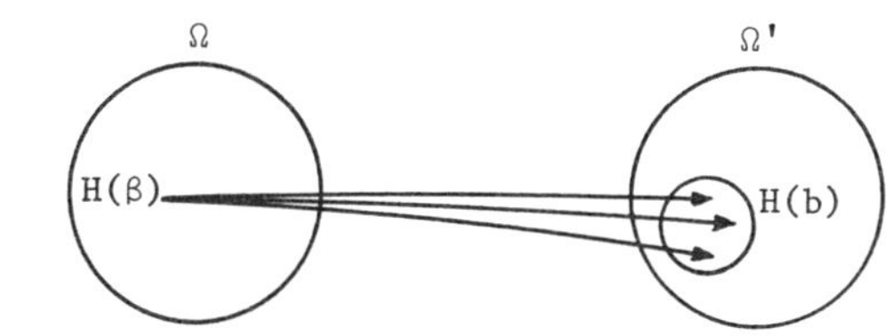

Fig. 2. Syntax, semantics, lexem

In the standard way of the "mathematical theory of communication" of Shannon, we can assume that two fields of probability are given on Ω and Ω' respectively, therefore we can characterize these latters by the entropies $H(\beta)$ and $H(b)$.

Definition 1. We shall say that the total amount of information which is contained in the NL: $\Omega \to \Omega'$ is $H(b) + H(\beta)$. □

The reader may question why do we not rather consider the expression

$$\begin{aligned} H(\beta,b) &= H(\beta) + H(b/\beta) \\ &= H(b) + H(\beta/b) \quad ? \end{aligned} \tag{4.1}$$

This is merely due to the fact that, in a first stage, Ω and Ω' are defined independently by the observer who so describes them without taking account of any relation they may involve.

Definition 2. We shall say that the amount of uncertainty which is involved by the NL: $\Omega \to \Omega'$ is $H(b) - H(\beta)$. □

Indeed, despite a word may have several lexems, one has generally $H(b) \geq H(\beta)$ for natural languages. This being so, when $H(b) = H(\beta)$, we have a somewhat of the special relativity. This analogy is not a mere coincidence, and we are allowed to expect more results by a systematic use of the relativistic physical framework to investigate general systems.

In this way, it is allowed to expect that the use of the general relativistic framework, suitably modified, should be of efficient use in general systems. □

REFERENCES

Jumarie, G.; *Int. J. Gen. Syst.* (G.B.), 5, 99-120, 1979.

Jumarie, G.; *Cybernetica* (Belg.), 22, 267-309, 1979.

Jumarie, G.; *Subjectivity, Information, Systems. A Unified Approach to Relativistic Cybernetics* (In French). Univers Publishing, Montreal; SIG Décision, Paris; Editions Vander, Bruxelles, 1980

ADDITIONAL REFERENCE

Jumarie, G.; An approach to Encoding in the Presence of Subjectivity. *Journal of Information and Optimization Sciences,* Vol. 1, No 2, pp. 166-186, 1980.

In this paper, the above model for subjectivity is applied to the standard communication chain "sender-channel-receiver" in which the receiver is a human observer who involves subjectivity, in the way that he has a prior partial knowledge about the content of the message. The concept of transinformation in the presence of subjectivity is defined, which is further applied to the problem of encoding in the presence of subjectivity. The resulting encoding is carried out by using the Faddeer equation, and comparison is done with standard results in the absence of subjectivity.

BRIDGING THE GAP BETWEEN NATURAL AND COMPUTER LANGUAGE

Douglas R. Skuce
Department of Computer Science
University of Ottawa
Ottawa, Ontario, Canada, K1N 6N5
November 1980

Abstract

Very few people possess the skill to express themselves concisely and unambiguously in natural language (NL). Modern society nevertheless can expect a continuing increase in demand for the production of NL text, for the concomitant comprehension of this information, and for converting NL information into computer languages (CLs). The semantic errors and confusion which usually result from the imprecise use of NL will therefore be multiplied, particularly in the latter case.

A solution to this problem is proposed, based on a language LESK (Language for Exactly Stating Knowledge), which is intermediate between NLs and CLs. LESK looks like NL, but employs simple mathematical, linguistic, and computer principles. An example in which LESK is used to clarify some of a university's academic rules is presented.

It is argued that both developed and developing nations could benefit greatly from widespread use of a lanugage like LESK.

Keywords

artificial language; natural language; artificial intelligence; database; knowledge bases; knowledge transfer; computer-aided instruction

Introduction

The explosive evolution of computer systems is making possible, and necessary, a great increase in the frequency of human-human comunication in natural language (NL) and human-machine communication in computer language (CL). However, most of the widely used CLs have their roots in the older conception of the computer as a numeric processor, programmed by computer specialists. Most of these languages are not suited to non-numeric applications, and the few that are so intended (e.g. LISP, SNOBOL, PROLOG) were designed only for computer specialists. The result is that still, most people do not communicate directly with machines and must remain "computer-illiterate". With declining skills in using NL, the quality of communication between people, and between people and machines, leaves much to be desired.

We suggest therefore that fresh approaches to both types of communication are necessary to permit a wide community of people to originate and transmit information quickly and, one would wish, accurately. The type of information we are most concerned with originates now as NL: terminology, statements, descriptions, specifications, regulations, laws, procedures, contracts, etc. If a human recipient has difficulty in understanding, the result is confusion and error. If the recipient believes he/she does understand, the chances are that the perceived meaning is not exactly what was intended.

On the other hand, to communicate such information to a machine, most people require the services of scribes who possess the knowledge of mysterious languages (CLs). These scribes are, fortunately, sworn to eternal vows of wisdom and infallibility, and hence transcribe correctly even the most poorly written NL into precise CL, accurately guessing the writer's intentions. Unfortunately, such scribes are very hard to find.

It would be desirable therefore to develop a language which is in some sense intermediate between NL and CL, so that it could be written, or at least understood, by a much wider group. It should be capable of expressing the basic conceptual structures underlying NL, in a syntax which is close to that of NL, and with a semantics which is not based solely on computer structure. Such a language would expedite both types of communication, but most immediately, would reduce the dependence upon scribes as translators from typical NL to CL. If a non-scribe wished to communicate using this language but could only read it (not write it), the scribe would make the translation as with older CLs, only the result would be understandable and hence verifiable to the originator, which is not the case with today's CLs. Perhaps the scribe could be a machine.

Such a language has been developed (Skuce 75-80). LESK (Language for Exactly Stating Knowledge) is based on simple linguistic, mathematical, and computer concepts so that it is very readable, yet its semantics is straightforward and precise. In its current implementation it looks like English, but many other NLs could be substituted, making automatic translation of LESK trivial. The user of LESK must understand its semantic bases, which requires a level of mathematical and computer science training comparable to that of a first-year engineering or business student.

We propose now to introduce LESK using simple examples and to discuss later some of its current and potential applications. We will also discuss its long-term implications for industrial and developing nations.

Describing Knowledge in LESK: An Example

University personnel constantly face a problem which is very typical of the kinds of communication problems LESK is intended to alleviate or solve: locating, understanding, and acting upon academic regulations. Some typical questions are:

1. (terminological) "Is winter term synonymous with spring term?" "How many kinds of student are there?" (academic categories); "What are the possible values for a mark in a course?" "How is did-not-write different from absent as a mark?"
2. (procedural) "Can a student use a course credit toward two degree programs ?"; "How does a student transfer from an engineering degree program to a science degree program ?"
3. (uncertain authority) "Who knows the rule about ...?"; "The calendar states X which seems to contradict this memo Y"; "Department A considers an X to be a Y, whereas Department B thinks an X is not a Y".

Were all such information defined precisely in LESK and available through interactive terminals which possessed a thorough ablility to answer questions deductively, most of these problems would disappear. The single computer-accessible source, termed a knowledge base, could as well be coupled to conventional databases to ensure that the databases conformed to the same rules. (This problem is becoming serious as independent databases proliferate.)

Let us now give a few examples of LESK-based knowledge involving typical concepts required to write or interpret academic regulations. LESK has three basic types of concept: <u>entities</u>, <u>statives</u> and <u>procedures</u>, corresponding approximately to noun phrases, stative sentences, and "recipes" in NL. (There is no NL analog to a procedure; LESK procedures remind one of recipes.) One introduces the desired terminology as one of the three concept types, either as sub-cases of other terms, or by specifying how terms are interrelated.

First, we introduce the entity PERSON (upper case denotes user terminology):

```
a PERSON:
  kinds (by SEX): MALE -, FEMALE -;
  - has 1 AGE A;
    0 < A < 150
end
```

This introduces the set of PERSONS, MALE PERSONS, FEMALE PERSONS, and the function SEX which partitions PERSONS. The function AGE is introduced on PERSONS. The symbol '-' denotes the definiendum. Next we introduce a subset of PERSONS:

```
a STUDENT:
  - is a PERSON;
  kinds: UNDERGRADUATE -, GRADUATE -;
  kinds: PART-TIME -, FULL-TIME -;
  has a unique STUDENT NUMBER;
  has a unique ACADEMIC RECORD;
  - has 1 or more DEGREE PROGRAMS
end
```

Thus anything true of PERSONS is also true of STUDENTS, and STUDENTS are further partitioned two different ways. Note that this declaration prevents a STUDENT from being both an UNDERGRAUATE in one DEGREE PROGRAM and a GRADUATE in another DEGREE PROGRAM. What is a DEGREE PROGRAM?

```
a DEGREE PROGRAM:
  kinds: UNDERGRADUATE -, GRADUATE -, SPECIAL -;
  - has 1 or more REQUIRED COURSES;
  - has 1 or more OPTIONAL COURSES;
  - has a number  called the REQUIRED NUMBER
                       of OPTIONAL COURSES;
  - has 1 DEPARTMENT D;
  - has 1 DEGREE TYPE;
  - is ADMINISTERED BY D
end
```

Here, for example, REQUIRED COURSES is introduced as a binary relation on DEGREE PROGRAM. Much of the basic declaration process consists of introducing set, relation and function names, and is very reminiscent of database structure declarations. Many of these latter could be automatically generated from the LESK declarations. The stative 'is ADMINISTERED BY' is also introduced. The reader is invited to interpret these statements in first-order predicate calculus; for example, the latter would become:

```
forall P: DEGPROG(P) => ADMINBY(P,DEPTOF(P))
```

The distinction between the following three concepts is a frequent source of confusion, although humans are quite adept at disambiguating the typical NL homonyms. The term 'course' is used with the following three meanings:

```
a COURSE:
  - is denoted by a COURSE name;
  - has a DEPARTMENT;
  - has a TITLE;
  - has a number called its CREDIT VALUE;
  - has 0 or more COURSE CREDITS called its PREREQUISITES;
  - has 1 or more TEACHERS
end

A COURSE OFFERING:
  consists of: a COURSE name CN,
               a SEMESTER IDENTIFIER;
  inherits properties from CN;
  - has a ROOM NUMBER;
  - has 1 or more MEETING TIMES;
  - has an ENROLLMENT list L;
  - has a number M called its MAXSIZE;
  the number N of STUDENTS in - =
  the length of L;
  N ≤ M
end

a COURSE CREDIT:
  consists of: a COURSE OFFERING,
               a MARK
end
```

Thus COURSE is the generic concept of courses in general, an instance of which might be "MATH 2343". A COURSE OFFERING is the scheduling and teaching of a COURSE in some SEMESTER in some ROOM. Note that COURSE OFFERINGS inherit all the statements made under COURSE. A COURSE CREDIT is what STUDENTS strive to accumulate. The 'consists of' primitive introduces an n-tuple used as a noun.

Statements may be given independently. The first of the following introduces a second stative 'is ENROLLED IN':

```
a STUDENT S is ENROLLED IN a COURSE OFFERING C
iff S is on the ENROLLMENT list of C

a MARK is 1 of:  A,B,C,D,E,F,DR,ABS,DNW,UNK,DEF

a PROFESSOR is a TEACHER

A PROFESSOR is TEACHING a COURSE C
iff there is a COURSE OFFERING CO such that
    P is a TEACHER of CO
```

Certain frequently needed concepts such as names, numbers and lists are provided as primitives in LESK, along with the logical quantifiers and operators, all with a NL syntax.

We turn now the the most difficult type of concept, the _procedure_. What makes procedures more complex is that they describe _change_. A procedure usually employs an imperative verb, and LESK borrows CL notation since NL does not (yet) have established terminology for computational notions. For example:

```
to ENROLL a STUDENT S IN a COURSE OFFERING C:
  let N = the number of STUDENTS in C
  if S has all the PREREQUISITES of C and
       N < the MAXSIZE of C
  then add S to the ENROLLMENT list of C;
       change N to N+1;
       succeed
  else print 'S CANNOT BE ENROLLED IN C';
       fail
end
```

The reader may be led to exclaim "but this is just COBOL!". COBOL simple converts existing CL constructs (circa 1960) into an English-like syntax. What is added here is a considerable number of linguistic and logical extensions which are automatically given a semantics which is not at all easy to translate into machine-level constructs. Also, a LESK knowledge base provides a deductive question-answering ability equivalent to that available from the underlying predicate calculus.

We trust that these few examples will have given the reader a reasonable appreciation for what LESK looks like and is suitable for. Many more examples can be found in (Skuce 75-80).

Current Applications and Status of LESK

LESK has been used on a wide variety of knowledge description tasks. For example, programming languages, neurophysiology, and the Canadian census have had certain of their concepts expressed precisely in LESK. The syntax, semantics, and computer implementation for entities and statives are quite straightforward, being based on first order predicate calculus. A compiler for LESK into an internal representation language called ARC (Algebra of Relational Composition) (Bainbridge and Skuce 80) has been implemented in the language PROLOG (Pereira, Pereira, and Warren 78) which was developed in artificial intelligence research on deductive systems. Also, Bainbridge has developed a question-answerer for ARC, written in PROLOG.

The exact form and method of implementing procedures in LESK is the subject of current research. Our goal is to employ only the few most straightforward features of conventional procedures which are compatible with LESK's predicate calculus basis. The above example then is only indicative of our current thinking on how one might combine these diverse and complex influences without excessive compromise.

Future Applications of LESK and Their Social Implications.

We may first consider the uses of LESK in the industrialized world in the next ten to twenty years. Let us assume that a reasonably complete implementation existed in a form which could be used on a wide variety of computer systems. (This could be achieved with ten or twenty person-years of effort.) We envisage applications at least as wide as current projections of the use of databases and of electronic word processing.

A second important application is in education. Virtually all teaching is still done using NL, either in textbooks or in lectures. Computer technology had promised to revolutionize teaching through computer aided instruction, although very few of us have ever experienced or even seen CAI. One big problem has been cost, but the rapidly dropping cost of hardware is quickly removing this impediment. What is left is the problem of transfering knowledge from existing NL sources into computers, at present a difficult problem due to the lack of adequate software, a proliferation of incompatible systems, and the imprecision of NL. A LESK system is as suited for CAI as it is for the other applications we have mentioned; there is no essential difference between, on one hand, the rules of chemistry, accounting or biology and, on the other hand, the rules of governments, business, etc. A CAI system based on LESK would have in addition the deductive properties LESK provides which are lacking in all current systems, and which are necessary to gain the student's respect as being "intelligent".

Turning now to developing countries, it has been well established that one of their major problems is knowledge transfer, particularly technical knowledge. Part of the problem is in the expense and delay of making available current information in printed form in their own languages. For LESK-based knowledge, an almost totally automatic translation system is a simple matter, as we have noted. Additionally, since the desired result is usually not a printed output but an interactive system, the cost and delay of publishing is eliminated. Advanced nations could possibly donate their obsolete computer systems, which no longer have any commercial value, but which would have enormous value to a country with very few computer systems.

Related Research

LESK is mainly a result of influences from artificial intelligence (AI) research, though other areas of computer science such as databases and programming languages have provided ideas. We will mention here only some of the more interesting and relevant work in AI. The question often arises "what is AI?". An easy answer is not possible, since it is hard to answer the question "what is intelligence?". The current best textbook in AI (Nilsson 80) takes the frequent expedient of simply giving examples of AI research, rather than trying to "define" AI. Computer systems have been developed which perform medical diagnosis at an expert level (better than most doctors) (Shortliffe 76); which have beaten the world's champion at backgammon (Berliner 80), which find chemical structures from mass spectrograms that the experts missed (Buchanan and Feigenbaum 78); which carry on conversations in English (Allen, J. and Perrault, R. 78); which have "hands" and "eyes" (Dodd and Rossol 79). In the latter two, performance is still far below human ability however.

Two other areas which are increasingly overlapping AI are database research (Wong and Mylopoulos 77) and computational linguistics. The work of Sowa (79) is particularly relevant to LESK.

We regret that an explanation of exactly how these various areas are related to LESK is not possible here. The reader may refer to Skuce (75-80) for amplification of these questions.

Concluding Remarks

Industrialized societies have progressed to the point where most university graduates (outside of the arts and letters) are being exposed to some conventional computer language, although few of them will make much use of this training. The reason for this unfortunate situation, like many in education, is that these languages are not appropriate to the information processing needs of most people: writing, reading, and trying to understand natural language information.

We argue therefore for a new dimension of literacy. Our institutions continue to fail to teach students to write their own languages well and continue to provide only barely minimal skills in using computer languages which are largely irrelevant to most student's needs. Instead, we advocate the widespread adoption of an artificial language which is a hybrid of natural and computer languages, and which most educated people could use constantly. Further, we have designed a language, LESK, to meet these needs, and have tested a major part of it on a variety of real examples. We have begun the implementation of computer systems which can "understand", for most practical purposes, statements and questions expressed in LESK. The completion of this task would not require an unusual investment.

The world-wide benefits of such a language seem undeniable. As a remarkable side effect, it is highly likely that by teaching people an artificial language such as LESK, their skills in using natural language would become considerably improved. The ultimate purpose of using LESK then, might be to obviate the very need for it!

Acknowledgements

The support of the Killam Research Foundation, and of the National Science and Engineering Research Council is gratefully acknowledged. Stewart Bainbridge has made many valuable technical suggestions concerning LESK, and Elyse Piquette contributed her editorial skills to this paper.

References

Allen, J. and Perrault, R. Participating in Dialogues: Understanding via Plan Deduction. Proc. of the Second National Conference of the Canadian Society for Computational Studies in Intelligence, Toronto.

Berliner, H. (80) Computer Backgammon. Scientific American, v. 324, no. 6.

Buchannan, B. G. and Feigenbaum, E. A. (78) Dendral and Meta-Dendral: Their Application Dimensions. Artificial Intelligence v. 11, nos. 1 and 2.

Dodd, G. and Rossol, L. (eds) (79) Computer Vision and Sensor-based Robots. Plenum Press, New York.

Nilsson, N. Principles of Artificial Intelligence. Tioga Publ. Co., Palo Alto.

Pereira, L, Pereira, F. and Warren, D. User's Guide to DECsystem-10 PROLOG. Dept. of Artificial Intelligence, Univ. of Edinburgh.

Shortliffe, E. (76) Computer-based Medical Consultations. Elsevier, New York.

Skuce, D. (75) An English-like Language for Qualitative Scientific Knowledge. Proc. of the Fourth International Joint Conference on Artificial Intelligence, Tbilisi.

Skuce, D. (77) Toward Communicating Scientific Knowledge Between Scientists and Machines. Ph.D. Dissertation, Dept. of Electrical Engineering, McGill University.

Skuce, D. (78) Describing Programming Language Concepts in LESK. Proc. of the Second National Conference of the Canadian Society for Computational Studies in Intelligence, Univ. of Toronto.

Skuce, D. (79) An Approach to Defining and Communicating the Conceptual Structure of Data. Dept. of Computer Science Tech. Report No. 79-05, Univ. of Ottawa.

Skuce, D. (80a) Knowledge Acquisition and Representation Using Logic, Set Theory and Natural Language Structures. Proc. of the Third National Conference of the Canadian Society for Computational Studies in Intelligence, Univ. of Victoria.

Skuce (80b) A Language for Exactly Stating Qualitative Generic Knowledge. submitted to: Journal of Computers in Biology and Medicine.

Sowa, J. (79) Semantics of Conceptual Graphs. Proc. of the 17th Annual Meeting of the Association for Computational Linguistics, La Jolla.

Wong, H. and Mylopoulos, J. Two Views of Data Semantics: A Survey of Data Models in Artificial Intelligence and Database Management. INFOR, v. 15, no. 3.

ON THE DEVELOPMENT OF A RECURSIVE MODEL OF WORD STRUCTURE IN THE ENGLISH LANGUAGE

K.S. O'Mara , W.M. Jaworski , and S. Klasa

Department of Computer Science
Concordia University
Montreal, Quebec, Canada

This work is dedicated to the memory of the late Professor K.V. Leung

ABSTRACT

This work presents a scheme of transition diagrams for finite automata which can synthesize english language words. To date this model has synthesized all words of size ≤ 5. The three transition diagram for all 2-letter words contained in Funk & Wagnalls Standard College Dictionary are given.

KEYWORDS

Orthographic perception, English Word Structure, Finite Automata, Natural Languages, Recursive Model, Transition Diagrams.

INTRODUCTION

Natural languages have a highly context-sensitive form. Chomsky (1956) has described this form as a type 0 language. The exact manner in which the human is able to interpret language has been of interest to psychologists since at least the last century (Pillsbury, 1887). Early studies sought to identify the perceptual orthographic clues that may underlie our ability to read (Woodsworth 1938). Modern psychologists have sought to determine the degree to which the recognition process involves whole words, letter clusters or single letters (Mewhort, 1974; Gibson and others, 1962). Conventional models of word recognition assume multiple levels of stimulus processing. These models, using the constructs of information theory, attempt to account for preceptual ability in terms of the redundancy of natural languages (Herdan, 1962: Pierce, 1965; Shannon, 1951). These paradigms of perception assume that reading involves learning a sophisticated guessing procedure (Rumelhart and Siple, 1974) or criterion bias (Broadbent, 1967) which allows us to perceive what sensory information alone is insufficient to determine (Adams, 1979). It would appear that it is possible to concisely specify rules for the synthesis and recognition of English words. These rules will be discussed in this paper.

METHODS

For our purpose we consider a dictionary to be a set of types or unique words found in a language.

The ultimate lexicon for the study of the english language is the Oxford English Dictionary (OED). While this dictionary is undoubtedly the most comprehensive single source listing of the English language it is neither complete nor unbiased to the British usage of the English language (Potter, 1975). For reasons outlined below the Funk & Wagnalls Standard College Dictionary (FW) is used as our source lexicon. This dictionary contains words found in the American usage of the english language.

In most dictionaries the set of words found in the language is ordered alphabetically. For the work described in this paper a dictionary in which the words or types were ordered by length or size was required. The task of re-ordering the set of words contained in an alphabetically sorted dictionary to form a dictionary ordered by word length is formidable. However, with the exception of: 1, 2 and 3-letter words, it is simpler to re-order the entire dictionary than to confirm the existance in the FW dictionary of every possible word of a given size. Mathematically there are 26^4 or 456,976 possible 4-letter words in the english language while the entire FW dictionary contains only about 152,000 entries. It is thus a simpler task to sort the entire FW dictionary than to look up all possible 4-letter words contained within it.

The principal reason for choosing the FW dictionary is that this source has a companion lexicon (CFW) in which entries are listed first by length and then alphabetically. The CFW was produced for the benefit of cross-word puzzle aficionados. It lists all 2, 3, 4, 5 and 6-letter entires contained in the FW dictionary. These listings have been censored to remove all phrases and words which were deemed obscene, scatological or racially offensive. However the CFW contains: foreign words as well as English; standard, informal and slang words in every part of speech and every tense, number and gender; variant spellings, abbreviations, acronyms, hyphenated words and prefixes and suffixes; geographical names and the first and last names of people.

The CFW differs significantly from other dictionaries. It contains only the name or index of the entry described in the FW dictionary. Also for the benefit of the crossword puzzle enthusiast it lists many permutations for each entry in the FW dictionary. In fact a given word of length N will be found $N!/((2*(N-2)!))$ times within the N-letter word list. As such one must delete all redundant entries from the CFW listings. The FW dictionary must then be consulted for each remaining entry in the CFW listings to determine if the cited letter-string or sequence is in fact listed in the FW dictionary as a part of speech in the english language. Each dictionary entry so established is called a valid english word (VEW). The lists of VEW contained in the CFW dictionary form dictionaries of 2, 3, 4, 5 and 6-letter VEW. It is from these derived dictionaries that the results presented in this paper are obtained.

OBSERVATIONS & RESULTS

1) All uni-letter VEW are vowels. They are: a,I,O.

2) All two-letter VEW listed in CFW contain at least one vowel. We shall consider these words to fall into three classes. The first case are VEW which are composed of a vowel prefixed to a consonant. The second case is the set of VEW composed of a vowel suffixed to a consonant. A third case is the set of VEW composed of two vowels. We shall denote these cases as: $\wedge\Upsilon$, $\Upsilon\wedge$, $\wedge\wedge$

where $\wedge \in \{a, e, i, o, u, y\}$ and

$\Upsilon \in \{b, c, d, f, g, h, i, j, k, l, m, n, p, q, r, s, t, v, w, x, z\}$.

Figures 1, 2, 3 are transition diagrams for finite automata which may be used to exclusively generate all 2-letter VEW of the forms: ∧Υ ; Υ∧ ; ∧∧, which are contained in the FW dictionary. In Fig. 1. we show that the set of 2-letter VEW {at, as, ar, ad, an, ah, ax, am} is obtainable from the initial state ⓐ. Similarily we see that the set of VEW of the form, ∧Υ, obtainable from the initial state ⓘ is {it, is, if, id, in}. From Fig. 2 we see that the set of VEW of the form, Υ∧, obtainable from the initial state ⓜ is {my, me, mi, ma, mu}. Similarily from the initial state ⓛ in Fig. 2 we may reach {la, li, lo}. From Fig. 3 we see that the set of 2-letter VEW of the form, ∧∧, obtainable from the initial state ⓐ is {ay, ae, ai, aa}, while from the initial state ⓨ we may reach {ya, ye}.

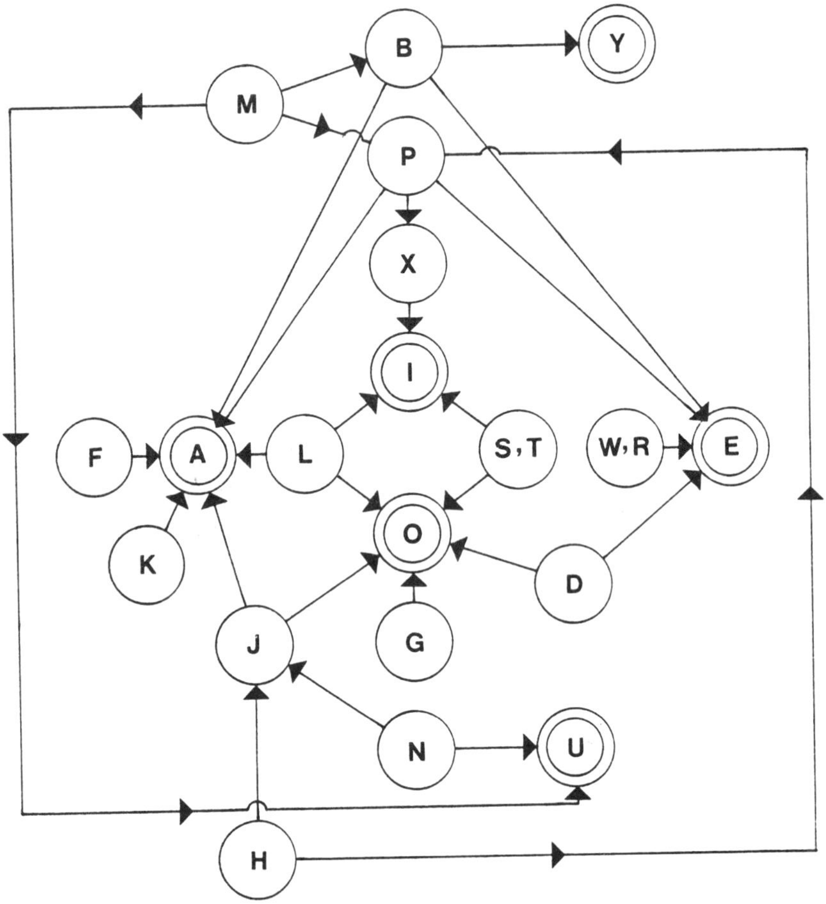

ALL CONSONANTS ARE INITIAL STATES

Figure 2

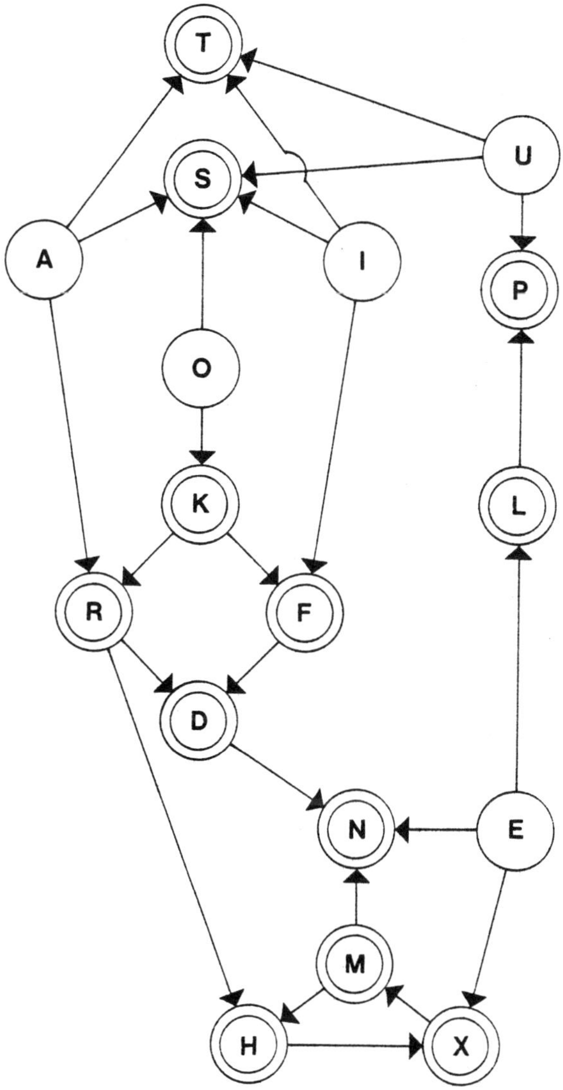

Figure 1

ALL VOWELS ARE INITIAL STATES

A complete list of the 73, 2-letter VEW obtainable from Figs. 1, 2. & 3 may be tabulated in matrix form in Fig. 4. In this association matrix A(i,j) the initial letter of all 2-letter VEW is specified by the row index i, while the second letter of each word is specified by the column index j. Thus A(7, 2) specifies the 2-letter VEW 'be'. All VEW contained in FW are demarcated by a √ all additional 2-letter VEW found listed in the OED are demarcated by an ε while those cited in the OED as obsolete, archaic or belonging to Middle English are denoted by a †. From this chart it may be clearly seen that 2-letter words of the form $\Upsilon\Upsilon$, have been prohibited throughout the history of the English language. The only living exception to this rule is the two letter injunction: 'sh'. In fact it would appear that all strings of the form Υ^N are prohibited from forming N-letter long VEW. It is interesting to note that most of the other possible $\Upsilon\Lambda, \Lambda\Upsilon, \Lambda\Lambda$ permutations have been valid english words at some point in the history of the language.

	A	E	I	O	U	Y	B	C	D	F	G	H	J	K	L	M	N	P	Q	R	S	T	V	W	X	Z
A	√	√	√		ε	√	ε	†	√	ε		√		†	†	√	√			√	√	√		†	√	
E	†	ε	†		†	√				ε		√			√	√	√	√		ε	†	†		†	√	
I		†		ε					√	√					†	†	√			†	√	√		†		
O		†		†	†	ε		†	√	√		√		√		√	√	†		√	√			†	√	
U				†					ε		†	ε				ε	†	√		ε	√	√				†
Y	√	√	ε	†	†				†	†				†	†		†			†	ε	†		†		
B	√	√	†	√	†	√																				
C	√	ε		†	†							†														
D	ε	√		√	ε	†																				
F	√	†				†																				
G	†			√		†																				
H	√	√	√	√	†	ε																				
J	√			√	†																					
K	√		†	ε	†	ε																				
L	√	†	√	√	†	†																				
M	√	√	√	ε	√	√																				
N	√	ε		√	√	†																				
P	√	√	√	†	ε	†																				
Q					†																					
R	†	√		†	†																					
S	†	†	√	√	†	†						ε														
T	ε	†	√	√	†																					
V	†	√		†		†																				
W	ε	ε	†	ε	†	†												†		†						
X	†		√																							
Z		†		†																						

Figure 4

Our results have shown that it is possible to generate the (425) 3-letter VEW contained in FW in terms of the 276 possible permutations specified by ϒΛ ,Λϒ, ΛΛ,. Seven transition diagrams of the forms ϒΛΛ ,ΛΛΛ, ϒΛϒ , ΛΛϒ, ϒϒΛ, ΛϒΛ, Λϒϒ are required for this task. They may be formed by prefixing either a vowel or a consonant to ϒΛ , Λϒ , ΛΛ , to yield the following partitions:

	ΛΛ	ϒΛ	Λϒ
Λ	ΛΛΛ	ΛϒΛ	ΛΛϒ
ϒ	ϒΛΛ	ϒϒΛ	ϒΛϒ

Alternatively they may be formed by suffixing either ϒ or Λ to ϒΛ, Λϒ ,ΛΛ as:

as

ΛΛ	ϒΛ	Λϒ	
ΛΛΛ	ϒΛΛ	ΛϒΛ	Λ
ΛΛϒ	ϒΛϒ	Λϒϒ	ϒ

VEW which may be formed from either a prefix or suffix operation are called multiradical words. Thus the multiradical VEW "pin" of type ϒΛϒ may be formed as ϒΛϒ←ϒ<: Λϒ (pin ← p + in) or ϒΛ: >ϒ → ϒΛϒ (pi + n → pin). Approximately 64% of 3-letter VEW are multiradical while the rest can be formed in only one way. For example the word "sky" of type ϒϒΛ can be legally formed only as ϒϒΛ ← ϒ<: ϒΛ (sky ← s + ky) in that ϒϒ is not a valid 2-letter base form.

In this recursive manner it is possible to describe all but two of the 2875 4-letter VEW found in FW. For this task 23 transition diagrams are needed.

Our preliminary results indicate that one may compute the relative frequency of use of each VEW obtainable from modified transition graphs by calculating the product of the positional probability of each letter within the word.

It would appear that in general all N-letter long English words may be specified with 2^N-1 transition diagrams. We are at present developing algorithms to test the validity and completeness of the productions of the type 3 grammar specified by this model on databases of larger words. This work would extend the grammar's rules to cover the problems of concatenation, hyphenation and hopefully syllabification.

ACKNOWLEDGEMENTS

The authors would like to thank William Gillespie and Kent Farrell for their many thought provoking discussions on this work. We would also like to thank Prof. C.Y. Suen for introducing us to this area of research and Prof. P.F. McCullach of McGill University's Classics Department for his help in coining the term multiradical. Partially supported by NSERC grant No. A9372.

REFERENCES

Adams, M. A. (1979). Models of word recognition. Cognition Psychology 11, 133-76.

Broadbent, D. E. (1976). Word frequency effect and response bias. Psychological Review 74, 1-15.

Chomsky, N. (1956). Three models for the description of language. IEEE Trans. On Information Theory 2, (3), 113-124.

Gibson, E. J., Pick, A., Osser, H., and Hammond, M. (1962). The role of grapheme-phoneme correspondence in the perception words. American Journal of Psychology 75, 554-570.

Herdan, G. (1962). *The Advanced Theory of Language As Choice and Chance*. Springer-Verlag, N.Y.

Mewhart, D. J. K. (1974). Accuracy and order of report in tachistoscopic identification. Canadian Journal of Psychology 28, 383-398.

Pierce, J. R. (1965). *Symbols, Signals and Noise*. Harper and Row Bros. Inc., N.Y.

Pillsbury, W. B. (1897). A study in apperception. American Journal of Psychology 8, 315-393.

Potter, S. (1975). *Changing English*. Andre Deutsch Ltd., London.

Rumelhart, D. E., and Siple, P. (1974). Process of recognizing tachistoscopically presented words. Psychological Review 81, 99-118.

Shannon, C. E. (1951). Prediction and entrophy of English language. The Bell System Technical Journal 38, 50-64.

Suen, C. Y. (1979). N-gram statistics for natural language understanding and text processing. IEEE Trans. on Pattern Recognition and Machine Intelligence PAMI-1, 2, 164-172.

Woodsworth, R. S. (1938). *Experimental Psychology*. Henry Holt and Co., N.Y.

Whorf, B. (1956). *Language, Thought and Reality*. M.I.T. Press.

Zipf, G. K. (1935). *The Psycho-Biology of Language*. Houghton-Mifflin, Boston.

Schwartz, E. I., and Landovitz, L. F. (1978). *Funk and Wagnalls Crossword Puzzle Word Finder*. The Stonesong Press, Grosset and Dunlap, Inc., N.Y.

Funk and Wagnalls Standard College Dictionary (1978). Canadian Edition, Fitzhenry and Witeside Ltd., Toronto.

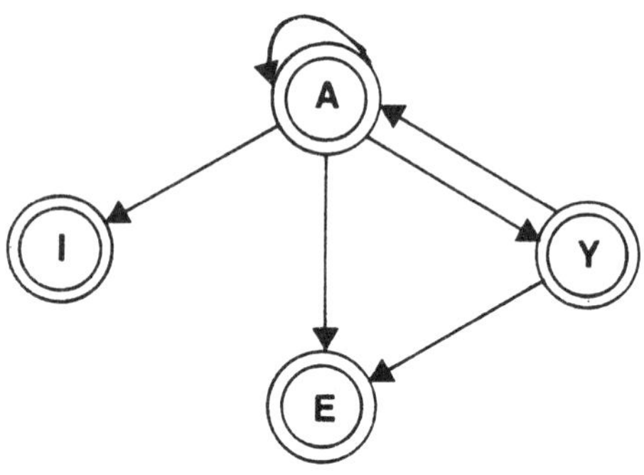

ALL STATES ARE INITIAL STATES

Figure 3

DESIGN ISSUES IN NATURAL LANGUAGE PROCESSING SYSTEMS

F. Oppacher

Department of Computer Science
Concordia University
Montreal, Canada

ABSTRACT

The aim of the work described here is to develop a programming approach for processing restricted but modularly extendible sublanguages of a natural language. Some theoretical assumptions underlying recent research efforts in computational linguistics are critically discussed. Unlike many recent natural language processing systems, the system presented in this paper emphasizes a fairly strict separation of syntax and semantics. A partial implementation which translates a small subset of English into French is described.

KEYWORDS

Natural language processing; automatic translation; transformational grammar; artificial intelligence; syntax; semantics.

DESIGN ISSUES

Recent work on natural language processing in artificial intelligence (AI) has begun to tackle some crucially important problems of intelligent language use that are largely ignored in theoretical linguistics. Foremost among these is the problem of devising representational media that are rich enough to encode conceptual and world knowledge in such a way as to facilitate the resolution of word sense and referential ambiguities. A related problem concerns the role of inference in understanding.

But whereas natural language processing systems (NLPS) have provided valuable insights into possible human conceptual mechanisms, they have not been extended to practically relevant domains. At the moment, the tremendous investment of time and effort by many talented workers in developing "understanding" programs can be justified only on theoretical grounds. (There are some doubtful exceptions to this claim, notably LSNLIS, one of the most robust systems. LSNLIS, the Lunar Sciences Natural Language Information System, was designed by Woods, Kaplan and Nash-Webber (1972) to enable lunar geologists to ask English questions about lunar rock samples from the Apollo mission. In Woods, Kaplan and Nash-Webber (1972) the system is reported to be quite successful in satisfying its prospective users: 78% of initial questions posed by geologists at a 1971 conference were processed correctly. Petrick (1977), in a critical evaluation of several well-known NLPS, notes that Woods has mentioned in talks that a much smaller percentage of follow-up questions could be processed correctly. Wolfe (1975) reports further disappointing experienceswith LSNLIS. Wolfe used the system to produce questions from a programming manual. He claims that 40% of the parsed sentenceswere assigned incorrect structures and that the program ran occasionally for over one hour before terminating unsuccessfully. All of this sounds quite discouraging, especially when one takes into account that LSNLIS is widely regarded as one of the most powerful systems, as one of few systems with potential serious users, and as the only system so far whose capacity to meet user needs has been assessed.)

It is tempting to suspect that the unimpressive performance of NLPS in practical applications is not entirely due to the complexities of natural language but, at least in part, to basic theoretical and methodological inadequacies in many of the presently fashionable approaches. This, at any rate, is the thesis of this paper. In the following, I shall outline an alternative view by contrasting it with two widely shared and, as I shall try to show, wrongheaded methodological assumptions, A1 and A2. The alternative view amounts to a rejection of both assumptions.

A1: Syntax is relatively unimportant. What little syntactic processing has to be done, is to be done under the control of semantic routines. A2: Natural language processing requires a single, homogeneous, "deep", language-free medium of representation.

It should be noted that I do not claim that everybody, or even anybody, subscribes to these assumptions exactly as stated but only that they are part of the "folklore" of the field, i.e. that many workers embrace views very similar to A1 and A2. A1 may be partly due to an overreactionto Chomsky's insistence that syntax is more important than semantics and can be pursued without appeal to semantic considerations. (Chomsky 1965, p. 75). It seems clear, on the contrary, that if syntactic constituents were not chosen to correspond closely to carriers of meaning, then syntax would be little more than a description of useless structure. But, while I agree with the proponents of A1 that syntax cannot stand on its own, I do not take the ultimate dependence of syntax on semantics as showing that the latter is sufficient by itself or that the former is unimportant.

It is precisely the dependence of syntax on semantics, i.e. the fact that appropiately chosen syntactic structures embody some semantic information, that makes a strong syntax crucial for language processing. It also seems quite clear that human understanding of text is greatly facilitated by syntactic preprocessing, although this is denied by some proponents of A1. For example, Schank (1975, p. 12) holds that syntax plays practically no role at all in understanding. This view can be readily refuted by noticing the difficulties people have when presented with text from which some syntactic redundancies have been removed. If if if if syntax is unimportant, semantics is primary, syntax does not facilitate understanding, they should be conflated, Schank is right, only if A1 is correct. Or is it?

To insist on a strong syntactic parse is not to claim, as does Chomsky, that people complete the parse before initiating any semantic routines, but only that categorizations which result from exploiting syntactic redundancies make the task of semantic interpretation much easier. The issue is partly a matter of trade-offs: if the semantics is very strong, one can get by with a relatively weak syntax, and, conversely, if the syntax is strong, then a relatively shallow semantics may suffice. Since syntax is better understood than semantics, a strong syntactic component seems desirable.

A prominent feature of programs based on A1 is their excessively complicated control structure. This makes such programs unintelligible to almost everybody and prevents their modular extension. Their complex structure is presumably due to the fact that the syntactic expertise is distributed throughout the system. There is some linguistic evidence for the desirability of integrating syntax and semantics, e.g. semantic information is often needed to remove structural ambiguities. But what the linguistic evidence shows is only that the syntactic parse must be capable to be occasionally affected by semantic routines, and not that it is desirable to integrate syntax and semantics to the point of conflating them. In the interests of improving modularity, it seems indicated to keep syntax and semantics as neatly separate as possible. A modular system should be designed in such a way that the question of how the program tackles a linguistic task may be taken to be settled, so that the programmer can concentrate on the content of semantic rules and need not be concerned with control issues or with the problem of having to identify the syntactic constituents to which the rules are relevant.

In the program described below, modularization is achieved by making the control structure largely transparent, i.e. by allowing it to reside in the parser. Put slightly differently, the phrase structure tree produced by the parser for a given sentence serves as the control structure which governs the semantic interpretation of that sentence. However, to eliminate uninterpretable parses,the syntactic component must be able to communicate with the semantic component. The extent to which this interaction disturbs the highly modular picture just sketched remains to be determined. Very little is known at present about how and to what extent syntax and semantics should cooperate. The system to be described is meant, among other things, to help study this question.

To summarize, I propose to replace A1 by A1':
A1': In order to make NLPS modular, the syntactic and semantic components should be kept largely distinct while allowing for some interleaving. A strong syntactic parse is useful because syntactic structure contains much semantic information.

According to A2, which is a basic tenet of many theories in AI and in cognitive psychology, the understanding process involves the extraction of abstract semantic features from the surface sentence. The former are then assigned to slots of pre-existing frame-like patterns. Semantic theories based on A2 are essentially ways of translating from a natural language into a "conceptual", internal "language of thought". This view of understanding is apparently confirmed by a large number of psychological experiments. For example, Anderson (1971) has found that people can often recall learned information on the basis of a cue that paraphrases the item studied. This would seem to suggest that meanings are stored in a conceptual medium and that people remember abstract meanings and not properties of lexical items : if the paraphrased cue is encoded into the conceptual representation, then it will match the representation of the studied information, given that the paraphrase is indeed meaning-preserving.

Such experiments do not, of course, say anything about any particular conceptual representation. They do not even confirm that there is some such representation unless one makes the further assumption that the learned material is recalled as easily with the paraphrased cues as with the literal cues. However, the latter assumption is empirically wrong: although the learned material is often recalled with paraphrased cues, it is recalled more often with literal ones. This fact, which has been established by Hayes-Roth and Hayes-Roth (1977) through an evaluation of many psychological experiments, undermines A2 and similar views: the postulation of a single, conceptual representation cannot account for the fact that people often employ more superficial representations.

Conceptual representations are often described as offering various kinds of efficiencies. Schank (1974, p. 10) among others, claims that conceptual representations facilitate inference making. In his system, inference rules are associated with primitive acts. Thus, the rules for a given act have to be written only once and need not be repeated for each word whose definition refers to that act. Moreover, attaching rules to low-level conceptual primitives makes inference more efficient, or so it is claimed, because the patterns occasioning inference are fully explicit in the representation.

However, it is patently false that decomposing all concepts into their primitive components increases the efficiency of inferencing. When everything is translated into the lowest level, then drawing inferences requires always that very large data structures are searched and matched. It would seem much more reasonable to initiate matches on lower-level structures only if more "superficial" matches, i.e. matches between high-level concepts or lexemes, fail. This precept is simply an instance of the eminently useful "heuristic": "Don't scratch where it doesn't itch".

To avoid scratching where it doesn't itch, i.e. to avoid drawing inferences always in terms of low-level primitives, is not just conducive to greater efficiency. The insistence on a single, primitive representation may well make inferencing programs on larger data structures entirely infeasible. This is so because symbolic pattern matching has the clique, subgraph isomorphism, and maximum subgraph problems as specific instances. The latter problems are all known to be NP-complete, i.e. their worst case times are at least exponential in the number of nodes. Since graphs representing the meanings of sentences or entire stories in conceptual representation systems would eventually, in non-toy applications, have to become quite large, pattern-matching and inference making would be impossible because of combinatorial explosion. Since inferencing is essential to natural language processing, and is done by pattern matching, it would seem that NLPS must deploy more superficial, higher-level representations. But paraphrasing and related linguistic tasks seem to require deep representations.

To summarize, I propose to replace A2 by A2':
A2': Adequate NLPS should be capable to deploy several levels of semantic and world knowledge representation. Inference should be done on "surfacy", high-level structures whenever possible.

In the last part of this paper I shall describe a program that is consistent with the alternative view characterized by A1' and A2'. However, the program will take only a few preliminary steps towards implementing the alternative view. The present, initial phase of implementation has two main goals: First, to create a modular programming environment to explore the usefulness of different representational media and, second, to study the trade-off between syntax and semantics, i.e. to find out how much mileage can be gotten out of a "surfacy" semantics, given a fairly strong syntax. These goals are realized by writing an almost fully transparent parser that generates context-sensitive parses and executes modular, production-like semantic rules. In this way some of the complexity of natural language processing is hidden away in the parser which functions essentially as the control structure for the entire system.

For practical reasons, I have tackled the problem of translating restricted sublanguages of a natural language into corresponding sublanguages of another natural language. The concept of a sublanguage is intuitively quite clear: the "language" of car repair manuals, of stock market reports, of programming manuals etc. would count as instances. For many such sublanguages it seems to be the case that the differences between sublanguages of a given natural language are greater than the differences between corresponding sublanguages in different natural languages. However, since I do not know how to define "sublanguage" precisely, no systematic use will be made of this concept. It serves only as a reminder that there are reasonably large, practically important, and relatively well-knit fragments of languages. Such fragments are characterized by powerful restrictions which can be exploited so as to make some of the tough problems of language processing quite a bit easier. In short, the program is not meant to be a computational theory of understanding but a tool to facilitate the processing of restricted but extendible sublanguages.

A PRELIMINARY IMPLEMENTATION

The parser chosen is similar to Earley's (1970) algorithm. Like the latter, it is an entirely general context free parser that works in time n^3 and space n^2 for ambiguous grammars. It differs from Earley's parser, however, in that it allows partially constructed phrase structure trees to be modified by the results of executing functions which inspect the immediate context of syntactic constituents. Its output is a phrase marker, or, if the analyzed sentence is structurally ambiguous, a list of several phrase markers. The latter are trees constructed by means of context free rules which are similar to the base rules in transformational grammars. There are, unfortunately, many aspects of a natural language that cannot be handled by means of an essentially context free grammar and restricted context sensitive tests.

To express various types of agreements, for example agreement in number (singular, plural) between auxiliaries or present tense verbs and nouns, or agreement in gender (male, female) between nouns and adjectives or between nouns and past tense verbs, we need rules which can operate on and modify entire trees. Agreements and similar grammatical constraints cannot easily be expressed by phrase structure rules because the constituents which should agree can be arbitrarily distant from one another. To handle such problems, the syntactic component contains transformational rules which can move subtrees around, delete and insert subtrees, and apply tests to entire trees.

The parser combines the context free and transformational rules as well as semantic rules in the following manner. Each grammatical rule consists of three parts. The first part is a context free rule in Chomsky normal form, i.e. a rule with at most two nonterminal symbols on its right hand side. This is an inessential restriction which is not required by the parser but which simplifies some of the tree modifying functions by guaranteeing that all trees are binary. The second part of a grammatical rule is a program that acts as a transformational or a semantic rule. For example, such a rule might copy relevant features from a controlling noun phrase onto a verb to achieve agreement. Some examples of semantic rules will be given below. The third part is a program that acts as a limited context sensitive test whose outcome may affect the parse path taken next.

Controlled by the context free rules and the context sensitive tests, the parser constructs a tree whose interior nodes are occupied by programs corresponding to transformational and semantic rules, and whose leaf nodes are occupied by the words in the surface string. The surface words are associated with dictionary entries and have various "semantic markers" such as AGENT, RECIPIENT, PHYSOBJ etc. attached to them. The processing of a sentence amounts to a postorder traversal of the parse tree. As can be seen from this very rough description, control issues involved in processing natural language text are indeed taken care of by the syntactic component. The desirability of this strategy has been argued in the previous section.

The current implementation, written in UT-LISP, translates a restricted sublanguage of English into French. There are many things it cannot do well and many more that it cannot do at all. For example, there is no difficulty in handling conjunctions consisting of complete sentences or complete noun groups as in "I love the girl and she loves the dog" and "The boy and the girl love the dog". But the program cannot handle incomplete conjunctions, i.e. those in which the connective occurs before the end of a conjunct is encountered, as in "I walk with and play with my dog". Adverbial phrases cannot be handled by the current version, and adverbs are treated as follows. Adverbs are classified as either sentence or predicate modifiers. A sentence modifying adverb Adv is one for which it makes sense to prefix a sentence S by "It is Adv-ly true that S". Thus, "Frequently he laughs", "He frequently laughs" and "He laughs frequently" would all be translated as having the form "Frequently he laughs". All other adverbs go after the modified verb in the translation. This approach simplifies the generation of French sentences but it makes the translations occasionally sound a bit stilted.

Structural ambiguities caused by a word occurring in several grammatical categories pose no problems. Word sense ambiguities are sometimes resolved by weak and superficial semantic rules as follows. Ambiguous words have disambiguating functions on their property lists. These functions pick semantic markers from the other words in a sentence and try to match a list of such markers against simple template patterns. For example, the lists of markers derived from "The flight takes four hours", "Joe takes the book", "Joe takes a chance" would match, respectively, the patterns (PROCESS takes TIME), (HUMAN takes PHYSOBJ), and (HUMAN takes ABSTROBJ).

SUMMARY

It has been argued that some basic assumptions underlying many current NLPS are linguistically or psychologically ill-founded. One of these assumptions, i.e. that syntactic processing should essentially be done only under the control of semantics, has been conjectured to lead to very complex programs. An alternative approach, based on a fairly strong separation of syntax and semantics, has been advocated. The alternative approach attempts to render the control structure of NLPS transparent by putting it into the syntactic component. Finally, a small, modular system, written in UT-LISP, that combines a context free parser with transformational and semantic rules, has been described.

REFERENCES

Anderson, R.C. (1971). Encoding processes in the storage and retrieval of sentences, J. Exp. Psych., 2, 338-340.

Chomsky, N. (1965). Aspects of the Theory of Syntax. MIT Press, Cambridge Massachusetts.

Chomsky, N., and H. Lasnik (1977). Filters and control. Linguistic Inquiry, 8, 425-504.

Earley, V.J. (1970). An efficient context-free parsing algorithm. Commum. ACM, 13, 94-102.

Hayes-Roth, B., and F. Hayes-Roth. (1977). The prominence of lexical information in memory representations of meaning. J. Verbal Learning and Verbal Behavior, 16, 119-136.

Petrick, S.R. (1977). In A. Zampolli (Ed.), Linguistic Structure Processing. North-Holland, Amsterdam, pp. 325-342.

Schank, R. (1974). Causality and Reasoning. Memoranda from the instituto per gli Studi semantici and cognitivi. Castagnola, Switzerland.

Schank, R. (1975). The conceptual approach to language processing. In R. Schank (Ed.), Conceptual Information Processing. North-Holland, Amsterdam, pp. 5-21.

Wolfe, J.H. (1975). An Aid to Independent Study Through Automatic Question Generation. NPRDC TR 76-18, AD-AD17 059, Navy Personnel Research and Development Center, San Diego, California.

Woods, W.A., R.M. Kaplan, and B. Nash-Webber (1972). Final Report. The Lunar Sciences Natural Language Information System: Final Report. BBN 2378, Bolt Beranek and Newman, Cambridge, Massachusetts.

```
    (init)
* * * (G1  ENTERED )

TYPE A SENTENCE !
* ? computers can't translate,
LES ORDINATEURS NE PEUVENT PAS TRADUIRE .

* ? i believe that he thinks that computers can't translate.
JE CROIS QU' IL PENSE QUE LES ORDINATEURS NE PEUVENT PAS TRADUIRE .

* ? he distrusted the girls.
IL SE MEFIAIT DES FILLES .

* ? he has distrusted the girls.
IL S' EST MEFIE DES FILLES .

* ? we have not remembered the girls.
NOUS NE NOUS SOMMES PAS SOUVENUS DES FILLES .

* ? the girls who are beautiful  have not remembered the old man.
LES FILLES QUI SONT BELLES NE SE SONT PAS SOUVENUES DU VIEUX HOMME .

* ? the girl whose house is big has not remembered the brown dog.
LA FILLE DONT LA MAISON EST GRANDE NE S' EST PAS SOUVENUE DU CHIEN BRUN .

* ? the  ugly old man by the red house doesn't have any friends.
L' HOMME LAID ET VIEUX A COTE DE LA MAISON ROUGE N' A PAS D' AMIS .

* ? the ugly boy and the very pretty girl are very happy.
LE GARCON LAID ET LA TRES JOLIE FILLE SONT TRES HEUREUX .

* ? the big happy red beautiful girl loves the ugly dog.
LA GRANDE FILLE HEUREUSE ET ROUGE ET BELLE AIME LE CHIEN LAID .

* ? men are big boys.
LES HOMMES SONT DES GRANDS GARCONS .

* ? ham is expensive.
LE JAMBON EST COUTEUX .

* ? i bought ham.
J' ACHETAIS DU JAMBON .

* ? men think that women are beautiful.
LES HOMMES PENSENT QUE LES FEMMES SONT BELLES .

* ? women don't think that men are beautiful.
LES FEMMES NE PENSENT PAS QUE LES HOMMES SONT BEAUX .

* ? the boy whose dog is happy loves a girl whose house is white.
LE GARCON DONT LE CHIEN EST HEUREUX AIME UNE FILLE DONT LA MAISON EST BLANCHE .

* ? he worked in the white house for the old man.
IL TRAVAILLAIT DANS LA MAISON BLANCHE POUR LE VIEUX HOMME .

* ? the boy whose house is red and the pretty girl love   dogs.
LE GARCON DONT LA MAISON EST ROUGE ET LA JOLIE FILLE AIMENT DES CHIENS .

* ? the boy whose girl whose dog was brown  is pretty  won't come.
LE GARCON DONT LA FILLE DONT LE CHIEN ETAIT BRUN EST JOLIE NE VIENDRA PAS .

* ? i have had ham.
J' AI EU DU JAMBON .

* ? i had had ham.
J' AVAIS EU DU JAMBON .

* ? she won't buy the house.
ELLE N' ACHETERA PAS LA MAISON .

* ? she won't buy houses.
ELLE N' ACHETERA PAS DE MAISONS .
```

MAN-COMPUTER COMMUNICATIONS BY VOICE

C. Y. Suen

Department of Computer Science
Concordia University
Montreal, Canada H3G 1M8

ABSTRACT

Voice is one of the most important media for man-to-man communication. With the advent of modern computer technology and continuous progress in the speech sciences and audiology, man-to-computer communication and vice versa by voice has become a reality. This paper reviews the metholodologies used in computer analysis and synthesis of speech signals, computational linguistics and the human factors involved. Machine performance and various applications will be discussed.

KEYWORDS

Man-computer communications; voice synthesis; computer voice; voice input; speech communication.

INTRODUCTION

Voice is one of the most important media in human communication. It is produced through the dynamic movements of the tongue, lips, jaw, velum, vocal cords and air passage through the vocal and nasal tracts. Speech is understood by human beings through an coustic analysis of the speech sounds and a linguistic analysis of the speech units, e.g. phonemes, syllables, words, and their combinations. Man-computer communication by voice is a most fascinating and challenging area of modern science and technology. It has a large number of potential applications for commercial, scientific, and industrial uses, e.g. voice answer-back systems, stock, bank, and telephone enquiries, airline and hotel reservations, computer assisted instruction, speech output for the blind, the deaf, and the aphasic, direct voice to text applications, etc. (Suen, 1979; Suen & de Mori, 1981) This paper describes the method used for two way communication between man and the computer, first computer production of voice to man and then computer recognition and understanding of human voice.

COMPUTER PRODUCTION OF VOICE

The computer voice is produced in two stages. In the first stage, the incoming text is analyzed and transcribed into phonetic strings of symbols. In the second stage, voice is produced through a piece of hardware (speech synthesizer) under the control of a computer, as depicted in Fig. 1.

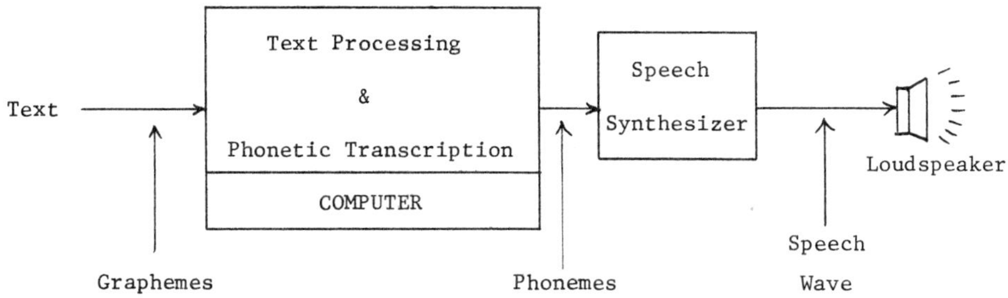

Fig. 1 Text-to-speech by computer.

The most difficult problem encountered in synthesizing English sounds is the inconsistency of English spelling (Suen & Jennings, 1978). Most current systems are based on letter-to-sound rules (see e.g. Elovitz and others, 1976; McIlroy, 1974) which give an approximate pronunciation of English words. A fairly intelligible voice has been obtained. However, accurate sounds can only be generated with reference to linguistic aspects other than spelling, e.g. syllable construction, syntax, semantics, and other prosodic features.

At Concordia University, we make use of both letter-to-sound rules and syllabic rules. The letter rules are organized by letter so that every letter of the alphabet has anywhere from 3 to 50 rules that operate on a grapheme and translate it into a distinct phoneme. The sequencing of phonemes is taken into consideration wherever possible, so that the number of rules associated with each grapheme is approximately correlated with the number of phonemically different environments for each letter. Thus, the number of rules necessary to pronounce each grapheme reflects the number of ways in which that grapheme can be pronounced. Examples of phonetic transcriptions of a few sentences by our computer are shown in Fig. 2.

Transcription of the sentence "Computers are useful":

Computers	(C) = /K/	(O) = /AA/	(M) = /M/
	(P) = /P/	(U) = /Y UW/	(T) = /T/
	(ER) = /ER/		
are	/AA R/		
useful	(U) = /Y UW/	(S) = /S/	(EFUL) = /F UH L/

Transcription of the sentence "Great people think alike".

Great	/G R EY T/		
people	/P IY P/	(L) = /L/	(E) = / /
think	(TH) = /TH/	(I) = /IH/	(NK) = /NX K/
alike	(AL) = /AX L/ (E) = / /	(I) = /AY/	(K) = /K/

Fig. 2 Examples of sentences transcribed by the computer.

In addition to the letter rules, we have syllabic rules. The first problem that we encountered when we began to work on stress came with the realization that the computer had no concept of syllable in its memory, and this being the case, there was no way of assigning stress because stress is based on both syllable make-up and the number of syllables in a word. Consequently, we operationally defined syllable number as the number of vowels or vowel clusters in a word. First the computer looks at the string of letters that makes up each word and assigns each letter either consonant or vowel status. Then, all the clustered consonants or vowels are grouped together to define the syllables.

The way speech is synthesized can be divided into two main categories:

1) Concatenation of phonemes, diphones or syllables(Suen and others, 1976, 1979)
2) Electronic networks and frequency generators (Linggard 1979).

The first category is the simplest one. It concatenates stored phonemes, diphones or syllables at the command of the computer. This method was used in the Spellex system (Suen and others 1976a, 1976b, 1976c) for the synthesis of the sounds of all letters and other symbols. It has also been used in a commercial answer-back system (Dixon and Maxey 1968; Allen 1977; Robinson 1979). The use of electronic networks is by far the most satisfactory and flexible method. Using available commercial synthesizers, synthesized voice has been produced at Concordia University and elsewhere.

MAN-COMPUTER COMMUNICATION BY VOICE

It would be very convenient to communicate with a computer which understands everything we say. Presently man-computer communication by voice may be divided into the following topics (Dixon and Martin 1979; Lea 1980; Newell and others,1973):

1) Voice recognition - recognizing the phonemes contained in a speech signal
2) Speaker recognition - recognizing the identity of a speaker
3) Speech comprehension - understanding the meaning of a spoken message.

To do the above,the computer must first of all be taught to do the following (de Mori 1979):

1) Recognize the phonemes contained in the speech signal
2) Assemble the phonemes to form syllables and words
3) Know the meanings of the words and their relations.

To date, we have had only limited success in this area. For a vocabulary below 30 words, current commercial and prototype systems in research laboratories can recognize almost perfectly all the words spoken by a selected group of trained speakers. However, the recognition rate deteriorates as the size of the vocabulary and the number of speakers increase. Speech is understood only when the sentences are constructed in a restricted manner using a modest vocabulary.

Most systems which recognize speech are characterized by the set-up shown in Fig.3. Speech is processed by the computer after an analog to digital conversion. Typical sampling rates are 8 to 20 kHz. A Fourier transform is commonly used to extract frequency information, energy concentration and distribution of the signal. Similarly, a filter bank is also used to extract the formants, frequency distribution, as well as time and energy information. The number of filters used defines the resolution of the spectrum with typical systems using from 20 to 45 filters. Phonemes are identified by comparing the extracted parameters with those obtained from a training process and stored in the computer. In speech comprehension systems, a linguistic analysis of the phoneme string is performed.

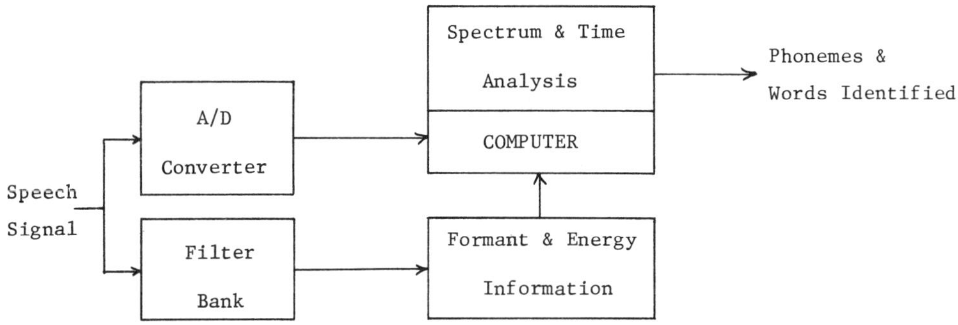

Fig. 3 Computer recognition of voice.

SPEECH AIDS FOR THE VISUALLY-HANDICAPPED

One area of man-computer communications by voice which of of special interest is in the creation of speech synthesizing aids for the handicapped. In this area, we have, over the past 10 years, developed a number of speech aids for the blind.

Apart from working on a reading machine for the blind, we have also developed a number of speech aids for the blind, e.g. the talking typewriter (Suen and others, 1976),the talking calculator (Suen and others, 1976b), and talking punch-card reader (Swail and others, 1977). These machines make use of a small digital speech generator for spelling invented by the author and his collaborator (Beddoes and Suen, 1979). Another device, the talking keypunch machine, has been constructed at Concordia University.

ACKNOWLEDGEMENT

This research was supported by a grant from the Department of Education of Quebec. The assistance and comments of Miss Tammi Rossman are gratefully acknowledged.

REFERENCES

Special Issue on Man-Machine Communication by Voice, Proc. IEEE, vol. 64, April 1976.

Suen, C. Y., "Recent advances in computer vision and computerized aids for the visually-handicapped," Invited paper, Proc. Computers in Ophthalmology, 1979.

Suen, C. Y. and de Mori, R. (1981). Computer Analysis and Perception of Visual and Auditory Signals, in press, CRC Press, Inc.

Suen, C. Y. and G. Jennings (1978). Computers and People, 27, 16-17, 25.

Elovitz, H. S., R. Johnson, A. McHugh and J. E. Shore (1976). IEEE Trans. Acous., Speech, and Signal Processing, 24, 446-459.

McIlroy, M. D. (1974), "Synthetic English speech by rule," Tech. Report, Bell Tel. Labs. Inc., Murray Hill.

Beddoes, M. P. and C. Y. Suen (1976). IEEE Trans. Biomedical Engineering, 23, 73-75.

Beddoes, M. P. and C. Y. Suen (1979). Generator for Spelled Speech and for Speech, Canadian Patent No. 261,200.

Dixon, N. R. and H. D. Maxey (1968). IEEE Trans. Audio and Electroacoustics, 16, 40-50.

Linggard, R. (1979). Computer & Digital Techniques.

Suen, C. Y., M. P. Beddoes, "Spellex - A system of speech aids for the visually-handicapped in vocational training," Proc. Conf. Systems & Devices for the Disabled, 1976a.

Suen, C. Y. and M. P. Beddoes (1976b). "Talking calculator of the Spellex system," Proc. 11th Int. Conf. Medical & Biological Engng.

Suen, C. Y., M. P. Beddoes, and J. C. Swail (1976c)."The Spellex system of speech aids for the blind in computer applications," Proc. AFIPS Conf., 45, 217-220.

Allen, J., "Synthesis of speech from unrestricted text," in A. Zampolli (ed.), Linguistic Structures Processing, North-Holland Pub. Co., Amsterdam, 1-30, 1977.

Robinson, A. L. (1979). IEEE Trans. Prof. Comm., 22, 159-165.

Sherwood, B. (1979). IEEE Spectrum, 16, 18-25.

Dixon, N. R. and T. B. Martin (Eds.) (1979). Automatic Speech and Speaker Recognition, IEEE Press.

Lea, W. A. (1980). Trends in Speech Recognition, Prentice-Hall.

de Mori, R. (1979). Signal Processing, 1, 95-124.

Kaplan, G., R. Reddy and Y. Kato (1980). IEEE Spectrum, 17, 22-29.

Newell, A. and others, (1973). Speech Understanding Systems, North-Holland Publishing Co., Amsterdam.

Rosenberg, A. E. and C. E. Schmidt (1979). Bell Syst. Tech. J., 58, 1797-1823.

Levinson, S. E. and K. L. Shipley (1980), Bell Syst. Tech. J., 59, 119-137.

Songo, D. C., S. I. Allen, P. S. Plexico, R. A. Mortford (1980). IEEE Spectrum, 17, 34-38.

Suen, C. Y., "A talking keypunch for blind programmers," in press, J. Visual Impairment & Blindness, 1981.

Swail, J. C. and C. Y. Suen, ACM SIGCAPH Newsletter, 26, 11-14, 1977.

Radhakrishnan, T. and C. Labrador, "On the design of an intelligent terminal for voice output in programming," Proc. 3rd Biennial Conf. Canadian Soc. Computational Studies of Intelligence, 327-331, 1980.

Webb, E., K. Menon and C. Y. Suen (1979). "A microcomputer based speech recognition system," Proc. 6th Man-Computer Comm. Conf., 41-48.

PRAGMATICS OF INFORMATION AND MAN-MACHINE COOPERATION

Ingbert Kupka

Fachbereich Informatik, Universität Hamburg
Schlueterstr. 70, D-2000 Hamburg 13

ABSTRACT

An approach towards a pragmatic definition of information is presented which allows a consistent view of transmitting, storing and processing of information. Based on these fundamentals communication and cooperation between man and machine, the communication behaviour of machines and the computer's role in social systems are discussed.

KEYWORDS

Information; context; data; communication; communication behaviour; dialog; man-machine communication; man-machine cooperation.

THE NATURE OF INFORMATION

Information and its various kinds of use play an important role in the human life of today: 'Information is bought, sold, stored, traded, exchanged and consumed in economic terms. It is treated both as a product and a service, as a commodity to be used in the process of attaining human goals and it is taking on the fundamental value aspects of money.' (Artandi, 1973, p. 242.) High-level techniques for transmitting, storing and processing of information have been developed in the last four decades. The fundamentals of information, however, are still unknown and the same holds for its impact on human work, social structures and on the cultural evolution. One cannot neglect the need for a definition of information which is more sufficient than that of the Mathematical Theory of Communication (Shannon, Weaver, 1949).

Information Models of Reality

Intuitively information is understood as knowledge about facts. Thus the structure of information seems to be at least as complicated as that of reality. On the other hand, whenever information is transmitted or processed, it is to be understood with respect to a special small part of reality. That means, when treating information about facts, one can assume that it is related to a limited part of the universe of facts. In order to exclude philosophical problems and subjects belonging to the physical sciences we presuppose that the facts to which some information refers only belong to a *model* of reality, which we call an *information model of reality*. The notion *fact* shall be applied to any detail of the model and it is assumed that the model contains concrete facts corresponding to physical things as well as abstract facts like notions, ideas or concepts. This is formally expressed by the demand that there exists a partial order 'is more general than' for facts called *abstraction relation*, symbolically

$$A \sqsubset B$$ for ' A is more general than B ' or ' B is more special than A'.

$A \sqsubseteq B$ means $A \sqsubset B$ or $A = B$.

A Semiotic Definition of Information

With respect to such an information model of reality and extending ideas of Klaus (1969) and Morris (1946) it is possible to define a concept of information. Klaus defines information as *equivalence class of signals* with the same meaning. Both, the signal as information carrier and its meaning are considered as facts of our information model.

Any collection of associations between facts as signals and facts as meanings is called a *context* if more general signals always are associated with the same or a more general meaning. We write

$$S \overset{C}{:} X$$ for ' S means X in the context C'.

This ternary relation is called *semiotic relation*.

We state: Let X be a fact.
The class of all pairs of facts U,V such that U means the fact X in the context V is called the information I(X) about the fact X, i.e.

$$I(X) = \{ \langle U,V \rangle ; \; U \overset{V}{:} X \}.$$

U is called the information carrier for I(X) with respect to the context V, i.e., U *represents* I(X). Thus informations are defined as equivalence classes like numbers are in mathematics. The abstraction relation induces a corresponding relation on informations which is denoted by the same symbol:

$$I(X) \sqsubseteq I(Y) \quad \text{if and only if} \quad X \sqsubseteq Y.$$

I(X) is called a *subinformation* of I(Y) and I(Y) is called a *superinformation* of I(X) in this case. Moving from I(X) to I(Y) is a *specialization* whereas moving from I(Y) to I(X) is an *abstraction*.

The abstraction relation determines the *logical aspect* of information. It makes no sense to distinguish in general between true and false information: Only with respect to a given fact X of a certain information model of reality an information I can be a correct information or a (correct) subinformation about this fact ($I = I(X)$ or $I \sqsubseteq I(X)$) otherwise one may call it an *incorrect information about X*.
Two informations I,J are called *consistent*, symbolically $I \sim J$, if they possess a common specialization (superinformation). If, additionally, there exists a most abstract common specialization, then this information is called the *union* of the informations I and J, symbolically $I \sqcup J$. Analogously, if I and J possess a most special common abstraction (subinformation), this is called their *intersection* and denoted by $I \sqcap J$.
In practice, the main operations to be performed on information are transmitting, storing and processing.

Transmitting Information

Transmission of information usually means the transfer of an information carrier, called message, representing the information with respect to a context which is common to sender and receiver. The Mathematical Theory of Communication refers to messages only. The actual message is considered as one selected from a set of possible messages. Information then is understood as a measure for the freedom of choice. It is important that the understanding of information introduced above can also be applied to the Mathematical Theory of Communication. If the alternatives of messages are substituted by similar alternatives of pairwise inconsistent informations, then a *measure of information* can be introduced as a mapping m from the set of informations into the set of real numbers such that

$$I \sqsubseteq J \quad \text{implies} \quad m(I) \leq m(J).$$

One can show that the functional dependence from the number of alternatives must be a logarithmic one.

Storing Information

For a general view of *information storing* it is assumed that the informations in question fulfill the following conditions:

(i) There exists a most abstract information. This corresponds to an empty data base.

(ii) Consistent informations can alwalys be (uniquely) united.

(iii) If I,J are arbitrary informations, then there exists a most special abstraction I' which is consistent with J. I' is denoted by I/J.

Thus I/J = I holds if $I \sim J$. We call I/J the *adaptation* of I by J.
Let I be the whole information stored in a data base and J an actual update information. Then the information of the updated data base is $I/J \sqcup J$.
An inconsistent state of a data base is characterized by the occurrence of inconsistent informations represented in the data base.

Information Processing

There exists no systematic correspondence between the input information and the output information of program-controlled computers in general. Such a correspondence exists, however, in the case of each special program. It can be described by a binary relation, called *information transformation*. It is defined as follows: (For a formal definition see Kupka, 1980).

The problem area defines an information model of reality with facts and corresponding informations for which the abstraction relation is defined. The informations are called *basic informations*. On a meta-level further informations M exist referring to specializations $I \sqsubseteq J$ of basic informations, such that by means of a monotonic *specialization operation* '∘' the information J can be constructed as $J = M \circ I$. The informations of the meta-level are called *meta-informations*. Here is a simple *example:*

basic informations: I = ' X hours and Y miles'; J = '3 hours and 180 miles'
meta-informations: M = { <X, 3>, <Y, 180> }.

Information transformations, denoted by the symbol '→' are now characterized by the property that for all basic informations A,B and all suitable metainformations M

$$A \rightarrow B \quad \text{implies} \quad M \circ A \rightarrow M \circ B.$$

As an example,

'X hours' → '60*X miles' implies '3 hours' → '180 miles'.

In order to show that algorithms can be expressed as information transformations, we present the Euclidean algorithm for the greatest common divisor (gcd):

Example: The meta-informations are composed of pairs <X,m> and <Y,n> with positive integers m, n. The transformation is derived from the following three rules:

gcd of X and Y *where* $X < Y$ → gcd of X and Y-X .
gcd of X and Y *where* $X > Y$ → gcd of X-Y and Y .
gcd of X and X → X .

A basic information 'gcd of m and n' is a specialization of 'gcd of X and Y *where* $X < Y$' if and only if $m < n$ holds.

Thus each step of a computing process is obtained by specializing one of the defining rules. *Programming* can be understood as building problem-specific information models and then constructing general transformation rules in this model.

PRAGMATIC ASPECTS OF INFORMATION

The interrelations between information and processes of the real world, including actions of human beings are mainly determined by the following pragmatic aspects:

- The immateriality of information.
- The necessity of representation.
- The context-dependence.
- The close relation between information and decision.

Immateriality of Information and Necessity of Representation

For every information that shall be stored, transmitted or processed its immateriality and the necessity of representing it are two sides of the same phenomenon.
If an a special application the fundamental problem of *copying information representations* is sufficiently solved - this has been done by biological evoluation processes as well as by human inventions, for instance, - then a *distribution* of information by distributing representations is possible. This facilitates access to information but impedes the protection of information and the cancellation of stored information. Unlike material things informations may be possessed as a whole by many owners. Hence a value in an economical sense usually is associated to *new informations* only. In a formal sense an information I is considered a new information with respect to another information J if I is not a subinformation of J.
There are no essential restrictions for *information representations*. Various kinds of them may be used for the same information. One takes full advantage of this by adapting the kind of physical representation to practical needs. Different media are taken for storing information and for high-rate transmission, for example. The recent progress of technical development just has led to a variety of possible physical representations with a high degree of miniaturization in the case of storing and high communication rates in the case of information transfer.
The structural characteristics of information can be strictly separated from the physical characteristics of their representations by the mediate constructions of abstract, i.e. *immaterial representations*. *Messages* in the case of information transmission and *data* in the case of storing or processing information are of this kind. They carry information and need themselves physical representations. Simple examples are sentences in a natural language or database entries. A useful *construction principle for data* as information carries is to represent composed informations by analogously composed data.

Context-Dependence

The essential property of a context is its role as a precondition of any representation. Whenever an information is to be distributed or transmitted, then the context already must have been distributed. With respect to a suitable context any information can be described completely and exactly. If a context is to be described under the same conditions, then a further adequate context is needed.
A *natural language* is a context which is widely spread and suitable for representing information occurring in a social community. Growing knowledge in special branches of science and techniques has led to *artificial technical languages*, each of which defines a context in a special area. For highly organized social systems with today's information technology the problem how to find a commonly understood context is unsolved, be it a language or a model.

Information and Decision

Information always results from decisions: Constructing concrete information representations as by talking, writing or modelling means a step-wise selection of alternatives. Thus decision processes are the *origin* of data or messages. Vice versa, the *purpose* of information is its use for decisions in real processes. If there is a class of processes which can be considered as repetitive runs of one abstract process, and if their dependence on the decision information is an algorithmic one, then a simulation of this class of processes by an information processing machine or automatic control of the real processes is possible. *Program- controlled information processing*, however, is much more general: All computation processes are repetitive runs of one universal computation process, which can be controlled by special programs as controlling information. Hence programs or information transformations are also informations.
If the logarithmic measure for informations is based on the number 2 then it indicates how many binary decisions correspond with an information.

COMMUNICATION AND COOPERATION BETWEEN MAN AND MACHINE

The class of information transformation which can be performed automatically, i.e. by applying a universal process controlled by a special program, is exactly determined by the notions of *computability* and *algorithm*. Since the algorithmic view refers to the internal functioning of the information processing machine only, it excludes the *communication behaviour* of the machine as well as that of its user. Application-oriented approaches of computer science like heuristic programming, non-procedural problem solving or interactive programming (see Kupka, Wilsing, 1980) and even organizational concepts of software production in practice require some kind of *trans-algorithmic view* which includes the pragmatic role of information, the form of communication and the kind of cooperation between all participants.
We analyze the main problems of this view by studying the phenomenon of *communication* in connection with machines and the more general subject of *man-machine cooperation*. Finally, some aspects of the role of *information and communication in the social environment* are discussed.

Communication Schemes

One of the earliest approaches to a communication theory as part of computer science is the thesis of Petri (1962) about 'Communication with Automata'. *Petri nets* as originally introduced there and nets of channels and agencies as a special interpretation of them (see Oberquelle, 1979) are excellent tools for modelling communication.

Figure 1 shows a simple net representation of communication. By refinement we get more complex representations with emphasis on details of special interest. We associate with each participant of the communication an *environment*, a *memory* containing stored information, a *communication channel* for sending and receiving messages and a context determining the informations represented by the messages.

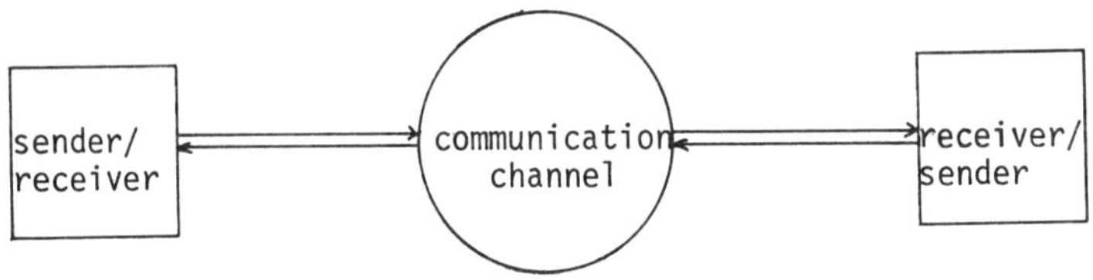

Fig. 1: Simple scheme of communication

Each participant is called *actor*. All his actions together, as illustrated in figure 2 are called the *communication behaviour* of the actor.

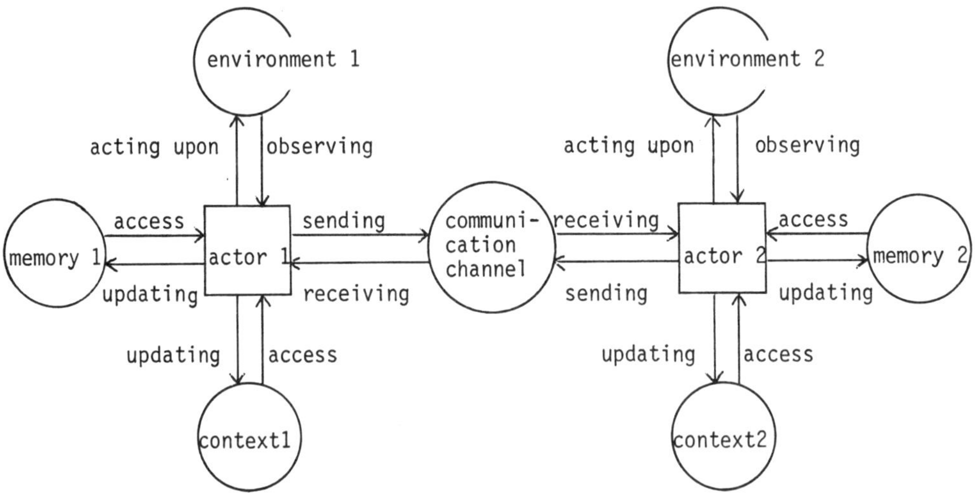

Fig. 2: Communication scheme for two actors

We consider communication as an *exchange of information between two or more actors.* Whether an exchange of messages is an exchange of information depends on the partial *coincidence of the contexts.* An information processing machine can be considered as an actor whose context can be described completely and exactly and whose communication behaviour, involving context changes, is algorithmic. Thus it is possible that the actual context of the machine is part of the user's context.

Man-Machine Cooperation

In practice communication cannot be separated from its *purpose*. We use the term *cooperation* instead of communication if the communication behaviour depends on some intention from which a *communication intention* can be derived. We discuss, in which sense a machine can have a communication intention.

Let a human actor have in mind an intention with respect to his environment. If in this special case his intended behaviour concerning observations, storing informations and affecting the environment is algorithmic, then he can *delegate* it as a whole or parts of it to a machine. This delegation includes the installation of a communication channel and the definition of an initial memory state, a context and a communication behaviour. A communication intention of the machine can then be derived as part of the human actor's intention. The situation is considerably changed if another human user communicates with the machine or if the original user - say some time later - has lost the primary intention or the relevant part of the context or both. Then a common context must be established again and then the machine works as if it has an own communication intention. The situation becomes still more complicated when the behaviour of the machine is changed by its user now. From a long-term point of view machines get access to increasing amounts of stored data and offer a variety of alternative kinds of behaviour, each represented by an algorithm. The more alternative actions are transferred to the machine, the greater is the complexity of the context for communicating with it and the difficulty to use ist. Man-machine cooperation takes place under participation of many users including those preparing the machine and over a long period of time. In the actual situation of a *dialog* between man and machine the user is confronted with a communication behaviour of the machine which is only partially known to him. An important improvement of the user interface in man-machine dialog (cf. Dehning, Essig, Maass, 1980) can only be obtained if a minimal common context is guaranteed allowing an exchange of information concerning the communication behaviour and the communication intentions of both actors.

Information and Communication in the Social Environment

The great impact of information and communication technology on the structure of social systems is an open secret today. The problem, however, is how to estimate and how to control this impact in the future. An important first step towards a control is to form a *public common understanding* of information and its role in the social community, of *individual and social rights with respect to information,* and of the role which computers really play. We should notice that the computer neither is a tool only with a well-known actual behaviour nor an actor with its own intention and intelligence but a medium which can take over intentions and communication facilities of many human actors.

REFERENCES

Artandi, S. (1973). Information concepts and their utility. *J.Am.Soc.Inf.Sci.,* 242-245.

Dehning,W., H.Essig, and S.Maass (1980). The adaptation of virtual man-computer interfaces to user requirements in dialog. Preliminary report, FB Informatik, Univ. Hamburg

Klaus, G. (1969). *Wörterbuch der Kybernetik.* Fischer, Frankfurt, Hamburg.

Kupka, I. (1980). Van Wijngaarden grammars as a special information processing model. *Math. Found. Comp. Sci.,* Proc.1980, Springer, Berlin, Heidelberg, New York, 387-401.

Kupka,I., and N.Wilsing (1980). *Conversational languages.* Wiley, Chichester.

Morris, Ch.W. (1946). *Signs, language, and behaviour.* Prentice-Hall, Englewood Cliffs, N.J.

Oberquelle,H. (1979) Nets as a tool in teaching and in terminology work. *Proc. of the Adv. Course on General Net Theory of Processes and Systems,* Springer, Berlin, Heidelberg, New York, 481-506.

Petri,C.A. (1962). Kummunikation mit Automaten. *Schriften d. Rhein.-Westf. Inst.f. Instrumentelle Math., 2,* Univ. Bonn.

Shannon, C.E., and W.Weaver (1949). *The mathematical theory of communication.* The University of Illinois Press.

ON THE USER INTERFACE OF COMPUTING SOFTWARE

Robert F. Ling

Department of Mathematical Sciences
Clemson University, Clemson, S.C. 29631

ABSTRACT

In principle, computing softwares are created for the problem-solving users and for the convenience of such users. In practice, software designers often neglect or fall short of this goal. In this article, some common weaknesses in existing software, particularly in the area of user interface, will be discussed. Reasons and possible remedies for the weaknesses are suggested.

KEYWORDS

Software design; man-machine interface; computer language; design philosophy.

INTRODUCTION

During the past quarter of a century, computing technology has advanced at an awesome pace. Computing hardware evolved through several generations in a span of two decades, increasing computing power a millionfold. The most widely used high-level programming languages of today (ALGOL, APL, BASIC, COBOL, FORTRAN, LISP, and PL/I) were all in existence by 1965. The past decade witnessed an explosive growth in the development of super-high-level software (special-purpose, discipline-oriented languages or application packages) built on existing high-level programming languages. For example, the number of publicly available software packages for statistical data analysis has increased from merely a few in 1970 to well over 50 by 1977.

Yet, amidst such powerful and often sophisticated technological advances, some essential aspects of computing, in the area of man-machine interface, seem to have been neglected or have not been able to keep pace. Certain "computerese" today are as primitive and uncivilized (or perhaps just unfriendly) as they were twenty-five years ago, and users today are no less frustrated, confused, or intimidated when they encounter such computerese than they did yesteryear. I shall consider some of the weaknesses in existing computing software, focussing on the area of user interface, and suggest some reasons and possible remedies for the weaknesses.

MAN-MACHINE INTERFACE

By "man-machine interface" or "user interface" I mean any medium, form, or content that pertains to, and is supposed to facilitate, the communication between Man and Machine. In particular, common tools of interface include languages for operating systems, programming languages, general-purpose and special-purpose applications packages, documentations, computer-generated diagnostics and error messages, and so on. Since, in principle, such tools are created for the problem-solving users and for their convenience, I take the position that an interface feature is unsatisfactory or undesirable if one or more of the following descriptions apply:

1. It is considered to be "unfriendly" by a sizeable group of users for whom the software is intended to serve.

2. It is unnecessarily cumbersome to use or its language is unnecessarily cryptic (as opposed to natural, in the man-to-man communication sense).

3. Its structure is unnecessarily rigid and inflexible, not allowing simpler, alternative modes of expression (that are logical or natural) by the user.

4. It is prone to induce human errors, because of its construct, its documentation, or whatever else.

An example of a software language to which all of the above descriptions apply is IBM's JCL (job control language). I know of a commercially available statistical program package whose documentation and usage are so obtuse that the designer of the package appears to be the only person who knows how to use it. The well-known, widely used programming language ANSI FORTRAN, which has not evolved much from its ancestor FORTRAN I (circa 1952), contains numerous unnecessary, inconvenient, and rather arbitrary rules and conventions. For example, one of the most common errors made by a novice FORTRAN programmer is the careless insertion of an extraneous comma in a DO statement, as in

```
DO 10, I = 1, 10
```

The extraneous comma causes no ambiguity, can be easily detected, yet it is treated as a "fatal error" in the attempted execution of a program rather than an optional delimiter in the syntax of the language or a "forgivable error".

The list of examples illustrating weaknesses in user interface in existing software can go on, but there is no need to. The point has been made. I am sure the reader can recall other examples of similar nature in his computing experience. Let us consider the questions "why?" and "what can be done?" regarding the existence and persistence of unfriendly software.

SOME OLD VIEWS THAT ARE DETRIMENTAL TO MAN-MACHINE INTERFACE

User interface problems are to a large extent caused by a combination of philosophical and practical viewpoints toward computing that should not have existed, but did; should have been obsolete, but are not. I shall paraphrase some such views:

1. That a user of a software should merely follow the rules set down by the designer or programmer of the software, and not to question "why this?" or "why not that?" because:

> The designer or the programmer generally knows so much more about computing than the user, he must know what's good for the user.

> Therefore, the user should trust and abide by the designer's judgment, and not question it.
>
> The designer or the programmer is doing the user a favor by creating the software. Since the user can get a task done much more easily or efficiently with the software than without, he should be grateful for what is available. All he has to do is to follow a set of rules prescribed by the designer. Criticisms are uncalled for.
>
> The programmer's job is to implement an algorithm to help Man to specify a task to be performed by Machine and to instruct Machine how the task is to be performed. So long as the programmer can document precisely the instructions for Man and Machine to follow, he is entitled to define the rules, which are to be treated as axioms.

There are numerous other similar views that treat the ordinary user as a second-class or third-class citizen in the world of computer users. A self-proclaimed elite group is thus formed, with memberships belonging to a small group of computer professionals and those users who are receptive or tolerant to the gibberish created by a few of the professionals. User interface in computing software suffers as a result. It is ironic that much of the best softwares today, not only in terms of user interface but also in terms of technical quality, are written by people who were not trained to be a computer scientist nor a computing professional.

2. That computing power should not be "wasted" on software features that are not essential to computation, i.e., features that primarily protect and benefit the "incompetent" users at a cost to the expert users.

Each of the following views has been expressed to me by one or more computer or computing experts, not in jest, but in all seriousness:

> High-level programming languages usurp computing power. Said expert wanted to harness all the raw computing power that is possible and not be burdened by layers of high-level software (including software for the operating system). Thus, in principle, he is opposed to all user interface software, with the exception of machine-language, of course.
>
> Friendly software wastes computing power. If a user is given the option to specify a task in any one of several ways (which makes it convenient for the user), then the computer will necessarily be less efficient in carrying out the task because extra computing time will be used to encode the task specification.
>
> Error detection codes and diagnostic messages waste computing power. Competent users rarely make errors. If and when they do, a simple code such as ERROR RFL123 should suffice. Most of the time a user should be able to figure out what caused the error without an actual text message. Other times, the Manual can be consulted.

What is at issue is the tradeoff of benefits between computational efficiency and user-convenience. The absurdity of the views stated above lies in the fact that some experts think that it is not worthwhile to use a few extra milliseconds of computing time even if it meant a potential saving of minutes or hours of frustration by the user. I expressed my view and sentiments on the issue elsewhere as follows:

To cite a specific example, suppose a program expects a real value to be input from a terminal. If the program is coded to accept the value in a particular floating-point format (i.e., with a specific number of digits before and after the decimal point), then little CPU execution time is used for this input phase of the task. However, the user is then required to remember precisely the form in which the input value must be typed. If the user is allowed to enter the value either as an integer or a floating-point number (in decimal form or in scientific notation), then a more elaborate input routine and additional CPU execution time would be needed for the task. Still more lines of code and more CPU time would be needed if the system is to check the validity of the user-given input value (i.e., whether it is numeric, as is required, and whether the value is within a permissible range so as not to cause overflow or underflow errors in machine arithmetic in subsequent computations) and provide appropriate diagnostics and on-line correction features. ... Of course the extra computation required by an interactive system to accommodate freeform input is negligible (and is unlikely to be detected by the user) and few would argue against freeform input as a waste of computing power. However, there are many other user-convenience features that can be accomplished at the expense of additional computations comparable to that of freeform input, yet many designers are reluctant to "waste" CPU time to make life easier for the user when these other user-convenience features are considered. In my opinion, much "unfriendly" computerese still exists today because such designers are grossly neglecting the human-interface features, and are defending themselves under the umbrella excuse of "optimizing computing power."

I hold a strong view that error-detecting codes, informative diagnostic messages, and codes for recovery from errors are all among the most important features in a computing software. The counter-argument of "wasting computing power" is a red-herring, both in theory and in practice.

MAN-MAN INTERFACE

Much of the weaknesses or ills in computing today resulted from Man's inability or unwillingness to communicate to Man. Ultimately, Man-Machine interface must be improved upon by better Man-Man interface. Space constraints in this article preclude me from suggesting a specific list of remedies. However, even if space were plentiful, specifics may not be necessary. It's the general attitude of some software designers and developers towards users and user-interface that requires modifications if not overhauling. Techniques of implementing interface features that would make myself and most other users happy should be child's play, in terms of technical feasibility, to most computer experts and software developers.

We (users of computers) have plenty of computing power at our disposal: Except for human interface problems, in a mere tick of a second-hand much computation can be done by a computer whose basic units of operations are measured in nanoseconds; we can communicate to the machine in one line of FORTRAN (or some other high-level programming language) what might have taken pages of machine-language codes; in one line of APL what might have taken pages of FORTRAN; in one word (in an applications package) what might have taken pages of APL ... we desperately need better communication among ourselves to better utilize what is existing by creating new or improved user-interface links within the network of man-machine communication links.

If Man[D/P] (designer of software/programmer) is not attuned to, or is insensitive to, Man[U]'s (user's) problem-solving needs and wishes and Man[U]'s imperfections (not always willing or able to read and absorb the intricacies contained in two shelves of technical manuals or a 500-page user's manual, not to mention that Man[U] makes occasional keypunch errors) then Man[D/P] would not be able to, or be unwilling to, communicate to Machine (via programming) what problems and potential errors of Man[U] to detect, nor what to communicate to Man[U] via error-handling capabilities, diagnostics and messages. Breakdowns in Man[D/P]-Man[U] communication precipitate breakdowns in Man[U]-Machine communication.

Lest my position be misunderstood, I shall close by stating that my article is not meant to be an indictment on all software designers and developers, not even the majority of them. I shall cite two software developers (organizations) as shining examples of developers for their attitude and accomplishments toward user interface. SAS, a statistical package, is highly user-oriented. Its personnel conducts an annual survey (with questionnaires sent to all of its known users) to find out the users' likes and dislikes as well as their priorities in the future enhancements of the package, in terms of both technical capabilities and user-interface features. Such Man[D/P]-Man[U] communication resulted in the continual growth of a friendly, high-technical-quality software. SPEAKEASY, a general-purpose, interactive system for scientific computing, is even more fundamental in its user-oriented philosophy. It is a very powerful high-level programming language (with APL-like features, developed independently of APL in the early 1960's) as well as a user-oriented language for problem-solving (with numerous advanced features in user-interface). Its technical features alone deserve recognition and high praise, but above all, its developers pride themselves in the fact that SPEAKEASY is an evolutionary system, designed by its users!

There are numerous other laudable software designers and developers who are sympathetic to the human cause, and the worst offenders of user-interface are in the minority. Unfortunately, some of the worst offenders (be it HAL or XYZ), because of their domination of market share in computing hardware or software, make their presence inescapable.

The main theme of this International Congress is: THE QUALITY OF LIFE AND HOW TO IMPROVE IT. It is my belief that the quality of life for the computing public will be greatly improved if only the present and future designers and developers of computing software will pay more heed to human-interface aspects of computing.

BUILDING USER INTERFACES WITH AN EXTENSIBLE LANGUAGE SYSTEM

Ronald E. Mills*

*School of Organization and Management, Yale University

ABSTRACT

Although matters of form and style are of indisputable importance to effective non-programmer computer use, the match between the user's knowledge and the computer's understanding of the user's needs is critical. We describe experiences with microcomputer-based software that allows the system builder to provide a collection of extensible languages matched to the technical capabilities of different groups of users.

INTRODUCTION

One view of human-machine interaction identifies a human, a computer and a communications channel between them. Much of the discussion of human-machine interaction focuses on the channel: its ease of use, ability to respond graphically or in color, its speed, error recovery abilities and the like. Equally important is the knowledge that must exist on either side of the channel: in the human's brain and in the computer's programs and data.

This knowledge can be roughly classified three ways:

1. Knowledge of the problem that the human and computer are jointly working on. Up to a point, adding problem-domain knowledge to either partner improves problem solving: the human makes "smarter" decisions and the computer can perform more functions.

2. Knowledge of the channel between them. A design objective is to make the necessary amount of this knowledge small. This translates into "ease of use" for the human and "ease of interfacing" for the computer.

3. Knowledge of the other partner. This area is our primary focus.

The machine obtains its knowledge of the abilities, needs, probable actions and shortcomings of its human user indirectly from the humans who have programmed it. The human obtains his knowledge of the abilities, needs (e.g., input format requirements), probable actions and shortcomings of the computer by training and experience. The training often comes (perhaps indirectly) from the same programmers who have provided the machine's understanding of the human.

To this process of developing mutual understanding, the human brings an extremely powerful tool: his lifelong experience communicating with and trying to understand other humans. This experience has taught him to expect that the intelligent entity with which he is communicating will possess the following attributes:

Consistency. If two messages are interpreted differently, then the difference should stem either from differences in the messages or differences in the context of the messages. Further, rules of message formation should not change unless the change reflects some meaning. Computers violate these principles when they use vastly different language styles for job control, file management, programming, text editing, etc.

Closure. Sets of facilities must seem complete. If it can add, subtract and multiply, then it is reasonable to expect it to divide.

Continuity. A small change in message should not generate a large unexpected change in behavior. Many database inquiry systems violate this when the cost to retrieve a variable further down in the storage hierarchy is enormously more expensive: the retrieval requests appear very similar to the user unless he understands the hierarchy.

The more the computer fails to meet these expectations, the more difficult and frustrating it is to use. At worst the human feels he is communicating with a malevolent moron, at best it feels like a benevolent, but somewhat alien, intelligence.

LANGUAGE

Language is central to human-human communication. Most human-machine communication also involves language: programming, job control, file management, database inquiry, statistical analysis, model development, etc. Knowledge of the other partner then involves language knowledge. When the human user is taught a problem-domain language that the computer is programmed to understand, and if the language is constructed with sufficient attention to issues of consistency, closure and continuity, then the human gains an "understanding" of the computer through his significant understanding of the problem domain and his experience with language in general.

To "understand" the human, the computer must be taught (i.e. programmed) how to recognize and react to messages in the language, which involves dealing with the concepts and structures of the problem domain. This programming is performed by humans, which gives rise to other human-machine interactions which themselves would benefit from organized language. The result is a hierarchy of languages.

HIERARCHY OF EXTENSIBLE LANGUAGES

At the top of the hierarchy, the end-user interacts with the machine in specific, problem oriented, languages. Often non-procedural, these languages use concepts, constructs and idioms which are comfortable to the user. Statements in such a language provide information or guidance to the machine in terms of the problem that it and the user are jointly trying to solve.

The machine's langauage understanding and problem solving ability may be, for the class of problems at hand, a special case of a more general, high level, methodological capability. Examples include database management, statistical analysis, and mathematical programming. If this capability is provided with a very high level language (e.g., the report generation language in a DBMS), then specification of the user accessible facilities is easier.

The general capability must be programmed, hopefully in a proper high level language. This, in turn, is implemented in a lower level systems programming or assembly language. Each language in the hierarchy is built by programming with the facilities provided by the next lower level. Each language provides the tools on which a higher level, more specific and problem oriented, language can be built.

A hierarchical language system can also be viewed as a mechanism for adding knowledge to the computer until it knows enough to communicate with the end user and help him solve his problem. Without programming, the compuer knows only the data structures and operations upon them provided by he hardware and operating system. The implementation of a high level language provides knowledge of a larger set of constructs (e.g., arrays, strings, lists, records) and more elaborate operations and control structures. Using this knowledge, an expert in some problem solving technique may build a system with enough knowledge to recognize and solve a large class of problems. An expert in the user's problem domain adds knowledge by particularizing the general system. The user finally brings his specific data and/or questions to be answered.

W.H.O. EXAMPLE

As an example of hierarchical languages, consider the following system built for the World Health Organization. The objective was to provide field analysts with access to a variety of LP-based population, disease incidence and resource allocation models so that the impact of health project funding decisions could be predicted. With an ALGOL-like programming language, a set of facilities for forming, naming, saving and accessing large, sparse, real matrices was built. That is, the computer's knowledge of real scalars and real arithmetic was increased to include real sparse matrices.

A language for using these facilities (along with those already on the machine) was provided. With this languae, a higher level language for forming LP coefficients and submitting them to a solution package (IBM's MPSX) was constructed. Then we built a language to allow one to describe symbolically the variables and relationships of a linear program: it resembled the mathematical notations commonly used by those who work with LPs. At this stage, the computer not only "knew" how to deal with LP coefficient matrices and the rules for using the solution package but also how to use symbolic relationships as a basis for building such matrices.

Several health system scientists expressed their own models in this linear modelling capability. Each expression generated a language in which a user of the model could provide to it further specifications and data. Thus we had the "Litsios Population Modelling Language" and the "Braun Population Modelling Language" and the "Deboeck Population Modelling Language", and so on. The end users, who were policy analysts, gathered data and "described" their countries in the modelling language of their choice. Thus we had the "Litsios Population Model of Sri Lanka" and the "Deboeck Population Model of Kenya" and so on. Neither the health system scientists nor the policy analysts needed to know much about computers other than the details of the special languages with which they communicated their special knowledge.

GENERAL ADVANTAGES

The hierarchical language approach has several benefits:

-- The computer's knowledge of the problem domain is built up as input is obtained from independent specialists. No single human needs to have detailed knowledge about both computers and the problem area.

-- The languages at all levels can be built with a consistent style, and facilities of lower level languages can be made available in upper level languages. The end user can therefore interact at whatever depth his own knowledge supports and his needs require without having to learn new communication skills.

-- Shortcomings in a system's closure and continuity properties can be traced to missing facilities at some particular level. As the missing facilities are added, the closure and continuity properties of all higher levels are improved. Further, lower level facilities provide debugging tools for higher level programs.

-- At the highest language levels, "throw away" programming is feasible. The cost of developing problem specific routines and languages, given all the supporting structure, is quite low compared to conventional programming.

IMPLEMENTATION

We have developed operational software organized into hierarchical interactive languages for DEC LSI-11 microcomputers and for IBM 370-type computers with time-sharing. At the base of the hierarchy on both machines is a language-defining mechanism which accepts syntax definitions in metalanguage and semantic definitions in any language known to the system. A structured assembler provides the base semantic language. Users of the LSI-11 environment have the advantage that the operating system and all of its utilities are built according to hierarchical language principles. With its fast video terminals, the micro also allows languages to handle "menu-driven" statement input. The IBM 370 environment offers the advantages of size and speed. A distributed system, using the LSI-11 as a user interface to the 370, is under development.

CONCLUSION

Computer language technology development in the last decade has been motivated in large part by the problems of software unreliability and the high cost of software maintenance. However, the need to prepare applications quickly and cheaply and the need to provide easy, but effective, access to the computer by non-programmers must also be recognized. Hierarchical language organization appears to offer some progress toward dealing with these needs.

SEMIOTICS IN THE REAL WORLD: MAKING INTERACTIVE COMPUTER SYSTEMS ACCESSIBLE TO IDEOGRAPHIC-LANGUAGE USERS

Ian H. Witten

Man-Machine Systems Laboratory, Department of Electrical Engineering Science, University of Essex, Colchester CO4 3SQ, U.K.

ABSTRACT

The relationship between signs that are used to communicate with an interactive computer system and the entities that are signified (like files, operating system interrupts, text-formatting commands) has grown in a chaotic and haphazard manner. This paper examines the practical problem of developing a front-end processor to make interactive computers accessible to users with different sign systems. Ideographic languages are selected for study because they show precisely how existing interactive systems depend on an alphabetic code, as well as having considerable commercial importance.

KEYWORDS

Semiotics; interactive computer systems; ideographic computing.

INTRODUCTION

Interactive computing provides a rich system of semiotics for man-computer communication. A little reflection easily identifies five of Jakobson's (1960) six linguistic functions in interactive dialogue. Most man-to-machine communication is imperative, and so takes place in the _conative mode_, because it is oriented towards the addressee, the computer. Commands usually contain information which qualifies the command (like filenames), and this implies use of the _referential mode_ which is oriented towards the context of the communication. Furthermore, much machine-to-man dialogue seems to be in the referential mode as well. It is commonplace in computing to use the _metalinguistic mode_ to alter the level at which the referent is interpreted, by, for example, placing the referent in quotes, prefixing it with an escape character, or writing it in an indirect command file for later execution. The _phatic mode_ of communication, which affirms, maintains, or halts communication, is apparent in the use of terminal interrupt characters, and in transactions whose purpose is simply to determine whether the communication channel is open following a compute-bound process. Interactive programs are frequently written to provide phatic ("I'm still here") information spontaneously by prompting the user when the execution of a command is complete. Some operating systems log a terminal out automatically if it is quiet for a long time; then, users may give phatic inputs to avoid this. Furthermore, they often "massage" files periodically by accessing them solely to fool the system into thinking that they are still in use and therefore should not be deleted in a periodic sweep-out. The *aesthetic mode*, defined by Jakobson as the relation between the message and itself, is used to some extent in text processing and document preparation, and also in program layout. The remaining function of communication, the *emotive mode*, concerns the relation between the message and its emitter, and seems to be absent in interactive dialogue with machines. Frustration caused by the lack of opportunity to use this mode is frequently manifested by terminal users thumping the keyboard and swearing at the VDU screen!

So much for generalities. The semiotics of interactive computing can be brought into sharper focus by examining the practical problem of developing a front-end processor which makes a machine accessible to a different community of language users. We are concerned here with the task of translating *artificial* language, and not with the much more difficult problem of translating *natural* language. Many differences in the way signs are processed, which are not always immediately apparent, must be considered. Translating an *ideographic* language like Chinese, rather than an alphabetic one, highlights these differences because a completely foreign (if I may use that word!) semiotic system is involved.

This paper addresses the problems associated with converting man-computer dialogues from an ideographic language to English, from the perspective of semiotics. It is assumed that the entire translation process takes place in a preprocessing computer, so that the host machine runs a completely unmodified English-based operating system. Technological background to the project is given by Witten and Ng (1980), who also discuss the advantages and drawbacks of the preprocessing approach as compared with specially-written ideographic operating systems.

SEMIOTIC CLASSIFICATION OF MAN-COMPUTER DIALOGUE

Much interactive dialogue involves text which is not interpreted or constrained by the system at all. For example, files can be edited or used for data-base retrieval, and comments can be included in programming languages, without being processed by the computer. This introduces a new function of communication which involves only storage for communication with oneself at a later time: we call it the *reflective* mode, because the purpose of communication is simply to be reflected back to the communicator.

When users speaking other alphabetic languages, like French or German, interact with English-based computer systems, relatively few problems arise. Reflective text remains unchanged. String decomposition by character position, as is common in older text editors, causes no difficulty. Although referential symbols need to conform with the system-specified syntax, this should not cause any more difficulty than it does in English. Often, foreign users are content to write conative symbols like keywords in English, because of its predominance in the computer world — international computer language standards usually use English, most programming-language primers are written in English, and so on. However, altering the keyword representations stored in the computer system is not a difficult task, at least in theory: they will appear as character strings in the source code and editing and recompiling it is all that is necessary. Hence the issue of language-translating preprocessors which is addressed in this paper does not arise for alphabetic-language users.

With ideographs, the situation is not so simple. Certainly an ideographic computer terminal could interface to a standard serial line, emitting ASCII code sequences which represent ideographs and accepting them to generate an ideograph on the screen. Then the operating system tables which store keywords internally could be adjusted to hold the ASCII sequences which represent the particular ideographs used as keywords. Although this copes correctly with conative signs, difficulties would still arise with referential and reflective ones. For example, string editing based on character positions would fail, unless the user was keenly aware of the internal representation of ideographs. It would be difficult to accomodate the syntactic constraints for referential signs. BASIC identifiers, for example, must contain at most two characters, the first being a letter and the second, if any, a digit. Resolution of these problems requires a more radical modification of the system software than just altering tables. Another, simpler, alternative is to leave the system unchanged and to try to incorporate the necessary intelligence in a preprocessor.

A further complication is raised by the _metalinguistic_ communication mode, for of course metalinguistic items become conative or referential when a certain level of processing is reached. To take a concrete example, the UNIX operating system makes much use of the character escape "\" to form control characters, and its effect can be delayed by further \'s if the interpretation is to take place at lower levels. Careful thought is necessary in the placement of \'s for interpretation to occur at the correct level. The treatment of metalinguistic input by the ideographic preprocessor is discussed below.

Let us summarize the coding requirements of the different kinds of symbols.

Reflective symbols constitute text that is not interpreted by the system. Certain special characters, which are interpreted at device driver level — like interrupt signals, rubouts, and line-erase characters — must not appear by accident in the translated text. The possibility of editing by character offset must be noted.

Conative symbols must agree exactly with internally-stored keywords. Hence they must be translated by the preprocessor, if operating-system modifications are forsworn.

Referential symbols are subject to syntactic constraints imposed by the system. Most systems insist that only _printing_ characters are used (not control characters), and in many cases only alphabetics and numerals are allowed (ie not "/", "!", etc.). Certain characters sometimes have special meaning in certain contexts. For example, many operating systems allow wild-card specification in filenames, either of a single character (often indicated by "?") or a string of characters (by "*"). Programming languages usually prohibit identifiers beginning with a digit, and some (eg BASIC) restrict their length to two characters.

CODING IDEOGRAPHS

For an ideographic terminal to interact with English computer systems, all ideographs must be converted to English pseudonyms. Desirable properties of the translation (discussed by Witten, 1980) are:

- pseudonyms should be short (for storage economy on the host computer);
- pseudonyms should be fixed-length (to enable string decomposition by character offset, and allow line justification in documents);
- given two strings of ideographs, it should be possible to tell from the translations if one is a substring of the other without false matches being caused by incorrect alignment (for contextual string matching);
- English inputs should be unlikely to masquerade as translated ideographs (for mixed-language work);
- translations should not violate any restrictions placed by the host computer on filenames, and should conform with the syntax of variable names in as many programming languages as possible (for referential signs).

The last requirement is rather stringent, because for most systems it rules out all characters except letters and digits. Furthermore, the fact that most languages prohibit identifiers which begin with a digit means that the standard 4-digit telecode (Walker and others, 1968) cannot be used.

A simple coding scheme meeting these requirements represents each of 27,000 ideographs by a 3-character string, with characters chosen from the digits and upper- and lower-case letters (Witten and Ng, 1980). A digit occupies the central position, flanked by alphabetics, to ensure unambiguous string matching, to minimize the likelihood of mistaking genuine English for ideographic translations, and to satisfy the name syntax of most programming languages.

This scheme suffices for reflective text and some referential symbols. Conative symbols must, of course, be translated by the preprocessor into the corresponding English command name. Referential symbols in upper-case-only languages, and BASIC, need to be handled separately.

DEALING WITH THE METALINGUISTIC MODE

A problem arises when text is entered in the reflective mode and subsequently used in the conative or referential modes. This happens, for example, when a command file is entered using the editor (as reflective text) and later interpreted by another subsystem which normally accepts input from the keyboard. The keyboard input typically contains a mixture of conative, referential, and reflective text (say keywords, filenames, and comments), and whenever a subsystem is being entered the preprocessor must obviously ensure that these elements are converted appropriately. However, when a previously-entered text file is used for input, the conversion cannot be done by the preprocessor because

— when the file was entered originally, the preprocessor could not know that it was destined for use by the particular subsystem as a command file,

and — when the subsystem reads the command file, its contents do not pass through the preprocessor and so there is no opportunity for conversion.

For a more concrete example, consider interacting with the BASIC language subsystem via an ideographic preprocessor. BASIC keywords are typed as ideographs, and converted by the preprocessor to English — PRINT, IF, GOTO, etc.. Furthermore, BASIC identifiers are also typed as ideographs, and must be converted to BASIC variable names — A0, A1, A2, etc. — through tables maintained by the preprocessor. Now if the BASIC program is SAVEd, what should the resulting computer file look like? If it is to be reLOADed by the BASIC subsystem, the keywords in the program should be in English form, so that they are recognized by BASIC on reloading. If, however, it is destined to be printed on a lineprinter by a standard system utility, it should have been SAVEd as text, with keywords represented as the reflective encoding of the ideographs which were typed originally.

For the preprocessor to work transparently, it must invoke mode-conversion programs automatically whenever the user issues certain commands (like SAVE a BASIC program). The virtue of our semiotic classification into reflective, referential, and conative signs is that it shows clearly the need for this.

IMPLEMENTATION AND CONCLUSIONS

Six important subsystems have been examined in some detail by Witten and Ng (1980). Perhaps the simplest, from the point of view of the ideographic preprocessor, is the LISP interactive language. The indivisibility of atoms and the generous maximum character length provided by most implementations encourages the use of the coding scheme presented earlier for all referential symbols. System-defined atoms are translated via a table in the preprocessor which associates the English name with its picturegram entry. Special characters like parentheses, full stops, and quotation marks which form part of the LISP syntax are all commonly used in Chinese text, and so there is no need to translate them. Even the atom decomposition functions EXPLODE and IMPLODE can be easily handled, by writing special ideographic EXPLODE and IMPLODE functions which take account of the structure of the codes.

More baroque languages give the preprocessor considerably greater difficulty. With BASIC, keywords are treated as in LISP. However, variable names must be identified and translated according to a special table which is built by the preprocessor at run time. The preprocessor translates the first-used ideographic identifier (whatever it may be) to "A0", the second to "A1", and so on, remembering the translations. Character strings are the most difficult items to translate, because BASIC contains string-processing functions such as LEN() (length of a string), LEFT$(), RIGHT$(), MID$() (substrings), and INSTR$() (string search). Since ideographs in strings are converted into three English characters each, the ideograph for LEN is translated into "(1/3)*LEN", and expressions like LEFT$(..., <expression>) into "LEFT$(..., 3*(<expression>))".

Some interesting system problems are posed by the operating system command level itself, the initialization of the preprocessor, and the mechanism which enters a new subsystem. Text editors, document preparation software, and interactive database management systems are relatively simple to handle, for the distinction between conative inputs (commands) and reflective text is fairly clear. In all cases in which information is saved on disk for later use, it is preserved in reflective form (in case it has to be printed); a special conversion operation is required for indirect input to each subsystem from a file. This conversion can be invoked by the preprocessor, transparently to the user.

Semiotics of man-computer dialogues can be classified in the same way as for human dialogue, except that the emotive mode is absent and a new mode appears, which we have called "reflective". This encompasses input whose sole purpose is to be reflected back to the communicator. Actually, these two modes could be equated, for emotive communication is to do with the relation between the message and its emitter, as of course is reflection. This would render the semiotic classification for computer dialogues identical with that for human ones. We prefer not to do so, however, for there is a difference in the purpose of emotive and reflective inputs: the former is intended to communicate the speaker's relationship with what he is saying to the hearer, whereas the latter communicates the message back to the speaker at a later time.

ACKNOWLEDGEMENTS

This research was supported by a grant from GWITS.

REFERENCES

Jakobson, R. (1960). Linguistics and poetics. In T. A. Sebeok (Ed.), _Style in language_, Wiley, New York. pp. 350-377.

Walker, G. L., S. Kuno, B. N. Smith, and R. B. Holt (1968). Chinese mathematical text analysis. _IEEE Trans. Eng. Writing Speech_, _EWS-11_, 118-128.

Witten, I.H. and Y. H. Ng (1980). An ideographic-language processor for accessing English-language computer systems. _Comput. J._, _in press_.

Witten, I.H. (1980). Translating interactive computer dialogues from ideographic to alphabetic languages. _Proc Coling 80_, Tokyo.

TOWARDS A PROGRAMMER'S ASSISTANT FOR NOVICE USERS

M.A. Bauer and D.M. Goldstein

Department of Computer Science

University of Western Ontario

London, Ontario, Canada N6A 5B9

ABSTRACT: In this paper the development of a programmer's assistant for novice users of the language BASIC is outlined. In particular, an overview of the repretation used to store information is presented as well as a description of the approach used to provide a conversational environment for such users. The paper includes several examples of the representation and dialogues.

Keywords: Programmer's Assistant, Knowledge Representation, Man-machine Dialogue, Semantic Nets, Natural Language.

1.0 INTRODUCTION

The main objective of this work was to develop an aid for a novice computer user. This programmer's assistant takes the form of software which is available to the user to answer questions about the programming language and help with the understanding of the syntax of the language by checking the syntax of the statements in the user's program. The user learns the programming language by being allowed to create a program and interrogate the assistant simultaneously. The user communicates with the assistant in a simple, restricted natural language dialogue. This provides an informal environment in which the user and the machine can carry on a conversation that could be similar to one between two humans. The programming language that this assistant currently deals with is BASIC. This choice seems reasonable since many first-year students and novice programmers are learning BASIC as an introductory language and it is based on an interactive environment which is conducive to a dialogue situation.

In this initial work a number of assumptions were made to limit the scope of the dialogue in order to get a handle on the problems involved in creating such an assistant. Firstly, if the assistant has notified the user that an error exists in a particular statement, the user must correct that specific error before he can correct any other or enter any new statements. The second assumption is that the user will always directly answer the question posed by the assistant. Thirdly, the assistant only allows the current BASIC statement to be edited, not any other statement. Information about the other statements is kept by the assistant, but presently no editing is allowed on them. Finally, to simplify the complexity of the dialogue, each BASIC statement entered by the user is treated as a "new" topic of conversation. This permits the assistant to limit the scope of the conversation both for efficiency and pedogogic reasons. Some motivation for this assumption will be presented in the discussion of the dialogue model.

The following is an example of part of a session that might occur between the assistant and a programmer. The programmer's responses are preceeded with a "P" and assistant's are preceeded with an "A". It is hoped this example will help to make the concept of a "programmer's assistant" clearer.

P: LET X=0

A: EACH BASIC STATEMENT MUST HAVE A STATEMENT NUMBER

P: GIVE IT NUMBER 10

P: 20 LET Y=

A: WHAT DO YOU WANT Y TO BE ASSIGNED?

P: HOW DO YOU TAKE THE SQUARE ROOT?

A: USE THE SQR FUNCTION.

.
.
.

A programmer's assistant with the capabilities outlined above can be thought of as consisting of four major components: language parser, sentence generator, knowledge base, dialogue model. The language parser and the sentence generator are the components that allow communication between the assistant and the novice user. The knowledge base contains the information the assistant requires about natural language and the programming language. A model for possible dialogues has been incorporated into the assistant to allow it to deal with the dialogue at more than simply a sentential level. This paper will present an overview of the representation being used to store knowledge required by the assistant and an overview of a model of dialogue that is being used to define possible dialogues.

2. Representation

2.1 Components of the Representation

For the programmer's assistant to be able to understand and respond to the student, it must have some *a priori* knowledge about natural language and the programming language BASIC. To allow the assistant fast retrieval, this information should be kept in an orderly, concise, uniform format. Semantic networks [Rumelhart and Levin 1975, Simmons 1973] and frames [Kuipers 1975, Minsky 1975, Winograd 1975] are two methods of representation that have been widely used in artificial intelligence applications. A combination of these methods is being used for the representation of knowledge required in this environment.

The objects used in the representation are <u>nodes</u>, <u>frames</u> and <u>items</u>. Links connecting these objects represent relationships, characteristics and properties of objects.

The idea of a node was taken from semantic networks and is used to represent concepts through a subset/superset hierarchy or indicate characteristics of a concept or object. This hierarchical structure is defined by <u>ISA-link</u> connections that join nodes to each other, while characteristics of a concept are defined by <u>CH-link</u> connections. Using these links, the assistant can represent the fact that a variable in BASIC could be used as an expression, or that FOR/NEXT loops have the characteristic that they must be properly nested inside one another.

A frame is used in this representation to define a concept or object represented by a node and contains a definition or description of the node. The frame allows for features that are common to a concept and ones particular to an observation to be stored and retrieved as a unit. Since there are basically two kinds of knowledge stored in this representation, two types of frames exist - <u>BDEF frames</u> and <u>VDEF</u>

frames. The links connecting a node to a frame are then BDEF, which is used for defining concepts within the programming language, and VDEF, which is used for defining concepts within natural language. These two types of frames are distinguished because they are used to define different classes of objects; their representation is the same.

The contents of a BDEF frame define the BASIC concept it is related to. It contains a description of the syntax of a statement, the components of a statement and any relationships between those components.

A VDEF frame is based on the notion of verb cases [Fillmore 1968, Rumelhart and Levin 1975]. When dealing with verb cases, references to components of a natural language sentence are made through the verb of the sentence. Since the sentences dealt with in this environment are imperatives (telling the assistant to do something) or interrogatives (asking the assistant a question) and have intentionally been kept short and simple, the primary parts of speech necessary are the verb, object, indirect object and in some cases a preposition and/or a locative. Hence a VDEF frame contains the actual verb, as well as references to the parts of speech in the sentence.

The primitive object of the representation is an item. There are several types of items which can be used to represent different types of information within frames. Firstly, the knowledge representation must contain general definitions, to allow for many possible variations of a single concept. To permit this, an item within a definition may be a variable item which can be instantiated with a particular object within an instance of a particular concept. To ensure that only proper instantiations occur, each of these items is linked to a node concept by a type (t) link, which defines the type of objects that may be used to instantiate that item. Secondly, in some cases only one instantiation can occur for a particular item - that is, the item is a constant or keyword. Rather than repeatedly instantiating such items, a symbol item is used, which contains the actual token. For example, keywords within the definition of a BASIC statement are defined by symbol items. Thirdly, complex constructs within a definition may be represented by frame items. Each of these is linked to a frame defining a particular component. Thus, frames may be nested in order to represent complex structures. Finally, for concepts which are better represented procedurally there are also procedure items. These procedure items (see Winograd 1975) reference a particular procedure that acts as a definition for a complex object. For example, a BASIC variable name can be a single alphabetic character A-Z, optionally followed by a single digit 0-9. Although this could be defined using variable and frame items, it is more easily and efficiently represented as a recognition procedure.

Figure 2.1 depicts a portion of the knowledge base showing the representation of a LET statement. Within the figures, nodes are represented by oval shapes, frames by rectangular shapes and items by small circles.

For use in VDEF frames, five more kinds of items exist. They are verb, object, indirect object, preposition and locative. These items are used in the verb definition (VDEF) frames to define the parts of speech related to a particular verb. The types of verbs used in this environment are basically action verbs - telling the assistant to do something. These actions involve obtaining information or performing a specific request. A procedure item is placed in the VDEF frame to perform the semantic "actions" of a particular verb. A segment from the knowledge base showing the representation of the verb MAKE is given in Figure 2.2.

2.2 Organization of Knowledge

As outlined above, the knowledge within the assistant is made up of both programming and natural language information. This knowledge is called the static knowledge

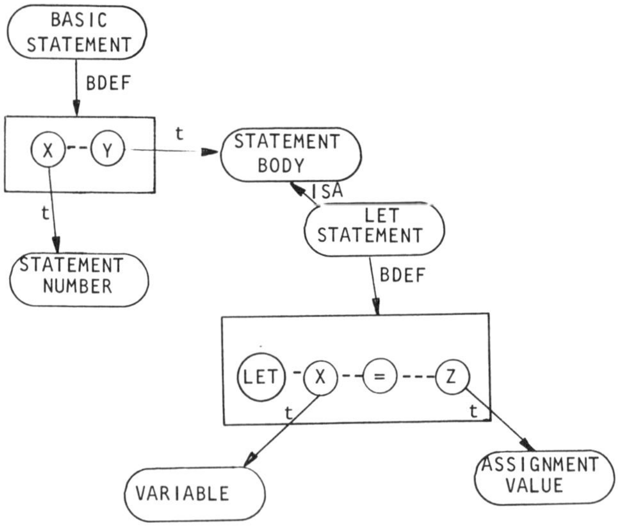

Figure 2.1: Representation of the LET Statement

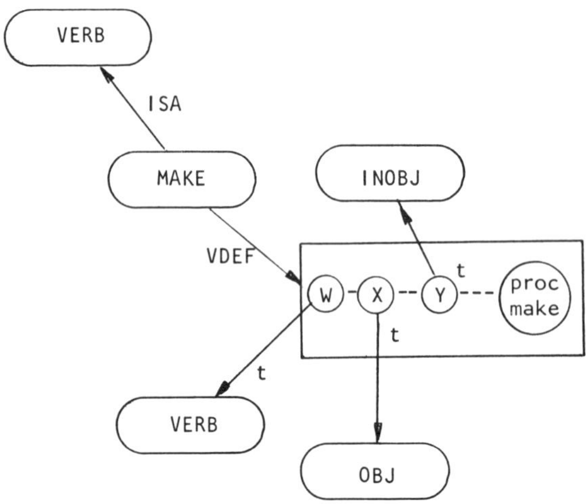

Figure 2.2: Representation of the verb MAKE

since it is present in the assistant at the outset of any session.

All relevant programming terms are contained in nodes and have a related definition in a BDEF frame. This is all the knowledge relating to the language BASIC that the assistant requires throughout a session. The detection and correction of errors is done using only the knowledge base since the syntax of any BASIC statement is contained in its definition. The assistant uses the static knowledge to answer any questions that are posed by the programmer.

The natural language knowledge is referenced through the verb of the sentence. Therefore each possible verb is represented by a node in the knowledge base and has an associated VDEF frame. The natural language knowledge is used by the assistant for both understanding the programmer's sentence and generating its own sentences.

As well as the static knowledge, the assistant must also keep information obtained during the dialogue - non-static knowledge. Some of this information is necessary so that the assistant can resolve pronominal references. The assistant must also remember the BASIC program that the user creates during a session to allow its retrieval. The responses of both the assistant and the programmer are stored as frames within the non-static knowledge base. Each of these frames is defined as an instance of the appropriate frame within the static knowledge. For example, the statement "10 LET X=0" would be represented as a frame, created from the frame defining a LET statement, and would be stored as an instance of that frame.

The initial representation of both assistant and user responses are kept in a section of the non-static knowledge base called the dialogue work area. The BASIC statement that is being discussed by the assistant remains in this work area until all the errors have been corrected and the next BASIC statement is entered by the user. Once a BASIC statement is corrected, it is transferred to a more permanent place within the non-static knowledge base, so it can be retrieved at a later time. As noted, this work is based upon the assumption that a new BASIC statement signifies a change of topic. This means that once a topic changes, the information relating to the previous dialogue is no longer pertinent and may be destroyed. This allows the dialogue work area to be continually reused.

3. Modelling Dialogues

3.1 Determining a Basis for Dialogues

To understand the model and how it works, one must first consider its development. Prior to actually creating a computational model of discourse, the actual structure of the dialogue needed to be considered. To that end the work of two linguists, J. McH. Sinclair and R.M. Coulthard [Sinclair and Coulthard 1975] has been used. Sinclair and Coulthard dealt with the discourse between a teacher and a class, which is similar to the dialogue between a programmer's assistant and a programmer. Following their approach, some examples of anticipated dialogue between an assistant and a programmer were generated and examined. A number of conclusions were drawn based upon these examples. Firstly, the programmer seemed to be controlling the direction of the conversation, that is, the assistant was usually reacting to a statement made by the programmer. Secondly, pronominal references contained in a sentence could usually be satisfied using the nouns from the preceding sentence. Thirdly, the topic of conversation was usually nested to a few levels, with the majority of the dialogue involving a single subtopic or none at all. This suggested that the current topic of the conversation in this kind of environment is usually dealt with before the topic changes. Thus, the current topic from the viewpoint of the assistant is simply the current BASIC statement. When the next BASIC statement is entered, it is treated as the new topic.

3.2 3.2 A Computational Model of Dialogue

Following the approach used by Sinclair and Coulthard, a grammar was developed from the example dialogues and the observations cited above. This grammar provided a definition and description of the possible dialogues that could occur between an assistant and a user. A computational model was then developed from the grammar. The computational model can be represented by a state diagram, which shows the flow of information that occurs during the operation of the assistant. The state diagram for the Model of Dialogue for the assistant is given below (Figure 3.1). The initial state of the model is S1, while S10 is the termination state.

There are two classes of states in the model - accepting and generating. In accepting states the assistant is poised to accept responses from the user, parse them via the language parser, interpret them and carry out the command if semantially sound. On the other hand, in generating states, the assistant generates a response by sending the appropriate information to the sentence generator. States S2, S4, S6 and S8 are all accepting states, while the generating states are S1, S3, S5, S7 and S9. As can be seen, these states alternate with each other. This is is an effect of the nature of the dialogue in this environment, that is, the user and programmer's assistant effectively take turns responding to one another.

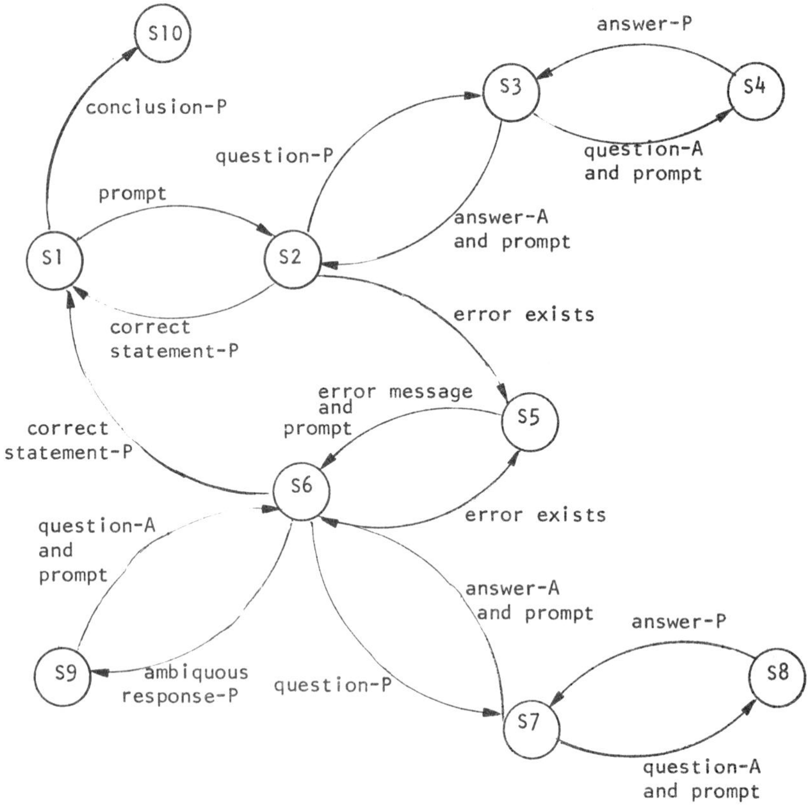

Figure 3.1 State Diagram of the Model of Dialogue

The arcs in the state diagram are labelled by the type of statement or condition that tells the assistant to move from one state to the next. The symbol "P" refers to an input by the programmer and an "A" signifies a response by the assistant. The assistant prompts the programmer to continue with the session, therefore some of the arcs are labelled with a "prompt". The label "error" refers to the fact that at one point the student has entered some information which is either incorrect, as in a statement with a syntax error, or information which cannot be interpreted, as in a sentence with an unknown verb. It may also involve a statement in which some errors have been corrected, but another error still exists within that statement. The arc "ambiguous" indicates that the correction to an error is ambiguous, that is, the correction may apply to one of several errors in a BASIC statement. The assistant at this point attempts to get the correction clarified.

It should be noted that certain sections seem to occur in more than one place of the state diagram. For example, the section consisting of states S2, S3, S4 is exactly the same as the section consisting of states S6, S7, S8. This "duplication" occurs because the assistant allows the user to ask questions at any time throughout the session. The distinctness of the two sections is necessary, because the assistant must return to the proper state within the context of the dialogue. For example, if the assistant was processing an error prior to a question posed by the user, it must resume processing the error when it has answered the question.

The Model of Dialogue provides the flow of control within the assistant. In the accepting states the assistant takes a response and decides from the type and content of the statement what to do and where to go next. In the generating states it creates a response and proceeds to a state where it expects a response of the appropriate type from the programmer.

4. A Partial Example

The following is an example of a few sentences that would be part of an actual session with the assistant along with their representation.

The assistant generates a prompt character in the state S1 and advances to state S2 where it waits to accept a statement from the programmer:

P: LET X=0

The parser determines that the user is trying to enter a BASIC statement since it recognises the word "LET". The parser then invokes a routine to store the information within the statement in the work area. In doing this, it discovers that the statement is incorrect - it cannot match the definition of a BASIC statement. In this case the error involves a missing statement number. The LET statement itself is acceptable and is stored. In addition, a frame is created describing the type of error. This is done by using the verb "missing" whose object references (dashed line in Figure 4.1) the item which has not been correctly instantiated. An error variable is then given the value of this frame. To flag this error a "?" is put in place of the statement number in the instantiated frame of this statement. This is depicted in Figure 4.1.

Since an error exists, the assistant moves to state S5 to generate an error message:

A: MISSING STATEMENT NUMBER?

By examining the error frame, the assistant determines that the item was actually missing and not simply misspelled. By following the t-link in the instantiated statement, the assistant discovers what was actually missing (a statement number). Knowing the above two facts, the assistant can generate the above error message. The assistant then transfers to state S6 to anticipate the correction from the programmer:

P: MAKE IT 10.

The user responds with a correction. The parserdetermines that the user's response is not a BASIC statement and evaluates it based upon the verb ("make"). Using the VDEF frame defining the case for the verb "make", the sentence is represented as a frame (depicted in Figure 4.1). To resolve the pronominal reference, the assistant examines the frame corresponding to its previous statement. Under the assumption that the user is trying to correct the error, the assistant associates the object of "make" ("it") with the object of "missing". This results in the missing item being made the referent of "it". The assistant then invokes the procedure that is associated with the verb "make". This procedure replaces the direct object by the indirect object and the statement number becomes instantiated to 10 . The assistant checks that this correction is valid by following the links in the representation and checking that the number "10" is a valid statement number. Since no further errors exist in this BASIC statement, the assistant returns to the state S1.

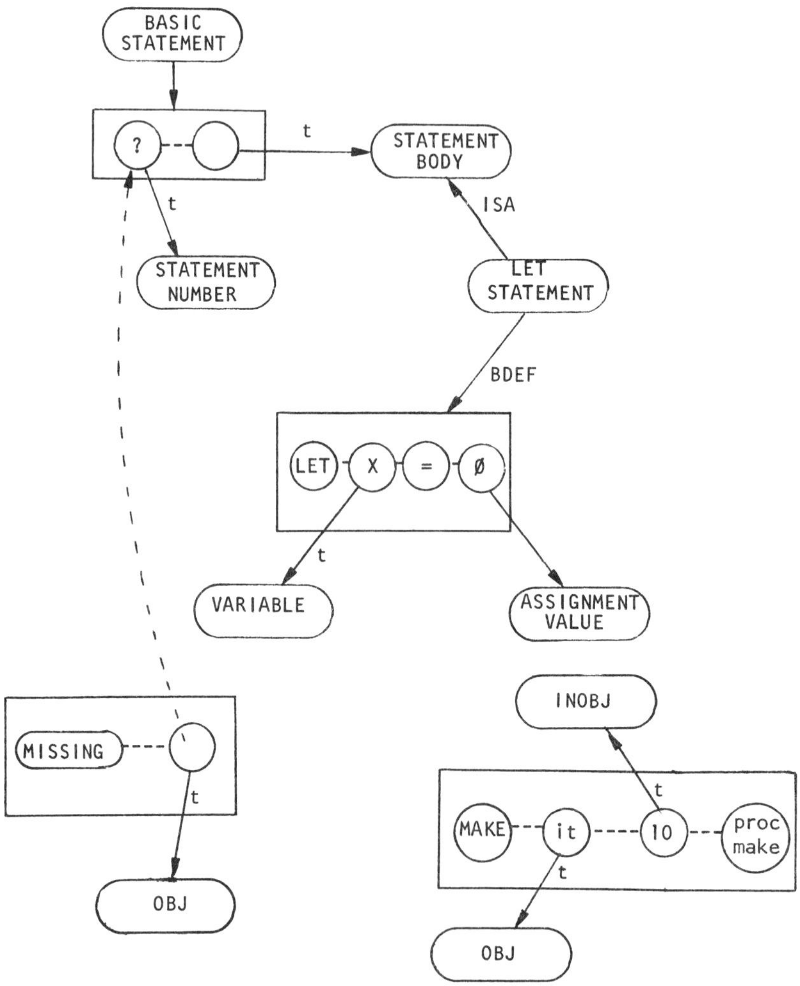

Figure 4.1: Representation of Portions of a Dialogue

This paper has discussed some of the concepts involved in developing an assistant for a novice user that can converse in natural language dialogue. A partial implementation of the assistant presented here exists and was implemented in the language POP-10, which is a version of POP-2 [Burstall et al 1971] written for a DECsystem-10.

In testing this assistant it was found that the model provided a good approach for dealing with the problems of how the assistant could understand statements within a dialogue environment. In particular the assistant can determine the context of a statement (that is, whether it is correcting a statement or answering a question) by "knowing" which state it is presently in. Although the specific model dealt with rather restricted dialogues, the general concepts introduced suggest a method for dealing with more complex dialogues.

The knowledge base representation used by the assistant is a network of nodes, frames and items. The network has allowed the assistant to represent both natural language and BASIC information. Both the pre-defined static knowledge and non-static information which is acquired through the dialogue can be stored in the same representation. This has permitted either type of information to be retrieved using the same procedures.

Only a simplified assistant has been considered, but it has been useful in demonstrating the concepts involved in the creation of a dialogue of system. Many features could be added to the assistant to enhance its operation. One enhancement would be to have the assistant allow the programmer to edit any BASIC statement in his program not only the last one entered. The assistant could also be made more flexible by removing the constraints that have been placed on the dialogue. Finally, a more challenging problem would involve extending the assistant to aid in detecting logic errors as well as syntax errors.

REFERENCES

Burstall, R.M., Collins, J.S., Poplestone, R.J. (1971). Programming in POP-2 Edinburgh: The Round Table and Edinburgh University Press.

Fillmore, C.J. (1968). The Case for Case. In E. Bach and R.T. Harms (Eds.), Universals in Linguistic Theory. New York: Holt, Rinehart and Winston.

Kuipers, B.J. (1975). A Frame for Frames: Representing Knowledge for Recognition. In D.G. Bobrow and A. Collins (Eds.), Representation and Understanding. New York: Academic Press Inc.

Minsky, M. (1975). A Framework for Representing Knowledge. In P. Winston (Ed.), The Psychology of Computer Vision. New York: McGraw-Hill.

Rumlehart, D.E., Levin, J.A. (1975). A Language Comprehension System. In D.A. Norman and D.E. Rumlehart Explorations in Cognition. San Francisco: W.H. Freeman and Company.

Simmons, R.F. (1973). Semantic Networks: Their Computation and Use for Understanding English Sentences. In R.C. Schank and K.M. Colby (Eds.), _Computer Models of Thought and Language_. San Francisco: W.H. Freeman and Company.

Sinclair, J. McH., Coulthard, R.M. (1975). _Towards an Analysis of Discourse_. London: Oxford University Press.

Winograd, T. (1975). Frame Representations and The Declarative - Procedural Controversy. In D.G. Bobrow and A. Collins (Eds.), _Representation and Understanding_. New York: Academic Press Inc.

MESSAGES FROM SYSTEMS TO USERS: A PROPOSAL FOR STANDARDIZATION

William F. Eddy

Department of Statistics, Carnegie-Mellon University, Pittsburgh, PA 15213

ABSTRACT

There is universal agreement among users that messages from computer systems are unnecessarily complicated. This is, of course, because the messages are created by and for system designers and not for users. Here, I suggest a standard for messages. The basic proposal is that the user, at entry to the system, be permitted to declare levels of expertise in computing and in the subject matter of the system, as well as levels of interest in these areas. The messages would then be restricted both in form and content to those particular levels. The proposal will be illustrated by a model of a statistical package.

KEYWORDS

Computer system; statistical package; error message; data analysis.

INTRODUCTION

Over the last five years or so a few statisticians have invested considerable time and money in an evaluation of existing statistical software. The fruits of this effort can be found in various papers in, for example, <u>The American Statistician</u>, <u>Computer Science and Statistics: Proceedings of the Xth Symposium on the Interface</u>, and <u>Proceedings of the Statistical Computing Section-American Statistical Association</u>. Of particular interest here is the paper by Francis, Heiberger, and Velleman (1975) which marks the formal beginning of this activity. They consider three categories of features of statistical program packages: user interface, statistical effectiveness and implementation. While some of their subcategories touch on system messages indirectly, nowhere do they explicitly consider the form or content of communications from systems to user. In a similar but not directly related paper, Ling (1980) has recently given an overview of design issues for interactive data analysis systems. He also touches indirectly on the topic of system messages, when discussing error detection and recovery, for example. However, he does not consider the form or content of the messages that are produced by the system. Here I will consider the general difficulties associated with messages from systems to users (for simplicity, in the context of a statistical package program or statistical system) and will make a specific easily implemented proposal for standardization. The ideas obviously apply to a system in any subject matter area.

BACKGROUND

Every computer user has, at sometime, received a message similar to

(1) ABC123Z PROGRAM INTERRUPT 12. PROGRAM COUNTER 54321.

(I have changed the form slightly to protect the guilty.) Imagine for the moment that this occurred while using a statistical system to perform a multiple regression analysis. If the user is a skilled programmer, expert on this particular machine, then the interrupt code 12 is instantly recognized as a "divide-by-zero" unless, due to faulty (human) memory, the user recognizes the 12 as an "underflow." If location 54321 is known, or later discovered, to be in the loop which accumulates the sum of cross-products of the regression variables then underflow is slightly disturbing to the programmer. On the other hand if 54321 refers to the statement that calculates the residual mean square then divide-by-zero is a really major statistical disaster. But in no case does the initial "ABC123Z" have any use without an extensive manual. Consequently, (1) contains some useless information and, worse, it is missing some useful information. The response depends not only on the level of computing expertise of the user but also the level of statistical expertise.

Recently most statistical system developers have become very conscious that messages like (1) are not only less than valuable to a system user but also can be actively harmful. Thus the latest "user-friendly" system fields all interrupts and masks off (from the user) those which the system designer considers unimportant. The attitude of the designer is: What they don't know won't hurt them. So when the system reaches the situation that resulted in (1) it either produces no message (in the underflow case) or (for the divide-by-zero case) the user gets

(2) NONPOSITIVE RESIDUAL DEGREES OF FREEDOM. COMMAND ABORTED.

(Again I have modified the form.) A skilled statistician instantly recognizes that a mistake has been made: there are as many independent variables as there are observations (or possibly doesn't recognize this because the system language has caused a serious case of <u>statisticii confusi</u>). If the user is not a skilled statistician then "command aborted" is hardly needed (because there was no output) and the initial portion of the message is less than useless. If the user is a skilled programmer then the transformation of (1) into (2) in a misguided attempt to be friendly has cut off the only useful information.

The time has come for systems to actually be friendly. Messages can and should be personalized. Statistical experts have a different world view than do computing experts and each has a different perspective than a novice at both. Each user should be permitted to specify to the system a level of expertise both in computing and in statistics and should be permitted to specify to the system a level of interest in both areas. These considerations will be more carefully delineated in the next section. Then, a specific proposal for standardization will be provided, and its possible implementation briefly discussed in the context of a statistical system.

BASIC PRINCIPLES

The goal of a genuinely friendly system should be to protect but not isolate the user. Thus masking interrupts is an actively harmful approach. If the actual cause of the interrupt is an underflow in the sum of cross-products loop the user should still be informed that something unexpected has happened even if no action may or will be taken. A basic principle for system designers is

(3) WHEN IN DOUBT, INFORM THE USER.

A rereading of the two forms in the previous section reveals an implicit assumption made by the system designers: the user can understand this message. I know expert statisticians who are unable (and unwilling) to decipher the message (1); and expert programmers who react likewise to (2). A truly friendly system operates on the principle

(4) THE USER KNOWS LESS THAN THE SYSTEM DESIGNER.

A natural extension of this principle is:

(5) THE USER KNOWS LESS ABOUT COMPUTING THAN ABOUT STATISTICS.

At its simplest this means that (2) is always preferred to (1). More generally it means that every message should convey statistical information in preference to computational information.

Many years ago (when I used a different brand of computer than I do now) the system, in an effort to be helpful, would produce, in addition to the message (1) a list of the contents of the accumulators, various system storage areas with cryptic names and best of all a core dump. These times have passed for me (and I hope for others) yet the famous dictum of Mies van der Rohe still applies:

(6) LESS IS MORE.

There are of course systems which carry this principle to ridiculous extremes; unless the user input is perfect there is no response from the system. An "improved" version of this is the system having very complex syntax which informs the user who errs even slightly: REENTER. This leads to another principle:

(7) LESS IS LESS.

At first glance (6) and (7) seem contradictory. However if they are applied simultaneously then the result is a message which is succinct but not laconic; the message has the necessary detail and does not have the unnecessary. Parenthetically I note that the two principles similar to (6) and (7): "More is less" and "More is more" are both very important but much less so, I think, than the two I have given.

A PROPOSAL

At this point it has become clear that there are two basic aspects to the content of messages: statistics and computing. Within each aspect there are two further dimensions: the importance of the message to the task at hand and the level of skill required to understand the message.

The framework of the proposal is simple. On entry to a statistical system the user declares a level of expertise in statistics (high or low). If the user claims a high level of statistical expertise then the user is entitled to declare a level of expertise in computing (high or low). Additionally the user is entitled to declare a level of interest in statistical and computational messages. The various possibilities I would allow are summarized in Table 1 below. My intention is that a default declaration would be "Novice". In any case the declaration determines not only the form of the messages but also the content.

Table 1

Aspect / User	Statistics		Computing	
	Skill	Interest	Skill	Interest
(A) System Designer	High	High	High	High
(B) Experienced Computational Data Analyst	High	Low	High	Low
(C) Experienced Data Analyst Previous User	High	Low	Low	Low
(D) Experienced Data Analyst New User	High	High	Low	Low
(E) Novice	Low	High	Low	Low

I would now like to consider the various possibilities, in turn, beginning with the highest levels.

(A) The system designer

This person should be highly skilled in both statistics and computing and should have a high level of interest in messages of all types from the system. In practice, the system designer seldom has the subject matter expertise that is ideally required. The messages for this user need to be very detailed and consequently should have a very structured form. In particular, a message for this user should have identification to indicate which program module and particular line of code is the source of the message. Also, if desired, messages for this user can contain information of the type displayed in (1) above; further, this user should have an option to display all the internal structures and storage areas at will. This user is already very well provided for in many existing package programs.

(B) The experienced computational data analyst

This person is also highly skilled in both statistics and computing but has little interest in messages from the system. Messages should be short and to the point. They should indicate the nature of the statistical activity and computational activity; thus, "Underflow in sum of cross-products loop" would be an acceptable sort of message.

The messages should not contain machine and system dependent information. However, if a manual is available it is permissible (but not desirable) to number the messages and provide a fuller explanation in the manual.

(C) The experienced data analyst/previous user

This person is an experienced data analyst and has experience using this particular system but does not have claim to general computing expertise. Consequently messages to this user should be short but the content must be restricted to statistical and not computational matters; thus, "Zero residual df" would be an appropriate message. In keeping with (5) above the message "divide-by-zero" would not be appropriate.

(D) The experienced data analyst/new user

This person is like the old hand except new to this system. The distinction is that this user wants and needs more detail in messages. Consequently the form of the message would much better be "Zero degrees of freedom in calculation of residual mean square."

(E) The novice

This person claims no expertise in computing, the use of this package, or even in the analysis of data. Consequently all messages must refer to causes and not consequences. Consequences require subject matter understanding. So the messages necessarily become slightly vaguer and use the simplest terms. Additionally, messages for this user may suggest actions to be taken, if such information won't cause further confusion. Because it is difficult to know how a novice will react to a message, the following should be considered as only the vaguest sort of example: "Zero degrees of freedom in calculation of residual mean square. This was possibly caused by too many independent variables being specified in the regression equation. Check to make sure that all the data has been used."

IMPLEMENTATION

The most obvious way to implement this proposal is to construct, for each situation, a distinct message for each type of user (five, in this case). The variants are all stored together in the source code, or preferably, in a data structure so they can be easily modified simultaneously. The message handling routine merely selects the appropriate one according to the declared levels of expertise and interest. This method is somewhat wasteful because some messages do not need to differentiate among capabilities of users (e.g., a mistyped number is a mistyped number); multiple copies are unnecessary. Storage space is an issue because I envision the message file having more lines of text then the source code for the system.

REFERENCES

Francis, I., R. M. Heiberger, and P. F. Velleman (1975). Criteria and considerations in the evaluation of statistical program packages. _Am. Stat._, _29_, 52-56.

Ling, R. F. (1980). General considerations on the design of an interactive system for data analysis. _Commun. ACM_, _23_, 147-154.

VIII. ARTIFICIAL INTELLIGENCE AND AUTOMATA THEORY

A MICROCOMPUTER BASED ARTIFICIAL INTELLIGENCE LABORATORY

Michael J. Magee

Associate Professor
Computer Science Department
University of Wyoming
Laramie, Wyoming 82071
USA

ABSTRACT

The design of a cost effective experimental artificial intelligence laboratory is described. Problems inherent in developing device control programs are minimized by using high-level language software to issue commands from a large mainframe to device control software that is resident on a microcomputer in the laboratory. Real world problems involving speech recognition, speech generation, image processing, and robotics are assigned to students using the laboratory.

INTRODUCTION

The integration of robotics into the computer science curriculum is of greater importance today than at any point in technological history. It is estimated that by the year 1985 almost 1/2 billion dollars will be invested in robotics research, development, and manufacturing [1]. This fifteen fold increase in demand over an eight year period will be spurred by the need to increase productivity as well as the desire to shift human workers to less tedious and more fulfilling tasks [2].

Collateral with the need to present robotic control concepts in an educational environment is the requirement that (artificially) intelligent programs be presented that demonstrate discovery, learning, and problem solving capabilities. The software associated with such control programs is often the emphasized portion of an artificial intelligence course in which simulated robots move about performing tasks in idealized and well-understood worlds. This modeling approach has been an effective tool to demonstrate the problem solving strategies that use, for example, resolution theorem provers (STRIPS) or procedural knowledge (Winograd) [3,4]. Such methods, while having impressive results in their respective domains, however, isolate the problem solver from "real world" difficulties. As Raphael [4] points out, simulated robots are less desirable than real robots for two primary reasons. First, the real world is too difficult to simulate if one desires a complete and accurate model. Second, it is difficult to know what to simulate due to real world uncertainties and errors that may occur.

It is the premise of this paper that a course in artificial intelligence should allow the student to experiment with the full range of control problems inherent in robotics, from high-level control software (the behavior of which is presumably deterministic) down to robotic hardware (which may have wheels that slip and control voltages that vary). For the general student environment the question thus arises, "How does one go about the task of developing a system that each student can manage?" As Hollis [5] notes, the software needed to totally control a robot is typically enormous. Our own experience verifies his conclusions to the point that it would be unreasonable to ask a single student to perform (in a single semester) all the programming necessary from microcontrol to high-level functions of the robotics system which will later be described. Furthermore, due to limited resources, namely that we have only one such system, the competition for the laboratory would prohibit more than a few students from using it. A final problem is that development of such a system can be rather expensive if one must start with little or no robotics hardware.

The advent of S-100 bus microcomputers has provided a nucleus around which to build a cost effective system that minimizes many of the problems associated with complex control software. There are currently many vendors supplying interfaces that digitize and display images, generate synthetic speech and music, digitize speech, and control robots (albeit that they are somewhat simple in nature). The interfaces are off-the-shelf items with little or no modifications required and are driven by 8080 or Z-80 software. A microcomputer system with several of these interfaces can be linked (using a serial or parallel I/O port) to a larger mainframe that controls the microsoftware. Control commands can then be issued from the mainframe using a high-level language (e.g. FORTRAN or LISP) to the microcomputer which, in turn, performs the low-level (device) control. It is the intent of this paper to outline the design of a currently existing system that provides a hands-on artificial intelligence laboratory.

System Design Criteria

A primary design goal for the system was to emphasize the control (cybernetic) aspects of the hardware components while minimizing the student's need to program at the machine code level. To accomplish this it was decided to use the microcomputer (an IMSAI 8080) for low-level device control only. A large mainframe (XEROX Sigma 7) issues commands through a serial I/O port to the IMSAI to perform one of the following tasks.

1. Digitize an image received from a vidicon (television camera).
2. Display (on a television monitor) an image in memory.
3. Generate synthetic speech.
4. Digitize (and recognize) a speech pattern input through a microphone.
5. Move a minirobot and monitor its tactile sensors.

In order to control the devices, the student uses a set of subprograms in a high-level language on the mainframe. This serves several purposes.

1. The student does not have to learn a new language to pursue the ultimate task of controlling the devices.

2. Many students are simultaneously able to perform a major part of the debugging of their programs using the host's time-sharing system instead of competing for time on the microcomputer.

3. Complex computations, such as linked list manipulations needed for natural language processing or trigonometric function evaluation for robot movement control, are reserved for the host which has a much more powerful instruction set.

4. Parallelism is possible by overlapping low-level control processing with high-level action planning. For example, the microcomputer can be generating speech while the mainframe is computing the next position for the robot.

5. Modularization of control tasks is possible. This is of primary importance since control of a robot is generally too formidable for one person to program.

THE HARDWARE CONFIGURATION

Figure 1 illustrates the fundamental configuration of the system. All items were purchased from the vendors noted with the total cost of the system (excluding CRT and modem) being less than $6,000. The only link between the microcomputer and the host is a 9600 baud serial I/O port. In essence, the mainframe views the microcomputer as it does any other timesharing terminal.

THE SOFTWARE CONFIGURATION

The software used to control the devices is delineated into two categories, that which resides on the microcomputer and that which resides on the mainframe. The two sets must constantly communicate with each other through the high-speed data link using a rather straight-forward convention of handshaking signals. In simplified form, this convention can be summarized as the following sequence:

1. The host sends a special signal for the microcomputer to execute a jump to the (microcomputer program) address that it will receive next.

2. The special control routine then waits for parameters that is may need. In the case of robot movement, for example, these would include speed, linear distance to be moved or angle to be turned, and action to be taken if an object is touched (stop or push it).

3. The microcomputer then causes the specified device to take the appropriate action. Upon completion of the action, the host is notified of the outcome. In the case of the robot, this would include the number of steps actually taken and the states of the tactile sensors. Control then reverts to the host.

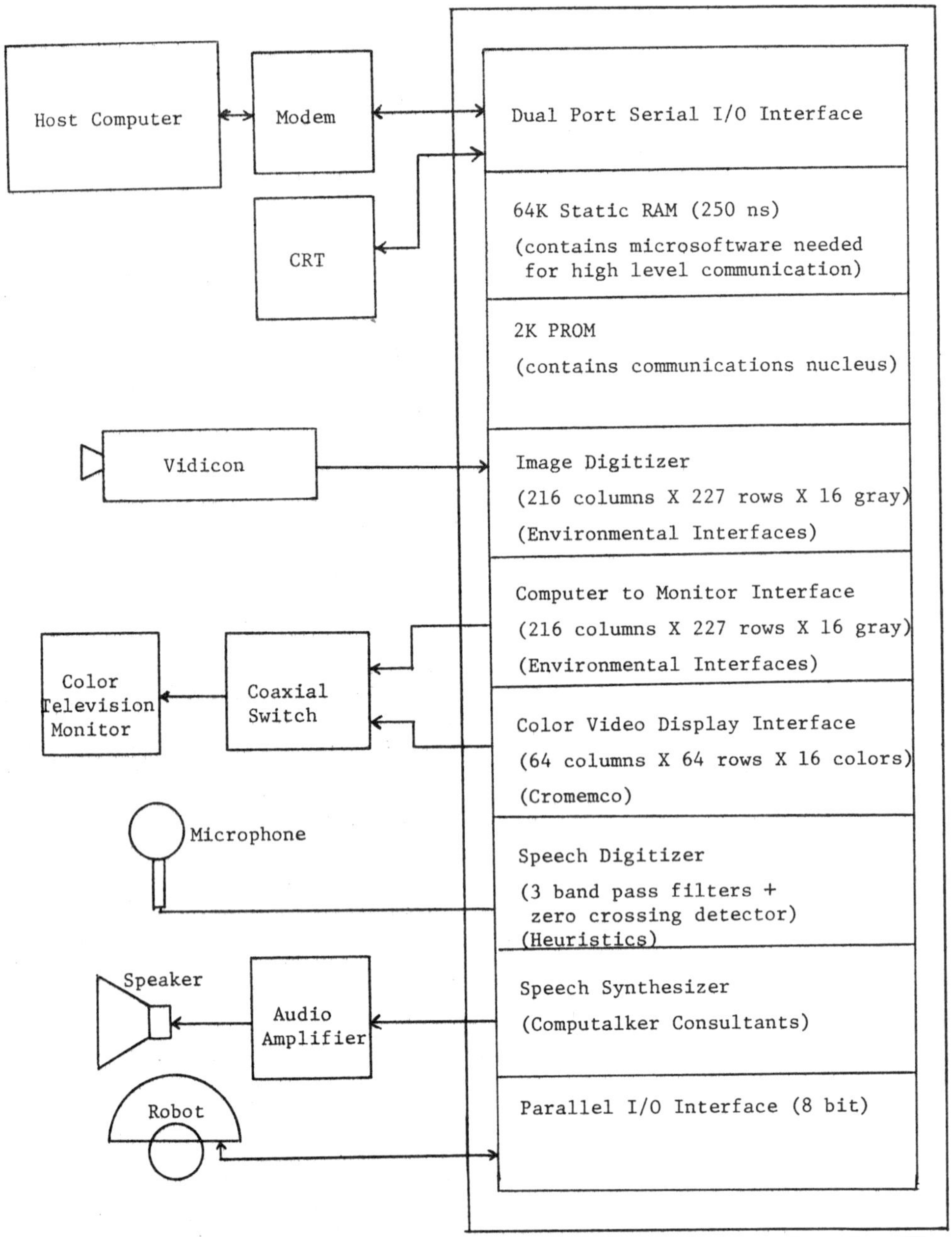

Hardware Configuration

Figure 1

With only minor exception, all of the actions taken by any of the microprocessor programs are followed by an acknowledgement or status readout of some kind. The type of response is of course dependent upon the nature of the specific device being controlled. Each of the responses is (bitwise) minimal to reduce delays caused by serial interfacing. These responses are summarized in the next several sections.

Speech Recognition

The host can request one of four actions for the speech software.

1. It can request the microcomputer to record the digital information in the next phrase spoken. The response signal is the index in the micro's local table of phrases where the new phrase resides.

2. The host can request the micro to forget a phrase (of a specified index), to which the micro simply responds that the phrase has been deleted.

3. The host can request the micro to forget all phrases, in which case the micro purges all recorded speech and is ready for a new training set.

4. The host can request the micro to recognize the next phrase input to the microphone. The micro software then waits until a phrase is spoken, performs a pattern match against the known library, and returns the index of the recognized phrase to the host.

A Speechlab digitizing interface (from Heuristics Inc.) is used for gathering the verbal input. The interface derives values from three band pass filters and a zero crossing detector [6]. Parameters extracted from the speech are either stored in a table (learning phase) or compared to values already in the table (recognition phase). Recognition is currently performed in the microcomputer using a modified version of a Chebychev difference measure program provided by Heuristics.

Speech Generation

When the host requests the microcomputer to "speak", it transmits a string of ARPABET phonetically spelled speech. The control software then generates a table in memory of control bytes to be sent to the speech synthesizer [7,8]. The microcomputer then sends control information to the synthesizer and acknowledges task completion to the host.

A Computalker CT-1 speech synthesizer in combination with CSR1 (Computalker Synthesis by Rule) software is used to generate spoken phrases. CSR1 translates the phonetic input string into a phonetic feature matrix which is scanned for combinations that must be adjusted to conform to English pronounciation. The matrix is then scanned by a parameter generator which causes the synthesizer to speak.

Visual Information Control

The host can request the microcomputer to alter the area in memory from which a direct memory access (DMA) interface is generating a low resolution color image or a higher resolution black/white image. The information sent is simply the memory location and the new content. The color interface is a Cromemco Dazzler which is capable of generating a 64 X 64 pixel matrix, each element of which can be one of 16 colors. Images generated by this interface are typically used to display crude maps of robot movement. The black/white interface is manufactured by Environmental Interfaces and generates a 227 X 216 pixel matrix of 16 gray levels. Its utility as a display unit is primarily for raw or enhanced images that have been digitized from a vidicon. Digitized visual information may also be transfered to the host.

Robot Control

The robot is sent commands to move in any one of six directions in two modes. The directions are forward, backward, sharp right and left turns, and wide right and left turns. The mode determines whether the control interface will cause the robot to stop when it strikes an object. The mainframe receives the result of the attempted action. The robot was purchased from Terrapin Inc. It has two bidirectionally current controlled (nonstepping) motors, a pen that may be raised or lowered, two LED nonsensory "eyes", and a two-tone beeper. In addition it has four orthogonally arranged touch sensors (microswitches). Microcomputer software commands the robot through a parallel I/O interface which, for the most part, was locally developed. Speed of the motors is controlled by issuing square waves of varying frequency and duration [9].

An Example of High-Level Control Software

One of the programs that illustrates the utility of the aforementioned software organization is controlling the robot using vocal input. In order to allow an arbitrary speaker to perform such control the following steps (in software) must be taken.

1. The microcomputer must purge all previously learned phrases.
2. A set of control phrases (e.g. "turn left", "move forward") is presented to the user. After each phrase is presented, the user must speak the phrase so that it can be learned. Typically, about 20 commands are learned.
3. The learning phase is followed by the command phase in which phrases that were learned are spoken, the index of the recognized phrase is returned, and the appropriate action is taken. Each action is accompanied by a phrase that is spoken by the synthesizer, which acknowledges the action taken (e.g. "I am turning left"). If the action involves robot movement, the color display map is updated.

The set of learned commands to which the robot responds include:

a. "affirmative" - used to affirm speech recognition table alteration commands.
b. "negative" - used to negate speech recognition table alteration commands.
c. "remember" - causes the next spoken phrase to be stored.
d. "relearn" - relearn a previously remembered phrase.
e. "forget" - purge a previously learned phrase from the phrase table.
f. "terminate" - stop the program.
g. "move forward" - move the robot forward.
h. "move backward" - move the robot backward.
i. "turn left" - turn the robot 90 degrees to the left.
j. "turn right" - turn the robot 90 degrees to the right.
k. "turn around" - turn the robot 180 degrees.
l. "backup" - retrace the last movement.
m. "go faster" - double the distance traveled in a single step.
n. "go slower" - halve the distance traveled in s single step.
o. "change color" - change the color of the robot trail on the display map.
p. "go home" - return to a previously specified location.

Advantages of the Modular Control Design

The mainframe high-level control subprograms are currently written in extended FORTRAN. These subprograms are collectively about 1500 lines of source. The structure of individual modules, however, makes programming of more complex tasks quite straightforward. For example, running a maze takes less than 100 statements. A second advantage is that the student may opt to replace any set of the control subroutines with a set of his or her own design as long as the communication protocols between the host and microcomputer are observed.

REFERENCES

1. "Industrial Robot Market to Reach $438 Million in 1985, F&S Says", Computer, Volume 12, Number 7, July, 1979, p. 102.
2. "Robots, Models, and Automation", Computer, Volume 12, Number 7, July, 1979, pp. 157-166.
3. P.H. Winston, Artificial Intelligence, Addison-Wesley, 1977, pp. 130-142, pp. 157-166.
4. B. Raphael, The Thinking Computer, W.H. Freeman and Co., 1976, pp. 249-288.
5. R. Hollis, "NEWT: A Mobile Cognitive Robot", Byte, Volume 2, Number 6, June, 1977, pp. 30-45.
6. B. Georgiou, "Give an Ear to Your Computer", Byte, Volume 3, Number 6, June, 1978, pp. 57-60.
7. D.L Rice, "Friends, Humans, and Country Robots: Lend Me Your Ears", Byte, Number 12, August, 1976.
8. D.L. Rice, "Hardware and Software for Speech Synthesis", Dr. Dobbs Journal of Computer Calisthenics and Orthodontia", Volume 1, April, 1976, pp. 6-8.

DESIGN AND EVALUATION OF AN IMAGE GENERATOR WITH INFORMATION EXCHANGE

Akira Ishikawa* and Hiroshi Mieno**

*Rutgers, The State University of New Jersey
**Tokyo University of Science

ABSTRACT

This paper deals with the preliminary design of an image generator with information exchange. An image generating process is considered as (1) retrieving information from a memory, (2) conceiving of their new combinations, (3) making novel judgments, (4) recording, and (5) making objective judgments. The proposed process is analyzed by use of the semigroup automata and stochastic processing theory.

KEYWORDS

Image creation, information exchange, intellectual development, image association, creativity analysis, value judgment.

INTRODUCTION

Lewin says that humans perceive and read their environment better as the development of their intelligence. In his view, then, knowledge space can be topologically constructed. It could generally be said that the development of knowledge can be made due, in part, to the enhancement of the intelligence through acquiring unknown knowledge or intelligence and, in part, to repetitive decompositions of the available intelligence space.

Modality of an intellectual development can also be shown by means of an algebraic machine, as is proposed by Hartmanis, in which synthesized knowledge is conjectured to be closed as a partial set of the knowledge space. Should this be granted, it implies that we may possibly design an accumulator of groups or semi-groups so that a new knowledge by other inputs can formulate other partial sets.

The human behavior model developed by Matsuda delineates the formation and conversion of images by human via a particular, cognizant, and meaningful symbol out of the sets of the signals as stimuli to be provided by one's action environment.

By an image we mean all the occupants in our brain at the usable time as part or whole of the meaningful memory filtered and stored. Some of them are well matured and will not evolve, while others have been and will be the outcome of two or more independent knowledges in the process of intellectual development. Note that a symbol encompasses semantic contents which emerge from the interaction between a set of signals, i.e., inputs, and the relevant object in a specific environment.

One of the present authors has designed the intellectual development process via the brainstorming approach that resulted in applying it to the computer-aided design and organization of a research and development group. In this process, a value judgment stage was incorporated posterior to the assessment as to whether or not a new image has been created.

Building upon previous studies including the ones as aforementioned, we have designed two intellectual development systems which permit information exchange in both serial and parallel components. By feeding new information successively, we simulate how the knowledge space would be enhanced and changed and how it would be varied as the norm of value judgment is applied to both systems.

ARRANGEMENTS

In preparation for establishing the model, we first of all will define basic terms that subsequently appear.

An image is defined as all the occupants in the brain, as defined in the previous section, which are not simply receipts of cognizant, external objects, but also the ones restructured through the possessor's filters.

Let two images at time t be i_1^t and i_2^t and their finite sets, $I_1^{t_1}$ and $I_2^{t_2}$, where

$$I_1^{t_1} \equiv \bigcup_{t=0}^{t_1} I_1^t, \quad I_2^{t_2} \equiv \bigcup_{t=0}^{t_2} I_2^t$$

Knowledge space are finite sets, $N_1^{t_1} \equiv \varphi(I_1^{t_1})$ and $N_2^{t_2} \varphi(I_2^{t_2})$ which correspond to I_1^{t1} and I_2^{t2}, respectively. Their finite sets, N_1^0, N_2^0 are called the initial knowledge set.

An image consists of four components as stated below:

1. $\varphi^{-1}(N_1^0)$, $\varphi^{-1}(N_2^0)$ retrieved randomly from N_1^0, N_2^0
2. i_1^{t-1}, i_2^{t-2}
3. i_2^{t-1}, i_1^{t-1} in other systems
4. Image generation via the association of β with α, or γ with α means that no image has yet to be created, i.e., $i_1^t \notin I_1^{t-1}$, $i_2^t \notin I_2^{t-1}$.

The norm of each system is subject to the value judgment inherent in each system. Consequently, the generation of an image is repeated S times (or S hours) so that it may meet the predetermined norm. The expression of an image at this point is $i_1^{t_1 t_2 \cdots\cdots t_s}$, $i_2^{t_1 t_2 \cdots\cdots t_s}$. Those sets which have met the norm, R_1, R_2 are expressed as $i_1^{t_1 t_2 \cdots\cdots t_s} \in (I_1^{t_1 t_2 \cdots\cdots t} \cap R_1)$, $i_2^{t_1 t_2 \cdots\cdots t_s} \in (I_2^{t_1 t_2 \cdots\cdots t} \cap R_2)$

Furthermore, a set of the whole knowledge, N, that has nothing to do with the system, and a set of the whole image, I, are denoted as follows: N_1, $N_2 \subset N$, i: $\in I$, $R_1 \cdot R_2 \subset I$.

A summary of the intellectual development system with notations stated above is shown in Fig. 1 (abbreviated)

Two intellectual development systems have been designed. One is represented by a group accumulator M_{g1}, which has itself inputs, state and outputs, partial group, G_1, of a group of image, G.

$$M_{G_1} = (G_1, G_1, G_1, \delta G_1, \lambda G_1)$$

δ_{G_1} and λ_{G_1} are mappings of state transition and output generation, respectively.

It is an independent system, whereby two subsystems do not exchange information with each other.

The other is an interdependent system in which two subsystems do exchange information. This system is further classified into serial and parallel systems. The former is serial in nature, as shown in Figure 2, and the output from Subsystem 1, posterior to the value judgment state, becomes the input for Subsystem 2. The latter is parallel in that the outputs of both Subsystems are synthesized and judged by the norm of R_2, succeeded by that of R_1, as shown in Figure 2 (abbreviated)

EFFICIENCY OF IMAGE CREATION AND REQUIRED TIME

First of all, let's consider the efficiency of image creation. In general, it can be said that the greater the number of trials we have to repeat, the less efficient the system is in terms of the generation of images within a given time span, assuming that we have an elucidated norm in our mind.

In other words, when we want to produce a number of products which may meet the predetermined norm, as trials increase, we would be able to obtain more qualified products, inasmuch as learning and creative effects continue. Accordingly, for simplicity, we herein define a measure of efficiency as a number of products created within a specified unit time that is the time needed for one trial.

Next, we consider the required time for the creation of one qualified product. Extraction of an image from the knowledge space commences with a null state and then is converted to the next stage that is equivalent to the image of an initial input. Then, the image already available is associated with a new input image extracted randomly from the knowledge space.

Now, let a set of images be an interval [a,b] on R', where $a<b$ and let $\{a,b\}$ an input image. For simplification, the creation of an image is assumed to find the midpoint between the state image and input image.

Let's also consider an event set, Ω , that creates an image, σ family of sets, $\mathcal{F}$, upon the generation of an image, and stochastic space (Ω, $\mathcal{F}$, P).

If [a,b] is to be a random variable, then the generation of an image may be explained by a binomial distribution and also Markov process, since one state is not irrelevant to the next state with the features of automata with a finite memory.

It is naturally considered that the segmentation of [a,b] into the minimum components and examination of each of them should be explored so that a required image may be included in it. The time required to find a new image is the basis of measurement.

The required time is subject to change, depending on the severity of the norm and also a number of norms appropriate to each subsystem. If the segmentation procedure is to be adopted, it may facilitate an examination and confirmation of a new image.

ILLUSTRATION

The object in this experiment is the stripe-patterns (vertical stripes). It is a set of varied colour stripes, Σ, ranging from yellow, Y, to blue, B, [Y,B]. The number of colour elements in a given stripe-pattern, $W(\in \Sigma) = 3 \sim 8$. (In the initial experiment, the number is three.) If the same colour is adjacent each other, the breadth of a single stripe becomes double.

Among Σ, let's assume that a finite set, σ, has minimum, cognizant components as follows:

$$\sigma \equiv \{Y^2, YG, G, GB, B^2\}$$

where $Y^2 \equiv Y \cdot Y$, $YG \equiv Y \cdot G$, $G \equiv Y \cdot B$, $GB \equiv G \cdot B$, $B^2 \equiv B \cdot B$

$Y^2 \equiv [Y^2, YG)$, $YG \equiv [YG, G)$, $GB \equiv (G, GB]$

$B^2 \equiv (GB, B^2]$

Accordingly, a finite set, $L \subset \Sigma^*$ (a set of Σ) is obtained so that it is a set with a complete lattice, $|W| = 3$, as shown in Fig. 3 (abbreviated)

The stripe-patterns mentioned above can be considered as a set of images. The association of one stripe-pattern with another is represented by mixed colours in accordance with the location of each colour.

In the case where W = 3, the number of elements associated with L becomes 125, and L is equal to a set of the whole images. When information is exchanged among elements within set L, the number of image generations in both systems with the change in the number of segmentations is shown in Fig. 4. In addition, the relationship between the number of colour segmentations and the summation of created images is shown in Fig. 5 (Figures 4 and 5 abbreviated)

In the process of running simulation, I_2^0 was set to be $\{(Y,Y,Y), (Y,B,B), (B,Y,Y)\}$.

There are two cases where $I_1^0 = I_2^0$ and $I_1^0 \neq I_2^0$, such that $I_1^0 = \{(B,B,B), (Y,B,B), (B,Y,Y)\}$. The summation of the number of images created in both systems during fifty trials (i.e., fifty unit periods) is shown in Table 1. These numbers shown in the numerator are the numbers of sum-set elements beyond the norm, while those at the denominator are the numbers of common-set elements. These numbers are the expected value after repeated trials. (Table 1 abbreviated)

Using the Table and Figures, the following findings can be inferred:

1. On the whole, in these interdependent systems, the parallel type is less efficient than the serial type. (There is a clear difference in the number of image generations, particularly when the norm is non-existent.)
2. No distinction is observed between the independent and interdependent systems, except for the parallel type.
3. However, it seems that the serial type with abundant knowledge and different norms has demonstrated the highest efficiency in terms of the number of sum-set elements.
4. As the number of colour segmentation increases, the summation of the number of images created increases at more than a proportionate rate.
5. As the number of location segmentation increases, the summation of the number of images created is generally enhanced, although a noticeable dispersion is

seen.

CONCLUDING REMARKS

In this paper, we have illustrated one of the simplest cases with the simplest form of information exchange. Certainly, numerous routes are open to be followed, in addition to the increasing number of trials and the increased number of colours in association with the type of the subject contained in this experiment.

In addition, we cannot disregard the conversion of one image to another in circumstances where no external stimuli are received. These issues may be related to deeply imbedded psychology, to graphic structure, or, on the surface, to the design and development of relational data bases.

Furthermore, a knowledge space should be semantically sensitive and responsive so that the same symbol (a word) may induce quite different images through the semantic manipulation. This would naturally necessitate the analysis of natural languages from the different perspectives. Through them, we can then explore semantically the conversion of more aggregative meaning--a phrase or sentence rather than just a few words--as the object of a creative manipulation in our intellectual development.

COORDINATION OF TWO ADJACENT SERVICE NETWORKS

Niv Ahituv*
Oded Berman**

* Faculty of Commerce and Business Administration, The University of British Columbia.

** Faculty of Management, The University of Calgary.

Abstract

Coordination between service networks (e.g., ambulance, courier) is quantitatively analyzed. The notion of coordination means that while each local government is allowed to maintain some degree of autonomy, it may also rent or render services to an adjacent municipality under certain conditions. Dispatching and render policies are discussed for two adjacent service networks.

1. Introduction

Careful design and efficient management of service networks is essential both in the public sector (e.g., ambulance) and the business sector (e.g., courier, maintenance services). Research in this area deals with problems of location [1,2,3], repositioning [3,4], and dispatching [5,6] policies. Most of the research focuses on issues related to an individual service network. In practice, there are situations where coordination between adjacent networks might be considered:

1. Each local government prefers to have a certain degree of autonomy in designing and operating its network; it primarily wishes to satisfy the needs of its local citizens. 2. Global optimum might imply some difficulties due to data gathering and complex computation, that can be avoided when considering several local policies and coordinating them. 3. Hiring unlocal services might be considered very seldom, thus an effort to solve a global model is not worthwhile. 4. A local government may wish to charge outsiders. Justification for this could be the decrease in quality of local services which can be calculated only after a local optimal policy is firstly computed, and deterioration due to external requests is then assessed.

2. Dispatching Policy for Two Separate Service Networks

The following analysis deals with two service networks. The ideas will be shown through an example, while the theoretical technicalities will only be noted in reference numbers.

Let N^1 be a service network illustrated in Figure 1. N^1 comprises of three nodes, 1, 2, and 3. A service unit (server) is positioned at node number 1 (indicated by a double circle). The figures near the links indicate travel times (or distances) between nodes; travel times are assumed to be deterministic. The calls for services are assumed to be Poisson distributed, at rate of $\lambda = 3$. The proportion of the calls in each node is given, denoted by h_i, ($h_i = 1/3$; i=1, 2, 3). Service times are expenontial distributed in a ratio of $\mu = 1$.

We can now use the Hypercube Model [7] to obtain some results pertaining to the analysis. Let $P(\hat{\hat{1}})$ denote the steady state probability that node 1 is occupied with an available server and $P(\hat{1})$ that node 1 is vacant. The steady states probabilities obtained as follows are: $P(\hat{\hat{1}}) = .25$; $P(\hat{1}) = .75$

For the purpose of clarification we will briefly explain how $P(\hat{\hat{1}})$ and $P(\hat{1})$ were obtained. (Later in this paper similar computations will be omitted). The first step is to set a dispatching policy, as exhibited in Figure 2. Since N^1 is a very simple network, the dispatching policy is rather simplistic: at state $\hat{\hat{1}}$ the server located at node 1 is dispatched to any calling node; at state $\hat{1}$ no server is dispatched.

Having the dispatching policy, a transition rates diagram can be drawn as in Figure 3. The nodes in Figure 3 represent states; the arrows represent transitions between states; the numbers near the arrows are the transition rates. The detailed balance equations for the system are:

$P(\hat{\hat{1}})(3) = P(\hat{1})(1)$; $P(\hat{\hat{1}}) + P(\hat{1}) = 1$

The solution for the above results in the values of the steady state probabilities. Assume that N^1 operates on its own so that the dispatcher cannot request any service from another network. Assume that R^1 is the 'cost' (penalty) of not providing a service upon receiving a call; R^1 is measured in time (or distance) units. The expected response time for N^1 (denoted by ERT_0^1 where the subscript 0 indicates 'no assistance' from adjacent networks) can be then calculated as follows (d(i,j) denotes the travel time between nodes i and j):

$$ERT_0^1 = P(\hat{\hat{1}}) \sum_{j=1}^{3} h_i d(1,j) + P(\hat{1})\ R^1 = (.25)\ (\frac{5}{3}) + P(\hat{1})R^1 = .4166 + .75\ R^1$$

Let N^2 be the service network illustrated in Figure 4 with two service units located at nodes 4 and 7. Again assume $\lambda = 3$, (the value of λ can be different for N^1 , N^2) $\mu = 1$ for the two servers and R^2 is the 'cost' of not providing a service upon receiving a call. Following the same calculations as for N^1, we get the steady state probability values for states in N^2 ;
$P(\hat{\hat{4}},\hat{\hat{7}}) = .1176$; $P(\hat{4},\hat{\hat{7}}) = .1764$; $P(\hat{\hat{4}},\hat{7}) = .1764$; $P(\hat{4},\hat{7}) = .5296$
Using the steady state probabilities, we can obtain the expected response time ERT_0^2 for N^2 :

$$ERT_0^2 = P(\hat{\hat{4}},\hat{\hat{7}}) \sum_{j=4}^{7} h_j \min\{d(4,j),\ d(7,j)\}$$

$$+ P(\hat{4},\hat{\hat{7}}) \sum_{j=4}^{7} h_j d(7,j) + P(\hat{\hat{4}},\hat{7}) \sum_{j=4}^{7} h_j d(4,j) + P(\hat{4},\hat{7})\ R^2 = .6762 + .5249R^2$$

3. Coordination Between Two Service Networks

Assume that N^1 and N^2 are connected via a link between nodes 3 and 4 whose travel time equals 5 units (Figure 5). We will assume that only N^1 is permitted to request service from N^2, and not vice versa, because N^1 does not wish to give away its sole server. Under this assumption, one could set a dispatching policy that would satisfy the dispatching constraint and, thereafter, solve the hypercube model for the unified network N. However, we assume that each network is administered by a different manager, who does not wish to obtain a global optimal solution but only to examine the implications of coordination on the local network. The manager perceives the adjacent network as a 'black box' which generates either 'demand' for or 'supply' of services. Thus, coordination is obtained by means of information transfer (negotiation) between managers. We will now describe the negotiation process.

Suppose the manager of N^1 (denote MN^1) wishes to call for N^2 services whenever N^1 is in state $(\hat{1})$. MN^1 notifies $P(\hat{1})$ to the manager of N^2 (MN^2). MN^2 views N^1 as an additional entity which generates demand at the rate of $\hat{\lambda} = \lambda P(\hat{1}) = (3)(.75) = 2.25$. MN^2 has to decide under which terms he or she will provide services to N^1. Suppose the decision is to respond only when N^2 is at state $(\hat{\hat{4}},\hat{\hat{7}})$, i.e., both servers are not busy. The steady state probabilities can be obtained by solving a set of linear equations, resulting in:

$P(\hat{\hat{4}},\hat{\hat{7}}) = .0707$; $P(\hat{4},\hat{\hat{7}}) = .2057$; $P(\hat{\hat{4}},\hat{7}) = .1659$; $P(\hat{4},\hat{7}) = .5575$.

Note that $P(\hat{4},\hat{7})$ has been increased from .5296 to .5575. This will affect the overall performance of N^2 as indicated by the expected response time (denote ERT_1^2). To calculate ERT_1^2 we have to take into account that nodes 1,2, and 3 are served by N^2 only when states $(\hat{1})$ and $(\hat{\hat{4}},\hat{\hat{7}})$ concur. The demand proportions (h_j) need to be scaled accordingly.

$$ERT_1^2 = P(\hat{\hat{4}},\hat{\hat{7}})\ [P(\hat{1}) \sum_{j=1}^{7} \frac{hj}{2} \min\{d(4,j),\ d(7,j)\} + (1 - P(\hat{1})) \sum_{j=4}^{7} h_j \min\{d(4,j), d(7,j)\}]$$

$$+ P(\hat{4},\hat{\hat{7}}) \sum_{j=4}^{7} h_j d(7,j) + P(\hat{\hat{4}},\hat{7}) \sum_{j=4}^{7} h_j d(4,j) + P(\hat{4},\hat{7})\ R^2 = .875 + .5575R^2$$

Denote $C_1^2 = ERT_1^2 - ERT_0^2 = .1988 + 0.3262\ R^2$

C_1^2 reflects the reduction in performance that N^2 has to bear due to services rendered to N^1. C_1^2 is spelled in time units and its transformation to cost (money-wise) is far from being straightforward. Hence, the process of charging for services appears to carry many difficulties. Instead, we would like to suggest a process that adheres to the notion of negotiation without having to set prices. Suppose MN^2 informs MN^1 that only at $\hat{P} = P(\hat{\hat{4}},\hat{\hat{7}}) = 0.0707$ of the time N^2 will respond to calls requested by N^1. MN^1 can now assess the improvement in performance due to services that will be rendered from N^2.

$$ERT_1^1 = P(\hat{\hat{1}}) \sum_{j=1}^{3} h_j d(1,j) + P(\hat{1})\ (1-\hat{P})\ R^1 + P(\hat{1})\ \hat{P}\ R^* \qquad (1)$$

where R^* is the 'cost' (in terms of distance or time) for services rendered from N^2. Assume for a moment that $R^* = 0$, than (1) results in $.4166 + .6969R'$

Denote $C_1^1 = ERT_0^1 - ERT_1^1 = .0531R^1$

C_1^1 reflects the maximum improvement gained by N^1 due to additional service provided by N^2. If one could obtain the monetary equivalent of C_1^1, its value would be the upper bound for the price MN^1 would be willing to pay for a service. Unfortunately, monetary utility as a function of response time is difficult to obtain, particularly for public services. However, we can obtain the initial values and the marginal changes in expected response times. Hence assuming for example that users in both N^1 and N^2 have linear utility functions (as a function of expected response time), we can confine ourselves to comparisons of C_1^2 and C_1^1. As a rule of thumb we may say that there are grounds for negotiation as long as $C_1^1 \geq C_1^2$. The exchange of probabilities and 'costs' between the managers of the networks is an exchange of information that can develop to negotiation as we will now portray.

Suppose MN^2 ranks various render policies according to the decrease of performance they cause to N^2. For example, rendering service at state $(\hat{\hat{4}},\hat{\hat{7}})$ is the least harmful; states $(\hat{\hat{\hat{4}}},\hat{\hat{\hat{7}}})$ or $(\hat{\hat{\hat{4}}},\hat{7})$ will rank second; states $(\hat{\hat{\hat{4}}},\hat{\hat{\hat{7}}})$ or $(\hat{4},\hat{\hat{7}})$ will be ranked third, etc.Obviously, each policy differently affects the performance of N^1. Thus, MN^1 may have his or her own rank ordering of the various offers given by MN^2. The two managers would reach an agreement if certain performance values in the two rank orderings coincides. This is demonstrated in Figure 6. Obviously, the values of R^1 and R^2 (the 'cost' of not responding to a call) would determine whether MN^1 and MN^2 can work out an agreement.

4. Some Concluding Remarks

The discussion in the paper has intended to raise some new ideas of coordination, rather than to be comprehensive. In fact, there are many topics still to be dealt with. For instance, both networks may hire or render services at time; both networks consider various response and request policies simultaneously; how can the model be expanded to fit any number of adjacent networks, rather than two? We hope that ensuing research will provide answers to these and to other pertinent questions.

REFERENCES

1. Handler, G.Y. and Mirchandani, P.B., Location on Networks, MIT Press,1979.
2. Berman, O., Larson, R.C., Odoni, A.R., "Development in Network Location With Mobile and Congested Facilities" to be published in the special issue on location in the European Journal of Operations Research.
3. Berman, O., "Repositioning of Distinguishable Urban Service Units on Networks", to be published in Computer and Operations Research.
4. Kolesar, P., Walker, W.E., "An Algorithm for the Dynamic Relocation of Fire Department Companies," Rand Institute R-1023-NYC, New York, 1972.
5. Carter, G.M., Chaiken, J.M. and Ignall, E. "Responses Areas for Two Emergency Units," Operations Research, 20, No. 3, May-June 1972, pp.571-594.
6. Jarvis, J.P., "Optimization in Stochastic Service Systems with Distinguishable Servers," IRP-TR-19-75, MIT O.R. Center, 1975.
7. Larson, R.C., "A Hypercube-Queuing Model for Facility Location and Redistricting in Urban Emergency Services," Computers and O.R. pp. 67-95 (1974).

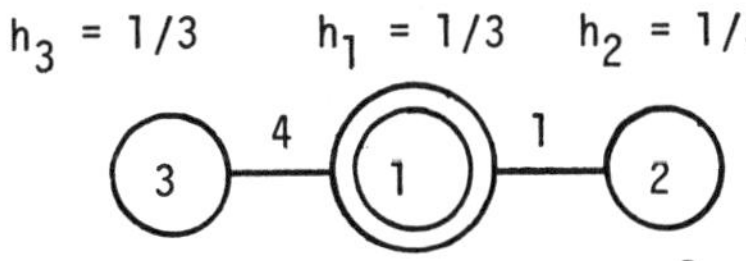

Figure 1: Service Network N^1

State \ Node	1	2	3
$\hat{\hat{1}}$	x	x	x
$\hat{1}$	-	-	-

Figure 2: A dispatching policy for N^1 (x denotes a dispatching while - denotes no dispatching)

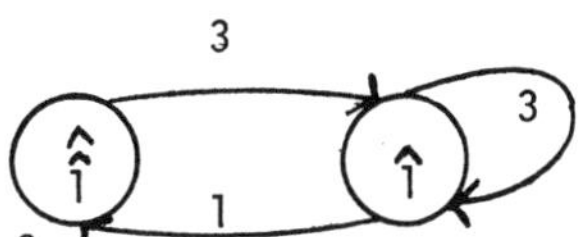

Fig: 3 The Transition Rates Diagram

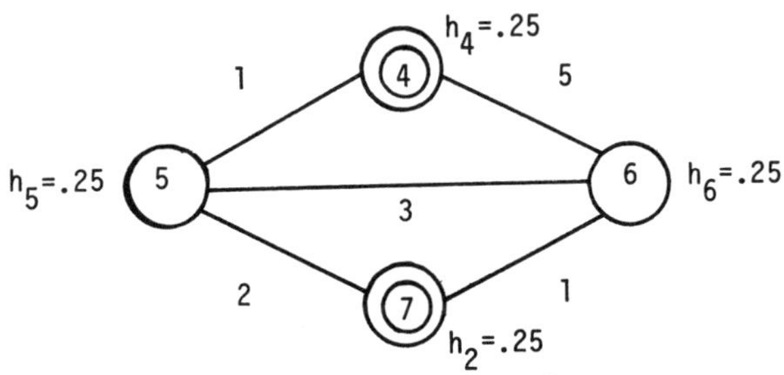

Figure 4: The Service Network N^2

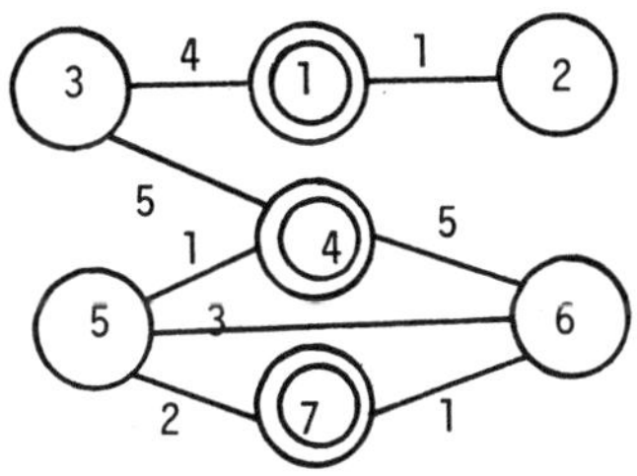

Fig 5:The Combined Network N.

States at which N^2 offers services	Decrease in N^2 performance (compared to no coordination)	Improvement in N^1 performance (compared to no coordination)
$(\hat{\hat{4}},\hat{\hat{7}})$	$0.1988 + 0.03265R^2$	$0.0531R^1$
$(\hat{\hat{4}},\hat{\hat{7}})$ or $(\hat{\hat{4}},\hat{7})$	$0.3556 + 0.08885R^2$	$0.13113R^1$
$(\hat{\hat{4}},\hat{\hat{7}})$ or $(\hat{\hat{4}},\hat{7})$ or $(\hat{4},\hat{\hat{7}})$	$0.7163 + 0.163R^2$	$0.234R^1$

Figure 6: Decrease and Improvement of Performance in Network N^2 and N^1.

KNOWLEDGE STRUCTURES BASED ON GENERAL AUTOMATA COMPARED WITH OTHER TYPES

Eldo C. Koenig

Computer Sciences Department
University of Wisconsin, 1210 W. Dayton St.
Madison, Wisconsin 53706, U.S.A.

ABSTRACT

There are three principal types of structures for representing knowledge. They are:
1) Semantic Nets
2) Conceptual Dependency Nets
3) General Automata Graphs

The ability of each of these to represent knowledge for each of the two kinds of knowledge, Declarative and Procedural, is examined, and knowledge contained in example sentences for the two kinds is structured and discussed.

KEYWORDS

Knowledge structures; general automata; semantic nets; conceptual dependency nets.

INTRODUCTION

Man receives knowledge and operates on the stored knowledge structures to perform in his environment. Knowledge that is stored is knowledge that is either received through natural language or is directly observed. The knowledge structures are operated upon to translate the knowledge into actions which may be observed externally. The actions may be either in the coded symbolic system of actions, i.e., spoken or written language, or in systems of actions involving manipulation and locomotion that exists also in lower animals.

Storing the meaning of sentences as knowledge structures involves more than a syntactical analysis as is demonstrated by the following two example sentences:

"John struck the table with a hammer."

"John struck the table with a glass top."

The sentences are syntactically identical but have different meanings. A distinction must somehow be made in the use of the hammer and the glass table top.

Man is also able to store the meaning of sentences involving interaction; for example:

"John gave a book to Jim."

An interaction takes place between John and Jim in the transfer of the book.

There are three principal types of knowledge structures aimed at storing the meaning of sentences. They are:

1) Semantic Nets
2) Conceptual Dependency Nets
3) General Automata Graphs.

The desire is to replace man by a computer in an interactive system and have it perform in the same manner. The ability of each of these three types of knowledge structures to represent knowledge for each of the two kinds, Declarative and Procedural is examined, and knowledge contained in example sentences for the two kinds is structured and discussed.

SEMANTIC NETS

The concept of semantic nets had its beginning with Quillian (1968). The elements of semantic nets are propositions. A proposition consists of a predicator and n arguments. The representation of a sentence as a proposition is enclosed by parentheses with the predicator written first followed by the argument list (Kintsch, 1974). Thre basic types of propositions are, adjective, noun, and verb propositions, and the predicator is an adjective, noun, verb, respectively. Examples are the following

"Mary is beautiful."	(BEAUTIFUL,MARY)
"Socrates is a man."	(MAN, SOCRATES)
"John struck the table."	(STRUCK,JOHN,TABLE)

For the second example, IS-A might have been used as a predicator. Then the representation would be (IS-A,SOCRATES,MAN).

The adjective and noun propositions are sometimes called appearance-propositions, coloring-propositions, location-propositions or size-propositions (Schubert, Goebel, Cercone, 1979). That is to say, they deal with declarative knowledge or non-procedural knowledge.

Links and nodes are common to all formalisms of semantic nets, and a semantic network is viewed as a directed labeled graph where the nodes represent concepts and the edges represent relationships between these concepts (Brachman, 1979; Lovesque and Mylopoulos, 1979). Simmons and Slocum (1972) represent all information in their semantic network by binary relations and Rumelhart et al. (1972) represent human conceptual knowledge as a labeled connected graph.

Consider semantic networks to represent the knowledge of the three sentences:

1) "Chris is a girl."
2) "John strikes the table."
3) "John gives the book to Jim."

Fig. 1a shows a semantic net representation of the knowledge of the first sentence following the notation of Schubert et al. (1979), and Fig. 1b shows the representation using the IS-A notation of Shapiro (1979). The knowledge of this first sentence is classified as declarative. Declarative knowledge is defined as the knowledge of "knowing that", or is knowledge of facts about the world (Norman and Rumelhart, 1975; Anderson, 1976). It is communicated verbally and is acquired suddenly.

The network diagrammed in Fig. 1c is for the knowledge of the second example sentence (Schubert et al, 1979). The concept "strikes" is used to relate the particular concepts "John" and "table". The knowledge relates to the procedural type knowledge. Procedural knowledge is defined as the knowledge "of knowing how" or is knowledge about how to do something (Norman and Rumelhart, 1975; Anderson, 1976).

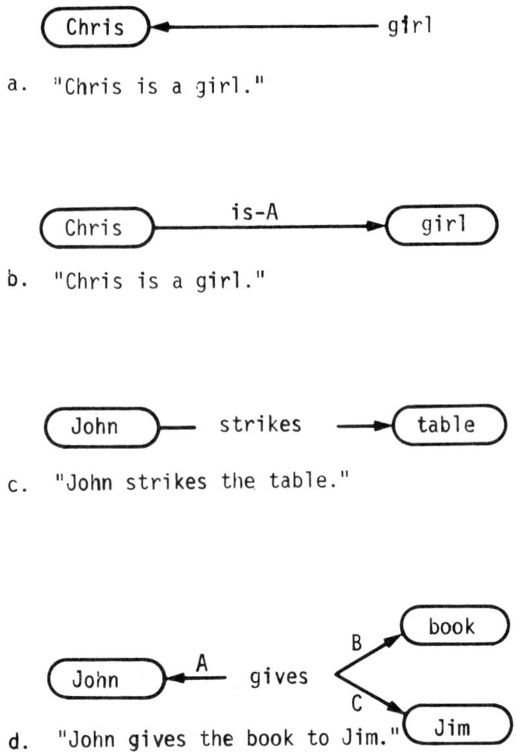

Fig. 1. Semantic net representation of knowledge of three example sentences.

Fig. 1d is a semantic network that describes the third example sentence, "John gives the book to Jim." The knowledge also relates to the procedural type knowledge, but it involves the interaction of two people, John and Jim. Argument order is given by the link lables A,B,C (Schubert et al., 1979).

CONCEPTUAL DEPENDENCY NETS

Schank presented a notion of conceptual dependency as a theory of natural language understanding (1972). The verb proposition is represented as an element of conceptual dependency nets, and therefore procedural knowledge is emphasized. The concept admits to only a small number of perhaps a dozen of very general primitive predicates (e.g. GRASP,INGEST,ATRANS) and cases (e.g., INSTRUMENT,RECIPIENT) and claims this set is adequate to represent the knowledge expressed in any natural language. The question may be asked whether these are truly primitive knowledge pieces (Brachman, 1979). For example, consider the two sentences

"John eats the food."
"John consumes the food."

The words "eat" and "consumes" are interpreted as the same ACT and are both described by the same primitive predicate INGEST. Schubert et al. (1979) argue that primitive representations are cumbersome and unnecessary.

Consider conceptual dependency nets for representing the three sentences previously analyzed for semantic nets:

1) "Chris is a girl."
2) "John strikes the table."
3) "John gives the book to Jim."

The first sentence does not contain any of the primitive acts and is not represented by any of the conceptual dependency nets for acts.

Fig. 2a shows the conceptual dependency network for the second sentence. John is recognized as the actor of a striking action in which the table is the conceptual object (Schank, 1975). But the action of striking is not considered conceptually primitive. Rather it consists of a PROPELling of an object X toward some goal, resulting in the physical contact of X and the goal. It is predicted that the table is the goal of the propelling action. The underlying ACT is PROPEL which requires, as do all the primitive ACTs, certain conceptual cases obligatorily. For PROPEL, those cases are ACTOR, OBJECT, DIRECTIVE (TO and FROM) in addition to TIME, LOCATION and INSTRUMENT cases, which are requirements of all acts. The cases OBJECT (O), INSTRUMENT (I) and LOCATION (L) remain unspecified, i.e., X,Z,Y are unspecified in Fig. 2a.

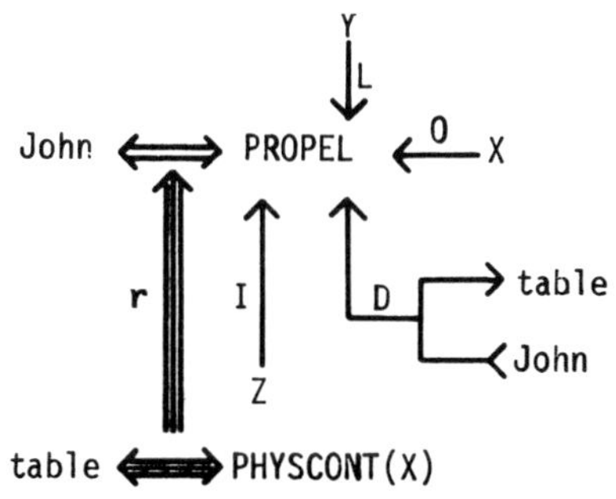

a. "John strikes the table."

b. "John gives the book to Jim."

Fig. 2. Conceptual dependency net representation for knowledge of the example sentences.

Fig. 2b is a conceptual dependency network that describes the third example sentence, "John gives the book to Jim.". The focus is the abstract transfer, and the book is changed by the ATRANS (one of the ACTs) from the donor in the RECIPIENT case (R) to the receiver. Instead of link labels A,B,C,---used in Semantic Nets (see Fig. 1d) we could have used labels suggestive of the role played by each argument in a proposition such as AGENT, OBJECT, RECIPIENT as was done by Schank. (Schubert et al., 1979). In fact, the role label is the notation used by Norman, Rumelhart (1975) in their networks.

It is apparent that the emphasis of conceptual dependency nets is upon procedural knowledge.

GENERAL AUTOMATA GRAPHS

Much of the knowledge of the world is about automata and histories of automata. The concept provides order and organization to the world and has been the subject of research of the author beginning over ten years ago. General automata graphs for structuring knowledge were introduced in the literature 1969 (Koenig, Frederick, 1969) and research has resulted in a continuing sequence of publications (Koenig, Frederick, 1970, 71, 72; Koenig, 1972, 73, 74, 75, 78, 79).

The concept of automata is not new (Chapuis, Droz, 1958). In the 17th century, lower animals were viewed as natural automata by Descartes, and mechanical devices reflecting the technology of the times were viewed as artificial automata. Leibnitz regarded clockwork as the model for all automata, and beginning the 19th century, both natural and artificial automata were studied in relation to energy. This century has witnessed a broad concept of automata viewed in total in its interaction with its environment.

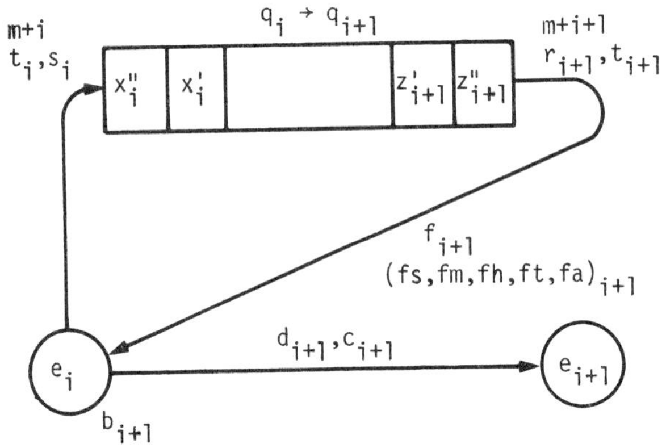

a. Diagram of operation of a general automaton.

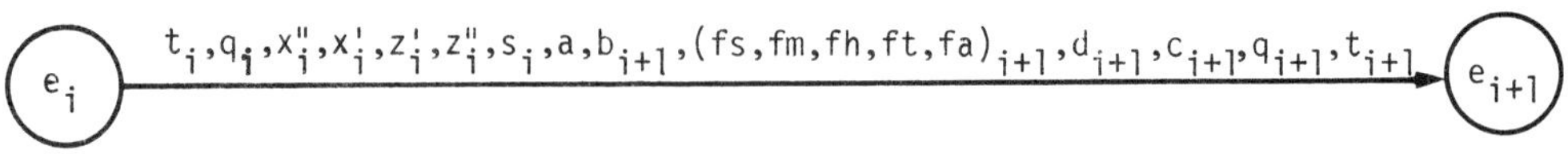

$(e_i, t_i, q_i, x''_i, x'_i, z'_i, z''_i, s_i, a, b_{i+1}, (fs,fm,fh,ft,fa)_{i+1}, d_{i+1}, c_{i+1} q_{i+1}, t_{i+1}, e_{i+1})$

b. Environment graph containing the information of a. and the corresponding graph tuple.

Fig. 3. General automaton operation.

The operation of a general automaton is diagrammed to operate as shown in Fig. 3a and is described as follows: Given a discrete moment $m + i$, at time t_i a general automaton a is accessing one and only one environment (stimulus location) e_i, receiving one and only one stimulus s_i, through one and only one auxiliary receptor x''_i and only one principal receptor x'_i, is in one and only one state q_i, and has one and only one time t_i associated with the moment $m + i$. Multiple accesses of a same environment during a period of consideration can be made only by the subject automaton. (For a system of automata, multiple accesses of a same environment can be made by different component automata of the system). Furthermore, during the next moment $m + (i+1)$, at time t_{i+1} the automaton a produces one and only one response r_{i+1} through one and only one principal effector z'_{i+1} and one and only one auxiliary effector z''_{i+1} based on the aforementioned conditions. The response r_{i+1} consists of a transformation response f_{i+1}, a spatial change response d_{i+1}, and a time change response c_{i+1}. The transformation response consists of responses interpreted as stimuli for sight fs_{i+1}, smell fm_{i+1}, hearing fh_{i+1}, touch ft_{i+1}, taste fa_{i+1}, and is recorded as a response b_{i+1} at the environment e_i accessed at the first moment $m + i$.

If the automaton is removed from the diagram of Fig. 3a and the information it contained is included in the label for the connecting line of two points e_i and e_{i+1}, Fig. 3b results. This is the basic automaton graph, called the environment graph for storing knowledge. There is an environment graph function, where $(e_i,t_i,q_i,x''_i,x'_i,z'_i,z''_i,s_i)$ is an element from the domain and $(b_{i+1},(fs,fm,fh,ft,fa)_{i+1},d_{i+1},c_{i+1},q_{i+1},t_{i+1},e_{i+1})$ is the image element. The requirement for single valuedness of the function must always be met. There are features like this resulting from the rigorous treatment of the problem of knowledge structures that should give a computer following these principles much greater power than the human for displaying intelligence.

The tuple of Fig. 3b is used for either storing knowledge of a sentence or storing knowledge that is directly observed. Consider the three example sentences previously analyzed:

1) "Chris is a girl."
2) "John strikes the table."
3) "John gives the book to Jim."

The first sentence is considered a class 4 sentence for this analysis (Koenig, 1973, 1974). The knowledge contained in this sentence is stored in the graph tuple of Fig. 3b as shown in Fig. 4a. The unknown or nonexistent values are indicated by ,-. The class of sentence identifies Chris as an automaton and the notation for Chris is (Chris,-,-,girl) which replaces a in the general graph tuple of Fig. 3b. The corresponding environment graph is also shown in Fig. 4a.

The second sentence is considered a class 1 type sentence (Koenig, 1973, 1974) and the knowledge contained in this sentence is stored in the graph tuple as shown in Fig. 4b. The corresponding environment graph is also shown. John is the automaton a, and the table is an environment e_i which supplies the stimulus s_i to John (see Figure 3a). Strikes is a transformation response of John which serves as both a sight stimulus and a hearing stimulus to an observer. Therefore, the automaton terms fs and fh have the value of strikes. If, for example, Mary is an observer, the following sentences are inferred: (Koenig, 1978)

"Mary hears John striking the table."
"Mary sees John striking the table."

Consider additions to the second example sentence to obtain the two sentences:
"John strikes the table with a hammer."
"John strikes the table with a glass top."
In automaton terms a distinction is made in the two sentences by the fact that "hammer" is an element of the set of auxiliary effectors while "glass top" is an element of the set of environments.

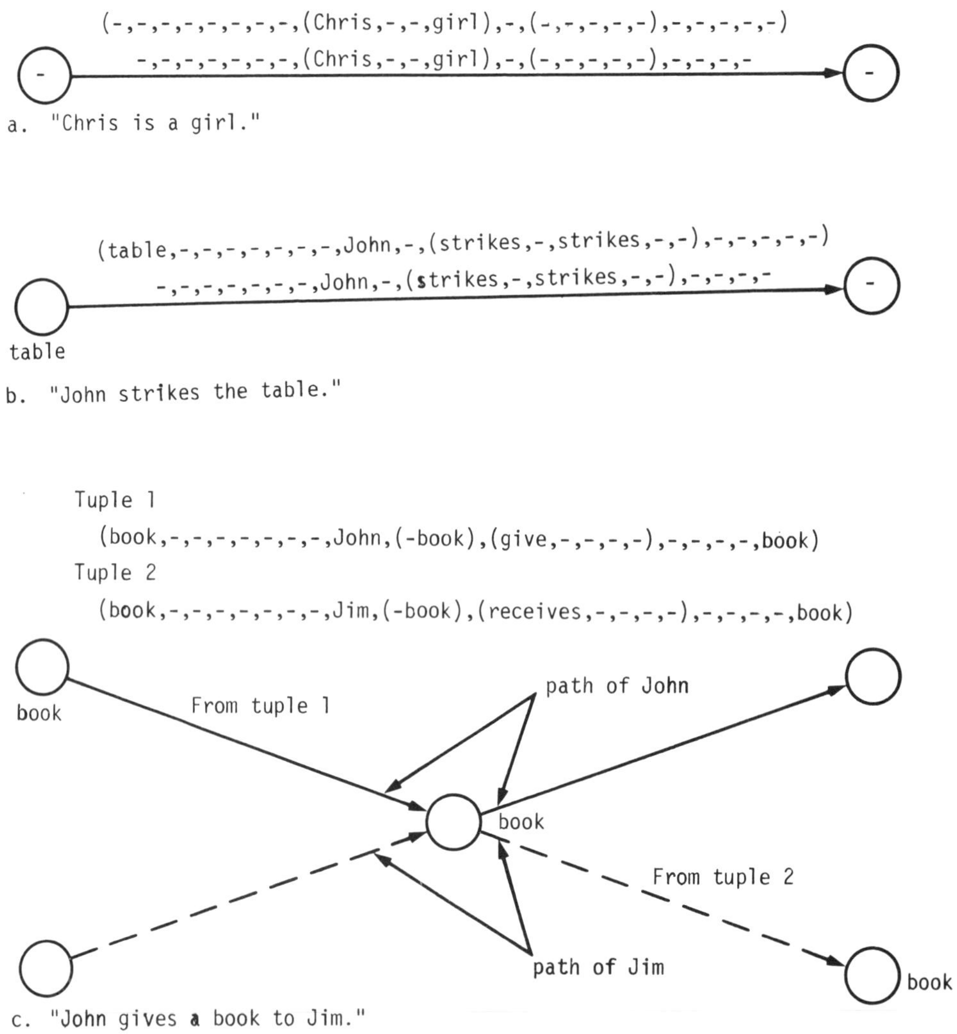

Fig. 4. Environment graph representation for the knowledge of the example sentences.

Sentence 3 contains knowledge about a system of automata where multiple accesses of a same environment are made by two component automata of the system (Koenig, Frederick, 1969, 1979). Two knowledge structures sharing a single point representing the environment are required to store the meaning of the sentence. A first, consisting of two points and an adjacent line, is required for John, the giver of the book, and a second, also consisting of two points and an adjacent line, is required by Jim, the receiver of the book (Koenig, 1979c). Referring to Fig. 3b, the two graph tuples (1) and (2) of Fig. 4c are established. The corresponding knowledge structure is also shown. The part of the knowledge structure with the solid lines pertains to John and the part with the broken lines pertains to Jim. A shared environment is required if a transfer of the book from John to Jim is to take place. An equivalent sentence is "Jim receives the book from John.".

Recall, example sentence 1 pertains to declarative knowledge. Declarative knowledge as used here in relation to automata may be said to aid in the identification of automata or objects, such as, the receptors and environments and is embedded in the knowledge structures with the aid of parentheses. Declarative knowledge in general is knowledge about histories of automata, i.e., knowledge about past recorded responses of automata. Procedural knowledge is knowledge about automata as they operate in their environments receiving stimuli and producing responses. Recall, example sentences 2 and 3 pertain to procedural knowledge.

Because the general automaton concept for representing knowledge is built upon order and organization of the world, it defines complete knowledge and makes for defined inferences. In the cases of semantic nets and conceptual dependency nets, we are not sure all possible inferences are included and we are not sure that the right inferences are made. In the case of conceptual dependency nets, all verbs map into a combination of ACTS and States, and references come from the ACTS and States (Schank, 1975). In the case of general automata graphs, all responses are verbs, and no attempt is made to categorize the responses into groups equivalent to ACTS. For the sentence "John strikes the table.", there is something (some stimulus) that caused John to strike the table. It is not clear that there is any allowance for this in the list of possible inferences for conceptual dependency nets. Recall that if Mary is an observer the following sentences could be inferred:

"Mary hears John striking the table."

"Mary sees John striking the table."

Again, it is not clear that there is any allowance for this in the list of possible inferences for conceptual dependency nets and semantic nets.

In the case of general automata graphs, knowledge can be obtained and structured directly by observation; i.e., through the use of transducers, knowledge of an operating automaton or automaton system can be entered directly into a knowledge structure. We need not first describe the knowledge in the form of a natural language with the human as an intermediate. Conceptual dependency nets and semantic nets deal only with the meaning of sentences.

CONCLUSION

Semantic nets, conceptual dependency nets, and general automata graphs are the basic types of structures for representing knowledge contained in sentences. In the case of general automata graphs, knowledge can also be obtained and structured directly through observations of operating automata and automata systems without first expressing it in the form of sentences.

Semantic nets and general automata graphs can represent both procedural knowledge and declarative knowledge. Procedural knowledge is represented by conceptual dependency nets. Declarative knowledge is embedded in the general automata graphs with the use of parentheses. Both procedural knowledge and declarative knowledge are represented by nodes and edges in the case of semantic nets.

The general automation concept for representing knowledge is built upon order and organization of the world and produces a well defined set of inferences. In the cases of semantic nets and conceptual dependency nets it is not clear that all possible inferences are included and that the correct inferences are made.

REFERENCES

Anderson, J. R. (1976). Language, Memory and Thought. Lawrence Erlbaum Assoc., Hillsdale, New Jersey.

Bobrow, D. G., and T. Winograd (1977). An overview of KRL, a knowledge representation language. Cognitive Science 1, No. 1, 3-46.

Brachman, R. J. (1979). On the epistemological status of semantic networks. In Associative Networks. N. V. Findler (Ed.), Academic Press, New York, pp. 3-50.

Chapuis, A., and E. Droz (1958). Automata - A Historical and Technological Study, (Trans. by Alec Reid), Editions du Griffon, Neuchatel (Switzerland).

Kintsch, W. (1974). The Representation of Meaning in Memory, Lawrence Erlbaum Assoc., Hillsdale, New Jersey.

Koenig, E. C. (1972). A general systems theory for studies in cybernetics and automation, (abridged). In Proc. Int. Congr. Cybernetics and Systems. London.

Koenig, E. C. (1972). Intelligence systems: Introduction to the modeling for the inherited ability for knowledge association. Proc. Brazilian Congr. for General Systems and Cybernetics.

Koenig, E. C. (1973). A language information theory: Obligatory relations, (abridged). Proc. 7th Int. Congr. Cybernetics, Belgium.

Koenig, E. C. (1973-74). A general systems theory for studies in cybernetics and automation. Kybernetes, Vol. 2, No. 4, pp. 217-223, and Vol. 3, No. 1, pp. 11-16.

Koenig, E. C. (1974). A language information theory: Obligatory relations. Cybernetica (Belgium). Vol. 17, No. 1, pp. 57-79.

Koenig, E. C. (1975). Cybernetic considerations for automata as reflected on a graph model. In Proc. 3rd Int. Congr. Cybernetics and Systems, Bucharest.

Koenig, E. C. (1978). Intelligent systems: Knowledge association and related deductive processes. Kybernetes, Vol. 7, No. 2, pp. 99-106.

Koenig, E. C. (1979b). Establishing valid arguments by computer and storing their meanings: A premiss of form A of Aristolelian logic. Kybernetes, Vol. 8, No. 4, pp. 299-303.

Koenig, E. C. (1979c). Knowledge structures for sentences describing systems of interactive automata involving words of "give", "receive" and "sell", "buy", Cybernetica (Belgium), Vol. XXII, 1, pp. 5-75.

Koenig, E. C., and T. J. Frederick (1969). Formal analysis for a general system of interactive automata. In Proc. 3rd Hawaii Int. Conf. System Science.

Koenig, E. C., and T. J. Frederick (1970). Formal analysis for a general automaton. In Progress in Cybernetics, Vol. 3, J. Rose, (Ed.), New York: Gordon and Breach, pp. 613-640.

Koenig, E. C., and T. J. Frederick (1970). A general system of interactive automata: Analysis for defining the graph model, (abridged). In Proc. 6th Int. Congr. Cybernetics, Belgium.

Koenig, E. C., and T. J. Frederick (1971). A general system of interactive automata: Analysis for defining the graph model. Cybernetica (Belgium), Vol. 14, No. 2, pp. 89-131.

Koenig, E. C., and T. J. Frederick (1972). Select properties of the graph model of a general automaton. Kybernetes, Vol. 1, No. 1, pp. 22-33.

Koenig E. C., J. Mason, and E. Jakubowski (1978). Extracting and storing the meaning of sentences by computer based on principles of general automata. Cybernetica (Belgium), Vol. 21, No. 1., pp. 45-60.

Koenig, E. C., and J. V. Schultz (1972-73). Analysis of a general logical discourse for man-machine interaction. Kybernetes Vol. 1, No. 4, 231-241; Vol. 2, No. 1, 27-36.

Koenig, E. F. (1976). Theory of graphs: Applications to automaton systems. Kybernetes.

Levesque, H., and Mylopoulos (1979). A procedural semantics for semantic networks. In Associative Networks, N. V. Findler (Ed.), Academic Press, New York, pp. 93-120.

Lindsay, R. (1973). In defense of ad hoc systems. In Computer Models of Thought and Language, R. C. Schank and K. M. Colby (Eds.), Freeman, San Francisco, California, pp. 372-395.

Norman, D. A., D. E. Rumelhart, and the LNR Research Group (1975). Explorations in Cognition, Freeman, San Francisco.

Quillian, M. R. (1968). Semantic memory. In Semantic Information Processing, M. Minsky (Ed.), MIT Press, Cambridge, Massachusetts, pp. 216-270.

Raphael, B. (1968). SIR: Semantic Information Retrieval. In Semantic Information Processing, M. Minsky (Ed.), MIT Press, Cambridge, Massachusetts, pp. 33-145.

Rumelhart, D. E., P. H. Lindsay, and D. A. Norman (1972). A process model for long-term memory. In Organization of Memory. E. Tulving and W. Donaldson (Eds.), Academic Press, New York, pp. 197-246.

Schank, R. C. (1972). Conceptual Dependency: A theory of natural language understanding, Cognitive Psychology 3, 552-631.

Schank, R. C. (1975). Conceptual Information Processing, North-Holland.

Schubert, L. K., R. G. Goebel, and N. J. Cercone (1979). Structure and Organization of a Semantic Net. In Associative Networks, N. V. Findler (Ed.), Academic Press, New York, pp. 122-175.

Simmons, R., and J. Slocum (1972). Generating English discourse from semantic networks. Comm. ACM 15, 10, 891-905.

FORMAL NEURONS TO SYNTHESIZE ARBITRARY PROBABILISTIC AUTOMATA

R.Moreno-Díaz, F.Hernández-Guarch and J.M.Limiñana

Dept.of Cybernetics and Basic Sciences. University College of Las Palmas. P.O.Box 550.Canary Islands. Spain.

ABSTRACT

As we show, a net of N probabilistic formal neurons in the McCulloch-Pitts sense is uncapable to synthesize an arbitrary probabilistic automaton of 2^N states, whereas organisms seem to be. The question arises: is there any physiological or logical mechanism missing? We advance the theoretical concept that there is interaction of efferents (axons), through their separating medium, in addition to the accepted interaction of afferents (dendrites). And also show that, for a particular type of efferent interaction, any arbitrary probabilistic automaton of 2^N states can be synthesized by a net of N modified formal neurons. Besides the logical and neurophysiological incidence of these results, they are also relevant to the general subject of cellular decomposition of probabilistic automata and probabilistic distributed computation.

KEYWORDS

Neural nets; formal neurons; interaction of axons; decomposition of automata;Markov chains; probabilistic computation.

INTRODUCTION

A relatively old problem in the theory of formal neurons (Moreno-Díaz and McCulloch, 1969) is the synthesis of arbitrary probabilistic automata by a minimal neural net with feedback, which originated from a number of logical and neurophysiological considerations (McCulloch et al,1969). The problem was not then solved because the limitations of probabilistic nets were not appropriately stated. Other later formulations of apparent solutions (Moreno-Díaz,1971 ; Santesmases et al.1977)are only partially valid for the same reason.

A recent attempt to obtain a modular dycotomic decomposition of probabilistic automata, for computer simulation, have finally inspired a correct generalization of the original net theory to cope with the problem. The basic concept in the formulation is that the firing of a neuron in a net depends not only upon the immediate past history of the whole net, but also upon the present states of firing of at least part of the rest of the neurons.

When this concept is represented by its neurophysiological counterpart, it means that a type of efferent (axons) interaction must take place. Therefore, as it shall become clear, the physiological candidate to embody the functional properties required by the generalization is an axon-axonal interaction of neurons through their separating medium.

A FORMULATION OF THE THEORY

A probabilistic neuron is a logical device having a number of boolean inputs and a single boolean output, such that for any possible input configuration, there is a probability for firing the neuron after a time delay (McCulloch,1959; Moreno-Diaz, 1971).

The possible output configurations for all the N neurons at each time are considered to be states of the net, so that the most general net, where any feedback is allowed, is a probabilistic automaton, and thus, for each configuration of the inputs external to the net, there is a state transition Markov matrix. But the inverse is not true, in general.

The simplest way to show this is to consider the degrees of freedom in each case. For a probabilistic net of N neurons there are $N2^N$ degrees of freedom for each external input configuration, whereas a probabilistic automaton of 2^N states, has, in general, $2^N(2^N-1)$ degrees of freedom in the same circumstances, which result from its probabilistic state transition matrix under the only restriction that each row sums 1.

Now, consider that there is interaction of efferents. The firing of a neuron at time t depends not only on the state of the net at time t-1, but also on the state of the net at time t, except for the digit in the place of the neuron being considered, since any degree of feedback is allowed. This results in a total of $N\cdot 2^{2N-1}$ probabilities for every external input configuration, but, obviously, they are not independent.

A set of independent probabilities for each formal neuron, which implies the modular or cellular decomposition of an arbitrary automaton, can be generated as follows. First, number the neurons of the net from N to 1 (say, left to right). Assume that each neuron is influenced by the axons of all the neurons at its right. Consider the Nth neuron, which is to be specified by 2^{2N-1} independent probabilities, one for each possible state at time t-1 combined with the 2^{N-1} possible configurations of the rest of the neuronal outputs at time t. Thus, the kth neuron is affected by efferent interaction from neurons k-1, k-2,...1, so that its firing is determined by 2^{N+k-1} independent probabilities.

The number of degrees of freedom of the net is $2^N(2^{N-1}+2^{N-2}+...+2+1)=2^N(2^N-1)$, which, as it should, is the number of degrees of freedom of an arbitrary probabilistic automaton for each external input. To include arbitrary external inputs, we proceed as indicated by Santesmases et al (1977). The synthesis is achieved as follows: let P_{ij} be the state transition matrix ($i,j=1,2..2^N$) and $p_i^\ell(k)$ ($i=1,2..2^N$; $\ell= 1,2..2^{k-1}$) the probabilities of firing the kth neuron, which are to be determined.

To find the values of $p_i^\ell(k)$, that is, the desired set of independent probabilities for each neuron, we proceed as follows. First, set for a first neuron:

$$p_i^1(1) = \sum_{j=1}^{2^{N-1}} P_{ij} \tag{1}$$

Naturally, the probability that said neuron does not fire is:

$$q_i^1(1) = 1-p_i^1(1) = \sum_{j=2^{N-1}+1}^{2^N} P_{ij} \tag{2}$$

For the second neuron, the firing probabilities for each state and external input configuration are two, corresponding to whether or not the first neuron fires. When the first neuron fires

$$p_i^1(1)\ p_i^1(2) = \sum_{j=1}^{2^{N-2}} P_{ij} \tag{3}$$

From (1) and (3):

$$p_i^1(2) = (\sum_{j=1}^{2^{N-2}} P_{ij})/(\sum_{j=1}^{2^{N-1}} P_{ij}) \tag{4}$$

The probability of not firing in this case is:

$$q_i^1(2) = (\sum_{j=2^{N-2}+1}^{2\cdot 2^{N-2}} P_{ij})/(\sum_{j=1}^{2^{N-1}} P_{ij}) \tag{5}$$

When the first neuron does not fire, one gets:

$$p_i^2(2)\ q_i^1(1) = \sum_{j=2^{N-1}+1}^{3\cdot 2^{N-2}} P_{ij} \tag{6}$$

and also:

$$q_i^2(2)\ q_i^1(1) = \sum_{j=3\cdot 2^{N-2}+1}^{4\cdot 2^{N-2}} P_{ij} = \sum_{j=3\cdot 2^{N-2}+1}^{2^N} P_{ij} \tag{7}$$

From (6) and (7), it follows that:

$$p_i^2(2) = (\sum_{j=2^{N-1}+1}^{3\cdot 2^{N-2}} P_{ij})/(\sum_{j=2^{N-1}+1}^{2^N} P_{ij}) \tag{8}$$

Also, from (7) and (2):

$$q_i^2(2) = (\sum_{j=3\cdot 2^{N-2}+1}^{2^N} P_{ij})/(\sum_{j=2^{N-1}+1}^{2^N} P_{ij}) \tag{9}$$

Probabilities given by (4), (5), (8) and (9) have a meaning only when the denominators are not zero. They are undetermined otherwise. Undeterminations are easyly interpreted since they correspond to events which are conditioned to other having a zero probability. Thus, they are practically not to be considered since they correspond to impossible events for the particular Markov chain to be synthesized.

The generalization of (1), (4) and (8) leads to the desired expressions for $p_i^{\ell}(k)$ and $q_i^{\ell}(k)$:

$$p_i^{\ell}(k)=(\sum_{j=1+(\ell-1)2^{N-k+1}}^{(2\ell-1)2^{N-k}} P_{ij})/(\sum_{j=1+(\ell-1)2^{N-k+1}}^{\ell\cdot 2^{N-k+1}} P_{ij}) \tag{10}$$

$$q_i^{\ell}(k) = 1-p_i^{\ell}(k) \tag{11}$$

Equation (10) shows that the event whose probability appears in the denominator always contains the one whose probability is in the numerator. Therefore, if the denominator is for the impossible event, so it is the numerator.

Equation (10) can be made clearer with an example. Assume a probability transition vector, p_{ij}:

$$P_{ij}=(0.5,\ 0.10,\ 0.15,\ 0.20,\ 0.25,\ 0.25,\ 0,\ 0)$$

which may correspond to the transition probabilities from the ith state to the different eight states in a probabilistic automaton, or even simply, to the distribution of a ramdom variable among a population. What is the corresponding three neuron net representation? From (10) and (11):

$$p_i^1(1) = 0.5 \qquad q_i^1(1) = 0.5 = 1-0.5$$

$$p_i^1(2) = \frac{0.15}{0.5} \qquad q_i^1(2) = \frac{0.35}{0.5} = 1-\frac{0.15}{0.5}$$

$$p_i^2(2) = \frac{0.5}{0.5} \qquad q_i^2(2) = \frac{0}{0.5} = 0 = 1-\frac{0.5}{0.5}$$

$$p_i^1(3) = \frac{0.05}{0.15} \qquad q_i^1(3) = \frac{0.10}{0.15} = 1-\frac{0.05}{0.15}$$

$$p_i^2(3) = \frac{0.15}{0.35} \qquad q_i^2(3) = \frac{0.20}{0.35} = 1-\frac{0.15}{0.35}$$

$$p_i^3(3) = \frac{0.25}{0.5} \qquad q_i^3(3) = \frac{0.25}{0.5} = 1-\frac{0.25}{0.5}$$

$$p_i^4(3) = \frac{0}{0} = \text{indet.} \qquad q_i^4(3) = \frac{0}{0} = \text{indet.}$$

In this example, it is clearly seen that, since the probability of not firing neuron two when neuron one fires is zero, the events conditioned to this ($p_i^4(3)$ and $q_i^4(3)$) are to be ruled out for meaningless.

CONCLUSION

What are the neurophysiological and epistemological incidences of our results? If we go back to the beginning of neural net theory (McCulloch and Pitts, 1943), pick up the transcendental points there and follow them thereafter, we end with an apparently simple conclusion: living machines seem to have built in the necessary mechanisms to cope with all of the most sophisticated artificial machines that we are able to create now, with a potentially minimal wiring effort. This could be interpreted in the sense that formal machines are logically behind the biological capabilities of living machines, though we need to create the former to understand the later.

Considering the theory of formal neurons by itself, two interesting additions appear historically to the logical proposal by McCulloch and Pitts in 1943. N original McCulloch-Pitts formal neurons cannot synthesize an arbitrary deterministic automaton of 2^N states (minimal synthesis).

After some neurophysiological evidences, Blum's interaction of afferents (dendrites) (Blum, 1961), led to the isomorphism of minimal neural nets with deterministic automata. Our contribution further shows that the interaction of efferents (axons) provides with the isomorphism of neural nets and arbitrary probabilistic automata.

In each case, constructive theorems have been demonstrated. Thus, this type of results are not only significant when we try to describe reality in formal terms, but have also provided with effective ways to the modular or cellular decomposition of arbitrary deterministic and probabilistic imaginary automata.

REFERENCES

Blum, M. (1961). Properties of a neuron with many inputs. In H.von Foester, G.Gopf (Eds.), Principles of Self Organization . Mc Millan-Pergamon. New York.

McCulloch, W.S., Pitts, W. (1943). A logical calculus of the ideas immanent in nervous activity. Bull. of Math.Biophy. 5, pp 115-133.

McCulloch, W.S. (1959). Agatha Tyche: of nervous nets - the lucky reckoners. Mechanization of Thought Processes, 10, Vol.II. Her Majesty's Stationary Office. London. pp 611-634.

McCulloch, W.S., Papert, S., Blum, M., da Fonseca, J.L., Moreno-Díaz,R. (1969). The fun of failures. Ann.New York Academy of Sciences. 152, 2. pp 963-968.

Moreno-Díaz, R., McCulloch, W.S. (1969). Circularities in nets and the concept of functional matrices. In L.Proctor (Ed). Biocybernetics of the Central Nervous System . Little and Brown. Boston. pp 145-150.

Moreno-Díaz, R. (1971) Deterministic and probabilistic neural nets with loops. Math. Biosciences. 11 pp 129-136.

Santesmases, J.G., Moreno-Díaz, R., Rubio Royo, F., Mira, J.(1977). Formal neuronal nets. In D.G. Lainiotis and N.S. Tzannes (Eds.) Aplications and Research in Information Systems and Sciences. Vol.II. Hemisphere. Washington-London, pp 490-495.

IX. INFORMATION RETRIEVAL

HOW PERSONAL COMPUTERS WILL AFFECT THE FUTURE OF LIBRARIES AND INFORMATION CENTERS

Eugene Garfield

Institute for Scientific Information®, 3501 Market St.
Philadelphia, Pennsylvania

ABSTRACT

The personal computer revolution will eventually change the way we receive and provide information services. This will describe PRIMATE™ (Personal Retrieval of Information by Microcomputer and Terminal Ensemble), a personal computing system designed to help individuals index their article reprint collections. With the addition of word processing software, users can prepare journal articles and other manuscripts using PRIMATE. Personal computers also have broader application. They could be used by special libraries for reprint collection indexing, as a terminal to a large data base, and to receive SDI services like ASCA® and ASCATOPICS®. Libraries or departments could even keep subject specific citation data bases on their system. Personal computers may be the missing link in the chain leading to the age of the electronic journal.

KEYWORDS

Personal computer, PRIMATE, reprint collection systems, electronic journal

It should come as no surprise that computers are extending our institutional and individual capabilities. We are now beyond the period of the computer revolution. We are already in the age of the personal computer and this will affect us in ways the old large computers did not (Winter & Stuart, 1979).

At ISI® we began a fascinating experiment a few years ago using personal computers. This experiment is primarily the brainchild of ISI's vice president of research Tony Cawkell. Tony recognized that scientists and scholars, like himself, were having difficulty managing burgeoning reprint collections. Since I and others at ISI also relied on Tony's reprint file we decided to use a microcomputer to provide better access to our collection.

Called PRIMATE, for Personal Retrieval of Information by Microcomputer and Terminal Ensemble, the hardware/software system Tony developed is still in development. Whether ISI markets hardware and/or software depends upon many factors. The microcomputer market is changing so rapidly it is difficult to focus on one particular piece of hardware lest it become obsolete overnight. The software we have already developed is remarkably simple to use. Our unique software will be able to serve a variety of purposes.

PRIMATE has generated genuine excitement from the moment we made its development public. I write a weekly essay which appears in Current Contents®. The response to my PRIMATE essay of July 17, 1978 (Garfield, 1978) was incredible. In fact, more people wrote or telephoned ISI about PRIMATE than any other idea I discussed before.

PRIMATE deals with a real problem almost every scholar faces. Gerry Jahoda (1970) recognized long ago that the task of organizing reprint collections is a widespread problem. Nor has it been an easy problem to solve. There are several basic systems in wide use. Most people file reprints alphabetically under the last name of the key author. Others file reprints in broad subject categories while others use a chronological system. All of these systems are inadequate when it comes to retrieving information by subject or other criteria. Clearly an author file is only useful if you can remember who worked on a particular subject. Filing by broad subject categories is very useful for teaching or writing reviews but you have to wade through a hundred reprints if you are looking for a specific piece of research information. The chronological arrangement is handy during the first six months to a year. That's because you may recall something you read within the last year. And this is not helpful if the file serves a lot of people.

So the average scientist has a large reprint collection to which there is no minimum access. That's not to say that there are not some vigilant individuals who keep elegant reprint collection systems. Perhaps one of the most elaborate systems I have heard of is the one established by Hans Selye (Selye & Ember, 1964) in Montreal.

PRIMATE, however, offers a reprieve for people with little time to index, sort, and arrange collections. As it was originally conceived, PRIMATE consisted of a microcomputer with a 64K-byte central processing unit, a 9-inch video display and keyboard combination, and two double-density floppy magnetic disks for mass storage.

Here is how data about a paper is entered into the PRIMATE system (Fig. 1).

Behavior Research Methods & Instrumentation
1979, Vol. 11 (2), 102-110

Teaching a computer to teach

RICHARD MILLWARD
Brown University, Providence, Rhode Island 02912

The long-range goal of computer-assisted instruction (CA) should be to construct computer programs that interact with a student as a human tutor does. In order to make computers act as intelligent tutors, we must consider three aspects of teaching: (1) A tutor makes use of many different knowledge sources. (2) A tutorial session is a kind of problem-solving situation in which the tutor is trying to solve the problem of imparting understanding to the student

Fig. 1. A typical paper to be entered into PRIMATE

By hitting a key labeled "Enter", the system tells you the serial number to be assigned to the document or reprint you wish to index. You simply write that number at the top of the reprint. After the serial number is assigned, you can enter more descriptive information about the reprint. In this case a paper falls into the general category of education (Fig. 2).

Serial no. 1254 EDCTN
Teaching a computer to teach.
Computer aided instruction (CAI).
SOPHIE system
Millward RM
Behav.Res.Meth.& Instr. 11(2),102-110
1979

Fig. 2. An entry into PRIMATE

While you could key the word "Education" you could simply use the mnemonic EDCTN. Then you would key in the title of the article followed by whatever descriptive words you believe will help retrieve the article later. (You can always add more entries later on.) The name of the author, journal, and pagination complete the entry. Obviously ISI would design a system that would enable you to retrieve by citation! When you are finished indexing the entries appear on the screen. If you are satisfied with them, a tap on a key puts the information into PRIMATE's memory where it is permanently stored.

In the first PRIMATE system you can use up to 300 characters, to characterize a document. This is more than adequate for most users. Several thousand such records may be stored on a single floppy disk. And when one disk is full another can be easily snapped into place. Eventually, advances in disk memory technology will allow even more data to be stored in PRIMATE.

To physically file the article or reprint, the user merely puts it in serial number order in a predesignated storage space -- a file drawer perhaps. In that way, finding the article to be retrieved when the time is right will be easy to do.

Retrieving an article that has been indexed and whose bibliographic information has already been stored in PRIMATE's memory is a simple matter. Suppose you wanted to retrieve the reprint on the "SOPHIE" system of instruction. All you would have to do is enter "SOPHIE" into this system and PRIMATE would find that article with the word "SOPHIE". If you didn't remember the name of the system but did recall that the article appeared in _Behavior Research Methods and Instrumentation_, you can key the journal name and PRIMATE will find and display all the articles you've indexed from this journal. You can then scan these citations until you find the right one. Or you might remember many things about the article: that the author's name was R.M. Millward, that it was an education article, that SOPHIE was the name of the system, that it was about Computer-Aided Instruction. With this information you could retrieve the bibliographic citation and serial number for just the article you wanted. PRIMATE will allow you to search up to five indexing terms at once.

With PRIMATE's sophisticated level of indexing the scholar will be able to use the system to search his or her reprint collection in other ways as well. To demonstrate, suppose you want to answer the question "in my reprint collection, do I have any articles written by G. Salton in 1979 on citation studies of the computer science literature?" You tap a key to indicate you wish to begin a search. On the display are five blank lines, on which you can type up to five word phrases, single words, or parts of words. To answer the question I just posed, you could enter: G. Salton; 79, citation studies; computer science. (Fig. 3)

SEARCH ARTICLES -- DISC; EG PERSONAL FILES

Search words 1. G. Salton
2. Citation Studies
3. Computer Science
4. 79
5.

Enter up to one word or word phrase per line.
Records searched; 709
Number of hits; 2

Fig. 3: A PRIMATE search display

You would then be able to retrieve the article: A Citation Study of Computer Science Literature by G. Salton and D. Bergmark which appeared in the IEEE Transactions on Professional Communication, Vol. PC-22, No. 3, September, 1979, and any others Salton may have written that year.

Since the price of PRIMATE would be in the same range as many word processors, we decided last year to incorporate this. Anyone who has ever attempted to put together a scholarly article can appreciate how word processing can be used to make this chore easier. Articles are usually revised many times before they are sent to a journal. They have to be edited, the parts have to be rearranged. and a million other details must be seen to before the paper can be submitted. In fact, word processing will allow a scientist or scholar to construct a journal article on-line -- edit it, manipulate it and have it printed out in camera ready copy.

Of course, when we first thought of PRIMATE we didn't conceive of all the other possible ways for using it that imaginative readers of Current Contents have since envisioned. Many of them recognized, as we did, that PRIMATE could "double" as a dumb terminal -- that is, a terminal that is on-line to another system such as Lockheed's Dialog, BRS -- you name it. So we decided that, to be practical, PRIMATE must have built-in data communications capability -- standard interfaces that allow PRIMATE to transmit and receive from local or remote distances connected to it by telephone lines. Just as large libraries dial up Lockheed to get SCISEARCH®, for example, an individual could do the same thing using PRIMATE in terminal mode.

Once systems like PRIMATE can support larger disk file capacities, librarians and individuals will be able to handle a lot of bibliographic applications. ISI will consider offering a unique SDI service called ASCAMATE. It will closely resemble two SDI services we already offer -- ASCA and ASCATOPICS . (Garfield & Sher, 1967)

If we can assure the accuracy of our telecommunications, we envision offering even more specialized services as time goes on. Picture a medical school library in need of a clinically oriented Science Citation Index® type of index. Such a subject specific index could be transmitted directly from us and received by the user's PRIMATE system.

It is easy to imagine that a PRIMATE-like system could become the necessary link in the first step towards the truly electronic journal. An author would write the article on the system, revise it, edit it, send it electronically to the editor of a journal having a compatible system. The editor can store the article in memory until ready to read it, display it on the CRT screen, read it, send it back for revisions or even send it to reviewers electronically. In the immediate future, the author will produce camera-ready copy on his or her system and transmit it to the publisher's electronic composing and printing system. Eventually, a journal may have electronic subscribers who want to receive the articles directly from the journal on their video displays.

Eventually such personal computing capability might lead to the partial elimination of the journal as we know it. Joshua Lederberg (1978) of Rockefeller University has broached this possibility in a recent article. He sees a time when journal articles will be submitted electronically to a vast number of reviewers in all disciplines, who may add information that became available since the appearance of the communication. Thus, many versions of the "same" article might exist -- the journal article will be a dynamic entity, changing and developing along with science and scholarship.

With a PRIMATE type of system the author might complete his or her version of the article and push a button. That button would send the article to some central storage area. When another scientist or scholar wanted to see what was new in his specialty or area of interest, he or she would request this information to be transmitted to his or her personal system. Upon reading the journal article, the individual would be free to comment, revise or send a message to the author. Certainly this system would require vigilance on the part of individuals whose own theories rest on continually updated scientific knowledge. However, it might actually advance the cause of scholarship because people in many disciplines will have access to a wide range of information and information exchange might more readily take place.

How will personal computers affect the future of libraries? The changes I've spoken about will take place gradually: do not expect the electronic journal to appear overnight. But within the next several years a specialized system to organize reprint collections should become available, whether from ISI or from some other manufacturer. It is very nearly a reality. And once scientists, scholars and special librarians begin to reap the benefits of computer technology, they will begin to demand more and more applications of the type I've described. That has been the case in almost every other area the computer has touched. The personal computer will become almost an extension of the scientist and librarian, the tool through which they will work more effectively.

REFERENCES

Garfield, E. and I. H. Sher (1967). ISI's experience with ASCA -- a selective dissemination system. _J. Chem. Docum._, _7_, 147-153.

Garfield, E. (1978). Introducing PRIMATE™ -- Personal Retrieval of Information by Microcomputer and Terminal. _Current Contents_, (29), 5-9.

Jahoda, G. (1970). _Information storage and retrieval systems for individual researchers_. Wiley-Interscience, New York.

Lederberg, J. (1978). Digital communications and the conduct of science: the new literacy. _Proc. I.E.E.E._, _66_, 1314-1319.

Selye, H. and G. Ember (1964). _Symbolic shorthand system for physiology and medicine_. Universite de Montreal, Montreal.

Winter, D. and D. Stewart (1979). Paper read at "_Towards 2001_": _Institute of Information Scientists Annual Conference, June 5-8, 1979_. Torquay, Devon, England.

ACCESSIBILITY OF LOCAL INFORMATION RETRIEVAL SYSTEMS

N. Howden* and K. Lee**

*Grad. Sch. of Lib. Sci., La. St. Univ., Baton Rouge, La.
**Dept. of Comp. Sci., La. St. Univ., Baton Rouge, La.

ABSTRACT

The authors discuss several existing local information retrieval systems and present criteria for developing an optimal system that can provide researchers with a personel supplement to complex informal and formal information systems.

KEYWORDS

Retrieval; FAMULUS; CFAMULUS; Scientific communication; local databases; personal files.

INTRODUCTION

Researchers in all fields from the humanities to the sciences appreciate the value of information and also the growing problems of gaining access to it. There are several types of information flowing to researchers including raw data, analyzed information, and conceptual knowledge. Each of these types of information showed a growth in quantity and complexity that threatened to stifle the scientific communication system of the 1960's and 1970's (Etzioni, 1971). Novel remedies have been tried including preprint distribution systems, research-in-progress systems, online retrieval systems, and current awareness systems. Information has been induced to flow faster and more selectively but also less formally. Lack of formality and overlap with formal systems has provided a source of concern that scientific information has not been properly assessed or synthesized (Etzioni, 1971).

Proper assessment in science is essential to the interaction between producers and users of information, for if new ideas are not examined and incorporated into the knowledge base, they die. Much of the producer/user evaluation and interaction process has been transferred to informal information processes such as the peer groups called invisible colleges. For either the formal communication systems or the informal communication processes we currently have, there are drawbacks. The formal system is very large. It often is saturated with too much information, and often requires some intermediary such as a librarian to make it work. Particularly hard to find, many useful documents are lumped together as "ephemeral", "gray literature", or as "fugitive literature" and excluded from formal communication systems. Informal systems are not consistent and operate on a private level, often too private for more than a few people to utilize.

One hopeful alternative that has entered the field to augment formal and informal systems is the local retrieval system. Such systems offer an assessment mechanism that supplements those found in formal (journal) communication. As Thomas Kuhn (1962) pointed out, local interactions between scientists have been important for providing informal assessment of research. Thus a researcher has one or more "trusted assessors" whose judgement he respects. It is realized that such relationships go well beyond friendly advice about publication, that they often extend to collaboration, consulting work, and barbershop quartets at the faculty club. Such local cooperation has grown to the development and use of computer programs to accomplish scientific tasks. One such programming effort is usually a local data file collected during a collaborative project. Files may contain data, a bibliography, or reference values used in research. Very often such files refer to fugitive literature or contain data accumulated informally through exchanges.

A number of retrieval systems for personal use by researchers have been identified by Jahoda (1970) and Mittman and Borman (1975) as well as by discussions in the Annual Review of Information Science and Technology. Most of the discussion and systems development to date however, focuses on how to construct such systems, where they should be standardized, or how efficient they are. An important aspect of any systems analysis seems overlooked in these instances and that is the question of purpose. Do the retrieval software packages that we base local systems on have the characteristics most beneficial to their function as elements in the scientific communication system? Mittman and Borman (1975), for example, tend to emphasize features more applicable to large formal retrieval systems such as online tutorials and a monitor log.

What features would make a local retrieval system more responsive to the role it might play in scientific communication? The authors' experience in the research environments at several universities and use of local systems led to the development of the following alternative list of essential criteria:

1. Easy input of data is important so that relatively untrained keyboard operators (including the researchers themselves) can do input.

2. Transportability to make data used at one location useful immediately if a researcher must relocate or just cooperate with a fellow researcher during a summer of research.

3. Use of online time-sharing, a service that is a reality at an increasing number of installations where it is difficult to find card reader equipment and where video terminals are ubiquitous.

4. Summary and report formats with a minimum of control commands to learn would simplify generation of vocabulary lists, KWIC indices, and other printouts.

5. Driver programs that make job control language transparent to the user would allow users to avoid retetitive housekeeping tasks.

6. Multiple file access making it possible to move from to file ought to be possible using either driver programs or a database management system.

7. Freedom to assign field labels should allow each user the flexibility necessary to express creativity both in using and applying the system.

8. Low cost or no cost access to the software would allow such a system to meet the goal of near universal availability to the community of researchers.

Have there been systems generated that meet all of the criteria for local systems and will interface well with the general scientific communication system? Several retrieval systems such as GIPSY, Mark IV, FAMULUS, and RIMS/RIQS are well known. Each, within the sphere it was designed for, performs some of our criterion tasks, but none, as we shall see, meet all the criteria. However, the authors will present a less well-known version of one of these systems that will soon meet all the criteria.

The GIPSY system and the Mark IV system are commercial packages that require substantial investment to lease or purchase. Neither is portable as a result of contractual restrictions, although versions are likely to be available for a number of different machines. Interfaces for easy data entry and program execution without use of JCL is not available, although user installations may certainly provide such interfaces. Flexibility in choosing field labels, number of fields, and output report formats is quite good in both cases.

RIMS/RIQS is a system constructed with the avowed intent of providing good access to local or personal files. Within its local environment it does that well, providing online access, ease of use in both storing and retrieving information, and providing maximum flexibility for users. It additionally accesses statistical software and allows stored data to be statistically analyzed. It is not at this point highly portable, being used primarily on CDC computers.

FAMULUS is a fairly ubiquitous set of FORTRAN programs that have been largely in the public domain since creation by the U.S. Forest Service. At least two major emendations to the basic programs have been made in what is termed the "Wisconsin version" and the "London version". A third version produced at the University of Kansas provides maximum flexibility for offline printout through a postprocessor package.

All of these systems require some basic knowledge of the computer on the part of the user and provide documentation of the programs that make up the software package. They all have in common the presence of an organization committed to maintaining them or making them available to users. Most of these systems are not generally available at minimal cost, nor are they necessarily user friendly to the unsophisticated user.

As the user and responsible agent for a set of FAMULUS programs, our working group has been able to gain substantial perspective on the needs expressed by our users and on means that are available to meet those needs. These needs are embodied in our list of criteria. Our experience has much in common with the experience of the originators of the FAMULUS package, especially with respect to scientific communication. As Yerke (1970) points out, "scientists do not learn of the most important developments pertaining to their work by visiting the library", instead they obtain information from their fellow workers, from phone calls, from "fugitive" newsletters and journals, and from colleagues met at conferences. Our differences are those imposed by time which has moved us from the original FAMULUS batch processing era of the 1950's-early 1960's into the present environment of time sharing, real time, interactive programming and a user population with rising expectations conditioned by the advent of large commercial database vendors and microcomputers. What validation can we provide for our asserted criteria for local systems? That our experience meshes or does not mesh with the model of scientific communication suggested by such people as Kuhn (1962) and Ziman (1968) will be the ultimate determinant of validity.

Conversational FAMULUS, or CFAMULUS is a version of the basic retrieval system created by adding PL/I programs for building and searching inverted index sequential files. Originally produced by another university, these programs were introduced into the local Louisiana State University environment in the fall of 1979.

Minor adaptations were made to fit the programs to the IBM 3033 computer but most of the programs, which came from an IBM 370, were compatible. A PL/I optimizing compiler library is the main support required from the system.

The environment for observing use of information resources has been unique in that little retrieval software was generally available beyond administrative use of Mark IV. Uses that have been found for CFAMULUS range from personal bibliography to library use for serials listings. One notable use has been files used for quick reference in the lab from a dumb terminal. The ubiquity of terminals now allows automated retrieval to be used for files that may never have been considered for automation before due to frequency of access or update. Many initial users were naive with respect to the computer so that we soon were forced to provide instructions in CFAMULUS for beginners. Observations of these initial uses prompted us to examine the need for system enhancement.

Our goals were established by a user group formed after the first few users became active. The primary goal established was to fully document the programs in order to provide a basis for both understanding the programs and for making any desired changes. Several changes were soon identified as priority items. Access to more complex sort routines for multiple field sorts and to allow "library" sorting would help the formal users of the system and as a byproduct give greater flexibility to the whole. High on the priority list has been the creation of driver routines to allow time sharing access to the full range of FAMULUS capabilities. This resulted from user comparison with CFAMULUS access. Further expansion is desired in the direction of providing capabilities included in the Wisconsin, London, and Kansas versions. Most pressing is a desire to expand the number of fields available to more than ten, although means have been found to obtain considerable mileage from those ten. CFAMULUS meets most of the other criteria for a suitable local retrieval system except for online thesaurus access and access to multiple files. These requirements are top priorities along with sorting as enhancements.

In summary, experience with the academic researcher shows that there is a need for information retrieval systems that supplement the information access capabilities of formal and informal communication systems. As extensions of the researcher's communication and evaluation abilities, local retrieval systems must be structured to provide maximum service with minimum investment in time or other resources.

REFERENCES

Etzioni, Amitai (1971). The need for quality fitters in information systems. Science 171, 3967.

Jahoda, Gerald (1970). Information Storage and Retrieval Systems for Individual Researchers. Wiley-Interscience, New York.

Kuhn, Thomas (1962). The Structure of Scientific Revolutions. U. of Chicago Press, Chicago.

Liman, John (1968). Public Knowledge: The Social Dimension of Science. Cambridge U. Press, London.

Mittman, Benjamin and Lorraine Borman (1975). Personalized Data Base Systems.

Yerke, Theodore B. (1970). Computer support of the researcher's own documentation. Datamation 16, 75-77.

SPRINT -A SYSTEM FOR INFORMATION RETRIEVAL IN LAW

R. Pavlov, V. Kiskinov, G. Angelova

Laboratory of Math. Linguistics, Institute of Mathematics with Computer Centre, Bulgarian Academy of Sciences, 1090 Sofia, P.O. Box 373 Bulgaria

ABSTRACT

The organization of modern society calls for an efficient operating with a large amount of law information. The foundations of the system SPRINT, designed at the Laboratory of Math. Linguistics, reflect the theoretical concepts about the structure of the socialist legal norm system. The formalized legal information is processed by the system SPRINT by using new techniques for information retrieval in law.

KEYWORDS

Mechanism of law regulation; legal norm system; legal relations; legal norm interactions; law information retrieval software.

INTRODUCTION

The organization of modern society calls for an efficient operating with a large amount of law information. One of the ways towards achieving this aim is the design of law information systems allowing to interpret the source material from various viewpoints.

The law information systems known to us are mainly of descriptive type. By means of a detailed dictionary design their authors try to achieve a more adequate correspondence between the needs of the users and the possibilities of the systems. The preliminary processing of the source legal material is usually reduced to retrieval of a set of keywords.

The system SPRINT was designed at the Laboratory of Mathematical Linguistics at the Institute of Mathematics with Computer Centre affiliated to the Bulgarian Academy of Sciences. The foundations of the system reflect the theoretical standpoints about the structure of the socialist legal norm system.

LEGAL-THEORETICAL STATEMENT OF THE PROBLEM

The mechanism of law regulation is a subject of systematic research. The mechanism of law regulation is viewed as consisting of the following four components: the legal normative system, the legal relations, the acts for realization of legal norms, and legal consciousness. The legal regulation of social relations is achieved by means of the dynamic interaction between the elements of the mechanism. A generalized system representation of these interactions has been conceived.

The system of legal norms is the basic component of the mechanism of law regulation. Its structure is being studied. The construction element is the legal norm with components, as follows: hypothesis and disposition (or sanction). The legal norm is represented as an ordered pair.

Functional differences divide legal norms into three main types: regulative, preventive and specialized. The differences in the functions justify the existence of some variations of the elements within the framework of the described structure.

The desired legal result is achieved by simultaneous or successive realization of a set of legal norms which are always interrelated. There is no norm manifesting its regulative effect independently, without interaction with the other elements of the system. The existence of various relations gives grounds to consider the set of legal norms as a system. In general, the relations can be classified as:
- relations of construction - an element of the structure of a legal norm participates as an element of the structure of another one;these are the most elementary relations in the system and they are scope-inclusive to a different degree;
- functional relations - the realization of one legal norm is a necessary condition for the function of another one. The accomplishment of the behaviour stipulated in the disposition generates one or several facts. Between certain of them and some of the facts described in the hypotheses of other norms there are relations of semantic equivalence or inclusion.

The relations are also subdivided according to the type of joined elements, their qualitative characteristics, the way two norms are connected and the exhaustiveness of connections.

Thus viewed, the system is a complex hierarchical formation of a set of elements (legal norms and relations between them). When regulating a particular social relation a part of this structure comes into action. Such an approach allows to study the chains of norms participating in legal regulation - "the legal path" for achieving the legal issues stipulated in the norms. The chains are a normative model of the actually arising and developing legal relations. By following the chains, a specific forecast of the possible developments of legal regulations is achieved. This is namely one of the main aims and information needs for both the law professional and the citizen incompetent in the related subject matter.

Due to the existence of relations of construction, as well as legal norms with alternative abstract and other elements, the chains grow into networks. As abstract structures they are equivalent to directed graphs. The corresponding mathematical apparatus allows to study the accessibility (along the chain), the coherence, cycles, flows,length, a certain (not necessarily numerical) value of their paths, and so on.

The norms and functional relations between them form chains that can be investigated by means of the apparatus of formal languages and grammars theory. Each chain can be considered as a word generated by some formal language, one terminal symbol of the language corresponding to each norm. A grammar has been found which describes the most general derivation rules between the elements of the normative system. Thus by the techniques of the theory of formal languages and grammars it is possible to describe the structural and technical-legal regularities in the legal norm system.

The presented orientation of research of the legal norm system is useful:
- for the creation of "forecasting" law information systems;
- as a technical legal tool used in legislative practice;
- in comparative law - for the comparison of formalstructures, etc.

EXPERIMENTAL REALIZATION

SPRINT is an experimental realization of the theoretical-legal views stated above. The system includes the possibilities of the already known to us law information systems and can answer the questions related to the specific informational needs of law professionals.

Due to the extraordinary specificity of the subject matter, SPRINT does not aim at drawing more complicated logical conclusions.

A law professional carries out a preliminary processing of the texts of laws. As a result the legal material is presented in structured form - legal norms, their elements and their more meaningful legal characteristics. Thus, the formalized legal information is processed by the system SPRINT by using new techniques for information retrieval in law.

The legal information used in the system SPRINT is organized in several files. The file of basic legal characteristics of the norms, the file of relations between norms and the file of keywords are used most.

SPRINT answers the following main groups of questions:
- by a given set of keywords it finds the normative text that contains them;
- for given norms, the system retrieves the text of factual content, the subject of the norm, the hypothesis and disposition (or sanction) of the norm, bibliography, legislative practice and basic legal features;
- for a given norm it retrieves its structural and functional relations and traces all possible input and output functional chains. In this way a specific legal forecast of the development of legal relations is accomplished;
- the system is able to retrieve a list of all normatively described rights, obligations, interdictions and sanctions about the subject for which data is available, as well as normative definitions of legal notions.

The system SPRINT is written in PL/I. The quality of the results from the experimental stage of the system allows to be reasonably optimistic about the elaboration of the theoretical-legal views on which SPRINT is based.

The authors foresee two trends for the development of the program realization of the system - increase and enrichment of the volume of legal material to be processed, and communication with the computer using a subset of natural language.

The authors believe that the system SPRINT will be useful not only as a tool for information retrieval but also as a legal advisor of the law professional. This effort to implement some theoretical viewpoints and elaborations of legal science is unique in Bulgaria for the present.

RETRIEVAL ASPECTS OF THE UNIVERSAL RELATION FOR ATTRIBUTE-ORIENTED QUERY LANGUAGES

A M. Tjoa* R.R. Wagner**

* Institut für Statistik und Informatik, University of Vienna, Vienna, Austria
** Institut für Informatik,University of Linz, Austria

ABSTRACT

Evaluating the data from the field of medicine, it always appears to be necessary to develop software-support which is user oriented and also easy to handle for the physician. The main task of this software-support is the manipulation of stored data and its transformation into standard interface forms of input of the evaluation program packages. The aim of the paper will be to describe a general system-architecture and the general concept of a query-language which makes it possible for the researcher (physician) to formulate in a simple way even complex problems for the purpose of preparing data from a data base.

KEYWORDS

Relational model; universal relation assumption; relational data base systems; total logical data independence; query-language; data-manipulation language.

INTRODUCTION

In medical research-projects such as epidemiological studies, a large amount of data are collected and stored in a database system for the purpose of analyzing these data by means of statistical methods. In such medical research-projects a great number of different questions are treated, whereby every question usually implicates the selection and statistical analysis of a part of the data stored in a data base system. This selection process means, that before analyzing data by statistical methods, the user must specify those data he needs for his evaluation by selcetion criteria. For the purpose of selection resp. manipulation of stored data, the computer scientist has to make a suitable software-support available.

It is the aim of this paper to propose and to describe a general system-architecture and the general concept of a query-language for such an software-support.

A basic concept of the query-language was introduced in Tjoa and Wagner (1979 a). In a next step an algorithm which is necessary for the implementation of the query-language was developed in Tjoa and Wagner (1979 b). A detail description of the language is given in Wagner (1979) and in Tjoa and Wagner (1980) it was shown that this query-language is more suitable for users like physicians than other known query- resp. data-manipulation-languages. Finally this language was implemented and integrated in 'test-data base system' called RELORD (Schmittner and colleagues, 1979).

GENERAL CONCEPT OF A DATA BASE SYSTEM

The proposed query-language is designed for a data base architecture in the sense of the ANSI/SPARC-proposal (Klug, Tsichritzis, 1979). In this data base architecture which is accepted by the data base designers, the views of data are divided into three levels:

- Conceptual level (logical view of all data stored in the data base)
- External level (logical user view of data)
- Internal level (description of the physical realization of the data)

The advantages of a database system designed in the sense of this architecture are:

- a high degree of flexibility
- logical and physical data-independence
- the ability to operate on different data structures
- access paths independence of data structures
- uniform retrieval time and (physical) update time

One of the most important advantage for the user consists in the logical and physical data-independence. That means that the user's knowledge of the internal data-organization is not necessary on the one hand. On the other hand the user is independent of any change of the logical structure of the data. The user has his own logical view of data, which is independent of the physical realization of these data and therefore this architecture makes it possible to communicate (i.e. the manipulation of stored data) with the data base system in a very easy manner.

The logical views of data (i.e. the conceptual level and the different external levels) are described by abstract models, the so called data models.

For the internal level a suitable model, which describes the physical organization of the data must be available.

The mapping between the different levels must be realized by a software package which is called the data base management system. For the user the knowledge of the data base management system is not necessary, every change of the data base management system, which is caused by a change of the data base system will be implemented by the so called data base administrator.

THE CONCEPTUAL LEVEL OF THE SYSTEM ARCHITECTURE

In the literature a lot of different data models are proposed and discussed for designing the conceptuel level of a data base system. For the system-architecture proposed in this paper Codd's relational model (Codd 1970, 1971a) was chosen, because this model is very flexible, more robust and more proven than other models and a performance study (Lochovsky and Tsichritzis 1976) has shown, that the relational model has a comparatively better user performance than other models.
In the data model of Codd (1970) the data are organized in the form of relations. It is important to distinguish between the relational schema of a relation which can be considered as a table with its attributes (see Fig. 1) and the realization of a relation which represents the content of the table (see Fig. 2).
In every relation there exist one or more attributes, which identify uniquely every tuple of a relation. Such attributes are called key-attributes of the relation R, all other attributes are called non-key-attributes.All rows of the table are distinct, the ordering of the rows and columns is inmaterial.

PATIENT	PATIENTNO	BIRTH_DAY	SEX	DIAGNOSIS

Fig. 1. Relational schema of the relation PATIENT.

PATIENT	PATIENTNO	BIRTH_DAY	SEX	DIAGNOSIS
	17 456 71	1932 04 03	f	D1
	13 367 71	1934 07 01	m	D2
	17 123 72	1924 03 02	m	D1
	15 245 72	1967 04 07	f	D3
	.. ..	....	.	..
	.. ..	....	.	..

Fig. 2. Example of a realization of the relation PATIENT

Using the relational model the conceptual level of the system-architecture is described by:

- RN which is a set of relation-names
- A which is a set of attribute names
- D which is a set of domains
- r: RN $\rightarrow$ P(A) a mapping which assigns for each relation-name the set of its attributes.
- d: A $\rightarrow$ D
- Σ_{RM} constraints-types used for the relational model.

THE EXTERNAL LEVEL OF THE SYSTEM ARCHITECTURE

In the last years many query-languages based on the relational data model were proposed and discussed in the literature (e.g. Chamberlin and Boyce 1974, Codd 1971b Zloof 1975). All these relational based query-languages turns out to be not very easy to use for a user like the physician, caused by a very formal notation and by the fact that the user's knowledge of the relations is necessary. In these convential relational query-languages the logical view of data (external level) consists of decomposed relations in third or fourth normal form. The demand of the user's knowledge of normalized resp. atomic relations means a kind of access-path. Although the user is independent from any physical realization of the data (i.e. physical data independence) in these conventinonal query-langauges there exists no total logical data independence. Therefore for the user few larger relations would be more desireable or in the optimal case if no knowledge of relations at all is required, i.e. only the knowledge of the attribute-names and the according domains is necessary.

In the effort to make a next step in the direction of a better user performance a concept of a relational based query-language was proposed (Tjoa and Wagner 1979), in which the knowledge of normalized resp. atomic relations is not necessary.

It is possible by means of an algorithm (Osborn, 1979 , Lozinskii, 1979 , Tjoa and Wagner, 1979) to produce from the normalized resp. atomic relations large joined relations. Each semantic reasonable interaction on a database is working within one relation. That means that for the a logical view of data these joined relations are the only relations to be considered. Of course, by using this concept it must be required that the attribute-names used in the database must be unique, i.e. every attribute of the database is identified uniquely by its name. So every joined relation fulfills the universal relation assumption (Beeri,Bernstein and Goodman, 1978).

The external level of the system-architecture proposed in this paper therefore consists of large joined relations (universal relation assumption).
Using the concept of the universal relation assumption the external level is described by:

- UR the universal relation
- A is the set of attributes of the universal relation
- D is the set of domains
- d: $A \rightarrow D$

The universal relation means that all relations in a data base are projections of a single relations. Let $R_1(A_1),\ldots,R_n(A_n)$ be the relations of the conceptual level of the system architecture, then it is assumed that a univeral relation UR(A) with $A = \bigcup_{1\le i\le n} A_i$ exists, such that $R_i = UR[A_i]$ for $1\le i\le n$.

THE INTERNAL LEVEL OF THE SYSTEM ARCHITECTURE

The internal level of the system-architecture proposed in this paper is represented by the concept of catalogue-relations, i.e. the relations of the conceptual level stored in the data base are described by relations. For the system three catalogue-relations are defined:

- Catalogue-relation 1: In this relation the semantic of the conceptual-relations are described.In this catalogue-relation the relationship between relations and attributes is stored.
- Catalogue-relation 2: In this relation attribute specific declarations are described.
- Catalogue-relation 3: In this relation declarations about the conceptual relations are defined (e.g. length of a relation, state of a relation, lockcode etc.).

A detail description of these catalogue-relations is given in Schmittner and colleagues (1979).

It should be noted that for the physical realization of the 'test-data base system' RELORD B-trees (Bayer and McCreight, 1972) are used. Since the RELORD-system was implemented in PL/1, the file management system was supported by regional-file-organizations.

THE CONCEPT OF THE QUERY-LANGUAGE

One of the main characteristics of the query-language which is designed for a system-architecture described in the chapters before consists in a very easy syntax. On the other hand the knowledge of the relations stored in the data base is not necessary, i.e. only the user's knowledge of the attribute-names and the domains belonging to it is necessary. The user-view for this query-language consists of only one large relation instead of many small relations.

In the effort to create a query-language which is very suitable for a user like the physician, the query-language was concepted as a dialogue-supported language. This dialogue-support means, that every step made by the user is controlled and if the user makes some mistakes he gets a system-message. In this message the user is informed about his mistake and he gets the possibility for correcting the input.

In addition every query is proved whether it is semantically unique. If the system locates that a query is not semantically unique, the user gets a message. An algorithm for test of uniqueness of a query is given in Tjoa and Wagner (1979b).

The output of a query is a relation which is specified by the user. This output represents the input-interface to any statistical analysis program. Of course by an interface-system the output-relation must be transformed into an adequate input form of such an evaluation program. Since very often only simple evaluations of the selected data are required, the user has the possibility of creating new variables. The creation of new variables means that the user can evaluate arithmetic expressions (in which some built-in-functions like mean, variance, sum, product are contained) and logical values by logical expressions. A detail description for the creation of new variables is given in Schmittner and colleagues (1979).

Using the concept of the universal relation assumption as external view the general structure of a user request by means of this query-languages is as follows:

- Definition of the attribute-names which are relevant for the specific user-request.
- Definition of the selection criteria for the specific user request by attribute/value-pairs.
- Definition of the output-attributes.

The query-processing with the proposed system-architecture can be operated in the following way:

- Specification of the relations which are relevant for the specific user-request.
- Creation of a join sequence for the attribute set specified by the user.
- Elimination of the tuples of the relations which do not fulfil the selection criteria specified by the user.
- Determination of the output-relation (i.e. a projections on the output-attributes specified by the user).

Of course it is not possible to describe this query-language in detail and therefore only the general structure of this language was shown. The query-language is described in Wagner (1979) and in Schmittner and colleagues (1979).

REFERENCES

Bayer, R., McCreight E. (1972). Organization and Maintenance of Large oredered Indexes. Acta Inf., 1, pp.173-189.

Beeri, C., Bernstein P.A.,Goodman N. (1978). A sophisticate's introduction to database normalization theory. Proc. Very Large Databases, Berlin.

Chamberlin, D.D., Boyce R.F. (1974). SEQUEL A structured english query language. IBM Research Report RJ 1200, San Jose.

Codd, E.F. (1970). A relational model of data for larged shared data banks. Commun.ACM,13, 6, pp.23-28.

Codd, E.F. (1971a). Further normalization of the database relational model. In Rustin (Ed.), Database Systems,Prentice Hall, Englewood Cliffs N.J.

Codd, E.F. (1971b). A data base sublanguage founded on the relational calculus. Proceedings of the ACM-SIGFIDET workshop, Data Description Access and Control.

Klug, A., Tsichritzis D. (1978). The ANSI/X3/SPARC DBMS framework report of the study on Database Management Systems. Inf.Systems, 3, 3.

Lochovski,F.H., Tsichritzis D. (1976). User performance considerations in DBMS selection. Technical Report Toronto.

Lozinskii, E.L. (1979). On Query-answering in relational data bases.Proceedings of the National Computer Conference, AFIPS-Press.

Osborn, S.(1979). Towards a universal relation interface.Proceedings Very Large Data Bases, Rio de Janeiro.

Schmittner,G.,Perr H., Perndl C. (1979). RELORD-Implementation of query-language and a relational data base system (in german). Techn.Rep.Univ.Linz.

Tjoa,A M., Wagner R.R. (1979). A relational based query-language for medical database systems (in german).In Ehlers C.Th., Klar R. (Eds.), Medizinische Informatik und Statistik, 16, Springer Verlag, Berlin,New York.

Tjoa,A M., Wagner R.R. (1979). Some more dependencies on the relational model and their relevance to queries on relational data bases. Proc.Euro IFIP Conference,North-Holland,Amsterdam.

Tjoa,A M., Wagner R.R. (1980). Some considerations on the relevance of relational data-manipulation-languages for medical research.Proceedings MEDINFO 80 Conference, North-Holland,Amsterdam.

Wagner,R.R. (1979). Query-by-attribute-characteristics.Techn.Rep.Univ.Linz.

Zloof,M.M. (1975).Query-by-example. Proceedings of the national Computer Conference AFIPS-Press.

NEED OF A RELIABLE MESSAGE IN BIBLIOINFORMATICS RETRIEVAL

A. Perales de Mercado

Seminario de investigaciones bibliotecológicas. Facultad de Filosofía y Letras. Universidad Nacional Autónoma de México

ABSTRACT

The application of informatics in the building of a library collection data base in developing countries, has two important aspects to be considered for success or not: 1) those against for a reliable biblioinformatics message; 2) those in favor for a reliable biblioinformatics message. Introduction to the problem. Situation in post industrial countries. Situation in developing countries. Actions to be taken in order to solve the problem.

KEYWORDS

Bibliographic control: crisis of descriptive cataloging; copy of old catalogs; reliable message; biblioinformatics retrieval; bibliographic quality; Information Science; errors; computer fraud; legislation; professionals boards; standards; users; planning information systems.

INTRODUCTION

The bibliographic control of library collections need of a reliable communication instrument with users and staff. This instrument is the card catalog of books invented by the French revolutionary era and carried out to its best expresion by the American librarianship.

The bibliographic elements written in the card body, represented the message between a transmisor and a receptor. All the history of the development of cataloging theory and practice tried to build a unified and reliable bibliographic message to the users.

That was the reason of the elaboration of cataloging codes according to the bibliographic materials of the moment.

In developed countries, the tradittional way of making library card catalogs became obsolete, when descriptive cataloging code was unable to solve all problems concerning to an increasing and diversifying bibliographic production. Added to this situation are economic problems of inflation, reduction of budgets and those of a closer library cooperation.

The crisis of descriptive cataloging in the libraries of United States caused impact not only in the same country but in all those depending on its codes and sales.

Most of Latin American countries are largely depending on American librarianship and they have not developed national cataloging codes, although they have different languages and other library problems.

Now with the influence of American library automation, some Latinamerican libraries have reached the economic point of using the benefits of biblioinformatics, copying the work already done in old catalogs. Are they going to have a realiable message in biblioinformatics retrieval? If so, what is the price to pay for this tremendous work?

IMPLEMENTATION OF BIBLIOINFORMATICS

The application of biblioinformatics to the main library card catalog and other catalogs has focused on the production of replicas in content to the card catalog to be replaced. As a result of this concern it is necessary to be considered some aspects not yet resolved: a) lack of information in material newly received at the library; b) incomplete catalog due to various reasons; c) an equality of cataloging due to standards, which may not represent the value of the material cataloged; d) to mantain the print bound mentality associated with 3 inch. by 5 inch. catalog cards; e) decisions to curtail fullness of cataloging description; f) partial catalog is available to users in library systems; g) lack of current information derived from printed lists that involve lag time in publication and distribution. In writting about these aspects, Brett Butler's experience was followed.

According to this distinguished author: "the data base cannot be used as a comprehensive managerial tool if it can not replace the necessary technical files that have supported library operations in the past"[1] In regard to limitations Butler pointed out: "integration of resources data bases with circulation systems is yet basically non existent"[2] Finally Spyers-Duran underlined: "Maintenance of bibliographic quality is primarly a professional concern. . . from a managerial viewpoint, automation has a tremendous impact on staffing a catalog department".[3]

If this is the biblioinformatics situation in a post industrial country, waht can be said in regard to developing countries, where Library Science has not yet achieved the necessary maturity and Information Science flourishes in government scientifique foreign resources data bases.

As an introduction to this problem it is important to remark the socioeconomic situation of developing countries with various grades of illiteracy, lack of food, demographic problems, poor sanitary condittions, insufficient housing, devaluation, inflation, developing commerce and communications, high public debt and so forth. Official prioritties are in regard to solve these priorities in order to avoid unemployment through a better agriculture, houssing, dressing, hygiene, education, energetics, industrialization and productivity.

If developing countries desire to improve their socioeconomic situation already mentioned, it is necessary to establish a National information system, with subsystems in each specific field. From rural areas to urban zones with proper instalations for good social communications. At the same time a deep renovation in state and federal library systems, school libraries, university informations networks, special libraries and national biblioinformatics networks.

A well planned biblioinformatics service, probably can solve the complexity of this concern. Care has to be taken in regard to the main topic of this paper NEED OF A RELIABLE MESSAGE IN BIBLIOINFORMATICS RETRIEVAL. It is not adviceable to copy old library card catalogs, plagued with errors as a consequence of a poor work done in the past and some times in the present too. They have not a reliable raw material to be used in a bibliographic information system. Due to many factors involved each part of the card body may have mistakes that can not be transfered to the network in a honest work.

It is important to admit that there are many aspects against a reliable biblioinformatics message in under-

---

[1] Butler, Brett "Collection and Resource data bases in bibliographic management" In: *Requiem for the card catalog. Management issues in automated cataloging.* Edited by Daniel Gore, Joseph Kimbrough and Peter Spyers-Duran. Westport, Conn., Greenwood Press, 1979. 200 p. 3-27 p.

[2] *Ibidem.*

[3] *Ibid.*

developed library card catalogs in regard to: human resources represented by an impersonal library bureaucracy, improvised executives and margined distinguished professionals, when existent. In regard to technical services, lack of standards, modifications to cataloging and classification tasks according to each new heads policies. Few libraries have a charging system in good condition, therefore there is not a bibliographic control: bibliographic material registered in the card catalog is not in the collection, items in the collection have no encryption in the card catalog. From a managerial viewpoint nobody knows how much was spent in errors and lost material. In such conditions biblioinformatics message can be an undesirable computer fraud.

As the same time there are many aspects in favor for a reliable biblioinformatics message in building a data base collection in developing countries. Par example: to enact a proper legislation in regard to library and information services, to establish professional boards in order to supervise professional work done in the bibliographic field, to consider the creation of National Standars for data encryption in the library and information fields with a conversion code for scientific and technological international network, to be conscious of the importance of taking in consideration user's needs, to look for an agressive administration wishing a high percentage of a reliable bibliographic information message at possible low costs when retrieving information, to think in the socioeconomic situation of the country when planning information systems.

CONCLUSIONS

Biblioinformatics is an outstanding finding in order to develop extension services for education and research.

Post industrial countries still have important biblioinformatics problems to be solved today and in the future.

Developing countries need to use this important tool to achieve a different quality of human life, taking in consideration their socioeconomic conditions.

The right use of this technology will probably give to developing countries a maturity in the socioeconomic field as well in Library and Information services.

REFERENCES

Bierman J. (1975). Automated alternatives to card catalogs. *j libr automat* 8, 277-298.

Fussler, H., and K. Kicher (1977). Contemporary issues in bibliographic control. *LQ* 47, 237-352.

Haas, W. J. (1977). Organizational structures to meet future bibliographic requirements. *LQ* 47, 274-284.

Kent, A. (1979). The online revolution in libraries. *AM LIB* 10, 339-342.

Martin, S.K. (1979). Upgrading brief and dirty data. *AM LIB* 10, 213-214.

Osborn, A. (1941). The crisis in cataloging. *LQ* 11, 393-411.

Thompson, J. (1979). The new catalogs and the unfinished evolution with increased automation and sharing in the last ten years, bibliographic processing has changed dramatically but catalogs for all is a dream still eluding our grasp. *AM LIB* 10, 357-360.

APPLICATION OF SYSTEMS APPROACH TO AUTOMATIC INTERPRETATION OF FREE TEXT DOCUMENTS

W. Delaney, L. Guerriero, C. Mero, E. Cannavò, E. Vaccari

Istituto di Fisica, Università di Bari, via Amendola 173
Bari, Italy

ABSTRACT

This paper presents results obtained in the free text analysis of medical documents using a software system whose philosophy is based on the formalization of the problem in terms of modeling concepts which implies a system's analysis of the physical system described by the medical documents. Our conclusion is that automatic content processing is possible for specific natural languages used to describe only one kind of physical or imaginary system. In this case, in fact, it is possible to base the analysis not only on natural language grammar and syntax but also on the meanings of semantically significant words, since each word will have a unique meaning when referred to a specific kind of system.

KEYWORDS

System modeling; free text analysis; language processing; documentation; radiological reports.

INTRODUCTION

Successful free text analysis presupposes computer access to the knowledge availablo to the human reader. Even though many suggestions have been made, the problem of WHICH information to furnish and HOW to formalize this information is essentially unresolved. Space limitations permit only brief references to the literature of the last few years where the increasing importance given to semantic considerations is evident, for example, in the conceptual dependency theory of Schank (Schank 1974 Schank 1975) and in the translator of Wilks (Wilks 1975). Both of these authors seek to define semantically primitive word classes and semantic relations between words. With respect to their approaches, we agree with the type of word classes used but it seems that a guiding principle determining the information to be associated with the words is lacking, implying the necessity to define an enormous set of semantic relations with the risk that, for any specific application, the set may be incomplete and will certainly contain superfluous relations.

Our approach (Vaccari 1975,Vaccari 1977) and Wilks's use similar techniques for the segmentation of text and for the synthesis of semantic structures; however they differ substantially at a conceptual level inasmuch as our approach assigns a fundamental role to the physical or imaginary context (the system object of the discourse, the SOD) in the organization of the word meanings and in the formalization of the purpose of the discourse, which we introduce as a fundamental consideration.

A SYSTEMIC APPROACH

In our approach language is seen as a means for constructing models of physical or imaginary systems for the purpose of transmitting finite quantities of information regarding specific aspects of such systems. For this purpose one uses linguistic expressions containing words of illimited semantic content (representing concepts corresponding to systemic entities) and words of limited semantic content (corresponding to systemic relations between such entities). In such an expression each word restricts the meaning of the others such that from a semantic point of view the expression can be considered to be a set of linguistic information units,LIU, of limited semantic content (representing detailed level concepts DLC corresponding to systemic information units, SIU, that is states, events and processes in physical or imaginary systems).

From the preceeding considerations it follows naturally that WHICH information to furnish and HOW to formalize them should be derived from an analysis of the system object of the document to be analyzed by the computer (the SOD).

Since they are not otherwise available to the computer one must explicitly furnish:
- -the systemic roles of the words: which systemic entities, or relations, they represent;
- -the types and structures of the SIU's which should be recognized in the analysis of the document (the operational definition of the purpose of the document and of the free text analysis).

FORMALIZATION

Having identified WHICH information to furnish, it remains to define HOW to formalize them.

Roles (in the SOD) are assigned to words by organizing the latter in a THESAURUS; the types and the structures of the SIU's are formalized in a SYSTEMIC SEMANTIC NETWORK which contains all the types of SIU's present in the document to be analized.

Operationally: considering the purpose of the document, one must identify which subsystems SS exist in the system SOD. Then the DLC must be defined in terms of the systemic entities: component C, attribute A, measure M interrelated by structural relations SR and by static and dynamic functional relations FRS and FRD respectively, so as to obtain the SIU's for the SOD and its SS's.

A state will correspond to one or more type 1 SIU's containing (1M,1A,(C,SR),FRS); an event will correspond to a type 2 SIU containing ((M),1A,(C,SR),FRD); and a process, or type 3 SIU will correspond to a set of type 1 and type 2 SIU's related by FRS's and FRD's.

The Thesaurus

Having identified the SIU's, the system analysis permits the identification of the specific system entities for the SOD and the organization of the words which may appear in the document in conceptual system classes CSC (C,A,M,R) subdivided into subclasses SCSC: active (CA) and passive (CP) components; action (AA) and characteristic (AC) attributes; measurements of the characteristics (MIS) and modes of the action (MOD); spatial (SRS) and gerarchical (SRG) structural relations; static FRS and dynamic (FRD) functional relations. This abstract classification scheme is rendered concrete forming lists of words (for each SCSC) which may appear in the document. This set of all these lists constitutes the thesaurus which organizes the word meanings with respect to roles in the SOD.

The Systemic Semantic Network

The systemic semantic network, SSN, is a set of SIU's; in the SSN each SIU is represented in the form of a binary tree in which the terminal nodes correspond to thesaurus lists belonging to the systemic class C,A,M and each non terminal node represents a list from the class of relations R identifying the sense in which the right sub-binary tree of the node specifies its left sub-binary tree.

Figure 1a illustrates the binary tree structure of a type 2 SIU (an event), the tree root represents the relation "performs the action of" . Each node is labeled by a number which symbolizes the list represented by that node; using that number one can locate in Fig.1b the name of the subclass containing the list and in Fig.1c a word which could belong to the list in a particular application.

a)

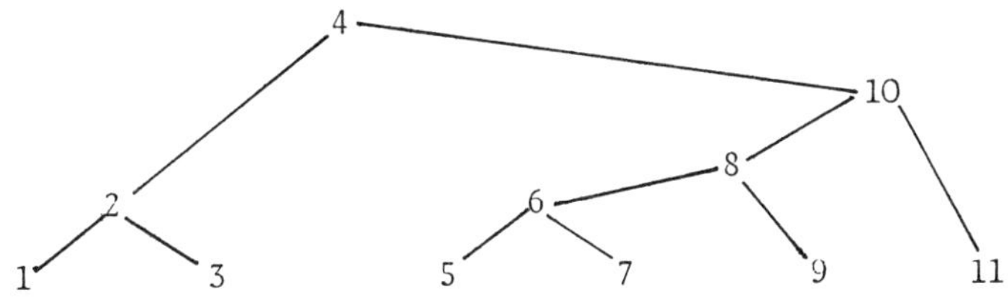

1 2 3 4 5 6 7 8 9 10 11

b) CA SR CA FR AA FR MOD FR CP FR CP

c) child (of) Mary play nicely with toys in park

Fig. 1. a) an SIU. b) subclasses containing node lists. c) words from the lists

When each list in the binary tree representing a SIU is replaced by a word from that list one obtains a binary tree called a normalized LIU (NLIU); the binary tree in Fig.1a with the words in Fig.1c at its nodes is an example of a NLIU.

The purpose of the free text analysis is to extract such NLIU's from the phrases of a document; an NLIU is extracted from a phrase by means of an analysis algorithm which organizes certain words contained in the phrase in the same way that their parent lists are organized in some SIU contained in the SSN. The NLIU exemplified in Fig.1 would be obtained for example from the phrase

Mary's child plays nicely with the toys in the park (1)

By inputing his SSN (and thesaurus) to the algorithm, the user specifies which kind of NLIU's he desires to receive in output. In the case in which his document contain phrases like (1), but he wants the algorithm to recognize and output simpler NLIU's than that shown in Fig.1c, he must input an SSN containing simpler SIU than that shown in Fig.1 (however more than one such simpler SIU will be necessary to extract all the information contained in a phrase like (1)).

The Analysis Algorithm

The goal of the analysis algorithm is to organize the words contained in each sentence of a document in one or more structures, called complete STRINGS, which satisfy the following criteria:

1. in a STRING words from the sentence are organized in a binary tree which mirrors the organization of their parent lists in some SIU of the SSN
2. a STRING is complete when it contains the essential elements of a NLIU. In general the user defines the essential elements of a NLIU by assigning appropriate values to a flag associated with each node of each SIU in the SSN.

Very synthetically the analysis of a sentence proceedes as follows:

1. using the thesaurus, each word in the sentence is classified according to the list to which it belongs
2. the sentence is devided in PARTS where each PART: begins with a syntactic operator, a linguistic element with limited semantical content, (except for the first PART of the sentence) and ends with the last word before the next syntactic operator
3. for each PART of the sentence an SIU in the SSN is found which contains all the parent lists of the words in that PART; the words in the PART are then organized into a STRING which mirrors the organization of their parent lists in the SIU
4. incomplete STRINGS are united to form more complete STRINGS; the unification operation is repeated until all STRINGS are complete.

APPLICATION : FREE TEXT ANALYSIS OF MEDICAL REPORTS

A software system, DARR (Vaccari, 1977) for the automatic classification of radiological reports has been realized in the framework of the approach described above.

In particular reports related to tomography of the ear were considered; this choice was motivated by the limited number of linguistic elements used (about 1000) and by the fact that such reports contain considerable more descriptive detail than that found in other type of radiological reports, which implies a level of linguistic complexity well suited to the testing of the software system.

A radiological report consists of a set of sentences whose information content corresponds to one or more observations of the radiologist (states of the system observed - the ear in our case). The sentences are the physical units of the report whereas the observations are the LIU's. In general the sentences are formally ambiguous; their constituent LIU's are unstructured and are not independent (they may have words in common).

The purpose of the free text analysis in our application is to transform the reports into a normalized language such that each observation is expressed by one NLIU (which corresponds to a physical unit in the normalized language); in this way the observations are formally independent and non ambiguous as necessary, for example, for computerized statistical research studies.

On the basis of an analysis of the SOD, a thesaurus organized according to the following classes and lists was obtained:

Class	Subclass	List
C	CP passive component	SA complex anatomic part SSA elementary " " GAP global physical aspect POS parts of SA's or SSA's
	AA pathological attr.	PC pathological findings P3 " conditions
A	AC physiological and anatomic character.	VIM,VIMP,VIMF,VIMD implicitly measured physiological and anatomic characteristics VNM,VNMP,VNMD physiological and anatomic characteristics not implicitly measured P1,P2 abnormal conditions NAC normal "
M	MIS measurements	MIS,MISP,MISD meas. relative to normal conditions ENT1,ENT2 absolute measurements MFD modifiers

The thesaurus also contains lists defining syntactic operators, one for punctuation marks and others (containing prepositions, conjunctions, etc.) for relations.

Figure 2 is a simplified representation of the SSN used in DARR. Each PATH from the first node (SA) to the last node (MFD) corresponds to an SIU. More precisely traversing such a PATH from SA towards MFD one encounters thesaurus list names corresponding to terminal nodes of an SIU; these "terminal lists" are encountered in the following order: first those belonging to class C, then those belonging to class A, and finally those belonging to class M.

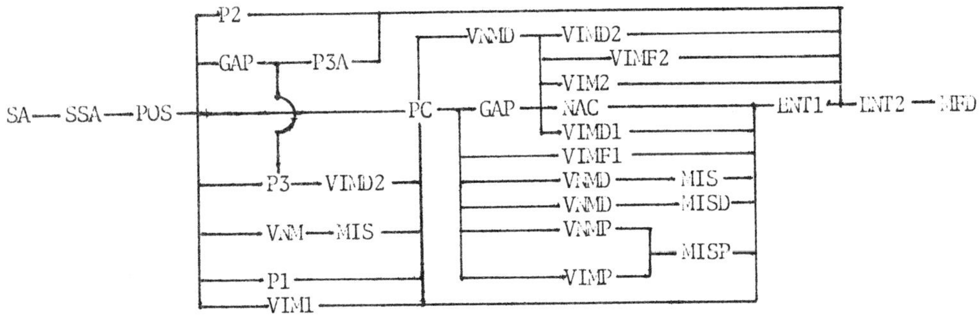

Fig. 2. The SSN used in DARR

To simplify the figure the relations are not shown, however (for the present application) they may be inferred as follows: in an SIU, consecutive terminal lists belonging to class C serve to localize the observation more precisely and consecutive terminal lists belonging to class A and M serve to specify more precisely the state being described.

Results

The DARR software system has been tested with 300 radiological reports written by two radiologist from the state hospital of Verona. The reports contain about 1600 different sentences, of which 60% contain type 1 LIU's, 29% contain type 2 LIU's and the remaining 1% contain type 3 LIU's.

The sentences containing the type 1 LIU's have been correctly analized by DARR using the SSN in Fig. 2 ; the modifications required for the analysis of the other sentences are presently being implemented.

REFERENCES

Schank, R., and C. Rieger (1974). Inference and the computer understanding of natural language. Art. Intelligence 5, 373-412

Schank, R. (1975). Conceptual information processing. North-Holland

Vaccari, E., W. Delaney, and S. Ventrella (1975). An approach to text analysis of medical reports. Proc. 28th ACEMB, New Orleans, Louisiana.

Vaccari, E., and W. Delaney (1977). Automated recognition and classification of independent observations contained in medical documents. Proc. COMPSAC, Whashington

Vaccari, E., W. Delaney, and A. Chiesa (1977). A free text analysis system for the automatic documentation of radiological reports. Meth. Inform. Med., vol 16, 3; 144-153

Wilks, Y. (1975). An intelligent analyser and understander of English. Comm. ACM, 18, 264-274.

SOME ASPECTS IN THE DESIGN OF AN AUTOMATED SYSTEM FOR TERMINOLOGY SERVICES "TERMSERVICE"

B. Nikolova, R. Pavlov

Laboratory of Math. Linguistics, Institute of Mathematics with Computer Centre, Bulgarian Academy of Sciences, 1090 Sofia, P.O. Box 373 Bulgaria

ABSTRACT

The paper sketches the general framework of an automated English-to-Bulgarian and Bulgarian-to-English system for terminology services in the field of mathematics, describing the motivation for the creation of the system, the type of services to be provided, the sort of persons using them, the methodology of the linguistic design and long-term aims and perspectives.

KEYWORDS

Applied computational linguistics; automated terminology services; terminology bank; parallel texts analysis; dictionary entry.

INTRODUCTION

Nobody working in a scientific or technical field can do without some information published in other languages. This implies that the need for translations is enormous, which means that for each language efforts must be made to compile the correct terminology.

As elsewhere, in our country mathematics professionals and translators of mathematical texts are asking to have the terminology relevant to their work made available in English. Moreover, the terminology itself both English and Bulgarian, needs to be standardized. A traditional dictionary (English-Bulgarian or Bulgarian-English) of mathematical terms does not exist in Bulgaria. The only technique which offers enormous storage capacity and fast retrieval of terminological information is computerized data processing. A well set up terminology bank offers ample opportunities to solve the difficulties met by the translator or mathematician in dealing with scientific texts.

To meet these needs, the Laboratory of Mathematical Linguistics at the Institute of Mathematics with Computer Centre affiliated to the Bulgarian Academy of Sciences has set itself a number of short, medium and long-term aims, designed to lead to the setting up of automated terminology services - the project TERMSERVICE.

FUNCTIONS AND USERS OF THE PROJECT

The types of services which will be provided by the system are, as follows:
- the system can be used as a computerized English-Bulgarian and Bulgarian-English dictionary of mathematical terms;
- the system can be incorporated in a machine translation project or be used in comprising the list of keywords and descriptors for information systems, as well as in systems for automated processing of English texts;
- the system can also be used as a base of a computer-aided instruction in English for students and specialists in mathematics;
- and, finally, the system provides sufficient linguistic information to conduct research on terminology, lexicography and lexicology.

When finalized, the system will serve the following groups of potential users:
- specialists in applied computational linguistics;
- specialists in information systems design;
- mathematicians;
- translators;
- linguists;
- students and lecturers.

METHODOLOGY OF THE LINGUISTIC DESIGN

The method used in the linguistic design of the system TERMSERVICE is parallel texts analysis. An original text in one language and its translation in another are termed as parallel texts. The task is to separate the items of translation in the original text, to find their equivalents in the target-language translation, to establish a one-to-one correspondence between them and fix the latter in a correspondence file. At present this is a task requiring much time from experienced linguists and programmers and involves considerable difficulties of linguistic and programming nature.

The idea of parallel texts analysis was taken up from presently existing automated systems for lexicographic services developed in the Soviet Union by the Centre for Translation but the procedure has been considerably improved towards further automation.

Preference is given to this method since it yields most reliable and genuine results about the state-of-the-art of terminology and allows to perform the lexicographical work mostly automatically.

STRUCTURE OF THE DICTIONARY ENTRY

The basic operative unit of the system TERMSERVICE is the dictionary entry. Headwords of dictionary entries are the English terms (simple or compound). The structure of a dictionary entry is organized according to the meaning or various meanings of a term in Bulgarian connected with the context relevant to it. In addition, each term is supplied with linguistic information which is an attempt to provide an extensive description of the corresponding lexical entry from a grammatical and syntactic viewpoint. Each term is also marked for its particular belonging to a subject field, the source it was taken from, spelling variants and standard abbreviations. Relationship to other entries, such as synonyms, antonyms, etc., is indicated by cross - references. The semantically relevant context is looked up automatically in the original English texts comprising the corpus or is a constructed context, but alway extracted from genuine original texts. Contextual examples (furnished with a translation in Bulgarian) have the advantage of giving each term with a more precise meaning than if it were isolated and serve to remove polysemy. Possible word combinations with the headword are also listed and stored separately. All this information is grouped in several information zones.

Since the system is not oriented only towards professional translators, general and specific terminology should be complemented by common-use lexems, their selection being motivated by frequency of usage.

Headwords are stored in conventional dictionary form. To assist the user's query, however, the project envisages a block of automatic reduction of inflected forms to standard form. Till that time the system will work with the aid of human pre-editing.

The project also envisages automatic recording of unsatisfied lookup requests for update purposes.

STAGES OF REALIZATION OF THE PROJECT

The project has several stages of realization - the design of subsystems for the terminology of various fields of mathematics - ALTERM for Algebra, LOGTERM for Mathematical Logic, ANTERM for Analysis, and so on - will end up in a general system covering mathematical terms as a whole - the system MATERM.

The long-term aim of the project is to providethe properly adapted software needed to create terminology banks for different scientific fields - Physics, Chemistry, Medicine, and so on.

The system will be used on-line with time sharing and should be operative in 1981.

A SYSTEMS APPROACH TO WORD PROCESSING

P.A.Land
School of Information Sciences, Canberra College of Advanced Education,
Belconnen, A.C.T. 2616, Australia.

ABSTRACT

Word processing directly affects the quality of work life and the job environment. A systems approach is used to analyse the goals, structure and function of word processing with a view to providing guidelines for organisations considering the introduction of such a system.

KEYWORDS

Word processing; office automation; sociotechnical systems.

INTRODUCTION

A communications assembly line is being introduced into the office and the technology which has enabled this is the word processor. Capitalisation per office worker has been low compared to that of factory workers (Zisman, 1979). Given the gains in productivity achieved by automating the factory, a similar effort is now being made in office communications. Tasks are being designed as part of a flow of activities to create an automated product. A technology has been developed to augment the productivity and capabilities of secretaries, and word processors are the machines in question. A supporting technology, including dictaphones for executives, is also used. "Specifically, WP seeks to channel the random and generalised work of traditional secretaries into two separate functions: . Typing [and] . All the other things a secretary does." (Kleinschrod, 1977). This point of view, presented by word processor vendors (Gilmore, 1979) and supported by office automation analysts coincided with the falling cost of computer hardware, the rising cost of secretarial salaries and the advent of microcomputers. It became cost-effective to introduce word processors (Stephens, 1979) in the 1970s and they have come into widespread use since then.

DEFINITION AND DESCRIPTION

Word processing is a part of office management which aims to increase the productivity and effectiveness of secretaries and executives. This is achieved by automating some or all of the process of document origination, preparation, printing, storage and distribution using special purpose equipment which has the ability to access supporting technical systems.

The characteristics of a word processor which distinguish it from more manual forms of document handling are

(a) editing of documents via a screen rather than retyping; for example, customising text, completing forms, revising lengthy reports;
(b) storage of documents on a disk for ease of retrieval;
(c) ability to merge lists with documents;
(d) speed and sophistication of some printers for presenting the final document.

The supporting technical equipment potentially enhances some of these functions. A shared disk or data base system avoids duplication (in stand-alone word processors) of lists and forms and provides a shared base for information retrieval; input can be provided via optical character recognition; output can be to fast printers with intelligent copiers, microform or photocopying devices and distribution can be achieved, via a network, using electronic mail. Further to this, general purpose data processing can be provided and the word processor can become a work station connected to this facility. One of the most recent studies incorporating many of these ideas is for a projected information system for the Australian Parliament (Aust. Parlt., 1980).

THE USER INTERFACE

Effectiveness of the technical equipment is both limited and facilitated by the user interface at the word processor screen. The user interface is the means of communication with the machine via typed instructions. Word processors have been accepted partly because these instructions have been designed for ease of use and are self-explanatory. Although they differ on each manufacturer's equipment they are designed for a special-purpose application, and present a coherent set of choices to the user. This feature distinguishes word processing from general purpose computers which, typically, use technical terms and cryptic machine-oriented codes rather than problem-oriented messages. To the extent that word processors are connected to other computer-based equipment the user is likely to experience an interface which is less sympathetic than that of the word processor. This limits its use and acceptability.

RELATIONSHIP WITH THE ORGANISATION

The introduction of word processing into an organisation is justified by the following benefits in operational terms;

(a) reduction in staff needed to perform the current workload or an increase in the workload completed by the current staff;
(b) improvement in product quality; and sometimes
(c) improvement in quality of work life due to more reliable and responsive document handling (Aust. Parlt., 1980).

In attaining these benefits executives, secretaries and managers are affected, and new procedures are required to allow a smooth flow of dictating, typing, correcting, revising, copying, filing and distribution of documents. Executives are expected to use dictaphones and standardised procedures for specifying the work to be performed. Secretarial work is extended to include the use of word processors, as a specialised job area or as an additional skill within secretarial work. The position of administrative secretary may be created in conjunction with the position of word processing typist. Office management may change from one of low visibility and little involvement with technology to one requiring understanding of information systems design, group psychology, telecommunications, EDP and office equipment selection, installation and management. Further to this, changes in working relationships will occur. Assuming that the allocation of secretarial support will now relate not to the status of the executive but to the work load within the office, some sharing of administrative support and of word processing typists is likely. This may also result in the appointment of supervisors of work groups of word processing typists and administrative support specialists.

There are also changes to the physical environment, especially for operators of word processing equipment. Seating, lighting and noise considerations are important as there seems to be no doubt that fatigue can occur from constant observation of VDU screens and from VDU flicker or glare, especially under poor lighting conditions (Rohlfs, 1979). Some industrial awards have been made to increase pay for specialist word processor operators doing this work exclusively (Watson, 1980).

WORD PROCESSING AS A SOCIOTECHNICAL SYSTEM

Given that word processing is a complex technical arrangement involving equipment and people and requiring the co-ordination of a set of task sequences to increase productivity and effectiveness in an office, it can be viewed as a system as shown in Fig. 1.

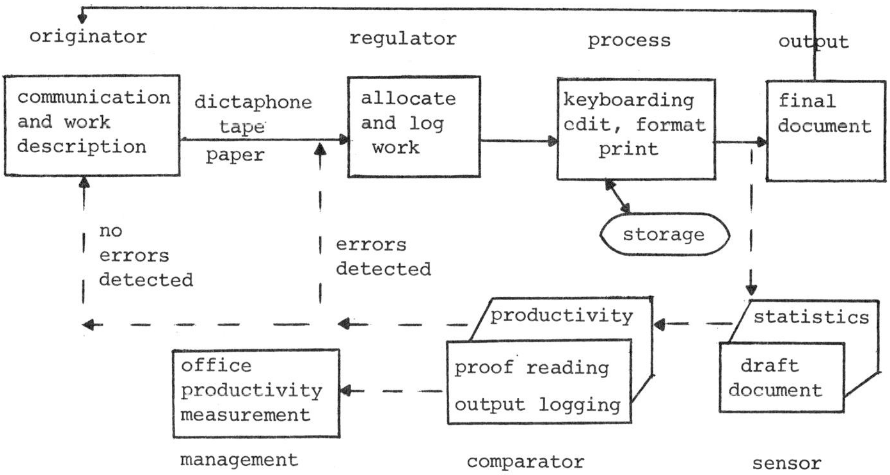

Fig.1 Word processing system

A document is originated in verbal or written form or by instructions for completing a standard form held on magnetic disk storage. Dictation or central recording equipment allows work to be originated at any time. New standards must be devised for specification of the work to ensure that the correct document is created, completed or edited. The supervisor logs the job and allocates tasks to word processor operators. The supervisor must balance workloads to ensure that varying priorities are met. The turnaround time will depend on how well a supervisor can balance work loads between operators and proof readers. It will also be affected by the number of printers available. Turnaround also depends on the reliability of word processors. If all keyboards are attached to one "shared logic" machine and this fails, no work can be done. A separate but supporting task is performed by an operator who guarantees that standard forms and documents currently being prepared or edited are available on storage. This requires a method for naming documents so that they can be identified by an originator, accessed by the operator and copied for security purposes in case any document is accidentally deleted.

Once the job has been allocated the document is transcribed or edited. This requires considerable skill and concentration. The operator must have mechanical aptitude and skills in editing, formatting, spelling and grammar. The operator must also know how to instruct the machine to perform these tasks which become complex when list manipulation and sorting is required.

After a draft document is produced, it is proof-read, corrected, and returned to the originator. The cycle is repeated until the originator distributes the document. Because it is easy to modify the document or reformat it, this may occur more often than with manual typewriters. At the final stage statistics are collected from the word processor and used in estimating productivity.

The structure of the system depends on management policy in the areas of task specialisation or task generalisation and electronic pacing or self pacing. A word processing system can be designed to fit into any of the four segments shown in Fig. 2.

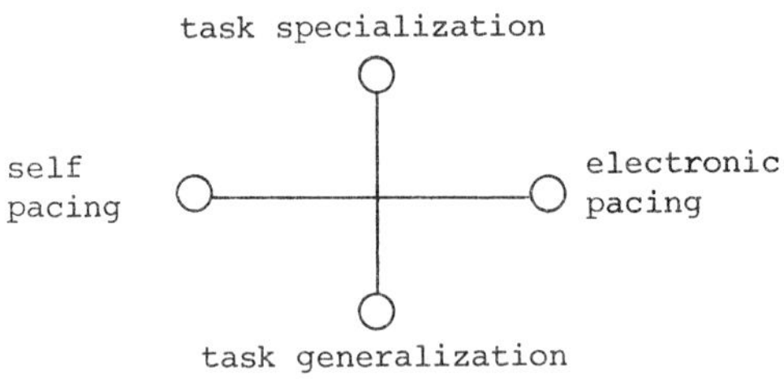

Fig.2 Sociotechnical system design considerations

Detailed studies have been made of the way executives and secretaries use their time in the office (Engel and others, 1979; Kleinschrod, 1977). On analysis, the conclusion drawn is that secretaries waste time through lack of task structure and control and that executives waste time by not delegating enough work because the appropriate support systems are not available. The suggestion is that secretarial productivity will improve through task structuring and specialised work supervision in combination with new technology. An example of the specialisation proposed is to divide administrative support tasks into the positions of supervisor, administrative secretary, receptionist and courier, and to organise typing tasks into the positions of word processing supervisor and word processor operator. Within these categories there are the grades of senior and junior. By introducing task specialisation and using support groups instead of generalist secretaries, savings are effected. Further, staff can be easily replaced. As far as executives are concerned, if they use dictaphones shorthand work is no longer required. Authority structures are also altered in the organisation's hierarchy as support staff are now supervised by specialist supervisors and not by executives. Task specialisation also results in changes in salary structures and working conditions (Watson, 1980). Examples of the latter are eye strain and postural fatigue which can be experienced with certain types of word processing arrangements (Rohlfs, 1979).

Task generalisation occurs when either executives or secretaries add the task of word processor operator to their current job descriptions. Within the literature, emphasis is placed on task specialisation as the way to gain control over secretarial time and hence to upgrade productivity. Task generalisation is seen as difficult to control and measure. However, a management philosophy which acknowledges morale as a job design consideration may find that task generalisation, in this context, results in increased productivity and morale. The nature of the task is challenging and most secretaries who use word procesors enjoy this work. Secretaries typically feel that mastering a word processor is a worthwhile skill requiring high mechanical aptitude and capacity for visual and spatial design. Some electronic pacing is taken for granted in many word processing systems, assuming the argument that costly technology should be used to full capacity. Part of the argument is that executives should use dictaphones. Dictaphones or central recording devices increase productivity by ensuring a smooth work flow with the word processor being a costly part of the system (in fact, the people are more expensive). An example of electronic pacing is the organising of a group of general secretaries' time to use the word processor continually. However, this is not a requirement of the equipment, but a matter of managerial policy. With electronic pacing, the technology itself may be seen as part of the authority structure of the office as it becomes perceived as the monitor. With self pacing, the word processor is seen as a tool to be utilised by the people involved, and to assist in the task to be performed. This may lead to a creative use of the equipment, when productivity is not the dominant design goal.

The literature is biased towards job specialisation and assumes electronic pacing; the resulting hierarchy has lead to claims of improved career paths for secretaries (Waterhouse, 1979). This may be the case in organisations with large installations but there is nothing inherent in the technology to determine that those secretaries will achieve the career positions and few operators could move up a career path in a small organisation. The reality for executives is some standardisation of their work, the requirement of using a dictaphone and the loss of a personal secretary.

PERFORMANCE MEASURES

Performance measurement can occur at several levels of system operation. The basic mechanical level is keystrokes per minute, lines of output, or documents printed per day. These can be measured by word processors themselves. However, they depend on the type of work done (original input, editing, reformatting) and do not reflect task variety. Higher levels of measurement such as document turnaround time, author satisfaction and volume of usage may be more appropriate. Further, an activity analysis of executives and secretaries both before and after introduction of a system and at opertional maturity, may be done. Once the system is stable, then quantitative productivity measures could be defined by calculating standard rates of work for original transcriptions and editing. In the final analysis, the aim is to achieve a designed system offering effectiveness and job satisfaction through task variety which will avoid the human and economic costs of absenteeism and high staff turnover as well as giving opportunities for development.

REFERENCES

Australian Parliament (1980). Planning and Design Study for the Information Systems and Information Services of the Parliament. Parliamentary paper 216.

Cooper, R., and M. Foster (1971). Sociotechnical Systems. Am.Psychol. 26, 467-474

Engel, G. H., J. Groppuso, R. A. Lowenstein, and W. G. Traub (1979). An office communications system. IBM Syst. J., 18, 402-431.

Gilmore, J. (1978). Office automation. Proceedings of the Digital Equipment Computer Users Society, 4,5, 1165-1182.

Kleinschrod, W. A. (1977). Management's Guide to Word Processing. The Dartnell Corporation, Chicago.

Rohlfs, S. (1979). User interface requirements. Infotech State of the Art Report Convergence : Computers, Communications and Office Automation, Vol. 2, Maidenhead, UK.

Stephens, W. J. (1978). Word Processors: Some Social and Industrial Implications of their Use in Australia. Department of Labour and Industry, Adelaide, South Australia.

Waterhouse, S. A. (1979). Word Processing Fundamentals. Canfield Press, San Francisco.

Watson, R. H. C. (1980). Public Service Arbitration Act 1920 Determination No.246 ... (Additional Word Processing Designations). Public Service Board, Australia.

Zisman, M. D. (1979). Integrated office automation systems. Infotech State of the Art Report Convergence: Computers, Communications and Office Automation, Vol. 2, Maidenhead, UK.

XI. PATTERN RECOGNITION AND SCENE ANALYSIS

AUTOMATIC INTERPRETATION OF SYMBOLIC DATA ON MAPS AND ENGINEERING DRAWINGS

J KITTLER
SRC Rutherford and Appleton Laboratories,
Chilton, Didcot, Oxfordshire OX11 0QX, U.K.

ABSTRACT

The problem of automatic recognition of symbolic data is considered in the framework of the more general task of automatic digitization and interpretation of maps and engineering drawings. The combined structural and decision theoretic approach which has been adopted to solve this problem is described.

KEYWORDS

Line drawing processing, pattern recognition, character recognition.

1. INTRODUCTION

The storing of line drawings such as maps and engineering drawings in a digital form has a number of advantages. It is an efficient way of recording information and making it easily accessible for many users. The digital form facilitates information transmission and if required, a subsequent redrawing of the line structured data on a digital plotter. The form is compatible with the current general trend of relying on computer memory devices in their capacity of an information recording medium to an increasingly greater extent. It allows a convenient retrieval of information and it is the only possible form that can be used in situations where the information serves as a reference data source for computer controlled or computer driven systems.

A line drawing can be transformed to a digital form simply by storing the binary outputs of an appropriate sensing device which correspond to the picture elements defined by the selected scanning resolution. However, the resulting raster representation of line structured data is very demanding in terms of the computer memory required. For instance, the storage of an alphanumeric symbol in a raster form would involve in the excess of 300 bits of information. If it was known, on the other hand, what symbol it was, it would be possible to represent it using a mere 7 or 8 bit data code. It is apparent that a considerable information compression can be achieved by interpreting the information in the ditigized line drawing in an intelligent manner. In the case of the line data this implies its modelling in terms of straight line segments or higher order curves while the interpretation of special symbols leads to the problem of automatic character recognition. Information compression is not the only justification for developing programs capable of automatic interpretation of line structured data. It is often the need for a representation with a high degree of flexibility, allowing modification, adaption and updating of the stored information, which calls for an intelligent approach to line drawing digitization.

In this paper we shall concentrate on the problem of interpeting symbolic characters on line drawings. The pattern recognition system developed for solving this problem is schematically represented in Fig. 1.

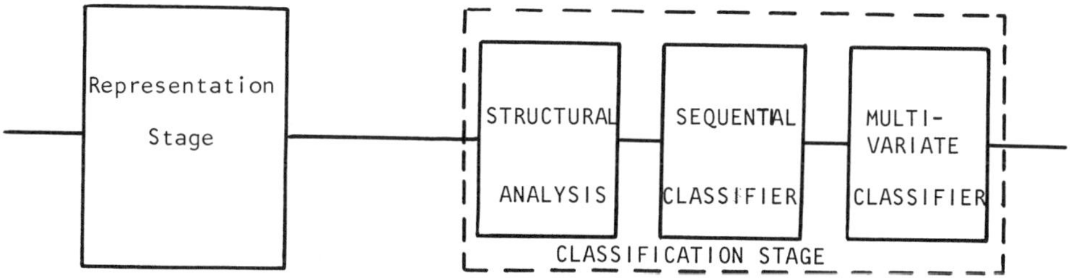

Fig. 1. Model of the recognition system

It comprises a pattern representation stage which transforms the symbolic characters into a form suitable for further processing and also carries out some rudimentary preprocessing tasks such as filtering. This stage is described in Section 2. The role of the second stage is to classify the input patterns. The classifier is implemented as a combined structural and statistical decision making system with three separate subsystems. The first element in the decision making chain performs structural analysis of the character. Its description is given in Section 3. The second subsystem delimits the set of possible candidate categories by a sequential analysis of special measurements extracted from the character. The stage is the subject of Section 4. Finally, Section 5 deals with the multivariate decision making stage which assigns symbolic characters to their respective classes. In this study only alphanumeric characters are considered as in the works surveyed by Suen (1978), but the approach is general enough to be applicable to any symbolic data.

2. DATA REPRESENTATION

The physical size of line drawings considered in the study (up to 1000mm x 800mm) and the resolution necessary to capture all the significant information (10 samples per mm) define the required number of raster elements per line and the number of lines per picture to provide a faithful raster representation of a drawing. In order to reduce this amount of information all the data is converted from the raster to a vector representation. This first involves thinning the captured image to one picture element width and then approximating the thinned lines by straight line segments. Only lines at angles 0,45,90 and 135 degrees are considered for fitting the data.

Each line is represented by its start and end point. Connected straight line segments are linked together to create entities each of which is characterised by a start node, end node and a set of point pairs defining the composite line segment between the nodes. The order of each node specifies the number of line segments connected to it.

A symbolic character is identified as a network of connected line segments which is not connected to any other network and fits a window of a prespecified size. Symbols thus extracted are represented by a number of nodes, their order, connecting segments and their geometries. Fig. 2 shows a few examples of the characters extracted from a map and represented in the form described above.

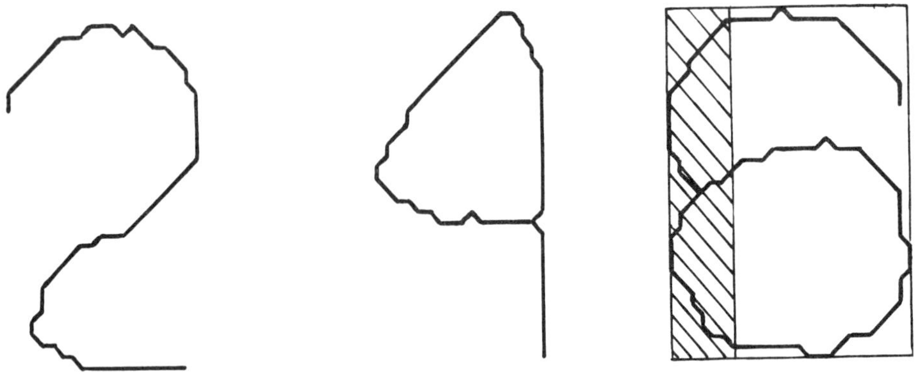

Fig. 2. Characters in the vector representation form

3. STRUCTURAL ANALYSIS

It is apparent from Fig. 2 that there is a difference in the structure of, say, characters 6 and 2. Character 6 has one node of order one and one node of order three while character 2 has two nodes of order one. It is, of course, advantageous to make use of this structural information for determining the class membership of each character. Unfortunately, no character has a unique structure. For instance, characters 4 and 6 in Fig. 2 have identical structures. The structural analysis (Deo, 1974; Pavlidis, 1977), therefore can be used only to reduce the number of possible candidate classes.

The possible structures of characters and their various fonts considered are shown in Table 1. According to their structure, characters can be divided into 10 categories. Each category is characterised by the number of nodes of each possible order between one and four. Note for instance that five different fonts of character four are catered for.

TABLE 1 CHARACTER STRUCTURES

CATEGORY	NUMBER OF NODES ORDER				CANDIDATE CLASSES
	1	2	3	4	
1	2				1,2,3,4,5,7,1,C,G,J,L,M,N,S,U,V,W,Z
2		1			O,D
3	1		1		4,6,9,P
4	1			1	4,R,Q,2,3
5	4			1	4,7,K,X
6	2		2		A,R
7			2		B,8
8	3		1		E,F,T,Y,3,4,2,G,J
9	4		2		H,K,I
10				1	8,B

Although in some cases the structural analysis can reduce the number of candidate classes very substantially, in others (eg. category 1 and 8 in Table 1) the number of possible hypothesis still remains too high and it need be further reduced before the input character is submitted to the final decision making stage. To this end sequential classification techniques are used to delimit the range of possible character categories. This approach is described in the following section.

4. SEQUENTIAL PRECLASSIFICATION

Let us consider again characters 4 and 6 in Fig. 2. Although their structure is identical there is additional information which allows us to differentiate between these symbols. First of all the topology of the nodes conveys very important discriminatory information which, for instance, in the case of structural category 3 is sufficient for ultimately identifying some of the elements in the category. In other categories additional measurements have to be made to allow discrimination between smaller subgroups. These include curvature measurements, angles of segments etc.

The sequential decision making process (Fu, 1968) of this second preclassification stage is illustrated in Fig. 3 on elements of category 1. Only node geometry information is used by the tests devised for the respective nodes of the decision tree. The sequential analysis assigns the unknown character to one of the six subgroups and the character is then passed on to the final decision making stage described in the following section.

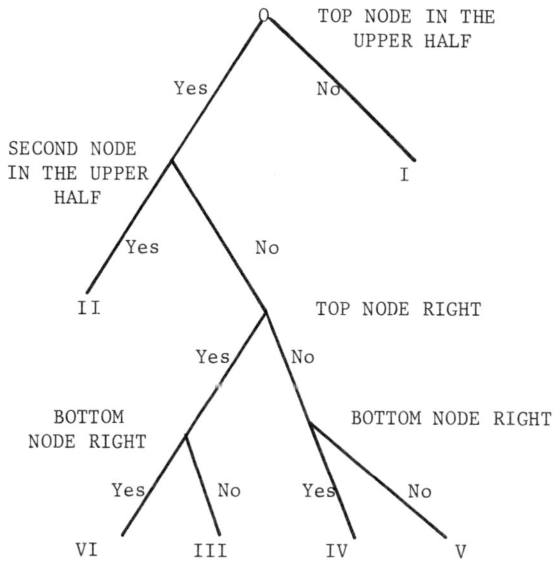

GROUPS	CANDIDATE CLASSES
I	M
II	U,V,W
III	N,S,5,1,4,J,G
IV	1,2,L,Z
V	3,7
VI	4,C,G

Fig. 3. Decision tree for category 1.

5. FINAL CLASSIFICATION STAGE

In the last stage of the decision making process each character is represented by an identical set of measurements. In particular, the measurements used are the lengths of the segments of a character falling into specific windows in the box enclosing it. An example of such a measurement is given in Fig. 2.

In total 16 measurements are extracted from each character. The measurements are considered as components of a pattern vector $\underline{x}$ representing the character, i.e.

$$\underline{x} = [x_1, x_2, \dots x_{16}]^T$$

It is assumed that in each class ω_i vector $\underline{x}$ is distributed normally with the mean $\underline{\mu}_i$ and the covariance matrix Σ_i. The distribution parameters are estimated from a set of learning patterns with known classification.

Suppose that the structural and sequential analysis assigns an unknown character to a subgroup containing classes ω_i, $i \varepsilon I$, where I_i is the set of indices of the candidate classes. Then the pattern $\underline{x}$ and, therefore, the corresponding character is assigned to that class ω_j which satisfies (Fukunaga,1972; Duda and Hart,1973)

$$P(\omega_j|\underline{x}) = \max_{i \varepsilon I_i} P(\omega_i|\underline{x})$$

$P(\omega_i|\underline{x})$ is the a posteriori probability of class ω_i given pattern $\underline{x}$. Assuming that a priori all classes in the group are equally likely, then this probability can be readily determined using the Bayes formula for calculating conditional probabilities.

The symbolic character recognition method described in the paper has been implemented in software. The system which is based on PDP 10 Computer can process handprinted alphanumeric characters at the rate of 10 characters per second. The size of the program is about 10k and additional 6k of computer memory is needed for storing the parameters of the statistical model of the classes. The system's rejection rate is less than 2%.

6. CONCLUSION

The paper describes a system for automatically interpreting alphanumeric characters on line drawings. The system is flexible enough to handle handprinted noncursive script of a wide range of fonts. It is based on a combined structural/statistical method of pattern recognition. Currently a microprocessor implementation of the system is being considered.

7. REFERENCES

Deo, N. (1974). Graph Theory with Applications to Engineering and Computer Science. Englewood Cliffs, N.J.

Duda, R.O., and P.E Hart (1973). Pattern Classification and Scene Analysis. Wiley, New York.

Fu, K.S. (1968). Sequential Methods in Pattern Recognition and Machine Learning. Academic, New York.

Fukunaga, K. (1972). Introduction to Statistical Pattern Recognition. Academic, New York.

Pavlidis, T. (1977). Structural Pattern Recognition. Springer, New Y ork.

Suen, C.Y., M. Berthod and S. Mori (1978). Advance in Recognition of Handprinted Characters. Proceedings 4 IJCPR, Kyoto, Japan.

VISUAL OBJECT ISOLATION AND RECOGNITION BY A CAMERA-COMPUTER SYSTEM

O.Santana, I. de la Nuez and J.Fortes.

E.T.S.I.I. and University College of Las Palmas
P.O.Box 550.Las Palmas de G.Canaria.Spain

ABSTRACT

An heuristic procedure, inspired by the dynamics of vertebrate retinal processing, have been developed to isolate an object from the visual image and to describe it. The isolation is performed after moving the desired object in, out or within the camera view. A description by a vector descriptor or by a convolution matrix is generated thereafter by the computer. These are subsequently used to recognize similar objects in new scenerys, to count them and to provide their coordinates to be, for example, manipulated.

KEYWORDS

Camera-computer; image processing; pattern recognition; object description; spatial convolution; pattern descriptors.

INTRODUCTION

A basic difficulty in artificial intelligence visual systems is usually to define an object and to detach it from the background of the visual image. This is partly so because, in many instances, object definition is tried on a single image (Hogg, 1977; Prager et al, 1977). Results obtained by simulation of some retinal basic processes in a camera-computer system (Rubio et al, 1980), have suggested that the difficulty in defining an object may be largely overcome by using several images.

Object isolation has been performed with ON-OFF like procedures by seeking the frame pixels where changes had occurred after the object of interest has moved in, out or within the camera view. The present intensity value of those points and/or the old intensity value are taken as definers of the object. The isolation task may be complicated by elaborated backgrounds, but, in general it performs reasonably well in a standard camera-computer system. The procedures are not, however, for analysis of changing images nor for motion detection (Tsotson, 1977; Jain et al, 1977).

The implementation of the procedures has been restricted to two dimensional objects (flat objects). Various processes are performed by the computer, as described below.

BASIC OPERATIONS

Frame Preprocessing.

The video data from a standard vidicon camera is digitazed and fed into the computer through the appropriate interface. The display is made by the computer printer, by means of a program in which the different levels of gray are represented by the densities of dots in the rectangle corresponding to each character. Thus each rectangle is to represent a pixel of the processed image, which is therefore capable of 6 x 6 + 1= 37 levels of gray. The resolution of the displayed images is also fixed by the printer in 128 x 116 pixels. This resolution has been estimated to be sufficient for the interpretation of results, though it is desirable to go to higher resolutions. Anyway, there is the facility to zoom into a smaller part of the camera raster under computer control, preserving the resolution, if detailled parts of the scene are to be examined.

Operations performed as preprocessing are the following:

Scaling of the data to be displayed. To optimally use the relatively low gray resolution of the display, a scaling is usually performed before a frame is printed.Two methods had been used. One is the asignation of the gray scale to the span of the intensities in the frame in a constant interval basis. This is not very efficient, so that an asignation based on a constant area is normally used.The area which is kept constant for each interval is the one in the histogram-number of pixels with a given intensity versus intensities. These histograms have also been found very useful in automatically setting thresholds for separating figures from a background. As illustrated in Fig.1, black figures in a lighter background, or viceversa, provide with histograms with a valley, and two clear peaks. When these figures are to be automatically detached from the background, all that the computer has to do is to find the valley and cut with the corresponding threshold.

Fig.1. Histogram for a picture of light objects on a dark background.

Integration of several frames. This is sometimes done on static images to improve the signal to noise ratio.

Thresholds setting. Thresholds can be stablished with are either fixed or a function of the image(adaptive thresholds).After a threshold, the resulting image can be displayed either by scaling the resting amplitude values or by a 0-36 value asignation. Apart from the thresholds implicit in spatial filtering operations, we use a threshold for eliminating progresive differences in background illumination. This is generated by adjusting the whole image to a three dimensional plane; the z values of the plane are then used to define the threshold.

Spatial filtering. Typical band and high pass filters of the centre periphery type (Laplacians) are also rutine in preparation of frames for further processing.

A typical sequence in the generation of what we call p-frame is the following: the initial data are cut by a threshold plane. Then, a band-pass spatial filtering is performed. Afterwards, a histogram determined threshold is applied. The resulting data is reduced to a two level frame thereafter. This sequence can be altered according to the nature of the images to be processed.

Object Isolation.

What is meant to be an object must be communicated to the computer. Its own visual system is used to do this, since the difference between two succesive frames from an unchanged scenary, after cutting by a threshold plane, is practically zero, except for some pure background noise. The noise can usually be eliminated by a histogram set threshold. Thus, to isolate an object, it is moved in or out the camera view, without overlapping other objects. The computer systematically performs the differences between two succesive frames; when there is a significant difference,it processes it to define the presuntive object, by cutting with a histogram set threshold. Thereafter, the computer generates a two level picture of the object.

Object Description.

Once the object is isolated, the computer generates a set of descriptors. Typical descriptors are heuristic, such as area, momenta, corners, holes, vertical and horizontal edges, orientation of edges and the usual. Now, these descriptors are, in general, too ad hoc to be satisfactory, so that we are trying a systematic generation of descriptors, based on normalized momenta that could yield the necessary invariances and the unique characterization at the limit. But this is still in the very preliminary stages. Anyway, the usual description give reasonable results in many practical cases, though they do not satisfy our global aspirations.

Another useful description, which is applicable just for translation is the mask generation. The computer can do this automatically when complicated backgrounds are present. The computer generates the mask, by producing a centre-periphery convolution matrix, where the centre extends within the profile of the object and it is excitatory while the periphery can be as drastically inhibitory as one wishes.

Object Recognition.

The recognition task is relatively simple once the appropriate set of descriptors is stablished. The computer does it in the old or in a new circumstance by providing with a normalized table about the probabilities that the detected object belongs or not to the object being seeked. New objects are defined on a cluster basis , which obviously may present real difficulties for particular p-frames. The generation of the table can be done in two basic ways, by using a property description or by a convolution. The resulting table is then cut by a threshold in the usual manner, to provide with specific results, for example the coordinates of the points where the similar presuntive objects are situated.

APLICATIONS.

The typical sequence in the generation of a p-frame is illustrated in Fig.2. Figure 2a is the printout of the original video data. Next, a threshold plane is generated. The result of cutting with said adaptive threshold is shown in Fig. 2b, where some background noise is still present.

A high-pass centre-periphery spatial filter, consisting of a 1 x 1 excitatory centre and a 3 x 3 inhibitory periphery is thereafter applied. Centre and periphery weights do not cancel. Positive values of the resulting filtered image are displayed, as shown in Fig. 2c.

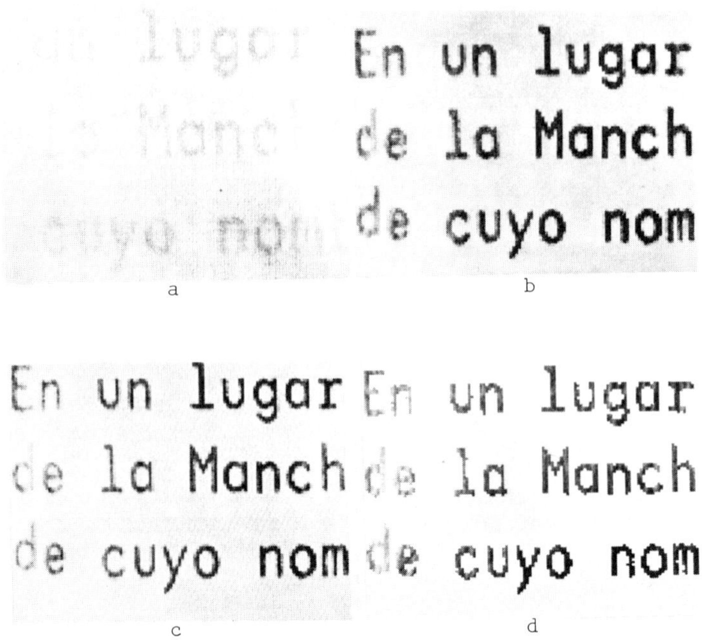

Fig. 2.- Illustration of the sequence for generating a p-frame.

Finally, a histogram set threshold is automatically applied to definitely detach the figures from the background. Figure 2d shows the result.

The use of descriptors to define objects has been performed in various images. One of these is that shown in Fig. 3. Typical descriptors generated to distinguish between screws and bolts are areas, centre of gravity and momenta of the clusters in the p-frame. Since there is a notorious difference between the two classes of objects to be separated, the above descriptors are sufficient for the task. The computer provides a printout of the coordinates of the centres of gravity, separately for screws and bolts.

The recognition of an object by template matching has been performed by letting the computer generate the corresponding convolution matrix. The process is illustrated in Fig. 4. Figure 4a shows the original video data. The object whose properties are to be communicated to the computer is taken away from the scene. The computer performs the difference between an old and a new frame, so that the desired object is detached, as illustrated in Fig.4b. After some preprocessing, a nitid profile is obtained, where the generation of the centre-periphery convolution matrix by the computer is automatically performed. This mask is convolved with a p-frame of the original, shown in Fig. 4c. The results of the convolution are thresholded and displayed, showing the position of similar objects, as illustrated in Fig.4d. Note that, since the convolution provides no invariance to rotations, only those objects with an orientation equal to the original are detected.

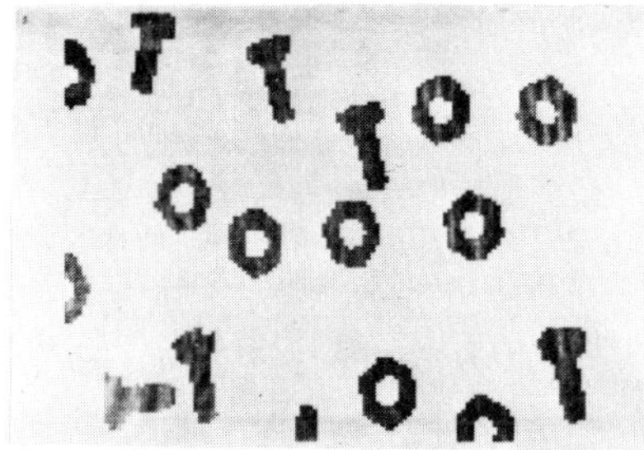

Fig.3.- p-frame for the computation of descriptors.

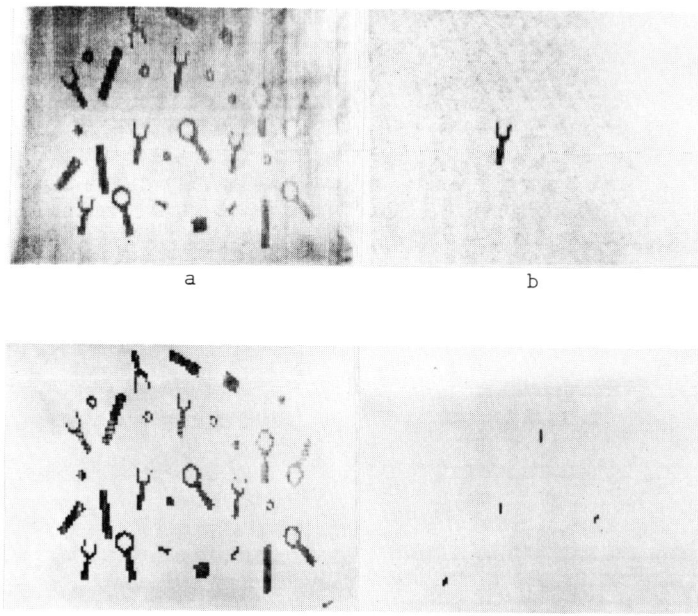

a b

c d

Fig. 4.- Generation and convolution of a mask.

ACKNOWLEDGEMENT

The authors manifest their gratitude to Prof. Roberto Moreno Díaz, Director of the Depts. of Physics and Cybernetics at the E.T.S.I.I. and the Colegio Universitario for his orientations and suggestions as well as for his teachings in the expressions.

REFERENCES

Hogg, D.C. (1977). A methodology for real time scene analysis. Proc. 5th. Int.Conf. on Artif.Intell. M.I.T., Cambridge M.A. p 627

Jain, R., Militzer, D., Nagel, H.H. (1977). Separating non stationary from stationary scene components in a sequence of real world TV-frames. Proc.5th.Int.Conf. on Artif.Intell. M.I.T., Cambridge, MA. pp 612-618.

Prager, J., Nagin, P., Kohler, R., Hanson, A., Riseman, E. (1977) Segmentation processes in vision systems. Proc. 5th.Int.Conf.on Artif.Intell. Cambridge MA. pp 642-645.

Rubio, E., Santana, O., Moreno-Díaz, R. (1980). Camera-computer simulation of retinal processes. This Congress.

Tsotsos, J.K.(1977). Some notes on motion understanding. Proc. 5th.Int.Conf.on Art.-Intell. Cambridge, MA. p 611.

A MICROCOMPUTER-BASED LETTER RECOGNIZER: DISCUSSION OF SOME PROBLEMS AND SOLUTIONS AND RESULTS OBTAINED WITH AN 8085

M. P. Beddoes and Wolfram Lunscher

Department of Electrical Engineering
University of British Columbia
Vancouver, B.C., Canada V6T 1W5

ABSTRACT

The main problems of prefiltering optical data and the decision problem itself are discussed. Solutions are outlined which may be incorporated in a microprocessor such as an Intel 8085.

A hardware recognizer has been built incorporating the principles discussed, and results with this machine are presented.

KEYWORDS

Optical character recognition; microprocessor data control; reading machines for the blind.

INTRODUCTION

The performance of the OCR is critically dependent on the recognizer being presented with single letters. A closed boundary must be drawn round a letter and all information inside the boundary can be declared useful. Other problems such as scaling (Shahidul-Hussein and co-workers, 1972), shift, tilt and even thickness of letters become simple to solve. Taxicab geometry (Gardner, 1980) is one method of visualizing boundaries and can be invoked in discussions of alternative boundaries, etc.

The other central problem is the choice of recognizer. In the literature, (Andrews, 1971; Beddoes, Kancia, George, 1974; Suen, 1979) transform methods are generally stated to be superior to non-transform methods. In our experiments we set out to test this assertion. We compared a Fourier transform recognizer (Beddoes, 1979) with a simple minimum distance template matcher. If registration between trained templates and incoming patterns is carefully maintained, the latter will out-perform the former by a small margin. With well produced print we obtained from 25 sets of one font an error rate .08%. The template method and the problem of registration will be discussed.

THE PROBLEM OF SEPARATING LETTERS

A main problem in prefiltering letter data is to isolate a letter from its neighbours. If a boundary could be drawn round it, the boundary would constitute a solution to the problem. A tracing technique suggested by Clemens (1965) is the basis of work reported here. The method obtained a boundary by local operations along an outside edge. It does not trace internal contours. Difficulties are encountered with very thin letters but these may be avoided by techniques described here. The method is based upon the following rules and equations which describe in detail the local operations.

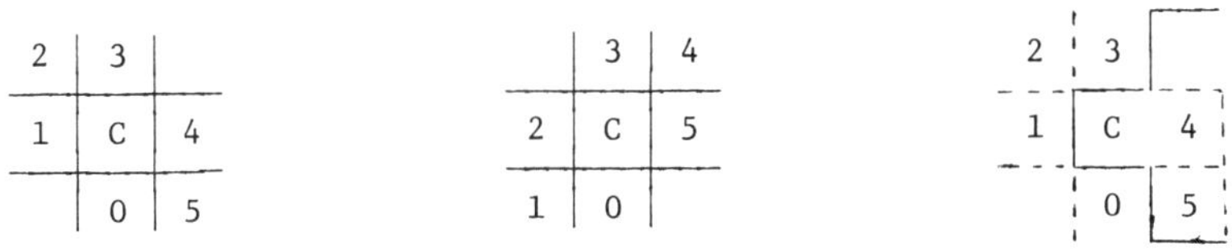

1a 1b 1c

Fig. 1: Clements hexagon boundary tracer.
Fig. 1a Basic hexagon of points
Fig. 1b Alternative basic hexagon (see text)
Fig. 1c Hexagon of 1a at one position in a pattern. The next centre will be point #4 and the trace will start from #3.

Refer to Fig. 1

1. The centre is placed at a black boundary (Fig. 1c).
2. A search commences from a white point (say, #1). The number increases modulo 6 until a black point is encountered.
3. This black square becomes the new centre (point 4).
4. A new starting point one higher modulo 6 than the old centre is taken (number 3).
5. Go back to step 2.

The local hexagon, Fig. 1a, is described on a single move basis by equations a.

$$hx_{k+1} = hx_k + hy_k - Py \tag{1a}$$

$$hy_{k+1} = px + Py - hx_k \tag{2a}$$

where px and py are the coordinates of the current centre, hx_k hy_k are a point (possibly the starting point) over white and hx_{k+1}, hy_{k+1} is the next point to be scanned. Equations 1a and 2a describe a clockwise search trace but the centre will move counter clockwise around convex curves. An opposite direction of travel is given by a counter clockwise trace which follows from the equations:

$$hx_{k+1} = hx_k - hy_k + py \tag{3}$$

$$hy_{k+1} = px - py + hx_k \tag{4}$$

Thin patterns cause trouble. Clemens found that the centre would be trapped on a 45° diagonal line of single picture elements. This difficulty is overcome by using a complementary set of points for the local trace and switching to the latter whenever a stationary centre is encountered. The complementary set corresponds to Fig. 1b and equations b) below.

$$hx_{k+1} = px - py + hy_x \tag{1b}$$

$$hy_{k+1} = px + hy_k - hx_k \tag{2b}$$

With "fat" characters, the centre travels around the outside boundary and only part of the video data is used producing useful data compression.

Figure 2 shows examples. The traces are produced by two points starting at the left of the pattern and travelling to the right with cc and ccw traces respectively given by equation sets 1,2 and 3,4.

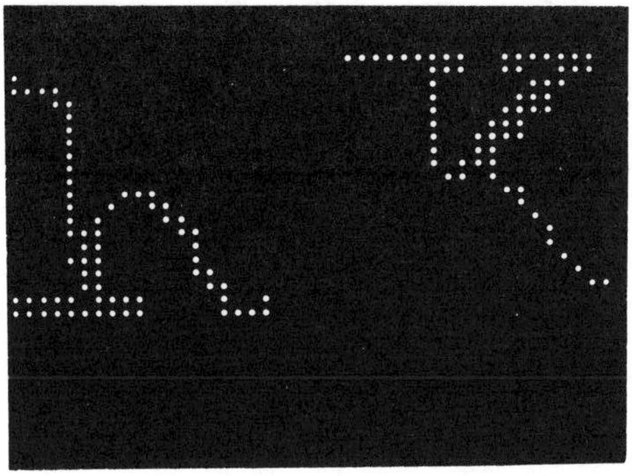

2a

2b

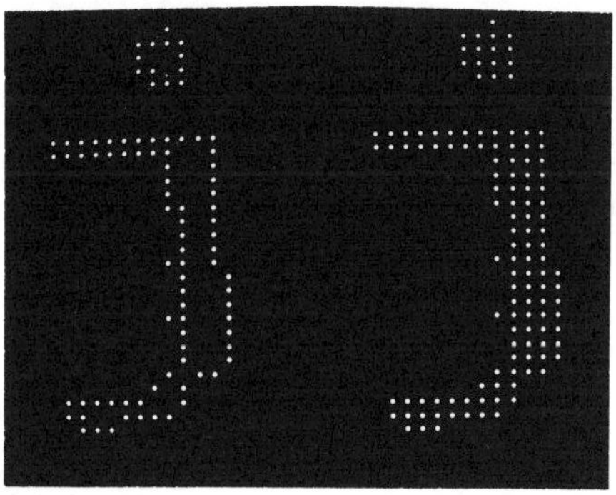

2c

Fig. 2. Examples of the modified Clements' contour trace.
Fig. 2a Left trace equations 1a,2a; right trace equations 3 and 4
Fig. 2b Two traces from 2a shows left; original pattern right, "K".
Fig. 2c Right original "j"; left shows two traces respectively around the jay and the dot.

Trouble with broken characters can be removed by scanning the same region several times removing at each scan information within any closed boundary (Fig. 2c). This is the root of the main disadvantage to the method. It is awkward with incomplete characters. The recognizer must include a scratch pad memory of at least one letter This dificiency can be removed at a cost which may be more apparent than real. In at least two other methods (Martelli 1976; Zahn, 1969) preprocessing all the video data is required. Operations may, however, be simple and fast. All boundaries, internal as well as external, are obtained. Comparisons with these methods is the subject of ongoing work.

The method outlined here is a basis for separating letters which are very close and for producing outer boundaries which may be used to remove excessive dependence on scaling.

COMMENTS ON A DISTANCE MATCH

A method to compare an input pattern suspected of being a letter is to compare the pattern with each of pretrained templates corresponding to a complete alphabet. The problems lie in two areas:

- using as simple a measure as possible in order to achieve high speed. The measure should be compatible with reasonable theory, say, a distance in hyperspace.
- registering the input pattern on the template.

The distance measure is based on the following considerations: An input pattern will be a binary variable in space, Zj = Z(j) where j = 1,2,--. L refers to points in pattern space. Zj is to be compared to the ith letter, Aij. The squared distance between the two, assuming they are registered properly, is

$$D_i^2 = \sum_{j=1}^{L} [Zj - Aij]^2$$

$$\triangleq \sum_{j=1}^{L} (Zj)^2 - 2\, l_i \tag{5}$$

where

$$l_i \triangleq \sum_{j=1}^{L} Zj\, Aij - \frac{1}{2} \sum_{j=1}^{L} Aij^2 \tag{6}$$

We note that $\Sigma(Zj)^2$, the first term in 5), is just the area of the input pattern and it will be fixed for a given pattern-search. Accordingly, monitor l_i only for the search i=1,2--52 and associated the match with the (l_i) max. The maximum l_i is given by setting Zj = Aij. Hence

$$l_{i\ max} = \frac{1}{2} \sum_{j=1}^{L} (Aij)^2 \tag{7}$$

Aij is a binary variable zero or one, and hence l_i, max is simply half the number of dots in the pattern i. This varies less than 2:1 from pattern to pattern.

The problem of relative displacement of input pattern and template has been approached using a centre of gravity method and nearest neighbours. The CG for a pattern is given by

$$\bar{x}^{11)} ; \bar{y}^{11)} = \left| \frac{\sum_{j=1}^{N} x_j\, n_j}{\sum_{j=1}^{N} n_j} \; ; \; \frac{\sum_{i=1}^{M} y_i\, n_i}{\sum_{i=1}^{M} n_i} \right| \tag{8}$$

where $n_i = n_j = 1$ if the i/j^{th} point is black.
$n_i = n_j = 0$ if the i/j^{th} point is white.
N=number of columns in the pattern matrix.
M=number of rows in the pattern matrix.

Because one must express CG as integer, unavoidable round off errors must occur. Recognition errors were made low by taking the best match from the CG and four nearest neighbours. This increased computation time by a factor of 5, but the error rate fell from about .2% to .08% and this was considered to be worthwhile.

In order to operate quickly, all operations, mainly these of equation 6 and equation 8, are realized in hardware. The trained characters reside in EPROM memory and are accessed using Direct Memory Access, or DMA. The total number of chips is 103. The microprocessor is required only to do housekeeping tasks: principal functions in the Computer/Recognizer and the prefilter are all realized in hardware.

CONCLUSIONS

The work described here is aimed at producing a simple reading machine for the blind. Two central problems are discussed. Letter separation is one problem and is solved by drawing boundaries around individual letters. The method developed is described, and experimental illustration of its operation is given (Fig. 2). The other problem is to choose an optimum recognizer. A template recognizer was found suitable for this application. A measure of "goodness of match" is developed (Equation 6). A modified centre-of-gravity method is used to align an input pattern to the trained templates.

The system has operated on a PDP 12 computer with 32K memory. Fortran programming was used. Good performance figures were obtained (.08% error rate on a single font) but processing time was unacceptable (5 seconds/letter). The system is now largely realized using dedicated hardware controlled by an 8085. Processing speed of up to 20 characters/sec. with error rate ∿.1% are obtained with well formed print from one font. Multi font recognition does not appear to be a difficult problem and informal experiments show that an increase in error rate of about two times may be expected, but this may be mitigated by using context.

ACKNOWLEDGEMENTS

We acknowledge the support from National Scientific and Engineering Research Council of Canada, and two grants from B.C. Health Care Research Foundation.

REFERENCES

Andrews, H.C. (1971). Multidimensional rotation in feature selection. IEEE Trans. Computers, C20, 1045-1051.

Beddoes, M.P., MacNeil, R. (1979). Recent advances in small computer techniques for identifying print information. Computers in Ophthalmology, 267-271.

Beddoes, M.P., Kancia, E., George, R.G. (1974). An optical character recognizer for a reading machine for the blind. 5 Canadian Medical and Biological Engineering Conference.

Clemens, J.K. (1965). Contour tracing techniques. Ph.D. Thesis, M.I.T.

Erhardt, V. (1977). Linear classifier for print recognition in a blind reading machine. Ph.D. Thesis, University Stuttgart.

Gardner, M. (1980). Mathematical games. Scientific American, 18.

Martelli, A. (1976). An application of heuristic search methods to edge and contour detection. Commun. ACM, Vol. 19, 73-83.

Shahidul-Hussein, A.B., Toussant, G.T., Donaldson, R.W. (1972). Results obtained using a simple character recognition procedure on Munsen's handprinted data. IEEE Trans. on Computer, C21, 201-205.

Suen, C. (1979). N-gram statistics for natural understanding and text processing. IEEE Trans. Pattern Analysis and Machine Intelligence, PAMI-1, 164-172.

Zahn, C.T. (1969). A formal description for two dimensional patterns. In J.K. Aggrawal and co-workers (Ed.), Computer Methods in Image Analysis, IEEE Press, 1977.

USE OF A RANGEFINDER FOR 3-DIMENSIONAL SCENE ANALYSIS[1]

Z. U. Islam and C. N. Shen[2]

Rensselaer Polytechnic Institute
Troy, New York 12181

ABSTRACT

This paper deals with the problem of 3-dimensional scene analysis using range measurements performed on the scene of interest with a laser range sensor. The 3-dimensional description of a scene from an analysis of its 2-dimensional projections is difficult in that depth information is lost in projection and also the recognition of objects in the resulting picture may be rendered very difficult by the ill effects of projection and ambient lighting such as, shadows, reflections, shading, and distortion. The scheme of range based scene analysis reported herein affords a 3-dimensional description of a scene in terms of various known objects in their true spatial positions and orientations with respect to the range sensor.

KEYWORDS

Three-dimensional scene analysis, edge detection, surface extraction.

INTRODUCTION

The description of a scene in terms of various solid bodies along with a specification of their true relative positions and orientations is called "Three-Dimensional Scene Analysis." Presently, the various approaches to this problem start by obtaining one or more pictures of the scene with a T.V. camera or similar device and then, recognize the objects by matching its images with a model image of the same object, obtained earlier during a modeling process [1], [2], [3]. There are, however, many drawbacks of using two-dimensional images for a three-dimensional description of the scenes, the notable ones being; the absence of depth information, and strong vulnerability to disturbing influences of ambient lighting which may cause serious errors in the final interpretation of the images [4], [5]. To afford a three-dimensional interpretation of a scene from its plane image, the idea of using multiple images from different vantage points has been utilized. The stereo T.V. technique [6], thus tries to obtain depth measurements of the scene indirectly by computation, from a

---

[1]Research sponsored by NASA Langley Research Center Grant No. NAG1-61.

[2]Professor, Electrical, Computer and Systems Engineering Department.

knowledge of the camera viewing geometries. To enable this, a correspondence must first be established between the images of a point obtained by different sensors. Much computational complexity is associated with this correspondence problem and even then the solution is not guaranteed.

In the recent years some attempts have been made to base scene analysis on depth information measured directly from the 3-dimensional scene. Thus, this principle has been utilized by Duda and coworkers [4], [5], to extract important features like planar surfaces, edges of solid bodies, etc., from the range information measured with a laser rangefinder. However, the particular feature extraction procedures used in that work, although being simple, have some limitations. Mainly, the assumptions of having surfaces largely horizontal or vertical, orthogonal faceted bodies and very little curved surfaces may not be very practical in the usual indoor scenes. Moreover the thresholding technique used for detecting edges of objects might fail to obtain satisfactory results when the actual changes are less than the threshold value.

The system described in this paper, in principle, uses a time of flight rangefinder similar to the one quoted above, to obtain range measurements of the workscene by scanning in the elevation and azimuthal directions. The feature extraction and scene description procedures, have been studied from a point of view of their utilization for space assembly tasks.

Advantages of the Rangefinding System of Scene Analysis

A number of important advantages were noted for the scene analysis system using range data over those using intensity information of a plane image.

The depth information is directly measured from the scene and is not extracted indirectly from the television or camera pictures, it is therefore more accurate. The procedure to calculate range from multiple T.V. pictures requires matching of corresponding regions in different pictures and involves lengthy and complex calculations.

The exact location and orientation of the various objects in the scene analyzed is calculated with respect to the robot's frame of reference, which can then plan its sequence of operations.

No lighting or illumination is required as the laser rangefinder provides its own beam to illuminate the scene. This is especially helpful in the space applications of robotics, where provision of lighting will be an additional burden.

Disturbing and interfering effects of ambient lighting such as shadows, reflections, shading, highlights, etc., that seriously impair the reliability of scene analysis procedures based on intensity, have no effect on the range based scene analysis.

The complete scheme of the rangefinding system of scene analysis is depicted in Fig. 1, and may be regarded as consisting of two major sections, the data acquisition and feature extraction section (blocks 1-7), and the object recognition/position and orientation determination section (blocks 8-13).

Coordinate Systems and the Scanning Scheme

The laser rangefinder of the robotic system is mounted on a mast which separates it sufficiently from the main body of the machine so that its field of view is not obstructed unduly by the manipulator arm movements. There are three separate coordinate systems involved in the whole process of laser sensor controlled movements of the robot viz, the coordinate system of the laser rangefinder in which the range measurements are performed, the coordinate system of the robotic machine in which the various manipulation tasks are defined and executed, and finally the coordinate system fixed in the object to be handled, whose orientation and position we wish to determine.

A spherical coordinate system, Fig. 2, is defined for the laser rangefinder in which the rangefinder itself is assumed to be at the origin of coordinates and its beam can define a cone of revolution, by rotating about the mast axis which we take as the z-axis. The y-axis for the system is taken along a general direction dictated by the robot's control system. This automatically fixes the x-axis for a right-handed system. The scene of interest is scanned using small increments of polar angle 'β' and azimuthal angle 'θ', and the range 'r' to each illuminated point is measured, Fig. 2. The range data is arranged in a matrix form with rows showing variation of angle θ and columns showing variation of angle β.

Rapid Estimation Scheme for Edge Detection

The rapid estimation scheme [7] (R.E.S.), detected the presence of horizontal and vertical edges in the workscene by processing the range data successively along each column and row of the range matrix. An edge was detected by a sudden change in the range and range slope using detection scheme with a decision tree. The result of R.E.S. is a collection of data points that belong to some edge of an object. These data points are widely spaced, depending on the data density that was imposed on the scanning scheme through the use of particular values of $\Delta\beta$ and $\Delta\theta$.

Slope Estimation Scheme

Range slopes are defined as the incremental changes in range for corresponding incremental changes in the pointing angles, i.e., range inpath slope = $\partial r/\partial\beta$ and range crosspath slope = $\partial r/\partial\theta$. The 2-dimensional smoothing algorithm [8] provides smoothed estimates of range and range slopes which are used for calculation of actual surface slopes.

Segmentation of Range Data into Plane Surfaces

A plane surface of an object, ideally, has constant slope along any two fixed orthogonal directions throughout its extent. Using this idea the observed data points of plane surfaces are grouped into various subsets, one for each surface, on the basis of their "surface slopes." These "surface slopes" are defined as the slopes shown by the surface of the object at the data points, Fig. 2. Reference directions for the orthogonal surface slopes with respect to the three coordinate planes are shown in Fig. 2. The segmentation of range data on the basis of surface slopes provides groups of connected data points that belong to one particular face of some observed object.

A Clustering Procedure for Extraction of Plane Surfaces

The problem of grouping range data points of different planar surfaces on the basis of their surface slopes becomes an application of clustering analysis [9], [10], in which:

1) The distinguishing feature vector is $X = [f_1, f_2]^T$, where f_1 and f_2 are the two orthogonal slopes with respect to one of the three coordinate planes of the laser rangefinder.

2) The number of clusters 'c', in other words the number of plane surfaces observed in the scene is not known initially.
3) An initial configuration or rough classification of feature vectors into clusters cannot be obtained by any simple procedure.
4) Feature vectors corresponding to parallel plane surfaces had to be discriminated.

Since an initial partitioning of data could not be found from the given problem specifications Agglomerative Hierarchical clustering [9] was used to obtain a reasonable initial partitioning. Moreover the whole data set is not treated at one time, but a recursive approach is adopted to cover the complete measurement space. Recursive approach was necessary for reducing computation, and differentiating between feature vectors of parallel planes.

To make the classification optimal, the familiar sum of squared errors criterion function [9], [10], is minimized. A least squares plane is then fitted to each of the clusters to describe the particular face of some object. The edges of the objects can then be calculated as the intersection of the plane faces.

Object Reconstruction and Recognition (Blocks 8-13)

Various edges and plane surfaces extracted above are then assembled into meaningful bodies using heuristic rules based on physical constraints of bodies. The strategy adopted for object identification is essentially a model based recognition process. Three-dimensional models of all the objects likely to be present in the workscene are first stored in the computer memory which consist of a specification of the lengths and spatial position of all the edges, the coordinates of all the vertices and the angles between their component edges. A coordinate system is then fixed in the model and is defined with respect to its vertices and edges observable from outside. The computer memory contains many stored models and an initial guess for the identity of the observed object is based on the lengths of the observed edges and the angles between them, at a particular vertex or face. A similar vertex or face is then searched in the model library and a list of candidate models is drawn. The viewing geometry of the laser rangefinder is then given by the position and orientation of its axes with respect to the coordinate system fixed in the object.

Using computer graphics techniques, for a 3-dimensionally stored model of some object, its view as seen from any direction, can be obtained. Thus by estimating the viewing geometry of the laser rangefinder for a particular observed object, we can generate a similar view of some stored model assuming it to be the correct one. The observed object can then be recognized by matching the observed and computer generated views.

CONCLUSIONS

A system for analysing scenes in case of space application of robotics has been described. The sensor assumed for performing measurements on the workscene is a time of flight rangefinder. Procedures have been studied for extracting important features of the scene like edges and surfaces of the objects, for describing individual objects by their edges, vertices and bounding surfaces, and finally for recognizing the previously known objects as observed in the scene by comparing the observed features with those of a model stored in the system memory. It is expected that the finally developed system will be able to recognize known objects in their true relative locations with respect to each other and with respect to the robot. The performance of the system will not be affected by variability in position and orientation of objects, and in the illumination conditions of the workscenes.

REFERENCES

[1] Evans, J.M., Jr. and Albas, J.S., "NBS/RIA Robotics Research Workshop", NBS Special Publication 500-29, April 1978.

[2] Johnsen, E.G. and Corliss, W.R., "Human Factors Applications in Teleoperator Design and Operation", John Wiley and Sons, New York, NY 1971.

[3] Young, J.R., "Robotics", John Wiley and Sons, New York 1973.

[4] Nitzan, D., Brain, A.E. and Duda, R.O., "The Measurement and Use of Registered Reflectance and Range Data in Scene Analysis", Proc. IEEE, Vol. 65, No. 2, 1975.

[5] Duda, R.O., Nitzan, D., and Barrett, P., "Use of Range and Reflectance to Find Planar Surface Regions", IEEE Trans. Pattern Analysis & Machine Intelligence, Vol. PAMI-1, No. 3, July 1979.

[6] Yakimovsky, Y. and Cunningham, R., "A System for Extracting Three Dimensional Measurements from a Stereo Pair of T.V. Cameras", Computer Graphics and Image Processing 7, 1978.

[7] Kim, C.S., Marynowski, R.C., and Shen, C.N., "Obstacle Detection Using Stabilizated Rapid Estimation Scheme with Modified Decision Tree", Proc. of JACC, October 1978.

[8] Kim, C.S. and Shen, C.N., "A Two Dimensional Recursive Smoothing Algorithm Using Polynomial Splines", IEEE Intl. Symposium on Info. Theory, 1980.

[9] Duda, R.O. and Hart, P.E., "Pattern Classification and Scene Analysis," John Wiley and Sons, New York 1973.

[10] Fukunaga, K.J., "Statistical Methods of Pattern Recognition," Academic Press, 1972.

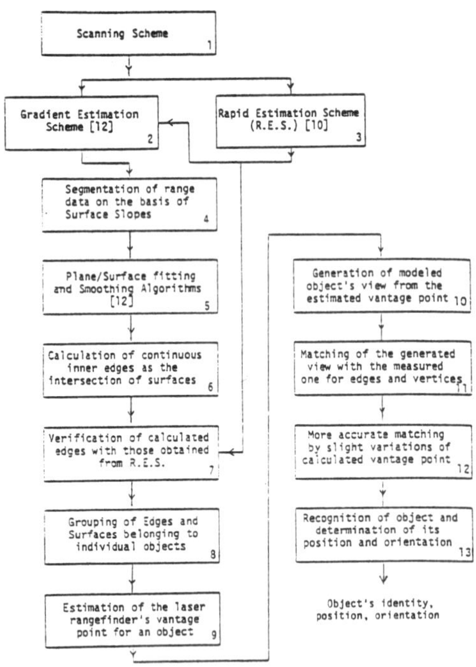

Fig. 1 Flow diagram of the scene analysis system

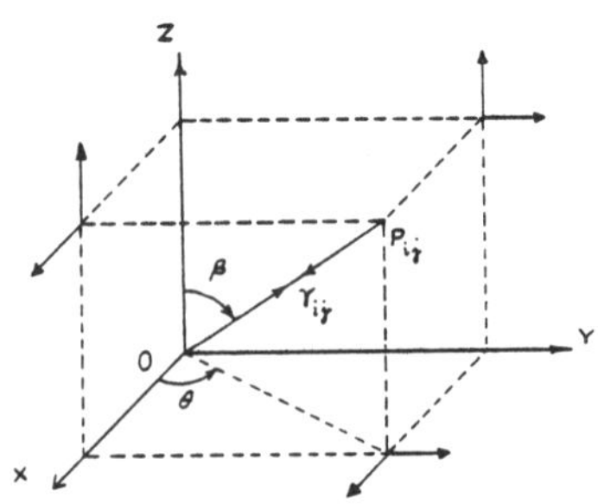

Fig. 2 Reference direction for the surface slopes

A Recursive Algorithm for Smoothing by Spline Functions*

C. S. Kim and C. N. Shen**
Rensselaer Polytechnic Institute
Troy, New York 12181

ABSTRACT

The problem dealt with in this paper is that of smoothing a set of noise corrupted measurements of an unknown signal. Here we interpret this smoothing problem as an optimal curve-fitting problem arising in approximation theory, and develop a recursive procedure for fitting smoothing splines. While the amount of computation needed for a non-recursive method increases rapidly with the number of data points, for this recursive algorithm it increases linearly with the number of data points. In addition, the algorithm is made adaptive to local data structure by estimating the parameter of an unknown original signal.

INTRODUCTION

By noting the fact that original signals such as visual scenes are in analog form, techniques were developed which reconstruct analog signals from discrete data by utilizing interpolation or approximating functions. Frequency domain interpretation of the interpolation process was reported in Ref. [1]. Also, B-spline interpolates [2-4] were used [5] in restoring a continuous signal from a set of digitized data. For 1-D noise corrupted data generated by unknown systems, Reinsch [6] utilized natural cubic splines [2-4] along with least squares constraints to solve the problem of curve plotting. Hou and Andrews [7] constructed continuous-discrete image model and utilized spline basis functions along with the least squares constraints for image restoration. Because of their nonrecursiveness, the algorithms in Refs. [6,7] are involved with complex computations and cannot be implemented on line. Recently, by using a reproducing kernel Hilbert space approach, Weinert [8,9] et al. developed a structural correspondence between spline interpolation and linear least-squares smoothing of a particular random process. From this structural correspondence, they developed recursive algorithms for spline smoothing and spline interpolation. In this paper, we take an algebraic approach to obtain a recursive algorithm for spline smoothing. In this procedure, we do not attempt to make a structural correspondence as Weinert did, rather we obtain a recursive algorithm directly from the structure of the spline smoothing problem. This approach will result in a simpler recursive algorithm. In addition, a similar technique can be utilized for the design of a 2-dimensional recursive estimation algorithm, which

* Research sponsored by NASA Langley Research Center Grant No. NAG1-61
** Professor - Electrical and Systems Engineering Department

will be considered elsewhere.

PROBLEM FORMULATION

From the viewpoint of approximation theory, when a set of discrete observation data is noise free, spline interpolation provides a means of optimally reconstructing an unknown original signal. When the observation data are corrupted by noises, and if the form of the original continuous signal is known, then we can use least squares estimation techniques to approximate the original signal. In this paper, we are dealing with a problem in which an unknown signal is approximated by smoothing splines from a set of noise corrupted observation data. Specifically, an unknown signal $f(\xi)$ is approximated by a polynomial spline $s(\xi)$ which minimizes the objective function;

$$J^* = \sum_{n=1}^{N} [s(\xi_n) - m_n]^T R_n^{-1} [s(\xi_n) - m_n] + \{ \sum_{n=2}^{N} \rho_n \int_{\xi_{n-1}}^{\xi_n} [s^k(\xi)]^2 \, d\xi \} \tag{1}$$

where

m_n is an observation data;

$m_n = f(\xi_n) + v_n$, for $n = 1, 2, \ldots, N$; (1a)

v_n is a white observation noise process with error covariance R_n;

$R_n \triangleq E\{v_n \cdot v_n^T\}$ (1b)

$\rho_n > 0$ is a smoothing parameter, and

$s^k(\xi)$ is the kth derivative of $s(\xi)$.

At this point, it is worth to note the physical role of the smoothing parameter ρ_n as follows: (a) when ρ_n becomes very small, $\rho_n \to 0^+$, the resultant approximating function will pass through each data point and becomes an interpolation function; (b) when ρ_n assumes a very large value, $\rho_n \to \infty$, minimization of the objective function in Eq. (1) corresponds to fitting a straight line to a data set using least squares criterion. Thus, it can be said that the smoothing parameter controls resolution in a tradeoff of the smoothness of the restored function.

A. Choice of an Approximating Function

As has been noted, it is desired to develop a recursive algorithm whose results are sufficiently close to those obtained by directly minimizing the global problem as given by the criterion in Eq. (1). Fundamental problems encountered in developing a recursive algorithm which generates approximating functions are:

(1) Feasibility of recursive structures.
(2) Feasibility of numerical calculations.

Regarding the first problem, it has been noted from Ref. [1-5] that some of the approximating functions such as polynomial splines and piecewise Hermite polynomials have finite support. That is, a resultant approximating function for one section is mostly affected by its neighboring data points. Thus, a recursive structure with one or more sample delays would result in sufficiently close results to nonrecursive ones.

For the second point, out of a certain set of functionals, an optimal solution to Eq. (1) is an L-spline [2-4]. L-spline is a piecewise polynomial of degree 2k-1, and has 2k-2 continuous derivatives in the region $[\xi_1, \xi_N]$. Here, we propose to restrict our approximating functions to piecewise Hermite polynomials [3] of degree 2k-1, which have k-1 continuous derivatives in the region $[\xi \ , \xi_N]$. Advantages in using piecewise Hermite polynomials are as follows:

(i) Define $x_i \triangleq [s(\xi), s'(\xi), \ldots, s^{k-1}(\xi)]^T \Big|_{\xi=\xi_i}$, $i=1, \ldots, N$, then a piecewise Hermite polynomial $s(\xi)$ is completely determined by x_i, $i=1, 2, \ldots N$. For the purpose of clarity in discussion, only the case of k=2 is treated in the following.

A piecewise cubic Hermite polynomial is represented as:

$$s(\xi) = \begin{cases} s_{1,2}(\xi) & \text{for} \quad \xi_1 \leq \xi \leq \xi_2 \\ \vdots & \vdots \\ s_{N-1,N}(\xi) & \text{for} \quad \xi_{N-1} \leq \xi \leq \xi_N \end{cases} \tag{2}$$

where $s_{k-1,k}(\xi) = [\phi_{k,1}(\xi)\psi_{k,1}(\xi)\phi_{k,0}(\xi)\ \psi_{k,0}(\xi)][x_k^T,\ x_{k-1}^T]^T$ (2a)

$$x_k = [s(\xi_k) \quad s'(\xi_k)]^T \quad k = 1, \ldots, N \tag{2b}$$

$$\phi_{k,1}(\xi) = (\xi-\xi_{k-1})^2[(\xi_k-\xi_{k-1}) + 2(\xi_k-\xi)]/(\xi_k-\xi_{k-1})^3 \tag{2c}$$

$$\psi_{k,1}(\xi) = (\xi-\xi_{k-1})^2(\xi-\xi_k)/(\xi_k-\xi_{k-1})^2 \tag{2d}$$

$$\phi_{k,0}(\xi) = (\xi_k-\xi)^2[(\xi_k-\xi_{k-1}) + 2(\xi-\xi_{k-1})]/(\xi_k-\xi_{k-1})^3, \text{ and} \tag{2e}$$

$$\psi_{k,0}(\xi) = (\xi-\xi_{k-1})(\xi_k-\xi)^2/(\xi_k-\xi_{k-1})^2 \tag{2f}$$

Thus, it becomes natural that the smoothing integral in Eq. (1) is expressed in terms of x_i's, i = 1, 2, ... N. With some manipulations in algebra the smoothing integral for k = 2 in Eq. (1) is represented in a quadratic form:

$$\int_{\xi_{n-1}}^{\xi_n} ||s''(\xi)||^2\, d\xi = (x_n - A^* x_{n-1})^T B^{-1}(x_n - A^* x_{n-1}) \quad \text{or,} \tag{3}$$

$$\int_{\xi_{n-1}}^{\xi_n} ||s''(\xi)||^2\, d\xi = \begin{bmatrix} x_{n-1} \\ x_n \end{bmatrix}^T \cdot \begin{bmatrix} C_{11} & C_{12} \\ C_{21} & C_{22} \end{bmatrix} \cdot \begin{bmatrix} x_{n-1} \\ x_n \end{bmatrix} \tag{3a}$$

where $x_n \triangleq [s(\xi),\ s'(\xi)]\Big|_{\xi=\xi_n}^T$ (4)

$$A^* = \begin{bmatrix} 1 & \Delta \\ 0 & 1 \end{bmatrix} \qquad B = \begin{bmatrix} \Delta^3/3, & \Delta^2/2 \\ \Delta^2/2, & \Delta \end{bmatrix} \tag{5a}$$

$$\Delta = \xi_n - \xi_{n-1}, \quad \text{for } n = 2, \ldots, N \tag{5b}$$

$$C_{11} \triangleq A^{*T} B^{-1} A^*, \qquad C_{12} \triangleq C_{21}{}^T \triangleq -A^{*T} B^{-1}, \tag{5c}$$

$$\text{and } C_{22} \triangleq B^{-1} \tag{5d}$$

(ii) Even though we restrict our approximating functions to piecewise Hermite polynomials, the one which minimizes the objective function in Eq. (1) becomes an L-spline. A simple example for the above discussions is shown.
Thus, the choice of a piecewise Hermite polynomial as our approximating function will also lead to an optimal solution.

A nonrecursive solution for the minimization of the objective function in Eq. (1) can be obtained by taking the gradient of J^* with respect to $[x_1, x_2, \ldots x_N]^T$ and setting it to zero. However, this approach will require solutions of a set of 2N simultaneous equations. To avoid this computational problem, we developed a recursive algorithm which requires inversion of 2 x 2 matrices only.

RECURSIVE ALGORITHM

Given a set of initial values for the mean $\hat{x}_1$ and its error covariance P_1, where $P_1 \triangleq E\{(x_1 - \hat{x}_1)(x_1 - \hat{x}_1)^T\}$, by using Eqs. (3) and (4) the objective function in Eq. (1) becomes

$$J_N = \sum_{n=2}^{N} [(Hx_n - m_n)^T R_n^{-1}(Hx_n - m_n)] + (x_1 - \hat{x}_1)^T P_1^{-1}(x_1 - \hat{x}_1) + \sum_{n=2}^{N} \rho_n (x_n - A^* x_{n-1})^T B^{-1}(x_n - A^* x_{n-1}) \tag{6}$$

where H = (1 0).

Let the solutions to the above optimization problem be $[\hat{x}_1^*, \hat{x}_2^*, \ldots, \hat{x}_N^*]$. If $\hat{x}_{p|q}$ is defined as the estimate of x_p obtained by minimizing Eq. (6) with N = q, then $\hat{x}_i^*$ can be written as $\hat{x}_{i|N}$. Here, it is proposed to approximate a nonrecursive solution $\hat{x}_i^* = \hat{x}_{i|N}$ by $\hat{x}_{i|i+\ell}$, where $\ell \geq 0$ and $i + \ell \leq N$. As has been mentioned before, due to a local base property of the polynomial splines, the smoothed estimate $\hat{x}_{i|i+\ell}$ would be sufficiently close to the nonrecursive solution $\hat{x}_i^*$ for $\ell = 1$, 2 or 3.

A. Filtering

From the definition, a filtered estimate $\hat{x}_{i|i}$ would be obtained by minimizing the objective function in Eq. (6) with N = i. By taking the gradient of J_i with respect to $[x_1, x_2, \ldots x_i]$, we have a set of simultaneous equations as follows:

$$\begin{pmatrix} G_j & \rho_j C_{12} & & & \underline{0} \\ \rho_j C_{21} & \rho_j C_{22} + \rho_{j+1} C_{11} + H^T R_{j+1}^{-1} H & \rho_{j+1} C_{12} & & \\ & +\rho_{j+1} C_{21} & \ddots & \rho_{k-2} C_{12} & \\ & & \rho_{k-2} C_{21} & \rho_{k-2} C_{22} + \rho_{k-1} C_{11} + H^T R_{k-1}^{-1} H & \rho_{k-1} C_{12} \\ \underline{0} & & & \rho_{k-1} C_{21} & \rho_{k-1} C_{22} - H^T R_k^{-1} H \end{pmatrix} \cdot \begin{pmatrix} x_j \\ x_{j+1} \\ \vdots \\ x_{k-1} \\ x_k \end{pmatrix} = \begin{pmatrix} d_j \\ H^T R_{j+1}^{-1} m_{j+1} \\ \vdots \\ H^T R_{k-1}^{-1} m_{k-1} \\ H^T R_k^{-1} m_k \end{pmatrix} \tag{7}$$

where j = 1 and k = i, and G_j and d_j are defined by

$$G_j \triangleq \rho_j C_{11} + E_j^{-1} \tag{7a}$$

$$d_j \triangleq E_j^{-1} \hat{x}_{j|j} \tag{7b}$$

$$\hat{x}_{j|j} = E_j [H^T R_j^{-1} m_j + \rho_{j-1} C_{21} G_{j-1}^{-1} d_{j-1}], \quad \hat{x}_{1|1} \triangleq \hat{x}_1 \tag{7c}$$

$$E_j^{-1} = \rho_{j-1} C_{22} + H^T R_j^{-1} H - (\rho_{j-1} C_{21}) G_{j-1}^{-1} (\rho_{j-1} C_{12}), \quad E_1 \triangleq P_1 \tag{7d}$$

It should be noted that the matrix on the left side of Eq. (7) is diagonally dominant and positive definite. Here we are interested in solving Eq. (7) for x_i. Thus, by eliminating the first four equations, i.e., the variable x_1, Eq. (7) is reduced to the same form as itself with $j = 2$, $k = i$ and $\hat{x}_{2|2}$ and E_2 are calculated by Eqs. (7c) and (7d). Noted that the quantity E_j defined by Eq. (7d) is called pseudo error covariance, because $P_{k|k} \triangleq E\{(x_k - \hat{x}_{k|k})(x_k - \hat{x}_{k|k})^T\}$ cannot be computed in a recursive manner directly.

By applying this reduction method repeatedly, the original equations in Eq. (7) are reduced to the same form as itself with $j = i-1$, $k = i$ and every $x_{j|j}$ and E_j are computed by Eqs. (7c) and (7d) recursively. Solving Eq. (7) with $j = i-1$, $k = i$ in terms of x_i, we obtain a recursive estimate algorithm as

$$\hat{x}_{i|i} = E_i [H^T R_i^{-1} m_i - \rho_{i-1} C_{21} G_{i-1}^{-1} d_{i-1}] \tag{8}$$

Equation (8) is rearranged as

$$\hat{x}_{i|i} = E_i H^T R_i^{-1} m_i + F_i \hat{x}_{i-1|i-1} \tag{9}$$

where $F_i = -\rho_{i-1} E_i C_{21} (\rho_{i-1} C_{11} + E_{i-1}^{-1})^{-1} E_{i-1}^{-1}$ (9a)

and E_i's are computed by Eq. (7d) recursively. Equation (9) above is the desired filtering equation which computes $\hat{x}_{i|i}$ from the previous estimate $\hat{x}_{i-1|i-1}$ and the present measurement m_i. From the viewpoint of smoothing spline, the recursive filtering algorithm can be interpreted as follows:

The estimate of x_i obtained by fitting cubic splines to the measurement data m_n, $n = 2, 3, \ldots$, with the initial values $\hat{x}_1$ and $E_1 = P_1$ is the same as the estimate of x_i obtained by fitting a cubic polynomial to the measurement m_i with the initial values at stage i-1, $\hat{x}_{i-1|i-1}$ and E_{i-1}. In fact, the above interpretation comes from the mathematical derivations in Eq. (7) through Eq. (9).

With reference to Eqs. (7) and (8), each iteration of the recursive filtering algorithm can be interpreted as fitting a cubic polynomial to the previous estimate $\hat{x}_{i-1|i-1}$ and the present measurement m_i in the region $[\xi_{i-1}, \xi_i]$.

B. Smoothing

A smoothed estimate $\hat{x}_{i|i+\ell}$ is defined as the estimate of x_i obtained by solving the minimization problem in Eq. (6) with $N = i + \ell$. In fact, this can be interpreted as fitting a polynomial spline to the first $i + \ell$ data and obtaining the function value and its derivative from the approximating function at the node i. In our formulation, this corresponds to solving the simultaneous equations of the same form as Eq. (7) with $j = 1$, $k = i+\ell$ and the quantities G_1 and d_1 are defined in Eqs. (7a) and (7b). By using the same reduction method as before, the result would be the same form as itself (Eq. (7) with $j = i$ and $k = i+\ell$).

For the case of a one-sample delay, $\hat{x}_{i|i+1}$ is obtained by solving the simultaneous equations in Eq. (7) with $j = i$, $k = i+1$, and which yields

$$\hat{x}_{i|i+1} = V_i E_i^{-1} \hat{x}_{i|i} + K_i m_{i+1} \tag{10}$$

where $$V_i = [-\rho_i C_{12} (\rho_i C_{22} + H^T R_{i+1}^{-1} H) \rho_i C_{21} + G_i]^{-1} \tag{10a}$$

$$K_i = -V_i \rho_i C_{12} (\rho_i C_{22} + H^T R_{i+1}^{-1} H)^{-1} H^T R_{i+1}^{-1} m_{i+1} \tag{10b}$$

and E_i is defined as before.

Equation (10) is the desired smoothing algorithm, in which the smoothed estimate

$\hat{x}_{i|i+1}$ is obtained by updating the filtered estimate $x_{i|i}$ with the measurement m_{i+1}. The smoothing procedure described above implies the following. The smoothed estimate of x_i obtained by fitting cubic splines to m_n, n = 2, 3, ... , i+ℓ with the initial values $\hat{x}_1$ and $E_1 = p_1$ is the same as the smoothed estimate of x_i obtained by fitting cubic splines to m_n, n = i+1, ..., i+ℓ with the filtered estimate $x_{i|i}$ and E_i.

A recursive procedure to obtain smoothed estimate $\hat{x}_{i|i+1}$ is summarized as follows:

(Part 1) obtain the filtered estimates $\hat{x}_{i|i}$, i = 2, ... , N, by using Eqs. (7d), (9) and (9a).
(Part 2) smoothed estimates $\hat{x}_{i|i+1}$, for i = 1, ... , N-1, are obtained by using Eqs. (10), (10a) and (10b).

As was mentioned earlier in this section, the smoothed estimate $\hat{x}_{i|i+\ell}$ is an approximation to the nonrecursive solution $\hat{x}_i^* = \hat{x}_{i|N}$. Thus, as the number of delays, ℓ, increase, we will get a better approximation to x_i^*. For ℓ = 2, 3, ... , only the smoothing part is modified by solving the simultaneous equations in Eq. (7) with k = i+ℓ. In fact, the smoothing algorithm developed by far is a fixed-lag smoothing algorithm, which is suitable for an on-line implementation. If the situation does not require an on-line implementation, we can also derive a fixed-interval smoothing algorithm with observation set $M = \{m_1, \ldots, m_N\}$. In this case, the resultant smoothed estimates become exactly the same as the nonrecursive estimates.

SIMULATION RESULTS

For a continuous signal:

$$f(\xi) = \sin(\xi) \tag{11}$$

measurements are obtained at discrete points:

$$m_i = f(\xi_i) + v_i, \quad i=1, \ldots, 100 \tag{12}$$

where v_i is white Gaussian measurement noise,

$$R_i = E\{v_i v_i^T\} = 0.000025, \text{ and } \Delta\xi = \xi_n - \xi_{n-1} = 2\pi/100 = 0.062832 \tag{13}$$

Function values and the first derivatives at discrete nodes are estimated from the measurements m_i, i=1, ... , 100 by the three schemes below:

(a) Difference quotients method.
(b) Recursive smoothing algorithm with ℓ=1: Eq. (10).
(c) Nonrecursive smoothing by cubic splines - as described in Ref. [6].

Table 1 shows the mean-square errors from the three schemes above.

Table 1: Mean-Square Errors

	Difference Quotients	Recursive Smoothing by using Eq. (10)	Nonrecursive Smoothing* by Cubic Splines, Ref. [6]
ε_0	2.5×10^{-5}	0.8×10^{-5}	0.57×10^{-5}
ε_1	4.30×10^{-1}	0.12×10^{-1}	0.1007×10^{-1}

*The computer program for this technique was provided by Prof. H. McLaughlin of the Mathematical Science Department at R.P.I.

where $\varepsilon_0 = \frac{1}{100}\sum_{i=1}^{100} (f(\xi_i) - \hat{x}_{i|i+1}(1))^2, \quad \varepsilon_1 = \frac{1}{100}\sum_{i=1}^{100} (f'(\xi_i) = \hat{x}_{i|i+1}(2))^2$ (14)

From Table 1, it is noted that both smoothing algorithms are successful in reducing the error in the estimated states. The error in the first derivative is decreased by more than 10 db. Moreover, Table 1 shows that the performances of the two smoothing schemes are comparable. However, it should be emphasized that the recursive smoothing algorithm developed in this paper is much simpler than the nonrecursive spline smoothing.

CONCLUSION

By taking an algebraic approach, a recursive smoothing algorithm was developed as an approximation to nonrecursive spline smoothing. Moreover, the scheme is made adaptive to local data structure by estimating the parameters of original continuous signals. Compared to the recursive smoothing algorithm suggested by Weinert [8,9], the smoothing algorithm in this paper is simpler in that the scheme is in a discrete form while the algorithm by Weinert contains integration procedure in the gain computation. Simulation result shows that the performance of the recursive smoothing algorithm is comparable to that of its nonrecursive counterpart [6]. In addition, the computational complexity with recursive smoothing algorithm is much less than its non-recursive counterpart. Also, recursive smoothing by splines can be implemented on-line. Two-dimensional recursive smoothing algorithm which constructs continuously differentiable surface to a set of discrete noisy data is being developed as an extension of this algorithm.

REFERENCES

1. Schafer, R. W., and L. R. Rabiner (1973). A digital signal processing approach to interpolation, Proc. IEEE, Vol. 61, No. 6.
2. Ahlberg, J. H., E. N. Nilson, and J. L. Walsh (1967). The theory of splines and their application. Academic Press, Inc.
3. Prenter, P. M. (1975). Splines and variational methods. John Wiley & Sons, Inc.
4. Greville, T. N. E. (1969). Theory and applications of spline functions. New York, Academic.
5. Hou, H. S., and H. C. Andrews (1978). Cubic splines for image interpolation and digital filtering. IEEE Trans. on Acoust., Speech, Signal Processing, Vol. ASSP-26.
6. Reinsch, C. H. (1967). Smoothing by spline functions. Numer. Math, Vol. 10, pp. 177-183.
7. Hou, H. S., and H. C. Andrews (1977). Least squares image restoration using spline basis functions. IEEE Trans. on Computers, Vol. C-26, pp. 856-873.
8. Weinert, H. L., and G. S. Sidhu (1978). A stochastic framework for recursive computation of spline functions: Part II, Smoothing splines. Journal of Optimization Theory and Applications, Vol. 30, No. 2.
9. Weinert, H. L., R. H. Byrd, and G. S. Sidhu (1980). A stochastic framework for recursive computation of spline functions: Part II, Smoothing splines. Journal of Optimization Theory and Applications, Vol. 30, No. 2.
10. Mendel, J. M. (1973). Discrete techniques of parameter estimation. New York, M. Dekker, Inc.

XII. SIMULATION METHODOLOGY

EXPERIMENTAL OPTIMIZATION BY SIMULATION TECHNIQUE

by

Alain CHECROUN[+] Motaz KHORSHID[++]

[+]Professor, Director of "CERIA", University of Paris, France

[++]Doctor of Engineering, Institute of National Planning, Cairo, Egypt

ABSTRACT

The purpose of this paper is two fold, (i) to propose an integrated framework for studying management systems by simulation and (ii) to evaluate the possibility of using simulation as an experimental optimization technique. A job shop model, which can be used to test both materials handling and dispatching rules, was developed in order to demonstrate the applicability of the proposed procedures.

KEYWORDS

Simulation, Experimentation, Optimum seeking methods, Multiple ranking procedures job shop simulation.

1 - INTRODUCTION

The attractive features that computer simulation offers, have encouraged operations researchers and statisticians to improve its practice through the use of various statistical techniques to analyse the obtained results [11,12,13,16,17, 18,19,23].

The commun conclusion of these studies indicate the need to consider two main problems when investigating simulated systems. First, the particular circumstances of simulation experiments that may lead to misinterpretation of results[1]. Second, the difficulty to achieve the assumptions of the statistical theory, as independence, normality, and homogeneity of variances. So, either we hope that the selected techniques are not influenced by assumptions violation, or we manipulate simulation runs to match them.

The purpose of this research is two fold. First, to propose an integrated framework for studyings complex systems by simulation. Second, to evaluate the possibility of using simulation to find the optimum solution. In order to demonstrate the applicability of the proposed procedures, we developed a model that can be used to test both materials handling and dispatching rules in a job shop production system.

As any statistical investigation, simulation user should consider the following steps when developing an experimental strategy :

i) The choice of a sampling plan that specify how each test run is to be executed an how to determine simulation run length.

ii) The development of an experimental design that will yield the desired information.

iii) The selection of a data analysis technique in order to reach some conclusion about the simulated system.

In section 2, we propose an integrated mathematical base for studying management systems by simulation. The proposed optimization techniques, are presented in section 3. In section 4, we explain briefly the simulation model that will be the setting of the study. Finally, the design and analysis of two experiments are discussed in section 5.

2 - MATHEMATICAL FORMULATION

In many simulation models, the process of interest apears as a stochastic process[2].

$$\{Y(t);-\infty\leqslant t\leqslant\infty\} \quad (1)$$

Since our research concerne discrete event digital simulations, we assume that during a small interval of time, the process shows little, if any change so that observing $Y(t)$ at periodic intervals result in no loss of information. In a more detailed manner, the sequence,

$$\{Y_t \text{ , } t = 1,2...,\infty\} \quad (2)$$

1-A list of these circumstances can be found in Fishman [11] page 262

2-We will consider only the stochastic simulation models as most management systems inevitably appear random to some degree in nature.

corresponds to the process $\{Y(t)\}$ at all integer values of the index t.

The index t may be the time ; for example Y_t may define the number of jobs in a production system at the instant t of simulation. It may simply denote order, for example Y_t may represent the waiting time for the tth customer to receive service in a queueing model.

In order to study several processes of interest, generated by different environmental conditions or input specifications, we should aquire a quantitative characterization of each of them.

The mean of the process ;

$$u = E\{Y_t\} \tag{3}$$

serves generally as the mathematical descriptor, and by definition, the variance[3]

$$\text{Var}\ (Y_t) = E\ \{(Y_t-u)^2\} \tag{4}$$

and the autocovariance function,

$$R_s = E\{(Y_t-u)(Y_{t+s}-u)\}\ ;\ s = 1,2,..., \tag{5}$$

Then regardless of the experimental objective and the type of simulated model, we should define a procedure for estimating "u" and for determining the accuracy of this estimator, i,e to select a sampling plan.

2.1. SAMPLING PLAN

In simulation experiments sample size can be increased by prolonging simulation run, or by conducting separate runs. Consequently, there is two sampling plans, the autocorrelated observations and the replicated runs.

i) Autocorrelated observations :

Let $\{Y_t, t\varepsilon n\}$ a time series of length n observed during the simulation run, the mean u can be estimated by :

$$\bar{Y} = n^{-1} \sum_{t=1}^{n} Y_t \tag{6}$$

where $\bar{Y}$ is called "simular response".

In order to determine the accuracy of $\bar{Y}$, we need to estimate its variance [12]:

$$\text{Var}\ (\bar{Y}) = n^{-1}\ \{\text{Var}\ (Y_t)+2 \sum_{s=1}^{n-1} (1-s/n)\ R_s\} \tag{7}$$

Where $\text{Var}(Y_t)$ and R_s are defined by (4),(5) and can be estimated by the formulas [18] :

$$\hat{\text{Var}}\ (Y_t)=(n-1)^{-1} \sum_{t=1}^{n} (Y_t-\bar{Y})^2 \tag{8}$$

---

[3] Assuming $\{Y_t\}$ a covariance stationary sequence {11,18}

$$\hat{R}_s = (n-s)^{-1} \sum_{t=1}^{n-s} (Y_t - \bar{Y})(Y_{t+s} - \bar{Y}) \; ; \tag{9}$$

$$s = 1,2,..,(n-1)$$

iii) Replicated runs :
We can generate independent observations in simulation experiment by repeating the run using different random numbers. Let

k = number of runs
n'= number of observations per run
n = total number of observations (=n'k)
$Y_{t,j}$= the observation generated at time t of run j.

The mean of each run is defined by :

$$Y_{n',j}=n'^{-1} \sum_{t=1}^{n'} Y_{t,j} \qquad ; \qquad j = 1,2,...,k \tag{10}$$

and then simular response is calculated by :

$$\bar{Y} = k^{-1} \sum_{j=1}^{k} Y_{n',j} \tag{11}$$

Since the sequence $\{Y_{n',j};j=1,2,...,k\}$ consist of k independent observations, Var $(\bar{Y})$ is given by :

$$\text{Var }(\bar{Y}) = \text{Var }(Y_{n,j})/k \tag{12}$$

and can be estimated by :

$$\hat{\text{Var}} \; (\bar{Y}) = S^2/k.$$

Where $$S^2 = (k-1)^{-1} \sum_{j=1}^{k} (Y_{n',j} - \bar{Y})^2 \tag{13}$$

2.2 SIMULAR RESPONSE FUNCTION :

When experimenter select a sampling plan, he can proceed to the study of the response $\bar{Y}$ as a function of the environmental conditions or the experimental factors $(x_1,x_2,...,x_p)$.

Factors are categorized as qualitative and quantitative. Examples of qualitative factors are policy specifications, such that alternative dispatching rules in job shop, or discrete environmental conditions. Quantitative factors are examplified by input parameters that can usually be thought as continious variates.

Since the stochastic features are spowned in the simulation model by incorporating the random number seed as an integrated part of input specifications, the response $\bar{Y}$ becomes a random variable because it is a transformation not only of the experimental factors $(x_1,x_2,...,x_p)$, but also of the randomly selected seed "r" [1]. This relation is defined by :

$$\bar{Y} = \phi(x_1, x_2, \ldots, x_p; r) = \phi(\vec{x}, r) \quad (14)$$

Although the random number seed "r" may be conceptually defined as a real number between o and 1, it could not be classified as quantitative factors, because $\bar{Y}$ will probably not be a continious function of it. Then it is unique among the other quantitative factors and expression (14) can be written :

$$\bar{Y} = \phi(x_1, x_2, \ldots, x_p) + \varepsilon(r) \quad (15)$$

where $\varepsilon(r)$ is a random effect dependent upon the random number seed .

Further, if we assume that $\varepsilon(r)$ is independent of experimental factors and has zero expectation, the expected simular response can be defined as :

$$E(\bar{Y}) = \phi(x_1, x_2, \ldots, x_p) = \phi(\vec{x}) \quad (16)$$

It is the nature of the unknown function $\phi(\vec{x})$, termed <u>simular response function</u> that we try to investigate by the simulation experiment.

In practical situations, any attempt to develop the exact form of $\phi(\vec{x})$ could not be justified from economical point of view. In addition, for many experimental purposes, it is unnecessary to consider the form of the true function. A flexible graduating function, will often be satisfactory to express the relation ship between $E(\bar{Y})$ and the p factors. Further more, many experimental strategies, divide the whole operability region of the factors space, into a number of smaller regions of immdiate interest. Withen these regions, the experimenter may feel it is reasonable to represent $\phi(\vec{x})$ by a known functional form, for example a polynomial, although he may know that such representation would be quite inadequate over the whole operability region.

As a result of the previous discussion, $\phi(\vec{x})$ may be approximated by :

$$E(\bar{Y}) \simeq f(x_1, x_2, \ldots, x_p\ ; \theta_1, \theta_2, \ldots, \theta_l) = f(\vec{x}, \vec{\theta}) \quad (17)$$

Where f is a known functional form indexed by some unknown vector $\vec{\theta}$.

The way with which we investigate the function $f(\vec{x}, \vec{\theta})$, in order to yield information about simulated system, depends on the experimental objectives. Accordingly we distinguish two types of experiments, Exploratory and Optimization.

2.3. EXPLORATORY EXPERIMENTS

If experimenter wishes to study the relative importance of the factors $\vec{x}$ as they affect the simular response $\bar{Y}$, he may use one of the following designs :

i) Screening designs

At the begining of investigation, specially with complicated simulation models, the experimenter may face the problem of so many factors. It may happen that not all the p factors are important but only a few, say p' factors. Therefore, he screen for them.

---

[1] A detailed discussion of this point can be found in Mihram [18]

The statistical literature contains many designs, for example, fractional factorial designs [4,8,17], random disigns [24], group screening designs [21], and super saturated designs [3]. The investigator has to select the design which fit his particular experimental situation.

ii) Designs for estimating parameters

When experimenter has a prior knowledge about the simulated system, either due to theoritical background or from previous investigations. He may assume that a particular functional form $f(\vec{x},\vec{\theta})$ is a good approximation of the true response function $\phi(\vec{x})$, in such a way that bias due to inadequacy of $f(\vec{x},\vec{\theta})$ to represent $\phi(\vec{x})$ can be neglected. So his goal will be to select an experimental plan to estimate the unknown parameters $\vec{\theta}$ with high accuracy. Two basic approaches were proposed to develop an experimental design, either to use a simple factorial or fractional factorial design [4,8,17], or to select a design based on a variance criterion as D - optimal designs [15].

iii) Designs for exploring response surface

When knowledge about simulated system is limited, the object is to approximate, withen a given region of factors space, $\phi(\vec{x})$ by some graduating function $f(\vec{x},\vec{\theta})$ which most closely represent the true simular response function.

Accordingly the following design requirements have to be considered :

a) The design should allow sequentialization so that designs of higher order can be developed with minimum loss of information.

b) The design should consider not only sampling variation but also bias error.

c) The design should allow a check to be made on the representational accuracy of the postulated model.

2.4. Optimization experiments

The purpose of these experiments is to find the combination of factor levels at which simular response function $\phi(\vec{x})$ is optimized.

A particular attention will be devoted, in the next section, to the explanation and the applicability of these experiments.

<u>To summarize</u>, any attempt to develop an experimental method for investigating management systems by simulation, should select a sampling plan which define an efficient procedure for estimating the variance of simular response $\bar{Y}$. The estimated variance measures the accuracy of results and then can be used to determine the appropriate run length. Having accomplished this task, an experimental strategy may be defined for investigating the inter-dependence between the simular response $\bar{Y}$ and the experimental factors.

3 - EXPERIMENTAL OPTIMIZATION TECHNIQUES

The choice of an experimental strategy that will yield an optimal solution depends on the type of factors in the simulation model. When all factors are quantitative, an optimum seeking routine can be used in order to find the combination of factor levels that optimize the response $\bar{Y}$. But the existence of some qualitative factors, as policy specification or operating rules, limit the search procedure to the choice between a number of experimental alternatives.

3.1. The search for an optimum combination of factor levels

When all factors are quantitatives, the investigator will wish to find in the smallest number of simulation runs, the Point $(x^{o}_{1}, x^{o}_{2}, \ldots x^{o}_{p})$, withen the factors space, at which $\phi(\vec{x})$ is a minimum or a maximum.

Since simular response function is not known in advance and is subjected to random variation, we think that the most reasonable strategy will be to fit a sequential program of investigation consisting of the following steps :

i) Divide the whole region of interest into a number of small subregions, so that we can explore adequately a small subregion with a moderate number of simulation runs.

ii) Use the results obtained in one subregion to move to a second in which simular response $\bar{Y}$ is better.

iii) Repeate the previous steps until the attainment of a near stationary region where no improvement in the simular response can be achieved.

iv) In this limited region, conduct a more detailed experiment in order to determine the local nature of the function. $\phi(\vec{x})$.

In the following sections we discuss briefly the two main elements of this sequential program, seeking a near stationary region and exploring it.

3.1.1. Seeking a near stationary region

When the starting conditions of simulation are fairly remoted from the stationary point, an optimum seeking technique will be needed to move rapidly through the factors space to a near stationary region.

Brooks [6] compared four optimum seeking methods, steepest ascent, univariate, factorial, and random search. He concluded that, when sequential investigation is possible, steepest ascent seems to be superior to the others, except in case of large number of factors, where random search is more efficient[5].

Recently, Smith [25] showed that random search should not necessarily be the search technique selected in practical situations even in case of so many factors and he recommended the use of the steepest ascent.

Since the steepest ascent method is explained in detail in Box and Wilson [5] and Davies [8], we just mention, here, some remarks that should be considered when applying the method to simulation experiment[6].

i) Since we use the error variance to test the adequacy of the fitted function and the significance of model parameters, an accurate estimate of the variance of $\bar{Y}$ is needed in order to avoid any wrong conclusion.

ii) As the steepest ascent method is affected by the size of the experimental error [5], we may try to reduce it, by selecting a minimum variance design (see section 2.3), by increasing simulation run length, and if possible, by using a variance reduction techniques.

---

[5]This is explained by the fact that, in random search algorithm, the number of experimental trials is not a function of the number of factors.

[6]The method will be explained using an example model in section.(5)

iii) If possible, provision should be made to estimate some of higher order coefficients that were not included in the postulated model. The study of these coefficients will provide some indication of whether the assumption that these terms can be ignored is a reasonable one or not.

3.1.2. Exploring the near stationary region

The experimenter may arrive at a near stationary region either as the result of successive application of steepest ascent method, or because he has already found it at the begining of his investigation. In either cases, only immediate neighbourhood need be explored to determine the local nature of response function $\phi(\vec{x})$ and this may be done without excessibely large number of experimental points.

Although many authers have ignored the exploration of near stationary region, and are only satisfied by finding the approximated optimum point, we think that it is an important step in case of simulation for the following reasons :

First, it should be remembered that because of random error and possible lack of fit between fitted equation and the true response $\phi(\vec{x})$, it must not be implied immediatly that the true surface has a maximum (or minimum) at the selected point So in practice further exploration and confirmatory runs should be performed arround the stationary point of the fitted surface.

Second, the discovery of factors dependence of a particular type may give us an idea about the cost of departure from the optimum point, if it was impossible to reach it in practice. For example, finding the direction of a stationary ridge means that we can know the different combinations of factor levels that optimize the response $\bar{Y}$. Then the choice between these alternatives can be decided according to the cost of each combination or according to an auxiliary response.

Two exploratory techniques are proposed in the statistical literature, Canonical analysis [5,8] and Ridge analysis [9]. The authors matched the two techniques in a single computer program in order to have more robust conclusions..This can be done by using canonical analysis to reveal the factor dependence withen the local stationary region, then using ridge analysis to evaluate the locus of the absolute maximum or minimum when augmenting the experimental region.

3.2. The choice between experimental alternatives

When simulation model contains qualitative factors, as managerial policies or operating rules, the search procedure will be reduced to the optimum choice between a number of experimental alternatives. More specifically, it is required to find the combination of factor levels corresponding to the best simular response $\bar{Y}$, such that the probability of correct selection (CS) is at least P^* , given that the difference Δ between the best and the next best simular response is at least Δ^*. This may be stated formally as

$$P_r(CS/\Delta \geqslant \Delta^*) \geqslant P^* \tag{18}$$

The previous formulation of the problem permit the use of one of the multiple ranking procedures [2,17,22,23]. Most of these methods assume normality, independence and commun known or unknown variances.

In practical simulation models, the distribution of the response $\bar{Y}$ is not Known, variances are not known and tend to differ, so either we manipulate simulation runs to meet these assumptions or we hope that the effect of their violation is negligible.

After consulting several multiple ranking procedures, the authers choosed three of them that seem to be attractive for simulation circumstances. The selected procedures are Bechhofer and Blumenthal [2], Paulson [22], and Sasser et al [23].

Bechhofer method is the only one extensivly tested for its sensitivity to assumptions violation, it is quite robust and relatively efficient [17]. Unfortunately, it cannot capitalize on favorable configurations of population means. Paulson's procedure give us the possibility to eliminate inferior populations, so it might be advantageous when comparing a large number of alternvatives.

The authors conducted a comparative study using two simulation models, a job shop production system [16] and a dynamic model of the firm [19]. The results of this study indicate that Paulson method is the most efficient method, specialy when the number of alternatives is large. Bechhofer method seems to be the best procedure from robustess point of view, but unfortunately it require large sample sizes when deeling with a large number of alternatives. Sasser method, which is a heuristic version of Bechhofer, is less efficient than Paulson procedure.

To conclude, the results of this emperical study[7], show that no ideal procedure, that consider the particular circumstances of simulation experiment, does exist in present time. We recommend Paulson method when experimenter is wary about computer time and the number of alternatives is sufficiently large. Otherwize, Bechhofer procedure seems to be the most robust one and it is relatively efficient.

4 - THE PROPOSED SIMULATION MODEL

Most researchs on job shop production systems using simulation technique assume that the major system constraint was machine availability [1,7,10]. Recently a particular attention has been devoted to the study of labor and machine limited production systems [14,20], where the performance of the shop was evaluated considering both dispatching and labor assignment rules.

Since, materials handling systems and transportation time between operations are ignored, till now, in most simulation models, the authors developed a model that can be used to test materials handling systems and rules, as well as the other operating polices. To our knowledge this is the first trial to incorporate materials handling system as an integrated part of a job shop simulation program.

In the following discussion the major characteristics of the model are briefly presented. For more detailed explanation, see Khorshid [16], chapters 2 and 3.

4.1. Job shop structure

The processing operations are carried out in m work centers. Each work center is characterized by a number of machines (C_i, $i=1,2,\ldots,m$), a machining queue for jobs waiting for service, and a handling queue for jobs to be handled between centers. The shop contains a number of materials handling equipment which are organized and assigned to work centers according to some handling rules.

4.2. Job characteristics

The time between successive arrivals of jobs in the shop is a random variable generated from certain probability density function a (.), with a mean interarrival time λ . Upon the arrival of a job, the sequence of work centers through which it must be routed are generated randomly using an $(m+1) \times (m+1)$ transition probability matrix P , with entries P_{ij} giving the probability of transition

---

[7]See also Kleijnen (17)

from work center i to work center j. The first row (i=o) represents entry into the system ,while the first column (j=o) represents departure from the shop. The choice of the elements of the matrix P determine the flow pattern in the tested shop which lie between the extreme cases, the pure job shop and the flow shop.

The time to process a job, as well as its handling time are random variables generated from a selected probability density function. Mean handling time depends on the distance between work centers and the used materials handling equipment.

4.3. Materials handling systems

The model can be used to test several handling systems, which vary according to the degree of centralized control exercised over the handling equipment. For on example, in case of completely centralized systems, each handling equipment, upon completing a transportation operation at a work center, is given its next assignment by central control. On the contrary, in case of decentralized systems, handling equipment return to central control for reassignment, only when there is no work remaining in the work center at which it was previously assigned.

4.4. Operating policies

Materials handling and dispatching decisions are required to operate the job shop. An operating policy consists of a specific combination of the following rules :

i) The rules governing the choice of a job from machining queue

ii) " " " " " " " " " handling queue

iii) The rules governing the assignment of handling equipment to different work centers.

iv) The rules determining when an equipment is eligible for reassignment to another work center.

4.5. Performance criteria

The efficiency of each operating policy is evaluated using a general performance measures , that can be easily transformed to a cost criteria in practical situations. Examples of these measures are, mean flow time per job, mean queue length, etc...

5 - EXPERIMENTATION

The hypothetical shop used in the experimentation consists of 6 work centers, each of them contains 2 identical machines. Four materials handling equipment are used to transport jobs between work centers. The job interarrival time, service time,and handling time are random variables generated from a negative exponential distribution. The means of these variables are adjusted so that average machine utilisation attain 85 % and handling equipment utilisation will be 95 %. The job routing is of the pure job shop type, and is generated using the transition probability matrix. A specific layout configuration is proposed in order to find estimates for distances between work centers.

5.1. First experiment

Finding the best operating policy:
The purpose of this experiment is to choose the operating policy that lead to the

best performance measure (simular response). Since out put of simulation experiment is stochastic in nature, a measure of the probability of correct selection is needed in order to justify the obtained results.

5.1.1. Experimental Factors

As it was indicated in section 4.4, an operating policy consist of a specific combination of dispatching and handling rules which we call, in statistical termonology, the experimental factors. The tested factors are[8] :

i) The rules governing the choice of a job from machining queue (x_1)
   1) Shortest - imminent - processing time rule ($x_1 = 1$)
   2) Minimum slack per operation rule ($x_1 = 2$)

ii) Rules governing the choice of a job from handling queue (x_2)
   1) Shortest handling time rule ($x_2 = 1$)
   2) Minimum slack per operation rule ($x_2 = 2$)

iii) Rules governing the assignment of handling equipment (x_3)
   1) assign available equipment to the work center whose handling queue contains the maximum number of jobs ($x_3 = 1$)
   2) assign available equipment to the work center whose queue contains the job which has been in the system the longest period of time ($x_3 = 2$).

iv) Rules determining when an equipment is eligible for reassignment (x_4)

   1) an equipment, upon completing a handling operation at a work center, is given its next assignment by central control ($x_4 = 1$)

   2) an equipment, upon completing a handling operation at a word center, stays at work center as long as there is any job in its queue ($x_4 = 2$).

Table 1, defines the 16 operating policies representing all possible combinations of the decision rules.

TABLE 1- DEFINITION OF POLICIES

Operating policies	Experimental factors (rules)				Simular response
	x_1	x_2	x_3	x_4	
1	1	1	1	1	0.4278
2	2	2	1	1	0.1832
3	1	2	1	1	0.2393
4	2	1	1	1	0.2178
5	1	1	2	1	0.2523
6	2	2	2	1	0.1814
7	1	2	2	1	0.2338
8	2	1	2	1	0.2046
9	1	1	1	2	0.3281
10	2	2	1	2	0.1916
11	1	2	1	2	0.2386
12	2	1	1	2	0.1938
13	1	1	2	2	0.3253
14	2	2	2	2	0.1979
15	1	2	2	2	0.2406
16	2	1	2	2	0.1985

[8]The choice of these rules and their definition are discussed in reference (16).

5.1.2. Simular response $\bar{Y}$:

The selected measure of performance is the mean delay ratio [10], which is defined for a particular job by the total service and handling time divided by the job flow time in the shop.

5.1.3. Sampling Plan :

At the begining of investigation, a study of the equilibrium conditions is conducted using some pilot runs. The results showed that the number of observations to be deleted, in order to reduce the bias caused by the starting conditions, was 500. On the other hand, since multiple ranking procedures require the generation of independent observations, we think that the best sampling procedure is to make a single simulation run per operating policy, to exclude 500 observations in the begining, and then to divide the run into a set of subruns in a way that the averages of the subruns would be uncorrelated. The method of Fishman [13] was used to determine the minimum number of observations per subrun. An iterative algorithm was incorporated in the main simulation program in order to estimate sequentially the variance of $\bar{Y}$ and then to terminate simulation run when desired accuracy is achieved.

5.1.4. Obtained results :

In section 3.2, we discussed multiple ranking procedures, and how they could be adopted to simulation circumstances. Table 2, shows the results of applying three of them, which were selected by the authors, to job shop simulation out put. All procedures choosed policy 1 as the best one, with probabilities of correct selection $P^* = 0.8, 0.95$ and different values of Δ^*. The results confirm also the superiority of Paulson method in case of large number of tested policies and favorable configurations of population means. This method requires a smaller sample sizes even with high probability of correct selection.

5.2. Second Experiment :

Finding optimum combination of factor levels :

In this experiment, we assume that simular response $\bar{Y}$ is related to a three quantitative factors by an unknown functional form. The purpose will be to find the values of the factors that optimize the simular response $\bar{Y}$.

TABLE 2. RESULTS OF MULTIPLE RANKING PROCEDURES

P^*	Δ^*	Bechhofer			Paulson			Heuristic		
		Sample size	Time in Sec.	Selected Policy	Sample size	Time in Sec.	Selected Policy	Sample size	Time in seconds	Selected Policy
0.80	0.01	11	1.1612	1	3	0.3855	1	7	0.4726	1
0.80	0.02	6	0.7800	1	2	0.3794	1	7	0.4818	1
0.80	0.03	4	0.6159	1	2	0.3784	1	7	0.4736	1
0.80	0.04	3	0.5571	1	2	0.3753	1	7	0.4731	1
0.95	0.01	17	1.769	1	6	0.4096	1	7	0.4690	1
0.95	0.02	8	0.9062	1	3	0.3820	1	7	0.4716	1
0.95	0.03	5	0.6840	1	2	0.3773	1	7	0.3912	1
0.95	0.04	4	0.6205	1	2	0.3773	1	7	0.4721	1

5.2.1. Experimental factors :

i) Due date control parameter (D1) :
When a job arrives at the shop, an approximated date for its completion has to be estimated. This due date depends on the shop utilisation factor and the number of operations to be performed on the job. We selected a simple and popular calculation formula, which is given by :

$$\text{Due Date} = r + D1.T$$

Where r is the arrival date and T is the expected processing and handling time for the job. $D1 \geqslant 1$ is a control parameter.

ii) Dispatching Parameter (U) :
Eilon et al [10] proposed a variant of the shortest Imminent processing time rule (SI), which they called SI* rule. It consists of the assignment of a float value F to each job in the machining queue, where F is defined by :

$$F = (D-r)-t_e - U$$

and
r = current date
t_e = expected time for the uncompleted operations on the job.
D = Due Date
U = a given control parameter.

Jobs with $F \leqslant 0$ are put in a priority queue, while those with $F > 0$ are put in a normal queue, the latter being processed only when the periority queue vanishes In each queue the SI rule applies.

The emperical study of Eilon et al [10] proved that this rule is superior to SI rule when selecting the optimum value of U.

ii) Handling parameter (UI) :

The previous rule is modified to deal with the choice of a job from the handling queue. In this case we calculate the float value by :

$$F = (D-r)-t_e - UI$$

Where D, r, t_e are previously defined, and UI is a given control parameter specified by the experimenter.
The rest of the rule is the same except that we use the shortest handling time rule instead of the SI rule.

<u>To summarize</u>, the purpose of this experiment is to find the values of the parameters (D 1, U, UI) that optimize the performance measure of a job shop characterized by a particular flow pattern and level of congestion. The selected measure is the mean flow time per job.

5.2.2. Seeking a near stationary point :

Since little is known about the orientation of the simular response function such that a good starting conditions could not be specified, it was thought that first order effects would be dominant, at least in the first stage of experimentation. Conséquently, a linear model is fitted using 2^3 factorial experiment. The levels chosen for the factors being those given in table 3.

TABLE 3 : FACTOR LEVELS FOR THE 2^3 EXPERIMENT

Factor levels in natural units	Factor levels in units of the design		Base Level	Unit
	-1	+1		
Due date parameter D 1	1.5	3.5	2.5	1
Dispatching parameter U	20	60	40	20
Handling parameter UI	20	60	40	20

Thus, denoting the variables measured in unites of the design by X_1, X_2 and X_3 respectively, then

$$X_1 = (D1-2.5)/1;\ X_2 = (U-40)/20\ ;\ X_3 = (UI-40)/20$$

and the linear model can be written.

$$E\ (\bar{Y}) = \theta_o + \theta_1 X_1 + \theta_2 X_2 + \theta_3 X_3$$

The following table shows the used factorial design and out put results[8]

Run	Factor Level			Mean flow time ($\bar{Y}$)
	X_1	X_2	X_3	
1	-1	-1	-1	66.8984
2	+1	-1	-1	88.0471
3	-1	+1	-1	66.7253
4	-1	-1	+1	67.6155
5	+1	+1	-1	84.9666
6	+1	-1	+1	94.1092
7	-1	+1	+1	65.4243
8	+1	+1	+1	74.1050

Since complete factorial experiment provides measures of linear and interaction effects, the following estimates were obtained :

$$\hat{\theta}_o = 75.986 \qquad \hat{\theta}_{12} = -2.589$$

$$\hat{\theta}_1 = 9.320 \qquad \hat{\theta}_{13} = -0.526$$

$$\hat{\theta}_2 = 3.181 \qquad \hat{\theta}_{13} = -2.367$$

$$\hat{\theta}_3 = 0.672$$

and the approximated standard error of the $\hat{\theta}$, based on the estimated variance of $\bar{Y}$ (see section 5.1.3) was 0.42.

[8]When number of factors is large, we can use a fractional factorial design which require a smaller number of experimental points 4

Two concluding remarks can be deduced from these results :

i) Factor X_3 has small effect compared with the other factors. Then, either the system is independent of the factor levels, or the unit adopted for the factor is relatively small. To guard against the second possibility, larger unit must be used in the next set of trials. If the factor continue to give small effect, it can be excluded from further investigation.

ii) Since interaction effects are not small compared to some first order effects, the starting condition is probably not too remoted from the stationary region. This suggest that we reduce the steepest descent path by a small increment. The calculation of the optimum path is shown in table 4.

TABLE 4. CALCULATION OF STEEPEST DESCENT PATH AND SUBSQUENT TRIALS

		X_1	X_2	X_3	$\bar{\bar{Y}}$
1) Base level		2.5	40	40	-
2) Unit		1	20	20	-
3) Estimated slop "$\hat{\theta}$"		9.320	-3.181	-0.672	-
4) Unit X $\hat{\theta}$		9.320	-63.622	-13.458	-
5) Changes in level per 5% change in X_1		0.466	-3.181	-0.672	-
6) Path of Steepest descent	0%	2.5	40	40	69.789
	5%	2.034	43.181	40.672	66.436
	10%	1.567	46.362	41.345	65.424
	15%	1.101	49.543	42.018	65.424

The results of table 4. show that no reduction of $\bar{Y}$ can be attained by the present path. So another cycle of steepest descent may be needed in order to improve the simular response.
When selecting factor levels for the next trials, we have to keep in mind that we are not too far from the optimum point.
Thus if further progress is to be possible, without taking second order effects into account, the best chance of success lay in reducing the units of the factors; table 5 indicate the new chosen levels.

TABLE 5. FACTOR LEVELS FOR SECOND GROUP OF TRIALS

	-1	1	Base level	Unit
Due date Parameter D1	1	2	1.5	0.5
Dispatching parameter U	36	56	46	10
Handling Parameter UI	16	66	41	25

A second factorial experiment was conducted, and the obtained estimates were :

$\hat{\theta}_o = 66.374$ $\qquad \hat{\theta}_{12} = -0.092$

$\hat{\theta}_1 = 0.950$ $\qquad \hat{\theta}_{13} = -1.222$

$\hat{\theta}_2 = -0.0496$ $\qquad \hat{\theta}_{23} = 0.106$

$\hat{\theta}_3 = -1.222$

We conclude from these estimates that a near stationary region is reached, since all first order effects are small compared with the standard error ($\hat{\sigma} = 1.19$); Further more, the effect of factor X_3, which was not significant in the first group of trials, became better than the effect of X_2. Thus neglecting it in the exploratory phase of experimentation may lead to misinterpretation of results.

5.2.3. Exploring the near stationary region :

In order to study the local nature of the response function, a central composite design [5] was used to fit the second order polynomial :

$$E(\bar{Y}) = \theta_o + \theta_1 X_1 + \theta_2 X_2 + \theta_3 X_3 + \theta_{11} X_1^2 + \theta_{22} X_2^2 + \theta_{33} X_3^2$$

$$+ \theta_{12} X_1 X_2 + \theta_{13} X_1 X_3 + \theta_{23} X_2 X_3$$

This design consist of 2^3 factorial experiment, 6 axial points, and one center point. It requires a reasonable number of trials and can benefit from results obtained from the first order design. On the assumption that a second order equation provides an adequate model, unbiased estimates were obtained for quadratique and linear effects.

Taking partial derivatives with respect to each factor, and solving the obtained system of linear equations, the levels of X_1, X_2 and X_3 corresponding to minimum flow time were as follows :

$$X_1^\circ = 0.781 \quad ; \quad X_2^\circ = 0.0445 \quad ; \quad X_3^\circ = 0.3153$$

Consequently, the minimum simular response $\bar{Y} = 64.930$

Transforming these values from design units to natural units, we get :

$$D1 = 1.11 \qquad U = 46.445 \qquad U_1 = 33.117$$

When the fitted surface was reduced to the canonical form, we obtained the following results :

i) the canonical form was

$$\bar{Y} - 64.93 = -0.2808\, x_1^2 + 0.3845\, x_2^2 + 1.1362\, x_3^2$$

ii) The new coordinates (x_1, x_2, x_3) for any points are given in terms of the old coordinates (X_1, X_2, X_3) by the following table :

	$(X_1 + 0.781)$	$(X_2 - 0.044)$	$(X_3 + 0.315)$
x_1	0.4972	- 0.0502	0.8661
x_2	0.0806	0.9966	0.0115
x_3	-0.8638	0.0640	0.4996

Since the coefficients of the canonical form are different in sign and each of them is not small compared with the others. This suggested that a minimax surface exist in the near optimal region. This conclusion was confirmed by applying the ridge analysis program to the fitted second order equation.

Figure 2, shows the values of the absolute maximum and absolute minimum of simular response $\bar{Y}$ as the radius R of the sphere, representing the experimental region, varies.

6. CONCLUSION

This research demonstrate the importance of using statistical techniques to design and analyse computer simulation experiments.

The problems concerning the choice of a sampling procedure and, the development of an experimental plan, were formulated considering the particular circumstances of simulation. This formulation can guid the investigator to select the most suitable statistical technique as a function of the desired objective.

The two applications of the proposed experimental techniques to a job shop production model showed that simulation can be used efficiently to find the optimum operating conditions. The essential advantage of these techniques is that they provide information about the precision of the estimated optimum, for example, the probability of correct selection of the best alternative, or the cost of departure from the stationary point, if it was impossible to reach it in practice.

REFERENCES

1. Ashour, S. and S.D. Waswani, "A GASP Simulation Study of Job Shop Scheduling", Simulation, Jan. (1972) pages 1-9.

2. Bechhofer, R.E; and S. Blumenthal, "A Sequential Multiple Decision Procedure for Selecting the Best One of Several Normal Populations with a Commun Unknown Variance, II : Monte Carlo Sampling Results, and New Computing Formulae", Biometrics, Vol. 18, March(1962) pages 52-67.

3. Booth, K.V.H., D.R. Cox "Some Systematic Supersaturated Designs "Technometrics, Vol. 4, (1962).

4. Box, G.E.P., and J.S. Hunter "The 2^{k-p} Fractional Factorial Designs, Part I, II" Technometrics, Vol 3, (1961) page 311.

5. Box, G.E.P. and K.B. Wilson "On the Experimental Attainment of Optimum Conditions" Journal of the Royal Statistical Society B, Vol. 13, (1951) pages 1-45

6. Brooks, S.H., "A Comparison of Maximum - Seeking Methods" Opns. Res. Vol. 7, (1959), pages 430-457.

7. Conway, R.W., W.L. Maxwell, and L.W. Miller, "Theory of Scheduling", Addison-Wesley Publishing Compagny (1967).

8. Davies, O.L., "Design and Analysis of Industrial Experiments", New-York : Hafner Publishing Company (1960).

9. Draper, N.R., "Ridge Analysis of Response Surfaces" Technometrics, Vol. 5, N° 4, (1963) pages 469-479.

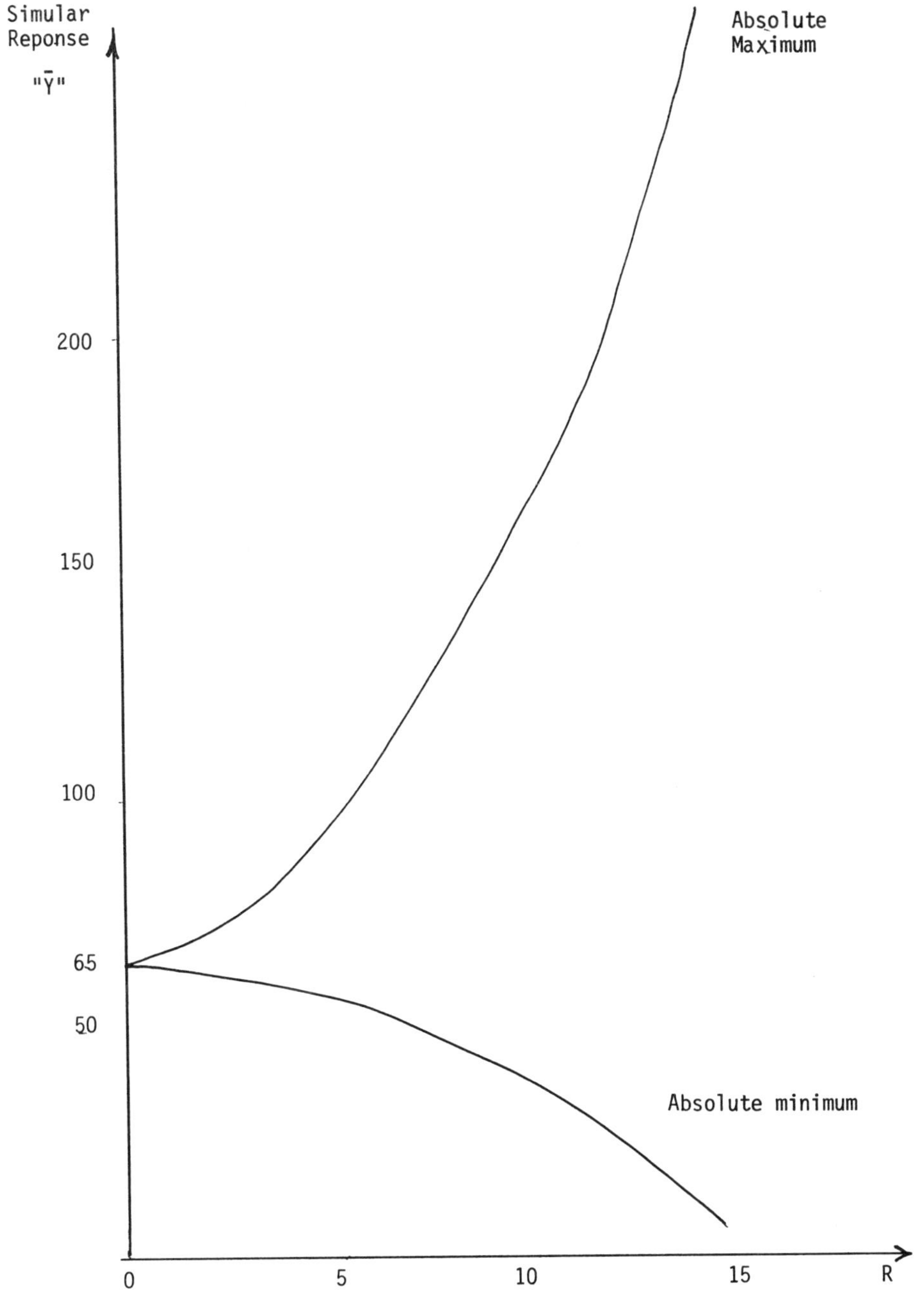

Figure 2 : Loci of stationary values as
R varies

10. Eilon, S. and I.G. Chowdhury, "Due Dates in Job Shop Scheduling" Int. Jour. of Prod. Res., Vo. 14, n°2, (1976) pages 223-227.

11. Fishman G.S., "Concepts and Methods in Discrete Even Digital Simulation" John Wiley & Son, Inc. (1973).

12. Fishman, G.S. "Problems in Statistical Analysis of Simulation Experiments : the Comparaison of Means and the Length of Sample Records". ACM Communications Vol. 10, (1967) page 94.

13. Fishman, G.S., "Grouping Observations in Digital Simulation" Management Science, Vol. 24, N°5, (1978) page 510.

14. Fryer, J.S. "Operating Policies in Mutiechelon Dual Constraint Jobs Shops"; Management Science; Vol. 19, N°9, May(1973) pages 1001-1012.

15. John, R.C. and Draper, N.R. "D.- Optimality for Regression Designs A Review", Technometrics, Vol. 17, n°1 (1975) page 15.

16. Khorshid, M.H., "Etude Experimentale de la Gestion d'Atelier par la Simulation", Thèse de Docteur Ingénieur, Univ. Paris-Dauphine (1978).

17. Kleijnen, J.P.C., "Statistical Techniques in Simulation" Marcel Dekker, New-York, Part II, (1975).

18. Mihram, G.A. "Simulation : Statistical Foundation and Methodology" Academic Press ; New-York, 1972.

19. Naylor, N.H. ; K. Wertz and T.Wonnacott, "Methods for Analysing Data from Computer Simulation Experiments" ACM. Comm, Vol. 10, n°11 (1967) pages 703-710.

20. Nelson, R.T. "Labor and Machine Limited Production Systems "Management Science, Vo. 13, n° 9, May (1967) page 648.

21. Patel, M.S. "Group Screening with More Than Two Stages", Technometrics, Vol. 4, (1962) page 209.

22. Paulson, E. "A Sequential Procedure for Selecting the Population with the Largest Mean from K Normal Populations" Ann. of Math. Stat. Vol. 35 (1964) page 174.

23. Sasser, W.E. ; D.S. Burdic ; D.A. Graham and T.H. Naylor "The Application of Sequential Sampling to Simulation" ACM. COMM., Vol. 13, N°5 (1970) page 287.

24. Satterthwaite, F.E. "Random Balance Experimentation" ; Technometrics Vol. 1, (1959) page 111.

25. Smith, D.E. "An Emperical Investigation of Optimum - Seeking in the Computer Simulation Situation", opns. Res. Vol. 21 (1973) page 475.

SIMBAS

Armando Guerra T.

Depto. de Biofisica, Escuela Nacional de Ciencias Biologicas, I.P.N., Mexico

ABSTRACT

An interactive program (SIMBAS) was developed with the primary goal of allowing an Ecology student with the minimal experience in either mathematics or computers to develop a system model and/or experiment with one of several models stored in disk. The program was written in BASIC for a WANG 2200-S with 12 Kbyts of core.

KEYWORDS

BASIC; computers; models; population dynamics; simulation

The application of mathematics to the modeling of biological phenomena has begun to play an important role in the field of biological sciences in the last years. However because of the sophistication of much of the mathematics involved few of the mathematical concepts and techniques have reached the undergraduated level of instruction. Despite this fact; it seems clear that the future development of many fields of biology will depend heavily upon the capabilities wich are provided by mathematical modeling. There are reasons to believe that mathematics will become a tool as important to biology as they have been to the physical sciences.

In many biological process it has been possible to analize and construct models. Unfortunatelly for the undergraduated most authors employ mathematical techniques which go beyond the typical mathematical background of the undergraduated and may be beyond the training experience of many faculty members.

The object of the program SIMBAS (SIMulation in BASic) is to provide a tool to introduce the student to the concept of dynamic modeling by the use of some simple mathematical techniques.

First the student must analize the system being studied and determine the basic components of the system and the variables on wich they depend. Second the student must stablish the mathematical relationships between the basic components of the system based on intuition; experience and experimental observation. Finally the mathematical relationships must be programed in the computer, and the computer will generate a set of results.

This results are then compared with the experimental data to evaluate the model and make the pertinent corrections (this process is called validation of the model).

Once the model has been validated it can be used to perform experiments to determine the behaviour of the system under a defferent set of parameters. It has been the aim of SIMBAS to help the student in this final steps of modeling.

SIMBAS can solve almost any problem which can be stablished as a set of first order differential equations with a set of initial conditions. Any set of this type can be expressed in the form:

$$y'(t)=f(t,y(t),p1,p2,p3,...,pn) \qquad (1)$$
$$y(0)=y0$$

where y(t)=(y1(t),y2(t),...,yn(t)) is the solution sought. y'(t)=dy/dt is the vector of first derivatives and y(0) is the vector of initial conditions. The quantities p1,p2,...,pn are the parameters which describe the relationship f between y'(t) and y(t); the independent variable t is equalled to time in many of the applications and will be called "time" from now on, even though it could represent some other quantity. the vector y(t) will often be called the "solution vector".
The techniques used to solve the equation (1) are the method of Euler and the classical 4th order Runge-Kutta method. It make it possible that the system of differential equations will not be restricted to some particular form (ie to be linear). These methods are not fool proof and the user must not expect it to solve incorrectly or ill-formulated problems.

The program consists of the main program and a series of subroutines. The main program can be divided in two parts the initialization and the execution-command loop. The initialization part allows the user to input a series of characteristics of the execution and of the model: the printer will be used?, are you asking for numerical output?, are you asking for graphycal output?, the data are included in your model?, what rule of integration will be used (Euler or Runge-Kutta)?.

Any or all of the values stablished in the initialization can be changed by the user in the command loop which consists of a command point where commands are accepted from the keyboard and then transffers to program segments which carry out the commands and then return control to the command point. The options that are available at command point are described later.

The subroutines in which the user can define his/her data are between the lines 150 and 270. In the lines 150-199 the user can write some comments about the model, the line 208 DATA contains the number of differential equations, the initial time, initial conditions, integration interval, time span for the simulation. The line 212 DATA contains the number of variables for graphycal output (if required). The line 216 DATA contains the print interval. The line 250 DATA contains the symbols for the graphycal output, in the lines 251-270 the user can define or ask the values of the parameters of the model. The subroutine that contains the mathematical relationship of the model most be writted between the lines 301-498 in the form: F(i)=parameter-i * Y(i), where parameter-i is one of the parameters of the model (eq. 1). The use of F() and Y() are mandatory.

Next the user has to define the scale of the graphycal output in the lines 804-813 in the form X(i)=Ki*Y(i) where Ki are the scaling factors and the use of X() is mandatory. In the lines 3000-3019 the user can define the variables for numerical output (T6 define the value of the independent variable).
Once the user has writted these lines the model can be loaded in SIMBAS and the user can run the model. The main program provides the interactive link between the user and the subroutines defined for the specific task requested at execution time. It consists of two parts: initialization and command loop. The initialization is carri-

ed out each time the program is started; it is the place where the user enter the values of the execution parameters and if necessary the model parametrs. The user can change any or all at execution time in the command loop segment.

The program asks the questions in such a way that the form of the expected response is obvious. The program first ask the user if he/she wants line printer output. If so the program gives the output in the line printer, this output provides a hard copy for future reference, if not, the output goes to the screen. The program next asks the user if she/he wants numerical output, if so the numerical output comes in the format XXX.XXXXE+XX adjusted for the number of variables in the model. Next the program asks the user if he/she wants graphycal output; in this form the user can select numerical output, graphycal output or both.

After the output form has been selected the program asks the user if the parameters of the model and of the program are in DATA statements (ie in a file); if this file exists the program asks the name of it and the program fetch it from the disk, if not the user must type it in. If the user selects graphycal output, the user can choose to obtain, simultaneously, numerical output of one of the variables of the model. Once the program has obtained all the information, it performs a check for errors and if the program detects some error it will be displayed on the screen with the apropiate message so the user can correct them. If no errors was detected the simulation begins.

The comand loop prints the option list and ask the user to select one of them. There are 18 options, three of them concerned with the running of the model, the rest of, them are for other tasks as you see in the following list: 1.- Line printer, 2.- Numerical output, 3.- Graphycal output, 4.- Integration rule, 5.- Initial time, 6.- Initial conditions, 7.- Integration interval, 8.- Time span for the simulation 9.- Print interval, 10.- Variables for graphycal output, 11.- Graphycal boundaries 12.- Variable for numerical output in the graphycal output, 13.- Continue, 14.- End program, 15.- Abort run, 16.- Reread parameters, 17.- Run again, 18.- Change values of variables or time. When the user selects his/her option the program executes this option and then returns control to the command point.

Once the program is running the user can atract the attention of the command point keying the S.F. 0, when the program recognizes this key returns control to the command point, print a message and the option list. All the dialogue concerned in the command loop is through the screen. After the completion of a run the program gives the user the option to end or continue. If she/he wants to continue the control returns to the command point.

The subroutines that carry out the integration resides between the lines 701-804. According with the rule selected by the user the program selects the Euler algorithm or the Runge-Kutta (4th order, fixed step) algorithm. The plotting subroutine resides between the lines 901-934, as the simulation proceeds the main program interrupts at the proper times to call the plotting subroutine and produce one line of output.

ON THE ESTIMATION OF A DISCRETE PARAMETER IN THE M/M/m/n QUEUEING SYSTEM

Jacob E. Samaan* and Derrick S. Tracy**

*Rehab Group, Inc., Falls Church, Virginia, U.S.A.
**University of Windsor, Windsor, Ontario, Canada.

ABSTRACT

In this paper, two estimators are proposed for estimating the number of servers m for the queueing model M/M/m/n with a finite waiting room with known capacity n. To investigate the properties of one of these estimators, the forward system of equations is written as

$$\frac{d}{dt}P(t) = AP(t)$$

where $P'(t) = [p_0(t),\ldots,p_{m+n}(t)]$ and $p_i(t)$ is the probability that the system is in state i at time t. Here A is a square matrix of order m+n+1, and it is necessary to investigate some properties of the eigenvalues and eigenvectors of A. Simulation is used to examine and compare the two estimators.

KEYWORDS

Discrete parameter estimation; queueing system; forward system of equations; eigenvalues; simulation.

INTRODUCTION

Most papers, in the statistical analysis of queues, have been concerned with the estimation of continuous parameters, namely, the arrival rate, service rate or mean waiting time for a single server or multiserver queueing system. In this case the theory of Billingsley (1961) is directly applicable. However, it may happen that the arrival and service rates of a many servers queueing system are known and estimators for the number of servers are required. Here the problem of estimating discrete parameters, which take on integral values only, should be investigated. In a previous paper, Samaan (1974) discusses the problem of estimating the number m of servers for the Markovian system M/M/m. In this paper the finite capacity Markovian system M/M/m/n is considered. It is assumed that the arrival rate λ, the service rate μ, as well as the capacity of the waiting room n, are known. Two estimates are proposed for estimating m.

MAXIMUM LIKELIHOOD ESTIMATION OF m

Assuming that we observe the arrival time and departure time of each customer up to time T, let a_i denote the total number of transitions from the state i to the state i+1; b_i denote the total number of transitions from the state i to the state i-1; γ_i denote the total time spent in the state i during the time of observation. Also let

$$\mu_i = i\mu \ , \quad 0 \le i \le m-1$$
$$= m\mu \ , \quad 0 \le i \le m+n \ .$$

The log-likelihood function of this sample is

$$L_T(m) = \sum_{i=0}^{m+n-1} a_i \ln\lambda + \sum_{i=1}^{m+n} b_i \ln\mu_i - \sum_{i=0}^{m+n} \gamma_i(\lambda+\mu_i)$$

$$= A\ln\lambda + B\ln\mu - \lambda T + C - \mu S$$

where

$$A = \sum_{i=0}^{m+n-1} a_i \ , \quad B = \sum_{i=1}^{m+n} b_i \ , \quad C = \sum_{i=1}^{m-1} b_i \ln i + \sum_{i=m}^{m+n} b_i \ln m \ , \quad S = \sum_{i=m}^{m-1} i\gamma_i + \sum_{i=m}^{m+n} m\gamma_i \ .$$

The maximum likelihood estimator of m is the value $\bar{m}$ which maximizes the function $L_T(m)$. By the theory of finite differences $\bar{m}$ must satisfy

$$\Delta L_T(\bar{m}) < 0 < \Delta L_T(\bar{m}-1)$$

where

$$\Delta L_T(m) = L_T(m+1) - L_T(m)$$
$$= \sum_{i=m+1}^{m+n} (b_i \ln \frac{m+1}{m} - \mu\gamma_i) \ .$$

Hence, the maximum likelihood estimator of m is the smallest value $\bar{m}$ which satisfies $\Delta L_T(m) < 0$, i.e.,

$$\sum_{i=m+1}^{m+n} (b_i \ln \frac{m+1}{m} - \mu\gamma_i) < 0 \ .$$

By a similar approach as in Samaan (1974), it is possible to prove that when $T \to \infty$, the function $L_T(m)$ has a unique maximum which occurs at the true value of the number of servers.

THE DIRECT METHOD FOR ESTIMATING m:

Let us observe the system for a period of time of length T and let

X_t = the number of customers in the system at time t,

τ_i = the first passage time to state i, and

$p_i(t)$ = the probability that the system is in state i at time t.

$$\hat{m}_T = \eta_T - n$$

is proposed as an estimator for m, where

$$\eta_T = \max_{0 \leq t \leq T} X_t \ .$$

In order to investigate the properties of this estimator, we investigate the probability

$$Q_m(t) = P(\hat{m}_t = m) = P(\eta_t = m+n).$$

We note that

$$Q_m(t) = P(\eta_t = m+n) = P(\tau_{m+n} \leq t) = P_{m+n}(t)$$

and the value of $Q_m(t)$ can be obtained by solving the sequence of forward equations, Saaty (1961),

$$p_0'(t) = -\lambda p_0(t) + \mu p_1(t) \ ,$$

$$p_i'(t) = -(\lambda+\mu_i)p_i(t) + \lambda p_{i-1}(t) + \mu_{i+1}p_{i+1}(t), \ 1 \leq i \leq m+n-2$$

$$p_{m+n-1}'(t) = -(\lambda+m\mu)p_{m+n-1}(t) + \lambda p_{m+n-2}(t)$$

$$p_{m+n}'(t) = \lambda p_{m+n-1}(t) \ .$$

This sequence can be written in matrix form as

$$\frac{d}{dt} P(t) = AP(t) \ , \qquad (1)$$

where

$$P(t) = [p_0(t), p_1(t), \ldots, p_{m+n}(t)]'$$

and A is a square matrix of order m+n+1 in the form

$$A = \begin{bmatrix} -\lambda & \mu_1 & & & & & & & \\ \lambda & -(\lambda+\mu_1) & \mu_2 & & & & & & \\ 0 & \lambda & -(\lambda+\mu_2) & \mu_3 & & & & & \\ & \cdots & \cdots & \cdots & & & & & \\ & & \cdots & \cdots & \cdots & & & & \\ & & & \lambda & -(\lambda+\mu_{m-1}) & \mu_m & & & \\ & & & & \lambda & -(\lambda+\mu_m) & \mu_{m+1} & & \\ & & & & & \cdots & \cdots & \cdots & \\ & & & & & \lambda & -(\lambda+\mu_m) & 0 & \\ & & & & & & \lambda & 0 & \end{bmatrix}$$

The solution of the system (1) is

$$p_{j-1}(t) = \sum_{i=1}^{m+n+1} c_i \, \alpha_j^{(i)} e^{w_i t} \ , \quad j = 1,2,\ldots,m+n+1 \qquad (2)$$

where c_i, $i=1,2,\ldots,m+n+1$, are arbitrary constants, and the w_i's are the eigenvalues of the matrix A, while $\alpha_j^{(i)}$ is the jth component of the ith eigenvector.

Assume that the system is in the state zero at $t = 0$, i.e.,

$$P(0) = (1,0,0,\ldots,0)' .$$

Then the arbitrary constants c_i are determined by

$$\sum_{i=1}^{m+n+1} c_i\alpha_1^{(i)} = 1,$$

$$\sum_{i=1}^{m+n+1} c_i\alpha_j^{(1)} = 0 \ , \ j = 2,3,\ldots,m+n+1 .$$

Now, since the eigenvalues of the matrix A are the solutions of the characteristic equation

$$|A-wI| = 0 ,$$

it is easy to show that one of the eigenvalues is $w_1=0$, and the others, $w_2,w_3,\ldots,w_{m+n+1}$, are all negative and distinct.

Using this property, the solution (2) can be written as

$$Q_m(t) = p_{m+n}(t) = c_1\alpha_{m+n+1}^{(1)} + \sum_{i=2}^{m+n+1} c_i\alpha_{m+n+1}^{(i)} e^{w_i t} . \tag{3}$$

Since $Q_m(t)$ is the probability of the first passage to the state m+n of a Markov chain, which is irreducible and recurrent,

$$Q_m(t) \to 1 \quad \text{as} \quad t \to \infty . \tag{4}$$

Also, since all the w_i's, $i=2,3,\ldots,m+n+1$, are negative,

$$\sum_{i=1}^{m+n+1} c_i\alpha_{m+n+1}^{(i)} e^{w_i t} \to 0 \quad \text{as} \quad t \to \infty . \tag{5}$$

From (3), (4) and (5), we have

$$c_1\alpha_{m+n+1}^{(1)} = 1 .$$

Hence, $Q_m(t)$ can be written as

$$Q_m(t) = 1 + \sum_{i=2}^{m+n+1} c_i\alpha_{m+n+1}^{(i)} e^{w_i t} ,$$

where constants c_i are determined from the relations

$$\sum_{i=1}^{m+n+1} c_i\alpha_1^{(i)} = 1 ,$$

$$\sum_{i=1}^{m+n+1} c_i\alpha_j^{(i)} = 0 , \quad j = 2,3,\ldots,m+n ,$$

$$\sum_{i=2}^{m+n+1} c_i\alpha_{m+n+1}^{(i)} = -1 .$$

PROPERTIES OF $\hat{m}_T$

Since

$$\lim_{t \to \infty} Q_m(t) = 1 ,$$

the estimator $\hat{m}_T$ is a consistent estimator. And consequently, the accuracy of $\hat{m}_T$ increases on increasing T .

Also, since $\hat{m}_T = \eta_T - n$, the accuracy of $\hat{m}_T$ increases when η_T , the maximum number of customers in the system during the time of observation, increases. And $\hat{m}_T$ takes the true value of m when at some $t \leq T$, all the servers are busy and the waiting room is full, i.e. $\eta_T = m+n$.

But when the utilization factor $\rho = \lambda/m\mu$ is large, the expected number of customers in the system is large and $\hat{m}_T$ is more accurate, see Table 2. By the same argument we see that the accuracy of $\hat{m}_T$ increases by decreasing m, see Table 3.

An interesting result is that the accuracy of $\hat{m}_T$ increases by increasing both λ and μ by the same ratio R, keeping ρ fixed. In fact, if λ and μ are replaced by $R\lambda$ and $R\mu$, then the matrix A is also replaced by RA for which the eigenvalues are $Rw_1, Rw_2, \ldots, Rw_{m+n+1}$. It is easy to see that this does not affect either the components of the eigenvectors or the constants $c_1, c_2, \ldots, c_{m+n+1}$. Here $Q_m(t)$ takes the form

$$Q_m(t) = 1 + \sum_{i=2}^{m+n+1} c_i \alpha_{m+n+1}^{(i)} e^{Rw_i t} ,$$

which shows that increasing λ and μ by the ratio R is equivalent to increasing t by the same ratio.

COMPARISON OF $\hat{m}_T$ AND $\bar{m}$

Since the analytical investigation of the properties of $\bar{m}$ seems to be difficult, we present here some numerical examples to compare m_T and $\bar{m}$.

For this purpose the system is simulated and a hundred samples are generated. Table 1 is the computer output listing the values of m_T, a_i, b_i and γ_i .

TABLE 1

$\lambda = 1.5$, $\mu = 1.0$, m = 4, n = 5, T = 100, LL = 8

I	A(I)	B(I)	T(I)	P(I)
0	29	0	15.326025	0.221040
1	58	28	35.715317	0.331560
2	32	57	25.259476	0.248670
3	24	31	11.916247	0.124335
4	16	24	8.036907	0.046626
5	5	16	2.733356	0.017485
6	2	5	1.372139	0.006557
7	1	2	0.261306	0.002459
8	0	1	0.035896	0.000922
9	0	0	0.0	0.000346

These are listed for $i = 1,2,\ldots,LL$ for one sample, where in the computer output, $LL \equiv \eta_T$, $A(I) \equiv a_i$, $B(I) \equiv b_i$, $T(I) \equiv \gamma_i$, and $P(I) \equiv p_i(t)$, the stationary probability that the system is in state i. Each of the random variables $\frac{a_i}{\lambda T}$, $\frac{b_{i+1}}{\lambda T}$ and $\frac{T_i}{T}$ converges stochastically to p_i.

Now let $\hat{Q}$, $\bar{Q}$, $\hat{p}$ and $\bar{p}$ be the estimated values of the probabilities $P(\hat{m}_T=m)$, $P(\bar{m}=m)$, $P(m-1 \le \hat{m}_T \le m+1)$ and $P(m-1 \le \bar{m} \le m+1)$ respectively, based on a hundred simulated samples. Thus

$$\hat{Q} = \frac{N_1}{100}, \quad \bar{Q} = \frac{N_2}{100}, \quad \hat{p} = \frac{N_3}{100} \quad \text{and} \quad \bar{p} = \frac{N_4}{100}$$

where N_1 is the number of times the estimator $\hat{m}_T$ has the true value m and N_3 the number of times $\hat{m}_T$ lies between $m - 1$ and $m + 1$ in the 100 samples. Similar definitions apply to N_2 and N_4.

To see the effect of increasing ρ on the accuracy of $\hat{m}_T$ and $\bar{m}$, Table 3 presents the values of $\hat{Q}$, $\bar{Q}$, $\hat{p}$ and $\bar{p}$ for m=3, n=5, T=100, for different values of ρ.

TABLE 2

$m = 3, \quad n = 5, \quad T = 100$

λ	μ	ρ	$\hat{Q}$	$\bar{Q}$	$\hat{p}$	$\bar{p}$
0.3	1.0	0.1	0.00	0.32	0.01	0.89
0.9	1.0	0.3	0.03	0.41	0.09	0.81
1.5	1.0	0.5	0.43	0.70	0.67	0.93
2.1	1.0	0.7	0.94	0.80	0.99	1.00
1.0	3.33	0.1	0.00	0.46	0.00	0.97
1.0	1.11	0.3	0.04	0.40	0.16	0.86
1.0	0.67	0.5	0.28	0.53	0.51	0.83
1.0	0.5	0.66	0.75	0.71	0.87	0.93

We notice that for large values of ρ, both $\hat{m}_T$ and $\bar{m}$ are satisfactory, while for small values of ρ, $\bar{m}$ is better.

Table 3 presents the estimated probabilities for $\hat{m}_T$ and $\bar{m}$ for λ=1.5, μ=1.0, n=5, T=100, for different values of m.

TABLE 3

$\lambda = 1.5, \quad \mu = 1.0, \quad n = 5, \quad T = 100$

m	$\hat{Q}$	$\bar{Q}$	$\hat{p}$	$\bar{p}$
2	0.97	0.91	1.00	1.00
3	0.41	0.69	0.63	0.87
4	0.66	0.35	0.20	0.80
5	0.02	0.32	0.03	0.81
10	0.00	0.01	0.00	0.01

In Table 4, we present the estimated probabilities when m = 3, for different values of n .

TABLE 4

$\lambda = 1.5, \quad \mu = 1.0, \quad m = 3, \quad T = 100$

n	$\hat{Q}$	$\bar{Q}$	$\hat{p}$	$\bar{p}$
2	0.99	0.55	1.00	0.72
3	0.94	0.68	1.00	0.90
5	0.48	0.76	0.67	0.87
7	0.11	0.71	0.18	0.86
9	0.04	0.74	0.08	0.91

REFERENCES

Billingsley, P. (1961). A Statistical Inference for Markov Processes. University of Chicago Press, Chicago.

Saaty, T. L. (1961). Some stochastic processes with absorbing barriers. J. Roy. Stat. Soc., Ser.B, 23, 316-334.

Samaan, J. E. (1974). On the maximum likelihood estimation of the number of servers for the M/M/m queueing system. Közlmények, 12, 29-51.

SIMULATING NEW MATERIALS PROCESSES REQUIRED FOR THE REALIZATION OF AN ALTERNATIVE ENERGY SOURCE

Alexander Kran

IBM East Fishkill Laboratories
Hopewell Junction, NY 12533
(914) 897-7142

ABSTRACT

An advanced software engineering methodology is being defined to permit interactive model definition and computer simulation. The method is being applied to solve complex problems in the field of new materials process developments. The scope of the problem requires complex computational capability and the ability to reproduce the behavior of a system as it operates in time, and where changes of state occur at discrete time intervals. In other words, there is complete compatability between a high-level, interpretive programming language and a discrete-event simulator.

The objective of this effort is to demonstrate the feasibility of low-cost, high-quality, single-crystal silicon. It is a key to a timely realization of an alternative energy source, achieved through direct conversion of solar to electrical energy.

KEY WORDS

Simulation, Modeling, APL, Discrete Event Simulation, Applied System Research, Industrial Cybernetics.

INTRODUCTION

Energy is the foundation of industrialized society. It is the prime mover of technology, and our progress is closely linked to its continued availability at affordable prices. All fossil fuels are finite in supply. We are already into the second half of this planet's liquid fossil fuel era, estimated to span a mere 100 years. These resources soon will be irretrievably depleted.

Solar energy is the only renewable energy source available to us in bountiful quantities, yet safe densities. Converting solar energy through the photovoltaic effect in semiconductors is particularly attractive, because without moving parts it provides direct conversion from free solar to valuable electrical energy. It is based on the absorption of solar photons and collection of these carriers by p-n junction. Since the photovoltaic solar cell is in principle not a heat engine, no thermal pollution occurs.

Photovoltaic solar energy conversion has potential for large-scale terrestrial applications. However, at this point a gating factor to producing significant amounts of electricity from solar cells is the expense of making sufficiently pure, crystal silicon for cell fabrication (1).

This paper describes an advanced simulation methodology for achieving practical solutions to complex problems encountered in large-scale silicon production. These are mostly technical problems discovered in the search for better materials, larger volumes, and tighter specifications in growing crystals. One goal is to achieve silicon material that also can be used in VLSI electronic applications. Computer simulation can at times allow researchers to explore means of achieving these goals without the time and expense of building experimental systems.

The method of solution may be analytical and/or numerical. Our approach combines analytical computer tools and simulation techniques. The results are practical application software systems for decision analysis, predictive modeling, experimentation, evaluation, and planning.

APPROACH TO COMPUTER SIMULATION

The advanced computer simulation methodology consists of an open-ended set of software systems, all coded in and linked by a high-level, interpretive programming language. APL has been chosen as the programming language because it is supported by our host system and has the broad capability for this effort. However, our longer-term objective is to couple simulation with reality by progressing from a computer and mathematical model of crystal growth to actual process control. Therefore, the concept must remain independent of host processor or programming language.

For low-cost applications, it has been shown that silicon can be crystallized directly from the melt in sheet form (2), or grown in cylindrical shape (3) by what is known as the Czochralski growth method. Simulated in Fig. 1, this is the most widely applied system in industry.

Therefore, the real system, or source of observable data, is a crystal grower designed to produce zero-defect silicon materials. Here polycrystalline silicon is melted in a crucible. Then a seed crystal is dipped into the rotating crucible and slowly (~10cm/hr) withdrawn, forming a single-crystal cylindrical ingot.

In simulating Czochralski crystal growth, a number of experimental frames are of practical interest. They are handled by different versions of the model, which may be created interactively. This is because performing experiments with the real system is expensive and time consuming. In addition, the model must be suitable for simulating the performance of next generation equipment, which has been designed to a set of performance objectives but is still untested as a manufacturing tool.

Two application software systems are described. The first supports decision analysis through interactive, discrete event simulation. It will be reviewed next.

Since the host system supports APL, it permits us to use APLGPSS, which is a GPSS (General Purpose Simulation System) look-alike residing in the APL workspace. With this situation, the simulation program can be readily decomposed so that functional program elements may be individually manipulated if necessary. The simulator divides the crystal-pulling process into a number of tasks as depicted in Fig. 2, a simplified version of the model logic flow. The process of crystal growth is represented as a transaction moving through the model.

Tasks to be completed include those performed by operators, and include cleaning the crystal puller and loading the polycrystalline silicon. These are strictly stochastic processes, characterized by the availability and arrival pattern of the technician and the service process which describes the activity to be performed. There are other tasks with stochastic characteristics that are also governed by the mathematics and physics of crystal growth. These are activities associated with polycrystalline silicon melt-down, seeding, neck and shoulder growth, constant diameter crystal body growth, and tail-off.

A probability of equipment or process failure is assigned to each task. The model determines if there is sufficient time for an entire growth cycle. For instance, in case of a two shift per day operation, many hours of production time can be wasted. Polycrystalline silicon melt-down, seeding, neck and shoulder growth is completed, and constant diameter crystal body growth commences if there is time. In addition, decision rules can determine, for instance, if a relatively short crystal turning poly is to be melted back or pulled out for its single-crystal portion. One growth cycle is complete once the crystal is cooled and removed from the furnace, and routine maintenance is performed. Random equipment failure based on statistics and decision rules which apply to crystal processing are also part of the model.

A valuable feature of this model is that a log of simulated crystal growth activity may be obtained as the model is executing. A brief example is shown in Table I. Time is kept three ways: a 24-hour clock, total elapsed time for the entire simulation period, and elapsed time between each crystal grown.

The second software system described here is a model designed to project the limits of technology and crystal growth equipment. Interactive executive functions help transact the analysis. Output data from this first software tool becomes compatible input data to the second one.

Using a cybernetic approach, manufacturing technology and process parameters are translated into economic and financial data by this model, establishing the baseline, or state-of-the-art. From this baseline, technological projections and sensitivity studies provide specific direction for subsequent plateaus in process development. This procedure is systematically iterated, analogous to a closed-loop feedback cycle.

Each production unit is described to the simulation system in terms of about 70 parameters within three categories: processing technology, resources (capital equipment, manpower, etc.), and raw materials.

Category 1 consists of processing-related parameters such as crystal growth rate and crystal diameter, which are available from the discrete-event simulation model.

Category 2 comprises direct cost items, such as crystal growth system cost. Also included are direct personnel to ensure efficient factory operation.

Category 3 contains polysilicon cost and services and supplies. This completes the list of specifically named items. The remainder--overhead, general and administrative expenses, and profit--are defined as percentages relating to other direct-cost items. Figure 3 shows how the production unit input data is arranged. The results from executing the production unit of the model are shown in Fig. 4. Output consists of the major contributing factors to silicon material cost, overhead cost, general and administrative expense, and profit. Annual revenue and material throughput also are computed in terms of value added.

RESULTS AND CONCLUSIONS

Present day simulation tools can provide answers to difficult questions regarding new materials processes required to realize alternative energy sources. In this effort, the simulation models have been run to explore means of improving silicon productivity and cost.

In general, this concept has proven valuable to decision makers in demonstrating the feasibility of low-cost solar power. It is aimed at providing guidance in the following areas:

1. Making decisions regarding short-term technological milestones, for example, the potential of rapid crystal growth.

2. Computer-assisted, interactive strategic planning, for example, to determine scenarios for technological directions in energy research.

3. Formulation of technological options for achieving future objectives, for example, to determine what sources of energy can be considered viable.

Simulating new materials processes can be a low-cost practical way of researching and ultimately achieving practical solar energy.

REFERENCES

Kran, A. (1980). The outlook for thin solar cells. _Proceedings of the Fourteenth IEEE Photovoltaic Specialists Conference_, San Diego, CA.

Schwuttke, G. H., Ciszek, T. F., Kran, A. (1978). Low cost silicon for solar energy conversion applications. IBM Journal of Research and Development, 22, 4.

Kran, A. (1978). System simulation and cybernetics affirm potential of low cost solar power. Proceeding of the Fourth International Congress on Cybernetics and Systems, Amsterdam, The Netherlands.

CLOCK	TIME ELAPSED		STATUS
00:01	1	0	SIMULATION STARTING
00:21	21	20	FURNACE CLEANED
00:21	21	0	TIME FOR GROWTH CYCLE?
00:21	21	0	•MELT-DOWN STARTED
02:13	133	112	•MELTING COMPLETE
03:00	180	47	•SEEDING,NECK,SHOULDER DONE
08:48	528	348	POLY GROWTH... TERMINATE RUN
08:48	528	0	•BODY GROWTH COMPLETED (58.00 CM)
09:20	560	32	•TAPER COMPLETED
10:43	643	83	••CRYSTAL COOLED « REMOVED
12:26	746	103	MAINTENANCE DONE... RESTART

TABLE 1. Log of Simulated Crystal Growth

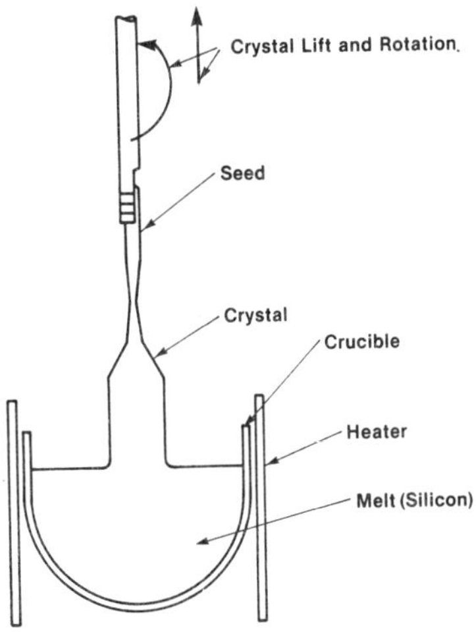

FIG. 1. Czochralski Crystal Growth

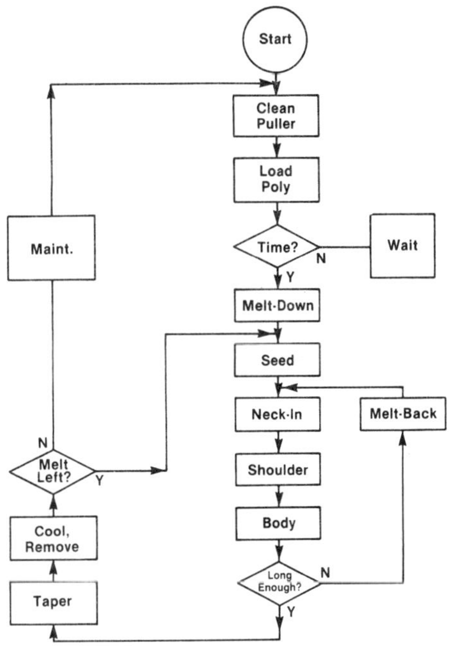

FIG. 2. Crystal Growth Model Logic Flow

PRODUCTION UNIT 1 'GROW ' INPUT PARAMETERS

	1980	1982	1984	1986
1. NUMBER OF MACHINES.......	1.00	1.00	1.00	1.00
2. CRYSTAL DIAMETER.......CM	10.00	12.00	12.00	12.00
4. GROWTH RATE ·/· OF MAX...	25.00	30.00	30.00	30.00
5. ACTUAL GROWTH RATE..CM/HR	9.83	10.77	10.77	10.77
6. MAXIMUM CHARGE.........KG	25.00	25.00	25.00	25.00
20. WAFER THICKNESS........MM	.85	.65	.63	.58
22. BATCH SIZE...............	1.00	2.00	3.00	4.00
25. YIELD LOSS............·/·	20.00	20.00	20.00	20.00
31. EQUIPMENT COST.......DOLS	250000.00	250000.00	250000.00	250000.00
32. EQUIPMENT LIFE......YEARS	7.00	7.00	7.00	7.00
33. INTEREST RATE.........·/·	10.00	10.00	10.00	10.00
34. ENERGY TO OPERATE......KW	60.00	60.00	60.00	60.00
35. EQUIPMENT AVAILABILITY·/·	65.00	70.00	72.00	75.00
37. AREA REQUIRED........SQ M	10.00	10.00	10.00	10.00
41. NUMBER OF SUPERVISORS....	.10	.10	.10	.10
42. ......AT A SALARY OF DOLS	25000.00	25000.00	25000.00	25000.00
43. NUMBER OF ENGINEERS......	.20	.20	.20	.20
44. ......AT A SALARY OF DOLS	20000.00	20000.00	20000.00	20000.00
45. NUMBER OF TECHNICIANS....	.33	.33	.33	.33
46. ......AT A SALARY OF DOLS	12500.00	12500.00	12500.00	12500.00
51. POLYSILICON COST..DOLS/KG	45.00	25.00	17.00	10.00
52. CRUCIBLE COST........DOLS	200.00	200.00	200.00	200.00
53. SERV/SUPPL/PARTS.....DOLS	55.00	60.00	60.00	60.00
54. POWER COST.......DOLS/KWH	.05	.05	.05	.05
61. OVERHEAD AS·/·OF PRSONNEL	100.00	100.00	100.00	100.00
62. ..AND AS·/·OF RAW MTL CST	10.00	10.00	10.00	10.00
63. GEN*ADM XP AS·/·OF DC+O/H	20.00	20.00	20.00	20.00
64. PROFIT BEFORE TAX.....·/·	15.00	15.00	15.00	15.00
65. WORKWEEK............HOURS	168.00	168.00	168.00	168.00
66. CONVERSION EFFICIENCY.·/·	14.00	15.00	16.00	18.00
67. NUMBER OF WEEKS PER YEAR.	50.00	50.00	50.00	50.00

FIG. 3. Production Unit Input Data

PRODUCTION UNIT 1 'GROW ' OUTPUT VALUES

	1980	1982	1984	1986
FROM PREVIOUS PU.....DOLS/SQ M	.00	.00	.00	.00
DIRECT COST..........DOLS/SQ M				
LABOR	11.44	5.14	4.84	4.28
EQUIPMENT CAPITAL RECOVERY	12.16	5.46	5.14	4.54
MATERIALS	98.15	41.43	27.31	14.79
SERVICES/SUPPLIES	31.49	13.08	9.90	7.83
SUBTOTAL:	153.24	65.11	47.18	31.44
OVERHEAD COST........DOLS/SQ M	24.41	10.59	8.56	6.54
MANUFACTURING COST...DOLS/SQ M	177.65	75.69	55.74	37.98
G*A EXPENSES.........DOLS/SQ M	35.53	15.14	11.15	7.60
TOTAL COST(THIS PU)..DOLS/SQ M	213.18	90.83	66.89	45.57
PROFIT...............DOLS/SQ M	31.98	13.62	10.03	6.84
SELLING PRICE........DOLS/SQ M	245.15	104.46	76.93	52.41
TOTAL SELLING PRICE..DOLS/SQ M	245.15	104.46	76.93	52.41
ENERGY CAPACITY COST (DOLS/KW)	1751.09	696.38	480.79	291.16

FIG. 4. Result of Simulating Production Unit

A SIMULATION MODEL FOR THE YUCATAN PENINSULA

Manfred Kudlek[+], Ingo Schwarz[x]

[+]Fachbereich Informatik, Universität Hamburg,
Schlüterstraße 70, D-2000 Hamburg 13, W Germany
[x]Wissenschaftszentrum Berlin, Internationales Institut für
Vergleichende Gesellschaftsforschung
Steinplatz 2, D-1000 Berlin 12, W Germany

ABSTRACT

A simulation model to study the future problems in food and water resources is considered. For this purpose the relatively small and isolated area of the Yucatán peninsula is chosen. In this model rural and urban population, food and water supply, as well as tourism and industry is studied. It turns out that a major for future problems is water resource.

KEYWORDS

Simulation, system dynamics, model, Yucatán, rural and urban population, tourism, food and water resource.

INTRODUCTION

Mexico in many aspects is a fast developing country. Her tourists are attracted by many different landscapes as well as by the extraordinary remains of her cultural history, and her still living Indian tradition. Since quite many years, and especially since the exploitation of vast oil fields in the region of the Gulf of Mexico, her economic development is of great interest also for the industrialized nations. Another factor of growth in which Mexico is at the top level in the world is her population increase. Indeed, 3 % is one of the highest factor of all nations. For this reason two major problems will arise in future, namely food and water supply which already are problems for some areas and groups of population. Looking upon the Yucatán peninsula, and having in mind the population and tourism development, water supply seems to be a severe long term problem.

Water resource management faces problems of increasing number and complexity. The scope of such problems has expanded enormously in recent years, particularly as a result of the rapid increase in population, the expansion of industry, the spread of urbanization, the emergence of new kinds of water use, and the increase in disposal of personal income. Irrigation also has become a key issue of water resource development on a worldwide level. In addition, water quality has deteriorated seriously having implications on water supply. Although relatively large volumes of water may be available, they often are suitable for rather limited ranges of use only.

These problems are shared by almost all nations, and the records of the recent UN World Water Conference offer the best evidence in this regard. The main purpose of this UN conference was the analysis of alternative courses of action which would assist the world in avoiding a water crisis of global dimensions by the end of this century. However, there is no global answer to questions concerning the outlook for water, or the likelihood, degree and consequences of an eventual water shortage. A continental or global picture of long-range expectations will necessarily consist of different local and regional outlooks with respect to foreseeable physical, technological and socio-economic patterns of water management.

To study such problems a relatively small and isolated area of Mexico was chosen, namely the Yucatán peninsula consisting of the federal states of Campeche, Yucatán, and Quintana Roo. For the model the method of system dynamics is used.

PROBLEM AREA

The Yucatán peninsula is a relatively flat karst region covering an area of about 140000 km^2 with an annual rain fall of 1200 mm. The population was .65 millions in the rural and .45 millions in the urban area in 1970.

There is a large part of Indian population, with a somewhat self-supporting agricultural economy. Small ejidos satisfy the daily need for food consisting mainly of corn and beans. There is also another form of agriculture in major plantations of henequen (sisal) with industrial manufacturing. Another rapidly increasing economical factor is tourism, especially along the Carribean coast, such as the complex of Cancun which might spread farther to the south, or the island of Cozumel.

There is a rather high increase of population by birthrate, being of the same amount as for the rest of Mexico, and there is also a rapid increase in the number of tourists. This has as a consequence an increase in demand for water, which usually is first satisfied in tourist centres and in towns, at the cost of the rural population.

As the ground of Yucatán is karst, after raining water directly passes through the soil into water reservoirs coming to surface only in cenotes, which are partially interconnected with each other. Only some part of the rain is filling cisterns. Wells with access to these reservoirs in open areas and tourist centres might very well decrease the water level over a long period to such an extent that rural population might be forced to move to urban areas. This problem will be simulated using the aggregated, heuristic method of system dynamics. Of course, with more information being available, refinements of such models will be possible.

REFERENCE MODE

According to the situation described, a reference mode shall provide a basic estimated behaviour of the system. It represents a specification of the topic of study and consists of graphical representations of the problems described by the model. The time span of such model developments are approximately 50 years, as demographic changes are relatively slow and develop over almost one generation's time. As figure 1 shows the expected development is some rapid increase and saturation in the urban population and the number of tourists, an increase up to a maximum and then a decrease of the rural population, and a growing decrease of the water resource.

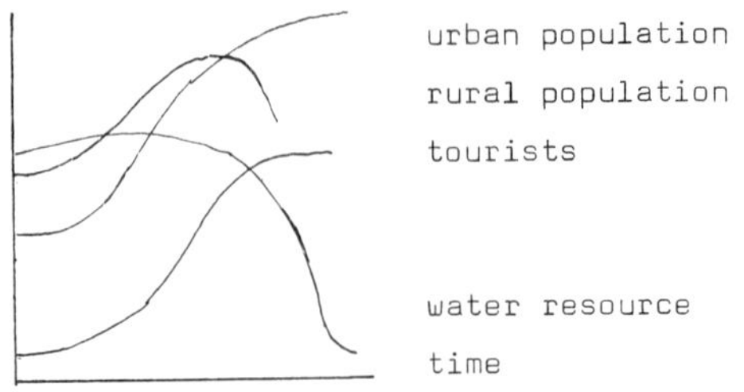

Fig. 1 Reference mode

SYSTEM STRUCTURE

Two populations are distinguished, namely the self-supporting rural population, and the urban one. The first one is rather dependent on their own agricultural production. Therefore the amount of agriculturally used and partially arable land together with its water supply is the most important system element for that part.

For the urban population jobs from industrial capital and from tourism are important. The labour market situation has some impact on migration. As capital investment in tourism and industry is mainly exogenous, an overall macro-economic model of Mexico itself would have to be incorporated. A link to the rural population is given by the food supply from that part of the population.

One can compare food and water needs with respect to population and tourism in order to correlate available and needed goods. These basic system elements and relations determine the following rough causal loop structure :

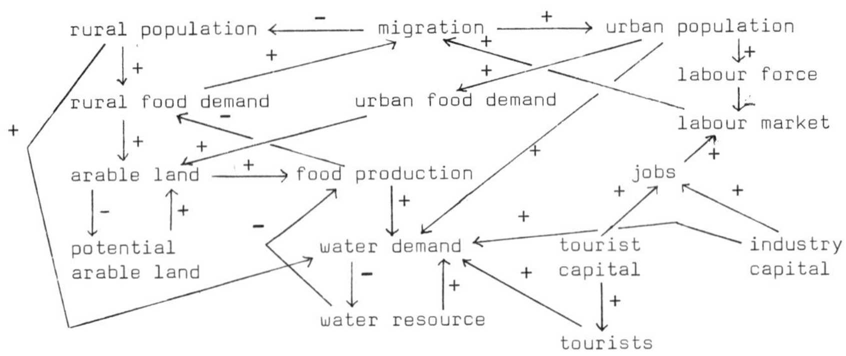

Fig. 2 Causal loop diagram

As may be seen from this rather aggregated structure, several negative loops are dominating. These loops are 'goal seeking' or stabilizing structures. If, on the other hand, there is a limiting element involved such as potential arable land or total water resource, the goal of water supply and agricultural production is limited. This in fact might cause overshoot and collapse situations if there are positive feedbacks involved in the model. One major positive feedback of this kind is the water, migration, urban population loop (figure 3).

If urban population (together with tourism) increases, water demand increases too. This in turn affects the water resources negatively. An increase in number of wells, or their deepening, close to urban areas causes a delayed decrease in the water level of the few cenotes in rural areas. This finally causes the rural population to migrate to urban areas. Here the cycle is closed and the effect becomes even worse.

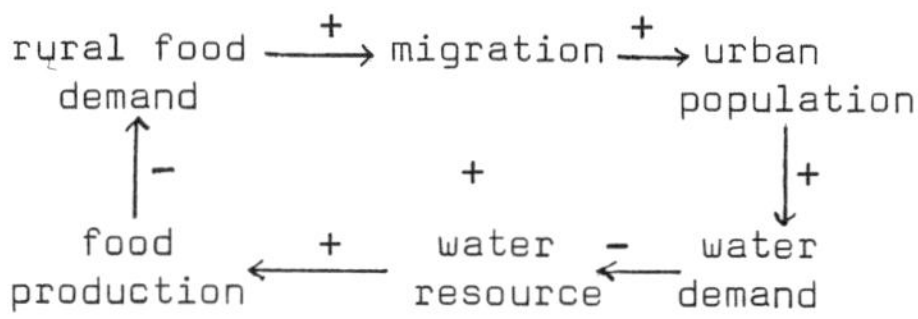

Fig. 3 The positive water resource-demography cycle

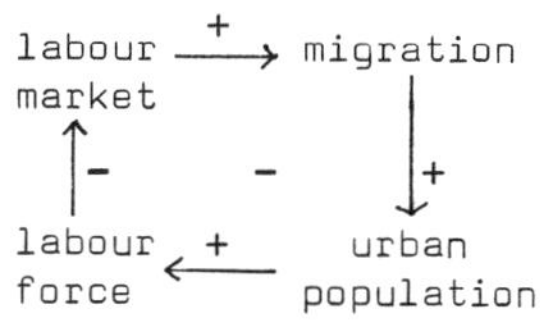

Fig. 4 The negative labour population cycle

The symptoms might be the refuge of rural population into urban areas which will lead to severe social problems there. There is no strong emigration from the Yucatán peninsula to other areas since population increases simultaneously in most other Mexican states.

Most of the self-regulating loops in the structure include the migration element which is coupled with the major positive feedback. Therefore the self-regulating effect will always be changed by the positive accelerating behaviour. The only balancing negative feedback mechanism is the urban population-labour force-labour market-migration loop (figure 4). This, however, might turn out to be too weak in order to have sufficient impact on the overall system's behaviour.

THE SIMULATION APPROACH

Within the frame of the economic development and the reactions by demographic changes due to scarce water resources, the model cannot be purely represented in an analytical form. A simulation approach will be necessary to provide quantitative information about the system's development during some time span. Taking the criteria mentioned above - long time horizons, a rather qualitative description of the system with small amounts of reasonable data, and delayed inter-connected structures - the simulation approach would be best suited by the system dynamics modelling technique.

System dynamics is a well-known and often used approach for modelling and analyzing complex dynamic systems. This modelling approach can be understood relatively easily. The models can be handled in a simple way by computer simulation, e.g. using SIMULA or DYNAMO. The simulation language DYNAMO stands in direct relation to the DYNAMO-flow diagram notation which provides a graphic representation of complex differential systems. The modelling approach is characterized by two major phases - conceptualization and representation — which contain several steps, most of which are relevant for any kind of modelling approach.

After defining the problem area and being familiar with the system under study - actually the most important steps of conceptualization - the dynamic behaviour of interest - the reference mode - have to be identified. This step includes as well the time horizon of the study and thereby the size of the time constants in processes to be explicitly represented in the model.

The system's boundary, as well as the basic mechanisms together with the selected important system elements determine a rough causal loop diagram. This diagram represents the basic relations among the selected elements. To get more detailed model assumptions, a DYNAMO-flow diagram has to be developed which then will give quantitative informations. It needs basic quantitative input data. The last step is the transformation of this flow diagram into a DYNAMO-program. Of course, any other suitable programming language or software system may be used for this step.

Instead of a DYNAMO-flow diagram a DYNAMO program is presented for one of the models. For some of the elements table functions are assumed, e.g. for investment in industry and tourism.

```
NOTE RURAL POPULATION
L POPL.K=POPL.J+DT*(BIRTHL.
X JK-DEATHL.JK-MIGI.JK)
R BIRTHL.KL=POPL.K*BRL
R DEATHL.KL=POPL.K*DRL
N POPL=POPLI
C POPLI=650000
C BRL=.05
C DRL=.02
NOTE URBAN POPULATION
L POPU.K=POPU.J+DT*(BIRTHU.
X JK-DEATHU.JK+MIGI.JK)
R BIRTHU.KL=POPU.K*BRU
R DEATHU.KL=POPU.K*DRU
N POPU=POPUI
C POPUI=450000
C BRU=.04
C DRU=.01
NOTE FOOD SECTOR
A FODML.K=POPL.K*AFDL
C AFDL=1
A FODMU.K=POPU.K*AFDU
C AFDU=2
A FPRO.K=ARLND.K*ADR
C ADR=60
A FGAP.K=FODML.K+FODMU.K*
X UFDMF-FPRO.K
C UFDMF=.2
NOTE WATER SECTOR
L WATDRN.K=WATDRN.J+DT*(
X RAIN.JK-WTDML.JK-WTDMU.JK
X -WTDMT.JK-WTDMI.JK-
X WTDMA.JK-WTEX.JK)
N WATDRN=WATDRI
C WATDRI=100
R RAIN.KL=RAINC
C RAINC=10
R WTDML.KL=POPL.K*WTUL
C WTUL=1E-6
R WTDMU.KL=POPU.K*WTUU
C WTUU=4E-6
R WTDMT.KL=POPT.K*WTUT
C WTUT=2E-5
R WTDMI.KL=ICAP.K*WTUI
C WTUI=.01
R WTDMA.KL=FPRO.K*WTUA
C WTUA=2E-6
R WTEX.KL=WTEXC
C WTEXC=.8
NOTE INDUSTRY SECTOR
L ICAP.K=ICAP.J+DT*(INVI.
X JK-DEPI.JK)
N ICAP=ICAPI
C ICAPI=150
R INVI.KL=TABHL(INVIT,
X TIME.K,0,30,5)
T INVIT=6/9/15/21/15/9/6
R DEPI.KL=ICAP.K*ATI
C ATI=.03
A ILF.K=ICAP.K*ICLR
C ICLR=100
NOTE TOURIST SECTOR
L TCAP.K=TCAP.J+DT*(INVT.
X JK-DEPT.JK)
N TCAP=TCAPI
C TCAPI=100
R INVT.KL=TABHL(INVTT,
X TIME.K,0,30,5)
T INVTT=5/10/15/20/15/10/5
R DEPT.KL=TCAP.K*ATT
C ATT=.04
A POPT.K=TCAP.K*TCF
C TCF=500
A TLF.K=TCAP.K*TCLR
C TCLR=50
NOTE AGRARIAN SECTOR
L ARLND.K=ARLND.J+DT*(
X NEWL.JK-DEVL.JK)
N ARLND=ARLNI
C ARLNI=14000
R DEVL.KL=ARLND.K*DEVF
C DEVF=.01
R NEWL.KL=CLIP(NEEDL.K,0,
X NOTLND.K,0)*CLIP(1,0,
X WATDRN.K,0)
A NEEDL.K=FGAP.K*CLIP(1,0,
X FGAP.K,0)/ADR
L TOTLND.K=TOTLND.J-DT*
X DEVL.JK
N TOTLND=TOTLNI
C TOTLNI=25000
A NOTLND.K=TOTLND.K-ARLND.K
NOTE MIGRATION SECTOR
A POPLF.K=POPU.K*LFP
C LFP=.25
A LF.K=ILF.K+TLF.K
A LFF.K=LF.K-POPLF.K
A MIGL.K=FGAP.K*MIGFL*
X CLIP(1,0,FGAP.K,0)*
X CLIP(0,1,NOTLND.K,0)
C MIGFL=1
A MIGU.K=LFF.K*LATT*
X CLIP(1,0,LFF.K,0)
C LATT=.5
R MIGI.KL=MIGL.K+MIGU.K
NOTE SPECIFICATIONS
SPEC DT=.1,LENGTH=30,
X PLTPER=1
PLOT POPL=L/POPU=U/POPT=T/
X WATDRN=W
PLOT ARLND=A/FPRO=F/FGAP=G/
X WATDRN=W
```

REFERENCES

Forrester, J. W. (1968). Principles of Systems. Cambridge, Mass.

Pugh, A. L., III (1970). DYNAMO II user's manual. Cambridge, Mass.

Schwarz, I. (1978). A generic regional model for interdisciplinary impact analysis. 2 nd International MAB-IUFRO Workshop on Tropical Rain Forest Ecosystem Research, (Djakarta), 244-259.

Doehring, D. O., and Butler, J. H. (1974). Hydrogeologic Constraints on Yucatán's Development. Science, 186, 591-595.

COMPUTER ANGST

Mark G. Graff* and Roger B. Johnson**

*Philip Morris Research Center, Richmond, Virginia
**Department of Comparative Literature, University of Southern Mississippi, Hattiesburg, Mississippi

ABSTRACT

Computer angst accounts for many widely-held doubts and reservations concerning computers in general and cybernetics research in particular. This unfocussed anxiety--which is rational, but not reasonable--is in part a result of the computer's early image and applications. Primarily, however, it results from the "grotesque" combination of the computer's anthropomorphic qualities and its electro-mechanical nature.

KEYWORDS

Fear of computers; aversion to computers; computers and society.

INTRODUCTION

Computers (considered as an institution) are distasteful--and threatening--to many people. It is not difficult to find examples of this distaste: *Time* and *Spiegel*, preachers and disc jockeys feel themselves on safe ground when they indicate a judicious reluctance to embrace this or that aspect of "the wired society".[1]

Objections are heard from those concerned about invasion of privacy, from people worried about losing their jobs to computers, from citizens who feel that computers should not "control" nuclear missiles or answer the telephone. Such issues abound. Many are substantive, and pose important questions. We will discuss them no further--but will argue, instead, that systematic consideration of such matters requires a more fundamental inventigation of human reaction to the computer itself. It is our thesis that computers provoke, in some humans, a reaction easily (inadequately) described in the word "angst" (dread). The pattern of recognition coincides with another, well-documented response: that is, human reaction to the grotesque.

[1]The term is, of course, James Martin's.

To make our point, we intend to take an indirect route, establishing that the computer is readily to be perceived as grotesque. After briefly discussing grotesquerie in the abstract, we'll cite computer traits which might qualify as grotesque. This, followed by a simple explication of Freudian angst, shows how computer angst could arise. Concluding, we will speculate about the impact of this reaction upon society in general, and the humanities in particular.

THE GROTESQUE

The study of the Grotesque is a long and honorable chapter in the history of European belles-lettres, being a subdiscipline of aesthetics. Although the term has had many applications, like other category tags, it has, since Renaissance art, indicated a melting or combining of otherwise imcompatible forms: animal and human, or flora and fauna. Historically, there is both an element of playfulness and an element of horror in this dissolution.

The perception of grotesquerie results, frequently, in confusion, for one effect of it is to dissolve the borders of our world, distorting the forms we suppose are natural.

> [That which is grotesque] challenges the conventional boundaries between male and female, sexed and sexless, animal and human, large and small, self and other, and consequently between reality and illusion, experience and fantasy, fact and myth.[2]

But human beings, although confused by the grotesque, are sometimes attracted to it, basically for three reasons:
1) it is often playful; 2) it is extraordinarily intricate: 3) it presents a similitude of truth--often, a disquieting one. These are also characteristics of many modern-day computing systems.

An analogous example is presented by the refined, complex, and disturbing novels of Kafka, wherein the prose structure itself reveals a "continuous onslaught of exactly rendered details, which is capable of no rational explanation, which forever eludes one's grasp, and to which one cannot get "accustomed".[3] The much-trumpeted "Information Revolution" has surely resulted in such onslaughts.

It is in this particular sense that computers--and, to a degree, player pianos--are grotesque, because they effect a dissolution of the borders between man and machine. It is the conjunction of the human and the non-human, in a kind of "mutilated anthropomorphism", that evokes computer angst.

COMPUTERS AS FREAKS

Many of the anthropomorphic qualities of computers are readily discernible: today's systems evince a wide range of identifiable personae. This seems natural; computers are, for the most part, designed by people, for people to use, for tasks which people have conceived (and, often, have performed). Thus, the "instructions" available to programmers, the "languages" designed to facilitate the use of these instructions, the algorithms used in passing information between people and computers--all these tend to be anthropomorphic. Computers in this sense reflect the humanity of their inventors, as others tools do.

[2]Fiedler, 1978, p. 24.

[3]Kayser, 1968, p. 147.

Moreover, many computer programs are specifically designed to imitate certain facets of human behavior. Such programs, Joseph Weizenbaum's ELIZA for example, which can mimic a Rogerian psychoanalyst, occasionally achieve a "suspension of disbelief", and the establishment of a strange kind of "relationship" between the program and its user. A disruption of this sort of exchange, due to a computer failure or program fault, is particularly disturbing.

It is also true, however, that computers differ from human beings in many ways not commonly understood.

They are, for example, temporally out-of-scale with human beings: computers can appear to be either amazingly fast or stunningly slow, depending upon the appropriateness and quality of a particular application. In a similar fashion, the "giant brains"[4] of the Forties and Fifties can now be thought of as either reduced to chip-size, or diffused over continents--such is the range of the computer's size and effect.

The traditional volatility of computer memories, necessitating recurrent "bootstrapping", is another non-human-like quality. The computer is, resultingly, a sort of eternal tabula rasa, continually foregetting and relearning, like some persevering amnesiac.

Perhaps most notably: the information manipulated by these machines is almost always imperceptible to humans. A disk pack upon which is stored Shakespearean sonnets, or baseball scores, is indiscernible from one filled with random sequences of "ones" and "zeros". Without specially-designed "interfaces", a person can't generally tell whether a computer is active at all!

ANGST

In his "twenty-fifth lecture", Sigmund Freud described "angst" as an emotion that "itself needs no description; everyone has personally experienced this sensation. . . at one time or another."[5]
It is in fact difficult to define. "Angst" is a German word usually translated as "anxiety"--but this is correct only in a narrow technical sense. A word with similar impact upon speakers of English is "dread."

Freud distinguished between "rational" and irrational" angst. Of the rational variety he said, "The occasion of it [angst], i.e. the objects and situations about which anxiety is felt, will obviously depend to a great extent upon the state of a person's knowledge and feeling of power regarding the outer world."[6]

This is in consonance with our assertion: the computer's grotesquerie contributes to a dissolution of important world-structuring boundaries, and threatens an invalidation of world-view. It is this threatened loss of control--dissolution--which results in computer angst.

---

[4]The term is Edmund Berkeley's.

[5]Freud, 1924, p. 400.

[6]Freud, 1924, p. 401.

The strength of the reaction is, of course, variable; it is obvious that knowledge tends to diminish angst. But knowledge of the processes and capabilities of the computer is not sufficient to allay this dread; required is a knowledge that can be translated into power, control over that part of the world which the computer seems to influence.

IMPACT OF COMPUTER ANGST

What impact does computer angst have upon society? Consider the following questions.

What accounts for the steady stream of fictitious computer villians, in books and movies such as "Colossus: The Forbin Project", "The Demon Seed", "2001: A Space Odyssey", and The Adolescence of P-1? What causes the prevalent computer-related uneasiness of middle-level managers which IBM, Honeywell, and M.I.T. have sought to relieve with "Know Your Computer" seminars? Why do so many artists ignore the computer as a medium for self-expression?

We assert that computer angst is at work here. Of course, its precise effect is difficult to ascertain.

Regarding the effect upon the humanities specifically, the evidence we have is chiefly anecdotal, and therefore diverse. Indeed, the impact is hardly to be measured in precise terms, because it is exhibited in negatives, a choice "not to", an avoidance of one avenue in the visualization of a project. In interviews with humanists, we encounter a vaguely recognized suspicion of the computer's "inappropriateness". Furthermore, to complicate this issue, angst is often accompanied by ignorance of the computer's use, for anything but "computing" in the narrowest sense.

What is the result? Retired persons and mid-career housewives, seeking continued education, glide toward art appreciation and French--almost never towards education in computers. We use the computer to teach more economically in some instances, but not radically better. Data for history are being stored, and occasionally organized, with the aid of computers,[7] but no resultant revolution of historiograhy seems to be on the horizon.

The impact of computer angst, then, is culturally to isolate computer activities, and to diminish the impact of computers themselves.

CONCLUSION

Although exacerbated by the computer's early image as a "giant brain" used by the military to control the terrible new weapons of the Forties and Fifties, computer angst is deeply rooted in the contrast between the electro-mechanical nature and anthropomorphic qualities of computers. This dissonance, which is amplified by the common personification of these machines, is sometimes perceived as grotesque; and it is in the wake of this perception that computer angst arises.

Of angst in general, Freud wrote, "One is tempted. . . to assert that the development of anxiety is never expedient."[8] Computer angst is surely inexpenient and costly; but more importantly, it stands in the way of cool, rational examination of the computer's proper role in our society, and of its effects upon the quality of life.

---

[7] Computerworld, 1980, p. 37.

[8] Freud, 1924, p. 402.

REFERENCES

__________(1980). "Henry VIII's Justices on the take? CPU checks." Computerworld, October 6, 1980, p. 37.

Fiedler, Leslie (1978). Freaks: Myths and Images of the Secret Self. Simon and Schuster, New York.

Freud, Sigmund (1924). A General Introduction to Psychoanalysis. New York: Washington Square Press, 1960. (A translation, by Joan Riviere, of the Boni & Liveright edition of 1924.)

Kayser, Wolfgang (1968). The Grotesque in Art and Literature. Translated by Ulrich Weisstein. P. Smith, Gloucester, Massachusetts.

USER SATISFACTION AND TELEPROCESSING

Dr. Kornel Terplan

Computer Sciences International
Germany

ABSTRACT

After outlining the high dependency of human day-to-day productivity on computing services, the measurability of human satisfaction will be discussed in greater detail. The major focus of the paper will be on introducing results of the latest technological development for making computing services more humane. Particular interest will be paid to discussing the effects achievable.

KEYWORDS:

System Modelling; Recent Advances in Computer Science; Improving Human Satisfaction.

INTRODUCTION

Computing became an essential part of our day-to-day functioning as well as the basis of corporate communication and transactions using advanced teleprocessing techniques. Data processing management should ensure that computer operations never become the bottleneck, but rather an effective tool that aids the process and progress of a corporation by providing information wherever required.

Computer Sciences have gone through a revolutionary development over the last 30 years. However, the human aspects of user satisfaction have not yet reached the proper consideration.

The primary objective of this presentation is to shed light on how computers and teleprocessing can improve human life and enhance use satisfaction. During my presentation, I will attempt to outline how computers are utilized at the present time and discuss how some of these services can be improved. After introducing the most important parameters of measuring user satisfaction, as response time, interactive throughput time, and availability, the Stimulus-Acknowledgment-Response-Model will be used for describing the system and user behaviour in teleprocessing applications.

PRESENT STATE OF THE ART

Today's teleprocessing activities can be characterized as follows:

CAPACITIY PLANNING - There are well proven techniques for the batch and partially proven for mixed (batch and interactive) environments. However, the majority of these methods does not take into consideration the dynamic behaviour of teleprocessing. Thus, user satisfaction cannot properly planned.

CONFIGURATION TOOLS - There exist only a few tools for teleprocessing systems design. The methods used are elementary and inadequate. The tools widely implemented try to optimize global system utilization and topology. However, user convenience at day-to-day work is not considered.

RESOURCE UTILIZATION - Performance parameters of central resources utilization are favoured against service level requirements. The strategy and tactic for improving systems efficiency are dealing with global tuning instead of responding to user requirements and offering better service than before.

DATA TRANSMISSION STATISTICS - Performance improvement should become an ongoing activity. Statistics on data transmission traffic by lines, concentrators and front-end-processors may help to adapt systems resources to user requirements.

REACTION TO SERVICE LEVEL DEGRADATION - At the present time service level degradation will be indicated by the user, instead of teleprocessing management. Thus, the duration between degradation and its elimination is longer than necessary.

HANDLING OF SYSTEM TROUBLES - Users should be alerted about systems breakdowns and their expected duration as soon as possible. Thus, users can adapt to unexpected conditions and can dynamically reschedule their activities. This kind of alerting information is still a very rare commodity.

Increasingly, the question of user/teleprocessing is becoming a key issue in the study of performance and objectives are being set for user service levels. The service levels, representing human satisfaction, consist of both qualitative and quantitative components. Fig. 1 shows a summary of performance parameters (GSA, 1978) that characterize the software development and systems operation processes. Also the classification in terms of process and product and effectivity and efficiency, respectively, are included. After careful consideration it can be recognized that central resources utilization profiles are favoured against service level requirements.

To properly manage to a set of service level objectives both qualitative and quantitative aspects must be considered. A set of quantitative measures must be selected and installed. Some means of identifying the qualitative fasters should also be established.

Summarizing the most important aspects of Fig. 1 the quantitative service level measures can be categorized as measures of timeliness, accuracy, availability and cost.

TIMELINESS - has to do with how closely the delivery time for a particular product can be met. It can range from a few seconds (response time) to hours (throughput time), even to months and years (in the case of a major development).

ACCURACY - is a measure of the errors that make their way through the system to the user. Usually, they are caused mostly by human mistakes, less by machine malfunction. The results are often reruns demanding additional resource usage. Also timeliness problems are usually involved. On the development side, accuracy problems generally relate to the failure of the development group to meet functional specifications. The result is redesign demanding additional resource usage and manpower.

AVAILABILITY - is preferred against reliability, that is fairly a technical measure, only. The user is not interested in percentage of uptime, MTBF and MTTR. He is only concerned about receiving service when required.

COST -is a sensitive criteria to the users. They tend not to accept excessive costs and/or when costs significantly vary from one period to the next. Charging users on the basis of an application related unit opposing resource based chargeback may help in solving problems in this area. In this case the user can control his data processing expenses.

TOOLS FOR MEASURING AND IMPROVING USER SATISFACTION

There are different levels for measuring service level parameters that may influence user satisfaction (Terplan, 1980). The measurement activities can be grouped as follows:

- technical analysis level
- technical control of the teleprocessing network
- performance analysis and
- performance control level.

Technical Analysis Level

The tools selected for this level are responsible for investigating the technical state of the components in the teleprocessing network once failures have been reperted. Most devices use the RS-232-interface for data collection.

Technical Control Level

The tools selected for this level are responsible for continuous technical control of the teleprocessing network. The majority of devices uses either the RS-232 interface or modems for data collection purposes. Using continuous technical control the response time to technical disturbances can be reduced. Thus, availability, accuracy and partially also timeliness can be improved.

Performance Analysis Level

The tools available in this category are primarily dealing with timeliness measures. The timeliness can be characterized by the response time of the system. Line load and front-end-processors throughput capabilities may influence timeliness to a great extent. There are different tools for collecting response time data: hardware monitors, software monitors (accounting), communication monitors. Figure 2 shows the data capturing locations in teleprocessing networks for various monitors.

Performance Control Level

There are three tools, that are capable of monitoring large and very large networks on continuous base.

NET/ALERT (Avant-Garde, 1980) - NET/ALERT is a tool for operations management. Its purpose is to continuously monitor a telecommunications system, display status of all critical elements of the system, alert operators to trouble conditions, record statistics for measurement and analysis, prepare performance reports for management.

MS-109 (Tesdata, 1980) - The MS-109 is a multi-component system designed to be flexible enough to satisfy a majority of teleprocessing network measurement requirements. It consists of components which interface to an AMS system, components which interface to the customer teleprocessing network, and components which interface both of these other categories. This allows measurement of the TP environment at the remote (end user) site in addition to the local (data center) site.

CTX-100 (Commtex, 1979) - CTX-100 has been developed for facilitating network management functions. These activities can be grouped for achieving tactical and strategic control. All measurement and evaluation steps are based on a central model, called SAR (Stimulus-Acknowledgement-Response). The system response will be accomplished in two steps. The "stimulus" is the input from the user; the "Response" is the output from the computer caused by the stimulus. The "Acknowledgement" is the output from the computer, or the network, which occurs after the stimulus, but before the response. The acknowledgement may convey some information to the user (e.g. stimulus was received and is being processed).

CONCLUSION

Using the latest technique introduced previously, computing operations will become more humanized. This major breakthrough will result in greater user satisfaction by:

- greater insight into network behaviour
- faster response for user inquiries
- early warning of potential bottlenecks
- automatic diagnosis by teleprocessing manager
- action instead of reaction
- matching response time with the internal information processing capabilities of the user.

After carefully examing present operational practices, it becomes apparent that the field of teleprocessing will go through drastic changes in the decade. If the human element - the user - is carefully considered while technological advancements are implemented, then they will yield not only greater user satisfaction but also a more efficient, a more profitable and effective tool to serve the needs of human resources-oriented management.

REFERENCES

1. TITN (1980), DPMS product overview, France
2. Tesdata Corp. (1980), AMS product overview, USA
3. Avant-Garde Inc. (1980), NET/ALERT product overview, USA
4. Commtex Inc. (1980), CTX-100 product overview, USA
5. Terplan, K. (1980): Measuring and improving the Performance of Teleprocessing Systems,
   ECOMA 80 London, Proceedings P. 101-108
6. GSA (1978), Management guidance for developing and installing an ADP performance management program, Washington, USA.

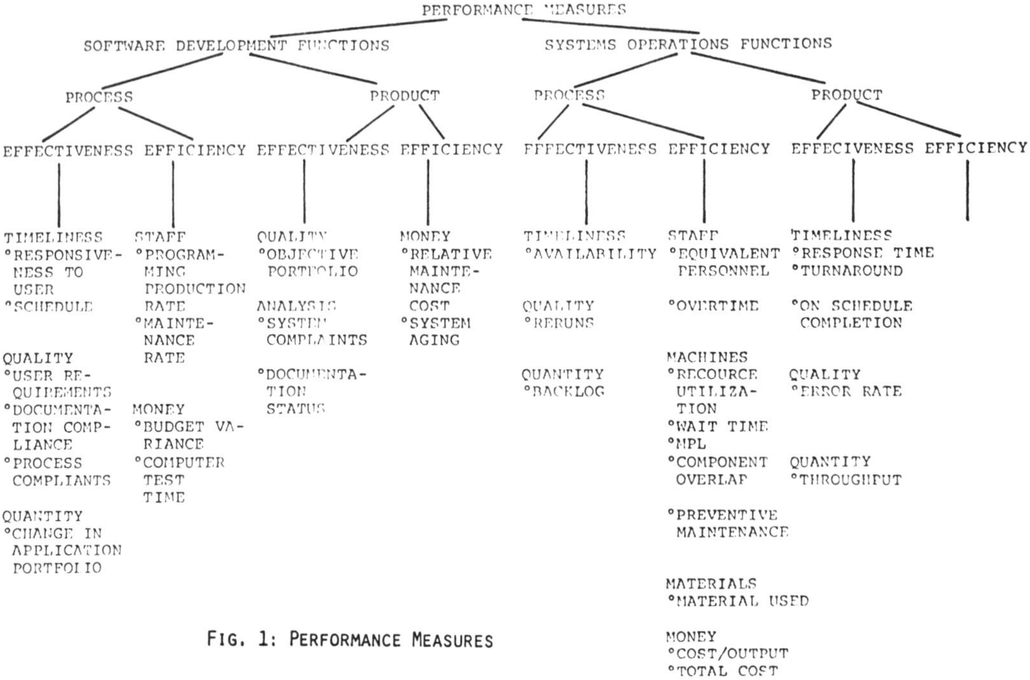

FIG. 1: PERFORMANCE MEASURES

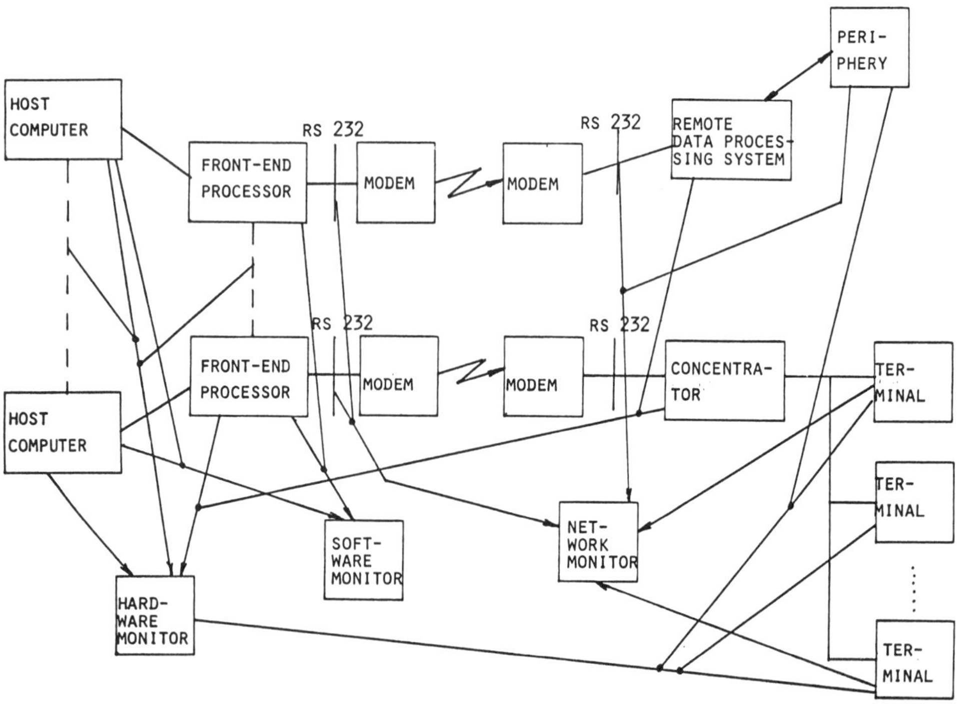

FIG. 2: DATA COLLECTION LOCATIONS

USER- AND ADMINISTRATOR- FRIENDLY ARCHITECTURE FOR INTERACTIVE SOFTWARE DEVELOPMENT

H. Kitagawa*, T. L. Kunii*, M. Azuma**, and Y. Mizuno**

*Department of Information Science, Faculty of Science, The University of Tokyo
7-3-1, Hongo, Bunkyo-ku, Tokyo 113, JAPAN

**Software Product Engineering Laboratory, Nippon Electric Company Ltd.
4-14-2, Shiba, Minato-ku, Tokyo 108, JAPAN

ABSTRACT

Human factors of an interactive software development tool FRIEND are discussed. FRIEND attains both the friendliness for tool users and the friendliness for tool administrators. These two types of friendliness are discussed and illustrated from the architectural viewpoint. The architecture is based on mappings between the user specification level and the database level.

KEYWORDS

Interactive software development tool; friendliness for tool users; friendliness for tool administrators; specification database; nested table data model.

1. INTRODUCTION

This paper discusses human factors of an interactive software development tool named FRIEND. FRIEND is a powerful aid for requirement analysis and software design. FRIEND supports creation, modification, verification, analysis, and maintenance of very large requirement and design specifications. The global architecture of FRIEND is depicted in Fig. 1. At graphic display terminals, system users who are actually software developers can interactively create a variety of requirement and design specifications including forms, graphs, and tables. Those specifications are verified and stored in the internal specification database in a unified way. The users can freely modify, retrieve, and analyze the specifications through the database management function and several utility programs. They can also easily get desirable hard-copied report documents.

The main feature of FRIEND is attainment of the following two types of friendliness.
(1) Friendliness for tool users. Tool users can specify and analyze requirements and designs from their own viewpoints in terms of a variety of human friendly representation forms such as forms, graphs, and tables.
(2) Friendliness for tool administrators. Tool administrators can manage internally stored specification data uniformly using only one type of data model to easily detect and remove redundancy and to preserve database integrity.

The database management function of FRIEND supports the nested table data model

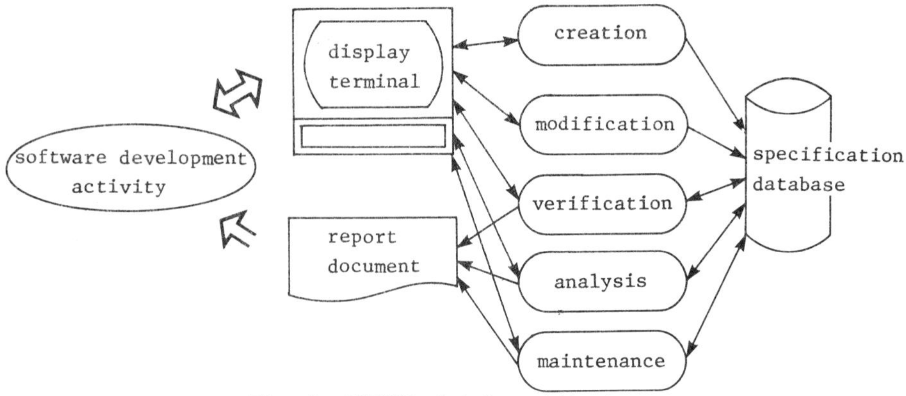

Fig. 1 FRIEND global architecture

(Kitagawa and Kunii, 1980a, 1980b). The nested table data model is a high-level, powerful, and flexible data model. The first friendliness for tool users is attained by the combination of flexibility of the nested table data model and direct handling of two-dimensional images such as forms, graphs, and tables through graphic display terminals. The use of the nested table data model is a key to the attainment of also the second friendliness for tool administrators.

Historically, a number of software development tools have been constructed and commercialized up to the present. Among them, SADT (Ross and Schoman, 1977; Ross, 1977) and ISDOS (Teichroew and Hershey, 1977) are most famous ones. However, the aforementioned two types of friendliness are attained in neither of them. In SADT, no database facility for uniformly managing specifications is provided. Therefore, the second friendliness is, of course, not achieved. In addition, the first friendliness is not fully attained, since SADT has no capability of supporting user views and analyzing specifications. By contrast, ISDOS has database management and report generation facilities. However, the user interface of ISDOS has very little in user friendliness. For example, users have to write specifications in a formal sequential language PSL. Specifications friendlily represented in two-dimensional forms , graphs, and tables cannot directly be handled. Moreover, users are not allowed to freely utilize data stored in the specification database, since their data usages are limited to those expressed by predefined PSA commands.

In section 2 of this paper, requirements for FRIEND are summarized. Those requirements are met in FRIEND by identifying two architectural levels, that is, the *user specification level* and the *database level*, and by achieving proper mappings between them. The downward mapping from the user specification level to the database level is explained in section 3. The converse upward mapping is illustrated in section 4. Section 5 is the conclusion.

2. REQUIREMENTS FOR FRIEND

The objective of FRIEND is a friendly support of creation, modification, verification, analysis, and maintenance of requirement and design specifications. At least the following three items are basic requirements for FRIEND to attain this objective.

Multiple User Views and Multiple Representation Forms

Software developers specify and analyze requirements and designs usually with certain views in their minds. Indeed user views are full of variety. For example, software architecture can be modeled from the viewpoints of functional decomposition, data flow, control flow, and so on. In addition, there is one more dimension of variety. That is, user friendly representation forms of specifications differ drastically depending on specified data types. For example, graphs fit to represent model structures, forms are suitable for specifying descriptive data, and tables are appropriate for listing homogeneous data. Therefore, FRIEND has to handle both the variety of user views and the variety of user friendly representation forms.

Analysis and Verification

To attain reliability of software systems, analysis and verification of their specifications are fruitful and have to be managed in FRIEND. As mentioned in the previous paragraph, requirements and designs are specified from various viewpoints and described in several representation forms. To analyze and verify the specifications, cross-references to those heterogeneous specifications are frequently necessary within the system. Therefore, to facilitate such system management, those heterogeneous specifications have to be managed homogeneously in the internal database of FRIEND.

Life Cycle Management

Software systems are tuned and evolved according to changes of their application and hardware environments. Requirement and design specifications are not only necessary for initially developing a software system. They are also useful for maintaining the system in the changing environments and for supporting design activities for systems of the next generation. To keep up with the tuning and evolution of the target system, FRIEND has to flexibly manage changes of specified data types, user views, user representation forms, and so on, and has to be readily tunable to ad hoc data usages and data analyses.

These requirements are met in FRIEND by setting two architectural levels, that is, the user specification level and the database level, and by achieving proper mappings between the two levels. (Fig. 2) At the user specification level,

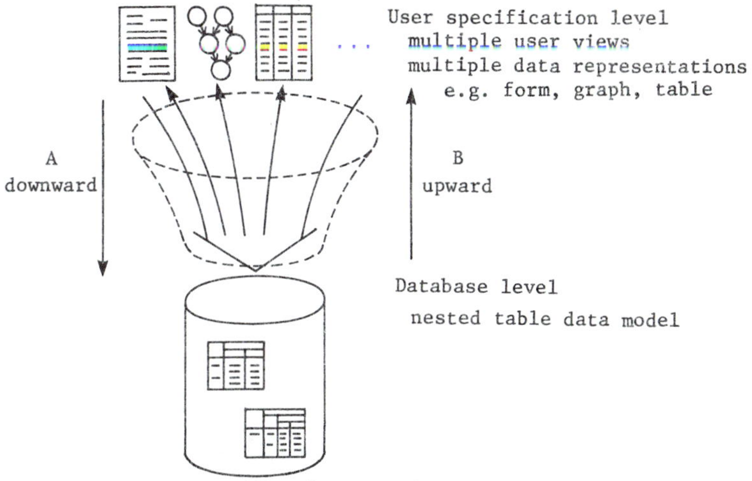

Fig. 2 FRIENDliness

multiple user views and multiple representation forms are supported. On the contrary, at the database level, such a variety of specifications are uniformly managed by means of one and only one data model, namely the nested table data model. The nested table data model is powerful enough to represent diversified specification data in a unified way, and is flexible enough to support multiple user views and to manage ad hoc data usages and analyses. In the nested table data model, data are represented in terms of nested tables. A nested table is a table with nests of columns and rows, and sample nested tables are shown in Fig. 3. Nests of columns and rows make it possible to directly represent both hierarchies of data and many to many relationships. Therefore, data described in the conventional data models (Sibley, 1976) (i.e., the relational model, the hierarchical model, and the network model) and data represented in forms, graphs, and tables can be directly and uniformly handled in the nested table data model. Thus, various objects at the user specification level can be readily mapped down to nested tables at the database level as indicated by arrow A in Fig. 2. On the other hand, the uniformly stored specification data can be freely retrieved and analyzed from several viewpoints and outputted in appropriate representation forms. In the nested table data model, a complete set of nested table operations is provided for manipulating nested tables. By means of them, nested tables fit for users' requirements of data usage can readily be defined through freely editing and joining nested tables stored in the specification database. Therefore, they facilitate the upward mapping complying with the users' requirements as indicated by arrow B in Fig. 2. The downward mapping and the upward mapping are explained with examples in the following section 3 and section 4, respectively.

FILE-SPECIFICATION-TABLE

FID	FNAME	ORG	REC-LENGTH	BLOCK-LENGTH	FIELD-SPECIFICATION			
					NAME	DESCRIPTION	PIC	KEY
F001	SALES FILE	S	30	600	OID	PURCHASE ORDER ID	9(5)	N
					DATE	ORDER DATE	9(6)	N
					CUID	CUSTOMER ID	X(4)	N
					COID	COMMODITY ID	X(8)	Y
					QTY	QUANTITY	9(5)	N
					FILLER	FILLER	X(2)	N
F002	INVENTORY MASTER FILE	S	50	600	COID	COMMODITY ID	X(8)	Y
					CNAME	COMMODITY NAME	X(25)	N
					PRICE	PRICE	9(6)	N
					QOH	QUANTITY ON HAND	9(7)	N
					FILLER	FILLER	X(4)	N
⋮	⋮	⋮	⋮	⋮	⋮	⋮	⋮	⋮

(a)

PROGRAM-SPECIFICATION-TABLE

PID	PNAME	CHART-SPEC	FILE-USAGE			SUBPROGRAM-ID
			FID	I/O-CLASS	ACCESS-MODE	
P001	SALES SUMMARY PROGRAM	P001C	F001	IN	S	P020
			F002	IN	S	P021
			F005	OUT	S	P105
			F101	OUT	S	
P002	INVENTORY UPDATE PROGRAM	P002C	F001	IN	S	P020
			F002	IO	S	P105
			F102	OUT	S	
⋮	⋮	⋮	⋮	⋮	⋮	⋮

(b)

CHART-SPECIFICATION-TABLE

CID	CNAME	CHART-SPEC						
		GRAPH-SPEC					SEMANTICS	
		ID	TYPE	ORIG	DESTN	SYMBOL	TYPE	ID
P001C	SALES SUMMARY GENERATION	1	NODE			1	PROGRAM	P001
		2	NODE			2	FILE	F001
		3	NODE			2	FILE	F002
		4	NODE			2	FILE	F005
		5	NODE			3	FILE	F101
		6	ARC	2	1	4	D-FLOW	D001
		7	ARC	3	1	4	D-FLOW	D002
		8	ARC	1	4	4	D-FLOW	D025
		9	ARC	1	5	4	D-FLOW	D026
⋮	⋮	⋮	⋮	⋮	⋮	⋮	⋮	⋮

(c)

Fig. 3 Nested tables

3. DOWNWARD MAPPING

As explained in the previous sections, requirement and design specifications expressed in several representation forms and specified from diversified viewpoints are supported at the user specification level of FRIEND, and they are uniformly mapped down to nested tables at the database level. Let us see some examples of the downward mapping. Suppose a form shown in Fig. 4 is necessary for file specification or has actually been used for manual documentation. This form can be directly mapped down to nested table FILE-SPECIFICATION-TABLE shown in Fig. 3(a), and data represented in this form are readily stored in this nested table. Data expressed in graphs and charts can also be represented in nested tables. For example, a simple chart shown in Fig. 5 are internally stored in a form of nested table CHART-SPECIFICATION-TABLE shown in Fig. 3(c).

FILE-SPECIFICATION		FID : F001	
FNAME : SALES FILE		ORG : S	
REC-LENGTH : 30	BLOCK-LENGTH : 600		
FIELD-SPECIFICTION			
NAME	DESCRIPTION	PIC	KEY
OID	PURCHASE ORDER ID	9(5)	N
DATE	ORDER DATE	9(6)	N
CUID	CUSTOMER ID	X(4)	N
COID	COMMODITY ID	X(8)	Y
QTY	QUANTITY	9(5)	N
FILLER	FILLER	X(2)	N

Fig. 4 File specification form

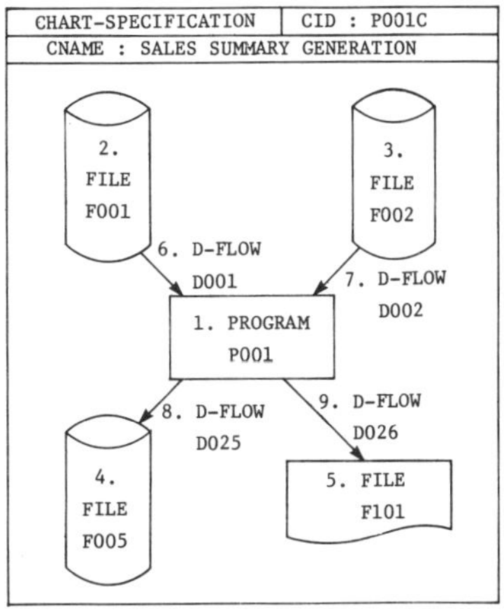

Fig. 5 Chart sheet

The above-mentioned mapping is automatically performed by system programs of FRIEND. Therefore, the users can interact with the system directly in terms of two-dimensional images such as forms and graphs through graphic display terminals. However, practical display screens have physical limitations of scale. In case manipulated two-dimensional images are larger than the screen, they may be decomposed into smaller parts and can be handled piece by piece by means of the window facility. Figure 6 shows screen images to be used for specifying files with the window facility in the prototype FRIEND system (Azuma and co-workers, 1980). In FRIEND, graphs can also be directly handled based on a new graph formalism named the R-graph theory. Its details are reported in Buchmann and Kunii (1979), Harada and Kunii (1979), and Kunii and Harada (1980).

FILE-SPECIFICATION-IMAGE 1

FID	:	F001
FNAME	:	SALES FILE
ORG	:	S
REC-LENGTH	:	30
BLOCK-LENGTH	:	600

NEXT IMAGE N

(a)

FILE-SPECIFICATION-IMABE 2

FIELD-SPECIFICATION

NAME	DESCRIPTION	PIC	KEY
OID	PURCHASE ORDER ID	9(5)	N
DATE	ORDER DATE	9(6)	N
CUID	COUSTOMER ID	X(4)	N
COID	COMMODITY ID	X(8)	Y
QTY	QUANTITY	9(5)	N
FILLER	FILLER	X(2)	N

NEXT IMAGE E

(b)

Fig. 6 Window facility

4. UPWARD MAPPING

In FRIEND, a variety of specifications are uniformly stored in the specification database as explained in the previous section. Conversely, FRIEND has to allow each user to retrieve and analyze the uniformly stored data from his own viewpoint and to output them in the most appropriate representation forms. This requirement is attained by deriving a nested table fit best for each application from stored nested tables using the nested table operations, and by outputing its contents in desirable representation forms by means of suitable utility programs. Here, two examples are given to illustrate the creation of such application dependent nested tables. However, details of the nested table operations are separated into a forthcoming paper.

Suppose a sample specification database consists of the three nested tables shown in Fig. 3. As the first example, let us define a report document on file usages in the registered programs. To do this, we have only to define a new nested table FILE-USAGE-TABLE by applying the nested table operations to nested table PROGRAM-SPECIFICATION-TABLE as shown in Fig. 7 and to select a proper utility program to print its contents in a desirable data layout. This report document is useful for analyzing ripple effects on programs caused by file modifications.

PROGRAM-SPECIFICATION-TABLE						
			FILE-USAGE			
PID	PNAME	CHART-SPEC	FID	I/O-CLASS	ACCESS-MODE	SUBPROGRAM-ID
P001	SALES SUMMARY PROGRAM	P001C	F001	IN	S	P020
			F002	IN	S	P021
			F005	OUT	S	
			F101	OUT	S	P105
P002	INVENTORY UPDATE PROGRAM	P002C	F001	IN	S	P020
			F002	IO	S	
			F102	OUT	[illegible]	P105

FILE-USAGE-TABLE						
	FILE-USAGE					
	PROGRAM				USAGE	
FID	PID	PNAME	CHART-SPEC	SUBPROGRAM-ID	I/O-CLASS	ACCESS-MODE
F001	P001	SALES SUMMARY PROGRAM	P001C	P020	IN	S
				P021		
				P105		
	P002	INVENTORY UPDATE PROGRAM	P002C	P020	IN	S
				P105		
	P015	INVOICE PRINT [illegible]	[illegible]			S

Fig. 7 Hierarchical reorganization

The second example is to define another report document describing detailed program specifications. For this purpose, a larger nested table DETAILED-PROGRAM-SPECIFICATION-TABLE embeding detailed information is derived as shown in Fig. 8.

PROGRAM-SPECIFICATION-TABLE						
			FILE-USAGE			
PID	PNAME	CHART-SPEC	FID	I/O-CLASS	ACCESS-MODE	SUBPROGRAM-ID
P001	SALES SUMMARY PROGRAM	P001C	F001	IN	S	P020
			F002	IN	S	P021
			F005	OUT	S	
			F101	[illegible]		

FILE-SPECIFICATION-TABLE								
					FIELD-SPECIFICATION			
FID	FNAME	ORG	REC-LENGTH	BLOCK-LENGTH	NAME	DESCRIPTION	PIC	KEY
F001	SALES FILE	S	30	600	OID	PURCHASE ORDER ID	9(5)	N
					DATE	ORDER DATE	9(6)	N
					CUID	CUSTOMER ID	X(4)	N
					COID	COMMODITY ID	X(8)	Y
					QTY		9(5)	N

DETAILED-PROGRAM-SPECIFICATION-TABLE													
			FILE-USAGE									SUBPROGRAM	
			FILE										
						FIELD-SPECIFICATION							
PID	PNAME	CHART-SPEC	FID	FNAME	ORG	NAME	DESCRIPTION	PIC	KEY	I/O-CLASS	ACCESS-MODE	PID	PNAME
P001	SALES SUMMARY PROGRAM	P001C	F001	SALES FILE	S	OID	PURCHASE ORDER ID	9(5)	N	IN	S	P020	SORT PROGRAM 1
						DATE	ORDER DATE	9(6)	N				
						CUID	CUSTOMER ID	X(4)	N				
						COID	COMMODITY ID	X(8)	Y				
						QTY	QUANTITY	9(5)	N				
						FILLER	FILLER	X(2)	N				
			[illegible]			COID	[illegible]	X(8)	Y				

Fig. 8 Join

Detailed file information and subprogram names are incorporated into this nested table. As demonstrated in this example, data included in two or more separate nested tables can readily be joined into one nested table.

5. CONCLUSION

In this paper, human factors of an interactive software development tool FRIEND were discussed. A special emphasis was laid on two types of friendliness attained in FRIEND. The first one was the friendliness for tool users. It was achieved by allowing the users to specify and analyze requirements and designs from their own viewpoints using friendly representation forms such as forms, graphs, and tables. The second one was the friendliness for tool administrators, and it was achieved by uniformly managing the internal specification database in terms of a powerful and flexible data model, namely the nested table data model. These two types of friendliness were discussed and illustrated from the architectural viewpoint. The architecture was based on mappings between the user specification level and the database level.

The University of Tokyo and Nippon Electric Company have been jointly developing FRIEND, and the prototype system has already been used for some experimental software development projects (Azuma and co-workers, 1980).

ACKNOWLEDGEMENT

This work was partially supported by grants from Special Research Project sponsored by the Ministry of Education, Science and Culture of the Government of JAPAN.

REFERENCES

Azuma, M., M. Takahashi, S. Kamiya, and K. Minomura (1980). Interactive software development tool: ISDT. Proc. 5th Int. Conf. Software Eng. (San Diego), To Appear.

Buchmann, A. P., and T. L. Kunii (1979). Evolutionally drawing formalization in an engineering database environment. Proc. IEEE Computer Soc. 3rd Int. Computer Software and Application Conf. (Chicago), 732-737.

Harada, M., and T. L. Kunii (1979). A design process formalization. Proc. IEEE Computer Soc. 3rd Int. Computer Software and Application Conf. (Chicago), 367-373.

Kitagawa, H., and T. L. Kunii (1980a). A language for office form processing (OFP) - with application to medical forms -. Proc. 3rd World Conf. Medical Informatics (Tokyo, Japan), 713-718.

Kitagawa, H., and T. L. Kunii (1980b). DNM: Dynamic nesting method - an aid for user friendly database application development -. Proc. 14th IBM Computer Sci. Symposium (Amagi, Japan), 434-455.

Kunii, T. L., and M. Harada (1980). SID: A system for interactive design. Proc. AFIPS National Computer Conf. (Anaheim), 49, 33-40.

Ross, D. T., and K. E. Schoman, Jr. (1977). Structured analysis for requirements definition. IEEE Trans. Software Eng. SE-3, 1, 6-15.

Ross, D. T. (1977). Structured analysis (SA): A language for communicating ideas. IEEE Trans. Software Eng. SE-3,1, 16-34.

Sibley, E. H. (Ed.) (1976). Special issue: data-base management systems. Comput. Surv., 8,1.

Teichroew, D., and E. A. Hershey (1977). PSL/PSA: A computer-aided technique for structured decomposition and analysis of information processing systems. IEEE Trans. Software Eng., SE-3, 1, 16-33.

COMPUTERESE AND THE INFORMATION BARRIER

Sandra Eveland

Linguistics Research Center
University of Texas at Austin
Austin, Texas 78712

ABSTRACT

As more scholars outside of the field of computer science begin to use the computer, the communication problem worsens. Scholars outside of technical disciplines especially encounter a jargon barrier that prevents them from getting the information necessary for completing their tasks. The reactions of these scholars to what they perceive as a hopeless mire of technical terminology is examined.

KEYWORDS

Jargon, computer terms, human factors, computers in the humanities, information.

Computers are a basic part of every American's life. Not only are virtually all financial transactions such as phone bills, pay checks and credit card shopping computerized, but the computer has also come into the home. There are microwave ovens, programmable tape decks and many toys driven by microprocessors. The bowling alley has a computer to keep score and set up pins, and cash registers everywhere are little computers. In addition, growing numbers of people are buying or building microprocessors for their own use. Now hobbyists can program home computers to turn off lights, keep income tax records or play blackjack and chess.

As the computer has become more important in our daily living, so has it become more widely used in all fields. As with all specialized machines that require training to operate, there are typically some of people who know the workings and quirks of each machine intimately, and some who don't. More and more non-users are discovering, however, that they could benefit from applying the computer to problems in a wide variety of fields. To greater and lesser degrees, these non-users have had to learn enough about the machine to use it. The degree of their success rests on how easily and thoroughly this new skill can be learned.

With this in mind, I decided to investigate the problems faced by a non-user who wants to become a user of computers. The obvious starting point seemed to be the area of communicating technical information to people with little background on which to build. I believed that new users would have varying degrees of difficulty in dealing with the complex terminology of the field. It also seemed that people in fields farthest removed from hard science would have the most problems. Thus, I set out to see if people outside the field encountered a jargon problem.

Because computing is a new field, very little work has been done by scholars on the problem of computing jargon. My primary source of information, therefore, was the users. I was able to interview in depth four scholars in the liberal arts. In addition to getting their reactions, I interviewed two representatives of the computation center whose job it is to help the public. I was also allowed to review the student evaluations for some of the short courses taught by the computation center staff. Most of these courses are designed for beginning users, so the comments of the participants, who are primarily graduate students or staff, are frequently pertinent.

In discussing the use of a computer as an aid to scholarly research, I encountered two types of users: those who hold the machine in awe and those who don't. The latter group of people had little or no trouble in learning how to use the machine, and in general, enjoyed the experience. One scientist, when asked if he had problems in learning how to compute, answered, "No, why should I?" Yet, at the other end of the scale, a linguist working on a dictionary that eventually will be stored in computerized form has been unable to learn even the simplest editing skills. He feels overwhelmed by the task, and stated that he might be able to learn the required procedures if he "spent twenty years locked away in a Tibetan monastery studying." From this, one can hypothesize that those individuals who are not afraid to try using a computer, and who are more aware of how they can benefit from using it are less bothered by the actual learning process. Also, the former are more likely to understand the limitations of the machine.

Many people are afraid of computers, so it is possible to discuss computer anxiety, just as one can discuss math anxiety. Most people first experienced computers through automatic billing of store accounts. This introduced the public to the computer error. We have all been told "the computer made a mistake," which is not true. Either a data entry clerk made a typographical error when entering data, or a programmer introduced a bug into a program that caused a computing error. Thus, the "blame-the-computer" syndrome is perpetuated by business refusing to admit human fallibility. (Aiken, 1980, p. 43) The result of this is that a lot of people have acceped the myth that computers are rigid, inflexible, have minds of their own, and can make one's life miserable. This kind of misrepresentation has been termed "cybercrud" by Nelson, who defines it as "putting things over on people by using the computer."

Television, films and the press have, in most instances, misinformed the public, too. For many years, we saw computer card sorters busily sorting cards, with the implication that this is what a computer looks like. When not showing us card sorters, television showed us ominous black boxes referred to in deferential terms. Hence, we came to accept that all computer output was valid and infallible.

These are some of the images that individuals carries with them as they decide to learn about computers. In some cases, these concepts are given credence by the computation centers. Schneider describes his first encounter with the world of computing as follows:

> I walked as in a French movie through an endless corridor to a doorway opening to another endless corridor, alone with the noise of my footsteps, until I reached at last two doorways. One of them was a closet. Upon opening the other I beheld an immense brightly lit glass case that appeared to contain the control room of the space ship Enterprise. Still I had not arrived. I found that the two doors opening into the glass case were locked, and, although I made motions to the space crew inside, they did not notice me, intent as they were on punching buttons and watching the messages appear at various stations.

Many people find this world hard to deal with for the simple reason that it is foreign to them. Dr. Jones, who has become an accomplished user of the DEC-10, says that before using a computer he had the usual stereotyped view of the machine as a monster. Once he became aware of the many applications, especially in his own field of linguistics, he became eager to learn the local system.

Part of the reason that computing is foreign is that the field has a highly specialized vocabulary. It is virtually impossible to use the machine without knowing the jargon. This fact is reflected by all of the interviews I conducted. Each of the people with whom I spoke felt the need for a glossary to help them decipher computer jargon. This also is a common thread in the evaluations of computation center short courses, as the following comments show:

> I appreciate the effort to avoid getting too technical or using too much computer "jargon."
>
> The jargon can be overwhelming - more reenforcement of computer jargon (maybe a list of terms) would help.

One participant had this to say about what was the the weakest part of the course: "assumption of mental comfort with the computer vocabulary".

The scholar working on the Gothic dictionary felt that not only is there a problem of different vocabularies, but of a completely different frame of reference as well. As he describes it, "If you learn a language, you must know the sounds that refer to a dog, but also know what a dog is." This particular scholar has a fear of machines in general, which he attributes to growing up in a less industrialized part of the world. He pointed out that he fears that he might somehow break or damage the computer, and therefore hesitates to use it. He also suggested that he has a greater respect for the value of objects than do most Americans. Interestingly, this fear of breaking the computer is a common reaction, as Mr. Able, manager of the computation center, points out in his interview. Also of significance is that the scholar hated math as a child and he associates computers with math. Interestingly, Dr. Baker, also of the computation center, points out that children are not afraid of computers as they don´t have preconceived notions about them.

Once people start using a computer, this fear is generally replaced by surprise at how ideosynchratic computer systems are. As Mr. Able pointed out, very few people are aware of the level of detail required. Not only are commands important, but punctuation is extremely significant. Dr. Charles, a linguist, describes the experience of using an editor as Kafkaesque. She says,

> You can do everything right except for 1 tiny detail (i.e. omitting a letter or space or using lc instead of caps) and nothing will work. People aren´t as picky as computers about details. There are rarely simple mistakes with people as there are with computers (e.g. for computers ^C) that will destroy hours of work.

Although it is possible to destroy one´s own work, it is almost impossible to accidentally destroy someone else´s, according to Mr. Able.

Dr. Jones, the most successful of the users whom I interviewed, kept stressing the game-like quality of computer interaction. He frequently mentions learning the rules, accepting the rules and going with the system as it is. This seems to have been a successful method for him. Apparently, the attitude of novices is significantly related to their success in learning a system. It has been shown that those with a negative view have more difficulty learning how to edit and generally make more mistakes. (Schneiderman, 1979, p. 10)

The universal concern of the scholars was the inability to get clear, concise information about how to use various systems. Most new users turn to the documentation provided by the computation center to get basic details. Everyone interviewed complained that these documents are inadequate for rank beginners.

Indeed, they are written by programmers for other programmers.

Programmers as a group do not generally have to deal with people. According to Dr. Baker, the job attracts introverts and allows introversion to be the norm. Thus, he says, one occasionally encounters a hard-core hacker that can't empathize with people. These people tend to be the ones who isolate themselves as high priests in the back room. Ben Schneider goes further in claiming that cases of nearly psychopathic withdrawal can be found wherever computers are. (Schneider, 1974, p.232) The point is again made in "The Hacker Papers", a serie of computer The point is again made in "The Hacker Papers", a series of computer mail messages recently puplished in a popular magazine. Although anti-social staff are the minority, such people are encountered with sufficient regularity to pose a problem to the new user.

At some point, the naive user will need to get information from someone who doesn't want to give it. From personal observation, one way for the holder of knowledge to discourage others from getting it is to "floor them with fancy footwork." By this I mean giving a high speed response filled with technical terms completely beyond the questioner's area of knowledge. The hoped for response is amazement and a belief that the questioner could never understand the subject. Therefore, the questioner will go away and never come back again. In doing this, the anti-social computing experts will have accomplished two things. First, they avoided extended social contact, and second, they reenforced their membership in the elite inner circle.

Interestingly, it is the use of jargon that allows the anti-social programmer to pull this off. Using jargon has been defined as making sounds for the purpose of announcing that one belongs to a certain class. Hence, it can be used for identification much as bird songs identify various species (Schneider, 1974, p.235). The analogy can be taken further by comparing such posing to the ritual displaying of plumage. Jargon serves to separate the initiated from the naive users, and hence is self-perpetuating. In addition, the programmer may not have an answer to the user's question, and rather than say, "I don't know," it is much more face-saving to hide behind jargon, which becomes a convenient substitute for knowledge.

This type of jargon is clearly "cybercrud." Another facet of the problem becomes apparent when people are quite willing to communicate the desired information, but are unable to do so. The majority of non-communicative people are of this type, rather than of the malicious type. There are two reasons for this type of communication failure. Many of these people have been associated with computing for so long that what is jargon for non-users is standard vocabulary for them. Thus, they cannot communicate when denied a large segment of their word pool. Hence, I continually heard complaints that the information provided was not sufficiently elementary.

In addition to the jargon problem, it is also apparent that some programmers do not have a good command of the English language. Their training is extremely technical. In addition, undergraduate programs typically encourage students to take math or engineering courses to satisfy minor requirements. There is no attempt to expose students to courses where the use of English is encouraged. It is possible to major in computer science and never have to write a paper outside of the basic English courses required of all undergraduates. This situation is decried by many, as the following passage indicates:

> It is ironic, even tragic, that a community which treasures and refines artificial languages (FORTRAN, PL/1, BASIC, and so forth) should set so little store by natural language that English becomes a more impoverished and weaker vehicle of communication amoung those who compute, while artificial languages become richer and more powerful. We can talk to our machines better than we can talk to ourselves. (Schneider, 1974, p. 235)

Part of the problem is that documentation produced is not adequate for people outside of the field.Part of the problem is in the subject matter. The information to be conveyed is so detailed and technical that communication can rarely be accomplished gracefully and easily, so one strives for efficiency. Thus, the new users are cut off from what should be their primary source of information. This is one area that could rather easily be improved. Even if there is insufficient staff at computation centers to provide individualized attention, elementary documentation could be made available and publicized.

Of the people I interviewed, all desired more clear documentation because they all found the available material too technical. Yet none of these people were aware that the computation center provides a free glossary of all terms used in published documentation. This is another facet of the communication problem. As Mr. Able pointed out, it is the segment of users who don't know where to go or the appropriate questions to ask, who "fall off the bottom." He further stated that it is these users who are likely to go away dissatisfied.

From interviewing various users and comparing their statements with what is written in the literature, I have found that there is a jargon surrounding computing and that this jargon presents a barrier to people unfamiliar with it in their attempts to use computers. It is also apparent to me that the jargon barrier will continue to exist in spite of attempts by the computation center staff to serve the public, because this has of necessity a low priority. The initial steps toward overcoming this problem will probably have to be taken by the new users. The first step in this is becoming aware of what computers can and cannot do. By learning their powers and limitations, people will come to see how they can benefit from using them. This in turn will motivate them to come to terms with the jargon and use it to convey what they want. This is already happening. There are already many terms, such as interface, flowchart and punched cards, that are in day-to-day use. These people should be encouraged to say, "Wait, I don't understand" when they need to. Perhaps when enough users demand information in terms that they can understand, some of the jargon that hampers their learning will disappear.

Craig, B. (no date). Computer Anxiety. Newsletter, Univ. Tennessee.

Fitz-enz, Jac. (1978) Who is the DP Professional? Datamation, 24, 124.

Gotlieb, C. C., and A. Borodin. (1973). Social Issues in Computing. Academic Press, New York.

Hewitt, H. J. J. (1977). MAYDAY: Computer Text Editing for non-computer People. LRC Basic Text-Processing Series, Austin, Texas.

Nelson, T. (1974). Computer Lib. Hugo's Book Service, Chicago.

Schneider, B. R. (1974). Travels in Computerland or Incompatibilities and Interfaces. Addison-Wesley Publishing Company, Reading, Mass.

Shneiderman, B. Human Factor Experiments in Designing Interactive Systems. Computer, 12, 9-20. (1979).

Weinberg, G. M. (1971). The Psychology of Computer Programming. Van Nostrand Reinhold Company, New York.

VIDEODISC: A NEW TECHNOLOGY FOR HUMAN FACTORS

Nancy H. McDonald
Department of Computer Science and Engineering
University of South Florida
Tampa, Florida

ABSTRACT

Designing computer systems to be useful tools for the general human population can be a difficult task. New technologies and innovations will surely provide assistance to this effort. One such recent development-videodisc technology-is discussed in terms of its potential human factors impact on two computer system application areas: Computer Assisted Instruction and Database Query Language.

KEYWORDS

Human factors in computer systems, human-machine interface, videodisc, computer assisted instruction, database query language.

INTRODUCTION

Anytime a human being <u>uses</u> an item to accomplish some activity, the human factors of the item may simplify or thwart the accomplishment. When using a hammer to secure a nail, if the grip of the hammer is made without concern for slippage, the nail may never be secured and the human using the hammer may be physically injured. This human factor concern must be especially keen when designing anything new that might invoke a bit of fear or anxiety at first. Surely, infant toys with their interesting sound, bright colors and large rounded shapes are meant to teach sensory awareness often for the first time. The human factors, in this case, are made so appealing that the accomplished activity is changed from a possible chore to a game.

When computers are being used by humans to accomplish particular activities -- whether it be playing pong, calculating nuclear power equations, maintaining customer records -- concern and consideration for the human factors of that computer system as a tool must be given equal import with concerns for the accuracy of the tool's functions. That is to say, if all the partial differential equations have been double and triple checked and the method for simultaneous equation solution is guaranteed, the calculation will be meaningless if the user cannot figure out how to enter the initial data -- or, if the user has no guidance in understanding the numbers produced as output.

The purpose of this paper is to describe some contributions videodisc technology is expected to make in the area of computer system human factors. Initially, we will discuss some general aspects of both computer system human factors and videodisc technology. Then, we will relate the issues by describing two research areas in which both topics may play very important roles: Computer Assisted Instruction (CAI) and Database Query Languages.

COMPUTER SYSTEM HUMAN FACTORS

As with any tool, the human factor aspects of a computer system occur wherever human and system meet, interface or interact. There are several dimensions that must be regarded when discussing such a complex tool as a computer system with respect to human factor concerns.

The computer system may be viewed (Fig. 1a) as a set of concentric subsystems; the center most subsystem being the physical hardware. A typical software nesting might entail (1) the operating system, the subsystem surrounding and closest to the hardware; (2) surrounded by a database management system; (3) followed by a detailed question and answering system; (4) encircled by a decision making system. Each system directs communications with and uses the subsystems nested within it, but not those surrounding itself.

A user classification can then be organized to coordinate with the nested computer subsystem scheme. As shown in Fig. 1a, the typical type of function performed by users interacting with the hardware involves designing the machine (hardware designing); users at the next level are generally developing software to allocate and utilize the hardware (system programming); the third level generally involves generalized application programming; the Q/A level might interface with a frequent user who was trained to interact with the system at that level; finally, the outer subsystem would most likely be used by someone who interacts with the machine occasionally and makes rather broad, general requests.

Figure 1b indicates one parameter of concern for the human user of this system (or subsystem) and one parameter of concern for the designer of the system (or subsystem). The deeper into the system the user's interactions go, the more knowledge he/she needs regarding computers in general and the specifics of the system at hand. The further out the designer goes in developing subsequent interfaces, the more care he/she must take with human factors because of the user's inexperiences with the system. If this inverse proportion of user knowledge to human factor design concern is not kept in balance, the psychological blocks of "boredom, panic, frustration, confusion or discomfort" (Foley, 1974) may thwart the user's activity.

Numerous computer system and subsystem designs have been devised to lessen the psychological effects just mentioned. Certainly the various interactive devices (e.g., keyboard, lightpen mouse, touch-panel, programming languages, query facilities, editors, etc.) reduce boredom, frustration and some discomfort. Presenting information in a variety of ways (e.g. verbally, graphically, interactively,) lessens certain feelings of panic and confusion. To these, we feel the addition of random accessible video/audio information will further enhance a computer system for the non-computer knowledgeable user.

VIDEODISC TECHNOLOGY

Within the last decade, a great deal of research and development has gone into the cost/effective fabrication of videodisc products. The general concept involves storing video and audio (i.e. analog) or digital information in a spiral pattern on a platter, shaped like a long-playing audio record. There are two general techniques used to playback the information on the disc -- the mechanical stylus method and the optical laser method. The first technique involves a stylus making contact with grooves on the disc as it revolves - very much like phonograph records. In the optical method, a laser is used to reflect from or transmit through pits on the disc's surface. See (Video, 1978: Velzel, 1978) for details. Both systems have capabilities for storing and randomly accessing information approximately 1 hour of video), slow motion, and fast forward. However, only the optical systems have been able to perfect the freeze-frame capability. Therefore, we will assume the optical technology when referring to videodisc efforts.

To provide further insight into videodisc technology's contributions to computer systems human factors, we present two examples of its use. First, we will describe the benefits CAI may derive: and then, we shall discuss a video-assisted graphic language for interacting with an inventory-type computerized database.

CAI-VIDEODISC HUMAN FACTORS

More than ten years ago, CAI was expected to revolutionize the educational process. The predictions envisioned individualized instruction that was more time efficient than conventional techniques and economically advantageous as well. While impressive CAI models have been developed by such projects as PLATO (Bitzer, 1970), LOGO, TICCIT (Sugarman, 1978 a&b) showing definite usefulness to the student, the time-efficiency and economy expectations with respect to the teacher have not been fulfilled. These hopes vanished because (1) most good pedagogy demands good graphics and would benefit from full-color motion picture capability which is still prohibitively expensive via computers alone; (2) translating effective teaching strategies, as well as student responses into appropriate data structures and commands for computer usage has resulted in typical author times of 50 hours per hour of CAI: (3) reluctance on the part of much of the teaching profession to get involved with CAI due to the preceeding two factors plus some job security anxieties.

With the advent of videodisc capabilities, instead of hours of laborious courseware design and typing in the information and commands - one need merely photograph all written and diagrammed material, any demonstrations, some lecture material, even exams. Because of the random accessibility, the ordering of information can be developed later and even individualized on the basis of dynamic evaluation of student performance.

At the University of South Florida, we are developing a Computer Aided Teachers System (CATS) to be used with the above description of teacher's input of courseware material. The CATS is a highly interactive, graphic-oriented interface facility which leads the teacher through the steps of labelling his/her information (stored on the videodisc) and sequencing the information for the student's usage. Through primarily menu-select and potentiometer-style interactions the teacher selects names for the material; assigns portions of the videodisc material to those names; and depicts the ordering(s) that the student may experience as he/she works through each lesson. There is very little typing and quite a lot of system provided assistance.

Figure 2 presents the visual image of the course design after the teacher's definition. At this point the teacher is directed to select the appropriate frame(s) of videodisc information to be presented. Then the teacher is lead through a sequencing operation to establish the order through which the student will encounter the material.

VIDEODISC QUERY LANGUAGE

In dealing with the typical query language, one notices such annoying features as:

1) An investment in training must be incurred to ask the simplest query.
2) Some queries are easy to state while others are more difficult than seemingly necessary.
3) Few languages make use of the actual data in the database.
4) It is often quite difficult to simply peruse the database with a vague query.

Psychological experiments studying human abilities to use query languages (Greenblatt, 1978; Brosey, 1978; Lochovsky, 1979; McDonald, 1975; Reisner, 1975; Thomas, 1975) have indicated certain features which are desirable for particular reasons. Most notable references for our graphical work are Thomas and Gould's study (Thomas, 1975) of the two-dimensional Query by Example language and this author's previous work (McDonald, 1975) comparing the graphic query language CUPID to a typical structured English query language. In both studies, a graphic approach to language design was shown to contribute to rapidly understanding and accurate user performance, and to a sense of user pleasure.

Our present research efforts, sponsored by the Defense Advanced Research Projects Agency, are combining the attributes of a two-dimensional language with videodisc technology to provide an unusual interface for an inventory database. This capability will utilize motion and still picture of actual data to provide the user with a perusal capability as well as spatial and visual cueing information.

The user will interact mostly via the naturally simple touch-panel devices. He/she will construct queries of the database by manipulating and positioning symbols with the assistance of the perusal capability for associating and identifying items. The system will have an accelerated scan facility for a "slot machine" effort to be used for subconscious/conscious associations. Also a split screen effect will allow various views of the data to be presented simultaneously. Figure 3 demonstrates the latter facility.

The user will have some flexibility and choice in the manner of query construction based on the query structures technique desired. The system will present several structure motifs (as exemplified in Fig. 4) for user selection. After one motif is selected, the system guides the user through query construction phase with accompanying video information.

The actual computerized database entails a relational model of the typical data along with additional spatial, video and associative information. The control of the data and the graphical interactions will be microprocessor based.

We expect this facility to reduce the training time investment needed with user selected structuring graphics; reduce the degree of difficulty of some queries via system interactions; and present actual data in perusal and "slot machine" modes to assist in understanding the data and constructing a query.

SUMMARY

This paper has presented an overview of the importance of human factors considerations for computer systems. While there have been numerous devices, strategies and notions developed to make the human-computer interface as effective for human use as possible, new technologies and developments may add benefits in this area here-to-fore not considered. To exemplify this, we discussed two research efforts applying videodisc technology. One project utilizes the videodisc as a random accessible storage media for a teacher's authoring system. For the purposes of

CAI authoring, the human factors may be drastically reduced so that the teacher's cost (time and energy) is more commensurate with the student's learing effort. The second project utilizes the video information to assist a user in understanding and formulating queries for an inventory database. In both projects, a primary goal is to bring the computer-tool to people with little or no knowledge of Computer Science, the teacher and the inventory manager. Both applications indicate that this new media may provide considerable human factors improvement for these specialized computer systems.

REFERENCES

Bitzer, D. L. et al, "The PLATO System and Science Education," CERL Report X-17, University of Illinois, Urbana, IL, August 1970.

Brosey, M. and B. Schneiderman, "Two Experimental Comparisons of Relational and Hierarchical Database Models, "Int. J. Man-Machine Studies 10, 1978, pp 625-637.

Foley, J. D. and Wallace, V. L. "The Art of Natural Graphic Man-Machine Conversation", Proceedings of the IEEE, Vol. 62, No. 4, April 1974.

Greenblatt, D. and Waxman, J. "A Study of Three Database Query Languages, "Databases: Improving Usability and Responsiveness, B. Schneiderman (ed.), Academic Press, New York, 1978, pp. 77-98.

Lochovsky, F. H., "Data Base Management System User Performance Variables", Ph.D. Thesis, University of Toronto, 1979.

McDonald, N. H., "CUPID: A Graphics Oriented Facility for Support of Non-Programmer Interactions with a Data Base", Ph.D. Thesis, University of California, Berkeley, 1975.

Reisner, P. et al, "Human Factors Evaluation of Two Data Base Query Languages: SQUARE and SEQUEL", Proc. AFIPS National Computer Conference, May 1975, Vol. 44 pp. 447-452.

Sugarman, R., "A Second Chance for Computer Aided Instruction", IEEE Spectrum, Vol. 15, No. 9, August, 1978, pp. 29.

Sugarman, R., "What's New Teacher? Ask the Computer" IEEE Spectrum Vol. 15, No. 9, September, 1978, pp. 44.

Thomas, J. C. and J. D. Gould, "A Psychological Study of Query by Example", Proc. AFIPS National Computer Conference, May 1975, Vol. 44, pp. 439-445.

Velzel, C. H. F., "Laser Beam Reading of Video Records", Applied Optics, Vol. 17, No. 13, July, 1978, pp. 2029.

"Video Disc", RCA Review, Vol. 39, No. 1, March 1978.

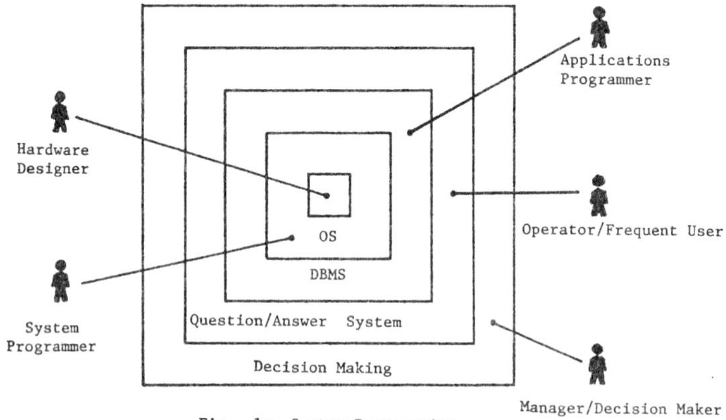

Fig. 1a System Perspective

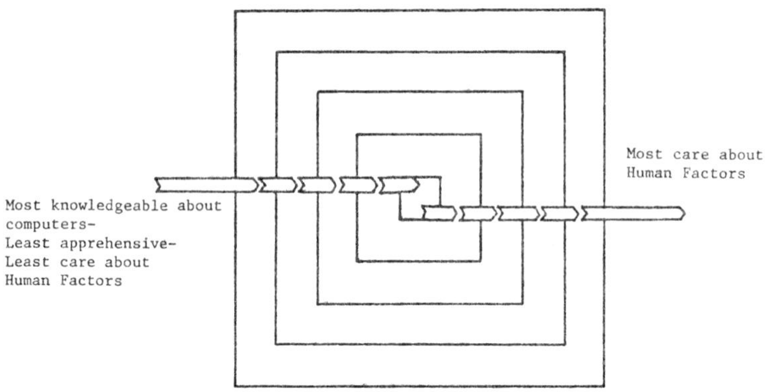

Fig. 1b Degree of Human Factors Concern

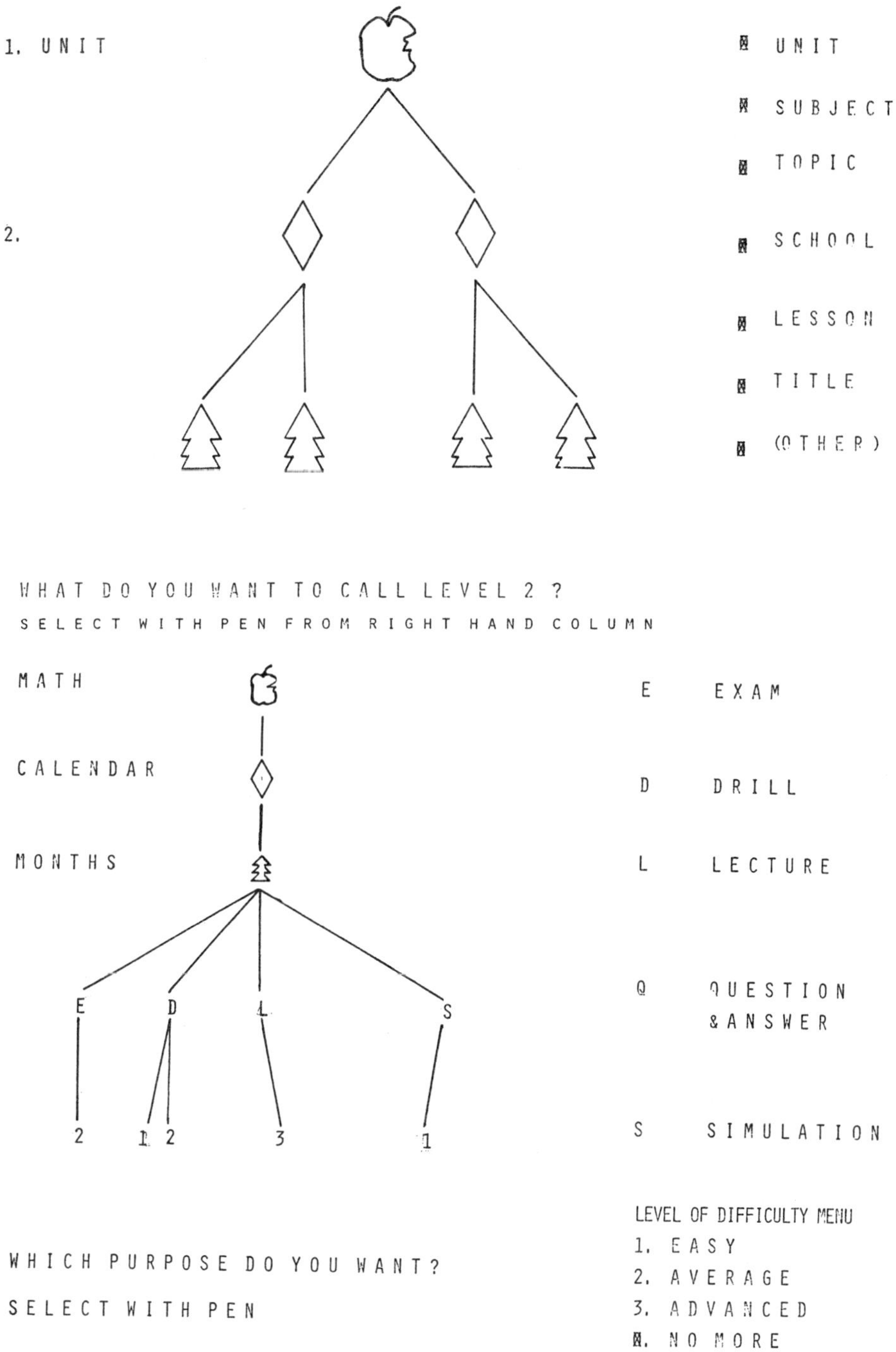

Fig. 2 CATS definition of courseware

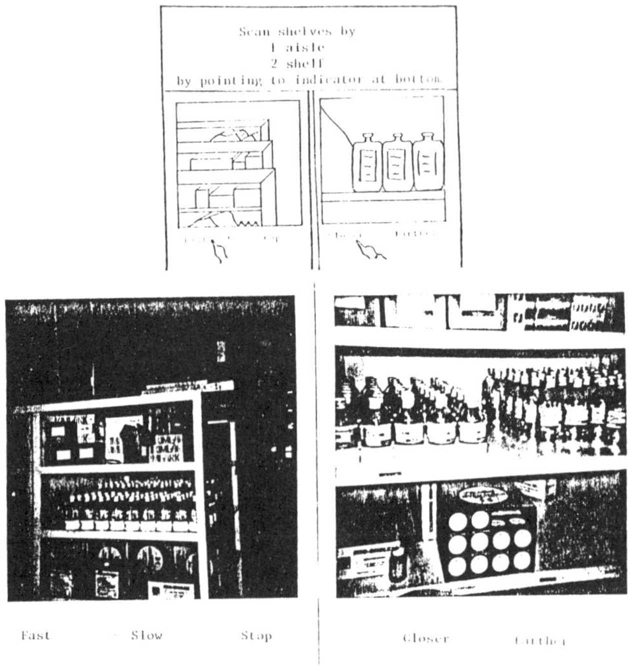

Fig. 3 Simultaneous views of data

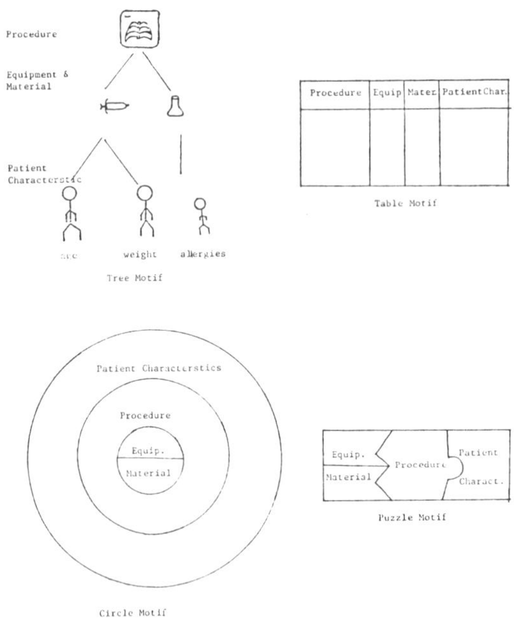

Fig. 4 Query motifs

Human Factors Issues in Office Information Systems

F.H. Lochovsky

Computer Systems Research Group
University of Toronto
Toronto, Canada
M5S 1A1

ABSTRACT

It is generally recognized that human factors issues are very important in the design and implementation of computer software. In the past, however, human factors issues were studied mainly after the fact so that their effect on software systems was often minimal. Office information systems are now just in their infancy. Because of their potential wide impact and large user community, human factors issues are of paramount importance in these systems. In this paper we outline those issues in office information systems that require human factors research.

KEYWORDS

Office information systems; human factors; data models; query languages; communication; system dynamics.

INTRODUCTION

When we design a computer system for use by people, we usually have two objectives in mind. First, we would like the computer system to fit the application and, second, we would like it to fit the people (end-users) who will use it. To meet the first objective, we study the interaction between the application and the computer system, noting the characteristics of each and trying to make them mesh. To meet the second objective, we study the characteristics of the interaction between the computer system and the end-users and try to make them mesh. This latter activity is called *human factors* (Lochovsky, 1978).

Office information systems are computer systems designed to bring "automation" to the office. They computerize such diverse facilities as word processing, electronic mail, forms processing, personal computing, data base management and office procedures. In addition, the goal is to provide all these facilities in one integrated system that will help the office worker in the performance of his/her job (Tsichritzis and Lochovsky, 1980).

In the past, reseachers in computer science have worked on human factors issues in many areas most notably in programming languages (Weinberg and Schulman, 1974) and data base management (Shneiderman, 1978). The emphasis has been on evaluating how easy the languages or systems are to use by untrained end-users. However, these evaluations are usually performed on "finished" systems. As a result, the impact of human factors research has been more of an expository nature than an integral part of the design process. (There have been some exceptions, e.g., (Reisner, 1977).)

We see the issues in designing office information systems that are acceptable to the intended user population as similar to those in data base management. The scope and complexity of the issues, however, are much greater. In data base management, the end-user population is usually composed of computer professionals. These people understand computers and are dedicated to their use. Thus, they can always be made to understand the "system" problems and be trained to overcome any idiosyncrasies of the system.

Office workers are not computer professionals. We cannot expect them to always be able to understand how the system functions or to provide them with extensive training to master hard to use facilities. We need to provide them with systems that are natural to them and are easy to learn to use and easy to use. In the design of office information systems, the resolution of human factors issues will be of paramount importance if the office is to be automated successfully. In the following sections we will outline the issues that require investigation in the design of office information systems.

DATA MODELS

Normally when there is data to be stored and processed, we like to organize the data in some way. To determine how best to organize the data, we analyze it to see if there is any pattern to the data occurrences or relationships among the data occurrences. We then abstract these patterns of occurrences and relationships into categories of data and categories of relationships among them. The ways in which we can construct these categories of data defines a *data model*.

In data base management, several data models have been developed for organizing computerized data (Tsichritzis and Lochovsky, 1981). For example, there are hierarchical data models (hierarchies of data), network data models (general interconnections of data) and relational data models (tables of data). These data models are very general to allow a wide range of applications to be represented. In addition, many of them are more oriented toward making it easier for the computer to process data than for representing the user's information requirements (Codd and Date, 1974).

We cannot expect that office workers will understand or find natural to use the data models used traditionally in data base management. But what kinds of data models are appropriate for office information systems? How does the worker in the office perceive and organize data? New data models have to be developed that are appropriate to the office and to the types of people that will be using office information systems. These data models will have to be much more flexible than data base management data models. The kinds of data handled by these data models, i.e., highly formatted data, are only one type of data found in an office. Data models for office information systems should also be able to handle non-formatted text data, picture data, voice data, notations to data (both text and voice), etc. Not only should they be able to handle these types of data, but the way in which they are represented in the data model should be easily understood and natural to the people that will be using this data. There is a need, therefore, when developing data models for office information systems to perform human factors testing to determine the suitability of the data model for the intended user population.

The notion of forms as a data model for office information systems has been proposed explicitly or implicitly by several researchers (Ellis and Nutt, 1980; Hammer *et al.*, 1977; Lefkovits, 1979; Tsichritzis, 1980a,b). Forms appear to be a natural candidate since they are familar to office workers and can represent much of the data that is captured and processed in an office. However, paper-based forms have many properties and it is important that the right ones be captured in a forms data model. In one case, a forms data model was developed that did not allow multiple-valued fields. It was discovered that this is a capability present in paper-based forms that users really needed and should have been included in the forms data model (Tsichritzis, 1980b). In data base management, data models often constrain the user to represent an application in ways that are not natural to him (Lochovsky, 1978). It is important that this same mistake not be made in office information systems.

Forms, while adequate for representing much of the data in an office, cannot represent all of it. More general data models need to be investigated that can capture additional information handling aspects of an office (Tsichritzis, 1980b). These data models need to be evaluated not only in terms of their power of representation, but also in terms of their acceptability to the end-users.

QUERY LANGUAGES

Once we have organized the data according to a data model, we would like to be able to manipulate it to access data or to modify it in some way. The operations that allow us to do this define a *query language*. In terms of office information systems, we are interested in query languages that can be used interactively and the characteristics that such query languages should have given the characteristics of the end-users.

Traditionally in data base management, query languages have been English-keyword oriented. That is, the user formulates requests in a restricted, English-like grammar where specific words (keywords) signal the start of specific parts of a request. Constructs in the language must appear in a fixed order and in a fixed format. In recent years other types of query languages have emerged that allow requests to be formulated by giving the system an example of the reply to a question (Zloof, 1977). For example, templates of tables can be displayed on a screen and the user fills in the table according to which columns he wants in the reply and the kinds of values desired in each column. Menu-based query languages and graphics-based query languages have also been proposed, but not used widely (McDonald and Stonebraker, 1976; Ellis and Nutt, 1980).

It is not clear which, if any, of these types of query languages are suitable for office information systems. Before we can answer this question, we first need to determine what types of querying facilities are required by office workers. Do they require the sophisticated facilities provided by data base management query languages or will simpler facilities do? Are the characteristics of current query languages suitable for the office? Can they be used by people eight hours a day in performing their job?

The facilities provided by query languages for data base management systems need to be general to handle diverse applications. In office information systems the query languages can be much more specialized. It would seem unlikely that English-keyword query languages would be suitable. They usually employ complex constructs that require some training to master (Reisner, 1977; Lochovsky, 1978). In addition, they require a good deal of input (typing) by the user and do not have particularly good interaction characteristics. By-example query languages may be appropriate if they can be tailored to the data model and the types of operations required in office information systems.

It is important not to be restricted by past experience in designing query languages for office information systems. Much more sophisticated technology is now available and it should be exploited wherever possible. Many types of technologies such as graphics, sound and colour can be combined to provide a very user-oriented query language (Herot, 1979). Research is needed to determine how best to combine these facilities to provide the types of capabilities required in office information systems.

COMMUNICATION

The objective of gathering and processing data is usually to communicate the results of the processing to people so that they can make decisions. Traditionally, these results are produced as printed reports and disseminated by hand to the appropriate individuals. With today's technology, electronic mail and networks can be used to disseminate information electronically. These capabilities have to be integrated into an office information system. It is important to study the characteristics of the communication processes in an office so that they can be supported in a natural way by office information systems.

People have many ways to communicate, written, verbal, visual, etc., and as many of these as possible should be supported by an office information system. To support these different modes of communication, different technologies will have to be integrated into a uniform communication capability. It is important that these facilities be provided in a way that is acceptable to the end-users. Studies are needed on the design of hardware and software components of the communications facility of an office information system.

As well as modes of communication, people also have different levels of communication, e.g., formal, informal and personal. An office information system should be able to distinguish between these different modes and be able to support them. The last level of communication may be particularly important to the acceptability of office information systems to the users. One may naturally be inclined to exclude this type of communication from the system since it is not "productive" to the organization. However, people have a need to communicate personally and if they cannot do it through the system, they will do it outside the system. Therefore, simply as an incentive to use the system, it may be important to support personal communication and even encourage it.

As well as determining how people communicate in an office, studies are required to determine how best to provide the means of communication within an office information system. It should be no more difficult to communicate with someone through an office information system than it is to do it via the telephone or by walking down the hall. These facilities need to be integrated with the data model and query language facilities so that their use is compatible with these other facilities.

INTERACTION

An office is not merely a place to do work. It is a social environment in which people must feel happy and useful if work is to be done at all. The introduction of office information systems could have a positive effect or a negative effect on this environment. A great deal depends on how the issues outlined in the previous sections are resolved. For this reason, it is important when dealing with these issues to always consider the larger issue of the office as a social environment. There are many interactions that take place within this environment, e.g., person to person, person to computer, etc., that need to be considered in designing office information systems.

The way in which the user interacts with the system, as much as the types of facilities provided will be a crucial factor in the system's acceptability. The interaction should be interesting and not tedious. The user should be able to interact with the system in different ways depending on his mood, the type of work to be done, etc. Above all, the user should always feel that he is the one who is in control of the interaction, not the system. Even though the system is always in control ultimately, it should make the user feel he is in control. The responsiveness and helpfulness of the system are factors that can contribute to a feeling of control.

The personal interactions in an office are as important as the business interactions. We have already indicated that office information systems should support personal communication between people. Thought should be given to providing recreational support. For example, the system could support games between it and people as well as between people. Such support could have an educational as well as a recreational value in that it can be a way for people to learn to use the system's facilities in a non-pressured mode of interaction.

SUMMARY

In the past systems were often designed without much thought given to human factors issues. Subsequent evaluations have shown the inappropriateness of this approach. Office information systems are in their infancy. Now is the time to consider the human factors issues involved in their design and in their introduction into the office.

In this paper we have outlined the areas in which there is a need for human factors research in office information systems. These areas, data models, query languages, communications and system-user interaction, relate to the interface between office information systems and the people who will use them. Unlike other computer systems, deficiencies in the human factors engineering of these systems will impact on a large user population. The human factors aspects of their design will be as important as the facilities they provide.

REFERENCES

Codd, E.F., and C.J. Date (1974). Interactive Support for Non-programmers: The Relational and Network Approaches. *Proc. ACM SIGMOD, Data Models: Data-Structure-Set versus Relational* (Rustin, R., ed.), 11-41.

Ellis, C.A., and G.J. Nutt (1980). Office Information Systems and Computer Science. *ACM Comput. Surv.* **12**, 27-60.

Hammer, M., W.G. Howe, V.J. Kruskal, and I. Wladawsky (1977). A Very High Level Programming Language for Data Processing Applications. *Commun. ACM* **20**, 832-840.

Herot, C.F. (1979). Spatial Management of Data. *Proc. ACM Int. Conf. Very Large Data Bases*, 16.

Lefkovits, H.C., *et al.* (1979). A Status Report on the Activities of the CODASYL End User Facilities Committee (EUFC). *SIGMOD Record* **10** (2 & 3).

Lochovsky, F.H. (1978). *Data Base Management System User Performance.* Tech. rep. CSRG-90, Computer Systems Research Group, Univ. of Toronto.

McDonald, N. and M.R. Stonebraker (1975). CUPID - The Friendly Query Language. *Proc. ACM Pacific 75 Regional Conf.*, 127-131.

Reisner, P. (1977). Use of Psychological Experimentation as an Aid to Development of a Query Language. *IEEE Trans. on Software Engineering* **3**, 218-229.

Shneiderman, B. (1978). Improving the Human Factors Aspect of Database Interactions. *ACM Trans. on Database Syst.* **3**, 417-439.

Tsichritzis, D.C. (1980a). OFS: An Integrated Form Management System. *Proc. ACM Int. Conf. Very Large Data Bases*, 161-166.

Tsichritzis, D.C. (1980b). *Form Management.* Submitted for publication.

Tsichritzis, D.C., and F.H. Lochovsky (1980). Office Information Systems: Challenge for the 80's. *Proc. of the IEEE* **68**, 1054-1059.

Tsichritzis, D.C., and, F.H. Lochovsky (1981). *Data Models.* Prentice-Hall (to appear).

Weinberg, G.M., and E.L. Schulman (1974). Goals and Performance in Computer Programming. *Human Factors* **16**, 70-77.

Zloof, M.M. (1977). Query-By-Example: A Data Base Language. *IBM Syst. J.* **16**, 324-343.

HUMAN PRODUCTION AND DETECTION OF ERRORS IN DATA SYSTEMS

I.B. Turksen

Department of Industrial Engineering
University of Toronto
M5S 1A4 Canada

ABSTRACT

The average outgoing quality of data is considered when the Type I and Type II errors are introduced into a data production, inspection and correction system. Production and inspection error rates are determined in an experimental study. A cost effectiveness model is proposed for a fractional sampling plan.

KEYWORDS

Human errors, production and inspection, data systems, fractional sampling

INTRODUCTION

The purpose of this paper is: first to summarize some of the known facts about the average outgoing quality of data in data production systems; second to present the results of an experimental study where the various error rates were determined; and third to propose a fractional sampling plan that suggests a trade off between the optimal cost and desired level of reliability.

A review of current literature reveals that human production detection and diagnosis of errors are investigated in the main in two overlapping areas: (i) human factors studies are essentially based on experimental data and attempt to assess the causes of human error in order to minimize these errors (Jacobson, 1952; McKenzie, 1957; Harris, 1966, 1968; Fox and Haslegrave, 1969; Carrol, 1969; etc.) (ii) quality control studies are basically concerned with the misclassification of production units (Minton, 1969, 1972; Collins, et. al., 1972, 1973, 1976; Case, et. al., 1973, 1975, 1977; Schmidt, et. al., 1974; Bennett, et. al., 1974; Hoag, et. al., 1975, etc.). The studies discuss the operating characteristic curves, average outgoing quality of data and the cost effectiveness of various sampling strategies.

Without digressing into the origins of human error (Senders, 1980; Norman, 1980) and without getting involved in a discussion of what constitutes an error (Myers, 1976), and being aware of the different domains of the control of human performance (Rasmussen, 1980), we present our work based on our earlier proposals, investigations, and experimental research (Turksen and Kul'ba, 1979; Raz, 1979; De Girolamo, et. al., 1980).

PRODUCTION AND INSPECTION

In Figure 1, we display a diagram of data-production activity. A coder, a keypuncher, a typist, etc., transfers N units of information from a source to a sink medium. An information unit could be a line of alphanumeric codes written on a source document, coding sheets, store receipts, etc. and key punched on a card or put on a CRT screen by keyboard entry. The content of the information, I, in a source medium is different from the content of the information I' in a sink medium, if at least one alphanumeric code is transferred erroneously from source to sink, i.e. a symbol is replaced by another. The decision-making is based, in general, on a function of information I' rather than the original information I. Hence the outgoing quality of the information could have far reaching implications.

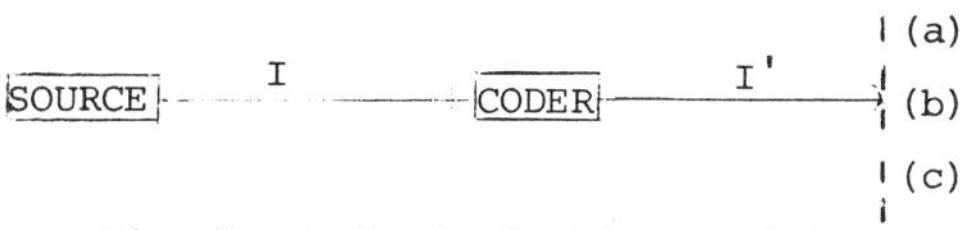

Fig. 1. Data Production Activity

(i) accept 100%
(ii) (a) screen 100%
(b) use an acceptance sampling plan

This creates a natural concern for the average quality, AOQ, of data. Several options are available to a controller of the quality of data: (i) accept the outgoing quality as it comes out of the coding process with information content I', i.e., 100% acceptance; or (ii) improve the outgoing quality by (a) a 100% screening or (b) an acceptance sampling plan.

It is known that banks, etc., implement a 100% screening and repeat this screening several times. In the implementation of an acceptance sampling plan, one could choose to use a single attribute sampling plan among various acceptance sampling plans. If a single attribute sampling plan is to be implemented, certain parameters are chosen first such as the acceptable quality level, AQL, the Type I error associated with the AQL, the unacceptable quality level, LTPD, and the Type II error associated with LTPD.

Then the sample size n and the acceptance number c are determined. A decision procedure for the single sampling plan is:

(i) inspect and correct the defectives in the sample;
(ii) if the acceptance number c is not exceeded then the batch is accepted;
(iii) if the acceptance number is exceeded, then the remainder of the batch is inspected and the observed defectives are corrected.

Inspection Without Errors. Suppose the probability that an information unit will be transferred incorrectly is denoted by p; and it is assumed to be constant for all units produced by the same person, so that the number of defective information units in a batch of N has a binomial distribution. If an acceptance sampling plan, with at most c defectives and a sample size of n, were to be used then the probability of accepting a batch without inspection error would be

$$P = \sum_{x=0}^{c} \binom{n}{x} p^{x}(1 - p)^{n - x} \tag{1}$$

and the average outgoing quality would be

$$Q = (1 - n/N)p\,P \tag{2}$$

Inspection and Correction With Errors. In reality (human) inspectors commit two types of errors. A Type I error is committed with probability α, whenever a non-defective information unit is classified as defective. And a Type II error is committed with probability β, whenever a defective information unit is classified as nondefective. In a similar acceptance sampling plan, the probability of acceptance would now be

$$P' = \sum_{x=0}^{c} \binom{n}{x} \{\alpha_1 + (1 - \alpha_1 - \beta_1)p\}^x \{\beta_1 + (1 - \alpha_1 - \beta_1)(1 - p)\}^{n - x} \qquad (1')$$

In addition, the (human) correctors commit two types of errors during the correction process. Given that a data unit is nondefective at the end of a production but that it is classified as defective in an inspection; then this data unit goes to a correction process. If, in the correction process, yet another error is committed, making the data unit defective, but the data unit is assumed non-defective, then a Type II error is committed. The probability that a Type II error is committed in correction is known as β_2. On the other hand, given that a data unit is defective at the end of a production and it is classified as defective in an inspection; then this data unit also goes to a correction process. Now if, in the correction process, an error is committed in changing this defective unit to yet another defective unit, then a Type I error is committed. The probability that a Type I error is committed in correction is known as α_2.

Minton (1972) states that the average outgoing quality would then be changed to

$$Q' = (1 - n/N)p\,P' + (n/N)\{\alpha_1 + (\beta_1 - \alpha_1)p\} + (1 - n/N)\{\alpha_2 + (\beta_2 - \alpha_2)p\}(1 - P') \qquad (2')$$

with the qualification that this formula "is correct only when a nondefective unit is classified as a defective and also changed to make it incorrect".

In summary, the effects of inspector and corrector errors are:

(1) The power of the operating characteristic, OC, curve to discriminate between good and poor quality batches is reduced and the OC curve shifts to the right. For a given fraction defective p, the probability of accepting the batch is increased from P to P' (see Figure 2).
(2) The shape of the AOQ curve is no longer unimodal. The AOQ curve initially increases, then decreases temporarily, and then increases directly with the incoming fraction defective. Inefficiency in correction of defective data units is the reason for this increase (see Figure 3).

EXPERIMENT

An experiment was conducted to estimate the parameters of a data production and inspection system in a laboratory environment. The experimental unit was a set of four computer programs picked randomly from a set of sixteen programs. Programs were written in GPSS language. Within each set, the four programs were placed in a random order. Two sets of programs were distributed to the subject randomly. Each subject key punched one set and inspected the other. In the production module, subjects used the IBM 1129 to keypunch cards by following (printout) program text as source document. The key punched cards were then listed in the form of a computer printout. In the inspection module, subjects inspected (matched) these listings with the source text. Later a perfect inspection was carried out using a computer comparison. An output data line in listing is <u>defective</u> if it differs from the source data line in at least one symbol. If a line is found to be non-defective but out of order, then one error is counted for every line it has to be moved to correct the sequence of lines. Similar recurrent errors throughout a listing are counted as independent.

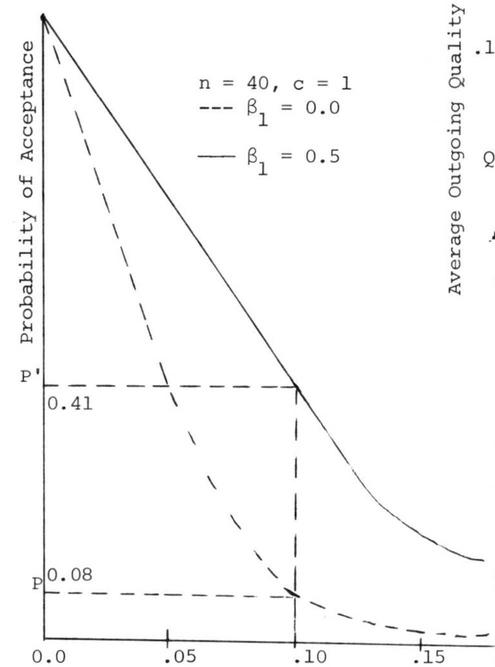

Fig. 2. Operating Charateristic Curves ($\alpha_1 = 0.0$)

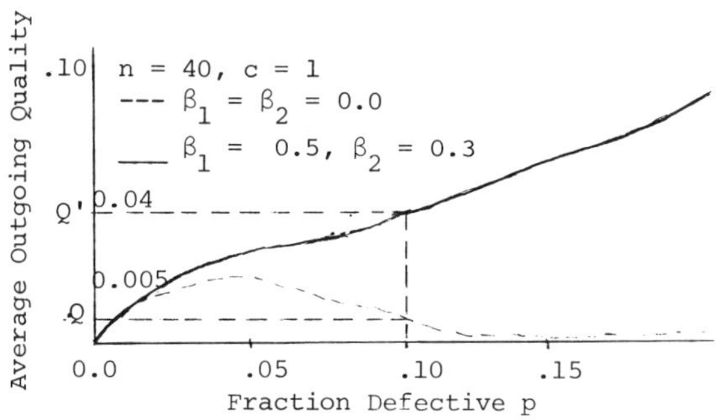

Fig. 3. Average Outgoing Quality Curves ($\alpha_1 = \alpha_2 = 0.0$)

A single factor, the subject's background and experience, was investigated and four groups of students participated in the experiment:

Group A. Non-engineering students, no keypunch experience, some typing experience, no knowledge of a computer language.

Group B. Non-engineering students, some keypunch experience, some knowledge of computer languages.

Group C. Engineering studnets, some keypunch experience, some knowledge of computer languages, no knowledge of GPSS language.

Group D. Engineering students with experience in both keypunch and GPSS. There were five students in each group twenty in all.

Production Errors. The production error rates for each subject and group is shown in Table 1. It is to be observed that the maximum error rate was 0.145 and the minimum error rate was 0.0096. The overall average was estimates as $\hat{p} = .0956$. The 95% confidence limits were $0.02595 < p < 0.2956$

The production error rate p was significantly different between groups (5%). Furthermore, there was a significant difference between the engineering and non-engineering students (5%). There was however no significant difference between the engineering students who had a knowledge of GPSS and those who did not (5%) and keypunch experience did not contribute significantly to the error rate p of the non-engineering students.

Inspection Errors. Not all subjects were able to complete the inspection process. Inspection error rates for those who were able to complete the inspection process are given in Table 2. The overall average of the Type I error was estimated as $\hat{\alpha}_1 = 0.0029$ with the 95% confidence interval as $0.00003211 < \alpha_1 < 0.2085$. The overall average of the Type II error was estimated at $\hat{\beta}_1 = 0.2882$ with the 95% confidence interval as $0.1224 < \beta_1 < 0.5404$.

The Type I error rate was significantly different between the two groups (5%). It is noted that none of the inspectors who are familiar with GPSS committed a Type I error, whereas a few of the students who were not familiar with the language committed Type I errors. It was also found that no significant difference existed with respect to Type II error rates.

TABLE 1 Estimates of the Production Error Rates

Students	GROUPS A	B	C	D
1	0.145*	0.107	0.0286	0.0534
2	0.061	0.0484	0.0471	0.157
3	0.107	0.111	0.0096**	0.0534
4	0.138	0.256	0.0822	0.107
5	0.168	0.106	0.0153	0.0929

TABLE 2 Estimates of the Inspection Error Rates

Subject	$\hat{e}_1$	$\hat{e}_2$
GROUP X - No Familiarity with G.P.S.S.		
1	0	0.1579
2	0	0.50
3	0.00862	0.208
4	0.008	0.20
5	0.025	0.385
6	0	0.0769
7	0	0.50
8	0.0183	0.409
TOTAL	0.00564	0.263
GROUP Y - Knowledge of G.P.S.S.		
1	0	0.25
2	0	0.571
3	0	0
4	0	0.267
5	0	0.25
6	0	0.571
7	0	0
TOTAL	0	0.3391
OVERALL TOTAL	0.0029	0.2882

Remark. Within these confidence limits of the production and inspection error rates, if we were to take $p = 0.10$, $\alpha_1 = 0.0$, $\beta_1 = 0.5$ and use a single inspection sampling plan with $n = 40$, $c = 1$, then we could read of the values of $P = .08$, $P' = .41$ and $Q = .005$, $Q' = .04$ from Figures 2 and 3. Thus, the inclusion of inspector errors would increase the probability of acceptance of the data processing by more than 30% and the average outgoing quality level of the single sampling plan would be poorer by about 3.5%. The overall effect of the single sampling plan would improve the quality of the output data by about 6%.

COST EFFECTIVENESS

With a slight departure from the cost models available in current literature, a fractional sampling model is proposed with the following cost structure:

$$C = \int_0^1 C(p)\ f(p)dp \tag{3}$$

where $C(p) = n\,C_i + n[(1-\beta_1)p + (1-p)\alpha_1]C_c + \{n[p\,\beta_1 + (1-p)\alpha_1\,\beta_2 + p(1-\beta_1)\alpha_2] + (N-n)p\}C_d$;

such that C(p) is the total cost per batch for any p, f(p) is the prior distribution of p, C_i is the inspection cost per unit, C_c is the cost of correction per unit, and C_d is the cost of defective per unit in the output, and all other terms are as defined previously.

For various distributions of p the analysis of this cost structure shows that the optimal policy is either not to inspect at all, $n = 0$, or to inspect 100%, $n = N$. For these two options, we have:

$n = 0$, $C(p) = Np$ and

$n = N$, $C(p) = NC_i + N\{(1-\beta_1)p + (1-p)\alpha_1]C_c + N[p\,\beta_1 + (1-p)\alpha_1\,\beta_2 + p(1-\beta_1)\alpha_2\}^i C_d$

In this formulation, the average outgoing quality of data is:

$$Q' = (n/N)[p\beta_1 + (1 - p)\alpha_1 \beta_2 + p(1 - \beta_1)\alpha_2] + (1 - n/N)p \qquad (4)$$

Thus, we have $Q' = p$ for $n = 0$ and $Q' = p\beta_1 + (1 - p)\alpha_1 \beta_2 + p(1 - \beta_1)\alpha_2$ for $n = N$. An apparent drawback of this fractional sampling plan is that cost optimality fixes the average outgoing quality. In turn if the average outgoing quality of the data is fixed to a desired level, from equation (4), we get a non-optimal n with a non-optimal cost value. Hence, if the average outgoing quality of data is an important factor, as it usually is, the cost optimality will have to be traded off with the requirements of the quality of data. With further analysis, it can be shown that the fractional sampling plan is always at least as good as a single sampling plan (see Figure 4).

For this analysis, one needs to take, for example, a cost formulation such as the one developed by Minton (1972), and first adapt it to our formulation with suitable changes and then compare it with the fractional sampling cost model developed here.

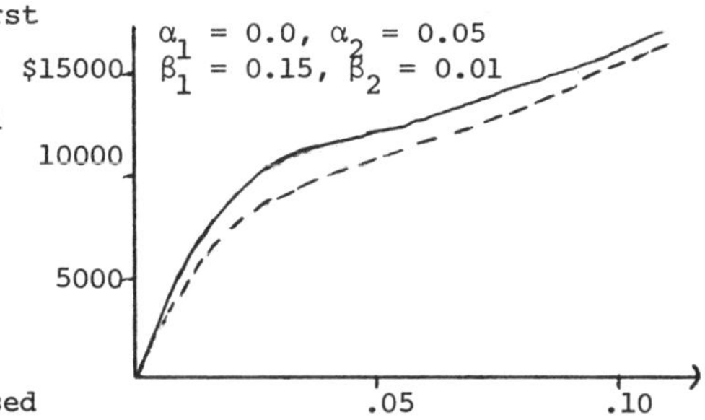

Fig. 4. Cost Comparison: $C_i = 1$, $C_c = 5$, $C_r = 100$, $N = 5000$

CONCLUSION

It has been shown that a judicious strategy could be selected among the alternatives of 100% acceptance, 100% screening and a fractional sampling based on cost of production, inspection and correction activities on the one hand and the required quality of data on the other by just using the proposed fractional sampling approach.

ACKNOWLEGMENTS

The author would like to express his gratitude to the Natural Science and Engineering Council of Canada for their support under operating grant No. A7698. Recognition is due to the late Professor B. Bernholtz and Professor J. Templeton for their suggestions. Thanks are due to T. Raz, R. Kelk, G. DeGirolamo and G. Ridout who assisted in the laboratory studies.

REFERENCES

Bennett, G.K. and Case, K.E. and Schmidt, J.W., (1974), The Economic Effects of Inspection Error on Attribute Sampling Plans, Naval Research Logistics Quarterly, 21, 431-443.

Carroll, J.M., (1969), Estimating Errors in the Inspection of Complex Products, AIIE Transactions, Vol. 1, No. 3, 229-235.

Case, K.E. and Bennett, G.K., (1977), The Economic Effect of Measurement Error on Variables Acceptance Sampling, Int.J.Prod.Res. 7, Vol. 15, No. 2, 117-128.

Case, K.E. and Bennett, G.K. and Schmidt, J.W., (1973), The Dodge CSP-1 Continuous Sampling Plan Under Inspection Error, AIIE Transactions, Vol. 5, No.3, 193-202.

Case, K.E. and Bennett, G.K. and Schmidt, J.W., (1975), The Effect of Inspection Error on Average Outgoing Quality, Journal of Quality Technology, Vol. 7, No. 1, 28-33

Collins, R.D. and Case, K.E., (1976), The Distribution of Observed Defectives in Attribute Acceptance Sampling Plans Under Inspection Error, AIIE Transactions, Vol. 8, No. 3, 375-378.

Collins R.D. and Case K.E. and Bennett G.K., (1972), The Effect of Inspection Accuracy in Statistical Quality Control, Proceedings the 23rd Annual Conference and Convention, American Institute of Industrial Engineers, Anaheim, California, p. 423-430.

Collins, R.D and Case K.E. and Bennett G.K., (1973), The Effects of Inspection Error on Single Sampling Inspection Plans, Int.J.Prod.Res., vol. II, no. 3, 289-298.

De Girolamo, G., Kelk, R.H., Ridout, G., 1980, Error Control in a Data Production and Inspection System at U. of T. Computer Centre, B.A.Sc. Thesis, Department of Industrial Engineering, University of Toronto.

Fox, J.G. and Haslegrave, C.M., (1969), Industrial Inspection Efficiency and Probability of a Defect Occurring, Ergonomics, vol. 12, no. 5, 713-721.

Harris, D., (1966), Effect of Equipment Complexity on Inspection Performance, Journal of Applied Psychology, vol. 50, no. 3, 236-237.

Harris, D.H., (1968), Effect of Defect Rate on Inspection Accuracy, Journal of Applied Psychology, vol. 52, no. 5, 377-379.

Harris, H.D. and Chaney, B.F., (1969), Human Factors in Quality Assurance, John Wilwy & Sons, Inc.

Hoag, L.L. and Hobbie, L.F., (1975), The Effect of Inspector Accuracy on the Type I and Type II Errors of Common Sampling Techniques, Journal of Quality Technology, vol. 7, no. 4, 157-164

Jacobson, H.J., (1952), A Study of Inspector Accuracy, Industrial Quality Control, 16-24.

McKenzie, R.M., (1958), On the Accuracy of Inspection, Ergonomics, vol. 1, p. 258-272.

Minton, G., (1969), Inspection and Correction Error in Data Processing, Journal of American Statistical Association, vol. 64, no. 328, 1256-1275.

Minton, G., (1972), Verification Error in Single Sampling Inspection Plans for Processing Survey Data, Journal of the American Statistical Association, vol. 67, no. 337, Applications Section, 46-54.

Norman, D.A., (1980), Post Freudian Slips, Psychology Today, April, 42-50.

Senders, J.W., (1980), Is There a Cure for Human Error, Psychology Today, April, 52-62.

Schmidt, J.W., Case, K.E., Bennet, G.K., (1974), The Choice of Variables Sampling Plans Using Cost Effective Criteria, AIIE Transactions, vol. 6, no. 3, 178-184.

Rasmussen, J., (1980), Some Trends in Man-Machine Interface Design for Industrial Process Plants, Riso M-2228, Riso National Laboratory, DK 4000, Roskilde, Denmark.

Raz, T., "A Model for Data Production Systems with Production and Inspection Errors" M.A.Sc., Thesis, 1979.

Turksen, I.B. and Kul'ba, V.V., (1979), A Model of Data Production and Error Control, Proceedings 1977 Spring Annual Conference AIIE, p. 1-10.

THE MAKING OF AN INSTRUCTIONAL PROGRAM WRITER

Ernest F. Garcia, Dean

School of Education
California State College
San Bernardino, California

ABSTRACT

This paper identifies knowledge and skills that writers of instructional programs for use in microcomputers should have. The role of the teacher in the production of instructional programs is emphasized. The need for training of program writers is compared to similar needs two decades ago when schools were involved with teaching machines and programmed learning.

KEYWORDS

Microcomputer, planning of instruction, implementation of instruction, programmers as technicians, program objectives, learning process, human development.

About two decades ago a great deal of excitement was generated in education in the United States by a phenomenon that was identified as the teaching machine and programmed instruction movement. The idea that teaching could be organized in self instructional modes and that teaching-learning events were reproducible, (Lumsdaine, 1963) caught the attention of educators who were under pressure to improve education in the aftermath of Sputnik I. Claims were made that the new auto-instructional methods were bound to revolutionize education and perhaps embodied the most important development since the emergence of the textbook.

If the teaching act could be defined as consisting of (a) the planning of instruction and (b) the implementation of instruction (Stolurow,1961) it was observed that the instructional events or steps were sequenced by the teacher, but that delivery or implementation of the program could be accomplished by use of a teaching machine or auto instructional device.

A proliferation of machines that were no more than dispensers or automatic page turners made it evident that programmed texts and other programmed materials delivered the instructional sequence and that most teaching machines were irrelevant. The important factor was the program, not the hardware. The exceptions, of course, were the computer-based teaching machine systems such as CLASS[1] at System Development Corporation in Santa Monica, California. The fact that computer-based machines allowed the student to interact with the program set them apart from those which merely dispensed a program.

The high cost of computer-based teaching machine systems made them impractical for widespread use and the state of the art, while showing great promise, was not at a point where wide application was feasible. In reality, the teaching machine and programmed instruction movement became the programmed instruction movement. The emphasis shifted to writing of programmed texts and other materials that did not require hardware for presentation. The writing of materials followed two major styles, a linear presentation attributed to B. F. Skinner and the intrinsic or branching style of Norman Crowder. The education literature of the early sixties contains many claims regarding the superiority of one method or the other. The controversy reached its highest level of absurdity when the linear advocates spelled "programed" while the branchers preferred "programmed"! Aside from silly controversies, the major problem that beset programmed instruction was that the quality of program organization and writing was suspect. There was a demand for programmed materials that resulted in hastily prepared materials written by people without proper training. The potential for making money from the new approach to instruction did not go unnoticed by publishers and soon-to-be publishers. By the early nineteen sixties, supermarkets had displays of programmed materials which informed the shopper that mathematics, history, languages, and anything else for that matter, could be learned in a few days through this new and exciting method. Shoddy writing and poor programming were not limited to materials sold in supermarkets. Programs bought by schools often did not represent the best in the state of the art, nor had they been thoroughly tested to determine if they were effective in helping the learner to attain the stated objectives. Efforts were made to train program writers through workshops sponsored by the Center for Programmed Instruction[2] as well as numerous college and university courses and institutes. Books and manuals for program writers emphasized principles of learning and programming as well as skill in writing (Markle, 1964; Stolurow, 1961).

A NEW NEED FOR PROGRAM WRITERS

The foregoing historical sketch serves as a comparison for a new problem that is emerging. The need for trained writers of instructional programs for use in microcomputers will become critical as the use of those devices continues to increase in the schools. The problem is not unlike the one presented above in some respects, but it is also very different because the instructional sequence must be prepared with computer language as an integral component. The expertise needed by an instructional program author is quite different from the know-

[1] Computer Laboratory for Automated School Systems, System Development Corporation, Santa Monica, CA (circa 1962)

[2] A non-profit center started by teachers to train programmers, produce and disseminate materials. In 1961 was located at 365 West End Avenue at 77th Street, New York.

ledge and skills needed by a computer programmer who writes programs for information retrieval, management of attendance data, or the matching of students with certain learning needs to the appropriate teaching materials available in the schools. The latter tasks are ones that any person trained in computer programming could accomplish without specific knowledge about learners or about teaching. As microcomputers come into broader use in the elementary and secondary schools, there will be an increasing demand for instructional programs. Who will do the programming? Companies that sell the hardware will want to have programs available in a variety of subject areas in order to demonstrate the advantages of computer-assisted instruction. It will be important to make critical evaluations of such programs to determine if knowledge of learners and sound pedagogy are in evidence and that the objectives of the program are measurable in terms of change in student behavior. Companies that produce hardware will probably not realize a great monetary return in the merchandising of software, yet the expense in hiring instructional program authors could become significant. As hardware costs continue to decline, human costs are increasing. The writing of instructional programs cannot be left to computer programmers who have little or no knowledge about learners and about teaching. It is likely that the writing of programs will become a task for teachers who will have to be trained in many of the same things as their programmed instruction colleagues in the past, and in author languages best suited for the programming tasks. Questions and arguments regarding teaching methodology seem to be of less importance now than they were during the programmed instruction-teaching machine era. The fact that a system with a microcomputer as the center can interact with the learner utilizing audio and visual support, makes the sequencing and branching of items more important than whether the program utilizes a linear format with remedial loops. Coulson (1962) established early in his research that instructional materials must be systematically organized and carefully prepared and that method variables are of little consequence when such preparation is not observed. What factors contribute to careful systematic preparation of instructional sequences?

SUBJECT MATTER CONTENT

A good starting point is to assure that the author has appropriate background in the subject to be programmed. While this statement might seem too obvious, one can point to instances where students have learned wrong concepts as quickly as accurate ones. A case in point were the children's books on jets and rockets that were authored after Sputnik I and did violence to Sir Isaac by insisting that rockets were propelled because their exhaust pushed against the air. Fortunately, students asked how such a device as a rocket could move in outer space where there was nothing to push against! Knowledge of the subject will help to assure that the material being presented to the learner is accurate. The programmer needs to accept the responsibility for accuracy of the subject matter content. Thus the first imperative is that the integrity of the subject matter must be maintained.

CLARITY OF PROGRAM OBJECTIVES

The author of a program should have a clear perception of what the learner is to be able to do upon completion of the instructional sequence. If the purpose of the program is to teach world geography, will the student be expected to identify land masses, describe their sizes and pertinent geological features? Is the learner expected to analyze the interdependence of geographic regions from an economic or political point of view? The difference between these objectives is substantial. One asks for recall of information while the other calls for analysis. The program author must have a way of structuring what is to be learned from simple recall to objectives that require a higher level of cognition. Bloom (1956) published a taxonomy that structures educational objectives into six categories that start with the lowest level of cognition designated as knowledge and progress through comprehension, application, analysis, synthesis, and finally the highest level according to Bloom, evaluation. The programmed texts of a few years ago seemed to concentrate too much on Bloom's lowest level of cognition and therefore presented facts to be recalled by the student as a measure of successful completion of the program. It has been said that there is nothing mere about a fact; however, program writers must challenge learners into higher levels of cognition. Unfortunately, education in all levels has dwelt too long within the confines of the knowledge category. The fourth grade social studies program and the college history course are often alike in the tasks required of the learner: remember something and give it back on the test! It is imperative that the author of an instructional program consider the objectives of the lesson in terms of what the student is expected to be able to do upon completion of the program (Gronlund, 1971). Objectives stated in behavioral terms help to focus on the organization of the instructional sequence to the extent that the author is forced to examine different levels of cognition in the program. Consequently, the learner is taken beyond recalling facts into understanding facts and principles (comprehension), applying concepts and principles to new situations (application), distinguishing between facts and inferences (analysis), and finally to learn to appraise, compare, conclude, on the basis of known data (evaluation). As authors go about the task of planning instruction for use in computer based systems, it is imperative that the objectives of the programs reflect clarity and specificity.

KNOWLEDGE OF THE LEARNING PROCESS

In addition to being able to write objectives that specify behaviors consistent with different categories of the cognitive hierarchy, the writer must also have knowledge of the learning process. Most teachers who would be the likely authors described in this paper have had a wide range of exposure to principles of learning. In preparation for program writing, there should be an extensive review of S-R approaches to learning, including operant conditioning, guided inquiry approaches, motivation, and response eliciting techniques such as cueing and prompting. Task analysis schemes (Gagne, 1977) can help to give an overall look at conditions of learning and to identify the tasks that must be performed in such diverse instances as simple S-R connections and problem solving. A review of learning principles and approaches with emphasis on application to teaching-learning situations will strengthen the writer's ability to organize instructional sequences that are productive for the learner.

KNOWLEDGE OF HUMAN DEVELOPMENT

Whether a person subscribes to the notion that readiness for a task can be taught (Bruner, 1960), or to a scheme of specific developmental stages which relegate learning to age spans, knowledge of human development is another imperative for program authors. Of particular interest should be the stages of cognitive development as devised by Jean Piaget. Between the stages of concrete and formal operations, there is a necessary transition into being able to cope with abstractions, yet teachers often present problems that require abstract thinking to learners who have not made the transition to the level of formal operations. An example of this practice was evident in the "new math" materials that emerged some two decades ago. The materials were developed by mathematicians largely in response to the "catch up" hysteria fomented by Sputnik I. When the materials were finally in place in the elementary school curriculum, they contained such things as set theory and Venn diagrams for six and seven year olds. Knowledge of child development was clearly not in evidence in such materials. At best, children in the pre-operational and concrete operational stages could memorize and recall facts about sets but seldom went beyond the knowledge level as described by Bloom (1957). Of course, knowledge of human development must not be limited to Bruner and Piaget. There are many other sources that can contribute to a program writer's effectiveness by presenting ideas on development that promote a sensitivity for the tasks that learners encounter at different stages.

WHO ARE THE AUTHORS?

A rather simple definition of software is: instructions that inform the computer what is to be done. In order to perform this task of informing, the programmer must use an author language that will allow the user to interact with the computer. Computer science programs in colleges and universities are producing increasing numbers of graduates who are skilled at using existing software and are able to write new programs for a variety of computer tasks. On the one hand there are skilled technicians who are able to provide the instructions that inform the computer what is to be done, but they may lack the knowledge of subject matter and principles of learning and development. On the other hand are the teachers who have the knowledge but lack the technical skill to "inform the computer". With wider use of microcomputers in schools, the need for a variety of programs for local use will grow. Teachers will want to make use of computers for specific tasks in their own classrooms. Schools will have to hire software technicians to team with the teachers in developing programs or there will have to be program writing training courses where teachers will acquire skills in using author languages. Teachers will also have to develop skill in preparing programs for systems that will include random access to audio and visual materials as part of the instructional sequence. Slides and film segments will be recorded on video discs and used as information or expository frames when the student needs additional information to make correct responses. There is a body of research in the use of films and other visuals which should prove of interest and benefit when effectiveness of programs is evaluated. Microcomputers will find their way into most aspects of our lives within a very short period of time. Factors such as size, cost, processing power, and versatility eliminate the barrier that existed when the early computer-based teaching machine systems were designed some two decades ago. It is now possible for even the most advanced technology to be available at a reasonable cost. Wise, Chan, and Yokely (1980) state that some social historians see the microcomputer as the third major cultural invention of the twentieth century, following the automobile and television. Ellis (1974) raises some critical questions about the use of the new technology. He sees the impact of the computer in education as forcing a critical look at education itself. The sequences of instruction and the processes that are automated will have to bear critical analysis in order to advance the art and science of teaching beyond doing the same old thing through a new device. Bushnell,(1964) writing on the computer as an instructional tool wrote:

> Automation requires a clear, operational statement of objectives to be accomplished by the system being automated. The desired student behavior must be more precisely defined, and research must be directed toward discovering optimal combinations of instructional techniques to produce these behaviors...a multitude of such practical problems must be solved if computers are to have widespread utility in education.

The agenda remains current.

REFERENCES

Bloom, B. S. (1956) ed., et al. _Taxonomy of Educational Objectives: Cognitive Domain_. David McKay, Inc., New York.

Bruner, J. S. (1960), _The Process of Education_, Vintage Books, New York, p.33.

Bushnell, D. D. (1964), _The Computer as an Instructional Tool: A Summary_, SP1554, System Development Corporation, Santa Monica, CA.

Coulson, J. E. (1962), A Computer-Based Laboratory for Research and Development in Education. _Programmed Learning and Computer-Based Instruction_, John Wiley and Sons, New York.

Ellis, E. B. (1974), _The Use and Misuse of Computers in Education_, McGraw-Hill, New York.

Gagne, R. M. (1977), _The Conditions of Learning_, Holt, Rinehart and Winston, New York.

Gronlund, N. E. (1971), _Stating Behavioral Objectives for Classroom Instruction_, The MacMillan Co., New York.

Lumsdaine, A. A. (1963), Instruments and Media of Instruction. _Hand-Book of Research in Teaching_, N. L. Gage, (Ed.), American Educational Research Association. Rand McNalley, Chicago, p.585.

Markle, S. M. (1964), _Good Frames and Bad_. John Wiley & Sons, New York.

Stolurow, L. M. (1961), _Teaching by Machine_, U. S. Department of Health, Education and Welfare, Washington, D.C., pp.4-5.

Wise, K. D., Kam, C. and Yokely, R. E., _Microcomputers: A Technology Forecast and Assessment to the Year 2000_. John Wiley and Sons, New York.

CAI SYSTEMS AND TEACHING MATERIAL PORTABILITY

M.Ferraris, V.Midoro, G.Olimpo

Istituto Tecnologie Didattiche, C.N.R., Genova (Italy)

ABSTRACT

The problem of portability of CAI material is discussed in connection with the structure of present and future CAI systems.

Most CAI systems existing today use the same computer for development and delivery; this approach will be less and less feasible in the future because of the increasing difference between development and delivery as to the demand of computer resources. In fact, while the work of course authors will be more and more computer assisted and resource consuming, student-computer dialogues will be carried on by inexpensive microcomputers.

Therefore, courseware portability will become a necessity not only for course delivery, but also for making the courseware utilization large enough so as to justify the high production investments.

In the last part of the paper COMES, a CAI system developed by the Istituto Tecnologie Didattiche of the Italian National Research Council, is described. COMES is a system oriented to courseware portability and naturally fits the trend toward separation between production and delivery expressed in the paper.

KEYWORDS

CAI, courseware, portability, microprocessors.

INTRODUCTION

Since the birth of computer assisted instruction in the fifties a great deal of different languages and systems were developed by companies and research groups to support the use of computers in education. These efforts were often independent and similar paths were followed anew several times; however, it is easy to recognize in all these systems two interesting common properties:

Property 1 - The same computer (or computers belonging to the same family) is used for both development and delivery of CAI courseware.

Property 2 - CAI courseware cannot be delivered on computers different from the one where it was developed.

These properties are by no means independent of each other. In fact, property 2 strictly implies property 1, while property 1 encourages property 2 in the sense

that a dogmatic acceptance of property 1 does not stimulate the research of techniques for producing portable CAI courseware.
The above mentioned properties induce several drawbacks in CAI applications.

Drawbacks coming from property 1

- The computer system used for both development and delivery has to be large enough in order to be able to fulfill the need for computer resources of the development system.
- A large computer is typically remote for most users; this increases the costs and reduce the total reliability of the system.
- A large computer is seldom wholly dedicated to CAI; this means that CAI delivery typically has to share a computer with other applications which exhibit different requirements mainly as to how the computer resources are managed.

Drawbacks coming from property 2

- All the drawbacks coming from property 1.
- Insufficient utilization of the human efforts required for production of CAI courseware. In fact, a large amount of courseware has been developed which is of high quality and very inetersting from the instructional standpoint, but is almost always bound to a particular computer family and cannot be easily transferred to other computers. This fact represents a waste of human resources because of both the multiple efforts for producing equivalent programs and the limited diffusion of CAI material due to hardware incompatibility.

From this short analysis we can state that courseware portability, i.e. the possibility to run CAI material on computers different from the one where it was developed is an extremely desirable property which represents one of the present day problems of CAI.
Some forecasts (1), (2), (3) predict that within one or two decades the problem will be completely overcome; in fact, hardware cost is rapidly decreasing and in a few years it will represent only a small fraction of the software cost. Therefore, the institutions, when buying a given package, will also be able to buy the hardware required for its delivery.
However, even in such a futuristic situation, the portability problem will play an important role in CAI, due to the need to use different computers for development and delivery of teaching material.
In fact, development and delivery are activities by no means homogeneous from the standpoint of their request of computer resources; and this disparity is bound to increase in the future since the need to support the author of CAI material will probably bring to some sort of "computer aided educational design".
This means that the development of CAI material will require computers of adequate power while delivery will be made by inexpensive microcomputers.
Therefore, suitable methodologies and tools will be needed for transferring CAI material from development systems to a variety of delivery (micro)systems.

FUTURE CAI SYSTEMS

In this section the different roles of computers in production and delivery of CAI material are further investigated and some previsions are made about future CAI systems.
Fig. 1 shows a possible structure for a CAI production system in the near future. Each box in the figure represents one of the phases of the production process and in each box some hints are given as to how the computer can assist that particular phase.
The production process starts with the instructional design, i.e. the design of the

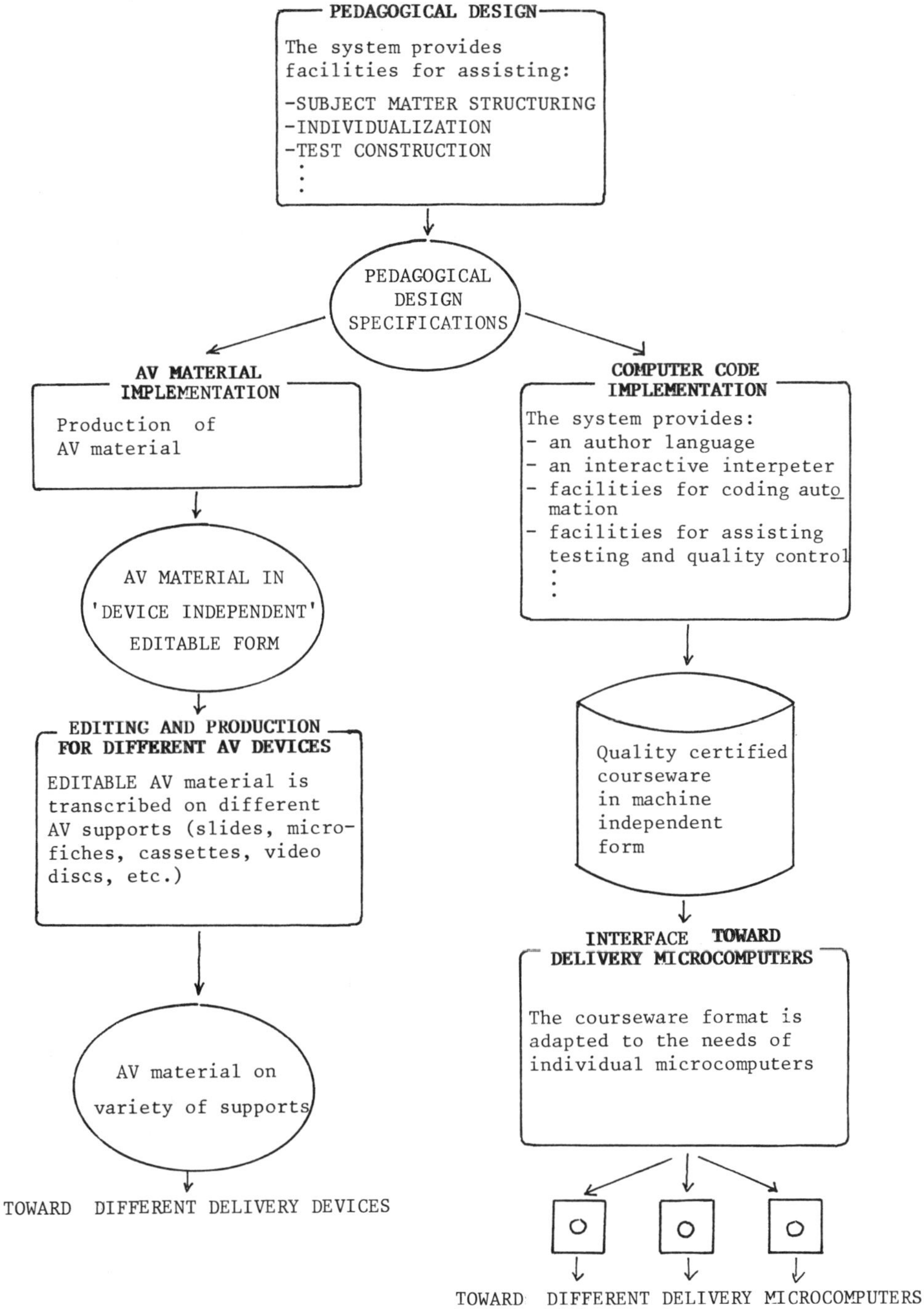

Fig. 1. Structure of a CAI production system.

process carried out without paying any attention to implementation considerations. Here the computer can play an important role assisting the course author in various activities such as subject matter structuring, individualization design, test items definition and so on.
Examples of computer support in these activities are already available (6), (10).

The instructional design phase yields a document which could be called 'instructional design specifications'. This document is the input to the implementation phase which in fig. 1 appears split into two subphases: computer code implementation and audiovisual material implementation. In fact, there is a trend even today, to integrate computer controlled AV devices into CAI delivery stations; and this trend is probably bound to increase with the advances in videodisc technology.

Technology

Here we will mostly focus our attention on the computer code implementation. Here the computer can provide not only a compiler for the author-defined courseware, but also a variety of interactive facilities, mainly for coding automation and for assisting testing and quality control; in fact, these activities are rather time consuming and error prone and exhibit several 'mechanical' aspects which can easily be dealt with by the computer.
The output of the implementation phase is the tested and quality certified courseware; of course, this courseware should not be dependent on any (micro)computer, but should be able to be run on a variety of delivery equipments.
Therefore, the output of the code implementation phase should be represented in an abstract, machine-independent form, easy to convert into a for usable on a particular microprocessor.
This conversion requires the rearranging of file structures, formats atc. for each given microprocessor by means of simple interface programs.
As far as delivery systems are concerned, we find that their functions are very simple and well-defined. Essentially they must provide an interpreter for the dialogues code, and occasionally they may be requested to communicate between each other, or with a larger computer for data exchange purposes. The latter function is very easy to accomplish even today since most microcomputers have built-in communication facilities.
The main purpose of this brief outline is to stress that there will be more and more difference between the resources needed by a CAI production environment and by a delivery system. This means that the use of different computers for production and delivery will become more and more of a necessity.
In fact, production environments will become more and more sophisticated, and resources consuming, and therefore justified only by a large market of CAI material; and such a market will be possible only by keeping delivery systems simple and inexpensive and by producing portable courseware.

THE APPROACH OF COMES

COMES is a new CAI system developed by the I.T.C. of the National NRC with two main purposes: first, to allow the separation of the phases of production and delivery, and second, to provide a tool for sharing CAI material at least on a national level.
Fig. 2 describes the basic structure of COMES.
In the development phase the author interacts with the facilities provided by the system (4), (5) and eventually the author-defined courseware is compiled into a form which is called 'portable code' (in the following it will be shortly referred to as PC).
The PC for a given instructional dialogue can be regarded as a program for an abstract processor having structure and instruction set strongly oriented to CAI applications; i.e. the functions associated with the instructions have been designed on the basis of the typical computational needs arising in CAI applications.
PC instructions and related spaces of memory are defined at a purely logical level;

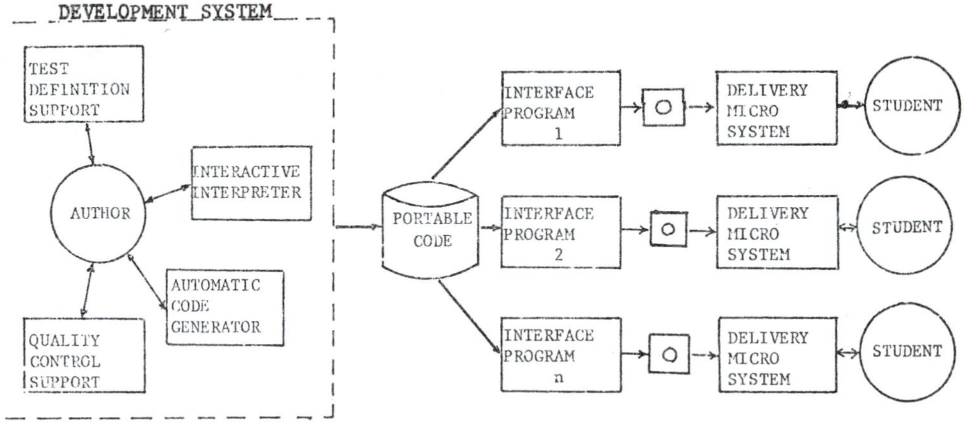

Fig. 2. The basic structure of COMES.

therefore, PC is independent of any particular machine, especially the one where it was produced. This holds also for graphic instructions and data which are independent of the model of the terminal used for development and delivery of graphic material.
Of course, courseware delivery requires that the PC interpreter is implemented on the delivery computer in order to create the CAI abstract processor on it.
At this point it should be clear that CAI material produced under COMES (in PC form) can be used on any computer system, provided that a PC interpreter running on it is available.
Of course, in most cases simple interface programs will be required to transform PC into an internal form suitable to the delivery system.
It is clear that in order to run COMES material on a new computer some effort is required to implement a new interpreter or modify an old one. This, however, should not be a serious obstacle, since the structure of the interpreter is very simple and full specifications are available for it (9). Therefore, it can easily be implemented on any machines [1]; and, when a machine is endowed with it, any material in PC form can run on it.
Finally, it is worth saying that this approach to portability does not imply that the author use a particular language for writing CAI material. What is actually required is that the language employed is translated by the author program in the portable PC form. It follows that changes and enhancements in the author system do not affect PC and the interpreters running on any machine.
Another interesting chance is that of implementing translators form different languages to PC in order to convert to portable form existing CAI material.

COMES, with its approach to portability, lends itself naturally to the separation of development and delivery discussed above, and naturally fits the trend expressed in Fig. 1. Besides, CAI material in PC form may be easily organized in 'CAI banks' typically resident on development computers. Banks can be accessed by different delivery computers of various types and sizes through communication links or via magnetic support.

Existing implementations of COMES are listed below:

---

(1) The only condition strictly required for the implementation of the interpreter is that random access to files be available.

Author System

a.1 version on an IBM 370/168 running under VM/CMS.

The system is presently operating and is being enhanced with facilities for automatic code generation and for assisting quality control.

Delivery System

d.1 version written in PL/1 running on IBM 370 under VM/CMS;
d.2 version written in BASIC running on an Olivetti P6060;
d.3 version written in UCSD Pascal running on an Apple II.

Version d.1 of the delivery program runs on the same computer used for development. Version d.2 and d.3 are examples of a standalone computer system receiving CAI material via floppy disc or communication lines.

REFERENCES

Bork, A. (1977). Instructional Systems. Educom Symposium, Minneapolis.

Bork, A. (1978). Transferability of Computer-based Learning Material. Internal Report, PCDP, University of California, Irvine.

Bork, A. (1978-79). Standalone Computer Systems - our Educational Future. Journal of Educational Technology Systems, Vol.7-3.

Ferraris, M. et al.(1979). COMES: Author Manual. Internal Report, Ist.Tecn.Didatt. CNR, Genoa.

Ferraris, M. et al.(1979). The Design of CAI Material in COMES. Internal Report, Ist.Tecn.Didatt.,CNR, Genoa.

Midoro, V. et al. (1980). A Methodology for a Computer Administered Testing. Proceed. of XIIth International Symposium on the Educational Technology, Poznan.

Olimpo, G. et al. (1978). COMES: A CAI System Oriented to Programming Languages. Proceed. ACM SIGSE/CSA Techn. Symposium on Computer Science Education, Detroit.

Olimpo et al. (1979). COMES: un Approccio alla Portabilità del Software Didattico. Rivista di Informatica, Vol. IX, Supplem. to n°3.

Olimpo et al. (1979). COMES System: Delivery Program Specification. Internal Report Ist. Tecn. Didattiche, CNR, Genoa.

Pask, G. (1976). Conversation Theory: Applications in Education and Epistemology. Elsevier.

FORESTS OF ORIENTED TREES AS METASTRUCTURES FOR A CAI SYSTEM

Ian R. Greenshields

Dept. of Computer Science,
Concordia University,
Montreal, P.Q. , Canada

ABSTRACT

A metastructure for a CAI system based on forests of oriented trees is developed from the notions of equivalence and order applied to the elements and partitions of a pedagogical data base. Routing through the forest is governed by a linear threshold decision making process.

KEYWORDS

CAI, tree structures, linear decision process, dynamic data structure

INTRODUCTION

The field encompassed by the term CAI can range from elementary systems designed to reinforce a learning situation through to large, complex structures aimed at being complete instructional systems in their own right (1). We propose here a metastructure biased towards CAI systems at the lower end of the complexity spectrum; the structure is designed to allow systematic evaluation of a student's performance over a question-answer environment and permit flexibility in routing students through a corpus of questions given the student's performance on prior questions. We will call such a corpus of material a pedagogical data base (PDB); this will be composed of elements (with associated answers) called units evaluatable in binary form. By means of successive linear weightings over subdivisions of the PDB the student's capability can be measured against some preset threshold upon which basis a decision to present the student with other portions of the PDB will be made. We introduce the notion of equivalent units by asserting that two units will be equivalent if they relate to a concept being taught (examined) at the same level of complexity and if they are independant of each other with respect to the understanding of the concept. We impose an order relation on the equivalence classes thus generated by saying that, for two equivalence classes A and B, A precedes B (B succeeds A) if a comprehension of the material in A is a necessary condition for a comprehension of the material in B. Such an equivalence class (and its associated linear weighting schema) can be represented by an oriented tree; the complete PDB will then comprise a forest of such trees. The entire structure of the forest and mechanism which routes students through the forest can be considered a metastructure imposed upon the PDB; a comparison with other meatstructures (e.g. (2)) indicates the simplicity of the construct.

METASTRUCTURE

Let $P=\{u_1,u_2,..,u_n\}$ be a PDB partitioned into equivalence classes $U_1,U_2,.U_m$ following the sense of equivalence described above. We order the U_i according to the notion of precedence outlined above; thus, U_i precedes U_j only if precedence holds. Decompose the U_i into subsets $V_{i,j}$ (the j'th subset of U_i). Then for every $u_i \in V_{i,j}$ we associate a terminal node $\tau(u_i)$ and a binary-valued function $f_B(u_i)$ such that $f_B(u_i)=1$ if theuser response to u_i is correct and $f_B(u_i)=0$ otherwise. Define a set of internal nodes $\Theta_I i$ and a relation on this set ($\sim$) with the property $\Theta_I i \sim \Theta_I j$ if $\Theta_I i$ is a direct predecessor node (father) of $\Theta_I j$. We associate with each internal node an integer valued constant w (the weight of the node) and an integer valued function $W(\Theta_I i)$ given by

$$W(\Theta_I^i) = w\sum_{\{\Theta_I^j:\Theta_I^i\sim\Theta_I^j\}} W(\Theta_I^j)$$

As a special case (subset) of the internal nodes we define the pre-terminal nodes $\Theta_p(V_{i,j})$ associated with the subsets $V_{i,j}$; they similarly have an integer valued weight constant w and an integer valued weight function given by:-

$$W(\Theta_p(V_{i,j})) = w\sum_{u_i\in V_{i,j}} f_B(u_i)$$

Following the definition of an oriented tree (3) as one in which the relative ordering of subtrees is unimportant, we can associate an oriented tree with each equivalence class U_i; such a structure is shown in fig. 1.

Fig. 1

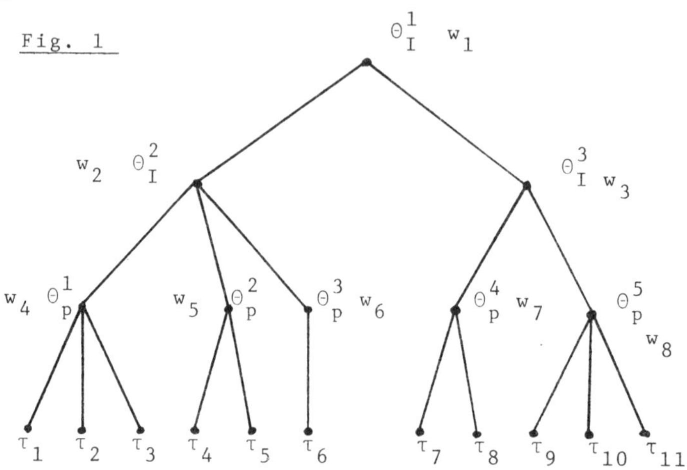

Note that the terminal (leaf) nodes of the tree are the τ-nodes associated with the units, while successive nodes represent various linear weightings over the subtrees until a final weighting is given by the root. Using the weightings beside the nodes of the tree and abbreviating $f_B(u_i)$ by u_i, the weighting for the example tree is clearly

$$W(\Theta_I 1) = w_1(w_2(w_4(u_1+u_2+u_3)+w_5(u_4+u_5)+w_6u_6)+w_3(w_7(u_7+u_8)+w_8(u_9+u_{10}+u_{11})))$$

Clearly the tree structure applies a system of linear weightings to sections of the PDB. A complete system of such oriented trees, each one corresponding to an equivalence class of the PDB, is a forest. Traversal through the forest is governed by comparing the student's performance (represented by the value of the weight function of the root node) against a preset threshold and choosing the next tree traversed on a success/failure basis. This is accomplished by introducing a new node (Δ-node) which in effect roots the forest to produce a semi-ordered tree (only the immediate sons of the Δ-node are ordered;these are roots of oriented trees). This situation is depicted in figure 2; note that leftsons precede rightsons following the concept of precedence over equivalence classes.

Fig. 2

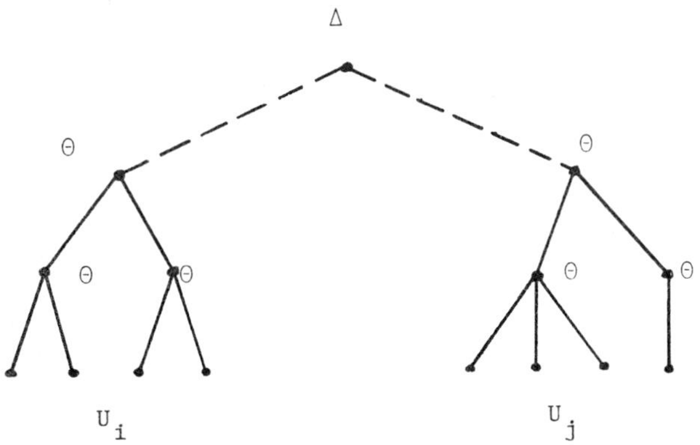

NODE STRUCTURE

We illustrate an implementation of such a forest using the data structures available in PASCAL (4,5). In order to avoid the introduction of defined pointer types, we proceed by defining records in the order PDB records, τ-node records, θ-node records and Δ-node records. Assume that the PDB is composed of records of type PDBRECORD such that each record contains both the unit and its affiliated answer; thus, in general, a PDB will be composed of records of the following type:-

```
type PDBRECORD =record QUESTION:QRECORD; ANSWER:ANSRECORD end;
```

τ-nodes will point into PDBRECORDs and maintain the result of the binary-valued function applied to the PDB record.Thus,

```
type TAUNODE= record  UNIT: ↑PDBRECORD; FBUI: 0..1 end;
```

In the consideration of the nodes to follow we omit certain ancillary definitions (for space reasons). Θ-nodes must, aside from storing their own weight w and the value of the weight function $W(\Theta)$ for use by the fathers of Θ-nodes, point either into other Θ-nodes or τ-nodes. Since pointer variables are bound to their defining type, we must distinguish Θ-nodes as having sons of type Θ or sons of type τ. The record structure therefore requires the use of variants. Since the number of sons a node can have must be bounded, we assume the existance of an integer constant MAXNODES which is the maximum number of sons to any node. Then a record structure representing a Θ-node can be expressed as:-

```
type THETANODE= record
     WEIGHTCONSTANT,WEIGHTFUNCTION: integer;
     case  MODE: (PRETERMINAL,INTERNAL) of
     PRETERMINAL: (SONPTR1: array[1..MAXNODES] of ↑TAUNODE);
     INTERNAL: (SONPTR2: array[1..MAXNODES] of ↑THETANODE);
end;
```

Clearly the scalar type MODE identifies the Θ-node as having sons of type τ (PRETERMINAL) or θ (INTERNAL). Δ-nodes maintain three blocks of information:- links, in the form of success and failure pointers per son (identifying which tree or (conceivably) which forest is traversed given the threshold determination)- necessitating the provision of pointers to both θ-nodes and Δ-nodes; thresholds, which are the integer values to be exceeded by the root weight function of a tree in order that a success be recorded and status information in the form of a success array whose i'th entry is set to 1 if the i'th son was traversed successfully (thus permitting re-entry to the tree without executing sons completed correctly). The Δ-node can be expressed as:

```
type DELTANODE= record
     THRESHOLDS: array[1..MAXNODES] of integer;
SPTRSTOTHETA,FPTRSTOTHETA: array[1..MAXNODES]of ↑THETANODE;
SPTRSTODELTA,FPTRSTODELTA: array[1..MAXNODES]of ↑DELTANODE;
     STATUS: array[1..MAXNODES] of 0..1;
end;
```

Traversal to a θ-node or a Δ-node from the i'th oriented tree is governed by inspecting the success/failure pointers in the arrays SPTRSTOTHETA/FPTRSTOTHETA indexed by i; if the i'th entry is non-nil then a θ-node is traversed else the success/failure pointer to a Δ-node is used (causing a new forest to be traversed). Traversal of the trees is performed postorder; thus the tree in fig. 1 is traversed as:-

$$\tau_1\tau_2\tau_3\Theta^1_p\tau_4\tau_5\Theta^2_p\tau_6\Theta^3_p\tau_7\tau_8\Theta^4_p\tau_9\tau_{10}\tau_{11}\Theta^5_p\Theta^2_I\Theta^3_I\Theta^1_I$$

EXAMPLE

An illustrative example is shown in figure3; here we assume the system is used to test a blind person's knowledge of the audio output of a blind reading aid. Thus the PDB comprises a set of audio vectors and answers in the form of strings of characters corresponding to the encoded word or character represented in the audio vector. The small PDB shown is composed of the set {a,e,i,o,f,m,r,s,t,so,if,at,or,me}. The equivalence classes are U_1={a,e,i,o,f,m,r,s,t} and U_2={so,if,at, or,me} with U_1 further decomposed into subsets $V_{1,1}$={a,e,i,o} and $V_{1,2}$={f,m,r,s,t}. $f_B(u_i)$ is 1 if the user correctly matches the output with the result string. Given the assumption that vowels are easier to discriminate than consonants, $V_{1,1}$ is given unit weighting

Fig. 3

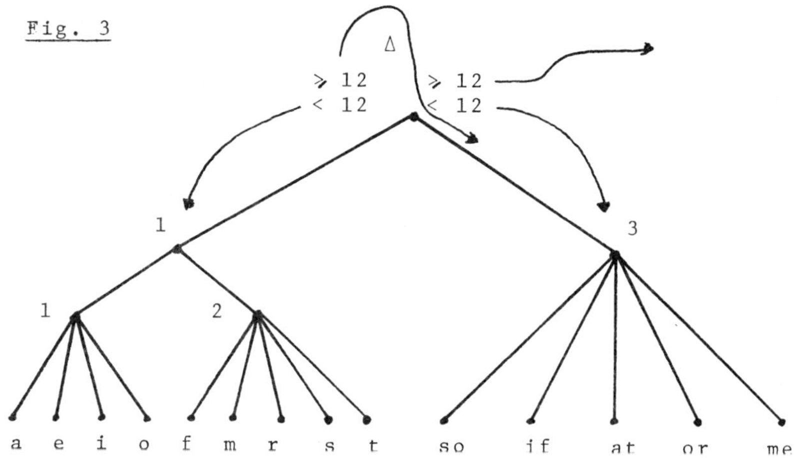

while $V_{1,2}$ is given a weight of 2. These are combined to weight U_2 at 3. The thresholds shown (>=12 for both classes) show both failure pointers returning to the failed tree; the success pointer associated with U_1 points to U_2 while the success pointer of U_2 points to a new forest.

LITERATURE CITED

(1):"Computer Assisted Instruction: Current Trends and Future Issues" J.A. Chambers and J.W. Sprecher, CACM vol23 no 6 June 1980
(2):"Transition diagram-based CAI/HELP systems" S. Feyock, Int. J. Man-Machine Studies(1977)9,399-413
(3):"The Art of Computer Programming: Fundamental Algorithms" D. Knuth, Addison Wesley,Mass. 1973
(4):"PASCAL User Manual and Report" K. Jensen and N. Wirth, Springer-Verlag, N.Y., 1974
(5):"Algorithms +Data Structures = Programs" N. Wirth, Prentice-Hall,N.J., 1976

TEACHING STATISTICS AND COMPUTERS

Walter W. Zwirner

The University of Calgary

ABSTRACT

Four years of intensive and experimental use of computers in the field of teaching statistics are summarized. The usefuleness of computers in seven areas of teaching are described. Experience indicates that some of these uses can be generalized to other fields of teaching. Wider use of computers is suggested and predicted.

KEYWORDS

teaching statistics; computers; data generator; sampling simulator; text transmittor; test administrator; numbers cruncher; calculator.

Computers slowly invade the classrooms at all levels of instruction. This "invasion" is perhaps slower than anticipated by some, nevertheless it finds most of us unprepared. We are suddently confronted with a powerful, and very often expensive tool.

The versatility of this device is perhaps nowhere easier to illustrate than in the area of teaching statistics. Experience has shown that the computer can be of valuable assistance in at least seven areas:

1. as a data generator;
2. useful for test construction;
3. useful in test administration;
4. helpful for text and information transmission;
5. useful numbers "cruncher";
6. assistance in test correction;
7. interactive calculator for classrooms.

Extensive use of different computer systems for the last four years has led to the strong belief that computers can be used as a didactive tool in teaching. Students at a graduate level, at which all this experience was developed, had little difficulty, although some frustration, in learning how to use computers. For most this meant learning the basic commands to "log" in and call on programs. Invariably some students find themselves intrigued and end up programming quite efficiently. A rather charming feature of the Multics system, part of the

Honeywell installation at the University of Calgary, is the communication available to users with other users. This mail system allows for immediate communication with all other users who are simultaneously using the system. All seven areas mentioned above are regularly used at different levels of statistics classes taught in Educational Psychology classes, this includes the standard introductory statistics course, a non-parametric statistics course, a course in multivariate statistics, and a course in factor analysis.

Data Generator.

The basis of statistical reasoning is the concept of randomness, in particular the collecting of random observations. Although statistics is founded on the observation of regularities of large numbers of observations, one of the misconceptions concerns the prediction of outcomes for small and medium size samples. The concept of errors and risks involved in making predictions can be developed intuitively through contemplating results of repeated sampling from the same population. Such samples are useful for both descriptive and inferential statistics. Textbook examples, although useful for exercises, nevertheless do not allow for simulation of the experiences which statisticians and actuarians collected over more than two hundred years. Students can receive some inkling of this process by considering large samples from the same population.

A random number generator was developed which can be used to generate data from a uniform, a normal, or a chi-square distribution, for use mainly in the introductory statistics course. Students can at any time ask for samples from these distributions, either with known or unknown parameters such as the mean, variance, degrees of freedom, minimum or maximum value. Students are then encouraged to find estimators for these samples and compare these to the optimum estimators developed by statisticians.

It was found that students developed a better understanding of the concepts of samples, estimators and populations when using the sampling program. Theoretically important concepts, such as the central limit theorem, can be developed by letting students sample repeatedly from a uniform distribution, arriving at a normal and then using these results to see the concept of a chi-square distribution developed.

Test Construction.

It is a commonly accepted truth that persons proceed at different speed in understanding and being able to find solutions to problems. Thus testing all students with the same test does not allow for differences in the speed with which students understand new concepts. Test construction for different students poses time problems if a faculty member has to develop each test anew. A solution to this problem exists in the case of statistics teaching since the main differences do not exist in the question as such, but in the data analyzed. The random number generator described above is used to "sample" data for each student, thus allowing students to write their own test, more like an analysis, when they feel ready. For administrative reasons deadlines are established for the completion of each test.

An added advantage of this method is the natural interaction between students and faculty in discussions concerning the test questions. I believe that a large amount of learning outside the classroom is encouraged through this method of testing. Another interesting development of statistical reasoning takes place when students compare their datasets, since they know that all data is sampled from the same population the distribution of sample statistics becomes clearer.

Test Administration.

"Links" between the instructor's workspace on the computer and that of all the students are easily established. These specialized workspaces include among others the test and information files. The test file includes all previous test questions as well as laboratory exercises, the information file can include solutions to "future" problems, as well as information concerning the use of the computer.

Test questions can be solved at any time that the student feels ready to answer the question. Many students have started to answer their questions also within the computer workspaces. Thus a certain amount of computer literacy is achieved through frequent use of the computers. This goal is not only valuable for statistics, but useful for teachers in general.

Text and Information Transmission.

A common problem with starting to study statistics has to do with the specialized language developed by statisticians. To confound the problem, different statisticians often differ in notation for some lesser used concepts. Theorems such as the central limit theorem, or the development of distributions, can be confusing if a student has to keep notes as well as follow arguments. It was found convenient and useful to store certain of these theorems in a permanent file with information concerning some widely used notational idiosyncracies. Students are told when such information is available, since these files are public, some again are able to study ahead, and find this an advantage.

Numbers Cruncher.

Allowing students to concentrate on the theory underlying statistical arguments should make them better users of statistical methods. The courses are all oriented toward understanding and away from learning how to perform calculations. Besides the standard statistical program packages such as SPSS, the BMD series and other custom tailored programs, we use interactive statistical programs based on APL. This particular language proved invaluable since it allows for the collecting and processing of rather complicated designs in class. Multivariate analysis as well as interactive graphing was performed and used to illustrate the results of different types of analysis. Lack of monitors at this time did not allow for the use of this technique in large classrooms. Thus tedious arithmetic could be done away with, and more time used to concentrate on the underlying models and theory.

Test Correction.

This part of the computer is only used as an assistance for the instructor at this stage. Flow diagrams of the solution to problems are stored and used to correct the answers.

It is conceivable and planned to use the computer for feedback to students on their progress in the solution of their problems. This would be a step in the direction of self directed learning. Although this step has been in the planning for some time, other problems seemed to be given priorities at the cost of this rather interesting problem.

Calculator.

Students have become to some extent used to working their problems on a computer instead of a desk, or pocket calculator. The two languages which were found most useful in this context are BASIC and APL. Since this is a preference determined by each student individually, not more than encouragement could be given.

Finally, after four years of extensive use of computers in the classroom it was found that students in time appreciate and become quite proficient in the use of computers. With the influx of computers into homes and elementary classrooms we can expect to find a more and more sophisticated group of students expecting and even demanding an efficient and versatile use of computers at the University level of instruction.

XV. COMPUTERS AND THE HUMANITIES

THE ESTATE OF THE ART: ASKING MORE OF COMPUTERS IN THE HUMANITIES IN THE EIGHTIES

Jeffrey F. Huntsman

Department of English, Indiana University, Bloomington, Indiana 47405, USA

ABSTRACT

This paper suggests that the greatest past failure of humanities scholars working with computers is one of the imagination--that, in other words, we have considerably underutilized even the existing hardware and software technology at our disposal. By way of illustration of projects with more ambitious but realistic goals, a modest project drawing on Artificial Intelligence procedures is sketched, a project that aims to replicate some of the essential features of the theory and practice of medieval English lexicographers.

KEYWORDS

Humanities applications; Artificial Intelligence; lexicology; lexicography; Middle Ages; England.

Since this paper is in part a kind of sermon, I might begin with a text, taken from Mortimer Taube's attack on machine translation and other aspects of Artificial Intelligence, *Computers and Common Sense: The Myth of Thinking Machines*:

> All noble things, says Spinoza, are as difficult as they are rare. So it is with respect to science and scientific advance in the computer and data processing field. Genuine advances are difficult and rare. And even more rarely are they the product of prophecy and the premature announcement of what someone expects to do but has not done (1961: 128).

Although I agree that "difficult and rare" indeed are major advances--most scientists (and humanists, for that matter) stay safely and unimaginatively well within the bounds of their paradigms--I disagree most strongly that speculation, even wishful thinking, is detrimental to progress. In fact, some of the most fruitful modellings of portions of the universe have been leaps in wonderment across chasms of theory where few solid facts exist. The analogical extension of the solar system model to the atom is still ringing currency in school classrooms, despite the fact that we now know it to be inaccurate to a distressing degree, and, as my colleague Philip Appleman has pointed out, Darwin's description of the process of evolution had to wait a century before we could understand even the rudiments of the evolutionary mechanism, the involved double helix and the genetic information it carries. I contend that with respect to the humanities, no less than to the sciences, Taube is dead wrong, and I mean to suggest that what we as virtual if not actual humanists exhibit is at least a failure of vision, if not a failure of courage. We are not limited, as Professor Logrippo suggested in a comment earlier in this conference, because we live in a "technological Stone Age," but because we have chosen to use our twentieth-century tools in nineteenth-century ways.

In the past decade, computers have gained a small but well-established place in the geographies of humanists. Publications like _Computers and the Humanities_ and _Computers and Medieval Data Processing_ list projects underway and completed that attest to the increasingly greater numbers of established humanists who are creating and using scholarly works like concordances, bibliographies, stylistic analyses, and editions--works produced with the aid of computers of all sizes. And if a few of our more traditionally-minded colleagues occasionally mutter neo-Luddite sentiments, many of our students today are taking the word-processing and mass storage capabilities of computers as much for granted as older scholars do their typewriters. Yet this general acceptance of high-speed data processing among many of us seems to have another, unfortunate appurtenance of the approaching maturity of the techniques: there are aspects of unconscious conservatism, of sameness, of a tendency to do old tasks in newer and more efficient ways ("A nanosecond saved is a nanosecond earned"--_Poor Tom's Almanac_) rather than to attempt new tasks. While all of these scholarly ancillae are of course very useful--and the current (accelerating) rate of improvement in both hardware and software promises that such tools and techniques will become increasingly efficient, precise, reliable, powerful, generally available, and cheap--they seem to be a kind of pleasant plateau, graced with laurels, on which we have rested too long. The State of the Art in these matters is peaceful enough for us coding contentedly in our dovish coteries, but it is stiffling nonetheless. Unless we expand the limits of what _we ourselves_ have chosen to do, unless we increase the _Estate_ of the Art, we run a danger of becoming like our programs--neat, perhaps even stylish in a dapper sort of way, but a trifle dull.

The alternative, I suggest, is tremendously more interesting, if only because we _can_ define, if not tasks totally new to the field, at least tasks approached in new and exciting ways. The humanities of course encompass more than language and literature, but those are the areas about which my ignorance is least acute, so I will speak of them. Literary theory (and to some extent literary criticism) is one area in which computer-wise humanists could formulate and test hypotheses whose falsification or affirmation could tell us much about how and why literature is written, read, and interpreted. I suspect we (and our students) will have to breach the wall around this area, for many of our more traditional colleagues are profoundly disturbed by the possibilities that their talents will be rendered obsolete. This fear is baseless, of course--none of us expects computers to relieve us of anything but the dogwork--but there is comfort in the warm and romantic mist that invests many humanities departments. So long as literary criticism and its counterparts are esoteric, the mysteries will remain enshrouded and the discipline protected. Thus we have book after book after book, each of which offers yet another so-called theory of literature, but few testable hypotheses and fewer replicable results are ever adduced. Testability is one of the criteria that separates a merely internally self-consistent system--a logic like mathematics or philosophy--from any system that attempts to deal with those intersubjectively congruent constructs we label "reality." Unfortunately, few literary "theories" floating about today meet this minimun standard of science--and I am here using the term _science_ in its classical sense of a _scientia_, a way and body of sharable knowledge, not in its more colloquial sense in which it is hierarchically preferred to systems of largely subjective interpretation like history or with those of cabalistic prognostication like sociology. Today literary theories seem to be in the position post-Logical Positivism was--some have retreated into personal speculations that deny testability (and thus are nonrational, although not necessarily irrational or even wrong), while others have trivialized what is considered important by looking only at the most easily testable (and therefore least interesting). Some, such as those outbreaks of structuralism that are not philosophical but descriptive in their intent have managed to commit both errors simultaneously. But literature--and our ways both of understanding it and of characterizing that understanding--may continue to provide the warranties of our humanity that we as a species demand, by increasing our awareness of how and why literature (or art or music or history or philosophy or any other intellectual endeavor, for that matter) moves us the way it does.

How might this ambitious (if not utopian) scheme be advanced? To begin with, as a profession I think we need to look well beyond the walls of our departments to the labor of workers in fields like psychology, epistemology, and Artificial Intelligence. In the past two decades, since the development of Artificial Intelligence as a discipline (one which obviously draws heavily on the findings of philosophy, psychology, linguistics, and cybernetics), researchers have discovered that the most successful modellings of the human mind are not those which postulate a small number of general principles but rather those which embody the ability to use a large amount of very diverse material in very different ways. Understanding is characterized, in a current concensus of many AI theorists, as the result of a process in which a system which performs TASKS that determine SETS OF KNOWLEDGE, which in turn prompt the system to modify the tasks so as to reconstitute the sets of knowledge, and so forth. Unlike earlier schemata for simulating intelligence, contemporary approaches emphasize procedural knowledge over declarative knowledge (e.g., "if X then do Y," not "all X is Y"). Further, some level of symbolic activity is requisite for any but the simplest tasks, and many tasks are better (meaning here both faster and more effectively) performed when several are done in parallel, not serial. (Here we might compare a common contemporary model of language, with its lexicon and constitutive rules, which characterize sets of knowledge, and a complex of other rules, which characterize tasks.) As Herbert Simon has said, "Knowledge without procedure is dumb, and procedure without knowledge is blind."

The complex programs to simulate intelligence which have been developed in this kind of theoretical environment are self-reflective, self-referencing, and self-modifying. They allow the investigation of more complicated, if narrowly defined problems (and here we might again compare the activities of linguists like Noam Chomsky). By way of illustration, I would like to conclude this brief presentation with a sketch of a simulation of an area in my own work which uses the insights of AI and its associated disciplines. The project is not done (indeed, except for the ideas which follow here, it is scacely begun in earnest), but I think it can be done and I especially think it should at least be tried. The particular area of research is medieval English lexicography, specifically involving the set of fourteenth- and fifteenth-century Latin-English translating dictionaries called the *Medulla grammatice* (found in 21 extant manuscripts, about half fragmentary). These texts are related in very complicated ways to dozens of other texts, but they still, taken together, constitute a reasonably coherent, if evolving body of linguistic knowledge current in England from ca. 1375 through 1480 and beyond.

Unlike grammarians, who aften discussed their professional motives and methods quite explicitly, medieval lexicographers seldom commented on the principles of their work; the exceptions, like Papias (c. 1053) and Johannes Balbus (1286) concerned themselves chiefly with mechanics such as the systems for alphabetizing the lemmata and indicating the grammatical information concerning gender, declension, conjugation, and so forth. About the conceptual structure of their works they had little to say beyond generalities like Papias' claim that his *Elementarium* surpassed mere glossaries, with their simple interpretations for words, by adding "definitions and notations according to rules." But medieval practice in lexicology and lexicography, like modern theory, recognized that words can be analyzed in differing ways and on several levels of abstractness and complexity; they may be given values in the form of statements of definition, equivalence, and similarity, which may not be interrelated in any comprehensive and principled way obvious to us. Despite the lack of an articulated and (apparently) unified system of definition, equivalence, and similarity, however, it seems clear that medieval lexicographers were proceeding according to some design, for their products show a consistency that cannot be ascribed merely to a tradition of copying and amalgamation. Tentatively, I have identified seven types of glossing procedures.

I. Definition
   A. Discrete hierarchies (genus-species systems)
   B. Conflated hierarchies (feature systems)
   C. Operations or recipies

II. Equivalence
   A. 1. Synonyms
      2. Translations
   B. Identifications
   C. Interpretations

My project has several dimensions:

1) the analysis of one full manuscript (Pepys MSS 2002′) of the dictionary, including lemmatizing and concording, so as to establish tables of synomyms, equivalents, etc., in a _discrimination net_ (that is, a flexible associational scheme that would provide directed linkings, like Edward Feigenbaum′s EPAM and John B. Smith′s RATSOL);

2) the development of algorithms for discovering the kind of gloss or definition used, beginning with the broad scheme outlined above;

3) with the iteration of the first two steps, the instruction of the program in Middle English, using the vocabulary already in the glosses of the Pepys manuscript; and

4) the generation and testing of new definitions (that is, definitions not already in the manuscript) first within the Pepys text, compared against the discrimination net, and second, compared against other mauscripts of the same dictionary;

5) the reformulation of the definition and equivalence schemata and subsequent iteration of step 4; and

6) the melding of additional _Medulla_ maunscripts into the data base and the recycling of the entire sequence of programs.

Clearly, there are several advantages of this scheme. By using the technique of _generation and testing_, the distinction between declarative knowledge and procedural knowledge is blurred--because with each interation of the procedure new knowledge becomes available to give the procedure a more precise, extensive, or profound scope for action--the system becomes self-reflective, as well as self-referencing, and thus can possibly replicate a process similar to that used by the English-speaking schoolboy who actually used the _Medulla_. Further, by testing against other manuscripts, it may be possible to replicate what processes a dictionary compiler used in making the work.

In this way, I would seek to discover the principles underlying medieval lexicography by developing a self-modifying computer model which, if successful, would discover the constants, parameters, and variables of medieval dictionary making. The translation procedures and lexical structures that have been continuing interests of researchers in computational linguistics and Artifical Intelligence should offer considerable illumination of the ways language users, including dictionary compilers, establish the conceptual mappings of the dimensions of lexical meaning and their variability in natural language. A model of medieval defining processes would thus be a heuristic device for developing and testing hypotheses about medieval lexicographic theory. Despite the fact that this project uses a method of modelling that I believe is new to the history of linguistics, it is clearly feasible.

I do not claim that my project will necessarily be interesting to anyone but myself, but I do hope its description, and the jeremiad that opened this paper, will be evocative, stimulating, challenging for some at least. As I see our work in the humanities during the next decade, the overriding limitation is not what we _can_ do but what we _wish_ to do. If there is any constant in today's computing world, it is that the technology is vastly underutilized by its human users. And in no field is that more true than in the humanities.

FRENCH SYMBOLIST LITERARY THEORY AND COMPUTER-AIDED ANALYSIS OF NOVELS

Paul A. Fortier

Department of French and Spanish, University of Manitoba
Winnipeg, Man., Canada

ABSTRACT

French symbolist theory provides the justification for applying information retrieval techniques to theme structures in novels. This approach was used to analyse three French novels. Results of the analysis indicate the applicability of this method to various types of literature.

KEYWORDS

Information retrieval; literature; French; symbolism; novel; theme.

INTRODUCTION

Translating Poe's (1927) "The Philosophy of Composition", Charles Baudelaire (1933) brought into the French symbolist literary tradition the idea that every aspect, even every word, in a literary text must contribute to the overall aesthetic effect of the work; nothing should be unrelated to this goal and nothing should be left to chance. Paul Valéry (1957, 1960) expanded and elaborated on this concept, making it a cornerstone of symbolist aesthetics in France. More recent scholars like Todorov (1970) and Ricardou (1971) have retained this concept. This idea can be applied to plot and characterisation with relative ease, but problems arise with themes and with atmosphere which is frequently created by thematic structures. Usually, one can work only with impressions about themes, but has verifiable facts about plot and characterisation.

In order to have precise and objective information about the frequency and distribution of themes, I had a computer analysis system developed by a professional programmer between 1971 and 1973 (Fortier, 1975). One part of the system builds a lemmatised list of all words in the text, showing their part of speech and location. Another part constructs an alphabetical list of words constantly related to a basic theme word like love or violence in French synonym dictionaries. Since both sets of lists are in alphabetical order and show the dictionary form plus the part of speech of words, they can be compared rapidly to produce information concerning the frequency and distribution of certain themes in a prepared text. This computer analysis system emulated the approach which I used working with pencil and paper on a keyword concordance to Céline's _Voyage au bout de la nuit_. Since

1973, I have been able to use the system to analyse Robbe-Grillet's la Jalousie and Gide's l'Immoraliste.

VOYAGE AU BOUT DE LA NUIT

Voyage au bout de la nuit (Céline, 1962) is a picaresque novel with philosophical overtones. The main character, Bardamu, joins the French army just before the First World War, and is subjected to wartime conditions both at the front and in Paris. He escapes to become a trader in the African jungle, but catches a fever and is sold as a galley slave on a vessel going to the United States. After a sojourn there, he returns to France, becomes a doctor and practices medicine in Rancy, a poor section of Paris. He ends up as director of a madhouse in Vigny, outside of Paris.

From a keyword concordance, I set up about one hundred frequency and distribution tables for various themes in this text. About forty of them were useful for interpretation. The rest were either not frequent enough, or not interesting enough to be useful. To give an example of the latter case, the theme of death is evoked about 500 times, but is concentrated in the war section of the novel, where Bardamu is obsessed by the fear of dying, or in chapters describing the death of certain secondary characters. This theme is not interesting because its distribution structure reveals information already obvious from a cursory reading of the text. Such is not always the case.

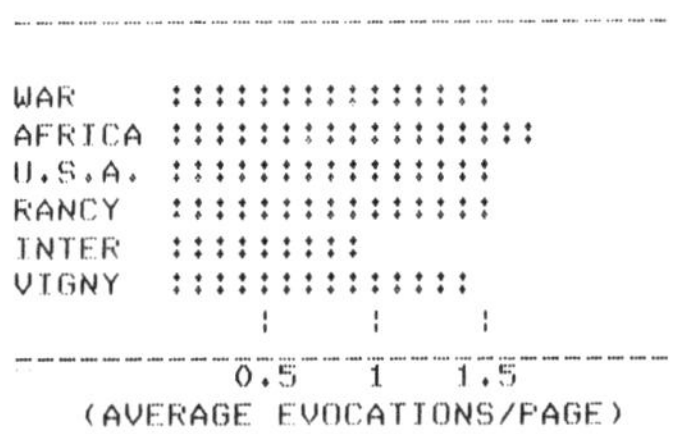

Fig. 1 The theme of violence in Voyage au bout de la nuit.

The theme of violence (see Fig. 1) appears most frequently, not in the war chapters of the novel, but in the section describing Africa. This fact is striking because war necessarily implies violence, but biological forces dominate in Africa. Violence, thus, is not primarily associated with an exceptional historical phenomenon --war--but with an integral aspect of the human condition. This structure of the theme explains in part the strong aura of pessimism which this novel generates.

Thematic clusters also help explain the novel. For example, in the war section, Bardamu describes almost lightheartedly the effect of a shellburst on two soldiers --one is decapitated, the other literally ripped open (20-21). But when he has to go to his regiment's improvised slaughterhouse, he becomes physically sick with disgust (24). It can be shown that the themes characterizing the description of the slaughterhouse echo those used in evoking conditions at the front. So Bardamu violently rejects a paradigm of war rather than a single incident. Similarly, he becomes beside himself with fear at the sight of a shooting gallery at a carnival (60-61). The description of the shooting gallery pulls together all the themes which this novel associates with war--both at the front and at the rear. So Bardamu's fear is profoundly justified--on the level of theme--although the other characters in the novel see it as cowardice or madness.

Towards the end of the novel, Bardamu hires a nurse, Sophie. His descriptions of her beauty contain strong overtones of revulsion (462-64), an ambivalence which one critic cites as an example of Céline's "fundamental perversity" (Beaujour, 1965). But on the level of theme, Sophie is intergrated into a myth of feminine beauty which attracted Bardamu to America--where he had suffered greatly. More than a hundred pages before the description of Sophie, Bardamu had debunked this myth (355-56), showing that it was related to the biological forces whose evil effects he had seen in Africa and in Parisian slums. So information about themes allows me to explain Bardamu's ambivalent attitude in terms of the structure of the novel, without needing to fall back on complexes which the author Céline might or might not have had.

Furthermore, I have been able to demonstrate (Fortier, 1974) that each section of this novel begins with a description which sets out the thematic content of what will develop in the following pages. This structure adds an overlay of inevitability to the novel's pessimism, and shows that this text, first published in 1932, is based on a reactionary world view which Céline did not express explicitly, on the political level, until 1936.

This, I think, is enough to show that precise analysis of themes provided the material for a nice interpretation of the novel. So in 1973, with a spanking new system to automate the routine aspects of theme analysis, and with high hopes, I turned my attention to Robbe-Grillet's (1957) la Jalousie, and hit a brick wall.

LA JALOUSIE

This novel (whose title can be translated as Jealousy) reflects the consciousness of an invisible narrator as he mulls over the circumstances surrounding a trip to town taken by his wife and an over-solicitous male neighbour. Because of car trouble, the pair had to spend a night in town, leaving the narrator alone on his plantation. The narrator recalls several times various incidents preceding or following the trip, but each time they change a bit. It is also impossible to fit these incidents into chronological order, because their order varies on the different occasions when they are recalled.

Computer processing allowed me rapidly to show the numerical importance of themes like uncertainty or the human body. But frequency and distribution tables proved of little help in analysing the novel because they showed only where evocations of the theme were concentrated, but gave no indication of plot or characterisation structures. So, I had only one term for comparison, when I needed two or three. For example, I could prove that the theme of travel is important in those places where the narrator is thinking of the trip to town which is the main incident of the novel, but nothing more. Computer analysis of theme distribution showed that theme structure is the basic organisational principle of the novel, but that is not new, nor does it lead to much in the line of interpretation.

Fortunately, a lemmatized concordance of the novel was a by-product of preparing it for theme analysis. It permitted me to trace each important theme, identify the places where it dominated, and more important, to relate it to other themes. I was thus able to interpret the text and show that most of the aspects of this novel produce a galling uncertainty, which resembles part of the emotional experience of jealousy. But I had to admit that for this text, my theme analysis system provided little that a lemmatised concordance would not have furnished. Since the expense of going from lemmatised text to analysable file, and the greater expense of recording themes for analysis, were not cost effective, I had to admit that this use of the system had led to a failure.

L'IMMORALISTE

Work on Gide's (1958) l'Immoraliste produced better results. Michel, the main character of this novel, contracts tuberculosis on his honeymoon in North Africa; he recovers his health, thanks to the devotion of his wife Marceline. The couple returns to France via Italy, but Marceline falls ill and Michel takes her to Switzerland to help her get better. But he can't stay put, and they leave for Italy, then North Africa again. There, worn out by travelling, Marceline dies.

One could interpret Michel's story as a case of almost pathological egotism, but this impression is not borne out by the distribution of first-person singular pronouns (see Fig. 2). Since Michel is a first-person narrator, these pronouns are frequent. Clearly, however, their frequency diminishes substantially, if not regularly, in the course of the novel. The language of this text shows that Michel speaks progressively less about himself--that he becomes less egotistical--as he tells his story.

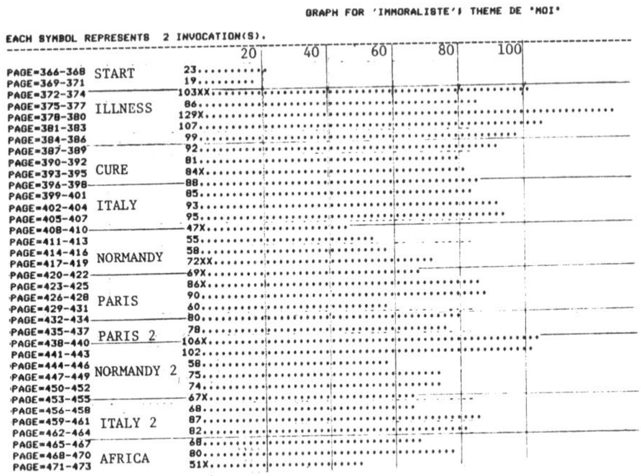

Fig. 2. First-person singular pronouns in l'Immoraliste.

A thematic cluster centred around illness suggests another interpretation (see Fig. 3). These themes are most frequently evoked at the beginning of the text when the narrator is quite sick. Their frequency drops off sharply as he gets better. But it goes up again--in the sections describing his cure, Italy, Normandy and Paris--because the narrator tends to associate themes like ugliness, boredom and order with the notion of illness. When these themes are important, sickness is always suggested as well, and the whole group's frequency rises.

In the second Paris section, Marceline has a miscarriage followed by serious complications. Her health improves later in Normandy and in Switzerland (the beginning of the section labelled ITALY 2), but declines during her trips through Italy and North Africa until she dies. The frequency of the sickness theme does not follow the same curve as Marceline's health, as might be expected. This anomaly can be explained in terms of another thematic cluster.

The theme of health is the nucleus of a group including beauty, life, warmth, water, strength and light (see Fig. 4). This group appears most frequently when Michel is getting over his tuberculosis. Its frequency rises as well when he describes the lush beauty of his farm in Normandy. Similarly, while Marceline is quite ill, Michel goes out into a splendid winter night in Paris, and this is reflected in the high frequency of themes associated with health. As Marceline's health improves in Normandy, the group's frequency rises, but it declines with the onset of bad weather and a small relapse by Marceline. Then, during their travels in Switzerland, Italy and North Africa, as Marceline's health declines, evocations of the thematic cluster around health increase. Michel is taking her to places where his health had improved, places which he associates with good health. He reveals his outlook by a strange remark: "Pourquoi tousse-t-elle par ce beau temps?" (465).

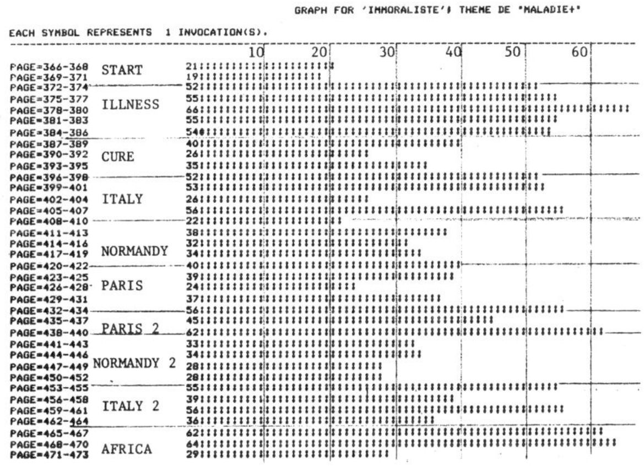

Fig. 3. Illness and associated themes in l'Immoraliste.

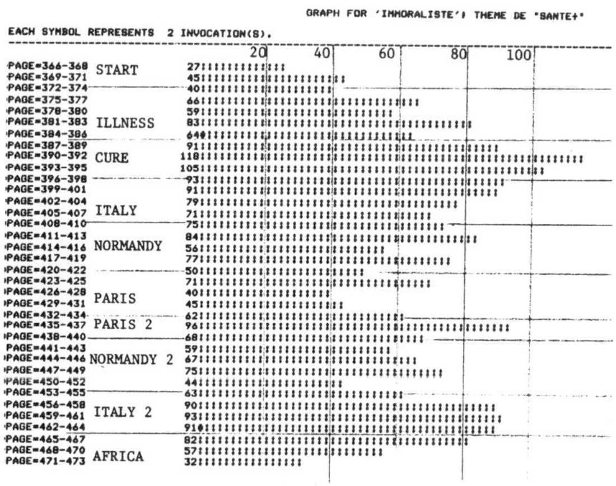

Fig. 4 Health and associated themes in l'Immoraliste.

Finally, as Marceline is clearly dying, the thematic cluster which linked good health with a certain environment is destroyed by the pressure of external events. In the last pages of the novel, Michel is grief-stricken and bewildered. He cannot understand how he could have acted as he did. He cannot deny the result of what he has done, but neither does he feel that his actions were wrong. And the reader is as mystified as Michel, unless he has paid close attention to theme development in this text.

For some time, I had the impression that the novel worked something like this. My theme analysis system allowed me to transform impressions into demonstrations. This, I think, shows rather clearly my system's usefulness.

CONCLUSION

So my approach has worked two times out of three. Should I give it a score of 67%, that is to say a "C" on the academic scale? I think not. It is a question of starting points. The theoretical underpinning for my theme analysis system is the concept that in a work of literature three elements--plot, characterisation, theme structures--work in harmony. In many recent novels, and in most poems, plot is practically non-existent. The same can be said about theme structures in many detective novels or adventure novels, and in most plays. My system cannot work effectively on these kinds of texts. That is to say that it is a special-purpose system fitted to a particular kind of literature, rather than an omnibus system applicable to all forms of writing. Given the reputation which many omnibus systems have acquired over the years, I am not entirely disappointed.

ACKNOWLEDGEMENTS

The work reported here has been funded by the following grants from the Canada Council: 69-418, S70-0735, S70-1561, S71-1933, S72-1650, W74-0453, S76-0734 and 451-77-520. The University of Manitoba has provided a postdoctoral fellowship, several research grants, and funds for travel.

REFERENCES

Baudelaire, C. (1933). Notes nouvelles sur E. A. Poe. In J. Crépet (Ed.), Nouvelles Histoires Extraordinaires. Conard, Paris, pp. v-xxiii.

Beaujour, M. (1965). Temps et substances dans "Voyage au bout de la nuit". Cahiers de l'Herne, 5, 178.

Céline, L.-F. (1962). H. Mondor (Ed.), Romans. Gallimard, Paris.

Fortier, P. A. (1974). La Vision prophétique: un procédé stylistique célinien. In J.-P. Dauphin (Ed.), Pour une poétique célinienne. Lettres Modernes, Paris, pp. 41-55.

Fortier, P. A., and J. C. McConnell (1975). THEME: A System for Computer-Aided Theme Searches of French Texts. University of Manitoba, Winnipeg.

Gide, A. (1958). Y. Davet and J.-J. Thierry (Ed.), Romans, Récits, Soties, Oeuvres lyriques. Gallimard, Paris.

Poe, E. A. (1927). R. B. Johnson (Ed.), The Poems and Three Essays on Poetry. Oxford University Press, Oxford, pp. 189-202.

Ricardou, J. (1971). Pour une théorie du nouveau roman. Seuil, Paris, pp. 59-88.

Robbe-Grillet, A. (1957). La Jalousie. Minuit, Paris.

Todorov, T. (1970). Introduction à la littérature fantastique. Seuil, Paris, pp. 97-106.

Valéry, P. (1957). J. Hytier (Ed.), Oeuvres I. Gallimard, Paris, pp. 769-84, 1314-39, 1359-78, 1412-15, 1438-43, 1456-91, 1786-88.

Valéry, P. (1960). J. Hytier (Ed.), Oeuvres II. Gallimard, Paris, pp. 1207-09.

VERBS: AN AFFORDABLE APPROACH TO COMPUTER-AIDED STUDIES OF NARRATIVE TECHNIQUES

B. B. Elias

French Department, University of British Columbia,
Vancouver, B.C., Canada V6T 1W5

ABSTRACT

This paper describes a method of computer-aided literary research which permits the analysis of numerous stylistic and technical aspects of a narrative which are outside the scope of concordance-based projects. For reasons of simplicity this method is based on the manual alphanumeric encoding of the verbs in a novel for the specific aspects to be studied. The verbs, being the linguistic and logical nucleus of a sentence, form a small yet totally representative data base, while manual encoding ensures maximum flexibility without the need for major computer programming.

KEYWORDS

Concordance-based data; frequency counts; manual alphanumeric encoding; prefix-suffix coding system; sequence patterns; verb-phrases.

The computer has finally begun to come into its own in the field of literary criticism. During the past few years scholars have been producing concordances of an ever-widening spectrum of works. At the same time these concordances have become increasingly more complex. Taking the French language as an example, where once the computer served to produce only word indexes, it can now be used to accomplish more exotic tasks such as those described in "The lemmatization of contemporary French" (Meunier, 1976) and "Computer-aided thematic analysis of French prose fiction" (Fortier, 1976). It is most encouraging to find that computer research of this kind is endowing literary studies with a renewed potential for precision and objectivity.

Yet each refinement of an 'ordinary' concordance program necessarily entails a major increase in its cost (in computer time and man-hours) and probably in its overall error factor as well. I support wholeheartedly all attempts to perfect methods of analyzing literary texts automatically. What I propose, however, is that consideration be given as well to methods of computer-based literary research that are accessible to students—let us say at the graduate level—and not only to faculty members with massive research grants and years to implement their projects.

One such affordable method is that which I developed over the past four years at the University of British Columbia to supply the necessary data for my doctoral thesis in French Literature. The topic of the dissertation was: "Etude du discours narratif dans Nord de Louis-Ferdinand Céline" (a study of the narrative techniques in the novel Nord by L.-F. Céline). The reasons for using the computer to study the literary work of this author were numerous. Louis-Ferdinand Céline is now generally recognized, with Marcel Proust, as one of the great stylists of the twentieth century French Novel. Although his vocabulary and themes had already come under more systematic study—thanks primarily to Paul Fortier's 1968 keyword concordance of Céline's first novel, Voyage au bout de la nuit—Céline's narrative techniques had long been neglected. It was all the more imperative to study the author's literary art from an objective point of view since his tendency to fragment the narrative makes it impossible to understand his narrative techniques fully by traditional means. Adapting the structural approach Gérard Genette (1972) had applied to Marcel Proust's A la recherche du temps perdu in his important essay entitled "Discours du récit," I therefore undertook to analyze the narrative structures in Nord with the help of special computer-produced data.

In order to avoid the excessive length of a concordance my method involves encoding only the verbs of a novel. In addition the encoding is alphanumeric, allowing for the use of easily recognizable abbreviations. The input text is thus drastically reduced—verbs account for less than 15% of the total number of words in a French novel—while still remaining quite 'readable' without machine decoding. Verbs represent the ideal basis for encoding narrative information. In fact the verb not only constitutes the grammatical and logical nucleus of a sentence, it is also highly representative of the text in its entirety. After all, every sentence contains at least one verb. Therefore no segment of the text can be neglected in arriving at the final results.

The advantage of choosing a limited data base of this kind is that it minimizes costs without any loss of efficiency. This in turn makes manual encoding entirely feasible—and indeed desirable—if one wants to analyze more than just a few different narrative elements. For while it may be possible to parse an input text automatically and to have the program distinguish between narration and direct or indirect speech, it would be totally impractical—if not impossible—to have it identify the more complex elements of the narrative such as indirect free speech ("style indirect libre" in French), stream of consciousness, iterative narration, flashbacks and different levels of narration (stories within stories, for example). The limitations affecting the role of the computer in this application are due to the fact that narrative techniques are difficult to define specifically a priori, depending as they do on the unique global context of each individual novel. Unlike lexical and thematic studies which can use computer-produced word and frequency lists as their starting points, the analysis of narrative techniques must start with a thorough reading of the text before the researcher can even hope to discover the various phenomena to be studied. Programming for the automatic recognition of narrative techniques would therefore have to start with finding all the different types of patterns existing within the text. Since the researcher is obliged to do this initial analysis of the text himself in any case, it is only practical that he encode his work manually. In this way he needs no complex programming, with its high computer-processing costs and lengthy verifications, and can proceed immediately with the 'busy work' of counting, tabulating and listing.

Another advantage of a manual encoding system is its high degree of adaptability. While automatic parsing, alone, would require a massive new program for each different language and for most individual styles, encoding the verbs by hand is easily adapted to any language and to any particular situation.

For my research on Nord I limited the coding to ten fields:

Field 1 = the page number
2 = the line number
3 = the chapter number

Field 4 = the type of speech
5 = the subject of the verb
6 = the verb tense
7 = negatives, interrogatives, etc.
8 = the levels of narration and of flashbacks
9 = the character speaking, iterative narration, etc.
10 = the verb-phrase.

The first three fields are numeric, the rest alphabetic. Given a simple hypothetical French narrative sequence such as:

> Le train partait.
> "Tu as pas amené ton pétard?" s'écria Georges deux fois.
> Mais n'avait-il pas dit qu'ils partaient en vacances?
> "T'es complètement dingue toi! C'était pour rire!"
> Georges allait regretter sa petite plaisanterie encore davantage quelques heures plus tard...

and given also that this sequence begins on line 4 of page 123 which is in chapter 5, I would encode the verbs as follows (they are underlined in the passage above):

1	2	3	4	5	6	7	8	9	10
123	04	05	NO	LEO	IMO	AFO	AOOO	OO	partait
123	05	05	DO	TUO	PCO	IPO	AOOO	GO	as = amené
123	05	05	VR	ILP	PSO	AFV	AOOO	OI	écria
123	06	05	IV	ILO	PQO	JNP	BFBA	OO	avait + dit
123	06	05	II	ISO	IMO	ITO	BFBA	OO	partaient
123	07	05	DO	TVD	PRO	AFO	AOOO	GO	es
123	07	05	DO	CEO	IMI	AFO	AOOO	GO	était = rire
123	08	05	NO	ILP	CPI	AFO	AFSA	OO	allait regretter

The first three fields are straightforward. They give in full the location of each verb in the text. The other fields contain abbreviations or codes relevant to the French language. These codes are sometimes arbitrary and would, in the case of a different researcher or a different language, most certainly change. Much more important is the fact that the coding system is based on the concept of prefixes and suffixes which gives it great flexibility despite the simplicity of the codes involved. (The letter "O" is used as the null set, in other words where there is no primary or secondary element to encode.) Thus, referring to the above example,

in field 4 (the type of speech)
(Prefixes): "D" = dialogue
"I" = indirect free speech
"N" = narrator's discourse
"V" = declarative verb (introducing the dialogue)
(Suffixes): "I" = second level indirect free speech
"R" = reflexive verb
"V" = second level declarative verb;

in field 5 (the subject of the verb)
(Prefixes): "CE" = "ce" or "c'"
"IL" = "il"
"IS" = "ils"
"LE" = "le"
"TU" = "tu"
"TV" = "t'" (colloquial elision)
(Suffixes): "D" = doubling of the subject pronoun
"P" = proper noun;

```
in field 6 (the verb tense)
   (Prefixes): "CP" = "conditionnel périphrastique"
               "IM" = imperfect
               "PC" = "passé composé"
               "PQ" = "plus-que-parfait"
               "PR" = present
               "PS" = "passé simple"
   (Suffix):   "I"  = infinitive;

in field 7 (negatives, interrogatives, etc.)
   (Prefixes): "AF" = affirmative
               "IP" = interrogative plus "pas"
               "IT" = interrogative
               "JN" = interrogative with inversion plus "ne"
   (Suffixes): "P"  = "pas"
               "V"  = inversion of subject and verb;

in field 8 (the levels of narration and of flashbacks)
   (Prefixes): "A"  = first level of narration
               "B"  = second level of narration
   (Roots):    "FB" = flashback
               "FS" = foreshadowing
   (Suffix):   "A"  = first level of flashback or foreshadowing;

in field 9 (the character speaking, iterative narration, etc.)
   (Prefix):   "G"  = Georges
   (Suffix):   "I"  = iterative narration;

in field 10 (the verb-phrase)
               "="  = a word separates the subject and the verb
               "+"  = the subject pronoun is between the verb and
                      the past participle.
```

It goes without saying that my encoding of the verbs in Nord was much more complete than this brief example would seem to indicate. In fact, I used several hundred different codes in all. The use of suffix codes allowed me to expand the scope of the information encoded beyond the limits of the general categories for each field. In this manner I was able to identify such secondary aspects of the narrative as ellipsis, summary, passive voice and interventions by the narrator.

The use of the prefix-suffix coding system has other benefits besides decreasing the number of separate codes required. By encoding complex phenomena in a cumulative and analytical manner my method facilitates the analysis of sequence patterns while at the same time ensuring the greatest possible uniformity of the results. In practical terms, whether one is encoding a simple finite verb or an entire verb-phrase, there is need for only one entry. For example, "n'avait-il pas dit" is coded as one compound tense, the pluperfect, instead of one imperfect tense ("avait") and one past participle ("dit"). With both compound tenses and verb sequences left intact, the codes can supply valuable information for grammatical and stylistic analyses which would otherwise be inaccessible. As for the individual elements such as auxiliaries, participles and infinitives that make up these complex grammatical structures, they can be identified by simply reducing the individual structures to their constituent parts. This method of encoding respects as well the structural unity of French double or triple negatives—a fact that is very important in studying novelists like Céline who use colloquial French in their writing.

The work involved in encoding the verbs of a novel is surprisingly low. In the case of Nord, I found that it took about 35 hours per 1000 lines to encode an unprepared text and to verify the keypunching. This figure would fall to at least 30 hours per 1000 lines if the data were entered from a terminal, and would be even less if one had access to a part of speech concordance of the work in question. These figures compare favourably with those of 20 hours per 1000 lines for doing an ordinary keyword concordance of a French text and of 95 hours per 1000 lines for producing a theme-searchable text file (Fortier, 1976).

Verification of the narrative data is quite simple as well. It suffices to scan for numbers or letters which were not supposed to be used, and to check any codes used only once or twice in all. The data thus becomes fully accessible almost immediately after encoding. In addition, the data is already meaningful even if only certain sections of the narrative have been encoded in this way. The initial stages of literary analysis need therefore not be delayed until the input text is complete. Instead, the formulation of hypotheses begins quite naturally during the encoding process itself. In the case of Céline, for example, I noticed that each of his characters seemed to use the French language differently. By singling out specific segments of dialogue, I was able to confirm this hypothesis quickly enough to adapt my coding system to this particular aspect of the narrative.

Once the encoding and verification have been completed, it is easy to do minor processing of the input data to make it more readable. For example, two entries appearing on the same line of text can be tagged automatically to indicate their order, or the abbreviations used can be translated into their complete form. Retrieval of the information is no less straightforward. Quotations and examples can be found by doing simple scans of the disc files. Even the most complicated listings and cross-tabulations can be produced by using packaged applications programs such as MIDAS and simple programs written in languages such as SNOBOL and PL/1. In other words, although the average researcher is likely to require some help from a computer programmer—in my case Lewis James of "U.B.C. Arts Computing" took care of the technical aspects of my work—the research will certainly not require the writing of any complicated programs. By encoding the verbs the researcher has already done the most difficult work. The rest is but scanning, matching and listing.

The usefulness of the output obtainable in this manner cannot be overestimated. From the input data one can produce not only alphabetical lists of all the verbs with their location and respective codes (invaluable for finding examples and quotations), but also all types of frequency counts. By using the computer it is possible as well to gain insight into the role of sequence patterns and transitions within the novel (something it would be impossible to do by hand). Naturally this and similar output is easily translatable into the form of graphs or histograms.

Another characteristic of this type of verb-oriented encoding is the versatility of its data. Since the verb-phrases are encoded for numerous factors it is possible to do very extensive cross-referencing in order to specify related aspects of the narrative or even to define new elements which have not been encoded before. In _Nord_, for example, the personal pronoun "vous" may refer to a character, the reader or the subject of a generalization. It was possible to distinguish these different uses automatically, simply by cross-referencing the subject pronoun in field 5 with the type of speech in field 4 (dialogue or narrator's discourse) and the level of narration given in field 8 (narrative or generalization).

The great versatility of the verb-coding system becomes all the more apparent when one realizes that in less than 20 letters per verb-phrase I was able to encode all the information necessary to do an exhaustive study of the narrative techniques in a novel. My thesis adopted the five categories of "order," "duration," "narrative frequency," "mood" and "voice" established by Gérard Genette (1972) as a basis for analyzing narrative structures. The first three chapters of the thesis dealt with the novel's time structures. The computer output gave me precise information on all flashbacks and foreshadowings, on the variations in the speed of the narration (descriptive pause, scene, summary, ellipsis) and on all aspects of repetitive narration. The fourth chapter analyzed the manner in which events and speech are represented in the novel. Verb lists and data on the number of verbs per line of text allowed me to single out and compare eventful passages, while cross-referenced frequency lists of the various types of speech allowed me to study this important aspect of the novel, and especially to show to what degree the author particularized the dialogue of each character. The final chapter dealt with the role of the narrator. There again the computer-generated data fulfilled the essential function of providing complete lists and frequency counts of the various aspects involved, thus giving me a clear notion of the real importance of the narrator in the novel.

The results of this type of computer-assisted research are most remarkable if the novel is complex. The fewer separate structural elements a novel contains, the less need there is for computer analysis. In the case of Nord, however, the researcher is faced not only with several different levels of language, but of narration as well. In addition, every aspect of the narrative is fragmented into tiny segments which make it impossible to get a realistic idea of the novel's narrative structures or even to undertake automatic parsing. Only by encoding the verbs by hand could I piece together the puzzle of Louis-Ferdinand Céline's narration.

My experience has convinced me that computer-generated data can be a very important adjunct to the study of narrative techniques in any language. By basing the input on the verbs alone one obtains a fully representative data base that is of manageable length (in the case of Nord the novel's 404 pages of small print were represented by only 19,417 verb phrases). This in turn makes manual encoding feasible, thereby opening up the field of computer-aided literary research to a wider spectrum of scholars. At a very low cost in computer time and man-hours one can produce results that are not only immediately useful, but that can serve as a data bank for subsequent studies as well. For example I have already received several requests from other Céline scholars for specific data on Nord. Even more important, however, is the fact that frequency-based data concerning narrative structures may finally offer researchers a set of objective criteria with which to compare different literary works. With this in mind I have begun to encode Céline's first novel, Voyage au bout de la nuit, using the same method and shall then utilize this data as the basis for publishing a systematic comparison of Voyage au bout de la nuit and Nord, the last novel to be published during his lifetime.

ACKNOWLEDGEMENT

I would like to thank the Social Sciences and Humanities Research Council of Canada for its generous support of my verb-based analysis of Voyage au bout de la nuit in the form of a postdoctoral fellowship (grant 456-800023). I am also grateful to the University of British Columbia for making its computer facilities available to me.

REFERENCES

Elias, B. B. (1980). Etude du discours narratif dans Nord de Louis-Ferdinand Céline. Doctoral dissertation, University of British Columbia.

Fortier, P. A., and J. C. McConnell (1976). Computer-aided thematic analysis of French prose fiction: II. Analysis of texts and preparation costs. In A. Jones and R. J. Churchhouse (Ed.), The Computer in Literary and Linguistic Studies, University of Wales Press, Cardiff. pp. 215-222.

Genette, G. (1972). Discours du récit. In Figures III, Editions du Seuil, Paris. pp. 65-282.

Meunier, J. G., S. Boisvert and F. M. Denis (1976). The lemmatization of contemporary French. In A. Jones and R. J. Churchhouse (Ed.), The Computer in Literary and Linguistic Studies, University of Wales Press, Cardiff. pp. 208-214.

THE IDEA BEHIND COMPUTER STYLISTICS

L. T. Milic

Cleveland State University
Department of English
Cleveland, Ohio 44115

ABSTRACT

Unlike traditional stylistics, computer stylistics can be free of the impediments to scientific inquiry introduced by the observer's deficiencies and the distortions created by adherence to hypotheses. No inquiry can be wholly free of prior analysis, but the most general results will be obtained when the experiment and its resultant data generate the hypothesis, as Newton insisted. The comparison of selected texts with carefully-constructed data bases constituting period norms can produce data which shed light on the state of the language during given periods and on the way that writers employed the stylistic resources of their language. Though such studies diverge from the generally-accepted idea of stylistics as an aspect of literary criticism, they have a wider field of application and a greater potential for social usefulness.

KEYWORDS

Stylistics; computer stylistics; quantitative stylistics; literary criticism; applied linguistics; linguistic data bases; stylistic theory.

TEXT

Stylistics has always been an empirical activity, dealing with phenomena which experience rather than theory has shown to deserve attention. Alchemists who heedlessly compounded chemicals they had little understanding of in order to "see what would happen" were truly empirical. They believed that their experiments were shaped by a theory (the hope that lead could be transmuted into gold.) It is this seeing what will happen that is the hallmark of empirical research, however dressed it may be in assumptions or hypotheses. It is not quackery if it presents reliable observation, so defined that the data may be properly interpreted by later researchers, as is the case for the astronomical records kept by the Babylonians and the Chinese. Perhaps ninety percent of the most useful findings in stylistics fall into this category. The instance of Mendenhall, nearly a century ago, offers a good starting-point.

Thomas Corwin Mendenhall was Professor of Physics at Ohio State University, eventually becoming the Superintendent of the U.S. Coast and Geodetic Survey. This self-educated man developed an interest in the question of Shakespeare's authorship of the plays. Mendenhall (1887) had counted word-lengths in the prose of Dickens, Thackeray, Mill and others in an attempt to develop a method of author identification as certain as the spectroscopic analysis of chemical compounds. Graphs of word-length could serve as "characteristic curves of composition." Two assumptions underlay this study: 1) each writer has a peculiar and unchanging vocabulary; 2) this peculiarity will be reflected in a consistent set of figures presented as a graph or curve. He found that groups of 5,000 words drawn from a given author produce graphs which are indeed consistent and generally differ from such groups drawn from other authors. He concludes: "[This method's] principal merits are, that it offers a means of investigating and displaying the mere mechanism of composition, and that it is purely mechanical in its application." Unfortunately, for <u>complete</u> reliability, he estimated that samples of 100,000 words would be needed.

This is a typical case of stylistic empiricism, despite the occasional references to theory. The study arises from an intuition for solving a problem by analogy to the method of a science. The conclusion is satisfying because the method is workable, mechanical and unbiased. No understanding of the linguistic process is referred to, no theory of language really even mentioned. Mendenhall believes that each writer has a peculiar vocabulary, though he has no theory to support this speculation. Later (1901) he presents counts of all the words in Shakespeare, in most of Bacon, Marlowe, Beaumont and Fletcher and others. He found that the figures for Shakespeare and Marlowe were identical!

Elsewhere in the Midwest, a professor of English was also at work on a scientific model. Sherman (1893) had noted the improvement in the teaching of chemistry which occurred when students were allowed to do the experiments themselves rather than watch them performed by a laboratory assistant. By analogy, he concluded that students of literature needed to feel the power of Shakespeare and Chaucer by examining the <u>elements</u> they have in common and compiling statistical tables exhibiting these differences and similarities. Sherman (1888) had already set himself the task of investigating English sentence-length. He counted thousands of sentences between Chaucer and such "recent" writers as Arnold, Pater and Lowell and concluded that sentences had grown shorter since Chaucer's time and that a constant sentence-rhythm can be revealed in any 300 sentences of a given author. In a second study (Sherman, 1892), he decided that he needed--not a theory but more data. After noting (but side-stepping) the problem of punctuation and its relation to sentence-form, he settled on a figure of 50 words as typical of the length of the English sentence up to Elizabethan times. Selecting five "modern" writers (De Quincey, Macaulay, Emerson and two New England preachers) he made counts of 500-800 sentences from each. To test the consistency of the average he counted over 2,000 sentences in De Quincey and 40,000 in Macaulay's <u>History of England</u>. He seemed satisfied with these averages, though he did not use any statistical tests to determine whether the dispersion was greater or less than might have been expected. He concluded: "The evidence seemed to indicate the operation of some kind of sentence-sense, some conception or ideal of form which, if it could have its will, would reduce all sentences to procrustean regularity...in this 23.43 [words-per-sentence average] was the resultant of the forces which had made Macaulay's literary character."

It is evident that Sherman was considerably in the dark about the nature or significance of his activities and propelled only by the pioneer's sense that over the next mountain range he will find--something. Mendenhall wanted to devise a method to identify Shakespeare's plays. Both he and Sherman operated on the analogy of science. Would either one have been advantaged by having a substantial body of theory behind him? Would their results or their interpretation of these results have been different if they had known what they were doing? And what kind of theory could have been of use to them?

A theory is a scheme or system which accounts for a set of data or phenomena. In this it is distinguished both from a law (like Grimm's or Verner's), which always holds true, and a hypothesis (like the Sapir-Whorf), which is merely a speculation and perhaps not provable. A sound theory should gather the relevant facts, satisfyingly relate them to each other and either predict new difficulties or suggest methods of experimental verification. Four areas of theory are applicable. In describing them I do not wish to claim that theory in any area has reached a stage of general acceptance. They are placed in descending order of reliability.

1. Theory of Language: the most general and could be said to include the others, for they are all aspects of the use of language. Strictly speaking, however, a theory of language explains the relation of languages to each other, the rules of syntax as they determine meaning, the mechanism of communication which uses the brains and senses of speaker and hearer, and the process of change in language. Evidently, items could be added to this list (e.g., the origin of language).
2. Theory of Rhetoric: this narrow field has always concerned itself exclusively with the relation between the conscious purposes of speakers and the linguistic structures best designed to generated specific effects in the audience. (Literary criticism was once located here but is now generally placed in the next category.)
3. Theory of Literature: although it claims its origin in Aristotle, this is an essentially new field, developed from the historical study of societies; it relates certain effects to specific genres, explores the parentage and filiation of types of literary works, attempts to establish model structures of meaning and wonders about the relation of literature to activities outside itself.
4. Theory of Style: the very existence of this area of inquiry has been questioned and its procedures regularly disparaged (Gray, 1969; Ellis, 1974), partly because of the extensive overlap between it and the three previous domains, which appear to make a theory of style superfluous; a minimum theory of style should deal with the reality of individual collections of linguistic resources ("idiolects" or "styles"), their unity and consistency, distinction from each other, acquisition and development.

It is possible that a theory of language could have provided Mendenhall with some information about constants and a theory of literature about the effect of genre on these constants. Sherman (apart from a course in statistics) could have used the information provided by a theory of rhetoric in considering sentence-length as an alternative option to effect predication and discerning the purpose of Macaulay's perhaps deliberate sentence-variation. If these pioneers had had such theoretical structures available to them, their conclusions might perhaps have been more interesting to their readers and more fruitful. The shape of their experiments would have been different: they might have concerned themselves with more complex stylistic features, which occur less often in a text, and therefore counted fewer instances. Would this have helped us, a century later? Unfortunately, the answer is no. Though we are far from solving the puzzle of style, we have made some progress in theory, especially of language. The theories of style available to Sherman and Mendenhall were an interesting mix of economics, aesthetics, correctness and old rhetoric. It is at least possible that their work would have been vitiated by ingenious but erroneous subtlety. As it is, the labor they contributed continues to be useful to those interested in the particular stylistic features they examined so thoroughly: word-length and sentence-length. The data that they amassed are still there for our use.

I must concede that I doubt the usefulness of much theory to the stylistician. Relevant theory at the present is little more than hypothesis and it is well known what harm *a priori* hypothesizing can do to the collecting of data. As Newton himself noted: "The best and safest method seems to be first to inquire diligently into the properties of things and to establish those properties by experiments and then to proceed slowly to hypotheses for the explanation. For hypotheses should be subservient in explaining the properties of things, not assumed in determining them." If the stylistician believes that the prose of writer A is better than that of B and that he can demonstrate this by revealing a greater quantity of features *x*, *y*, or *z* in A's work than in B's, he may be influenced to define *x*, *y*, and *z* so as to accord with A's practice and exclude all of B's uncertain or ambiguous instances. Whenever judgment is required, bias is possible and if present perverts the results. To avoid bias and maintain scientific objectivity, could one proceed by excluding *a priori* commitments? The attractiveness of such a proceeding is lamentably undermined by the fact that the very decision to examine any stylistic feature is an *a priori* decision that this feature is of consequence.

A *stylistic feature* may be any of a large number of aspects of an author's work which has been selected for special examination, usually because of its prominence or its significance. Probably the number of possible stylistic features is limitless, bounded only by the number of units in a text or canon and their permutations multiplied by the number of kinds of units that may be found. This extensive array in practice usually comes down to a much smaller number which may for convenience be grouped into four classes.

The basis of the classification is "unsophisticated recognition," the ability, that is, of a naive reader (such as a computer) to recognize the structures referred to with a minimum of instruction. The first three classes proceed away from easy computer ability. The fourth class belongs in a different set but is added here for completeness.

I *Formal*. This is the class of shapes (letters, words, sentences) with tangible demarcations. The features in this class include length of word and sentence, frequency of letters, words, affixes, punctuation marks, vocabulary items, combinations and collocations of units and positions of items (e.g., sentence-final).

II *Categorial*. These are the decisions that computers can make if carefully instructed, though never with entire accuracy, with respect to syllables, word-classes, syntax, morphology, parallelism and other simple rhetorical arrangements.

III *Cognitive*. These are the judgments that usually require human attention, including such distinctions as abstract/concrete, such matters as imagery, rhythm and the notional/semantic aspects of language.

IV *Derivative*. This class depends on quantities provided by the three preceding classes and consists of second-order relations of which the writer is almost certainly unaware: standard deviation of the sentence-length, vocabulary richness (K factor), type-token ratio, gap recurrence, letter frequency rank-order.

A recent group of books on style may serve to illustrate the significance of these categories. O'Donnell (1970), attempts to demonstrate Stephen Crane's authorship of an unfinished work entitled "The O'Ruddy" and gives a list of "fourteen seemingly mundane, unexciting variables" (stylistic features) on which his pilot study is based. Three (colon, impersonal construction, introductory word) belong in Class I, the other eleven (sentence type, personal construction, introductory clause, parenthetical remarks, types of participle, adverbs, split verb form and unbalanced structure) in Class II. As the experiment was designed to identify an author, the features which showed no predictive power were dropped and others put in their place. The ultimate list of seventeen variables came also to include items in Class III (color words, metaphors) and in Class IV (verb-adjective ratio, sentence-verb ratio).

A monograph on Orwell's essay (Ringbom, 1973) examines series (Class II), punctuation and sentence-length (I), favorite words (I and IV), words of perception, qualification and categorization (III) and certain miscellaneous words (I). A study of Virginia Woolf's *To the Lighthouse* (Leaska, 1970) provides tables showing the number of sentences (I), clause types (II), content words (I), Latinate words (II), types of verbals and verbs (II), types of nouns (I, II, III), function words (I), syllables (II) and lines (I) spoken by each character in the novel.

The narrower the task (attribution: O'Donnell; specification of characters: Leaska), the greater the number of Class I and II variables. The more we move into interpretation, the greater the number from Class II and III, especially those that are _ad hoc_, observed to be peculiarities of the author. I contend that the far-out stylistic feature ("Types of deviant sequence in hypotactic complexes," "References to _ear_ and _knee_ in eight history plays") is ultimately the least significant, in the statistical sense. One gathers that such features do not represent anything typical or meaningful in the text, the genre or the period, but that they simply fall out of the fabric when it is shaken. They are more accidents of vision at the service of a hypothesis than genuine observations of the corpus of the writer.

I have noted (Milic, 1972) that a hypothesis may be an obstacle to a stylistician because it focusses his attention on the features he would willingly discover while blinding him to those for which he has no use. The conclusion follows that a proper stylistic analysis would involve an examination of all Class I features in a text and as many of Class II as could be reliably elicited. Eliminating _ad hoc_ stylistic features and _a priori_ hypothesizing would of course narrow the scope of stylistics, which hitherto has always been taken to have some necessary relation to the interpretation of texts. I am convinced that this is an error which has hopelessly complicated the definition of style and confused the work of stylistics. Interpretation belongs properly within the domain of literary criticism (theory of rhetoric or theory of literature). A proper theory of style, as suggested above, could be limited to a consideration of the development of an individual's acquisition and use of the resources of his native language. Its practical side would then require a complete description of this individual's _style_. Such a description might ultimately furnish some materials which correlated with the literary effects for which a writer may be well known, but this correlation is neither important nor reliable.

As an example, one might consider Swift's tendency to create long lists (catalogues) of units in bizarre order. Some may associate this tendency with the qualities for which Swift's style is famous (irony, clarity, simplicity, force), but there is no way of knowing whether there is a connection. No one has tried an experimental study based on the responses of readers which might help to determine actual perceptions. Moreover, it is unlikely that the qualities relate to the features in any manner that could be useful in literary study. If both (features and qualities) are there and we create a link between them, we must be aware that it is our creation. On the other hand, we know that catalogues similar to Swift's are characteristic of persons afflicted with certain kinds of schizophrenia. It seems at least as logical to draw the conclusion that Swift was driven to the use of this feature as a consequence of the madness that eventually overcame him and that he had very little conscious control over it. Any lengthy consideration of such a rhetorical device (Class II feature) brings one inevitably closer to the solitary writer bent over his writing table than to the domain of aesthetics.

The paradox into which this line of argument leads is that writers--by definition craftsmen with words--are to be treated as if their production was no more under conscious control than the ravings of certified lunatics. It also moves in the direction of the writer and his psychology and away from the product and its exegetical requirements. This should not be altogether startling since it is the process rather than the product that is the human miracle. The practical result of accepting this line of reasoning is the need to study the features of the language of an individual in a very general way.

A compilation such as the Brown Corpus is founded upon no ingenious hypothesis, except perhaps in the choice of samples, although it is based on certain theoretical concepts: linguistic norms are important, genre will determine certain quantities, certain types of composition belong together. The analyses included in the book describing the Corpus (Kucera and Francis, 1967) and based on the million-word text are very general and belong in Class I and in Class IV, including both initial data and derivative statistical features: hapax legomena, type-token ratios, words of very high frequency and their parameters of distribution, word-length, sentence-length with parameters of variety, homogeneity and the like. The interesting aspect of this collection is that its lack of emphasis on theory of style has made it, with all its limitations and shortcomings, a major reference tool for the study of style. Other Class I features that can be derived from a basic alphabetical word-list include the size-ordered word-list and the reverse-alphabetical list. Needless to say, the Brown Corpus is not a study of style, but an accumulation of linguistic constants with respect to a theory of literature based on genre. It is a background for studies of style, specifically mid-Century American.

A more obviously stylistic study of a single author is the important concordance to Shakespeare (Spevack, 1968). Its extensive detail about the text provides all the necessary data for an enormous variety of studies and conclusions. To be sure, it is richer in positional information (what words go together) than in strictly analytic data (how many words begin with a) but everything can be retrieved from what is here. With it any question about Shakespeare's language can be answered. The only thing missing is an Elizabethan Corpus to provide a background for Shakespeare's individuality.

The best instruments for the study of style are a general computer program, which will indifferently compile all the necessary data, and a corpus of the period under consideration to serve as a norm. At the moment (November 1980) a number of computer packages exist which can be used in computer stylistics (e.g., EYEBALL, COCOA), but no systematic corpora exist for the study of earlier periods except my own Augustan Prose Sample (available) and Century of Prose Corpus (in progress). The Oxford Archive is accumulating all available English data bases and will in time be able to fill the need described here.

REFERENCES

Ellis, J.M. (1974). The Theory of Literary Criticism. Berkeley. Chap. 6.

Gray, B. (1969). Style: The Problem and Its Solution. The Hague.

Kucera, H. and W.N. Francis (1967). Computational Analysis of Present-Day American English. Providence, R.I.

Leaska, M. (1970). Virginia Woolf's Lighthouse. New York.

Mendenhall, T.C. (1887). Characteristic curves of composition. Science, 9, 237-249.

Mendenhall, T.C. (1901). A mechanical solution of a literary problem. Popular Science Monthly, 60, 97-105.

Milic, L.T. (1972). Autocoding in computational stylistics. In B.J. Kachru and H.F.W. Stahlke (Eds.), Current Trends in Stylistics. Edmonton. p. 264.

O'Donnell, B. (1970). An Analysis of Prose Style to Determine Authorship. The Hague. pp. 43-4.

Ringbom, H. (1973). George Orwell As Essayist. Abo.

Sherman, L.A. (1888). Some observations upon the sentence lengths in English prose, University of Nebraska Studies, I, 119-130.

Sherman, L.A. (1892). On certain facts and principles in the development of form in literature, University of Nebraska Studies, I, 353.

Sherman, L.A. (1893). Analytics of Literature. Ginn, Boston. p. v-xvi.

Spevack, M. (1968). A Complete and Systematic Concordance to the Works of Shakespeare. 6 vols. Olms, Hildesheim.

STATISTICAL SOFTWARE TOOLS FOR COMPUTER-ASSISTED ANALYSIS OF MUSIC.

by Luigi Logrippo and Bernard Stepien
Department of Computer Science
University of Ottawa
Ottawa, Ont.
Canada K1N 9B4

ABSTRACT

Statistical methods such as cluster analysis and factor analysis are being used in order to quantitatively investigate the stylistic characters of musical compositions. The tools employed are simply adaptations of off-the-shelf software packages. An example of utilization of cluster analysis for the study of modal structures in Inuit music is presented.

KEYWORDS

Statistics; music; cluster analysis; computers; inuit; eskimo; modality; statistic analysis; data analysis.

1. INTRODUCTION

Computers are being used to man's advantage in very many different fields. It is only slowly, however, that they are gaining recognition as valuable tools for research in the humanities. Factors which contain this development are the scarce familiarity of researchers in the humanities with quantitative methods, costs incurred in securing computer time and the aid of programmers, and (last but not least) the researcher's concern that the use of a mechanical tool may cause him to lose control over his conclusions. In this paper, we will show that some of these difficulties are less serious than many people may suspect, since widely available software packages, whose utilization is well-established in other disciplines, can be relatively easily adapted to the needs of research in the humanities. Moreover, some of these packages are flexible enough that they can be adapted to serve the researcher's needs in a variety of ways.

We will discuss here the experience we have gained in using one of these packages in the framework of an ethnomusicological project, involving the stylistic analysis of Inuit monodies (i.e., unaccompanied songs). Over one-hundred monodies were placed in a computer data base, and analysed by using a variety of packages. The most useful of these has been BMDP, the biomedical statistical package of the Health Sciences Computing Facility of the University of California (Los Angeles). With some additions to the input and output functions, we have been able to use BMDP in order to perform a wide range of experimentations. The parts of the package used so far are the cluster analysis and factor analysis programs, however other programs seem equally promising, and we are planning to experiment with them in the future. In this paper, we will only discuss some aspects of cluster analysis.

Cluster analysis attempts to classify a set of entities according to their similarity from the point of view of some predefined set of characters, or "indicators". For example, if the set of entities being considered is a set of monodies, and the set of indicators is the set of notes used in each monody, cluster analysis will group together monodies using similar sets of notes, a very useful thing from the musicological point of view (and also a slow chore to do by hand).

In our research, cluster analysis is being used in order to classify the monodies from several points of view. We will limit ourselves to the presentation of an example of utilization of cluster analysis to the classification of songs by scales (i.e., notes used), and intervals. The results are immediately useful to a musicologist attempting to classify the melodies by modes, for which scales and intervals are one of the most important characters of identification. Other classification criteria that have been investigated include melodic contours, rhythmic contours, and rhythmic patterns. Earlier results of this research were presented in Pelinski (1980).

The songs analyzed are Inuit personal ("a-ja-jai") songs, some collected by Pelinski (1980) in the settlements of Rankin Inlet and Eskimo Point, others collected by Hauser (1977) in Thule. These are the three "corpora" we will be discussing below.

Work related to the one discussed here was previously reported by Wenker (1977) and Morando (1979). The first author presented a computer-assisted statistical analysis of scales and intervals in North-American folk music, without however using cluster analysis or factor analysis. The second author used factor analysis in order to uncover statistical relationships between the musical styles of several classical composers.

2. CLUSTER ANALYSIS

The technique used was the standard clustering technique developed in the sixties by various researchers such as Sneath and Sokal (1973). As mentioned above, the specific implementation used was BMDP (Dixon and Brown, 1979), in its version of November 1979. We refer the reader to the abundant literature existing on this topic. It is however important to summarize here the two major steps of this automatic classification method which, because of the complex computations involved, is only possible in practice with the use of a computer. The clustering algorithm proceeds in two steps:

1) First it computes a coefficient of similarity between the entities in consideration, in our case, the songs. There is a variety of methods for computing such coefficients in classical statistics, some of them well-known as far back as the 19th century. For the research discussed here we used the method of the Chi Square because it was the most appropriate for our problem: this choice was both recommended by the literature and verified empirically. In this paper, the coefficient of similarity will be called "distance".

2) The distances are then used in the second step: the amalgamation process. This step will group the songs together in "clusters", starting with those that are mutually most similar, and continuing with progressively less similar ones. The results of the amalgamation process are shown in

several forms, of which we will discuss the following two:

- The amalgamation tree, which shows the order in which the songs were "clustered" together, and therefore implicitly their distances.
- A table, that we shall call "sorted indicator table", which shows the songs sorted by similarity, together with their indicators. This table is not provided by the standard BMDP package.

First, we used the method on each corpus of songs individually, to understand the internal organization of the songs of a given Inuit settlement. Subsequently, we mixed the three corpora (Rankin Inlet, Eskimo Point and Thule) to verify the influence of geographic dispersion on the Inuit singing style. Finally, we tried to correlate classifications according to different indicators: i.e. tonal systems vs. interval systems etc.. For reasons of space, we will limit our discussion to the Rankin Inlet corpus. The term "density" will be used to denote the proportion of notes or intervals in a song. The density of notes will be called "tonal density".

3. CLASSIFICATION OF SONGS BY TONAL DENSITY

In order to classify the songs by tonal density, we proceeded as follows. First, we ran on the encoded songs (Hickey and others, 1979) a simple frequency count program (not part of BMDP). This gave us a table of percentages of time values for each note in each song. In rearranged form, this is Table 1. On this table, we ran BMDP, and this gave us the amalgamation tree of Fig. 1. Another simple program (also not part of BMDP) rearranged the table of percentages according to the results of Fig. 1, and this yielded the sorted indicator table shown in Table 1.

Next, we explain how these outputs are to be read. In Fig. 1, the numbers at the top are song numbers, sorted in such a way as to group together similar songs as much as possible. The numbers at the left are "amalgamation distances", i.e. the distances at which the songs or groups of songs were clustered together. For example, from Fig. 1 it appears that songs 19 and 38 are very similar from the point of view of tonal density, since they were clustered together at the small distance of 1.085. On the other hand, songs 34 and 36, which are quite dissimilar (having been clustered together at the fairly large distance of 12.993), are also very dissimilar from the other songs in the corpus, since they join the closest cluster, the one consisting of all the songs appearing at the right side of the figure, only at the large distance of 24.067. The tonal density sorted indicator table (Table 1) shows song numbers and song authors at the left. For each song and each note, it also shows what is the percentage of time value of the note in the song (these percentages are the "indicators" discussed above). For example, note F accounts for 10.5% of the total played length of song 1, whose author is Kavik. Notes are represented according to the external representation discussed in Hickey and others (1979), i.e. -C is middle C, etc.. Note G appears always very frequently because all songs were transposed so that G corresponds to the main recitation tone. This table gives us a clear picture of the tonal relationships among songs.

With Fig. 1 and Table 1 in hand, the classification of the songs by tonal density does not present difficulties (see Table 2). Songs 34, 36, 11, and 28 are clearly tonally "unusual" songs, or "outliers", according to statistical terminology. Songs 31 and 17 are not classified as outliers because they will also appear together in the interval density analysis, and therefore deserve to belong to a class of their own. Further, we detect a fairly large class of songs (class 1), characterized by the use of notes F, G, A, and (with the exception of two songs), C. This class does not seem to have any particularly interesting subclasses, with the exception of (29,27), two songs by Tiktak, which exhibit a low percentage of A, and will reappear together at the interval analysis stage. Other subclasses appear, but they are either small, or fail to be characterized by the strong occurrence (or non-occurrence) of certain notes. In fact, the internal structure of class 1 seems to be due mainly to variations in frequency, rather than to note replacements. It is quite possible that a larger sample would reveal a somewhat more interesting picture.

In class 3, characterized by notes E, G, A, and +D, the substructure is more sharply defined, giving raise to the two subclasses 3' (using C), and 3" (using B and D instead). Class 4 is clearly the most strongly characterized of all, and in fact it will reappear intact in the interval analysis.

Looking at Table 1 from a more global point of view, one discovers a yet deeper organization that encompasses the whole set of songs. Schematically, this is shown in Fig. 3. There are four notes that appear in the greater majority of songs, and these are D, G, A, C, and +D. Class 1 is characterized by a strong F, class 2 is characterized by the replacement of A sharp for A, class 3 is characterized by a strong E, and class 4 is characterized by a strong F sharp and the absence of D and +D. Within class 3 we have the following substructure: class 3' is characterized by the absence of D, and class 3" by the replacement of B for C.

4. CLASSIFICATION OF SONGS BY INTERVAL DENSITY

The basic information here is given by Fig. 2 (Interval Density Amalgamation Tree), and Table 3 (Interval Density Sorted Indicator Table). Fig. 2 is to be read as Fig. 1. Table 3 is similar to Table 1, however the indicators here are percentages of intervals. Intervals are represented by the number of semitones they contain, e.g. -2 is the descending tone, 0 is the prime interval, and 4 (also written +4) is the ascending major third. So descending tones are 13.5% of all the intervals in song 1 by Kavik. Note that what is shown here is only the central and most important part of the table, which is too large to fit in the page.

According to the method used for the classification by tonal density, we obtain Table 4, which shows the classification by interval density. Note that Table 4 contains more classes than Table 2, as is to be expected because of the fact that there are many more intervals than notes.

Again, a global look at Fig. 2 reveals some interesting structural facts, as shown by Fig. 6. The analysis of Fig. 6 is less clear-cut than the one of Fig. 5, because of the greater complexity of the picture. For example, class 2 represents a sort of "transition" between class 1 and class 3 (it is shown as related to class 3 in Fig. 2). The modal analysis to be given in the next section will show that class 2 consists of the two subclasses (17,31) and (38,19), of which only the first relates to class 1 (because of the -4 and +4).

5. COMBINED TONAL/INTERVAL DENSITY ANALYSIS

It is now natural to attempt a combination of the previous two analyses. Songs that have both a similar tonal density and a similar interval density will then be classified together in this step. The resulting classes of songs can be said to identify "modes", in the sense that two songs belonging to the same mode will have similar tonal and interval structures (the authors are aware of the fact that the usual definition of modes in musicology does not only involve the distribution of notes and intervals, the only aspects considered here). A study of the modal system of the Eskimo Point corpus, carried out with a different method not depending on the use of the computer, was presented in Pelinski (1980).

We hope that the reader will believe that it is by pure coincidence that we end up with eight modes, just as many as the medieval church modes. These are shown in Table 5. Eight songs (i.e. about a quarter of the total), do not belong to any of these eight modes. As we have mentioned already, this analysis helps in finding some interesting subclasses of the tonal classes: such are (38,19), two very similar songs, (27,29), the mentioned two songs by Tiktak, and (15,35,23), a subclass of tonal class 3" characterized by the presence of +D.

6. CONCLUSIONS

We have presented a computer-assisted method for classifying melodies. Time and space constraints are not allowing us to complete this paper with its logical conclusion, which is the musicological analysis of the results. However, our experience has shown that the computer-assisted analysis is a valuable starting point for further musicological research.

Even though our analysis could, in principle, have been done by hand, in practice this would have been extremely laborious and error-prone. The same software packages could be used equally well in order to classify hundreds or thousands of monodies, even though for such quantities some additional processing would be necessary, in order to further compress the information.

It is interesting to note that cluster analysis is a common tool in research areas in the social and health sciences, as well as in Engineering, Physics, and many others. The authors believe that similar methods will one day be common tools in musicological research. They are quite aware, however, that in order for this to become a reality, it will be necessary to somewhat "customize" the software packages in order to meet the musicologist's specific needs, a task that may require much experimentation and research.

ACKNOWLEDGMENT

Research reported here was funded in part by the Social Sciences and Humanities Research Council of Canada. The authors are grateful to Ramon Pelinski for having motivated this research. They are also indebted to Ugo Berni Canani and George White for many stimulating discussions.

REFERENCES

Dixon, W.J., and Brown, M.B. (Ed.) (1979). BMDP-79, Biomedical Computer Programs P-Series. University of California Press, Berkeley, Los Angeles.

Hauser, M. (1977). Formal structure in polar Eskimo drumsongs. Ethnomusicology, XXI, 1, pp. 33-53.

Hickey, P.J., Logrippo, L., Pelinski, R., and Strong, E.C. (1979) Internal and external representation of musical notation for an ethnomusicological project. Proceedings of the Tenth Ontario Universities Computing Conference, University of Ottawa.

Morando, B. (1979). Construction et analyse de nuages associes a des parallelipedes de correspondance: une contribution a l'analyse musicale. These de 3eme cycle, Universite Paris VI.

Pelinski, R. (1980). Cinq etudes sur la musique des inuit du caribou. Presse de l'Universite de Montreal. In print.

Sneath, P.H.A., and Sokal, R.R. (1973) Numerical Taxonomy. W.H. Freeman and Co., San Francisco.

Wenker, J. (1977). A computer-aided analysis of Canadian folksongs. In: Lusignan, S., and North, J.S. (Ed.) Proceedings of the Third International Conference on Computing in the Humanities, The University of Waterloo Press.

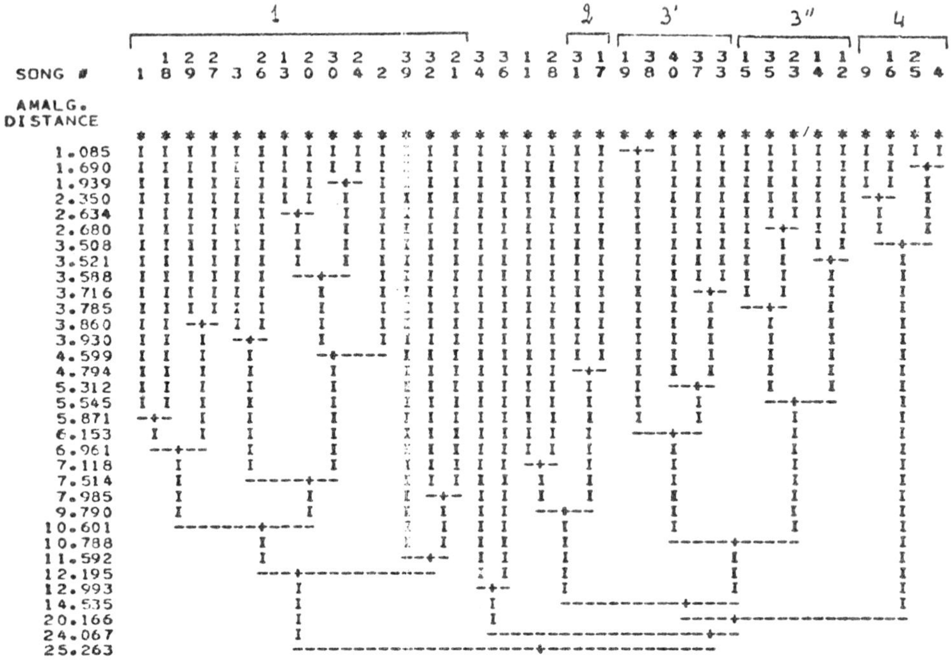

FIGURE 1. RANKIN INLET TONAL DENSITY AMALGAMATION TREE

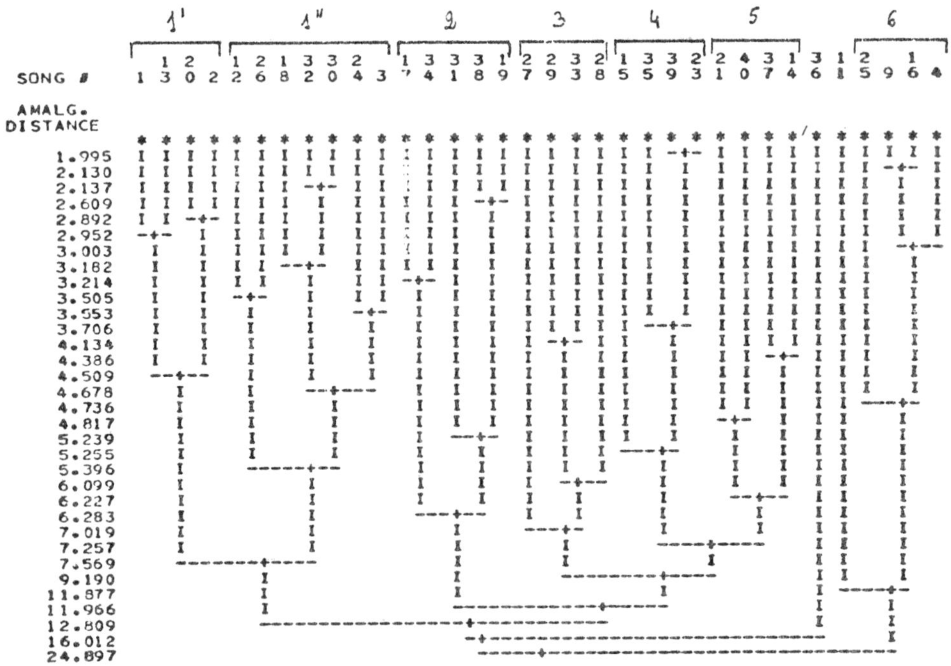

FIGURE 2. RANKIN INLET INTERVAL DENSITY AMALGAMATION TREE

PROPORTIONS OF NOTES USED FOR RANKIN INLET ON CLUSTERED SONGS

SONG	#	COMPOSER		-C	-C#	D	D#	E	F	F#	G	G#	A	A#	B	C	C#	+D	+D#	+E
RK	1	KAVIK							10.5		73.3		1.3	12.3						
RK	18	ULUJUK					3.4		25.8		59.2		11.6							
RK	29	TIKTAK		3.5		1.6			13.9	4	66.0	1.4	6	2.8	4	9.4				
RK	27	TIKTAK							20.6		68.8		6		4.1	5.9				
RK	3	KAVIK				1.6		1.3	8.0		51.4		15.9		6.7	6.4		8.5		3
RK	26	KOLIT				2.8		5	13.4		46.5		28.3			1.9		6.6		
RK	13	TAUTUNNG	1						16.3		58.1		16.9			8.7				
RK	20	MIRANAA							14.4		60.1		10.6	3.7		9.6		1.6		
RK	30	MANILAK							7.7		65.2		11.7			13.2		2.2		
RK	24	USSAK				1.6			7.3		62.9		17.3			8.5		2.4		
RK	2	KAVIK				2.4		4	17.7		42.1		21.9			15.6				
RK	39	T.KIMALA		2.8		7.5		5	59.4		13.5		11.2			5.1				
RK	32	ANGUTIQT							11.0		39.3		10.0	22.5		15.4		1.9		
RK	21	NILAULA							4.0		21.0		49.9		1.7	16.4		3.4		
RK	34	KASIGIAK		8.9		3		65.2			19.2		6.4							
RK	36	MCCASKIL						3.2			6.3					48.7		33.3		8.5
RK	11	PANIGONIA								12.2	55.9	2.7	2.7		18.5	8.1				
RK	28	TIKTAK				23.8			7		63.5		2.1		9.9					
RK	31	ANGUTIQT	2			13.9					68.6			12.2		2.9		2.5		
RK	17	ULUJUK				5		5.4			62.9			17.6		7.3		5.1		
RK	19	TAUTUNNG		4				7.5	2		68.5		5.5			15.3		2.6		
RK	38	T.KIMAL	3'					10.2			65.4		6.5			15.5		2.4		
RK	40	T.KIMAL						15.3			46.4		25.6		4.2	5.4		1.7		1.5
RK	37	KIMALAR				4.4		8.2			49.1		9.5			13.4		12.3		3.1
RK	33	KASIGIAK		1.3		4.2		8.5			57.2		15.1	9	9	7.8		4.0		
RK	15	ANARUAK				5.6		13.4			50.6		9.1		11.1			6.9		3.2
RK	35	TAPARTI	3''			6.2		3.1			61.8		7.4		15.4			6.2		
RK	23	USSAK				2.6		7.7			55.8		15.3		15.1			3.6		
RK	14	TAUTUNNG			1.9	5.6		14.8			59.9		13.0		4.9					
RK	12	TAUTUNNG				4.8		8.3	1.7	3	49.7		24.8		10.4					
RK	9	PANIGONIAK								37.5	53.9		7.8			7				
RK	16	ULUJUK	4					2.0		25.2	60.4		11.2			1.2				
RK	25	USSAK								15.8	76.2		8.1							
RK	4	IRKOOTI								21.9	69.1		7.5			1.5				

TABLE 1. RANKIN INLET TONAL DENSITY SORTED INDICATOR TABLE

PROPORTIONS OF INTERVALS USED FOR RANKIN INLET ON CLUSTERED SONGS

SONG #	COMPOSER		-8	-7	-6	-5	-4	-3	-2	-1	0	1	2	3	4	5	6	7	8
RK 1	KAVIK							5.6	13.5		61.8		14.6	3.4					
RK 13	TAUTUNNG	1'					4.9	4.9	7.4		59.0		18.9	4.9					
RK 20	MIRANAA					1.4	4.1	4.1	12.9		54.4	7	13.6	4.8	4.1				
RK 2	KAVIK					7	6.7	8.1	11.4		43.0		17.4	6.0	3.4				
RK 12	TAUTUNNG				1.6	2.4		8	14.6		47.2	8	26.8	3.3					
RK 26	KOLIT					2.5	5.0	1.3	13.8		48.4		25.2	1.9		1.9			
RK 18	ULUJUK						6.7		10.4		60.6		21.8		5				
RK 32	ANGUTIQT	1"		1.1		3.8	4.9		7.7		57.9		19.1	3.8	1.6				
RK 30	MANILAK			2.0			5.2		6.5		58.2		20.9	4.6	2.6				
RK 24	USSAK			4.0		1.5			9.5		63.2		16.4	3.0		2.5			
RK 3	KAVIK			1.8	4	2.7	2.2	9	5.8		58.3	1.8	16.6	4.5	3.1	9			
RK 17	ULUJUK		5				5.5	16.0	1.8		51.6		7.3	13.7	1.4	1.8			
RK 34	KASIGIAK						6.6	10.5	6.1		53.6		6.1	10.5	6.6				
RK 31	ANGUTIQT	2				3.8	1.9	5.7			73.3		3.8	7.6		3.8			
RK 38	T.KIMAL		3.3			7		7.9	3.9		56.6		9.9	17.8					
RK 19	TAUTUNNG		1.8			6		5.3	7.1		57.6		10.6	12.9	6	2.9			
RK 27	TIKTAK						4.2		15.8	5.0	56.3		12.6			5.0			
RK 29	TIKTAK	3	5	2.7		4.5	5	3.6	10.8	9	52.3	2.3	8.1	1.4	5	4.1		3.2	
RK 33	KASIGIAK		2.0	2.0		5.4		4.1	7.4		51.4	1.4	11.5	4.7		10.1			
RK 28	TIKTAK			7		11.9	1.3		2.6		63.6		3.3		4.0	11.3			
RK 15	ANARUAK			1.4		4.2			14.1		38.0		11.3	4.2	7.0	12.7			
RK 35	TAPARTI	4		2.5			8	2.5	9.3		55.1		10.2	5.9	5.9	3.4			
RK 39	T.KIMALA		7	2.1		2.8		2.1	9.1		46.2		16.8	9.8	1.4	4.9			
RK 23	USSAK			2.1		2.1		2.1	9.6		54.1		13.0	6.2	2.7	5.5			
RK 21	NILAULA					1.6		3.2	14.4	2.4	45.6	3.2	17.6	8.8	8				
RK 40	T.KIMAL	5	3.9	1.1		8.3		6	9.4	6	43.9	1.1	11.7	13.3	5.6	6			
RK 37	KIMALAR					4.4	2.2	4.4	5.8		46.7		14.6	10.2	2.2	5.1			
RK 14	TAUTUNNG			3.0	3.0	1.0		8.0	6.0		43.0	3.0	15.0	11.0		4.0			
RK 36	MCCASKIL					7.8			32.5		16.9		32.5			2.6			
RK 11	PANIGONIA					1.9	11.5	3.8		13.5	36.5	17.3			15.4				
RK 25	USSAK								3.4	9.7	73.7	9.7	3.4						
RK 9	PANIGONIAK	6						6.9	5.2	10.3	44.8	17.2	5.2	8.6		1.7			
RK 16	ULUJUK							9.7	4.1	9.7	49.8	12.4	8.3	6.0					
RK 4	IRKOOTI							7.4	2.9	6.3	64.6	8.6	5.1	5.1					

TABLE 3. RANKIN INLET INTERVAL DENSITY SORTED INDICATOR TABLE

CLASS		CHARACTERISTIC NOTES		INCLUDES SONGS
1		F,G,A,[C]		1,18,29,27,3,26,13,20,30,24,2,39,32,21
2		[D],G,A#,C,+D		31,17
3	3'	E,G,A	C,+D	19,38,40,37,33
	3"		B,D,[+D]	15,35,23,14,12
4		F#,G,A,[C]		9,16,25,4
OUTLIERS				34,36,11,28

TABLE 2. CLASSIFICATION OF SONGS BY TONAL DENSITY

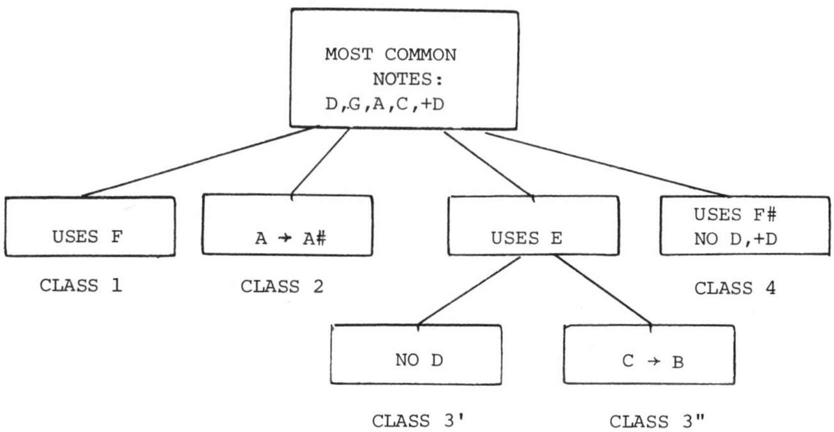

FIG. 3

CLASS		CHARACTERISTIC INTERVALS		INCLUDES SONGS
1	1'	[-4],-2,0,2,3	-3	1,13,20,2
	1"			12,26,18,32,30,24,3
2		-3,[-2],0,2,3		17,34,31,38,19
3		[-7],[-5],-2,0,2,5		27,29,33,28
4		-7 ,[-5],[-3],-2,0,2,3,4,5		15,35,39,23
5		-5,-3-2,0,[1],2,3,[4],[5]		21,40,37,14
6		[-3],-2,-1,0 ,1,2,[3]		25,9,16,4
OUTLIERS				36,11

TABLE 4. CLASSIFICATION OF SONGS BY INTERVAL DENSITY

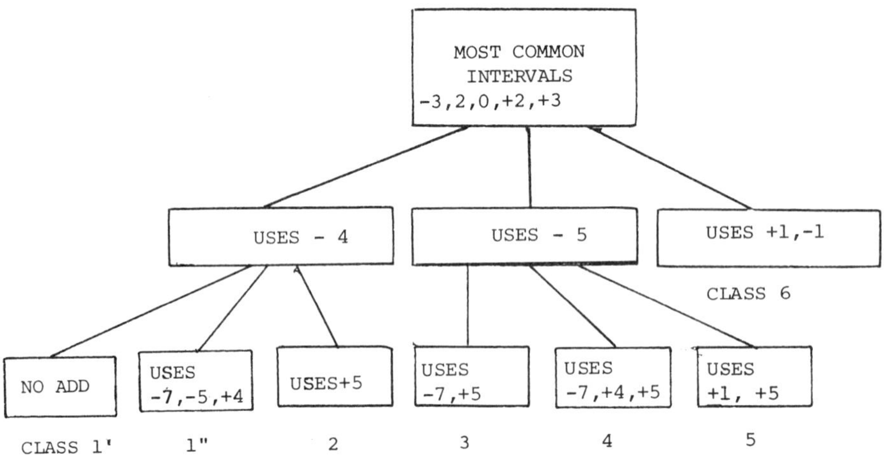

FIG. 4

MODE	Tonal Class	Intval Class	SONGS
1	1	1'	1, 13, 20, 2
2	1	1"	26, 18, 32, 30, 24, 3
3	1	3	27, 29
4	2	2	31, 17
5	3'	2	38, 19
6	3	5	40, 37, 14
7	3"	4	15, 35, 23
8	4	6	9, 16, 25, 4
OUT-LIERS			21, 39, 33, 12, 34, 36, 11, 28

TABLE 5. SONG CLASSIFICATION BY "MODES"

COMPUTERIZED CONTENT ANALYSIS OF NARRATIVE IN ANNUAL REPORTS: A COMPUTER IN THE HUMANITIES LOOKS AT THE CORPORATE WORD

Robert L. Oakman
Department of English
University of South Carolina
Columbia, South Carolina 29208

ABSTRACT

Two colleagues in Business Administration and I are experimenting with a computerized content analysis system--Howard P. Iker's WORDS package--to investigate narrative in financial reports of American corporations. For example, do narrative portions of corporate annual reports reflect or obscure the quantitative information scrutinized by auditors in the financial statements? Discussion includes theoretical assumptions, computer methods, and preliminary results of this research--textual analysis, including grammatical and statistical components, and isolation of "theme" clusters of related words.

KEYWORDS

Content analysis; corporate annual reports; auditing; narrative analysis; word-frequency analysis; WORDS computer package.

INTRODUCTION

In recent years the U.S. Securities and Exchange Commission has begun to emphasize "truth in packaging" in the various kinds of information sent out to investors and stockholders in corporate annual reports. Gradually the responsibilities of auditing firms have evolved beyond the strictly regulated analysis of the financial statement according to generally accepted accounting principles. Moving beyond these traditional concerns, auditors, in order to be "responsible," are now required to make at least a cursory judgment about obligatory narrative summaries of the company's financial statement to see if they directly conflict with the financial statement itself.

Besides this summary, an annual report contains other kinds of information in narrative form directly related to the finances of the company for the use of stockholders and potential investors--usually a president's letter and sometimes a general discussion of operations. Unlike the narrative financial summary, this information is optional, discretionary, unaudited, and unregulated. Of particular interest to investors may be discussion of future plans and financial goals; often management wishes to disclose information selectively which best presents its point of view. Recent accounting research suggests that management does manipulate the information reported. It seems clear that the concept of information inductance, "that the behavior of an information source may be affected by anticipated reactions to the information being sent," (Ingram and Frazier, 1980, p. 1) is a subtle form of propaganda often at work in narrative disclosures to stockholders.

CONTENT ANALYSIS

Given this background, in 1979 two faculty colleagues in the College of Business Administration, Robert W. Ingram and Katherine B. Frazier, asked my advice about the possibilities of applying computerized content analysis to the narrative portions of corporate annual reports. Their initial objectives were two-fold: to evaluate a method for representing narrative elements of corporate annual reports in a systematic, quantitative way and, more importantly, to see if the narrative sections, especially those that are unaudited, differ for firms with poorer financial performance than for those with better performance. More basically, do the narratives reflect or obscure the quantitative information examined by the auditors in the financial statements? If there is evidence of discrepancy between the audited and unaudited materials, perhaps the auditing profession needs to reconsider its methods of making judgments about narrative disclosure within the reports.

What was needed was a method to measure the characteristics of narrative disclosures, especially those that are unaudited and not well suited to current auditing practice. Content analysis, a method for identifying the manifest content of narratives through the systematic identification of specified characteristics within them, seemed an appropriate choice to try to ferret out important themes and concepts within the documents. Computer-related content analysis schemes can be broken down into two general types: those beginning with a pre-determined list of words, a dictionary of expected terms, that has been coded into content categories and those involving no *a priori* decisions before analysis about frequency or meaning of the words that may be found in the documents.

We decided to base our investigations on a series of content analysis programs of the second kind without a predetermined dictionary, collectively called the WORDS package and developed by Howard P. Iker and his associates at the University of Rochester School of Medicine (Iker and Klein, 1974; Iker, 1974). The attractions of Iker's methods for the business analysis were several. First, its search for content is empirical; it starts with no pre-determined dictionary based on *a priori* assumptions. It ferrets out the significant content words from the documents themselves on the basis of their frequency and interrelationship with each other.

In its current form, the WORDS system for content analysis consists of a series of sophisticated modular computer programs written in PL/I that can be called by the user in a sequence that he chooses. Essentially, WORDS incorporates components to accomplish two basic kinds of operations on text: linguistic analysis of words and sentences to create a list of most frequent content words and statistical analysis applied to these words. According to the assumptions underlying WORDS, a pool of the most frequent content words that occur in association with each other, excluding function words like prepositions and conjunctions, will contain the most significant thematic information in the documents. In order to isolate these words and analyze them with statistical rigor, the user typically needs to arrange the WORDS programs to carry out three functions: grammatical analysis of the text, choosing the associationally "rich" subset of most frequent content words, and deciding which of several, optional statistical methods he desires for the analysis.

As a humanist concerned with the automation of language analysis, I was especially concerned with the linguistic components of WORDS in recommending it for this business application. To begin with, WORDS starts with the documents themselves in encoded form, with very little pre-editing. In our texts, the only major changes made in the literal transcripts of the narratives within the annual reports dealt with company names and numerical material, primarily dates and financial figures incorporated in the prose itself. We replaced company names with the generic term "company" and numbers with a symbol to generalize information of these kinds over the several companies and still make it available for tabulation.

After the texts are split into words and stored along with their original locations in the source documents, a parsing program begins to analyze the words into classes, omitting function words and merging words with the same stem together, the process of lemmatization, to allow different forms of a verb like *ask (asked, asks, asking)* to be considered as one word in the later frequency counts. The program to assign word classes finds some words, including the common function words, in a small dictionary; identifies others by means of morphological endings that signal parts of speech (*-ing* as verb, *-ly* as adverb, with exceptions in the dictionary); and resolves ambiguities of words tentatively assigned to several classes by means of frame analysis. A word such as *like*, which can be noun, verb, or preposition, must be resolved within context in this way, since the content meaning of *like* as noun or verb needs to be counted, whereas the function word *like*, a preposition, can be excluded.

After the parsing and lemmatization processes, WORDS is ready to sum up word frequencies and to proceed to choose a set of up to 215 high-frequency terms for statistical analysis. The limit of 215 words is determined by a concern for economy in computing time in the calculations of inter-correlation matrices among the words and the need to keep the number of content themes resulting from factor analysis down to a reasonable size for ease of interpretation. With factor analysis and varimax rotation, the computer produces a set of "factors," interpretable sets of word groups chosen because of their observed statistical association within the larger set. For example, instead of 215 words, whose relationships are difficult to categorize, factor analysis reduces them to a small set of factors that account for most of the variance in the original set and can be more easily analyzed in their place. Finally, the computer prints out the word groups with their statistical properties for the user's perusal and further analysis.

MATERIALS FOR STUDY AND PRELIMINARY RESULTS

Initially we tried content analysis with WORDS in a sample of annual reports for 1978 for thirty-four corporations involved in steel and other metal manufacturing. After the linguistic analysis yielded frequency tables for all the content words, we simply chose the most frequent 215 words from each set of narratives for factor analysis. For the president's letters, the twenty resulting factors varied from one factor with 40 words accounting for 5.85 per cent of all significant variation to one with only 12 words accounting for only 1.72 per cent of all significant variation. For the narrative summaries, the twenty derived factors varied from one containing over a hundred words (17.7 per cent of significant variation) to others with four and five words (accounting for a minimum of significant variation, .37 per cent and .45 per cent respectively). Using 2.0 per cent variation as a cutoff point left twelve worthwhile factors for both kinds of narrative, accounting for 45.5 and 65.8 per cent of cumulative significant variation in the letters and summaries, respectively.

Closer analysis of some of the factors related to the documents themselves is more suggestive of the results than a statistical summary. Among four annual reports picked at random for this description, only one gets a high enough Z-score among the twelve more important word groups in each document category to be considered statistically significant. The long letter from the president emphasizes positive, active verbs; favorable nouns and adjectives; and a future orientation, along with words associated with the business itself--all linked in factor 4, containing the following words among 35 with factor loadings above the cutoff point (.30):

invest	high	equity
expand	second	industry
remain	over	material
grow	overall	stockholders
base	past	million
demand	future	sales
have	period	economy
include	present	earnings
increase	history	steel
		management
		capacity

The upbeat tone of the letter is set in the first sentence, containing many of the important terms: "NVF's sales in 1978 were the highest in our seventy-five year history and net earnings, second only to thoseof 1974, were almost double NVF's previous second best results in 1977." We note that the subheadings within the letter reinforce the very positive tone: "A decade of achievement," "Growth through investments," "1978--a springboard to future success," and "Outlook for 1979." Factor 12, important in the short one-page financial summary, contains fourteen significant terms (factor loadings above .30), organized in lists similar to those above:

cause	part	coal
decrease	primary	volume
offset	high	selling
increase	operating	profits
	segment	steel

The more formulaic nature of such narrative disclosures, signaled by the higher percentage of significant variation accounted for by the twelve top factors in the summaries, is evident here in two typical sentences embodying much of this vocabulary:

> The 21% increase in steel sales in 1978 was caused by higher industry prices on steel products coupled with a 9.6% increase in Sharon's steel product shipments. . . . The 35% decrease in coal sales was primarily the result of a 33% decrease in shipments caused by the lengthy coal strike that shut down Sharon's coal mining operations through the first quarter of 1978.

Examination of the whole report reveals that substantial gains in overall profits make the positive vocabulary of increase related to steel much more obvious than that of decrease, which occurs more than once only in the sentence above, where coal losses are explained away by a crippling strike. Compared to the president's letter with its oratorical appeal, the financial disclosure is properly explanatory and mainly, if optimistically, factual.

Two of the three other reports examined against the results from WORDS offer nothing of substance worthy of detailed discussion. The president's letter of the last company does have a Z-score of 2.01 on factor 11 with seventeen terms, among them the following:

elect	over	vice
serve	under	president
add	approximate	opportunity
represent	large	economy
report	late	common
		business

Compared to the finances of the first company discussed previously, profits in this case are up along with inflation, but stock prices are not. The president's emphasis is on growth opportunity from restructured management reflected in the three words, vice and president and elect. The long four-page financial summary is very high on factor 6 (Z-score of 5.22), with 31 words, mostly general terms expected in a business discussion; the top five scorers are representative: products, markets, employ, report, program. Profit and selling are words in this factor shared with the first company, but improve has replaced the heavy reliance on increase and decrease, neither word significant enough to fall within the factor group of associated words. More expansive in its narrative presentation, this more subdued but forward-looking summary fits a company with fewer positive results to report than the first one.

CRITIQUE AND FUTURE PLANS

What kinds of successes do these preliminary results suggest about the use of WORDS for content analysis of text like corporate annual reports? Given the reasonable lengths of the narratives--a long one runs five or six pages--a careful reader probably could have picked out the most important words for each one, perhaps on the basis of raw frequencies; in fact, WORDS starts with the initial assumption that the frequency of content words within a corpus is a key to their importance. On the other hand, the reader would probably not have noticed the emphasis on the improvement in one company's financial summary, as opposed to an emphasis on increase, since the opening paragraph contains five forms of increase as noun and verb to only one form of improve. A loading of .307 for improve places it above the cutoff point of .30 for statistical inclusion in the factor, whereas increase at .251 falls below the significance level. In other words, factor analysis allows making subtler distinctions based on accepted statistical principles than even careful readers are likely to discern.

Another advantage of using the document-generated vocabulary means that words are associated through frequencies and factor analysis without predetermined classifications. With the business narratives, a common set of economic and financial terms can be expected in every document; yet even the word lists given for two company's financial summaries showed different emphases within this overlapping vocabulary. In sum, by means of an empirically derived, scientifically objective methodology, WORDS in these early experiments has revealed significant content, represented by the factors of associated words and sanctioned by the validity of generally accepted statistical techniques.

Work in hypothesis testing is probably the most interesting and potentially the most powerful and influential part of the ongoing analysis. We want to examine the relationships between the numerical information in the audited financial statement; the narrative financial summary examined also by the auditors; and the unregulated, discretionary information in the president's letters. Are the narrative disclosures consistent with the content of the audited financial report? To represent the "content" of the financial statement, we are using a variety of financial ratios such as stock market performance, profitability, and financial solvency that are commonly cited in cross-comparisons of corporations and widely adopted by investors and market analysts for decision making. When these are factor-analyzed and grouped for an industry and for each company, the statistical method of canonical correlation can be carried out on both sets of factors, numerical and word-based, for both kinds of narratives to show the maximum amount of relationship between the two sets of factors. In other words, canonical correlation can suggest how close the financial data for many companies correlate with the narrative summaries of this information.

Although at present only one or two computer tests of canonical correlation have been carried out, the results already suggest the weakness of the hypothesis that the president's letters vary systematically with a firm's performance portrayed in the financial statement. As might be expected, there seems to be a closer, statistically verifiable correlation between the financial statements and the narrative summaries. If these trends can be substantiated over many documents for a variety of industries, auditors and investors may wish to consider the relative importance, seriousness, and truthfulness of these narrative portions of annual reports as measures of financial performance.

As a humanist and novice in the realm of business analysis, questions like these are beyond my expertise and can best be handled by my colleagues in Business Administration. On the other hand, our joint experiences in seeking a systematic, objective, empirical measure of content in the narratives of corporate annual reports have already been enlightening to both parties and even suggestive for our upcoming work. With the WORDS package, we seem to have found a method with much flexibility for different kinds of text analyses and, more importantly, with adequate foundations in widely accepted linguistic and statistical techniques adaptable to computer processing. We are happy with the suggestiveness of tentative results already produced. As the outsider primarily concerned from day to day with the interpretation of language materials, central to the humanities, I personally am pleased that this computer method seems to validate the kind of subjective judgments an intelligent reader might make about the corporate word. In fact, it probably will open more potential avenues of inquiry than he can see by traditional narrative analysis. And, more generally, it will be making a real, humanistic contribution to the understanding of language usage in the larger world outside the confines of the university.

ACKNOWLEDGMENTS

I want to thank Robert W. Ingram and Katherine B. Frazier of the College of Business Administration and Tim White of the Social and Behavioral Sciences Laboratory at the University of South Caroilna for their gracious and helpful assistance with this research. This project has been funded by a grant from the Peat, Marwick, Mitchell Foundation through its Research Opportunities in Auditing program. The views expressed herein are those of the author and do not necessarily reflect the views of the Peat, Marwick, Mitchell Foundation.

REFERENCES

Iker, H.P. (1974). SELECT: A computer program to identify associationally rich words for content analysis. I. Statistical results. Computers and the Humanities, 8, 313-319.

Iker, H.P., and R.H. Klein (1974). WORDS: A computer system for the analysis of content. Behavior Research Methods and Instrumentation, 6, 430-438.

Ingram, R.W., and K.B. Frazier (1980). Environmental performance and corporate disclosure. Journal of Accounting Research, 18, forthcoming in Autumn issue.

A TYPOLOGY OF MELODIC CONTOURS IMPLEMENTED BY COMPUTER

Ramon Pelinski

University of Montréal

ABSTRACT

In this study, a technique for establishing melodic contour typologies is investigated.

KEYWORDS

Melodic contour, numerical taxonomy, clustering, typology, Eskimo music.

INTRODUCTION

Melodic contour has been used in comparative musicological analysis as a taxonomic criterion for the determination of structural relations between different cultures or cultural sub-groups (Adams 1976:179-215).
This idea is founded upon the following assumptions:

a. Melodic contour is a configuration of pitches which establishes the relation of similitude or dissimilitude between the songs of a given corpus or of given corpora.
b. The predominance of types of melodic contours is an index of the stylistic organization of the songs.
c. Melodic contour can be associated with the symbolization of contents (feelings, cultural categories,etc) external to the sound material.

While contour analysis has been practiced on a "manual" basis, we propose a technique whose novelty resides in the application of certain procedures of numerical taxonomy (Sneath,Sokal 1973) implemented by computer.
The samples under consideration are taken from three corpora of personal songs of the Cariboo Eskimo from the Hudson Bay (Rankin Inlet and Eskimo Point) and the Ammassalik-Eskimo from Greenland (Thule).

The rpoposed technique includes the following operations:

1. <u>Standarization of the length of the songs</u> : The reduction of the length of the songs to a standard length of 100 pitches, which is required due to the comparative aims of the analysis.

2.Decimal segmentation: Each standarized song is segmented into 10 sections of equal length; each section is caracterized by the pitch which is most distant from the tonal center G, common to all songs.

3.Clustering of the songs: The standarized songs are clustered according to the shortest Euclidean distance (coefficient of similarity).
Steps 1,2 and 3 are represented by a diagramme in which the horizontal axe is the tonal center G; the segments are caracterized by their melodic peaks

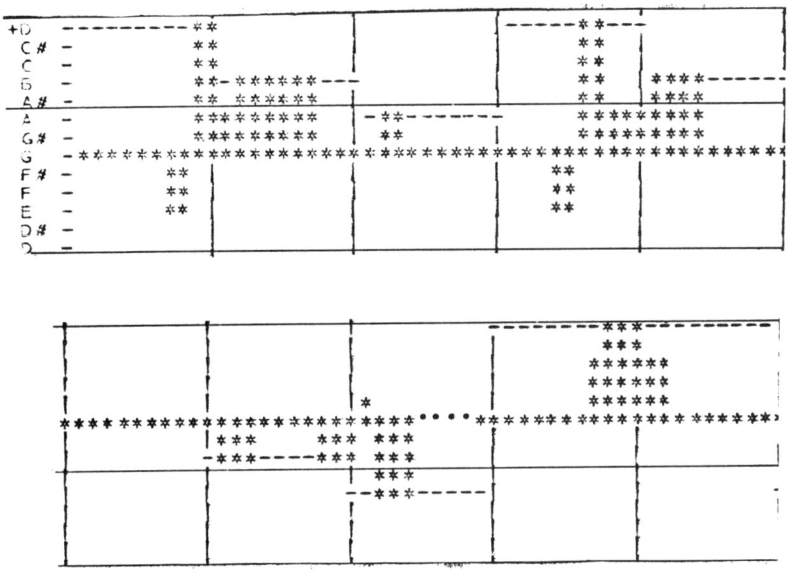

Fig.1 : Contour representation of a song after standarization and decimal segmentation

4.Minimum spanning tree: After establishing the similarity coefficient, a minimum spanning tree may be constructed. It visually represents the clustering (grouping) of the songs of the the three samples according to their degree of similarity (s. Fig.2)

5.Typology of contours: The minimum spanning tree is the point of departure for establishing the contour types of a determined sample. A typology presuposes cutting points or limits between the types. However, th ese limits are not explicit in the minimum spanning tree. Indeed, we ourselves have to decide about the degree of generality of the typology by establishing pertinent limits between the members of the tree. In practice, we have isolated sub-trees or types, considering as affinity groups or typological unities those contours which are adjacent to a given song. In Fig. 3 the numbers design ate the members of the sample; the simple alphabetic letters designate the coefficient of similarity, while RI, TH and EP designate the samples respectively:Rankin Inlet, Thule and Eskimo Point.

The use of the minimum spanning tree shows an important characteristic of our procedure: the isolated sub-trees or typological unities are not hermetic but

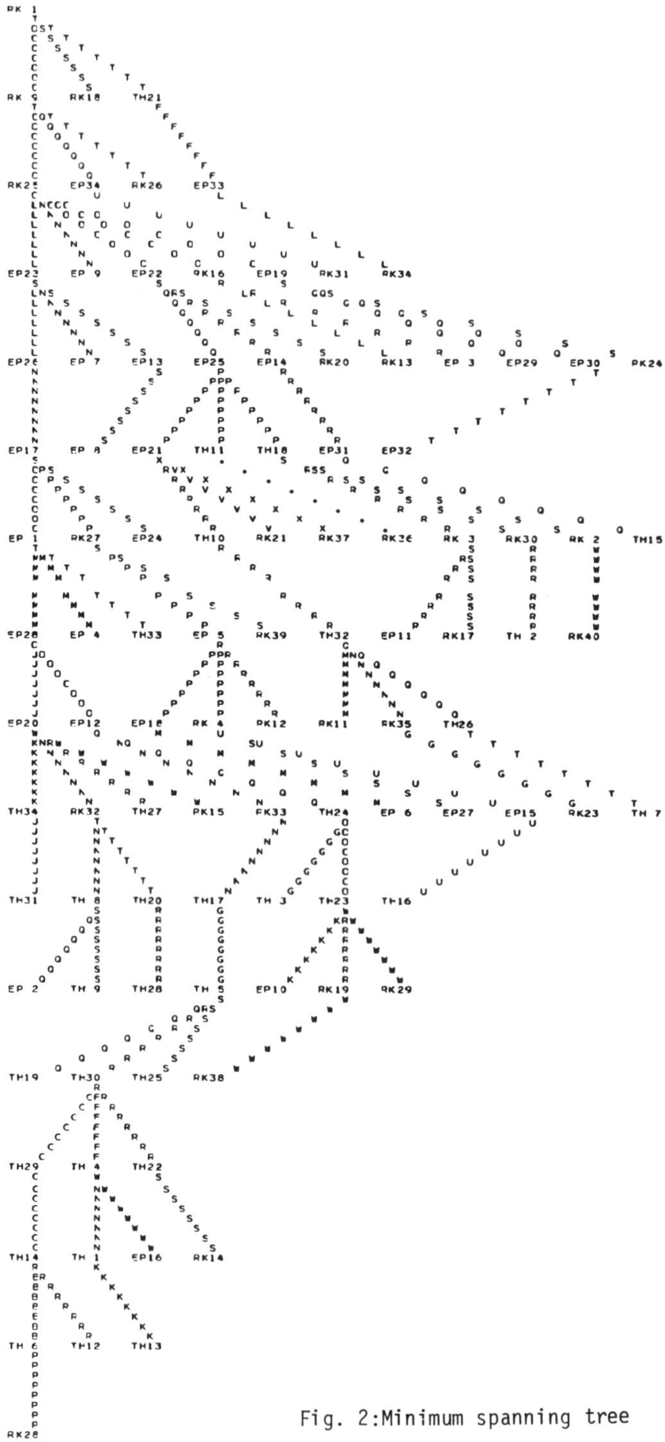

Fig. 2:Minimum spanning tree

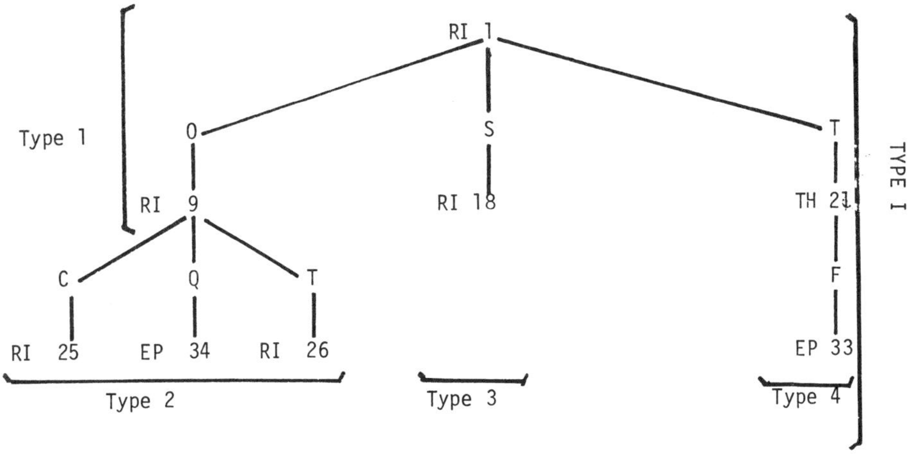

Fig. 3 : Typology of contours

rather oper types whose members can simultaneously belong to different types according to their respective degree of similarity. If we want to obtain a more synthetic typology, we can extend the limits which separate the memebers of the tree. The generality of a typology depends, indeed, on the number of borders one is willing to delimit (Levy 1979:26).

6. Calculation of the contour type (phenotype) : The melodic contour of a determined type is obtained by calculating the following elements:
a. the average of the coefficient of similarity between the single contours of the same sub-tree.
b. the standard deviation of the single contours around the average contour (Fig.4_)

7. Genotype of the melodic contours: If operations a. and b. are applied to the whole sample of songs, we obtain the genotype of the melodic contours which musically characterize the sample under consideration (Fig. 5).

8. Phenogramme of melodic contours: The phenogramme constitutes another way of representing the distances of amalgamation of the contours and their clustering according to the measure of their Euclidean distances: to the minor distances of junction correspond the more closely related contours (Fig. 6).

9. Correlational analysis of variables: The minimum spanning tree and the phenogramme are useful tools for establishing relationships between melodic contour and other variables such as: the appurtenance of the singer to a determined cultural group, the tonal density (number of pitches) of a song, its rhytmic organization, etc.

CONCLUSION

The advantages of the proposed technique made evident by this study include the following:
a. We can measure with mathematical precision not only the contours but also the degree of similarity between them.

b. The construction of an open typology does justice to the structural fluidity of a determined sample of songs. Indeed, the taxonomic binarism is overcome by the gradation of the relations of affinity.

c. The operations required by the technique are explicit; they can be reproduced and their results confirmed or infirmed. It may be not necessary to talk about this advantage among informaticians. However, explicitness of procedures can still be considered a taboo in certain hermeneutic musicological tendencies.

d. If we leave it up to the computer to accomplish those tasks which only computers can perform, we can vastly increase our enjoyment of those tasks that only human can perform (Cook 1967:104).

SONG # 1	0.0	1.0	5.0	5.0	-1.0	-3.0	-2.0	-3.0	-2.0	-3.0
SONG # 17	-1.0	-1.0	5.0	5.0	5.0	-2.0	-2.0	-2.0	-2.0	-1.0
SONG # 28	4.0	2.0	4.0	2.0	0.0	-3.0	-1.0	-1.0	-1.0	-1.0
SONG # 4	-3.0	-3.0	4.0	4.0	-3.0	-4.0	-3.0	-1.0	-3.0	-2.0
AVERAGE SONG	0.0	-0.3	4.5	4.0	0.3	-3.0	-2.0	-1.8	-2.0	-1.8
STD.DEVIATION	2.5	1.9	0.5	1.2	2.9	0.7	0.7	0.8	0.7	0.8

	SONG # 1	SONG # 17	SONG # 28	SONG # 4	AVERAGE SONG
DISTANCES		6.856	6.164	6.245	6.421

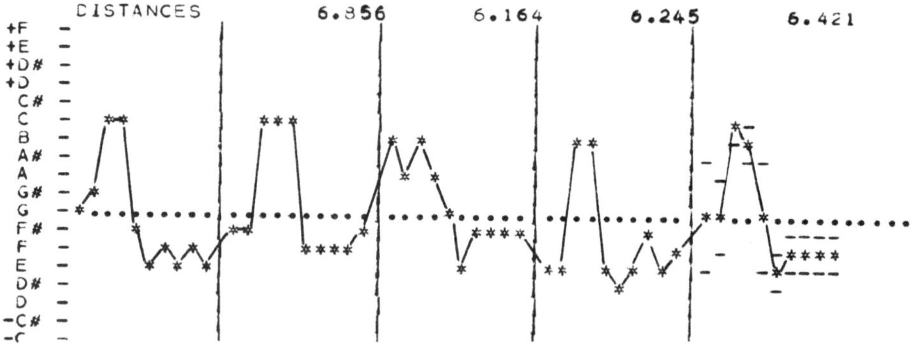

Fig. 4 : Phenotype of a group of songs from Rankin Inlet

ARCHETYPE SONG	2.1	2.4	0.7	1.8	1.6	-0.1	0.4	-0.4	0.8	-0.4
STD.DEV ARCHETYPE	3.3	4.0	3.6	3.7	3.6	4.7	4.5	4.5	3.9	2.3

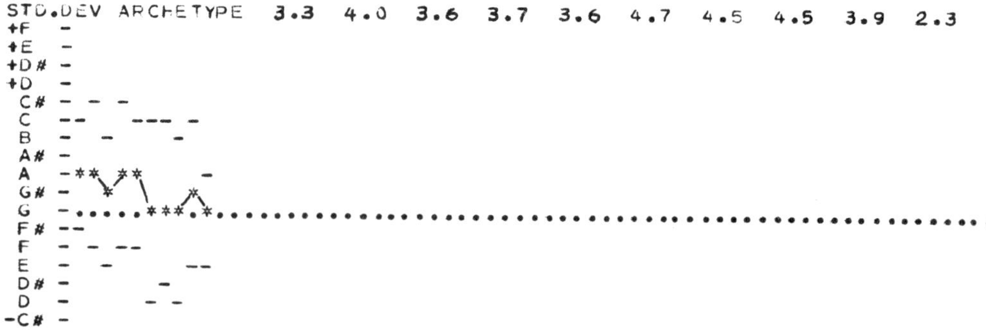

Fig.5: Genotype of the sample of Rankin Inlet

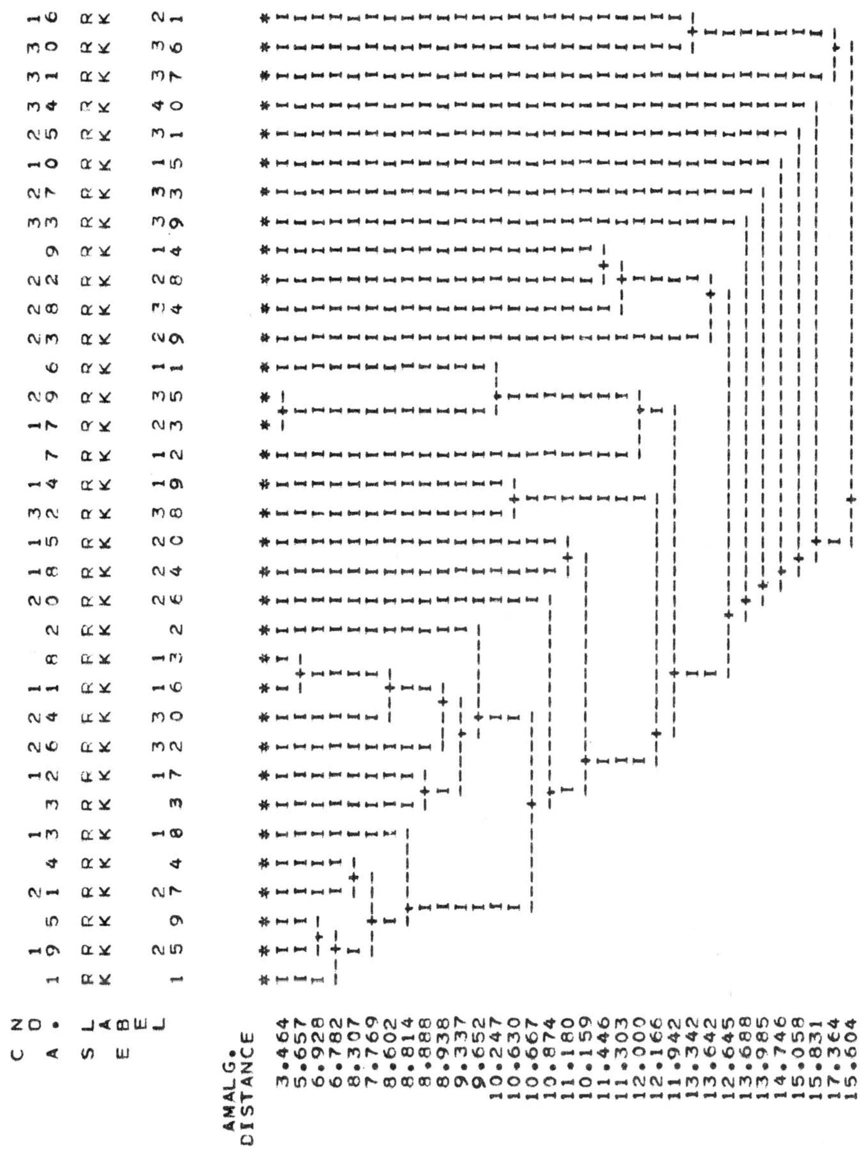

Fig. 6:Phenogramme of the melodic contours of the sample of Rankin Inlet.

REFERENCES

Adams,Ch.R. (1976).Melodic contour typology. Ethnomusicology,XX,2,179-215.

Cook,Ch.C. (1967). An Introduction to the Information Processing Capabilities of the Computer. Computer Applications in music.,G.Lefkoff,ed. W.Virginia U.P.

Levy ,M.L. (1979). Comprendre les statistiques.Paris Seuil.

Sneath,P.M.A., Sokal,R.R. (1973).Numerical Taxonomy. W.H.Freeman.

STATISTICS AND THE COMPUTER IN FORMULA ANALYSIS OF SERBO-CROATIAN HEROIC VERSE

Rudy S. Spraycar

Department of English
Louisiana State University
Baton Rouge, Louisiana 70803

ABSTRACT

The "oral-formulaic theory" of Milman Parry and Albert Bates Lord has influenced hundreds of stylistic studies of Homeric and medieval epic verse. Lord's original quantitative analysis of a short sample of oral folk poetry recorded in the field in Yugoslavia has assumed undue importance, obscuring Parry's and Lord's sensitive account of the complex phenomenon of repetitive poetic diction. No statistical basis has been established for a distinct quantitative threshold of formula density, above which only the work of an oral poet can reach. Rigorous analysis of significant amounts of data requires the development of automatic lemmatization for Serbo-Croatian, and perhaps of a new mathematical model as well.

KEYWORDS

"Oral-formulaic theory"; automatic lemmatization; Serbo-Croatian; oral poetry; statistics and literary style.

INTRODUCTION

Humanists have made great strides in adopting scientific methods for their research since the poet W.H. Auden (1946) wrote, "Thou shalt not sit/ With statisticians," and C. P. Snow (1959) bemoaned the formation of "two cultures," literary and scientific; much of this progress has been precipitated by the computer. This trend is not without its dangers, however. Serious problems can arise when humanists borrow rudimentary quantitative tools and methodologies developed by scientists; if such researchers in the humanities lack the methodological rigor to interpret their data soundly, they run the risk of offering only apparently elegant simplicity in their solutions to humanistic problems.

THE "ORAL-FORMULAIC THEORY"

An example of a humanistic endeavor whose quantitative methods long escaped rigorous methodological scrutiny is the "oral-formulaic theory." Advanced by Milman Parry (1930) and further articulated by Albert Bates Lord (1960), the hypothesis accounts for the repetitive or "formulaic" style of Homeric and, later, medieval epics on the basis of their oral composition.

Wishing initially to account for the ubiquity of such familiar Homeric epithets as "rosy-fingered dawn" and "grey-eyed Athena," which he had so rigorously analyzed (1928), Parry posited a "traditional style" in which the oral poet worked. Composing the poem anew in each performance without the writer's leisure to choose one word at a time, the oral-traditional bard was compelled by the necessity to get on with the poem to make use of ready-made phrases. Seeking an analogous, living oral poetic tradition to serve as a "laboratory" for his hypothesis, Parry (1933) turned to the Southslavic heroic poetry of Yugoslavia. The works of these illiterate, oral poets displayed stylistic features, including formulaic diction, similar to those of Homeric verse. Albert Bates Lord (1960) extended and completed Parry's comparative study of Serbo-Croatian oral poetry.

Though the arguments of both Parry and Lord are complex and intricate, many of their followers, particularly those interested primarily in medieval epic, have fixed upon a single, quantitative criterion for discriminating between oral and written verse. Although Lord has repeatedly (1974, 1976) bemoaned this trend, many scholars have concentrated exclusively on his original study (1960) of fifteen lines of Serbo-Croatian oral folk poetry recorded in the field in Yugoslavia. Lord analyzed his sample by determining which of these fifteen lines (and/or thirty half-lines) recurred in a larger corpus of about 12,000 other lines of the same oral poet's work. (This discussion will be restricted to the figures for "formulas" that are half-lines repeated verbatim.) Of the thirty half-lines in the fifteen-line sample, 63% recur verbatim (word-for-word, disregarding differences in inflection) in the course of 12,000 other lines of verse.

The figure of 60%, or in some cases 40%, "formulas" came to be regarded as a "threshold," above which poetry of uncertain origin could be regarded as essentially oral in composition. Literally hundreds of other studies were based on this single analysis, as can be seen in the bibliographies, no longer current, of Haymes (1973) and Lord (1974). Yet, as Bruce A. Beatie (1979) has observed, no scholar of the "oral-formulaic" school attempted to test or even to replicate this study.[1]

To be sure, Lord was aware that the formulaic style of such poetry as medieval epic could be the result of literate imitation; a "control" (he does not call it that) is needed. He offers (1968) the results of John Lindow's analysis of thirty lines of literate poetry from Andrija Kačić-Miošić's Razgovor Ugodni Naroda Slovinskoga (The Pleasant Discourse of the Illyrian People) (Matić, 1964). Lindow found the poetry of Kačić, a literate monk avowedly imitating oral verse, only 27% "straight formula," as opposed to Lord's earlier (1960) figure of 63% for authentic oral poetry. However, Lindow searched for repetitions of the half-lines in his thirty-line sample through only 3,208 other lines of literate poetry, despite Lord's original consultation of some 12,000 lines in seeking recurrences of phrases of oral poetry.

REPLICATION OF ANALYSIS OF CONTROL

The fourfold difference in the number of lines consulted seems to account for the difference that Lord reports (1960, 1968) between oral and written verse. Using a computer-generated concordance, I have re-analyzed Lord's and Lindow's original thirty-line sample of Kačić's literate poetry, consulting a full 12,000 other lines, as Lord did with oral poetry; I found Kačić to be 65% "formula," compared with 63% for Lord's oral sample. In fact, since Lord's oral sample was only fifteen lines long, it is striking that in my analysis the first fifteen lines of Lord and Lindow's literate sample are 77% formula, or substantially more formulaic than the oral sample.

---

[1]Beatie seems unaware of my initial critique of the approach (Spraycar, 1976).

SAMPLE SIZE

As Joseph J. Duggan has argued (1973), part of the problem lies with the shortness of the sample; this is easy to demonstrate. Table 1 analyzes all possible fifteen-line samples in the first fifty lines of poem twenty-six in the 1759 edition of Kacic's Razgovor (Matic, 1964).

TABLE 1 Percentage Formularity of Successive Fifteen-Line Samples

Lines	Non-Unique Half-Lines*	Percentage
1-15	23	77%
2-16	24	80%
3-17	24	80%
4-18	23	77%
5-19	22	73%
6-20	21	70%
7-21	22	73%
8-22	21	70%
9-23	20	67%
10-24	20	67%
11-25	20	67%
12-26	20	67%
13-27	20	67%
14-28	20	67%
15-29	19	63%
16-30	18	60%
17-31	18	60%
18-32	16	53%
19-33	16	53%
20-34	15	50%
21-35	17	57%
22-36	18	60%
23-37	19	63%
24-38	21	70%
25-39	21	70%
26-40	20	67%
27-41	20	67%
28-42	21	70%
29-43	20	67%
30-44	21	70%
31-45	22	73%
32-46	21	70%
33-47	23	77%
34-48	21	70%
35-49	23	77%
36-50	23	77%

*Lines repeated elsewhere in a range of 12,000 lines.

The percentage of verbatim half-line formulas ranges from a low of 50% to a high of 80%, averaging 68% (compared with Lord's figure of 63%). While any one of these samples of literate poetry would cast doubt on the idea that only the oral poet can attain so high a degree of formularity, the substantial fluctuation from sample to sample testifies further to the dubiousness of this approach.

INCREASED SAMPLE SIZE

Following Duggan's suggestion (1973) that the entire referent be taken as a sample,

I have developed a computerized method of lemmatization to disregard Serbo-Croatian inflectional variation automatically in analyzing samples of about 12,000 lines of poetry (Spraycar, 1980). Other problems remain to be worked out, however. While some of the blame for the problematic character of Lord's findings can be blamed on his sample size, even setting the sample equal to all the hemistichs of a given text as Duggan does will not enable one to compare the percentage formularity of two texts of unequal length without further adjustments by way of statistical analysis.

Table 2 shows how increasing the number of lines analyzed increases the percentage of formulas quite dramatically and in a relatively uniform progression; this occurs even when one keeps the sample size and the referent equal, analyzing every line in the sample but comparing it only with the other lines in that sample. On the other hand, samples of widely varying length, even though they are from the same poem, vary substantially in formularity as a function of the size of the referent used.

TABLE 2 Repeated Whole Lines as a Function of Sample Size

Number of Lines	Number of Repeated Whole Lines*	Percentage
100	2	2%
500	31	6%
1000	86	9%
5000	816	16%
10000	2489	25%
15000	4105	27%
20000	6115	31%

*Not to be compared with the much higher figures for half-lines in Table 1.

Consequently, I must suggest that Lord's judgement was correct when he chose to examine 12,000 lines of oral Serbo-Croatian verse--roughly the length of a single Homeric epic. We need a synthesis of Lord's judgement and the statistical concern of Duggan and others. Moreover, I know of no studies in this area that have employed even the most rudimentary statistical measures of validity.

CONCLUSION

The work cannot stop here. If the relatively unsophisticated quantitative methodologies outlined above are inadequate, statistical sophistication alone will not offer a new theory to explain a phenomenon that remains inadequately accounted for. Perhaps the mathematical modelling offered by cybernetics and systems approaches that has already borne fruit for linguistic researchers will eventually provide the key to this problem as well.

ACKNOWLEDGEMENTS

The completion of the research reported in this paper was facilitated greatly by my tenure during summer and fall 1979 of a research fellowship granted by the American Council of Learned Societies, and by released time from the Louisiana State University English Department in spring 1980. An earlier version of part of this paper was presented to the 1980 meeting of the Société Rencesvals, American-Canadian Branch, Berkeley, June 1980. For their helpful comments on early drafts, I am indebted to Professors John Fischer, Edward C. Schweitzer, Rosan Jordan, and Jack Winner of Louisiana State University; I am grateful for their suggestions.

REFERENCES

Auden, W. H. (1946). Under Which Lyre: A Reactionary Tract for the Times (Phi Beta

Kappa Poem, Harvard, 1946). In E. Mendelson (Ed.), Collected Poems, Random House, New York. p. 262.

Beatie, B. A. (1979). Measurement and the Study of Literature. Computers and the Humanities, 13, 185-194.

Duggan, J. J. (1973). The Song of Roland: Formulaic Style and Poetic Craft. University of California Press, Berkeley and Los Angeles.

Haymes, E. R. (1973). A Bibliography of Studies Relating to Parry's and Lord's Oral Theory. Publications of the Milman Parry Collection, Harvard University, Cambridge, Mass.

Lord, A. B. (1960). The Singer of Tales. Harvard University Press, Cambridge, Mass.

Lord, A. B. (1968). Homer as Oral Poet. Harvard Studies in Classical Philology, 72, 1-46.

Lord, A. B. (1974). Perspectives on Recent Work on Oral Literature. Forum for Modern Language Studies, 10, 187-210. Rpt. in J. J. Duggan (Ed.), Oral Literature: Seven Essays. Barnes and Noble, New York.

Lord, A. B. (1976). Discussion of J. A. Russo, Is 'Oral' or 'Aural' Composition the Cause of Homer's Formulaic Style? In B. A. Stolz and R. S. Shannon (Eds.), Oral Literature and the Formula. Center for the Coordination of Ancient and Modern Studies, University of Michigan, Ann Arbor.

Matić, T. (Ed.) (1964). Djela Andrije Kačića Miošića. Jugoslavenska Akademija Znznosti i Umjetnosti, Zagreb.

Parry, M. (1928). L'Épithète traditionnelle dans Homère: Essai sur un problème de style homérique. Société Éditrice Les Belles Lettres, Paris.

Parry, M. (1930). Studies in the Epic Technique of Oral Verse-Making. I. Homer and Homeric Style. Harvard Studies in Classical Philology, 41, 73-147.

Parry, M. (1933). Whole Formulaic Verses in Greek and Southslavic Heroic Song. Transactions of The American Philological Association, 64, 179-197.

Snow, C. P. (1959). The Two Cultures and the Scientific Revolution. Cambridge University Press, New York.

Spraycar, R. (1976). La Chanson de Roland: An Oral Poem? Olifant: A Publication of the Société Rencesvals, American-Canadian Branch, 4, No. 1 (October), 63-74.

Spraycar, R. (1980). Automatic Lemmatization in Serbo-Croatian. Association for Literary and Linguistic Computing Journal. Accepted for publication.

THE ROLE OF LANGUAGE IN THE NEW COMMUNICATION TECHNOLOGY

Henry Kučera

Brown University
Providence, R.I., U.S.A.

ABSTRACT

The new communication technology, which relies primarily on "written" language as its medium, will require from all its users a disciplined and, eventually, quite sophisticated use of language skills. The understanding of linguistic performance in various types of communication will thus become important. This paper reports on the analysis of the various styles represented in the one-million-word corpus of present day American English, known as the Brown Corpus. Information is presented on sentence length, on sentence complexity, and on the occurrence of active and passive constructions in the various Corpus genres.

KEYWORDS

Natural language processing; automatic parsing; performance grammars; sentence complexity; semantic mapping; stylistics.

INTRODUCTION

Advances in communication technology are bound to affect the status and development of the principal medium of communication, human language. But the specific effect will obviously depend on the role that certain forms or aspects of language play within the particular system of information dissemination. The invention of movable type had a long-range beneficial effect on the process of codification of languages, establishment of "standard" linguistic codes and the investigation and study of grammars - not to speak of the general cultural and political effects of the spreading literacy that it brought about. The impact of the major advances in communication technology of the last hundred years, however, has probably been far less beneficial. The telephone, radio and, especially, television rely on the spoken language and do not necessarily enhance formal language skills. The case of television, where the spoken language is often only an ancillary medium to the visual message, is particularly complex and controversial. It is not my purpose here to debate the question of whether television has contributed significantly to the problem of "functional illiteracy" that seems to plague even economically highly developed societies. But even a cursory observation of the form and function of language in television makes it hard to imagine how this medium could possibly have a positive effect on higher-level language skills.

Be it as it may, the future presents an entirely different picture. The current developments in communication technology, as well as those that are on the horizon, are based on computer-directed editing, formatting, transmission and storage of information, the primary medium of which is the written (or printed) language. The various word-processing and text-editing systems, electronic mail, direct access links to library-like data banks from which one will be able to retrieve a particular segment on one's own terminal, the home-based query systems connected to a central processor, all of these, of course, will rely primarily on the "written" form of language and thus require from its users a rather high level of linguistic sophistication. It is important to emphasize, in this connection, that it would be quite erroneous to consider the "written" form of language to be simply a recoding of the spoken form into a more permanent medium. Both the theoretical studies of linguists interested in the differences between spoken and written language (e.g. Vachek, 1959; Francis, 1974) and the empirical research of spoken and written corpora of languages show that we are dealing here with overlapping but yet different codes of communication, each with its properties significantly influenced by the different capacities of the communicational channel involved. Written corpora, in general, consist of well-formed sentences which are parseable - certainly by humans and, to a considerable extent, automatically - thus providing at least some prospect of computer-based understanding of such texts. In the spoken samples, on the other hand, fragmentary utterances, false starts, redundant repetitions and hesitation phenomena are extremely common, as is an overwhelming reliance on various kinds of supra-segmental features and indicators, primarily intonation (cf. Svartvik and Quirk, 1980). All of this clearly makes any direct automatic parsing of such texts a rather hopeless proposition.

There is actually an important lesson in this: Even if, in the late 1980's perhaps, the technology of direct *voice* communication with machines becomes sufficiently well developed for general use, formal language skills will still be required from all those who will wish to use this technology. As a matter of fact, the formal properties of such voice-based communication systems will need to be rather strictly defined and enforced; it is clear that - as one moves beyond the elementary technology of recognizing individual digits and a few key words or primitive commands to a really meaningful machine recognition of continuous speech - the need for a disciplined voice input which adheres quite rigorously to a standard language structure and its grammar becomes imperative. Thus those who may wish to use the new technology - be it one based on the written language or one requiring a disciplined spoken input - will need to master the "new" literacy as more and more complex systems become available. One can only speculate about the social consequences of such a development; it is clear that the burden on the educational system will not decrease but rather present a whole new set of challenges.

ANALYSIS OF LINGUISTIC PERFORMANCE

The future - if one views it with at least some optimism - thus suggests an important and as yet unfulfilled need in language study, namely the description of linguistic structures that are actually used in some well-defined (primarily written) form of communication, and the development of what might be called "performance grammars" which make reasonable predictions about actual language use. It is with the above goals in mind that my analysis of the one-million-word corpus of present-day American English has proceeded. Because of space limitations, I shall be able to describe here in detail only one aspect of this analysis, the distribution of active and passive sentences in different genres of English, a distribution that bears on the relation between syntactic structures and semantic function.

The one-million-word corpus was originally assembled by W.N. Francis and Henry Kučera at Brown University in the 1960's and is thus frequently referred to as the Brown Corpus (Kučera and Francis, 1967). More recently, the compilers of the Brown Corpus have completed a grammatical annotation of the data base. The entire one million words of the Corpus have been "tagged", with each word given a specific grammatical symbol. The "tagging" procedure, which was semiautomatic, assigned to each running word an unambiguous symbol based on a taxonomy of 82 grammatical categories (Francis and Kučera, 1979). The basic principle of our tagging is a largely expanded set of grammatical word classes, supplemented by morphological information (singular vs. plural, tense of verbs, etc.) and some syntactic coding. Because of the system of grammatical annotations used, the retrieval of various types of syntactic information can now be accomplished algorithmically. Our current research is concerned both with automatic parsing of the annotated text and with the study of linguistic performance.

If we disregard headlines and other headings (of chapters, sections, etc.), the Brown Corpus contains 54,724 sentences, with the mean sentence length of 18.49 words. However, both sentence length and sentence structure vary greatly among the 15 genres of writing represented in the Corpus. In investigating linguistic performance, it is useful to focus on the two main subdivisions of the Brown Corpus, "informative" prose and "imaginative" prose. In the Brown Corpus, the term *Informative Prose* is applied to all those samples that have been selected from nonfiction sources. This section is divided into nine genres, ranging from newspaper selections of various kinds to learned and scientific writing. There are altogether 374 samples of Informative Prose in the Corpus; with each sample being approximately 2,000 words long, this part of the Corpus consists of 755,010 words. *Imaginative Prose*, on the other hand, includes samples taken from a variety of fiction sources and is represented by six genres. There are 126 samples of Imaginative Prose, again of about 2,000 words each, accounting for 256,955 words. The entire Corpus thus consists of 500 samples of texts and contains 1,011,865 running words (word tokens), not counting headlines and other headings.

All genres of the Informative Prose portion have a higher mean sentence length than any of the genres in the Imaginative Prose section of the Corpus, pointing to a stylistic difference between the two sections. But sentence length is only an elementary parameter in the study of actual linguistic performance and of style. Much more important, in this respect, is the investigation of sentence structure and sentence complexity. I have presented the results of a sentence complexity study of the Brown Corpus in a previous paper (Kučera, 1980) where I define sentence complexity essentially as the number of predictions per sentence. While the details cannot be reproduced here, Table 1 gives a general summary of my results. (INFO refers to the Corpus subset of informative prose, while IMAG is the abbreviation of the section of imaginative prose).

TABLE 1 Sentence Length and Sentence Complexity

Measure	INFO	IMAG	CORPUS
Words/Sentence	21.12	13.55	18.48
Predic./Sentence	2.80	2.38	2.65
Words/Predication	7.54	5.69	6.97

While Table 1 simply confirms that sentence length is highly genre dependent, it also shows that the predication-per-sentence figure is not directly correlated with sentence length. A closer look at the individual genres represented within each of the two major sections of the Corpus provides further insight into these differences. While the average sentence length in *all* genres of INFO is above the Corpus mean length and, conversely, *all* genres of IMAG are below the Corpus mean, the same neat distributional division does not obtain with regard to the number of predications. Here we find that within INFO, there is one genre (Skills and Hobbies) which is below the Corpus mean in the number of predications, and two newspaper genres (Press, reportage and Press, reviews) which are exactly at the mean. In IMAG, on the other hand, the predication/sentence figure of one genre (Humor) is well above the Corpus mean. The words-per-predication figures show, in essence, that the number of words needed to express a predication is considerably smaller in those styles of writing in which sentences tend to be shorter. This implies an interesting conclusion about the overall structure of sentences in INFO as compared to IMAG. Since, aside from the verbal groups, the other major constituents of a sentence are the nominal groups (i.e. noun phrases), the statistics presented in Table 1 clearly suggest that nominal groups in INFO generally tend to be longer (and, in some sense, thus more complex) than those in IMAG. Both cognitive and automatic parsing of texts of the informational kind will thus put greater demands on noun-phrase processing.

The investigation of the distribution of active vs. passive sentences in the Brown Corpus presents further interesting results about genre differences as well as processing strategies in natural language comprehension. If we adopt a simple definition of a passive predication, i.e. as any predication, finite or non-finite, that includes a passive participle, the differences between genres become readily apparent. Table 2 provides a detailed display of the actual number of predications and of the percentage of active and passive predications for each genre of the Brown Corpus.

TABLE 2 Active and Passive Predications

Genre	Number of Actives	Number of Passives	Percent of Actives	Percent of Passives
A. Press, report.	8924	2283	79.63%	20.37%
B. Press, edit.	5979	1498	79.97%	20.03%
C. Press, review	3554	871	80.32%	19.68%
D. Religion	3754	929	80.16%	19.84%
E. Skills	7793	2270	77.44%	22.56%
F. Pop. lore	10656	2839	78.96%	21.04%
G. Belles lett.	16699	4243	79.74%	20.26%
H. Miscel.	4917	2212	68.97%	31.03%
J. Learned	14591	6080	70.59%	29.41%
K. Fiction, gen.	8654	1520	85.06%	14.94%
L. Mystery/Det.	7599	1167	86.69%	13.31%
M. Science fiction	1754	318	84.65%	15.35%
N. Adv./Western	9224	1301	87.64%	12.36%
P. Romance	9323	1378	87.12%	12.88%
R. Humor	2478	479	83.80%	16.20%
CORPUS	115899	29388	79.77%	20.23%

Table 2 shows that all genres of IMAG are well below the Corpus mean in their average number of passive predications. Interestingly enough, the converse is not quite the case; although most genres of INFO are above the Corpus mean in their number of passive predications, three are below the Corpus mean (B. Press, editorials, C. Press, reviews and D. Religion) and two (A. Press, reportage and G. Belles lettres) are only very slightly above the Corpus mean.

Comparing the two major subdivisions, however, the basic distribution of active/passive predications emerges in a striking manner, as Table 3 shows.

TABLE 3 Summary of Two Major Subsections

Group Type	Number of Actives	Number of Passives	Percent of Actives	Percent of Passives
INFO	76867	23225	76.80%	23.20%
IMAG	39032	6163	86.36%	13.64%
CORPUS	115899	29388	79.77%	20.23%

The distribution of active vs. passive predications in the performance study of various genres has, of course, direct consequences in any consideration of cognitive processing of language utterances as well as in designing possible strategies for natural language comprehension by computers. This last issue is clearly of some importance since technological advances that allow access to tremendous amounts of stored information will also require some automatic selection procedures for retrieving that particular information whose *content* is relevant to a given task. Thus a semantically oriented analysis of the language data bases will become highly desirable. The active-passive distribution of predications has clear relevance for semantic processing strategies. But a word of caution is necessary here. It can be plausibly argued, of course, that - in general - the mapping of semantic relations is more direct in syntactically active sentences than in passive ones: the surface subject of an active sentence is much more likely to correspond to the semantic agent than is the case in passive sentences where agency will remain either unexpressed or will be expressed by other types of phrases (such as the *by* phrase in English). Thus the active sentence expresses on the surface the underlying thematic relations between the predicate and its arguments more directly than the passive sentence. In traditional linguistic terminology, the "logical subject" of the active sentence coincides with the surface subject, something which is not the case in passive constructions. However, the relation between the set of syntactic concepts, such as "logical subject", and semantic ones, such as "semantic agent", is not isomorphic. There are many predicates with which the "logical subject" may serve other functions than that of agency, quite commonly that of a beneficiary or recipient (as in *Tom got a good grade in this course*, or *This situation requires special care*) or as instrument *(This key will open the back door)* as well as others. Nevertheless, the mapping between the semantic function and syntactic structures is bound to be more direct, at least statistically, in data bases in which passive constructions have a low probability of occurrence (such as fiction) than in those where they are relatively common (such as scientific writing, official documents,etc).

To summarize then, we can describe the syntactic style of Informative Prose, compared to Imaginative Prose, by at least these three characteristics: longer sentences, more complex nominal structures, and a larger proportion of passive predications. In a previous paper (Kučera, 1980), I have also identified a fourth parameter, namely a rather large proportion of non-finite predications (i.e. gerunds and infinitival complements) in informative prose. In contrast to this, the texts of Imaginative Prose exhibit shorter sentences, a significantly smaller number of word tokens per predication (pointing to less complex nominal groups), and a smaller percentage of passive predications, as well as a relatively smaller number of non-finite verbal constructions. The research which we are now conducting with the Brown Corpus should provide us with further insights into the syntactic structure of English texts and their stylistic properties, as well as into problems of automatic parsing in general.

REFERENCES

Francis, W.N. (1974). Language, speech, and writing. *Spelling Progress Bulletin, XIV, 2,* 2-6.

Francis, W.N. and H. Kučera (1979). *Manual of Information to Accompany the Standard Corpus of Present-Day American English.* Department of Linguistics, Brown University, Providence.

Kučera, H. (1980). Computational analysis of predicational structures in English. *Proceedings of the 8th International Conference on Computational Linguistics.* Tokyo. pp. 32-37.

Kučera, H. and W.N. Francis (1967). *Computational Analysis of Present-Day American English.* Brown University Press, Providence.

Svartvik, J. and R. Quirk (1980). *A Corpus of English Conversation.* CWK Gleerup Publishers, Lund.

Vachek, J. (1959). Two chapters on written English. *Brno Studies in English, 1,* 7-38.

TRENDS IN SOVIET CYBERNETICS

Ladislav Matejka

University of Michigan, Ann Arbor

ABSTRACT

The initial stages of Soviet cybernetics and its application to the humanities were markedly influenced by the role of mathematicians and logicians with a natural penchant for abstract detachment from variables and for lucid presentation of complex problems. Gradually, however, the scientific rigor imposed on the humanities has given away to the language of social science or, more specifically, to the historians primarily concerned with the problem of cultural relativism. The combination of the formalization and reductionism of the exact sicences, on the one hand, and a central focus of attention on the traditional objects of study of the humanities, on the other, constitute the basis for the most important differential tendencies which are traced and discussed in the proposed paper: its aim is to provide a concise historical survey and to discuss the potentials and outlooks.

INTRODUCTION

For a long time, modern logic, semiotics and structural linguistics, were banned in the Soviet Union as ideological deviations incompatible with Marxism. However, in the 1950's, when it became apparent that the technological advances of data processing devices were intrinsically related to theoretical achievements in those fields, the provincial guardians of the official party line loosened their grip to allow the Soviet Union to catch up with the West in the development of the computational sciences and their application to industry, the exploration of the universe and, by implication, also to modern warfare.[1] The term "cybernetics," promoted by Norbert Wiener from the M.I.T.,[2] turned into the Russian *kibernetika* and swiftly became a household word used in both the hard and soft sciences from economics to musicology. It provided a shelter not only for the Marxist theoreticians, concerned with the applicability of things and thoughts, but also for scholars primarily interested in knowledge regardless of its applicability. The first issue of *Kibernetiku na službu kommunizmu* [Cybernetics Called to the Service of Communism], published in 1961 includes not only programmatic articles on the study of messages as a means of controlling machinery and society, but also articles about cybernetics in biology, medicine, psychology and even jurisprudence. It was in such company that V. V. Ivanov and S. K. Šaumjan published their outline of structural linguistics in which they drew attention to Hjelmslev's dualistic doctrine, close to Šaumjan's logical bent, and to the traditions of *Opojaz,* the school of Russian formal method, close to Ivanov's predilections. The article in its setting is characteristic for the early stages of Russian cybernetics which tried to encompass the most diverse ingredients, sometimes mutually exclusive in nature. The two authors have since that time developed in rather different directions: Šaumjan remained faithful to the radical dualism of the Copenhagen school and pursued its interpretation of Saussure's legacy, while Ivanov has continued in his intellectual growth to take into account the progressive traditions of Russian linguistics and poetics whether cultivated in the Soviet Union or abroad.

---

[1]Cf. L. Matejka, "On the First Russian Prologomena to Semiotics" in V. N. Vološinov's *Marxism and the Philosophy of Language,* New York, 1973, 161-174; "Cybernetics" in L. S. Graham's *Science and Philosophy in the Soviet Union,* New York, 1972, 324-354.

[2]Norbert Wiener, *The Human Use of Human Beings,* New York, 1954, 15.

1. Engendered during a period of precipitate "catching up" with Western scientific and technological advances, the Soviet cybernetic movement developed with amazing speed and force thanks not only to the fortunate conjunction of scientific and political circumstances at the time but also to the availability of a cadre of exceptionally qualified and talented scholars to serve as its originators. Although always subject to the vicissitudes of a strongly ambivalent attitude in its regard on the part of Soviet officialdom, the movement gained a steadily increasing, enthusiastic following in Soviet intellectual and academic circles and, as a result, has undergone what might be described as an "explosive" theoretical and methodological evolution, drawing into its orbit a vast variety of fields of knowledge and producing a staggering number of scholarly works. Publication abroad, particularly of works in translation and works written in Western languages, plus participation by some of the movement's members in international conferences, eventually secured an even broader base of interest and influence, above all among members of parallel movements in France, Italy, Germany, England and the United States.

As elsewhere, also in the Soviet Union it was the science of linguistics that served as the matrix for the formation and development of the structural approach, and it was the conception of language as a system of sings that led to expanded semiotic dimensions of study. However, not linguistics in and of itself alone was responsible for this development but rather coupled with the optimistic outlooks of cybernetics. It was precisely cybernetics that opened a broad avenue of access for the influence of other modern scientific disciplines such as mathematical logic and, above all, information theory. Indeed, such was its fascination for the Soviet scholars that the idea of cybernetics instantly brought forth virtually a whole new intellectual era infused with an extreme degree of "scientific optimism." As Ann Shukman describes it: "Not only did cybernetics promise modernity and a better future, it also held out the prospect of the triumph of rationality in the man-made brain, and, with it, proof of the omnipotence of human reason; there was talk of poetry by computer, of the 'formalization of the human intuition'. . . It seemed that all fields of human activity would at last be opened to rational inquiry; and this with the approval of the authorities."[3]

2. In general, cybernetics and information theory, two newly coined terms of American science, have been used in the Soviet Union far more enthusiastically than the term *semiotic(s),* coined in the 17th century by the English empiricist John Locke and redefined in the 19th century by the American pragmatist Charles Sanders Peirce. The doctrine of sign with its ancient philosophical roots and its susceptibility of dualistic interpretation (*viz.* something material may stand for something spiritual) has never been wholeheartedly embraced by the conservative Marxists apprehensive of the possible semiotic reinterpretation of the Marxist differentiation between the base and the superstructure. It was the concept of sign that made life difficult in the 1930's for the Russian followers of Ferdinand de Saussure ("sosurianstvo") and also for the Neo-marxist school of M. M. Baxtin, ambitious to propagate semiotics in the name of Marxism.[4] In the 1950's, the spectacular success of cybernetics and information theory naturally prepared the ground for a more comprehensive understanding of the nature of sign. Finally, in 1962 the Academy of Sciences of the USSR gave its blessing to semiotics by undertaking a role in organizing a Symposium on Structural Studies of Sign Systems in Moscow. The official organizers of the Symposium were two branches of the Academy: the Scientific Council on Cybernetics and the Institute of Slavic Studies. The Symposium had five sections, covering natural languages as systems, written sign systems and deciphering, non-verbal sign communication, semiotic modeling, art as semiotic system and structural-mathematical investigation of literature. The introduction to the Proceedings provided a definition of semiotics and, at the same time, introduced the interpretation of Saussure by both the Copenhagen and Prague schools, representing, in fact, two mutually exclusive views.

3. The Russian interest in semiotics that revived in the 1960's was nurtured both by the retrieval of the domestic tradition of the late 1920's and by freer access to Western sources such as the Vienna school of modern logic (Carnap and his application of logical calculi in empirical sciences), American behavioral semiotics (Charles Morris and his semiotic Organon of Sciences) and its linguistic counterpart (Bloomfield's descriptivism and especially Pike's structural theory of human behavior). The Soviet contributions to semiotics reflect–often through a critical prism–the terminological and conceptual problems of the

[3] Ann Shukman, *Literature and Semiotics (A Study of the Writings of Yu. Lotman)*, Amsterdam-New York-Oxford, 1977.
[4] Cf. V. N. Vološinov, *Marxism and the Philosophy of Language*, tr. by L. Matejka and I. R. Titunik, New York, 1973.

Western semioticians, including the unsettled differentiation between signal and sign, on the one hand, and sign and symbol, on the other, the threefold distinction of syntax, semantics and pragmatics, and *etic* and *emic* standpoints. In his observations about a text as a signal, A. M. Pjatigorskij, a linguistically trained scholar of Buddhism, refers to Charles Morris' *Foundation of a Theory of Sign* and Kenneth Pike's behavioristic *Language and Life,* but his attitude is ideologically critical. He sharply distinguishes between the physical embodiment of a cultural text and its pragmatic bonds with its users, whether regarded as abstract constructs or concrete participants in time and space. In fact, Pjatigorskij insists that a text does not necessarily imply an inherent bond with any concrete verbal language (as illustrated by ideographic and certain hieroglyphic texts). At the same time, he keeps textual and non-textual signals apart and, by implication, excludes zoosemiotics–a subject not too popular with Soviet semioticians. In his study, Pjatigorskij tests the empirical potentials, as well as the limits, of semiotics. Among the problems attracting his particular attention is that of the relation of the text to the domain of aesthetics or, more specifically, to a given cultural tradition. His question is whether artistic intention is embodied in the signal itself and, consequently, whether the text of verbal art as an artifact should be juxtaposed to the texts of non-art. In subsequent studies, Pjatigorskij abandoned the use of the term signal and embraced the position of the Russian mathematician P. A. Florenskij (1882-1943), seeing sign-ness and thing-ness so closely bound together in their nature that "you could not get a razor-blade between them."[5]

4. Appropriately enough, in the early phase of the Soviet movement problems of machine translation or "automatic language processing" occupied the focal point of research. As was true also in the West, the success of these efforts lay not nearly so much in their practical results as in the theoretical elaborations which, while stimulated by the needs to adapt human language to technology, considerably transcended the original "applied" purposes. Intensive investigations into the theory of formal language modelling led to a fruitful rapprochement between structural-transformational and comparative-historical linguistics and, thereby, set the stage for the whole future course of the movement's development. According to D. Segal, a role of crucial importance in this entire process was placed by the polymath linguist V. V. Ivanov:

> Ivanov's early works present the picture of structural linguistics not as an abstract deductive science, but as a complex structure of models which strive to represent the inherent complexity of language as system of systems. This view of structural linguistics became accepted by many linguists, especially by specialists in comparative linguistics and structural typology. It also served as a cornerstone for the development of structuralism as a scientific method for humanities. The view of structuralism as a complex scientific trend which draws its methods and arguments both from deductive and inductive approaches and which strives to elicit the structure inherent in the data became decisive in the forming of the ideas of Soviet semiotics.[6]

V. V. Ivanov, A. Zaliznjak, and V. Toporov, have applied the formidable combination of their knowledge and talents to elaborate a program for the structural-typological study of modelling systems that broadly anticipates the whole future course of the Soviet structural-semiotic movement. The authors have chosen to tie their discussion to the particular topic of religion, seeing religion as a sign system with minimal abstraction and maximal modelling capacity. But it is clear that religion functions here as an expeditious point of reference for the construction of a general program including among other things the problem of the relationship between instrument of study and object of study and the special complications that arise when both instrument and object are human beings. The author's high concern for scientific objectivity leads them to posit special value for unconscious data and a special role in eliciting and interpreting such data for psychoanalysis.[7]

---

[5] Cf. A. M. Pjatigorskij, "O nekotoryx teoretičeskix predposylkax semiotiki," *Sbornik statej–1973,* pp. 187-191; Cf. "On Some Theoretical Presuppositions of Semiotics," *Semiotica,* 1974, 185-188.

[6] *Aspects of Structuralism in Soviet Philology,* Papers on Poetics and Semiotics, 2, Department of Poetics and Comparative Literature, Tel-Aviv University, Tel-Aviv, 1974.

[7] A. A. Zaliznjak, V. V. Ivanov, V. N. Toporov, "O vozmožnosti strukturnotipologičeskogo izučenija nekotoryx modelirujuščix semiotičeskix sistem," *Strukturnaja tipologija jazykov,* Moscow, 1962, 134-143.

The basic flowchart of Claude Shannon's Information Theory has excited the visionary fancy of both the engineers and the philologists in the Soviet Union. V. V. Ivanov and V. N. Toporov have elaborated it into a model for inductive-deductive textology, applicable not only to diverse types of existing texts but also for the reconstruction of texts which have not been preserved and exist only potentially. In this way the linguistic reconstruction of theoretical protosystems has been supplemented by a program for the reconstruction of the prototexts themselves. Clearly, the optimistic zeal which characterized the 19th century comparatists has found in Shannon's scheme a new impetus promising to improve August Scheicher's famous reconstruction of the Indo-European protofabel by means of new, sophisticated methods of internal projection of the present into the past. It is surely symptomatic that one of Toporov's daring journeys into the Indo-European past was dedicated to Schleicher's memory and published in the Tartu *Trudy,* 4 (1969) on the one hundred year anniversary of his death.[8] The introduction to Ivanov-Toporov's book *Semiotic Modeling Systems of Slavic Languages* (Moscow, 1965), outlined a general framework for a number of theoretical as well as practical problems including deciphering of texts in unknown languages or secret codes. The linguistic methods appear here in complex interaction with the social, as well as the natural, sciences, coordinated in the spirit of empirical rationalism for the noble cause of a modern semiotic encyclopedia.

5. The initial stages of Soviet cybernetics and its application to the humanities were markedly influenced by the role of mathematicians and logicians with a natural penchant for abstract detachment from variables and for lucid presentation of complex problems. Gradually, however, the scientific rigor imposed on the humanities has given away to the language of social science or, more specifically, to the historian primarily concerned with the problem of cultural relativism. In the course of this development, the Soviet trend ceased to parallel French "structuralism" and its antihistorical, "Kantian" posture characteristic for Claude Lévi-Strauss and his followers. The main role in this gradual change has been played by Jurij Lotman, the chief architect of Tartu semiotics of culture and ardent advocate of the conceptual difference between the primary and secondary modelling systems in semiotics. One of the first—and, as it happens, the most persistent and "loaded"—issues was that of the relationship of the new structural poetics to the Russian formal method of the 1920's. The scientific ambitions of structural poetics and direct references to the *Opojaz* by certain leading representatives aroused suspicions that structuralism was a rebirth of the despised "formalism" of the past. Argument and explanation on the past of the new school became necessary, and its spokesmen typically took one of two possible tacks: either to prove, on the basis of their opponents' concept of formalism, that structural poetics fundamentally differed from the formal method; or to prove that Soviet structuralism had honorable native roots in improperly understood and appreciated structural aspects of the Russian formalist movement, as well as of other phenomena of early Soviet intellectual history.

The first of these tacks was taken by Ju. Lotman in one of his earliest published contributions to the structural-semiotic trend.[9] Using the assumption that formalism means the neglect of content in favor of form and the dominance of linguistics in the study of verbal art, the author bears down on two main points: the necessity and, indeed, overriding importance of semantic (ideological) analysis because of the special organization of content in a literary work of art and the crucial difference, in precisely that regard, between the concepts of structure in literary science and linguistics.

Lotman's paper "Dynamic Model of Semiotic System," published in 1974, characteristically rejects Saussure's radical dualism separating semiotics systems as static entities from their history (i.e. their dynamism). In his criticism of Saussure, Lotman embraces the Prague school's insistence that Saussure's static semiotic system is a mere fiction and that the dynamism of history has to be projected into the study of semiotic systems rather than artificially kept aside. Although Lotman generously quotes Roman Jakobson, most of the quotations are from Jakobson's Prague years rather than from his subsequent search for universality of distinctive features in human languages and for poetic invariables of mankind. It is in the name of history, cultural relativism and Hegelian dialectics that Lotman challenges the concept of poetics as an integral part of linguistics, the very fundament of Jakobson's recent search for grammar of poetry and poetry of grammar. In sharp contradistinction to Jakobson, Lotman sees the

---

[8] V. N. Toporov, "K rekonstrukcii indo-evropejskogo rituala i ritual'nopoètičeskix formul (na materiale zagovorov), *Trudy,* 4 (1969), 306-344.

[9] Ju. M. Lotman, "O razgraničenii lingvističeskogo i literaturovedčeskogo ponjatija struktury," *Voprosy jazykoznanija,* 3, 1963, 44-52.

language of verbal art as a special secondary modeling system which is superimposed on the natural language and does not belong to the domain of linguistic studies. Lotman's paper is just one of an entire series of studies indicating that the development of Tartu semiotics of culture conceptually deviates from the development of Jakobson's poetics seen as an integral part of linguistics.

6. At the same time that "the logic of machine translation research led Soviet structuralists to an understanding of the overwhelming role played in linguistic structure by semantics, i.e. that part of the linguistic edifice which is based on relationships with other semiotics systems,"[11] the same understanding created a marked methodological bias toward procedures of formalization and reductionism. Indeed, precisely formalization and reductionism became the cardinal issues in a debate between the structuralists and their opponents that was held on the pages of the journal *Voprosy literatury* throughout the 1960's. The various opponents shared the common objection to structuralism that whereas formalization and reductionism are in accord with the requirements of the exact sciences, their application to literary study or other humanistic disciplines risks minimizing or misconstruing or wholly obliterating precisely those qualities that constitute the distinctive nature of the object of study in the humanities. The structuralists' position contended, contrariwise, that formalization and reductionism are essential cognitive instruments in any field of knowledge, literary study or other humanistic discipline being no less than any other kind amenable to their application, and that, far from limiting or distorting the nature of the objects of study in the humanities, the structural approach opens scientific access to the study of the immense richness and complexity of the humanistic domains and reveals their vital interconnectedness with other kinds and levels of human knowledge. Moreover, V. V. Ivanov noted in his contribution that instead of "reducing" the human to the machine, as so many of the opponents of structuralism seemed to believe it would, the structuralist method is in a position to use the machine as a way of measuring and defining the stupendous magnitude of human creativity.

While the defenders of exact methods were unimpeachably correct in laying claim to the use and value of advanced theories from the exact sciences in application to literary study or other humanistic fields, and while the unprejudiced eye easily detects the party-line prejudice and sophistry that underlie much of their opponents' argument, it is evident that formalization and reductionism did and do present certain dangers. This was to some extent acknowledged by Ivanov's complaint about the "vague and loose" application of terms by A. Žolkovskij and Ju. Ščeglov.[12] But the full implications here surely also include the fact that abstract, formal schemes, such, for instance, as mathematical set theory or the principle of opposition (so successfully applied in the distinctive features theory), can be manipulated to provide "scientific description" of anything whatsoever. The justification and productiveness of such schemes when knowledgeably and substantively applied are undeniably, indeed provably, high, but their potential for abuse is certainly also high. Enthusiasm generated out of scientific optimism, coupled with innovatory élan and group allegiance, make this potential abuse very real. It can hardly do the Soviet movement's true magnitude any detriment to suggest that its wide appeal among students and its astounding prolificness have to some degree been conditioned by such abuse. The same may to some degree also be reflected in the fact that almost from the movement's inception its hypotheses have been applied to an extraordinary diversity of topics and materials.[13] However, the abuse of which formalization and reductionism are susceptible is a risk inherent and ineluctable for an intellectual movement which is nomothetic, universalist and pancyclopedic as the Soviet cybernetic movement has been.

[10] Jurij M. Lotman, "Dinamičeskaja model' semiotičeskoj sistemy," *Institut russkogo jazyka*, Moscow, 1974.

[11] D. Segal, *op. cit.*, p. 17.

[12] V. V. Ivanov, "O primenenii točnyx metodov v literaturovedenii," *Voprosy literatury*, 11, 1967, 115-126.

[13] For instance, topics have included parapsychology, personology, hallucinogenics, numerology, alchemy, cartomancy, sexuality, etiquette and so on; the argument that such topics supplied relatively simple sign systems and that therein lay their appeal and advantage for study is no doubt true enough. The time and space spread of the materials the Soviet semioticians have used is phenomenal, ranging–often in one and the same article or book–from prehistoric to contemporary times and literally from around the world, the Soviet scholars evincing special propensity toward what used to be called the "exotic" cultures.

XVI. MATHEMATICAL PROGRAMMING

NONLINEAR PROGRAMMING: A SURVEY OF ITS USE IN IMPROVING THE QUALITY OF LIFE[1]

Louis G. Birta
Department of Computer Science
University of Ottawa
Ottawa, Ontario K1N 9B4
CANADA

ABSTRACT

Nonlinear programming methods have been extensively applied in solving a broad spectrum of real-world problems. The intent of this paper is to examine some typical applications which can be characterized as having a potential impact on improving the quality of life. A brief review of currently used methods for solving the nonlinear programming problem is also provided.

KEYWORDS

Nonlinear programming applications; nonlinear programming methods.

INTRODUCTION

The nonlinear programming problem is concerned with locating the minimizing (maximizing) argument of a given criterion or objective function, normally subject to constraints which place restrictions on the set of permissible solution candidates. A wide variety of real-world problems can be cast into such a framework. The technological sector, in particular, has produced many important applications; e.g. in engineering design, industrial process control and aerospace systems, and surveys of these are available; e.g. Avriel and Dembo (1979) and Rauch (1980). Applications outside this sector are increasing and the intent of this paper is to examine some representative examples from a variety of such non-technological areas. These applications can be viewed as having some relevance to improving the quality of life inasmuch as they have a potential impact on issues which are of general popular concern.

The paper is divided into two parts. In the first part a brief overview of contemporary methods for solving nonlinear programming problems is given in order to provide a better perspective for the discussion of applications which appears in part 2.

---

[1] This research was supported in part by grant number A8109 from the National Sciences and Engineering Research Council, Canada.

SOLUTION METHODS FOR THE NONLINEAR PROGRAMMING PROBLEM

The nonlinear programming problem (NPP) can be stated as:

$$\text{minimize } f(x)$$
$$\text{subject to } \begin{cases} c_i(x)=0 & , i\varepsilon E \\ c_i(x)\geq 0 & , i\varepsilon I \end{cases}$$

where E and I are respectively index sets for the equality and inequality constraints associated with the problem. We take x to be an n-vector and denote by x* the solution to the NPP. Solution algorithms for the NPP normally presume certain differentiability conditions on f and the c_i and a typical assumption in this regard is that these functions are twice continuously differentiable.

Depending on the specific nature of the functions f and c_i, a variety of specialized versions of the problem can be identified; e.g. the linear, the quadratic and the geometric programming problems. There specialized versions are not considered in this discussion. However, a specialization which is of particular significance, is the unconstrained problem (UP) which corresponds to the case where the c_i functions are absent (sets I and E are empty). The importance of this case stems from the fact that methods for solving the NPP are invariably based on a transformation procedure that leads to the requirement of solving a sequence of unconstrained problems (in the ideal case, this sequence contains only one member). The need for effective methods to handle the UP is therefore of crucial importance and the discussion below first considers this particular case.

The Unconstrained Problem

Many heuristic methods for the UP have been proposed over the years and these methods continue to enjoy some popularity. Included here are the simplex method of Nelder and Mead (1965), the pattern search method of Hooke and Jeeves (1961) and random search methods; e.g. Price (1979).

The steepest descent and Newton-Raphson methods can be regarded as the predecessors of most contemporary methods for handling the UP. Although these methods, in their "pure" form, are rarely used in practice, the principles underlying their operation have played an important role in guiding the evolution of contempory methods. These fall into two broad categories; namely, the conjugate gradient methods and the variable metric (or quasi-Newton) methods. A closed related class of methods, called B-conjugate, has also been proposed (Powell (1975)).

The original procedure in the class of conjugate gradient algorithms was proposed by Fletcher and Reeves (1965) and is as follows (we use $g_k=f_x(x_k)$):

(i) choose x_0 (the priming guess), let $p_o=-g_o$ and set k=o

(ii) let $x_{k+1}=x_k+\alpha^*p_k$ where $f(x_k+\alpha^*p_k)=\min_{\alpha} f(x_k+\alpha p_k)$

(iii) set $p_{k+1}=-g_{k+1}+\beta_k\, p_k$ where $\beta_k=g_{k+1}^T g_{k+1}/g_k^T\, g_k$

(iv) check a termination condition and if not satisfied, repeat from step (ii) with k replaced with k+1.

The significant feature of this procedure is that when applied to the quadratic function $f(x)=\frac{1}{2} x^TAx+b^Tx+c$ (with A positive definite) it will locate the minimizing argument in at most n steps. Variations to the procedure based on an alternate specification for β_k have been proposed; e.g. Polack and Ribière (1969) suggest $\beta_k=g_{k+1}^T \gamma_k/g_k^T g_k$ $(\gamma_k=g_{k+1}-g_k)$ while Sorenson (1969) recommends $\beta_k=g_{k+1}^T \gamma_k/p_k^T \gamma_k$.

The quasi-Newton or variable metric methods can be viewed as an evolution of the Newton-Raphson method. The general procedure is as follows:

(i) choose x_o (the priming guess) and a value for the n×n symmetrix matrix H_o (normally the identity matrix); set k=o

(ii) using a suitable value for α_k, set $x_{k+1}=x_k-\alpha_k H_k g_k$

(iii) update the value of H_k by replacing it with H_{k+1}

(iv) check a termination condition and if not satisfied, repeat from step (ii) with k replaced with k+1.

In much of the original work with the quasi-Newton methods, the α_k in step (ii) was chosen so that $f(x_k+\alpha_k s_k)=\min_\alpha f(x_k+\alpha s_k)$, $(s_k=-H_k g_k)$. The circumvention of this linear search (or line search) subproblem has received much attention; e.g. the work of Davidon (1975), Powell (1976), Goldfarb (1977) and Shanno and Phua (1977).

The two most successful up-dating formulae for H_k in step (iii) are the DFP formula (Davidon (1959); Fletcher and Powell (1963)):

$$H_{k+1} = H_k + \frac{\delta_k \delta_k^T}{\delta_k^T \gamma_k} - \frac{(H_k \gamma_k)(H_k \gamma_k)^T}{\gamma_k^T H_k \gamma_k}$$

where $\delta_k=x_{k+1}-x_k$, and the BFGS formula (Broydan (1970); Fletcher (1970); Goldfarb (1970); Shanno (1970)):

$$H_{k+1} = H_k + \frac{1}{\delta_k^T \gamma_k} \left[(1 + \frac{\gamma_k^T H_k \gamma_k}{\delta_k^T \gamma_k}) \delta_k \delta_k^T - ((H_k \gamma_k)\delta_k^T + \delta_k(H_k \gamma_k)^T)\right]$$

A comprehensive discussion of many important features of the quasi-Newton methods can be found in Dennis and Moré (1977).

The Constrained Problem

With some notable exceptions, contemporary methods for solving the constrained problem are formulated around a set of sufficiency conditions which apply at the solution x*. A typical set of such conditions (based on the assumption that f and the c_i are twice continuously differentiable) is as follows:

(i) There exist unique parameters (Lagrange multipliers) $\lambda_i^* \geq 0$ such that $L_x(x^*,\lambda^*)=0$ where $L(x,\lambda)=f(x) - \sum_{i\varepsilon I^*} \lambda_i c_i(x)$ (the Lagrangian function) and $I^*=\{i: c_i(x^*)=0\}$.

(ii) There exists $\sigma>0$ such that $w^T G(x^*,\lambda^*)w \geq \sigma w^T w$ for every w for which $w^T \nabla c_i(x^*)=0$, $i\varepsilon I^*$ where $G = \frac{\partial^2 L}{\partial x^2}$ and $\nabla c_i = \frac{\partial c_i}{\partial x}$.

One of the early methods for handling the constrained problem is based on the use of penalty functions; e.g.

$$\phi(x,r) = f(x) + r\{\sum_{i\varepsilon E} c_i^2(x) + \sum_{i\varepsilon I} [c_i(x)]_-^2\}, \quad r>0$$

where $[c_i(x)]_- = \min(0, c_i(x))$. The approach here is to add to the objective function a positive penalty for constraint violation. For any particular value of the penalty parameter r; e.q. $r=r_k$, the minimization of $\phi(x,r_k)$ yields a result $\hat{x}(r_k)$ and by successively increasing the value of r in a sequence of subproblems, it is plausible to assume that $\hat{x}(r_k)\to x^*$ as $r_k \to \infty$. In practice, the parameter r cannot be allowed to actually assume an infinite value, thus upon termination of the procedure, the final $\hat{x}(r_k)$ value will technically violate the constraints. This situation is avoided in an alternate approach which uses a barrier function transformation. A typical function of this type (for the case where E is empty) is:

$$\phi(x,r) = f(x) + \frac{1}{r}\sum_{i\varepsilon I} 1/c_i(x)$$

The procedure here is again to successively locate the minimizer of $\phi(x,r)$ as the parameter r takes on a sequence of predetermined positive values of increasing magnitude. The advantage in this approach is that the sequence of $\hat{x}(r_k)$ values satisfy the constraints; i.e., are feasible, provided that the initializing guess on the first step is feasible. Much of the basis for the penalty and barrier function transformations was formulated by Fiacco and McCormick (1968) and a comprehensive review of salient features can be found in Ryan (1974).

Hestenes (1969) proposed the "method of multipliers" as a means of avoiding the inherent difficulties with the above transformations which arise from the need to assign excessively large values to the penalty parameter. This approach (for the case where I is empty) focuses on the function:

$$\phi(x,\lambda,r) = f(x) - \sum_{i\varepsilon E} \lambda_i c_i(x) + r\sum_{i\varepsilon E} c_i^2(x), \quad r>0$$

which has subsequently become known as the augmented Lagrangian function. The associated procedure also involves two levels of iteration. In the inner iteration, an unconstrained minimization procedure is used to locate $\hat{x}(\lambda^k,r_k)$ which minimizes $\phi(x,\lambda^k,r_k)$ where (λ^k,r_k) is fixed. Then (λ^k,r_k) is updated and the procedure repeated. In order that the sequence of $\hat{x}(\lambda^k,r_k)$ converge to x^*, it is necessary that the λ^k up-dates converge to λ^*, the vector of Lagrange multipliers at x^*. The significant feature here is that this can be achieved without requiring $r_k\to\infty$; it is only necessary (under mild assumptions) that r_k be greater than some limiting value $\bar{r}$. Methods for adjusting (λ^k,r_k) are described and evaluated by Fletcher (1975) who also shows that the method of multipliers is equivalent to a proposal made by Powell (1969). Extension of the concept to include inequality constraints has been studied; e.g. Rockaffellar (1973) and Fletcher (1975). Various aspects of the method are discussed by Bertsekas (1976) and Powell (1978).

Another method which avoids the ill-conditioning problems inherent in the classical penalty function approach is based on the use of "exact" penalty functions (Pietrzykowski, 1969). The underlying objective here is to create a function whose minimizing argument is x^*, thereby making it possible to locate the solution of the original problem with a single unconstrained minimization.

The most common such function is:

$$\phi(x,r) = f(x) + r\left(\sum_{i\varepsilon E} |c_i(x)| - \sum_{i\varepsilon I} [c_i(x)]_-\right)$$

and in this case it can be shown (e.g. Howe, 1973), that the desired correspondence between x* and the minimizer of ϕ, holds for any r value above some threshold; i.e. the value of r need not approach infinity.

There are nevertheless, two sources of difficulty with the approach. The first arises from the inherent non-differentiability of $\phi(x,r)$ at the minimizing argument which can severely tax the performance of standard unconstrained minimization methods, and the second is concerned with the determination of the threshold value for the parameter, r.

The reduced gradient and gradient projection methods fall into a category which is quite distinct from the various penalty function approaches outlined above. The underlying concept in these methods is to generate and move on a sequence of directions which simultaneously reduce the criterion function and maintain feasibility of the up-dated estimate. These methods have their origins in algorithms for the linearly constrained problem where they are reasonably straightforward to implement, (Wolfe, 1967).

A basic feature of these methods is to use equality constraints to express some subset of the problem variables in terms of the remaining variables. Since only this latter group of variables is "independent", the dimensionality of the problem is reduced. When the constraints are nonlinear, reasonable movement in a locally feasible direction will normally lead to a violation of some constraints and an issue in such methods is how to best restore feasibility. The Generalized Reduced Gradient (GRG) method, as outlined by Abadie and Carpentier (1969) is the original procedure in this class of methods. A comprehensive discussion of these methods is given by Sargent (1974).

APPLICATIONS

Most of the applications described below are concerned either with model refinement or plan formulation. The objective in the first category is to establish values for a set of parameters within a model of a process or system which cause the model's behavior to best approximate reference data (typically obtained from the "real world"). In the case of plan formulation, the objective is to develope a strategy (in terms of embedded parameters) which, in some predefined sense, optimizes the evolution of a given model of a process or system over a specified time interval.

Agriculture

Day and Sparling (1977) provide an extensive non-technical survey of applications of optimization concepts in agriculture and resource use, from the perspective of economics. A recent paper by Rovinsky et al. (1980) examines a specific problem in agricultural economics; namely the problem of determining the optimal allocation of means of production among alternative commodities under various types of cooperation by producers. The problem is formulated as a nonconcave quadratic programming problem and a solution algorithm is developed. The authors note that the approach has been successfully applied in a cotton production problem.

Scarborough and Hunt (1977) consider a somewhat different cost minimization problem. It focuses on the observation that there is associated with a given field pattern (traversal route over a field) a best set of parameters with which to operate the machines required to work the field; e.g. ploughing speed. By using these best values, an optimal cost can be associated with any given field pattern and hence a means for ranking alternative patterns is obtained. Constraints arise from physical limitations of commercially available equipment. The problem is solved using the classical penalty function method.

The important problem of protein specification in an optimal least-cost diet is examined by Smith (1974). This aspect of the diet specification problem is difficult because the amount of protein required is dependent on the "quality" of the protein and this inter-relationship introduces a nonlinearity into an otherwise linear mathematical model. Allowing both these attributes to be available in an optimization procedure yields a far better solution to the underlying problem of how to best allocate agricultural land in a developing country (the paper is specifically concerned with Nigeria). Because of the predominantly linear nature of the problem, solutions are obtained using a linear programming procedure that is adapted to linearize the nonlinear constraints.

A particularly important area which will undoubtedly receive increasing attention in the future is that of agricultural pest management. The work of Varadarajan and Tummala (1980) addresses this problem but from an optimal control formulation which is beyond the scope of this survey.

The Environment

The paper by Marchuk (1976) provides a general discussion of the use of optimization in the allocation of industrial sites having regard to minimization of environmental pollution. The specific case of aerosal pollution is considered, however the presentation is at a conceptual, rather than applied, level.

Ito and Sakawa (1979) consider the problem of formulating an optimal strategy for constructing public sewage treatment systems over a 20 year planning period. The mathematical model developed and used by the authors in the study pertains to the Yodo River basin in Japan. The objective function incorporates both construction and operating costs for the projected sewage treatment plants. Inequality constraints arise from the need to satisfy various environmental standards relating to water pollutants. The GRG method is used to solve the problem.

A similar study is reported by McNamara (1976). It is concerned with the design of efficient water quality management systems which specifically allow for the combination of different types of pollution abatement techniques. The main objective is to control BOD (biochemical oxygen demand) in a river channel and the approach is applied to the upper Hudson River. The problem is formulated and solved as a geometric programming problem. A study with similar objectives which deals with the Deleware Estuary can be found in Graves et al. (1972).

Water Resources

Johnston and Pilgrim (1976) describe a model refinement problem that is concerned with determining best values for a set of 8 parameters in a rainfall-runoff model called a Broughton model. The reference data used in the criteria function is observed rain water run-off in a specific watershed. The optimization is unconstrained and the authors use both the simplex method and the DFP variable metric method for carrying out the search. A major conclusion of the study is that the problem is exceedingly hard and the authors remark that "a true optimum set of values was not found over 2 years of full-time work concentrated on one watershed, although many apparent optimum sets were readily obtained".

A planning application in the water resources area is reported by Young and Pisano (1970). It is concerned with determining the least-cost mix of alternatives to satisfy projected water demands within an urban region. Typical water sources considered are surface water, desalination, wells, electrodialysis and waste treatment. The criterion function used in the study incorporates process costs, transport costs and various economic factors. Constraints arise from the inherent production limits of the various sources and the need to satisfy projected demand. A gradient projection technique is used to generate a solution for a specific case study (the James River, Virginia, region).

Another planning application is described by Peters et al (1978). The model used is distinctive unasmuch as it incorporates the notions of recourse actions and chance-constraints to account for uncertainty of water inflows into the region under investigation (in this case the Karum River and its tributaries in Khuzestan, Iran). The objective in the study is to determine the monthly water release policies from three reservoirs in the system, as well as distribution policies in order to maximize expected benefits over the planning period (one year). The solution procedure used by the authors combines both linear and non-linear programming routines (the latter based on the augmented Lagrangian method) to handle different aspects of the problem.

Energy (Electric Power Generation)

Nonlinear programming methods have been extensively used as planning tools in the electric power generation industry. A comprehensive survey of typical applications; e.g. economic dispatch, generation planning, maintenance scheduling etc., is given by Sasson and Merrill (1974). Studies in the generation planning area have been recently reported by Juseret (1978) and Hanscon et al (1980). The problem considered by Juseret is concerned with the timing and the type of future power generation plants. The formulation involves linear constraints on the decision variables and a gradient projection type of solution method is employed. The objective in the study by Hanscon et al is to determine weekly outflows from the larger reservoirs in the Hydro-Quebec hydro electric system in order to most economically meet projected weekly energy demands. The planning period runs over a one to two year interval. The solution algorithm used is of the reduced gradient type.

Gagnon et al (1974) describe a program designed to assist in planning the development and operation of systems of power generating plants; e.g. the timing of new installations and maintenance scheduling. The program has been applied in genuinely "large" problems; e.g. 6000 variables, 4000 linear equalities, 11,000 linear and nonlinear inequalities, formulated in the context of the Columbia River basin. A hybrid solution method is used which combines a classical penalty function transformation with a variation on the GRG method.

Medicine and Biology

The applications in this group are all in the category of model refinement problems. In the study described by Pendrys (1979) the model in question relates to the behavior of digoxin in a patient undergoing digoxin treatment for congestive heart failure. The objective is to "tune" parameters in a general model in order to tailor the model's behavior to a specific patient. This is achieved by minimizing a norm of the difference between the model behavior and clinical data obtained from the patient. The problem is unconstrained. One of the important uses of such a patient dependent model would be in assisting a physician to establish a proper continuing dosage level of digoxin. This is of considerable importance inasmuch as there is a 30% mortality rate among patients who develop an excessive level. Both the conjugate gradient and DFP methods are used in the study.

A similarly oriented study is reported by Saffer et al (1976), except in this case a model of Rose Bengal (an effective tracer for studing biliary function) transport through the hepatobiliary system is investigated. The model is formulated by the authors as a 4 state Markov chain and the study is concerned with determining best values for 4 entries in the transition probability matrix for the model. A least squares minimization between the model prediction and clinical data provides the basis for the parameter adjustment. The pattern search method (Hooke & Jeeves, 1961) is used as the basic optimization procedure, although a Gauss step is carried out in the final phase to more accurately locate the minimum.

The problem of tracer movement in biological systems is also treated by Davis and Ottaway (1972). The model in this study involves 4 first order differential equations containing 8 parameters and relates to the metabolism of I - labelled Throxine in perfused rat liver. Best parameter values are obtained by minimizing the deviation between experimental data and model prediction. The relative merits of 3 different unconstrained minimization methods are compared; namely, the conjugate gradient, the DFP and the simplex (Nelder and Mead, 1965) methods.

The paper by Cornacchio (1975) discusses the effectiveness of the conjugate gradient and DFP methods in solving some other biological model refinement problems. In particular, a class of models of ionic conduction across biomembranes is considered.

The Economy

A substantial amount of literature exists on the application of nonlinear programming methods in the optimization of econometric models. These applications (of the plan formulation category) are essentially discrete time optimal control problems where the objective is to determine best values for a set of control ("instrument") variables over some sequence to time intervals, subject to the constraints imposed by an econometric model. The criterion function normally is a measure of the deviation of certain critical variables from a prespecified "target" path. Such problems are generally "large" in the sense that many (often hundreds) of variables, equations and inequalities are involved in the model. Much of the recent work is documented in the Annals of Economic and Social Measurement; e.g. Bock et al (1977-78) and Caine (1977-78). An excellent synopsis of this particular application area can be found in Lasdon and Waren (1980).

A particular example is the study by Haurie et al (1974) who use the method of multipliers to determine optimal economic policies over a 4 year horizon. This is carried out within the framework of an econometric model of Canada having modest size (39 equations, both linear and nonlinear). Their criterion function incorporates a number of factors which include unemployment rate and balance of payments and the two control variables over which the optimization takes place are government investment and the exchange rate between the Canadian and U.S. dollars.

Applications of nonlinear programming methods to model refinement problems in economics have also been reported. Two examples of this type are given in the papers of Dagum (1977) and Singh and Maddala (1976). In both cases the objective is to find values for parameters in an income distribution model which results in a best fit to available data within a specific country.

CONCLUDING REMARKS

The above survey is intended only to be representative of some of the nonliner programming applications outside the technological domain. It is by no means exhaustive, especially since new applications are continually being reported. Additional applications can be found in the survey by Lasdon and Waren (1980). A recent synopsis of software issues is available in Waren and Lasdon (1979).

REFERENCES

Abadie, J. and J. Carpentier. (1969). Generalization of the Wolfe reduced gradient method to the case of nonlinear constraints. In R. Fletcher (Ed.) Optimization, Academic Press. pp. 37-47.

Avriel, M. and R. S. Dembo. (1979). (Eds.). Engineering Optimization, Mathematical Programming Study 11. North Holland, Amsterdam.

Beale, E. M. L. (1972). A derivation of conjugate gradients. In F. A. Lootsma (Ed.), Numerical Methods for Non-Linear Optimization, Academic Press. pp. 39-43.

Bertsekas, D. P. (1976). Multiplier methods: a survey. Automatica. 12, 133-145.

Bock, G., V. Wulfingen and P. Pauly. (1977-78). An optimization approach in multiple target problems using inequality constraints: The case against weighted criterion functions. Annals Econ. Sco. Meas. 6, 613-631.

Broyden, C. G. (1967). Quasi Newton methods and their application to function minimization. Math. of Comp. 21, 368-381.

Broyden, C. G. (1970). The convengence of a class of double-rank minimization algorithms 2. The new algorithm. J. Inst. Math. Appl., 6, 222-231.

Broyden, C. G. (1970). The convergence of a class of double-rank minimization algorithms. J. Inst. Math. Appl. 6, 76-90.

Caine, R. (1977-78). Restrictions on the control vector in economic optimization problems. Annals. Econ. Sco. Meas. 6, 493-501.

Cornacchio. J. V. (1975). Conjugate gradient and variable metric search algorithms: applications to computer modelling of biological processes. Conf. on Engineering in Medicine and Biology. Vol. 17.

Dagum, C. (1977). A new model of personal income distribution: specification and estimation. Economie Appliquee. 30, 413-436.

Davidon, W. C. (1959). Variable metric method for minimization. Atomic Energy Commission, Reseach and Development Report No. ANL-5990,(Revised).

Davidon, W. C. (1975). Optimally conditioned optimization algorithms without line searches. Math. Prog., 9, 1-30.

Davis, R. H. and J. H. Ottaway. (1972). Application of optimization procedures to tracer kinetic data. Math. Biosci. 13, 265-282.

Day, R. H. and E. Sparling. (1977). Optimization models in agriculture and resource economics. In G. G. Judge, R. H. Day, S. R. Johnson, G. C. Rausser and L. R. Martin. (Eds.). A Survey of Agricultural Economics Literature, 2, Univ. of Minn. Press. 2, 93-127.

Dennis, J. E., and J. J. More. (1977). Quasi-Newton methods, motivations and theory. SIAM Review, 19, 46-89.

Dixon, L. C. W. (1975). Conjugate gradient algorithms: quadratic termination without linear searches. J. Inst. Math. Appl., 15, 9-18.

Fiacco, A. V. and G. P. McCormick. (1968). Nonlinear Programming: Sequential Unconstrained Minimization Techniques. Wiley, New York.

Fletcher, R. (1970). A new approach to variable metric algorithms. Computer J., 13, 317-322.

Fletcher, R. (1975). An ideal penalty function for constrained optimization. J. Inst. Math. App. 15, 319-342.

Fletcher, R., and A. P. McCann. (1969). Acceleration techniques for nonlinear programming. In R. Fletcher (Ed.), Optimization, Academic Press.

Fletcher, R. and M. J. D. Powell. (1963). A rapidly convergent descent method for minimization. Computer J., 6, 163-168.

Fletcher, R. and C. M. Reeves. (1964). Function minimization by conjugate gradients. Computer J., 7, 149-154.

Gagnon, C. R., R. H. Hicks, S. L. S. Jacoby and J. S. Kowalik. (1974). A nonlinear programming approach to a very large hydroelectric system optimization. Math. Prog. 6, 28-41.

Goldfarb, D. (1970). A family of variable metric methods derived by variational means. Math. of Comp., 24, 23-26.

Goldfarb, D. (1977). Generating conjugate directions without line searches using factorized variable metric up-date formulas. Math. Prog., 13, 94-110.

Hanscom, M. A, L. Lafond, L. Lasdon and G. Pronovost. (1980). Modelling and resolution of the medium term energy generation planning problem for a large hydro-electric system. Mangt. Sc. 26, 659-668.

Haurie, A., A. Jacquet and A. van Peeterssen. (1974). Optimal control of a nonlinear econometric model of Canada using the method of multipliers. Int. J. Systems Sci. 5, 1077-1095.

Hestenes, M. R. (1969). Multiplier and gradient methods. J. Opt. Theory and Appl. 4, 303-320.

Hooke, R. and T. A. Jeeves. (1961). Direct search solution for numerical and statistical problems. J. ACM. 8, 212-229.

Howe, S. P. (1973). New conditions for exactness of a simple penalty function. SIAM J. on Control. 11, 378-381.

Huang, H. Y. (1970). Unified approach to quadratically convergent algorithms for function minimization. J. Opt. Theory and Appl., 5, 405-423.

Ito, K. and M. Sakawo. (1979). A nonlinear optimization model for water treatment of a river basin. Int. J. Systems Sci. 10, 1065-1080.

Johnston, P. R. and D. H. Pilgrim. (1976). Parameter optimization for watershed models. Water Resour. Res. 12, 477-486.

Juseret, R. (1978). Long term optimization of electrical system generation by convex programming. In M. L. Balinski and C. Lemarechal (Eds.) Mathematical Programming Study 9. North Holland, Amsterdam. pp. 186-195.

Lasdon, L. S. and A. D. Waren. (1980). A survey of nonlinear programming applications. Working Paper 80-01, Computer and Information Sciences Dept., Cleveland State University, Ohio.

Marchuk, G. I. (1976). Environment and some problems of optimization In Lecture Notes in Computer Science £40, Springer-Verlag. 13-30.

McNamara, J. R. (1976). An optimization model for water quality management Water Resour. Res. 12, 125-134.

Nazareth, L. (1977). A family of variable metric updates. Math. Prog. 12, 157-172.

Nelder, J. A. and R. Mead. (1965). A simplex method for function minimization. Comp. J. 7, 308-313.

Pendrys, J. P. (1979). Application of unconstrained optimization techniques to computer-assisted digoxin theraphy. Phd thesis, School of Advanced Technology, SUNY, Binghampton.

Peters, R. J., K. C. Chu and M. Jamshidi. (1978). Optimal operation of a water resources system by stochastic programming. In M. L. Balinski and C. Lemarechal. (Eds.). Math. Prog. Study 9, North-Holland, 152-175.

Pietrzykowski, T. (1969). An exact potential method for constrained maxima. SIAM J. Num. Anal. 6, pp. 299-304.

Polak, E., and G. Ribiere. (1969). Note sur la convergence de methode de direction conjugees. Rev. Fr. Inform. Oper. 3, 35-43.

Powell, M. J. D. (1969). A method for nonlinear constraints in minimization problems. In R. Fletcher (Ed.), Optimization, Academic Press. pp. 283-298.

Powell, M. J. D. (1975). A view of unconstrained minimization algorithms that do not require derivatives. ACM Trans. on Math. Software 1, 97-107.
Powell, M. J. D. (1976). Some global convergence properties of a variable metric algorithm for minimization without exact line searches. R. W. Cottle and C. E. Lemke (Eds.), Nonlinear Programming, SIAM-AMSA Proc. Vol. IX, American Math. Soc., Providence, R.I., 53-72.
Powell, M. J. D. (1977). Restart procedures for the conjugate gradient method. Math. Prog. 12, 241-254.
Powell, M. J. D. (1978). Algorithms for nonlinear constraints that use Lagrangian functions. Math. Prog. 14, 224-248.
Price, W. L. (1979). Controlled random search procedure for global optimization. Computer J. 20, 367-70.
Rauch, H. E. (1980). (Ed.). Control Applications of Nonlinear Programming, Pergamon Press.
Rockafellar, R. T. (1973). A dual appraoch to solving nonlinear programming problems by unconstrained optimization. Math. Prog. 5, 354-373.
Rosen, J. B. (1960). The gradient projection method for nonlinear programming, Part I, linear constraints. SIAM J. 8, 181-217.
Rovinsky, R. B, C. A. Shoemaker and M. J. Todd. (1980). Determining optimal use of resources among regional producers under differing levels of co-operation Operations Research 28, 859-866.
Ryan, D. M. (1974). Penalty and barrier functions. In P. E. Gill and W. Murray (Eds.), Numerical Methods for Constrained Optimization. Academic Press. pp. 175-190.
Saffer, S. I., C. E. Mize, V. N. Bhat and S. A. Szygenda. (1976). Use of nonlinear programming and stochastic modelling in the medical evaluation of normal-abnormal liver function. IEEE Trans. Bio. Eng. BME-23, 200-207.
Sargent, R. W. H. (1974). Reduced-gradient and projection methods fornonlinear programming. In P. E. Gill and W. Murray. (Eds.) Numerical Methods for Constrained Optimization. Academic Press. pp 149-174.
Sasson, M.S. and H. M. Merrill. (1974). Some applications of optiminization techniques to power systems problems. Proc. IEEE, 62, 959-972.
Shanno. D. F. (1970). Conditioning of quasi-Newton methods for function minimization. Math. of Comp. 24, 647-651.
Shanno, D. F. and K. H. Phua. (1977). Inexact step-lengths and quasi-Newton methods. B. Gilchrist (Ed.), Information Processing 77, North Holland. Proc. IFIP Congress 77, Toronto Aug. 8-12, 1977 , pp. 189-193.
Singh, S. K. and G. S. Maddala. (1976). A function for size distribution of incomes. Econometrica. 44, 963-970.
Smith, V. E. (1974). A diet model with protein quality variable. Mangmt. Sci. 20, 971-980.
Sorenson, H. W. (1969). Comparison of some conjugate direction procedures for function minimization. J. Franklin Inst. 288, 421-441.
Varadarajan, R. V. and L. Tummala. (1980). Systems analysis techniques for environmental problems - applications to agroecosystems. Proc. Int. Symp. on Sys. Anal. and Sim., Berlin, Sept. 1-5, 1980.
Waren, A. D and L. S. Lasdon. (1979). The status of nonlinear programming software. Operations Res. 27, 431-456.
Wolfe, P. (1967). Methods for linear constraints. In J. Abadie (Ed.) Nonlinear Programming. North Holland. pp 120-131.
Young, G. K. and M. A. Pisano. (1970). Nonlinear programming applied to regional water resource planning. Water Resour. Res. 6, 32-42.

A MIXED-INTEGER NONLINEAR PROGRAMMING MODEL WELL SUITED FOR USE IN THE ANALYSIS AND CONTROL OF ENERGY SYSTEMS

Dudley V. Goetschel
Amerock Corporation
Rockford, IL. 61101

ABSTRACT

Given a total energy plant system or subsystem consisting of a variety of different component types, and given the energy demands imposed on that system for a chosen time period, an important problem contributing to the conservation of energy and related costs is the determination of optimal operating strategies for that system. Presented in this paper is a model for this problem as well as a brief discussion of nonlinear programming algorithms.

KEYWORDS

Energy systems modeling; systems analysis; optimization; mixed-integer nonlinear programming; variable metric.

INTRODUCTION

With the availability of data describing the operating characteristics and performance of the various component types used in energy conversion systems, and analytical component models derived from this data, one can study the performance of a given system. In particular, given a set of demand profiles, one can determine the optimal operating rules for a system over a specified time period. The understanding of sound techniques for the modeling and solving of this optimization problem must occur before one can adequately examine the potential for savings in the energy consumption of an existing plant, or determine the optimal configuration of components for a new plant.

The first step one usually takes in studying a system is to decide how to conceptually view the system. Once it is decided which components and component interactions are to be included in the model, a mixed-integer nonlinear programming problem can be formulated which has as a solution the optimal operating levels (or operating rules in general) for each component.

THE MODEL

For this discussion, a component in a system is simply defined to be one or more physical pieces of equipment, thought of as a unit and modeled as such, which takes one or more forms of energy as input, "processes" the energy, and produces one or more output energy forms. We designate one of the output energy forms as primary. This is the type of energy the component was designed to produce and the demand imposed on the component for this energy is defined to be the component's operating level. The quantities which need to be modeled at the component level are 1) the energy input quantities required to produce the primary output; 2) the "secondary" output energy forms produced as a result of operating the component at the level dictated by the primary output; 3) properties of the energy forms which are dependent on the performance of the component; and 4) upper and lower bounds on the operating level of the component. All of these quantities can be model· as functions of 1) the primary output (the operating level); 2) the properties associated with the input and output energy forms which influence the performance of the component; and 3) environmental conditions. A complete set of data should be available for each component which contains observed measurements of each quantity being modeled with corresponding values for the operating level,energy properties and environmental conditions which influence the dependent variables. A linear or nonlinear least squares analysis can then be performed to yield nonlinear (or linear) functions which can be used for the calculation of the dependent variables. We can assume that within the valid operating range, these functions will be at least one time continuously differentiable and in most cases low-degree polynomial models are acceptable. The appeal of the polynomial form is the ease with which one can supply analytic expression for first-order derivatives used by the nonlinear programming algorithm during the optimization. A discussion of component modeling using multiple regression, with an example, can be found in (Schultz, Cornwell, Aldridge and Goetschel,1979)

Using the identification of the physical energy flow for each component, one can view a system as in Fig. 1 for example.

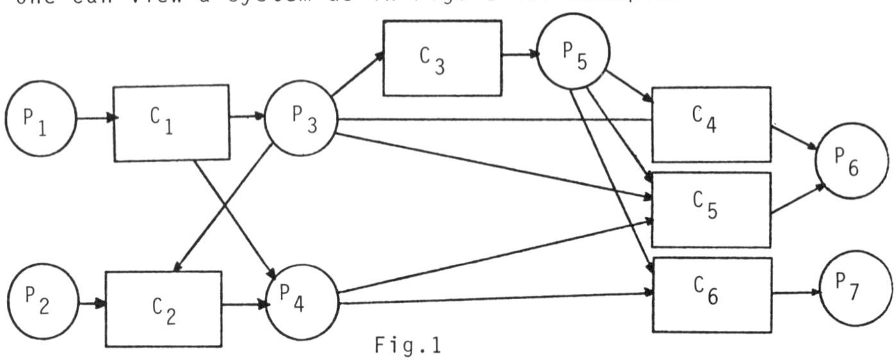

Fig.1

In Fig.1 the rectangles represent physical components while the circles are used to conceptually represent the energy types used, transferred or supplied by the system. The circles can be thought of as consolidation points and all energy quantities corresponding to arcs leading to a given circle should be of the same type and be characterized by the same properties. If the values of a certain property of energy flowing from more than one component to a consolidation point are unequal, capabilities must be provided to determine the value of that property to used in other component calculations. Let x_k^j be the primary energy output flowing from C_k to P_j and let v_i^j represent the ith physical property of energy associated with P_j. Unless a components operation is fixed at a certain level, the x_k^j are all treated as variables in the optimization. The v_i^j can be constant, a variable in the optimization or be represented as a function of parameters depend- on the operation of surrounding components. Denote by u_{ik}^j the ith physical property of the energy type represented by P_j used in or determined by computations associated with component k. Further, let (x_k,u_k) be the row-vector consisting of the operating level and physical energy properties included as independent variables in models associated with component k. Then we can let

$f_k^j(x_k,u_k)$ = the physical input energy required by component k from P_j to produce the energy output x_k;

$g_k^j(x_k,u_k)$ = the available physical output energy flowing from component k to P_j resulting from the operation of the component at the level x_k;

$\hat{c}_k(u_k)$ = the operating capacity of component k;

$\hat{a}_k(u_k)$ = the minimum amount of energy component k is allowed to produce before the component is turned off.

And if we define the sets J_1, J_2 and J_3 to contain the indices corresponding to the resource consolidation points, the internal consolidation points and the external demand consolidation points respectively, we can formulate the problem of determining optimal component operating rules for a fixed set of demands and environmental conditions as follows:

$$\text{P1 minimize} \quad \sum_j \sum_k \alpha_j \, f_k^j(x_k,u_k), \qquad j \in J_1 \tag{1}$$

$$\text{subject to} \quad -(\sum_k f_k^j(x_k,u_k)) + r_j \geq 0 \qquad \text{for } j \in J_1 \tag{2}$$

$$\sum_k x_k^j + \sum g_k^j(x_k,u_k) - d_j \geq 0 \qquad j \in J_3 \tag{3}$$

$$\sum_k x_k^j + \sum g_k^j(x_k,u_k) - (\sum f_k^j(x_k,u_k) + d_j) \geq 0 \qquad j \in J_2 \tag{4}$$

$$\hat{c}_k(u_k) \geq x_k \geq \hat{a}_k(u_k) \tag{5}$$

$$\left.\begin{aligned} U_i^j &\geq v_i^j \geq L_i^j \\ U_i^j &\geq u_i^j \geq L_i^j \end{aligned}\right\} \tag{6}$$

where the α_j are weights attached to the various resources, the r_j represent the available resource levels, the d_j represent demand levels and the U_i^j and L_i^j are used to bound the energy properties for the consolidation points. In addition to the above constraints, it may also be necessary to include several equality constraints to assure that energy properties assume correct values. This model is discussed in more detail in (Goetschel,1979).

In solving the above problem, however, we are assuming the demand values are such that the optimal solution occurs at a point where all components are operating at a positive level. Since we are allowing the demand levels to be arbitrarily small, there is usually a need to allow components to be turned off. The problem is therefore, inherently, a mixed-integer nonlinear programming problem. We can let

$$\delta_k = \begin{cases} 0 & \text{if component } k \text{ is off} \\ 1 & \text{if component } k \text{ is on} \end{cases} \qquad \text{for } k = 1,\ldots,n$$

and multiply the appropriate terms in the problem by δ_k to represent the activity (or inactivity) of component k.

It should be noted that in this type of problem, many, if not the majority of the integer variables, can be eliminated for a given set of demands via an exploitation of the system structure. One may be able to predict accurately which components will be operating simply by knowing the level of the various demands. The merits in having the capability to solve P1 are limited though, since one is usually interested in examining the performance of a system over an extended time period. This of course increases the number of discrete and continuous variables. However, even though problems solved over an extended time period can be very large, they are also very sparce and starting points for the optimization algorithm are easily generated.

Once the individual components have been modeled and an overall system has been constructed, one can decide on an optimization approach. Branch and bound methods can be used for moderately sized problems to handle the 0-1 variables directly via use of Special Ordered Sets (in particular, Linked Ordered Sets) introduced by Beale and Tomlin (1970). The author believes, however, that the greatest potential for success in solving these problems with adequate efficiency and accuracy lies within the use of nonlinear programming methods. A study was performed by Goetschel and Kalita (1979) in which a variable metric algorithm was successfully applied to determine solutions for a small example system. Variable metric algorithms for constrained nonlinear programming arrive at a solution to a given problem by solving a sequence of quadratic programming subproblems. A quadratic approximation to the Lagrangian function associated with the problem is minimized in these subproblems subject to linearizations of the constraints. The solutions to these subproblems define search directions used by the algorithm to decrease the objective function on each iteration. The variable metric (or quasi-Newton) characteristic of these methods stems from the updating done on each iteration to the approximation of the Lagrangian Hessian matrix used in the quadratic approximation. For information relating to the evolution of quadratic programming based algorithms, one should see (Wilson,1963),(Robinson,1972) and

(Rosen and Kreuser,1972). Variable metric type methods are discussed by Murray (1969) and Wright (1976) while a general survey of available software for solving nonlinear programming problems can be found in (Wright,1978).

CONCLUSION

The appeal of a variable metric approach arises from two observations. The first is that most energy plant components operate in a very predictable fashion. In fact, in most cases many of the functions used to model component performance are linear, quadratic or cubic. This suggests that the linear approximations to the constraints will probably be very good during the iterative process. Secondly, the constraint approximations are based on the constraint gradients and the resulting constraint matrix used when solving the quadratic programming subproblem for any iteration is very sparce and assumes a standard, block-diagonal structure (with possibly a few coupling constraints). Treatment of large-scale sparce problems is discussed in (Gill and Murray,1974). Both of these observations suggest increased reliability and efficiency.

REFERENCES

Beale, E. M. L. and J. A. Tomlin (1970). Special facilities in a general mathematical programming system for nonconvex problems using ordered sets of variables. Proceedings of the Fifth International Conference on Operations Research, (J. Lawrence,Ed), pp.447-454.

Gill, P. E. and Murray, W. (1974) Methods for large-scale linearly constrained problems. Numerical Methods for Constrained Optimization (P. E. Gill and W. Murray,eds.), Academic Press, New York, pp.29-66.

Goetschel, D. V. and J. K. Kalita (1979). Two experiments involving the implementation and use of nonlinear optimization software on a minicomputer system. Argonne National Laboratory, Applied Mathematics Division, TM-331, pp.12-33.

Murray, W. (1969). An algorithm for constrained optimization.Optimization (R. Fletcher,ed.), Academic Press, New York, pp.244-258.

Robinson, S.M.(1972). A quadratically convergent algorithm for general nonlinear programming problems. Mathematical Programming 3, pp.145-156.

Rosen, J. B. and J. Kreuser (1971). A gradient projection algorithm for nonlinear constraints. Numerical Methods for Nonlinear Optimization, (F.A.Lootsma,ed.), Academic Press, New York, pp.297-300.

Schultz, H., L. W. Cornwell, N.E. Aldridge, and D. V. Goetschel (1979). Component modeling using multiple regression. Proceedings of the Tenth Annual Pittsburgh Conference on Modeling and Simulation, Vol.10, Part 3, pp.849-855.

Wilson, R. B. (1963). A simplified method for convex programming. Ph.D thesis, Harvard University, Cambridge, MA.

Wright, M. H. (1976). Numerical methods for nonlinearly constrained optimization. Stanford Linear Accelerator Center, Report 197, Stanford, CA.

Wright, M. H. (1978). A survey of available software for nonlinearly constrained optimization. Stanford University Systems Optimization Laboratory, Report 78-4.

SOLVING A CAPITAL/TECHNOLOGY CONSTRAINT PROBLEM BY IMPLICIT ENUMERATION*

H. T. Lau

Computer Science Department
Vanderbilt University
Nashville, TN 37235
U. S. A.

ABSTRACT

Nowadays, the telephone network planning development is mainly concentrated in digital technology. One of the problems is to make proper decisions to change individual traditional analog switching office to a new digital terminal. This capital budget exercise is formulated as a zero-one integer programming problem. (A special case of it gives rise to the technology availability constraint problem.) In 1965, Balas introduced the additive algorithm to solve the general class of zero-one integer programming problems. The principle of the algorithm is based on an implicit enumeration technique. Since then numerous investigators had made refinements and extensions to the basic Balasian procedure. We give a variation of the Balas algorithm to solve our capital/technology constraint problem.

KEYWORDS

Network planning; zero-one integer programming; branch-and-bound method.

INTRODUCTION

Nowadays, the telephone network planning development is mainly concentrated in digital technology. One of the problems is to make proper decisions to change an individual traditional analog switching office to a new digital terminal. This capital budgeting exercise can be formulated as a zero-one linear programming problem [15] : a capital budgeting problem. A special case of it gives rise to the technology availability constraint problem. In 1965, Balas introduced the additive algorithm [1] to solve this class of problems which arise frequently in practice. Since then numerous investigators had made refinements and extensions to the Balasian [see 2, 3, 4, 5, 6, 7, 8, 9, 10, 12, 13, 14 for example]. Patterson [11] reported that the surrogate constraint concept developed by Geoffrion [5, 6] and the dynamic origin concept of Salkin and Spielberg [10] are especially effective.

*Part of the work was done while the author was a scientific staff member at Bell Northern Research, Canada.

The basic Balasian procedure uses the branch and bound strategy. In this paper, the telephone network planning problem will be formulated. The general branch and bound method will be reviewed. A variation of the Balasian algorithm is then applied to solve our problem.

PROBLEM FORMULATION

Given n analog switching offices with T time periods (which usually correspond to years) mapped onto P decision periods, $P \le T$, it is required to determine the decision period for each office to be digitized based on the costs of equipments and maintenance. Let

w_i : decision variable for keeping the office i to be in analog state throughout time periods T;
$CA_i(t)$: cost in time period t of making decision w_i;
A_i : Present Worth Annual Cost (PWAC) of the cash flows $CA_i(t)$;
$y_i(k)$: decision variable for changing the analog office i into a digital remote (RLM) in decision period k;
$CR_i(k,t)$: cost in time period t of making decision $y_i(k)$;
$R_i(k)$: PWAC of $CR_i(k,t)$;
$z_i(k)$: decision variable for changing analog office i into a digital switch (DMS) in decision period k;
$CD_i(k,t)$: cost in time period t of making decision $z_i(k)$;
$D_i(k)$: PWAC of $CD_i(k,t)$;
$C(t)$: budget constraint for time period t.

To satisfy the budget constraint for each time period :

$$\sum_{i=1}^{n} \{ CA_i(t) \cdot w_i + \sum_{k=1}^{P} [CR_i(k,t) \cdot y_i(k) + CD_i(k,t) \cdot z_i(k)]\} \le C(t)$$

for $t=1,2,\ldots,T$.

Each office can only be in one of the analog, remote or digital state in decision period k :

$$w_i + \sum_{k=1}^{P} [y_i(k) + z_i(k)] = 1 \qquad i=1,2,\ldots,n. \tag{1}$$

Let A_i, $R_i(k)$, $D_i(k)$ be the global costs of maintaining and/or changing an office to analog, remote or digital state. The objective is to minimize these costs. The decision variables take the value zero or one. The problem can now be stated as follows :

$$\begin{aligned}
\text{minimize} \quad & \sum_{i=1}^{n} \{ A_i w_i + \sum_{k=1}^{P} [R_i(k) \cdot y_i(k) + D_i(k) \cdot z_i(k)]\} \\
\text{subject to} \quad & \sum_{i=1}^{n} \{ CA_i(t) \cdot w_i + \sum_{k=1}^{P} [CR_i(k,t) \cdot y_i(k) + CD_i(k,t) \cdot z_i(k)]\} \le C(t) \\
& \qquad\qquad \text{for } t=1,2,\ldots,T, \\
& w_i + \sum_{k=1}^{P} [y_i(k) + z_i(k)] = 1 \qquad i=1,2,\ldots,n, \\
& w_i,\ y_i(k),\ z_i(k) = 0 \text{ or } 1 \qquad i=1,2,\ldots,n,\ k=1,2,\ldots,P.
\end{aligned} \tag{2}$$

Let us consider a variation of (2). If the coefficients w_i are zero and the other coefficients are interpreted as follows :

$C(t)$: number of digital lines manufactured in time period t;
$CR_i(k,t)$: number of digital lines used in making decision $y_i(k)$ in time t;
$CD_i(k,t)$: number of digital lines used in making decision $z_i(k)$ in time t;

then (2) becomes a technology availability constraint problem. We shall discuss primarily on (2) since the above variation is a special case of it.
By changing the variables and making the substitution of $K = 2P + 1$, it is not difficult to see that (2) can be transformed into the following standard formulation :

$$\begin{aligned} &\text{minimum} \quad \sum_{i=1}^{n}\sum_{j=1}^{K} c_{ij}x_{ij} \\ &\text{subject to} \quad \sum_{i=1}^{n}\sum_{j=1}^{K} a_{ij}(t)\cdot x_{ij} \geq b(t) \qquad t=1,2,\ldots,T \\ &\qquad\qquad \sum_{j=1}^{K} x_{ij} = 1 \qquad i=1,2,\ldots,n \\ &\qquad\qquad x_{ij} = 0 \text{ or } 1 \qquad i=1,2,\ldots,n \\ &\qquad\qquad\qquad\qquad\qquad j=1,2,\ldots,K. \end{aligned} \tag{3}$$

In the next section we shall present a variation of the Balasian additive algorithm to solve (3). Basically it is a branch and bound method.

BRANCH AND BOUND METHOD

To our present knowledge, to order to solve any integer programming problem, there is no general method which can work faster than exhausting all the feasible solutions. The most intelligent trick is to avoid complete enumeration and be able to arrive at conclusion by only partial examination of all possibilities. A technique to achieve this is called Branch and Bound. The method consists of two basic procedures. Branching is the process of partitioning a large problem into smaller subproblems, and bounding is the process of calculating a lower bound on the solution of each subproblem generated in the branching process. The braching process is carried out by an enumeration tree in which an original problem is replaced by two or more new problems which are mutually exclusive and exhaustive smaller subproblems of the original.

Consider the coefficients c_{ij} in the objective function in (3). We can divide them into n groups : the i-th group is $(c_{i1}, c_{i2}, \ldots, c_{iK})$.

In order to speed up the bounding procedure, we first find the minimum element of each of these groups. The smallest element in the i-th group will be denoted by $\bar{c}_i$. Suppose $\bar{c}_i$ is the v-th element within the i-th group of the c_{ij}, then for each $\bar{c}_i$ the coefficients in the constraints of (3) will be denoted by $a_{iv}(t)$, $t=1,2,\ldots,T$.

Similarly, the $a_{ij}(t)$ can be divided into nT groups : T large groups each of which contains n subgroups. The i-th subgroup within the t-th large group is $(a_{i1}(t), a_{i2}(t), \ldots, a_{iK}(t))$.

We find the maximum element for each of these nT subgroups. The largest element in the i-th subgroup within the t-th large group will be denoted by $a_{iu}(t)$. This preprocessing step is done once and for all.

To solve (3), each solution candidate can be characterized by a permutation of integers 0 and 1. In (3) since there are nK variables each permutation has nK integers, each of which is either 0 or 1. because of constraints (1), each node in the enumeration tree has exactly K sons. There are altogether n levels in the tree and number of digits in each node is a multiple of K.

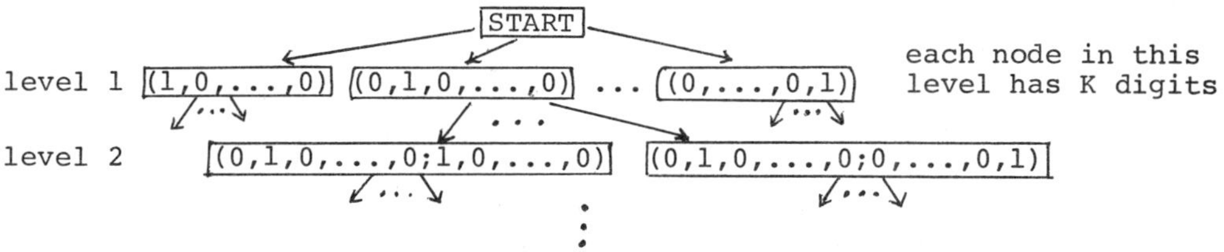

Fig. 1. Enumeration tree for system (3); each node in level i has iK digits and there are n levels.

At the outset, we need an initial upper bound Z for the solution of (3). A convenient one is $Z = \infty$. In the special case of (2), a rough guess is obtained : with no digital conversion, set $Z = \sum_{i=1}^{n} A_i$.

Now the following algorithm suggests itself.

Step 1. [Preprocessing and Initialization.]
Find the numbers $\bar{c}_i$, $a_{iv}(t)$ and $a_{iu}(t)$, for i=1,2,...,n.
Set $Z = \infty$.
The initial K nodes (Fig. 1, level 1) are put in a list.

Step 2. [Select.]
The most recently generated node S is selected.

Step 3. [Branch.]
A son s of S is generated.

Step 4. [Bound.]
If s is a leaf then do the following :
calculate the value Z' of the objective function given by s; if Z' is less than Z then replace Z by Z'.
If s is not a leaf then do the following : (suppose at level q)

a) if $\sum_{i=1}^{q} \sum_{j=1}^{K} a_{ij}(t) \cdot x_{ij} + \sum_{i=q+1}^{n} a_{iu}(t) < b(t)$

for any t=1,2,...,T

then s is fathomed and goto Step 5;

b) compute a lower bound $L = \sum_{i=1}^{q} \sum_{j=1}^{K} c_{ij} x_{ij} + \sum_{i=q+1}^{n} \bar{c}_i$

if $L \geq Z$ then s is fathomed and goto Step 5;

c) if $\sum_{i=1}^{q} \sum_{j=1}^{K} a_{ij}(t) \cdot x_{ij} + \sum_{i=q+1}^{n} a_{iv}(t) \geq b(t)$

for all t=1,2,...,T

then s is fathomed, otherwise add s to the list and goto Step 5;

if $L < Z$ then set $Z = L$ and store this solution as the current best solution.

Step 5. [Branching finished ?]
If all the sons of S are already generated then delete S from the list and proceed to Step 6; otherwise return to Step 3.

Step 6. [Terminate ?]
If the list is empty then the optimal solution is found and is stored in Z; otherwise return to Step 2.

REFERENCES

[1] Balas, E., "An additive algorithm for solving linear programs with 0-1 variables," Oper. Res. (USA), 13, 4, 1965, 517-546.

[2] Driebeek, N., "An algorithm for the solution of mixed integer programming problems," Man. Sci. (USA), 12, 7, 1966, 576-587.

[3] Fleischmann, B., "Computational experience with the algorithm of Balas," Oper. Res. (USA), 15, 1, 1967, 153-155.

[4] Geoffrion, A. M., "Integer programming by implicit enumeration and Balas' method," SIAM Review (USA), 9, 2, 1967, 178-190.

[5] Geoffrion, A. M., "An improved implicit enumeration approach for integer programming," Oper. Res. (USA), 17, 3, 1969, 137-151.

[6] Geoffrion, A. M., and A. B. Nelson, "User's instructions for 0-1 integer linear programming code RIP30C," Memorandum RM-5657-PR, May 1968, RAND Corporation.

[7] Glover, F., "A multiphase-dual algorithm for the 0-1 integer programming problem," Oper. Res. (USA), 13, 6, 1965, 879-925.

[8] Holcomb, B. D., "Zero-one integer programming with heuristics," IBM contributed program (360D-15.2.011), 1968.

[9] Lemke, C., and K. Spielberg, "Direct search algorithm for 0-1 and mixed integer programming," Opers. Res. (USA), 15, 5, 1967, 892-914.

[10] Lemke, C., and K. Spielberg, "DZIP1, Direct search 0-1 integer programming 1," IBM contributed program (360D-15.2.001), 1968.

[11] Patterson, J. H., "Balasian-based enumeration procedures : a study in computational efficiency," Bicentennial Conference on Mathematical Programming, Gaithersburg, 1976.

[12] Peterson, C., "Computational experience with variants of the Balas algorithm applied to the selection of R & D projects," Man. Sci. (USA), 13, 9, 1967, 736-745.

[13] Salkin, H. M., and K. Spielberg, "DZLP, Adaptive binary programming," IBM contributed program (360L-15.001), June 1969.

[14] Trotter, L. E., Jr., "User's instructions for the integer programming code ENUMBER8," Mathematics Research Center, University of Wisconsin, Madison, Technical Summary Report #1347, December 1973.

[15] Tsui, M., private communication.

IDENTIFICATION OF STOCHASTIC PARAMETER INEQUALITIES

Dobrin Burev*
*Base for Automation of Scientific Experiment, Bulgarian Academy of Sciences, LENIN Blvd 72, Bulgaria

ABSTRACT

This paper discusses a method of parameter identification of stochastic inequalities with the help of experimental observations. The problem is defined in the class of problems of mathematical programming. There is demonstrated a problem, realized on a computer for the parameter identification of stochastic inequalities, defining the growth kinetics of the kind of tumour cell populations.

KEYWORDS

Parameter identification, stochastic inequalities, mathematical programming, adaptive algorithms, non-stationary structural indeterminacy.

INTRODUCTION

The mathematical modeling in the contemporary practical problems is usually done in conditions of indeterminacy. This indeterminacy is parametric, if the values of a finite number of the parameters are unknown; it is a structural one, if the functional model form is unfamiliar and a probabilistic one, if the probability distributions of the observable and (or) unobservable variables (the noise) are not familiar. It is well known, that the mathematical models construction in the form of differential equations under conditions of stationary structural indeterminacy is connected with many methodological and computational difficulties, concerning the approximation problems. These difficulties become practically insurmountable in conditions of non-stationary structural indeterminacy, when the functional model type changes in an unfamiliar manner in previously unknown time moments. The necessity of estimating and solving in real time imposes some additional heavy restrictions on the permissible volume of comcomputations. This paper suggests the application of mathematical models in the form of bilateral stochastic differential inequalities instead of the use of differential equations in modeling and control of a class of plants with non-stationary structural indeterminacy, which satisfies the following assumption: for every given class of functions L_f there exists a stationary, bounded and closed in the n-dimensional phase plant space, compact range D, that performs the following: a) all the bounds of D can be described with the help of functions, precise up to unknown parameters; b) all the possible trajectories of the plant with non-stationary structural indeterminacy belong to (lie in) the range D. In this way the modeling of the class of plants discussed, whose behaviour can be described by unfamiliar differential equations, altering their functional form in time in an unknown manner, is done with the aid of bilateral differential inequalities with parametric indeterminacy.

FORMULATION OF THE PROBLEM

The mathematical model of the plant

$$dx/dt = F_t(x,u,t), \; x(0)=x_0 \, , \tag{1}$$

$$x = (x_1(t)\ldots x_n(t))' \, , \; u = (u_1(t)\ldots u_m(t))' \, ,$$

is synthesized in the form:

$$f_1(x,u,t,C_1) \leq dx/dt \leq f_2(x,u,t,C_2), \tag{2}$$

where u is a control variables vector, x is an observable random vector of the state variables, t is time, $F_t(.)$ is unfamiliar vector-function, changing its type in time, f_1 and f_2 are vector-functions, defined in the class of functions L_f, C_1 and C_2 are sets of unknown parameters. The differential inequalities (2) are replaced by their discrete analogs:

$$x^{(k-1)} + \Delta t.f_1(x^{(k-1)},u^{(k-1)},k-1,C_1) \leq x^{(k)} \leq x^{(k-1)} + \Delta t.f_2(x^{(k-1)},u^{(k-1)},k-1,C_2), \text{ or} \tag{3}$$
$$\varphi_1(x^{(k-1)},u^{(k-1)},k-1,C_1) \leq x^{(k)} \leq \varphi_2(x^{(k-1)},u^{(k-1)},k-1,C_2),$$

where φ_1 and φ_2 are known vector-functions, k is the time discretization moment, Δt is the discretization interval.

The parameter identification problem of (3) is defined in the class of stochastic nonlinear programming problems with the use of the observable values of x and u :

$$E_{x,u}\left\{\left[\varphi_{1i}(x^{(k-1)},u^{(k-1)},k-1,C_{1i}) - \varphi_{2i}(x^{(k-1)}, u^{(k-1)},k-1,C_{2i})\right]^2\right\} \rightarrow \min_{C_{1i},C_{2i}} \, , \tag{4}$$

the following constraints being satisfied:

$$E_{x,u}\left\{x_i^{(k)} - \varphi_{2i}(x^{(k-1)}, u^{(k-1)},k-1,C_{1i})\right\} \leq 0, \tag{5}$$

$$E_{x,u}\left\{\varphi_{1i}(x^{(k-1)},u^{(k-1)},k-1,C_{1i}) - x_i^{(k)}\right\} \leq 0, \tag{6}$$

$$i = 1,n \; ; \; k = 1,2,\ldots \, .$$

METHODS AND ALGORITHMS

The problem (4,5,6) can be written in the class of nonlinear programming problems:

$$W_o = \frac{1}{K}\sum_{k=1}^{K}\left[\varphi_{1i}(k,C_{1i}) - \varphi_{2i}(k,C_{2i})\right]^2 \rightarrow \min_{C_{1i},C_{2i}}, \qquad (4a)$$

$$W_1 = \sum_{k=1}^{K} He^1(i,k).(\varphi_{1i}(k,C_{1i}) - x_i^{(k)}) = 0, \qquad (5a)$$

$$W_2 = \sum_{k=1}^{K} He^2(i,k).(x_i^{(k)} - \varphi_{2i}(k,C_{2i})) = 0, \qquad (6a)$$

where $\varphi_i(k,C) = \varphi_i(x^{(k-1)},u^{(k-1)},k-1,C)$;

$$He^1(i,k) = \begin{cases} 0, \text{ if } \varphi_{1i}(k,C_{1i}) - x_i^{(k)} \leq 0 \\ 1, \text{ if } \varphi_{1i}(k,C_{1i}) - x_i^{(k)} > 0 \end{cases}, \qquad (7)$$

$$He^2(i,k) = \begin{cases} 0, \text{ if } x_i^{(k)} - \varphi_{2i}(k,C_{2i}) \leq 0 \\ 1, \text{ if } x_i^{(k)} - \varphi_{2i}(k,C_{2i}) > 0 \end{cases}, \qquad (8)$$

$i = 1,n$; $k=1,K$.

The problem (4a,5a,6a,7,8) can be solved, applying the nonlinear programming methods and algorithms, that do not use derivatives (Himmelblau,1972).

For the adaptive identification of (3) in real time, the current observations of the values of x and u are used. The estimation of C is done through recursive algorithms, realizing the stochastic approximation method (Tsypkin, 1968):

$$C_{1i}^{(k)} = C_{1i}^{(k-1)} + \left\{G_C^{(k)}.\left[\varphi_{1i}(k,C_{1i}^{(k-1)}) - \varphi_{2i}(k, C_{2i}^{(k-1)})\right] - \lambda_{1i}^{(k)}\right\}.(\partial\varphi_{1i}(k,C_{1i}^{(k-1)})/\partial C_{1i}^{(k-1)}), \qquad (9)$$

$$\lambda_{1i}^{(k)} = \max\left\{0, \lambda_{1i}^{(k-1)} + G_\lambda^{(k)}.(\varphi_{1i}(k,C_{1i}^{(k-1)}) - x_i^{(k)})\right\}, \qquad (10)$$

$$\lambda_{2i}^{(k)} = \max\left\{0, \lambda_{2i}^{(k-1)} + G_\lambda^{(k)}.(x_i^{(k)} - \varphi_{2i}(k, C_{2i}^{(k-1)})\right\}, \ \lambda_{1i}^{(o)}=\lambda_{2i}^{(o)}=0, \ i=1,n; l=1,2 \qquad (11)$$

The constants G_C and G_λ satisfy the conditions:

$$0 < G^{(k)} < 1 \ ; \ \sum_{k=1}^{\infty} G^{(k)} = \infty \ ; \ \sum_{k=1}^{\infty} (G^{(k)})^2 < \infty \ . \qquad (12)$$

Boundedness of $\partial \varphi_{li} / \partial C_{li}$, l=1,2 , i=1,n is also assumed.

In the cases, when parameter stationarity of $\varphi_1(C_1)$ and $\varphi_2(C_2)$ is supposed, the criterion for cancelling of the adaptive algorithms (9,10,11), consists in the simultaneous fulfillment of the conditions, given below:

$$\lambda_1^{(q)} = \lambda_2^{(q)} = 0; \ \max_{\ell} \ \max_{i} \left| C_{li}^{(q)} - C_{li}^{(q-1)} \right| \leq \varepsilon \qquad (13)$$

$\forall q = k_o, k_o + K_1$, where ε and K_1 are definite positive numbers. In the cases, when C_1 and C_2 are unfamiliar time functions, the identification of C_1 and C_2 in real time is a continuous process.

APPLICATION IN THE CONTROL PROBLEMS

The solution of the parameter problem for optimal control of the plant, described by the differential inequalities of type (2) is considered:

$$J_o(B) = J_o(x,u,t) \to \min_B \ , \ J_o(B) \text{ - unimodal}, \qquad (14)$$

$$f_1(x,u,t,C_1) \leq dx/dt \leq f_2(x,u,t,C_2), \ x(0)=x_o \ , \qquad (15)$$

$$u = R(x,t,B) \ , \qquad (16)$$

$$J_c(x,u,t) \leq 0, \ c=1,N_1 \ , \ 0 \leq t \leq t_f . \qquad (17)$$

The functions $J_c(.)$, $c=0,N_1$, $f_1(C_1)$, $f_2(C_2)$, $R(B)$ and parameters C_1, C_2, x_o, t_f are accepted to be known. The parameters B of the feedback law are not familiar.

The following symbols are introduced:

$z_1 = z_1(B,t,x_o,t_o=0)$ is the solution of the system:

$$dx/dt = f_1(x,R(x,t,B),t,C_1), \ x(0)=x_o, \ t \in [0,t_f];$$

$z_2 = z_2(B,t,x_o,t_o=0)$ is the solution of the system:

$$dx/dt = f_2(x,R(x,t,B),t,C_2), \ x(0)=x_o, \ t \in [0,t_f].$$

Then the optimal control problem (14-17) is defined as a non-linear programming problem:

$$\max\{J_o(z_1(B),R(z_1(B),B,t),t) \ , \ J_o(z_2(B),R(z_2(B),B,t),t)\} \to \min_B, \qquad (18)$$

$$J_c(z_1(B),R(z_1(B),B,t),t) \leq 0, \quad c = 1,N_1 \tag{19}$$

$$J_c(z_2(B),R(z_2(B),t,B),t) \leq 0, \quad c = 1,N_1 , \tag{20}$$

In order to solve problem (18-20) the software package PROPTICON (Dimitrov,1979) may be used in the integrating of systems for determination of z_1 and z_2, and the computing of J_c, $c=0,N_1$ (18-20) with current values of B.

In the case of adaptive control the following non-linear programming problem is solved at every k-th iteration (k=0,1,...):

$$\sum_{j=1}^{m} (u_j^{(k)} - u_j^{(k-1)})^2 \rightarrow \min_{u_j^{(k)},\ j=1,m} , \tag{21}$$

$$\varphi_1(x^{(k)},u^{(k)},k,C_1^{(k)}) - d^{(k+1)} \leq 0, \tag{22}$$

$$-\varphi_2(x^{(k)},u^{(k)},k,C_2^{(k)}) + d^{(k+1)} \leq 0, \tag{23}$$

$$J_c(x^{(k)},u^{(k)},k) \leq 0, \quad c=1,N_1 , \tag{24}$$

where vector $d^{(k+1)}$ is constructed with the aid of the desired values of x, $C_1(k)$ and $C_2(k)$ are determined by the algorithm (9-11). The initial point of the procedure, solving (21-24), is u(k,0)=u(k-1).

ON THE MODELING OF THE KINETICS OF GROWTH OF NOT-TREATED TUMOUR SPO IN A MOUSE

A model of the type, given below, is synthesized:

$$\varphi_1(t,C_1) = (C_{11}.\mathrm{arctg}(t+C_{12})+C_{13}.\mathrm{arccotg}(t+C_{14})).t + C_{15} \leq$$

$$\leq \ln V \leq (C_{21}.\mathrm{arctg}(t+C_{22})+C_{23}\mathrm{arccotg}(t+C_{24})).t+C_{25} =$$

$$\varphi_2(t,C_2) , \tag{25}$$

where V is the tumour volume in μm^3, t denotes the days after tumour transplantation. The choice of functions (25) is dictated by the facility to use one and the same functions φ_1 and φ_2, modeling the growth of not-treated and treated tumours, with and without recidives. The parameter identification problem of (25), having in mind 13 observations on 6 animals (K=78), is defined in the form (4a,5a,6a,7,8). The experimental data (Table 1) are results from the investigations of G.Grashev and K.Christov, from the Institute of onkology, Medical academy,Sofia. Parameters of the program FLEXI (Himmelblau, 1972), used for identification of C_1 and C_2 (25) are: CONVER = 10^{-4}, SIZE=2, NX=10, NC=2,NIC=0. Initial values of the parameters, being identified:

TABLE 1 Experimental data

t → / V μm³	25	32	35	38	41	45	49	52
No 1	25.6	93.6	156.8	230.4	281.6	289.0	420.0	595.35
No 2	60.0	180.0	356.4	440.0	691.2	520.0	617.4	1620.0
No 3	44.8	176.4	281.6	388.8	480.0	487.5	540.0	921.6
No 4	35.2	100.8	204.8	291.6	324.0	361.0	485.1	748.8
No 5	25.6	93.6	230.6	260.1	440.0	580.8	1176.0	1303.56
No 6	60.0	144.0	256.0	317.8	560.0	560.0	864.0	1149.2

t → / V μm³	55	61	66	70	75
No 1	750.20	979.2	1530.0	2196.40	3680.0
No 2	2196.40	2980.8	4320.0	6681.6	7142.4
No 3	1440.0	2592.0	3520.0	5078.4	8164.8
No 4	1350.0	1740.8	2743.6	2743.6	3840.0
No 5	1843.6	3528.0	4057.0	4452.8	6000.0
No 6	1530.0	2312.0	4000.0	5033.6	6997.0

$C_{11}^{(o)} = C_{12}^{(o)} = C_{24}^{(o)} = C_{25}^{(o)} = 1$; $C_{13}^{(o)}=C_{14}^{(o)}=C_{15}^{(o)}=C_{21}^{(o)}=C_{22}^{(o)}=C_{23}^{(o)}= 0.$

Obtained optimal values of C: $C_{11} = 0.06495285$, $C_{12} = 4.738107$, $C_{13} = 0.6709097$, $C_{14} = -5.573483$, $C_{15} =-0.1166938$, $C_{21} = 0.05205575$, $C_{22} = -1.177827$, $C_{23} = -0.017890$, $C_{24} =-0.7104388$, $C_{25} = 3.268661$.
Values of the object function and the constraints: $W_1 = W_2 = 0$; $W_o = 2.6505$.

CONCLUSION

The paper discusses the problem of mathematical modeling and control of plants, characterized by non-stationary structural indeterminacy. This problem is formulated as parameter identification of bilateral stochastic differential inequalities, defined with accuracy up to unfamiliar parameters.

REFERENCES

Himmelblau, D. M. (1972). Applied Nonlinear Programming. McGraw-Hill Book Company, London.

Tsypkin, Ya. Z. (1968). Adaptation and Learning in Automatic Control Systems. Nauka, Moskow, in Russian .

Dimitrov, D. K. (1979). Ph.D. Dissertation Thesis. Institute of Industrial Cybernetics and Robotics, Bulgarian Academy of Sciences, Sofia, in Bulgarian .

XVII. MODELING AND CONTROL OF LARGE SCALE SYSTEMS

LARGE-SCALE SYSTEMS: STRUCTURE, STABILITY, RELIABILITY

D. D. Šiljak
School of Engineering
University of Santa Clara
Santa Clara, California, U.S.A.

SUMMARY

This paper presents an overview of the progress report describing the results obtained at the Department of Electrical Engineering and Computer Science, University of Santa Clara, during three years of research sponsored by the Division of Electric Energy Systems of the Department of Energy. The broad objective of the research has been to study the relationship between structure, stability and reliability of large scale systems via the decomposition principle, with the theoretical developments directed towards analysis and control of interconnected power systems. The central new concept formulated in the research is that of *generalized decompositions*. The concept has been introduced to partition models of power systems into overlapping subsystems. Such unorthodox decompositions have been necessary in order to obtain the subsystems which represent physical entities. In this way, the subsequent analysis can take advantage of the special structural features of the models under consideration, and produce results with significant conceptual implications. Decentralized control strategies with overlapping information sets have been developed, which can be made compatible with the subsystems obtained by a generalized decomposition. The most attractive property of the overlapping decentralized control schemes is that they are robust with respect to structural perturbations of the interconnected system, and can serve as a basis for building fault-tolerant control schemes with inherent dynamic reliability enhancement using the new multiple controller configurations.

The generalized decompositions have been applied to the transient stability of multimachine power systems, with subsystem sharing the reference machine. An overlapping decentralized control scheme has been used for a two-area automatic generation control, where each area represents a subsystem with the tie-lines as overlapping parts. The results obtained via the "piece-by-piece" approach using the generalized decomposition, compare favorably with the standard "one-shot" methods and designs of interconnected power systems.

The plan of the paper is as follows:

In Section 1, we describe the three major tasks of the research:

(I) Generalized Decompositions

(II) Overlapping Decentralized Control

(III) Reliable Control Systems

We also survey some of the related results which support the development in the three major tasks. In Section 2, we describe several directions for future research, with the main thrust towards a comprehensive theory of fault-tolerant control. Section 3 contains the complete list of papers and conference reports written under the contract with the Department of Energy.

1. RESEARCH DESCRIPTION

1.1 Research Tasks

A system is considered large if either for conceptual or computational reasons it cannot be analyzed as a whole, but has to be decomposed into a collection of smaller interconnected subsystems. A common recipe is to divide the system into disjoint subsystems on the basis of a physical insight and intuition or a mathematical artifice, solve the subproblems independently, and finally combine the solutions of the subproblems to get the solution of the overall problem. This approach is successful and produces meaningful results if (and only if) the subsystems are selected properly.

Even the best results available at present in the area of large-scale dynamic systems, are not suitable for consideration of interconnected power systems, because of the restrictive notion of "disjoint subsystems". The reference machine in a multimachine system appears as a natural part of each subsystem when transient stability should be determined. In automatic generation (AGC), the tie-lines are common parts which are shared among the areas, and any meaningful decomposition of (AGC) should recognize this fact. For these reasons, the major objective of the research has been:

(I) To develop a sufficiently broad theoretical basis and specific procedures for *generalized decompositions* of dynamic systems, with emphasis on transient stability analysis of multimachine systems.

(II) To formulate and study *overlapping decentralized control* and produce feedback schemes for automatic generation control of interconnected power systems.

(III) To analyze the relationship between structure and stability in order to develop new methods for stability analysis and decentralized stabilization of large-scale systems under structural perturbation. The ultimate goal is to invent new schemes for *reliable control* of interconnected systems using multiple control systems.

The three tasks are now described separately by referring to the relevant material in publications.

1.2 Task I: Generalized Decompositions

After three years of intensive research, the generalized decompositions of dynamic systems is now a rather well established area of research. An *inclusion principle* has been formulated in [7], which serves as a basis for generalized decompositions of linear systems using algebraic means. When physically identified subsystems share certain parts of the overall system, they can be stacked on the top of each other to form an *expansion* of the overall system which is a system representation in a space of larger dimension. Conditions are provided under which the expanded system includes the original system, in which case, it is possible to extract from the expansion the essential information about the behavior of the original system. Since in the expanded space the overlapping subsystems appear as disjoint, the behavior of the expansion can be studied using standard disjoint decompositions.

In stability analysis of large-scale systems using the concept of vector Liapunov functions, applications of the generalized decompositions have been particularly effective. It has been shown in [7] that overlapping decompositions of linear systems can establish stability where orthodox disjoint decompositions fail. Since the inclusion principle has been extended to nonlinear systems [8], it has been possible to duplicate the same result for the nonlinear differential equations of the Lotka-Volterra type. Subsequently, the inclusion principle has been extended in [13] to consider not only the overlapping decompositions by repeating the states of the original system, but also to consider the situations in which the original and expanded states are linear combinations of each other. This added freedom in the choice of decompositions has been shown to be useful in establishing stability where both disjoint and overlapping decompositions cannot.

A natural decomposition of multimachine power systems is an overlapping decomposition with the reference machine common to all the subsystems. A "piece-by-piece" stability analysis can be carried out using the vector Liapunov functions. To each subsystem a scalar Liapunov function is assigned, and stability of the overall system is established from stability of the subsystems and the constraints on the interconnections among the subsystems using the appropriate construction of a vector Liapunov function. Since the appeal of this approach depends on the ease by which subsystem functions are chosen, a pair-wise decomposition was used at first [1, 16], whereby each machine is paired with the reference machine to form an overlapping subsystem. In this way, subsystems turned out to be second-order systems of Lur'e-Postnikov type, for which ready-made Liapunov functions exist. Obviously, this approach produced overly conservative estimates of the stability regions simply because, in reality, the power systems are invariably composed of weakly connected groups of tightly coupled machines with more than two machines in each group. A pair-wise decomposition cuts across the strong interconnections violating the coherency aspect of the system, and results into relatively poor stability region estimates if any at all.

A new decomposition scheme has been proposed recently [19, 22], whereby a multimachine system is partitioned into interconnected subsystems which may have more than two machines. Using such a generalized decomposition, one can include the strong couplings inside the subsystems, which enhance stability of subsystems, while exposing only the weak couplings as subsystem interconnections. The decomposition produces multivariable Lur'e-Postnikov systems for which a systematic construction of Liapunov functions is available via Moore-Anderson version of the Popov-Yakubovich-Kalman method. The results of the "piece-by-piece" method compare favorably with standard "one-shot" technique of Willems to the extent that the saving in computer time outweighs the conservativeness inherent in the stability region estimate obtained via vector Liapunov functions. Furthermore, the decomposition procedure decreases the liabilities of errors due to handling a large amount of data, which is what all "one-shot" techniques do. Further extensions of the decomposition approach should exploit this fact and introduce more refined models of the subsystems without obscuring the intrinsic structural characteristics of the multimachine interconnections.

1.3 Task II: Overlapping Decentralized Control

For various conceptual and practical reasons, the control problems in systems composed of interconnected subsystems are decentralized in nature. It is common to assume that each subsystem is controlled independently on the basis of its own performance criterion and locally provided information. The informational restriction is essential and rules out the standard optimization procedures. The situation is further complicated by the fact that imperfect knowledge of the model of interactions is invariably a part of the control problem. No one knows at present if the limitations of the existing design methods for decentralized control and estimation can be removed by developing better procedures, or if these limitations can be attributed to the lack of precise axiomatic definition of the optimization problem. These facts are well recognized in the literature (for references, see [5] and [17]).

To bypass these difficulties, we formulate a decentralized control problem as a problem of interconnecting subsystems which had been optimized by local feedback.

The underlying idea is to treat the interconnections as perturbation terms and determine suboptimality of a locally optimal control law when applied to the overall interconnected system. The measure of suboptimality is the size of performance bounds of the interconnected subsystems relative to the performance of the disconnected optimal subsystems used as reference. Turning this approach the other way, it is possible to investigate suboptimality of the decentralized control which arises from the globally optimal control of the overall interconnected system when the feedback connections between subsystems are removed to satisfy the information structure constraint.

Most of the results available for decentralized control of large interconnected systems whether they are of optimal or suboptimal type, are obtained under the assumption that the actions of the individual controllers are carried out using disjoint information sets; that is, there is no sharing of information among the individual control agents. There is, however, a large class of models which are composed of overlapping subsystems and the sharing of information among the controllers is essential. A notable case is the problem of Automatic Generation Control (AGC) for interconnected power systems. In this context, it is natural to use the models for each individual power area and apply the concept of autonomous local control to each area and the associated tie-lines. This results in a *decentralized control scheme with overlapping information sets* [15].

At present, very little is known about how to formulate decentralized control strategies based upon overlapping information sets. Despite the appearance of a number of semi-rigorous resolutions of certain practical situations, the general aspects of the problem have remained obscure. The inclusion principle can be extended to systems with inputs and outputs [7] to provide a theoretical basis and necessary tools for a systematic and unifying approach to this problem. Again, the underlying idea is to enlarge the state space of the original system so that the overlapping subsystems appear as disjoint. This can be done in such a way that with respect to the enlarged space, each subsystem has its own distinct input. A stabilizing decentralized control can then be designed in the expanded space using the standard schemes of [5], and subsequently contracted for implementation in the original system as shown in [7].

To address the optimization problem in the AGC context, it has been necessary to extend existing framework to include expansions and contractions of performance indices associated with individual power areas. This has been accomplished in [15], where it has been shown how the suboptimality criteria of [5] can be used in the design of decentralized control for overlapping interconnected systems. An Application of the criteria is outlined in [15] using a 19-th order model for load-frequency control of a two-area interconnected power system.

1.4 Task III: Reliable Control Systems

The objective of this task is to provide a mathematical framework for building reliable control using less reliable controllers. The new ingredient in the design is the multiple control system which provides the necessary redundancy for reliability enhancement in control systems subject to controller failures. The proposed reliability design is based upon the decentralized control schemes used so far for synthesizing reliable control systems subject to perturbations in the plant interconnection structure.

Due to increasing complexity of interconnected power systems, reliability of control has become an essential requirement in the design. The desired degree of reliability cannot be achieved merely by diligent application of standard control engineering practice, that is, by using high-quality components in otherwise standard optimization schemes. For this reason, there has been a considerable effort in constructing control schemes with some form of built-in reliability enhancement. The major objective has been to synthesize a control structure so that the system performs satisfactorily under faulty conditions.

It has already been demonstrated that with regard to reliability, the decentralized control schemes are superior to their centralized counterparts. By now, there is a considerable number of results (for example, see [5] and [9]), which show explicitly the high reliability of decentralized control strategies for large systems when subject to structural perturbations of interconnections among the subsystems, as well as the robustness [24] of such strategies to wide range of nonlinearities in subsystem couplings. Reliability of decentralized control schemes, however, is low with regard to structural perturbations in feedback interconnections and controller failures. If a local controller breaks down, it is entirely feasible that the whole system may do the same. Replacement of a faulty controller by a standby, or disconnection of the corresponding subsystem for the purpose of preventing the system breakdown may be either impossible or undesirable due to design constraints. From the point of view of reliability theory, the decentralized control schemes are series connections of controllers, which have no redundancy with regard to control capability, and a failure of any one controller may cause a system breakdown.

A considerable research effort under the contract has been devoted to construction of control schemes which have a necessary redundancy with respect to controller failures. It has been shown in [4] that this can be done in a natural way by using a new concept of *multiple control systems*. In terms of reliability theory, a multiple control case represents a parallel connection and, thus, a structural counterpart of reliability related to controller failures. It has been demonstrated in [4] that, in principle, it is possible to build control schemes by multiple control systems, which use arbitrarily poor controllers but have arbitrarily high reliability. It is strongly believed that the movement of control technology in the direction of modularization and use of microprocessors reinforces the attractivity of the application of multiple control systems to large interconnected systems.

1.5 Related Results

In addition to the results described under the three major tasks, a number of supplemental developments have been pursued, which broaden the base of the research. They also serve to re-enforce the main results and point out the new directions of research.

It is a well-known fact that for decentralized stabilization of linear systems, controllability of both the subsystems and the overall system is not sufficient. In note [3], it has been shown that connectedness of the system digraph is neither necessary nor sufficient supplement for decentralized stabilizability.

In [6], a decentralized scheme is proposed for stabilization of a class of linear time-varying interconnected systems. For this special class, the only stabilizability condition is the uniform complete controllability of the subsystems. The class of such systems and the stabilizability conditions are further broadened in [9], [10], [14], and [24].

Absolute stability of interconnected Lur'e-Postnikov systems, which appear in the multimachine systems, have been considered in [11] using similar techniques as in [1]. Another such variation was analyzed in [21], where decentralized stabilizability was established via output control.

A distinct effort has been devoted to the use of graph-theoretic concepts in the analysis of interconnected systems. In [12], such concepts have been used to decompose a large control system into block-triangular interconnection of input reachable subsystems. Stabilization of the system has been thereby reduced to stabilization of a number of smaller subsystems. The special structural properties of the obtained subsystems can be exploited to further simplify the stabilization problem. Connective stabilizability of large scale systems composed of interconnected subsystems, has been considered in [18] and [20]. Both analytic and graph-theoretic conditions are provided in terms of the interconnection structure of the system.

The problem of vulnerability of dynamic systems has been also formulated and partially resolved in the context of directed graphs [23]. A system is considered vulnerable if its structural properties such as input (output) reachability or structural controllability (observability) is destroyed by a perturbation characterized as removal of a line, or set of lines, from the corresponding graph. Graph-theoretic procedures have been developed in [23] to identify the minimal sets of lines which are essential for preserving the structural properties of the system. These results complement those of reliability [4], and can be used as a point of departure in the direction of reliability analysis of controllability (observability) whereby the probability of such properties to hold is determined in terms of the survival probabilities of the interconnecting links of the system.

Finally, special attention is devoted to robustness of suboptimal control. The reference [24] devoted solely to this problem introduced the gain and phase margin as measures of robustness of suboptimal linear-quadratic regulators. It has been shown that the suboptimal control retains the infinite gain margin of the corresponding optimal system, but that the phase margin and gain reduction tolerance depend on the degree of suboptimality of the nominal optimal control law. This establishes the degree of suboptimality introduced in [5], as an index of both the system performance regarding the optimality criterion and the robustness to plant parameter uncertainties and distortions of the optimal control law. It has also been shown that the suboptimal closed-loop systems remain stable despite insertions of memoryless nonlinear gains inside individual feedback loops, thus raising further the confidence in suboptimal designs of linear-quadratic regulators for interconnected systems.

2. FUTURE RESEARCH

After three years of intensive research, the concept of generalized decompositions and decentralized control with overlapping information sets is now a well established framework for studies of certain basic problems in the context of complex dynamic systems. The future research should exploit the advantages of this new development and come up with important results concerning our ability to control reliably large power systems.

Some of the directions in future research can be summarized as follows:

(i) Broadening further of the inclusion principle to consider imperfect contractions and expansions of dynamic systems using the notion of approximately invariant subspaces. This should open up a real possibility of using the principle in model reduction of power systems.

(ii) Falut tolerant control schemes should be strengthened by studies of redundancy allocations via the classical reliability theory. Dynamic output controllers should replace the state-injection controller configurations. Reliable regulators should be built via overlapping estimation and control concept. These efforts should produce a systematic design of control and estimation schemes for interconnected systems with a built-in dynamic reliability enhancement.

(iii) Graph-theoretic analysis of dynamic systems should be pursued further to produce computer algorithms for vulnerability assessments of large scale control systems. The enormous simplifications achieved by the topological investigations should pay-off in the structuring of interconnected power systems for reliable performance.

When completed, the above research plan should result in a comprehensive and sufficiently general study of the intricate relationship between complexity and stability in large-scale dynamic systems which can be used as realistic models of interconnected power systems. We intend to pursue this plan with vigor and as rapidly as possible in order to achieve a satisfactory degree of completeness and utility of the end product.

3. PUBLICATIONS

1. "On Transient Stability of Multimachine Power Systems", Lj. B. Jocić, M. Ribbens-Pavella, and D. D. Šiljak. IEEE Transactions, AC-23(1978), 325-332.

2. "Overlapping Decentralized Control", D. D. Šiljak. Handbook of Large Scale Systems Engineering Applications, M. G. Singh and A. Titli (eds.). North-Holland, Amsterdam, 1978, pp. 145-166.

3. "Counterexamples to Fessas' Conjecture", M. Ikeda and D. D. Šiljak, IEEE Transactions, AC-24(1979), 670.

4. "Reliable Control Using Multiple Control Systems", D. D. Šiljak. International Journal of Control, 31(1980), 303-329.

5. "Suboptimality of Decentralized Stochastic Control and Estimation", R. Krtolica and D. D. Šiljak. IEEE Transactions, AC-25(1980), 76-83.

6. "Decentralized Stabilization of Linear Time-Varying Systems", M. Ikeda and D. D. Šiljak. IEEE Transactions, AC-25(1980), 106-107.

7. "Overlapping Decompositions, Expansions and Contractions of Dynamic Systems", M. Ikeda and D. D. Šiljak. Large Scale Systems, 1(1980), 29-38.

8. "Lotka-Volterra Equations: Decomposition, Stability, and Structure", M. Ikeda and D. D. Šiljak. Journal of Mathematical Biology, 9(1980), 65-83.

9. "On Decentralized Stabilization of Interconnected Systems", M. E. Sezer and Ö. Hüseyin. Automatica, 16(1980), 205-209.

10. "On Decentrally Stabilizable Large-Scale Systems", M. Ikeda and D. D. Šiljak. Automatica, 16(1980), 331-334.

11. "Local Stability of Composite Systems: Frequency Domain Condition and Estimate of the Domain of Attraction", M. Saeki, M. Araki, and B. Kondo. IEEE Transactions, AC-25(1980), 936-940.

12. "On Structural Decomposition and Stabilization of Large-Scale Control Systems", M. E. Sezer and D. D. Šiljak. IEEE Transactions, AC-26 (to appear in April 1981).

13. "Generalized Decompositions of Dynamic Systems and Vector Liapunov Functions", M. Ikeda and D. D. Šiljak. IEEE Transactions, AC-26 (to appear in the Bellman Special Issue, 1981).

14. "Decentralized Stabilization of Large-Scale Systems with Time-Delay", M. Ikeda and D. D. Šiljak. Large Scale Systems (to appear).

15. "Suboptimal Decentralized Control of Systems with Overlapping Information Sets", M. Ikeda, D. D. Šiljak, and D. E. White. Journal of Optimization Theory and Applications (to appear in June 1981).

16. "Decomposition and Stability of Multimachine Power Systems", Lj. B. Jocić and D. D. Šiljak. Proceedings of the 7th IFAC Congress, Helsinki, Finland, 1978, pp. 21-26.

17. "On Decentralized Control and Estimation", D. D. Šiljak, Proceedings of the 7th IFAC Congress, Helsinki, Finland, pp. 1849-1856.

18. "Structural Decomposition and Stabilization of Large-Scale Control Systems", M. E. Sezer and D. D. Šiljak. Proceedings of the 13th Asilomar Conference on Circuits, Systems, and Computers, Pacific Grove, California, 1979, 176-181.

19. "Generalized Decompositions for Transient Stability Analysis of Multimachine Power Systems", M. Araki, M. Mohsen-Metwally, and D. D. Šiljak. Proceedings of the Joint Automatic Control Conference, San Francisco, California, Paper No. TA3-B, 1980.

20. "Stabilization of Large-Scale Systems using Graph-Theoretic Concepts", Proceedings of the Joint Automatic Control Conference, San Francisco, California, Paper No. WAI-D, 1980

21. "Robust Decentralized Control Using Output Feedback", Ö. Hüseyin, M. E. Sezer, and D. D. Šiljak. Proceedings of the 18th Annual Allerton Conference on Communication, Control, and Computing, University of Illinois, Monticello, Illinois, 1980 (to appear).

22. "Generalized Decompositions and Stability of Nonlinear Systems", M. Ikeda and D. D. Šiljak. Proceedings of the 18th Annual Allerton Conference on Communication, Control, and Computing, University of Illinois, Monticello, Illinois, 1980 (to appear).

23. "Vulnerability of Dynamic Systems", V. Pichai, M. E. Sezer, and D. D. Šiljak. Proceedings of the 19th IEEE Conference on Decision and Control, Albuquerque, New Mexico, 1980 (to appear).

24. "Robustness of Suboptimal Control: Gain and Phase Margin", M. E. Sezer and D. D. Šiljak. Proceedings of the 19th IEEE Conference on Decision and Control, Albuquerque, New Mexico, 1980 (to appear).

STABILITY OF DISTRIBUTED PARAMETER SYSTEMS WITH FINITE DIMENSIONAL COMPENSATORS VIA SINGULAR PERTURBATION METHODS

Mark J. Balas

Associate Professor
Electrical, Computer, & Systems Engr. Dept.
Rensselaer Polytechnic Institute
Troy, NY 12181 USA

ABSTRACT

Using a singular perturbation formulation of the linear time-invariant distributed parameter system, we develop a method to design finite-dimensional feedback compensators of any fixed order which will stabilize the infinite-dimensional distributed parameter system. The synthesis conditions are given entirely in terms of a finite-dimensional reduced-order model; the stability result depends on an infinite-dimensional version of the Klimushchev-Krasovskii lemma developed in [10].

1.0 INTRODUCTION

Many engineering systems exhibit a distributed parameter nature and must be described by partial differential equations. Examples of such distributed parameter systems (DPS) include heat diffusion & chemical processes, wave propagation, & mechanically flexible structures. Various aspects of the control of DPS have been considered in, for example, [1] - [5]; our experience in DPS has been shaped by applications in large aerospace structures [6].

The state spaces for DPS have infinite dimension; so, at best, reduced-order models must be used in controller synthesis. However, the closed-loop stability of the infinite dimensional DPS with a finite dimensional feedback controller becomes a fundamental issue. The synthesis of finite dimensional controllers for DPS and the analysis of their closed-loop stability by singular and regular perturbation techniques have been our main areas of emphasis [7]; this theory has been developed with flexible structures and other highly oscillatory DPS applications in mind.

Even in large-scale, lumped parameter systems, such as electric power distribution networks, it is necessary to perform model reduction and reduced-order controller synthesis and to analyze closed-loop stability. The use of asymptotic methods, especially singular perturbations, has been very successful in this regard (e.g. [8]). We have extended certain of these singular perturbations methods for DPS to provide estimates of stability in an infinite dimensional setting [9] - [10] and applied them to mechanically flexible structures [11]. In this paper we will use these singular perturbation results to analyze the general finite dimensional compensator for linear DPS.

We consider linear, time-invariant DPS with the form:

$$\begin{cases} v_t = Av + Bf, \quad v(0) = v_0 \\ y = Cv \end{cases} \tag{1.1}$$

where the state $v(t)$ is in a Hilbert space H (called the state space) for all $t \geq 0$; the state space has inner product $(\cdot,\cdot)$ and associated norm $||\cdot||$. The control vector f and the observation vector y have dimensions M & P, respectively, which denote the number of (independent) actuators and sensors in use; thus, rank B=M and rank C=P. The operator A is an unbounded differential operator with domain D(A) dense in H and A generates a C_0 semigroup U(t) on H. Many practical DPS can be stated in this form [7].

2.0 REDUCED-ORDER MODELING OF DPS: A SINGULAR PERTURBATIONS FORMULATION

Since the state space H of the DPS in (1.1) is infinite dimensional, we must obtain a reduced-order model (ROM) upon which to base the finite dimensional controller design. In general, this is done by selecting a finite dimensional subspace. H_N (with dim. $H_N = N < \infty$) contained in D(A). This subspace H_N is the ROM subspace; its complement H_R is the Residual subspace, and together they decompose the state space:

$$H = H_N \oplus H_R \tag{2.1}$$

We define the (not necessarily orthogonal) projection operators P_N and P_R and let $v_N = P_N v$ and $v_R = P_R v$. These decompose the state $v = v_N + v_R$ and the DPS (1.1).

In this paper, we will make two basic assumptions on model reduction processes

(1) the subspaces H_N, H_R are reducing subspaces (i.e. they are both A-invariant);

(2) the subspaces H_N, H_R can be chosen to give a singular perturbations formulation with small parameter $\varepsilon \geq 0$ and an exponentially-stable residual subsystem.

These assumptions yield the decomposed DPS (1.1) in the following form:

$$\frac{\partial v_N}{\partial t} = A_N v_N + B_N f \tag{2.2a}$$

$$\varepsilon \frac{\partial v_R}{\partial t} = A_R v_R + B_R f \tag{2.2b}$$

$$y = C_N v_N + C_R v_R \tag{2.2c}$$

where $A_N = P_N A P_N$, $B_N = P_N B$, $C_N = CP_N$, etc. and A_R generates the C_0 semigroup $U_R(t) = P_R U(t) P_R$ with exponential growth condition:

$$\| U_R(t) \| < K_R \, e^{-\sigma_R t}, \quad t \geq 0 \tag{2.3}$$

where $K_R \geq 1$ and $\sigma_R > 0$. All operators except A_R are bounded. The terms $B_R f$ & $C_R v_R$ are called, respectively, control & observation spillover; they represent the interconnections through which the controller can affect the residual subsystem.

The ROM of the DPS, in this formation, is obtained by setting $\varepsilon=0$:

$$\begin{cases} \frac{\partial v_N}{\partial t} = A_N v_N + B_N f \\ y = C_N v_N + D_N f \end{cases} \tag{2.4}$$

where $D_N = - C_R A_R^{-1} B_R$ & A_R^{-1} is a bounded operator due to (2.3). The ROM is a finite dimensional system and the parameters (A_N,B_N,C_N,D_N) may be identified with matrices by choosing a basis in H_N; this will be done wherever necessary. We shall assume, henceforth, that the ROM is controllable and observable (easily verifiable conditions exist for finite dimensional systems). The ROM is completely determined by the choice of the subspaces H_N & H_R.

3.0 FINITE DIMENSIONAL COMPENSATORS FOR DPS

The form of the finite dimensional feedback compensators used here will be the following:

$$\begin{cases} f = L_{11}y + L_{12}z \\ \dot{z} = L_{21}y + L_{22}z \end{cases} \tag{3.1}$$

where the compensator state z has dim. $z = \alpha \leq N \leq \infty$ and $L_{11},L_{12},L_{21},L_{22}$ are matrices of appropriate sizes. We say the compensator is output feedback when $\alpha=0$. The order α of the compensator is assumed to be fixed at some acceptable value which reflects the available capacity of the on-line computer being used to implement (3.1). In [10], the compensator order was $\alpha = N$.

The compensator design is synthesized as though the ROM (2.4) were the full DPS (1.1), i.e. as though $\varepsilon=0$. Let

$$F = \bar{A}_N + \bar{B}_N \bar{L}\, \bar{C}_N \tag{3.2}$$

where

$$\bar{A}_N = \begin{bmatrix} A_N & 0 \\ 0 & 0 \end{bmatrix}, \quad \bar{B}_N = \begin{bmatrix} B_N & 0 \\ 0 & I_\alpha \end{bmatrix}, \quad \bar{C}_N = \begin{bmatrix} C_N & 0 \\ 0 & I_\alpha \end{bmatrix}, \ \& \ \bar{L} = \begin{bmatrix} \bar{L}_{11} & \bar{L}_{12} \\ \bar{L}_{21} & \bar{L}_{22} \end{bmatrix}$$

with

$$\bar{L}_{11} = (I_M - L_{11} D_N)^{-1} L_{11}$$

$$\bar{L}_{12} = (I_M - L_{11} D_N)^{-1} L_{12}$$

$$\bar{L}_{21} = L_{21}(I_P + D_N \bar{L}_{11})$$

& $$\bar{L}_{22} = L_{22} + L_{21} D_N \bar{L}_{12}.$$

Let the composite state $q_N = \begin{bmatrix} v_N \\ z \end{bmatrix}$ in $H_N \times R^\alpha$; from (2.4) & (3.1), this satisfies (for $\varepsilon=0$):

$$\frac{\partial q_N}{\partial t} = F q_N \tag{3.3}$$

The following theorem gives the conditions under which a stable design can be synthesized:

Theorem 3.1: If the ROM (A_N,B_N,C_N,D_N) is controllable and observable and

$$M + P + \alpha \geqslant N + 1 \tag{3.4}$$

then $\bar{L}$ may be chosen so that F given by (3.2) has any desired pole locations in the complex plane.

The proof follows easily from results in [12] or [13] and depends on the finite dimensionality of the closed-loop system when $\varepsilon=0$. Since the gains $\bar{L}$ can be synthesized, the compensator gains $L_{11}, L_{12}, L_{21}, L_{22}$ can be found; however, we must assume that $\| D_N \|$ is sufficiently small to make both $(I_M - L_{11} D_N)^{-1}$ & $(I_P + D_N \bar{L}_{11})^{-1}$ exist.

4.0 CLOSED-LOOP STABILITY FOR THE DPS WITH A FINITE-DIMENSIONAL COMPENSATOR

Let ε be such that $0 < \varepsilon \leq \varepsilon_0$, and consider the closed-loop system consisting of the infinite-dimensional DPS (2.2) and the finite-dimensional compensator (3.1). Let the closed-loop state w in $\check{H} = H_N \times R^{\alpha} \times H_R$ be given by

$$w = \begin{bmatrix} w_N \\ w_R \end{bmatrix}$$

where $w_N = \begin{bmatrix} v_N \\ z \end{bmatrix}$ is in $H_N \times R^{\alpha}$ & $w_R = v_R$ in H_R. The norm on $\check{H}$ is defined by

$$\|w\| = (\|v_N\|^2 + \|z\|^2 + \|v_R\|^2)^{1/2} \tag{4.1}$$

From (2.2) & (3.1), the closed-loop state satisfies:

$$\frac{\partial w_N}{\partial t} = H_{11} w_N + H_{12} w_R \tag{4.2a}$$

$$\varepsilon \frac{\partial w_R}{\partial t} = H_{21} w_N + H_{22} w_R \tag{4.2b}$$

where

$$H_{11} = \begin{bmatrix} A_N + B_N L_{11} C_N & B_N L_{12} \\ L_{21} C_N & L_{22} \end{bmatrix}$$

$$H_{12} = \begin{bmatrix} B_N L_{11} C_R \\ L_{21} C_R \end{bmatrix}$$

$$H_{21} = [B_R L_{11} C_N \quad B_R L_{12}]$$

$$H_{22} = A_R + B_R L_{11} C_R$$

But, (4.2) is in the form of the infinite dimensional Klimushchev-Krasovskii lemma whose proof appears in the revised version of [10]; estimates on the upper bound ε_0 are presented there, as well. From the infinite-dimensional K-K lemma, we obtain the following closed-loop stability result:

Theorem 4.1: If H_{22} generates an exponentially stable C_0 semigroup and the finite rank operator

$$\bar{H} = H_{11} - H_{12} H_{22}^{-1} H_{21} \tag{4.3}$$

has all its poles in the open left half-plane, then there exists positive ε_0 (whose value can be estimated from the model data) such that, for all $0 < \varepsilon \leq \varepsilon_0$, (4.2) is stable.

This theorem says that under certain conditions the finite dimensional controller based on the ROM (ε=0) will continue to stabilize the infinite dimensional DPS when ε is "small" but non-zero. The "smallness" required of ε can be estimated from bounds on the control and observation spillover terms, the constants K_R, σ_R in (2.3), and the ROM data.

The following lemma greatly simplifies Theo. 4.1:

Lemma 4.2:
$$\bar{H} = F = \bar{A}_N + \bar{B}_N \bar{L} \bar{C}_N \quad (4.4)$$

The proof appears in the Appendix. Consequently, Theo. 4.1 may be restated as our main result:

Theorem 4.3: If the stable compensator synthesis conditions of Theo. 3.1 are satisfied for the ROM (2.4) and if the control and observation spillover coefficients $\beta = \|B_R\|$ & $\Gamma = \|C_R\|$ are sufficiently small that

$$\frac{\sigma_R}{K_R} > \beta\, \Gamma\, \|L_{11}\| \quad (4.5)$$

where K_R & σ_R appear in (2.3), then there exists positive ε_0 such that for all $0 < \varepsilon \leqslant \varepsilon_0$ the same finite-dimensional compensator (3.1) stabilizes the infinite-dimensional DPS (1.1).

This follows from Lemma 4.2 and use of the semigroup-generator perturbation result ([14], Theo. 10.9) applied to

$$H_{22} = A_R + B_R L_{11} C_R$$

because $B_R L_{11} C_R$ is a bounded operator.

5.0 CONCLUSIONS

Our main result (Theo. 4.3) gives conditions under which a general finite-dimensional compensator, based on a singular perturbations reduced-order model, will stabilize an infinite dimensional (linear) distributed parameter system. This is an extremely useful result since all practical feedback compensators must be finite dimensional in order to be implementable with on-line digital computers. The result is valid for large-scale lumped-parameter systems, as well.

The most difficult assumption to satisfy is the choice of reduced-order subspace H_N to achieve a singularly perturbed formulation (2.2) of the distributed parameter system (1.1); the other assumptions are reasonably easy to satisfy. Modal methods have worked for mechanically flexible structures, e.g. [11]. A general discussion of this modeling difficulty is given in [15]; it is a fundamental problem in large-scale or distributed parameter systems and should not be overlooked.

ACKNOWLEDGEMENT

This work was supported by NASA Contract No. NAS-9-16053.

1. Lions, J.L. (1972). Some Aspects of the Optimal Control of DPS, SIAM, Philadelphia, PA.
2. Ray, W.H., and Lainiotis, D., eds., (1978). DPS: Identification, Estimation, & Control, M. Dekker, NY.
3. Russell, D., (1978). Controllability & Stabilizability Theory for Linear PDE: Recent Progress & Open Questions, SIAM Review, 20, 371-388.
4. Curtain, R., and Pritchard, A., (1978). Infinite Dimensional Systems Theory, Springer, NY.
5. Aziz, A., et.al. eds., (1977). Control Theory of Systems Governed by PDE, Academic Press, NY.
6. Balas, M., Some Trends in Large Space Structure Control Theory: Fondest Hopes, Wildest Dreams, IEEE Trans. Autom. Control (to appear).
7. Balas, M., (1980). Toward A (More) Practical Theory of Distributed Parameter System Control, Proc. Int'l. Conf. on Computers & Circuits, Port Chester, NY.
8. Kokotović, P., O'Malley, R., and Sannuti, P., (1976). Singular Perturbations & Order Reduction in Control Theory: An Overview, Automatica, 12, 123-132.
9. Balas, M., Singular Perturbations & Model Reduction Methods For Large Scale & Distributed Parameter Systems, submitted to J. Franklin Inst.
10. Balas, M., (1980). Reduced-Order Feedback Control of DPS Via Singular Perturbations Methods, Princeton Conf. on Information Sciences & Systems, Princeton, NJ; revised version submitted to SIAM J. Contr. & Opt.
11. Balas, M., (1980). Closed-Loop Stability of Large Space Structures Via Singular & Regular Perturbation Techniques: New Results, IEEE Control & Decision Conf., Albuquerque, N.M.
12. Kimura, H., (1975). Pole Assignment by Gain Output Feedback, IEEE Trans. Autom. Control, AC-20, 509-516.
13. Davison, E., and Wang, S. On Pole Assignment in Linear Multivariable Systems Using Output Feedback, Ibid, 516-518.
14. Curtain, R., and Pritchard, A., (1977). Functional Analysis in Modern Applied Mathematics, Academic Press, NY.
15. Kokotović, P., (1980). Subsystems, Time-Scales, & Multi-Modeling, IFAC Symp. on Large Scale Systems, Toulouse, France.

APPENDIX: PROOF OF LEMMA 4.2

$$\bar{H} = H_{11} - H_{12} H_{22}^{-1} H_{21}$$

$$= \begin{bmatrix} A_N + B_N Z L_{11} C_N & B_N Z L_{12} \\ L_{21}(I_P - C_R H_{22}^{-1} B_R L_{11}) C_N & L_{22} - L_{21} C_R H_{22}^{-1} B_R L_{12} \end{bmatrix}$$

where $Z = I_M - L_{11} C_R H_{22}^{-1} B_R$

& $H_{22} = A_R + B_R L_{11} C_R$.

$$F = \bar{A}_N + \bar{B}_N \bar{L}\, \bar{C}_N$$

$$= \begin{bmatrix} A_N + B_N \bar{L}_{11} C_N & B_N \bar{L}_{12} \\ \bar{L}_{21} C_N & \bar{L}_{22} \end{bmatrix}$$

where $\bar{L}_{11} = (I_M - L_{11}D_N)^{-1} L_{11}$

$\bar{L}_{12} = (I_M - L_{11}D_N)^{-1} L_{12}$

$\bar{L}_{21} = L_{21}(I_P + D_N\bar{L}_{11})$

$\bar{L}_{22} = L_{22} + L_{21}D_N\bar{L}_{12}$

& $D_N = - C_R A_R^{-1} B_R$

Therefore, $\bar{H} = F$ if and only if

(a) $Z = (I_M - L_{11}D_N)^{-1}$ and

(b) $C_R H_{22}^{-1} B_R = - D_N(I_M - L_{11}D_N)^{-1}$

However, (a) → (b):

$$- D_N(I_M - L_{11}D_N)^{-1} = - D_N Z = - D_N + D_N L_{11} C_R H_{22}^{-1} B_R$$
$$= C_R A_R^{-1}(I_R - B_R L_{11} C_R H_{22}^{-1}) B_R$$
$$= C_R A_R^{-1}(H_{22}H_{22}^{-1} - B_R L_{11} C_R H_{22}^{-1}) B_R$$
$$= C_R A_R^{-1}(H_{22} - B_R L_{11} C_R) H_{22}^{-1} B_R$$
$$= C_R A_R^{-1}(A_R) H_{22}^{-1} B_R = C_R H_{22}^{-1} B_R$$

Thus it remains to prove (a) which we recast as:

(a') $(I_M + QW)^{-1} = I_M - Q(WQ + I_R)^{-1}W$

where $Q = L_{11}C_R$ and $W = A_R^{-1} B_R$ are both bounded (finite rank) operators.

Since $(WQ + I_R)^{-1} = (I_R + A_R^{-1} B_R L_{11} C_R)^{-1} = H_{22}^{-1} A_R$ and H_{22}^{-1} is assumed to exist for small enough spillover bounds, therefore

$$Z = I_M - Q(WQ + I_R)^{-1} W$$

where the required inverse exists. Thus, to prove (a) we need only prove (a'); however, this follows directly:

$$(I_M + QW)\,(I_M - Q(WQ + I_R)^{-1}W)$$
$$= I_M + QW - Q(WQ + I_R)^{-1} W - Q[WQ(WQ + I_R)^{-1}]W$$
$$= I_M + QW - Q(WQ + I_R)^{-1} W - Q[I_R - (WQ + I_R)^{-1}]W$$
$$= I_M + QW - Q(WQ + I_R)^{-1} W - QW + Q(WQ + I_R)^{-1} W = I_M$$
$$\therefore\ (I_M + QW)^{-1} = I_M - Q(WQ + I_R)^{-1} W$$

and (a') is proved.

A SYSTEM ENGINEERING METHODOLOGY FOR EVALUATING THE BEHAVIOUR OF LARGE SCALE POWER SYSTEMS UNDER SMALL PERTURBATIONS

M. Eremia, H. Crişciu, and A. Gheorghe

Dept. of Electrical Engineering, Polytechnic Institute of Bucharest, Bucharest, Romania

ABSTRACT

A methodology is given for evaluating the behaviour of power systems under small perturbations, using analysis techniques based on the modal method - as eigenvalues localisations, and matrix sensitivity - and on linear programming methods. Besides the criterion of stability, the method permits the identification of the most important parameters for mantaining the stability and to establish critical oscillation frequencies in the system. The optimal magnitude of the automatic regulators parameters in given range is also established.

KEYWORDS

Power systems analysis; modal techniques; linear programming; general systems.

INTRODUCTION

Power system, as a part of general economic and social system may be taken as a complex system made up of sub-systems with interactions and reactions. Placement and complexity of power systems, in general, and especially of electric power systems lead to development of computers, and introduction of cybernetics as a superior step in system organizing. Electric power system ooncept is defined as a model of making of a functional assembly in order to produce, transmit, deliver and use of electric power (Fig. 1). For practical reasons it is necessary to consider electric power systems stability under small perturbations due to internal or external causes, which in turn affect the reliability of the large system under consideration. Aleatory behaviour of the consumers lead to power systems perturbations and finally one can consider that the latter can loose its stability with great technical and economic consequences. The continuity in operation of the generator groups interconnected depends upon the automatic adjustement of the equipment performances, which influence the dynamic processes in the system.

THE METHOD'S ALGORITHM

The mathematical model of electrical power system in dynamic operation to small perturbations is similar to the mathematical model of an automatic system in which the input values are those that can be modified by the means of voltage and speed regulators.

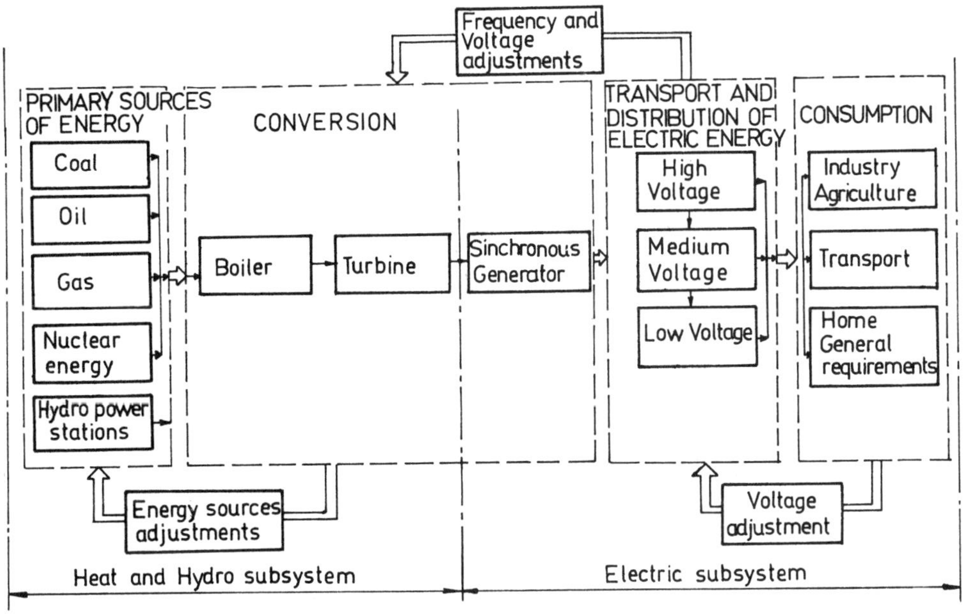

Fig. 1. Cybernetic model of an energetic system.

The system engineering methodology for evaluating the behaviour of large scale power systems in presence of small perturbations includes the following steps:

Step 1 : Representation of the operating status of the power system by a time - invariant smoth linear system:

$$\dot{X} = A \cdot X + B \cdot u(t) \qquad (1)$$

Step 2 : Transformation of the initial system into a free system by using a complet feed-back:

$$\dot{X} = A \cdot X \qquad (2)$$

where A is the coefficient matrix.

Step 3 : Application of modal analysis on the A coefficient matrix-a calculation technique of the automatic control theory - allows the determination of the dynamic answer in small amplitude perturbations, without integration of the non-linear system of equations. Developing in Taylor series of equations round of an operation point ($\Delta\dot{X} = A \cdot \Delta X$), the first term is taken into account and stability is evaluated-by calculation of eigenvalues of A matrix - in the field for which the results of the linearized equations are valable.

Step 4 : Choice of evaluating the critical parameters by means of the sensitivity matrix. From the list of the eigenvalues are selected only those which are close to the imaginary axis.

Step 5 : System eigenvalues layout for different values of critical parameters as given at step 4.

Step 6 : Calculation of the damping factor and the establishment of frequences of the existent oscillations in the analysed system in order to choose the possibilities of attenuation and to avoid the defective operation of the protective systems.

Step 7 : Establishment of optimal strategy from the stability point of view, using the Simplex linear programming method.

To bring into an unitary form the equations of the different component elements of the power system (2) and to build the matrix coefficient A, there were proposed several method (Anderson, 1973; Laughton, 1966; Undrill, 1968; Van Ness, 1968; Yu, 1972). One of these which utilizes submatrices corresponding to different basic blocks which describe the operation of the system was presented by the Eremia and Potolea (1976), and by Eremia and Gheorghe (1978). As the dynamic behaviour of the analyzed system depends upon the localisation of the eigenvalues of the matrix A in the complex plane, it is necessary to study the sensitivity coefficients of the eigenvalues to change the different elements of the matrix A. A simple method is to consider small variations to some parameters of the coefficient matrix, to recalculate the eigenvalues, and to use a subroutine for drawing the results, thus obtaining the roots loci.

THE CRITICAL PARAMETERS IDENTIFICATION

The study of the sensitivity coefficients of the eigenvalues and of the eigenvectors permits the determination of critical presumtive parameters of the system. The general expression of the coefficients or matrices of sensitivity of the order I is :

$$[S_{ik}] = \frac{\partial \lambda_i}{\partial \alpha_k} = \left[\frac{\{[\partial [A] / \partial \alpha_k](X_i, V_i)\}}{\{(X_i), (V_i)\}} \right] \tag{3}$$

in which λ_i are the eigenvalues of the matrix A, and (X_i), (V_i) are the eigenvectors of the matrix A, and of the matrix A transposed, respectively. Interpretation of the results is made by comparing the values of sensitivity coefficients. The better the obtained value, the greater the influence of the respective parameter upon the dynamic behaviour.

PARAMETERS OPTIMIZATION

It is known that for a given operation point, a way of stability improving is to move to the left the complex graphical representation of the critical eigenvalues. This leads to the minimisation of the function :

$$f(\lambda_i, c_i) = \sum_i |c_i + \lambda_i| = \min! \quad (i = 1, \ldots, n) \tag{4}$$

with the restrictions :

$$\alpha_j^{min} \leqslant \alpha_j \leqslant \alpha_j^{max} \quad (j = 1, \ldots, m) \tag{5}$$

where λ_i is the eigenvalue i of the matrix A, in function of the parameter α_j , c_i - the limit of the eigenvalue, n - the number of eigenvalues, and m - the number of system parameters. We shall use the method of simplex linear programming. Taking into consideration the first terms from the Taylor series development of (4) round an operation point, and gruping the terms we have :

$$f(\alpha_1^{p+1}, \alpha_2^{p+1}, \ldots, \alpha_m^{p+1}) = f(\alpha_1^p, \alpha_2^p, \ldots, \alpha_m^p) - \sum_j \frac{\partial f}{\partial \alpha_j} \alpha_j^p + \sum_j \frac{\partial f}{\partial \alpha_j} \alpha_j^{p+1} \tag{6}$$

Knowing the parameter values α_j^p at the operation point p, we try to find new (optimum) parameters α_j^{p+1}, which correspond to another operation point (p+1). These parameters are obtained by minimization of the function ϕ, taking into account the restrictions (5). This function may be written as :

$$\phi = \sum_j^m A_j \alpha_j^{p+1} \tag{7}$$

where

$$A_j = \frac{\partial f}{\partial \alpha_j} = \frac{\partial|\lambda_1|}{\partial \alpha_j} + \frac{\partial|\lambda_2|}{\partial \alpha_j} + \cdots + \frac{\partial|\lambda_n|}{\partial \alpha_j} \tag{8}$$

The terms $\partial|\lambda_i|/\partial\alpha_j$ represent the coefficient of the sensitivity matrix $[S_{ik}]$, which were determined above. Using the Simplex algorithm, we obtain the optimal values $\alpha_1^{p+1},\ldots,\alpha_m^{p+1}$ which minimalize the function ϕ, as well as the function f (λ_i, c_i) given by (4).

APPLIED STUDIES ON A COMPLEX POWER NETWORK STRUCTURE

The modal analysis has been applied to the study of an electrical system as represented in Fig. 2. The data of the synchronous machines and the network are calculated in the relative units as related to 114o MVA and 4oo kV.

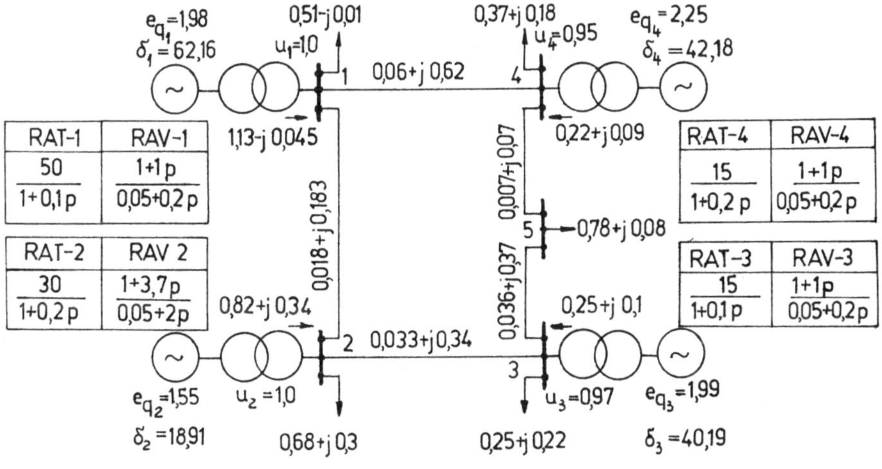

Fig. 2 An electrical power system (RAT and RAV are the transfer functions' of the voltage and speed regulator).

As critical parameters there were identified first of all T_{r1} - time constant of the voltage regulator RAT-1 - and with a small weight τ_1, τ_3, τ_4 (the promptitude of the speed regulators RAV). Using the localizing technique of the T_{r1} of the eigenvalue in the complex plane (Fig. 3) it is given their sliding mode for some values of T_{r1} , larger and smaller then the reference one. It is remarked that the system remains stable for the entire range of values (o, ol and ls).

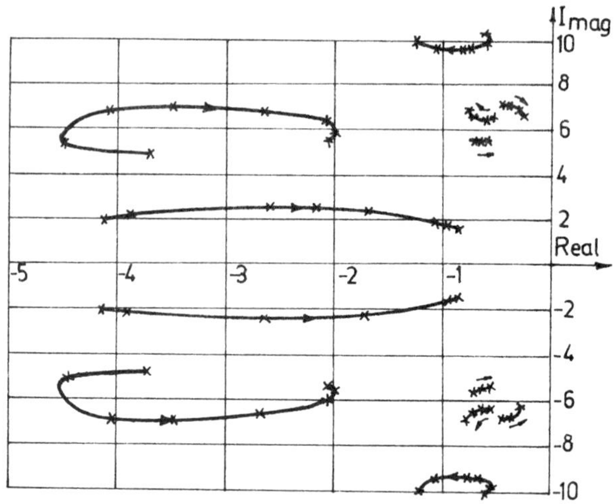

Fig. 3 Computation resultats

Critical oscillatory frequency in the system for this value range decrease from 1,lo2 Hz for T_{rl} = o,ols to o,lo6 Hz for T_{rl}= ls. The parameters optimisation for the restrictions imposed are given in Table 1.

TABLE 1 The Parameters Optimisation

		T_{rl}	τ_1	τ_3	τ_4
Initial Parameter		o,1	o,2	o,2	o,2
Restriction	maximum	o,o5	o,2	o,2	o,2
	minimum	o,3	2	2	2
Optimised Parameter		o,o5	2	2	2

REFERENCES

Anderson, J. H., D. J. Leffen, and V. M. Paina (1973). IEEE Winter Meeting, N. Y.
Eremia, M., and E. Potolea (1976). Bul. ISPE (Bucharest), 3-4, 49-71.
Eremia, M., and A. Gheorghe (1978). International Congress of Cybernetics and General Systems, Amsterdam
Kundur, P., and P. L. Dandeno (1975). PSCC, Cambridge, U. K.
Laughton, M., A. (1966). Proc. IEE, 113, 325-336.
Undrill, J. M. (1969). IEEE, Trans. PAS, 87, 835-844.
Van Ness, J. E., J. M. Boyle, and F. P. Imad (1965). IEEE Trans. Autom. Control AC-lo, 3o8-315.
Yu, Y. N., A. M. Moussa (1972). IEEE Trans. PAS, 91.
Zein El-Din, H. M., R. T. H. Alden (1977). PICA Toronto, 186-193.

STABILITY ANALYSIS OF ECOLOGICAL SYSTEMS

Helen A. Ryaciotaki-Boussalis, and Steven D. Martinez
Department of Electrical and Computer Engineering, Box 3-0
New Mexico State University
Las Cruces, New Mexico 88003

ABSTRACT

The stability of multivariable ecological systems is examined. The components (or subcommunities) of the composite system (or entire community) are allowed to display non-linear characteristics. The stability analysis of the large-scale system is performed in the Lyapunov sense.

INTRODUCTION

In recent publications (Davenport, 1973; Gardner, 1970; May, 1972, MacArthur, 1970; Siljak, 1974; Siljak, 1975), the analysis of large-scale ecological systems as pertaining to density dependent species has been based on computer simulation. These simulations have shown that an increase in the level of interaction between subcommunities leads to sudden instability in the community independent of the interconnection pattern. Linear models of n predator-prey situations have also been developed to demonstrate the complexity versus stability problem (MacArthur, 1970). Although the above techniques were useful in determining the complexity versus stability problem, they lack the mathematical generality necessary to predict stability in untried situations.

Siljak (1974) has stated that the complexity versus stability problem is appropriately formulated in the framework of the general competitive-cooperative equilibrium analysis.

The present paper examines the stability of systems involving nonlinear time-varying interactions. The stability analysis of the ecological system is based on the construction of a Lyapunov function for the entire community from the individual subcommunity Lyapunov functions. The resulting stability criteria are sufficient conditions for global stability and are presented in forms which are well suited for digital computation.

NOTATION AND DEFINITIONS

Denote by R^n the n-dimensional Euclidean space of real n-vectors. The norm $|X|$ of a n-vector $X \in R^n$ is taken to be the Euclidean norm.

$$|X| = (X^T X)^{1/2} = (\sum_{i=1}^{n} x_i{}^2)^{1/2}.$$

Let H_h denote the spherical neighborhood of the origin

$$H_h = \{ X \varepsilon \mathcal{R}^n : |X| \leq h \}$$

and let H_{h,t_0} denote the half-cylindrical neighborhood

$$H_{h,t_0} = \{ (X,t) \varepsilon \mathcal{R}^{n+1} : |X| \leq h, \ t \geq t_0 \}$$

of the t-axis in the motion space where h is some finite constant.

Definition 1. An n-dimensional function F(X,t) of the n-dimensional vector X and t is said to belong to class $\mathcal{B}$ in H_{h,t_0} if for all $(X,t) \varepsilon H_{h,t_0}$ the following conditions are satisfied:

- F(X,t) is continuous in X and t, and
- has continuous first partial derivatives with respect to the components X_i of X.

Consider the system of differential equations

$$\dot{X} = F(X,t) \ , \quad F(0,t) \equiv 0. \tag{1}$$

Definition 2. The solution X = 0 is exponentially stable if there exist two constants $\alpha > 0$ and $\beta > 0$, independent of the initial values, such that for sufficiently small initial values the following inequality holds:

$$| X(t; X_0, t_0) | \leq \beta |X_0| \exp [-\alpha(t-t_0)]. \tag{2}$$

If (2) is satisfied for all initial values in a region H_{h,t_0}, then (1) is said to be exponentially stable in $H_{h_0,\tau}$. If (2) is satisfied for all initial conditions then (1) is said to be exponentially stable in the large.

Results which yield conditions for stability and instability in the sense of the above definitions involve the existence of functions V(X,t) which satisfy any of Lyapunov's stability and instability theorems and are called Lyapunov functions. Let a positive definite function V(X,t) be differentiable function of X and t. Then the total derivative of V(X,t) with respect to t along the solutions of equation (1) is given by

$$\dot{V} \equiv \frac{\partial V(X,t)}{\partial t} + [\nabla V(X,t)]^T \dot{X}$$

where

$$\nabla V(X,t) = [\frac{\partial V}{\partial X_1} \ \frac{\partial V}{\partial X_2} \ . \ . \ . \frac{\partial V}{\partial X_n}]^T. \tag{3}$$

If the system (1) is exponentially stable, then under certain conditions it is possible to define a Lyapunov function which satisfies certain estimates which can be used to establish a lower bound on the rate of decay of the system solution. These concepts are stated in the following theorem, the proof of which is given by Ryaciotaki-Boussalis (1979).

Theorem 1. Consider the n-dimensional system of differential equations (1).

(i) If there exist a positive definite scalar function $V(X,t)$ and a set of positive constants c_1, c_2, and c_3 satisfying (a) and (b) of the following relations

$$\begin{aligned} &(a) \quad c_1|X|^2 \leq V(X,t) \leq c_2|X|^2 \\ &(b) \quad \dot{V}(X,t) \leq -c_3 V(X,t) \\ &(c) \quad |\nabla V| \leq c_4|X| \end{aligned} \tag{4}$$

for all $(X,t) \in H_{h,\tau}$, $\tau > 0$, then the system (1) is exponentially stable for all $X_0 \in H_{h_0}$.

(ii) If $F(X,t) \in \mathcal{B}$ in $H_{h,\tau}$, $\tau \geq 0$, and if (1) is exponentially stable for all $X_0 \in H_{h_0}$, then there exist a positive definite function $V(X,t)$ and associated positive constants satisfying (4) for all $(X,t) \in H_{h,\tau}$.

SYSTEM STABILITY

Consider a system S described by

$$\dot{X} = F(X,t) + AG(X,t). \tag{5}$$

Let the i-th component be given by equations of the form

$$\dot{X}_i = F_i(X_i,t) + A_i G_i(X_i,t) \tag{6}$$

where $i=1,2,\ldots,N$, $X \in \mathcal{R}^n$, $n = \sum_{i=1}^{N} n_i$, and $X_i \in \mathcal{R}^{n_i}$.

Theorem 2. Consider the ecological system (5) composed of subcommunities of the form (6). If the following conditions hold:

(i) For each component (6) there exist Lyapunov functions $V_i(X_i,t)$ and positive constants c_{i1}, c_{i2}, c_{i3}, and c_{i4} satisfying constraints as given in (4);

(ii) there are positive constants $\gamma_i > 0$ such that

$$|G_i(X_i,t)| \leq \gamma_i \; |X_i| \; ; \tag{7}$$

(iii) there exists a parameter vector $K = [k_1, k_2, \ldots, k_N]^T$ with positive entries such that $c_1 > 0$, $c_2 > 0$, and $c_3 > 0$ where

$$\begin{aligned} c_1 &= \min_i \; (k_i c_{i1}) \\ c_2 &= \max_i \; (k_i c_{i2}) \\ c_3 &= \min \; (k_i c_{i3} - k_i \gamma_i \frac{c_{i4}}{c_{i2}} |A_i|) \end{aligned} \tag{8}$$

then the interconnected system (5) is exponentially stable.

Proof: Let

$$V(X,t) = \sum_{i=1}^{N} k_i V_i(X_i,t) \tag{9}$$

be a Lyapunov function for system (5). Then from (4a) and (4b) one gets

$$\sum_{i=1}^{N} k_i c_{i1} |X_i|^2 \leq V(X,t) \leq \sum_{i=1}^{N} k_i c_{i2} |X_i|^2$$

or $$c_1 |X|^2 \leq V(X,t) \leq c_2 |X|^2$$

where $$c_1 = \min_i \{ k_i c_{i1} \} > 0 , \quad c_2 = \max_i \{ k_i c_{i2} \} > 0. \tag{10}$$

The gradient of (9) is given as

$$|\nabla V(X,t)| = \left(\sum_{i=1}^{N} |\nabla V_i(X_i,t)|^2 \right)^{1/2}$$

or $$|\nabla V(X,t)| \leq \left(\sum_{i=1}^{N} (k_i c_{i4} |X_i|^2 \right)^{1/2} \leq c_4 |X|$$

where $$c_4 = \max_i \{ k_i c_{i4} \} > 0. \tag{11}$$

The total derivative of (9) with respect to system (5) is

$$\dot{V}_S(X,t) = \sum_{i=1}^{N} \{ k_i (\frac{\partial V_i}{\partial t} + (\nabla V_i)^T F_i(X_i,t)) + k_i (\nabla V_i)^T A_i G_i(X_i,t) \}$$

$$\leq \sum_{i=1}^{N} k_i (\dot{V}_i(X_i,t) + |\nabla V_i|^T \; |A_i G_i(X_i,t)|. \tag{12}$$

Concequently applying (4b),(4c), and (7) into (12) yields

$$\dot{V}_S(X,t) \leq \sum_{i=1}^{N} k_i (-c_{i3} V_i(X_i,t) + \gamma_i c_{i4} |X_i|^2 |A_i|)$$

$$\leq \sum_{i=1}^{N} -k_i (\; (c_{i3} - \gamma_i \frac{c_{i4}}{c_{i2}} |A_i| \;) \; V_i(X_i,t) \;)$$

or $$V(X,t) \leq -c_3 V(X,t)$$

where c_3 is given by (8). If $c_3 > 0$ Theorem 1 is satisfied and the ecological system (5) is exponentially stable.

Theorem 3. Consider the interconnected system (5) with components as in (6). If the following conditions are satisfied:

(i) For each subcommunity (6) there exist Lyapunov functions $V_i(X_i,t)$

and positive constants c_{i1}, c_{i2}, c_{i3}, and c_{i4} satisfying constraints as given in (4);

(ii) $(\nabla V_i)^T A_i G_i(X_i,t) \leq -c_{i4} V_i(X_i,t)$; (13)

(iii) there exists a parameter vector $K = [k_1, k_2, \ldots, k_N]^T$ with positive entries such that $c_1 > 0$, $c_2 > 0$, and $c_3 > 0$ where

$$c_1 = \min_i (k_i c_{i1})$$
$$c_2 = \max_i (k_i c_{i2}) \tag{14}$$
$$c_3 = \min_i (k_i (c_{i3} + c_{i4}))$$

then the composite system (5) is exponentially stable.

Proof: Let

$$V(X,t) = \sum_{i=1}^{N} k_i V_i(X_i,t) \tag{15}$$

be a Lyapunov function for the ecological system (5). The constants c_1, c_2, and c_4 are given by (10) and (11) and are positive. Taking the total derivative of (15) with respect to system (5), one obtains

$$\dot{V}(X,t) \leq \sum_{i=1}^{N} k_i [\dot{V}_i(X_i,t) + (\nabla V_i)^T A_i G_i(X_i,t)] \tag{16}$$
$$\leq -\sum_{i=1}^{N} k_i [c_{i3} V_i(X_i,t) + c_{i4} V_i(X_i,t)] \leq -c_3 V(X,t)$$

where c_3 is given by (14). If $c_3 > 0$ then V(X,t) satisfies Theorem 1 and the ecological system (5) is exponentially stable.

CONCLUTIONS

Lyapunov-type functions are utilized to obtain sufficient conditions for exponential stability for a class of ecological systems. The results are obtained in terms of subcommunity characteristics and interconnection properties, and are presented in forms well suited for computational purposes.

ACKNOWLEDGMENTS

The authors would like to thank Dr. O.A. Davenport for his useful comments on the subject.

REFERENCES

Davenport, O. A. (1973). Structuring and Modeling Insect Ecosystems. Ph.D. Dissertation, New Mexico State University, Las Cruces, New Mexico.

Gardner, M. R., and W. R. Ashby. (1970). Connectedness of Large Dynamical Systems: Critical Values for Stability. Nature 228, p. 784.

May, R. M. (1972). Will a Large Complex System be Stable? Nature 238, pp. 413-414.

MacArthur, R. H. (1970). Species Packing and Competitive Equilibrium for Many Species. Theor. Popul. Biol. 1, pp. 1-11.

Ryaciotaki-Boussalis, H. A. (1979). Stability of Large-Scale Systems. Ph.D. Dissertation, New Mexico State University, Las Cruces, New Mexico.

Siljak, D. D., (1974). Connective Stability of Complex Ecosystems. Nature 249, p.280.

Siljak, D. D., (1975). When is a Complex Ecosystem Stable? Mathematical Biosciences 25, pp. 25-50.

COMPUTATIONAL METHODS FOR DISTRIBUTED PARAMETER SYSTEMS: IDENTIFICATION AND CONTROL

H. T. Banks
Lefschetz Center for Dynamical Systems
Division of Applied Mathematics
Brown University
Providence, R.I. 02912

ABSTRACT

Approximation methods for functional and partial differential equations are discussed in the context of parameter identification and optimal control problems. Theoretical and numerical results are presented.

KEYWORDS

Parameter identification; optimal control; delay systems; distributed parameter systems; computational techniques.

SUMMARY

In this lecture we discuss our recent efforts on development of approximation schemes for linear and nonlinear infinite-dimensional state systems. Included in our considerations are functional and partial differential equations such as

$$\dot{x}(t) = f(q_1, x^t, x(t), x(t-\tau_1), \ldots, x(t-\tau_\nu), u(t)) \tag{1}$$

(here x^t denotes the function $x^t(\theta) = x(t+\theta)$, $-r \leq \theta \leq 0$),

$$v_{tt} = q_1 v_{xx} + q_2 v_t + q_3 v + f(q_4, t, x, v), \tag{2}$$

and

$$v_t = q_1 (p v_x)_x + q_2 v + f(q_4, t, x, v), \tag{3}$$

where appropriate boundary and/or initial conditions (possibly also depending on unknown parameters q_i and control functions u) are chosen for (1), (2) and (3). Our approach entails rewriting these systems as abstract evolution equations

$$\dot{z}(t) = \mathcal{A}(q) z(t) + F(q, t, z(t), u(t))$$

in appropriately chosen Hilbert spaces $Z(q)$ and then approximating this system in finite-dimensional subspaces $Z^N(q)$ (e.g., a Ritz-Galerkin approach). The vector parameter q is, in general, an unknown parameter to be determined from "observations" of the systems. For linear systems the proofs of convergence and establishment of error estimates are facilitated by results (Trotter-Kato type approximation theorems) from linear semigroup theory.

The methods discussed for (1) are based on the general (arbitrary-order) spline approximation theory for linear and nonlinear functional differential equations developed in (Banks and Kappel, 1979) and (Banks, 1980) along with the general ideas for optimal control of nonlinear delay systems given in (Banks, 1979). As we shall illustrate with numerical examples, our methods can also be used to identify the delays τ_i in (1) from data. The underlying theory for such problems has been developed in (Banks, Burns and Cliff, 1980) and (Banks and Daniel, 1980).

For a class of systems of which (2) and (3) are special cases, we discuss the general methods developed in (Banks and Kunisch, 1980a; Banks and Kunisch, 1980b). In this situation the theoretical results are established by employing the Trotter-Kato type approximation ideas along with the perturbation techniques used in (Banks, 1979). Numerical findings for schemes based on modal (eigenfunction) and cubic spline approximation subspaces Z^N are presented.

ACKNOWLEDGEMENT

The research reported has been carried out during the past several years and has been supported in part by continuing research grants from the National Science Foundation (NSF-MCS7905774), the Air Force Office of Scientific Research (AF-AFOSR-76-3092C), and the U.S. Army Research Office (ARO-DAAG29-79-C-0161). Two of my graduate students, Patti Daniel and Jim Crowley, have contributed in a significant way to the numerical investigations and I would like to express my appreciation for their continuing efforts on these problems.

REFERENCES

Banks, H.T. (1979). Approximation of nonlinear functional differential equation control systems. J. Optimiz. Theory and Appl., 29, 383-408.

Banks, H.T., and Kappel, F. (1979). Spline approximations for functional differential equations. J. Differential Eqns., 34, 496-522.

Banks, H.T. (1980). Identification of nonlinear delay systems using spline methods, Proc. Int'l. Conf. on Nonlinear Phenomena in the Math. Sciences (V. Lakshmikantham, ed.), Academic Press, to appear.

Banks, H.T., Burns, J.A., and Cliff, E.M. (1980). Parameter estimation and identification for systems with delays, SIAM J. Control and Optimiz., to appear.

Banks, H.T., and Daniel, P. (1980). Estimation of delays and other parameters in nonlinear functional differential equations, to appear.

Banks, H.T., and Kunisch, K. (1980a). Parameter estimation techniques for nonlinear distributed-parameter systems, Proc. Int'l Conf. on Nonlinear Phenomena in the Math. Sciences (V. Lakshmikantham, ed.), Academic Press, to appear.

Banks, H.T. and Kunisch, K. (1980b). An approximation theory for nonlinear partial differential equations with applications to identification and control, to appear.

A COMPENSATION METHOD FOR SLOWLY TIME-VARYING DISCRETE DYNAMIC SYSTEMS

M. de la Sen* and S. Dormido**
Dpt. de Automática e Informática, Fac. C. Físicas

*Universidad de Bilbao, Aptdo. 644. Lejona (Vizcaya), Spain

**Universidad Complutense. Madrid-3, Spain

ABSTRACT

A solution is proposed to compensate slowly time-varying discrete dynamic systems using an additional input which is applied during an adequate sampling period. The compensation is achieved from interval to interval, and therefore the method may be applied in real time. For this purpose we introduce the concept of adaptive reference model in the sampling period, whose matrices A_k and B_k are calculated at the nominal parameter vector of the transfer function, but in general its sampling intervals are not periodic. Finally, some examples are shown that experimentally verify the theoretical results previously obtained.

KEYWORDS

Adaptive sampling, discrete dynamic systems, adaptive reference model, compensation to plant parameter variation, compensation in real time.

INTRODUCTION

In many practical cases, alterations in the environment induce undesirable changes in the vector which parameterizes the continuous transfer function of a discrete dynamic system. In this situation, the sampling period may be used as a general compensator in the problem of compensating perturbations in the parameters of the system (Dormido, de la Sen and Mellado, 1979). This is of special relevance when a digital controller is designed in order to satisfy a set of specification given a priori. Under these conditions, a change in the parameters, keeping the sampling period constant, normally leads to design the compensator again and therefore increasing the cost.

The purpose of this paper is to give a method to compensate the undesirable changes indicated above by using an adaptation in the sampling period from interval to interval and an additional external input. The only and very reasonable condition assumed on the parameter vector is that it is slowly time-varying and, therefore, it may be considered constant over at least one sampling interval.

GENERAL EXPRESSION FOR THE TRANSMISSION OF THE PERTURBATIONS IN THE PARAMETERS OF A SAMPLED DATA SYSTEM.

Let the linear discrete time system be: $\underline{x}(k+1) = A\,\underline{x}(k) + B\,\underline{r}(k)$ (1)
where $\underline{x}(k)$ and $\underline{r}(k)$ are the state n-vector and the external input p-vector in the sampling instant kT. A and B are constant real matrices of appropriate dimensions. If there are perturbations in the parameters of the system, they are slow enough not to have influence on the last sampling.

Let the perturbed system in both plant parameter and period described by:

$$\hat{\underline{x}}(t_{k+1}) = \hat{A}_k\,\hat{\underline{x}}(t_k) + \hat{B}_k\,r(t_k) \tag{2}$$

$$\text{where } \hat{A}_k = A + \Delta A_k \qquad \hat{B}_k = B + \Delta B_k \tag{3}$$

$\Delta(.)$ denotes the variation in the corresponding matrix in the last sampling interval.

In equation (3) it must be taken into account that:

$$\Delta A_k = \delta A_k + \delta^2 A_k + \delta^3 A_k + \ldots.$$
$$\Delta B_k = \delta B_k + \delta^2 B_k + \delta^3 B_k + \ldots.$$

where $\delta^i A_k (\delta^i B_k)$ is the i-order variation for the increment $\Delta A_k (\Delta B_k)$ in the (k+1)th sampling interval.

By considering only the variations of first order for the increments $\Delta(.)$ and neglecting the high order terms, Eq. (2) becomes:

$$\hat{\underline{x}}(t_n) \cong \sum_{j=0}^{n} A^j \delta A_{n-j-1} A^{n-j-1}\,\hat{\underline{x}}(t_0) + \Big[\sum_{k=0}^{n-3} A^{n-k-1}\,\delta B_k + (A^{n-k-1} + \delta A_{n-1}\,A^{n-k-2})B + \sum_{j=1}^{n-k-2} A^j \delta A_{n-j-1} A^{n-j-k-3}B\Big]\,\underline{r}(t_k) + (A\delta B_{n-2} + (A+\delta A_{n-1})B)\underline{r}(t_{n-2}) + (\delta B_{n-1} + B)\underline{r}(t_{n-1}) \tag{4}$$

where $A^j \equiv 0$ if $j < 0$.

The equation which gives the error between the states of the perturbed systems in the parameters and the nominal one at the new sampling instants can be easily obtained as follows. Let $\tilde{\underline{x}}(t_k)$ be the error at the instant t_k, and $\underline{x}(t_k)$ the state of the original system at the new sampling instants, and $\hat{\underline{x}}(t_k)$ the state of the actual system at the t_k-th instant.

$$\tilde{\underline{x}}(t_k) = \hat{\underline{x}}(t_k) - \underline{x}(t_k) \tag{5}$$

where

$$\tilde{A}_k = \tilde{A}(\underline{p}_k, T_k) = \hat{A}(\underline{p}_k, T_k) - A(\underline{p}, T_k) = \hat{A}_k - A_k \tag{6}$$

$$\tilde{B}_k = \tilde{B}(\underline{p}_k, T_k) = \hat{B}(\underline{p}_k, T_k) - B(\underline{p}, T_k) = \hat{B}_k - B_k$$

where $\underline{p}_k$ is the value of the vector of parameter at the instant t_k, $\underline{p}$ its nominal value and $T_k = t_{k+1} - t_k$.

From (5) and (6) it is easily followed that:

$$\tilde{\underline{x}}(t_{k+1}) = \tilde{A}(\underline{p}_k,T_k)\underline{x}(t_k) + \hat{A}(\underline{p}_k,T_k)\tilde{\underline{x}}(t_k) + \tilde{B}(\underline{p}_k,T_k)\underline{r}(t_k) \tag{7}$$

Applying (7) iteratively leads to

$$\tilde{\underline{x}}(t_{k+1}) = \sum_{j=0}^{k} \prod_{i=j+1}^{k} (\hat{A}(\underline{p}_i,T_i))\tilde{A}(\underline{p}_j,T_j)\underline{x}(t_j) + \prod_{i=0}^{k} (\hat{A}(\underline{p}_i,T_i))\tilde{\underline{x}}(t_0) +$$

$$+ \sum_{j=0}^{k} \prod_{i=j+1}^{k} (\hat{A}(\underline{p}_i, T_i)) \tilde{B}(\underline{p}_j, T_j) \underline{r}(t_j) \quad (8)$$

DYNAMIC CALCULATION OF THE SAMPLING INTERVAL BY EMPLOYING AN "ADAPTIVE REFERENCE MODEL IN THE SAMPLING PERIOD".

In Eq.(5), $\underline{x}(t_k)$ can be considered as the state at $t = t_k$ of an adaptive reference model in the sampling period for the actual system. This is because the matrices (6) are evaluated at the new instant with the nominal parameters (Landau,1974,1976)

$$\underline{x}(t_{k+1}) = A(\underline{p},T_k)\underline{x}(t_k) + B(\underline{p},t_k)\underline{r}(t_k) = A_k\underline{x}(t_k) + B_k\underline{r}(t_k) \quad (9)$$

In the following it is assumed that the expressions (6) are derivable at the nominal values of both the vector parameter and period. In a first order approximation they can be written as:

$$\tilde{A}(\underline{p}_k, T_k) \simeq (\partial A(\underline{p}, T_k)/\partial \underline{p}) \cdot \Delta \underline{p}_k$$

$$\tilde{B}(\underline{p}_k, T_k) \simeq (\partial B(\underline{p}, T_k)/\partial \underline{p}) \cdot \Delta \underline{p}_k$$

Since our aim is to compensate the variations of the parameter vector by using proper variations at the sampling period, the modified state $\hat{\underline{x}}(t_k)$ and the state $\underline{x}(t_k)$ of the adaptive reference model in the sampling period must be very similar. The increment ΔT_{k+1} is obtained minimizing in each interval the following cost function:

$$J_{k,k+1} = \tilde{\underline{x}}^t(t_{k+1})Q_k \tilde{\underline{x}}(t_{k+1}) + \lambda_k(\Delta T_{k+1})^2 \quad (10)$$

where Q_k is an nxn positive semidefinite matrix, λ_k is a non-negative real number and the superscript t denotes transposition. The first term in (10) is a non-decreasing function of the error that weighs large errors much more heavily than small errors. The second term weighs the cost of varying the sampling interval. After some calculations it is deduced

$$\Delta T_{k+1} = -(\tilde{\underline{x}}_1(t_k)^t Q_k \tilde{\underline{x}}_2(t_k)) \;/\; (\tilde{\underline{x}}_2(t_k)^t Q_k \tilde{\underline{x}}_2(t_k) + \lambda_k) \text{ , where} \quad (11)$$

$$\tilde{\underline{x}}_1(t_{k+1}) = \frac{\partial A(\underline{p},T)}{\partial \underline{p}} \Delta\underline{p}_k . \underline{x}(t_k) + \hat{A}(\underline{p}_k,T).\tilde{\underline{x}}(t_k) + \frac{\partial B(\underline{p},T)}{\partial \underline{p}} \Delta\underline{p}_k . \underline{r}(t_k)$$

$$\tilde{\underline{x}}_2(t_{k+1}) = \frac{\partial^2 A(\underline{p},T)}{\partial \underline{p}.\partial T} \Delta\underline{p}_k . \underline{x}(t_k) + \frac{\partial A(\underline{p}_k,T)}{\partial T} \tilde{\underline{x}}(t_k) + \frac{\partial^2 B(\underline{p},T)}{\partial \underline{p}.\partial T} \Delta\underline{p}_k . \underline{r}(t_k) \quad (12)$$

APPLICATION OF AN EXTERNAL ADDITIONAL INPUT

Now a solution is proposed to compensate slowly time-varying discrete dynamic systems using an additional input which is applied for an adequate sampling period. As in the previous case, the compensation is achieved from interval to interval. Consider the state equations:

$$\hat{\underline{x}}(t_{k+1}) = \hat{A}_k \hat{\underline{x}}(t_k) + \hat{\underline{b}}_k r(t_k) + \underline{1}_k v(t_k) \quad (13)$$

$$\underline{x}(t_{k+1}) = A_k \underline{x}(t_k) + \underline{b}_k r(t_k) \quad (14)$$

Eq. (13) describes the compensated system (the compensator is achieved modifying the sampling interval and applying an external additional input $v(t_k)$ simultaneously). Without loss of generality the external input $r(t_k)$ and $v(t_k)$ are assumed to be scalar. Eq. (14) describes the adaptive reference model in the sampling period, whose matrices A_k and $\underline{b}_k$ are calculated at the nominal parameter vector.
We choose as a cost function to achieve the compensation:

$$J_{k,k+1} = \underline{\tilde{x}}^t(t_{k+1})Q_k\, \underline{\tilde{x}}(t_{k+1}) + \lambda_k\, v(t_k)^2 \tag{15}$$

where Q_k and λ_k are a symmetrical positive definite matrix and a non-negative scalar respectively. From (5), (13), (14) and (15) we obtain:

$$J_{k,k+1} = \underline{\hat{x}}^t(t_k)\tilde{A}_k^t Q_k \tilde{A}_k\, \underline{\hat{x}}(t_k) + \underline{\tilde{x}}^t(t_k)\, A_k^t\, Q_k A_k\, \underline{\tilde{x}}^t(t_k) + \underline{\tilde{b}}_k^t Q_k \underline{\tilde{b}}_k (r(t_k))^2 +$$

$$+ (\underline{1}_k^t Q_k \underline{1}_k + \lambda_k)(v(t_k))^2 + 2(\underline{\tilde{x}}^t(t_k)A_k^t Q_k \tilde{A}_k \underline{\hat{x}}(t_k) + \underline{\hat{x}}^t(t_k)\tilde{A}_k^t Q_k \underline{\tilde{b}}_k r(t_k) + \underline{\hat{x}}^t(t_k)\tilde{A}_k^t Q_k \underline{1}_k v(t_k) +$$

$$+ \underline{\tilde{x}}^t(t_k)A_k^t\, Q_k \underline{\tilde{b}}_k\, r(t_k) + \underline{\tilde{x}}^t(t_k)A_k^t Q_k\, \underline{1}_k\, v(t_k) + \underline{\tilde{b}}_k^t\, Q_k\, \underline{1}_k\, r(t_k)v(t_k)) \tag{16}$$

where $\tilde{A}_k = \hat{A}_k - A_k$ and $\underline{\tilde{b}}_k = \underline{\hat{b}}_k - \underline{b}_k \quad k = 1,2,\ldots$

The optimum $v(t_k)$, with no restrictions upon it, would be obtained by direct calculation of the gradient of $J_{k,k+1}$ with respect to $v(t_k)$ from (16). That yields:

$$v(t_k)^* = -(\underline{1}_k^t Q_k \underline{1}_k + \lambda_k)^{-1}\ \underline{1}_k^t Q_k (\tilde{A}_k \underline{\hat{x}}(t_k) + A_k \underline{\tilde{x}}(t_k) + \underline{\tilde{b}}_k r(t_k)) \tag{17}$$

From (16) and (17) we obtain:

$$J^*_{k,k+1} = \underline{\hat{x}}^t(t_k)\tilde{A}_k^t P_k A_k \underline{\hat{x}}(t_k) + \underline{\tilde{x}}^t(t_k)A_k^t P_k A_k \underline{x}(t_k) + \underline{\tilde{b}}_k^t\ P_k \underline{\tilde{b}}_k (r(t_k))^2 +$$

$$+ 2(\underline{\tilde{x}}^t(t_k)A_k^t P_k \tilde{A}_k \underline{\hat{x}}(t_k) + \underline{\tilde{x}}^t(t_k)A_k^t P_k \underline{\tilde{b}}_k r(t_k) + \underline{\hat{x}}^t(t_k)\tilde{A}_k^t P_k \underline{\tilde{b}}_k r(t_k)) \tag{18}$$

where $P_k = Q_k(I - \underline{1}_k(\underline{1}_k^t\ Q_k\ \underline{1}_k + \lambda_k)^{-1}\ \underline{1}_k^t\ Q_k)$, $k = 0,1,\ldots$ and I is the (nxn) identity matrix.

An optimization process in the sampling interval is used to improve the compensation. It must be noted from (17) that $v(t_k)$ is dependent on the corresponding sampling interval and that $\underline{1}_k^t\ Q_k\ \underline{1}_k + \lambda_k \neq 0$ for all k must be imposed on Q_k and λ_k as a previous condition.

<u>Example of Application</u>. The method proposed above has been used to compensate a free ($u_k = 0$, for all k) second order discrete system, whose plant matrix is in the real canonical form with a pair of complex eigen-values -0.1 ± i 0.05.

$$\underline{x}(k+1) = \begin{pmatrix} -.1 + .05\,i & 0 \\ 0 & -.1 - .05\,i \end{pmatrix} \underline{x}(k)$$

The imaginary part is assumed to be constant and the real one linearly increasing with time with slope .01 ($a(t) = -.1 + .01\,t$). Q_k and λ_k are taken as the identity matrix and zero respectively. The initial state is $\underline{x}(0) = (1,1)^t$, and the nominal sampling interval is assumed to be $T_0 = 1$ second.

For simplicity in the calculations, $\underline{1}_k = (1,1)^t$ for all k, that is to say the compensation is achieved by adding the state variables and the additional input $v(t_k)$. In Eq. (17) estimations for $\underline{\hat{x}}(t_k)$ and $\underline{\tilde{x}}(t_k)$ based on the previous measures $\underline{\hat{x}}(t_{k-1})$ and $\underline{\tilde{x}}(t_{k-1})$, and an evaluation of the parameter vector at the k-1 sampling instant have been used. In this way, this method may be applied in real time and there is no compensation over the first and second sampling intervals. Table 1 shows the simulation results for the state variables x_1 and x_2 in the nominal perturbed and compensated systems for a time interval $[0,5]$. The maximum variation in the sampling interval has been restricted to ± 20 % of its nominal value.

CONCLUSION

In this paper, a method to compensate discrete systems whose continous transfer functions have not constant parameters due to undesirable perturbations has been given. It has been shown that if the variations in the parameters above are slowly time-varying, an additional input may be applied during an adequate sampling interval to achieve the compensation.

TABLE 1

TIME	NOMINAL SYSTEM		PERTURBED SYSTEM		COMPENSATED SYSTEM	
	x_1	x_2	x_1	x_2	x_1	x_2
0.0	1.000	1.000	1.000	1.000	1.000	1.000
0.4	0.9703	0.9511	0.9711	0.9519	0.9711	0.9519
0.8	0.9323	0.9138	0.9353	0.9168	0.9353	0.9168
1.2	0.9368	0.8321	0.9453	0.8382	0.9453	0.8382
1.6	0.9175	0.7813	0.9293	0.7914	0.9293	0.7914
2.0	0.8964	0.7329	0.9145	0.7477	0.9145	0.7477
2.4	0.8751	0.6868	0.9007	0.7069	0.8874	0.6936
2.8	0.8539	0.6429	0.8880	0.6685	0.8747	0.6554
3.2	0.8326	0.6012	0.8763	0.6328	0.8427	0.5992
3.6	0.8113	0.5615	0.8656	0.5991	0.8321	0.5661
4.0	0.7901	0.5238	0.8559	0.5674	0.8020	0.5145
4.4	0.7691	0.4880	0.8472	0.5376	0.7934	0.4861
4.8	0.7481	0.4540	0.8395	0.5094	0.7858	0.4592
5.0	0.7377	0.4376	0.8360	0.4959	0.7477	0.4113

REFERENCES

Dormido, S., de la Sen, M. and Mellado, M. (1979). Sensitivity compensation and adaptive sampling. Proceedings INFO II, Vol. 2, 37-44. Patras (Greece).

Landau, I.D. (1974) Automatica, 10, 353-379.

Landau, I.D. (1976) IEEE Trans. Autom. Control, 21, 194-202.

DESCRIBING FUNCTION METHODS FOR HIGH-ORDER HIGHLY NONLINEAR SYSTEMS

James H. Taylor

School of Mechanical and
Aerospace Engineering
Oklahoma State University
Stillwater, Oklahoma 74078

ABSTRACT

Applying the modern systems approach to studying the "quality of life" often entails developing and using complicated dynamic models (high-order, highly nonlinear differential equations). Dealing with mathematical models of this sort can be difficult. In some instances, however, these models can be effectively studied using recent advances in describing function theory. These methods are described and illustrated in this paper.

KEYWORDS

Large scale systems; highly nonlinear systems; describing function methods; limit cycle analysis; statistical performance analysis; nonlinear covariance analysis.

INTRODUCTION

The complexity of problems facing modern society is rapidly outstripping man's capability to manage them on the basis of "intuition" and the old "rules of thumb" that have been successful in the past. The systems theoretic answer to this difficulty is to develop analytical models (usually dynamic) that account for the complexity of the system, and study the resulting model behavior via analytical methods or simulation. Analysis and simulation are the usual bases for both validating the model (determining if the model is realistic) and for problem solving (determining how to control the system or modify the system so that it behaves more appropriately).

Many models that result from such activity are differential equations that are of high order (having many state variables) and nonlinear. In some circumstances, the nonlinearity can be neglected, and well-established linear system analysis and design methods can be used for predicting system behavior and for problem solving. Often, however, nonlinearity cannot be neglected and the systems engineer feels forced to rely solely on simulation as a tool for performance evaluation and design.

The last decade has seen substantial progress in extending quasi-linearization or <u>describing function techniques</u> to permit the analytic study of high-order highly nonlinear dynamic systems models. Three areas are especially noteworthy:

- <u>Limit Cycle Analysis</u> - Until recently, nonlinear systems that exhibit limit cycle (oscillatory) behavior could not be analyzed unless they were low-order or contained only one dominant nonlinearity. New extensions of sinusoidal-input describing function (SIDF) theory (Taylor, 1975, 1980; Hannebrink, Lee, Weinstock and Hedrick, 1977) permits systems of any order, with any number of nonlinear effects, to be treated analytically.

- Sinusoidal Input Response Analysis - The same generalized SIDF approach allows the direct determination of the response of complicated nonlinear systems to sinusoidal inputs (Taylor and Mohan, 1980).

- Random Input Response Analysis - Many systems are driven by random inputs (e.g. first-order Markov processes or white noise) or have random parameters. The response or performance of such a system is generally described statistically (e.g. rms vibration levels may provide a measure of ride quality in a train or bus). The statistical behavior of high-order nonlinear systems can now be determined by an analytic technique (Kazakov, 1965; Gelb and Warren, 1973) that combines covariance analysis with random-input describing functions (RIDF's); this method has been proven to be much more effective than monte carlo simulation in many instances.

All of the relevant describing function techniques are outlined in this paper.

QUASILINEARIZATION/DESCRIBING FUNCTION METHODS

DF Definitions

The basic idea of the describing function (DF) approach for studying nonlinear system behavior is to replace each system nonlinearity with a linear term whose "gain" is a function of "input amplitude", where the type of input signal is assumed in advance; this concept is dealt with very thoroughly in Gelb and Vander Velde (1968) and Atherton (1975). In this paper, two cases are considered:

Sinusoidal-Input Describing Functions (SIDF's)

$$f(b + a \sin \omega t) \cong f_o(a,b) + n_s(a,b) \cdot a \sin \omega t \tag{1}$$

Random-Input Describing Functions (RIDF's)

$$f(m + r) \cong \hat{f}(m,\sigma) + n_r(m,\sigma) \cdot r, \qquad r \sim N(0,\sigma) \tag{2}$$

[m is the mean value of the input, r is a zero-mean gaussian random variable with standard deviation or rms value σ.]

The DF elements (f_o, n_s or $\hat{f}$, n_r) are mathematically formulated to minimize the approximation error in (1) and (2). The usefulness of DF methods lies in the subsequent treatment of the resulting quasi-linear model using linear system analytic techniques, which are well established and usually very straightforward to apply. The power of DF methods is derived from the amplitude-dependence of the DF elements, which accounts for one of the basic effects of nonlinearity. Standard linearization (small-signal or Taylor series linearization) fails to capture this essential property of nonlinear phenomena. A wide variety of DF's, both sinusoidal- and random-input, are catalogued in Gelb and VanderVelde (1968) and Atherton (1975), so we will not consider that aspect of DF theory further.

Limit Cycle Analysis

To illustrate the fundamentals of limit cycle analysis, consider the following autonomous nonlinear differential equation:

$$\dddot{x} + 2\ddot{x} + [2\dot{x} + \mathrm{sgn}(\dot{x})] + [3x + x^3] = 0 \tag{3}$$

The two nonlinearities are quasilinearized as follows: Assume that $x \cong a \sin \omega t$, implying that $\dot{x} \cong a\omega \cos \omega t$. Then (Atherton, 1975)

$$x^3 \cong \frac{3}{4} a^2 \cdot a \sin \omega t$$

$$\mathrm{sgn}\,(\dot{x}) \cong \frac{4}{\pi a \omega} \cdot a\omega \cos \omega t$$

The quasilinearized system then has the characteristic equation

$$s^3 + 2s^2 + \alpha s + \beta = 0, \quad \begin{cases} \alpha = 2 + \dfrac{4}{\pi a \omega} \\ \beta = 3 + \dfrac{3}{4} a^2 \end{cases} \tag{4}$$

Two roots of this equation are in the imaginary axis -- as demanded by the limit cycle assumption -- if

$$2\alpha = \beta \tag{5}$$

and the corresponding natural frequency is

$$\omega = \sqrt{\beta/2} \tag{6}$$

Combining (4) to (6) gives the limit cycle condition to be

$$2\left[\frac{3}{2}\left(1 + \frac{a^2}{4}\right)\right]^{1/2} + \frac{4}{\pi a} = \left[\frac{3}{2}\left(1 + \frac{a^2}{4}\right)\right]^{3/2} \tag{7}$$

If a value or values of a can be found that satisfies (7), then SIDF theory predicts that a limit cycle of amplitude a and frequency ω given by (6) exists; the single solution is $a = 1.63$, $\omega = 1.58$.

This example illustrates the following most general limit cycle conditions (Taylor, 1975): Given

$$\dot{\underline{x}} = \underline{f}(\underline{x}, \underline{u}) \tag{8}$$

when $\underline{x}$ is an n-dimensional state vector and $\underline{u}$ is an m-dimensional input vector. Assuming that $\underline{u}$ is a vector of constants, denoted $\underline{u}_o$, it is desired to determine if (8) may exhibit LC behavior.

As before, we assume that the state variables are nearly sinusoidal,

$$\underline{x} \cong \underline{x}_c + \mathrm{Re}[\underline{a} \exp\,(j\omega t)] \tag{9}$$

where $\underline{a}$ is a complex amplitude vector and $\underline{x}_c$ is the state vector center value (which is not a singularity, or solution to $\underline{f}(\underline{x}_o, \underline{u}_o) = \underline{0}$ unless the nonlinearities satisfy certain stringent symmetry conditions with respect to $\underline{x}_o$). Then we again neglect higher harmonics, to make the approximation

$$\underline{f}(\underline{x},\underline{u}_o) \cong \underline{f}_{DF}(\underline{x}_c,\underline{a},\underline{u}_o) + \mathrm{Re}[F_{DF}(\underline{x}_c,\underline{a},\underline{u}_o)\ \underline{a}\ \exp(j\omega t)] \qquad (10)$$

The real vector $\underline{f}_{DF}$ and the matrix F_{DF} are obtained by taking the fourier expansions of the elements of $\underline{f}(\underline{x}_c + \mathrm{Re}\ \underline{a}\ \exp(j\omega t),\ \underline{u}_o)$ as illustrated above, and provide the quasi-linear or describing function representation of the nonlinear dynamic relation. The assumed limit cycle exists for $\underline{u} = \underline{u}_o$ if $\underline{x}_c$ and $\underline{a}$ can be found so that

$$\begin{aligned} &\text{(i)} \quad \underline{f}_{DF}(\underline{x}_c,\underline{a},\underline{u}_o) = 0 \\ &\text{(ii)} \quad [j\omega I - F_{DF}(\underline{x}_c,\underline{a},\underline{u}_o)]\underline{a} = 0,\ \underline{a} \neq \underline{0} \end{aligned} \qquad (11)$$

(F_{DF} has a pair of pure imaginary eigenvalues, and $\underline{a}$ is the corresponding eigenvector.)

The nonlinear algebraic equations (11) are often difficult to solve. An iterative method, based on successive approximation, can be used successfully for more complicated problems such as the highly complex aircraft performance analysis problem (9 state variables) described in Taylor (1980).

Frequency Response Analysis

The system model is the same as (8), except $\underline{u}$ is a vector of sinusoidal inputs, $\underline{u} = \underline{u}_o + \mathrm{Re}(\underline{c}\ e^{j\omega t})$. Then $\underline{x}(t)$ has the form indicated in (9), and we quasilinearize $\underline{f}(\underline{x},\underline{u})$ as follows:

$$\begin{aligned} &\underline{f}(\underline{x}_c + \mathrm{Re}(\underline{a}\ e^{j\omega t}),\ \underline{u}_o + \mathrm{Re}(\underline{c}\ e^{j\omega t})) \\ &\quad \cong \underline{f}_o(\underline{x}_c,\ \underline{a},\ \underline{u}_o,\ \underline{c}) + \mathrm{Re}[F_{DF}(\underline{x}_c,\ \underline{a},\ \underline{u}_o,\ \underline{c})\ \underline{a}\ e^{j\omega t}] \\ &\quad + \mathrm{Re}[G_{DF}(\underline{x}_c,\ \underline{a},\ \underline{u}_o,\ \underline{c})\ \underline{c}\ e^{j\omega t}] \end{aligned} \qquad (12)$$

Applying the same conditions of harmonic balance that underlie the limit cycle condition (11), it is possible to solve for $\underline{x}_c$ and $\underline{a}$ using

$$\begin{aligned} &\underline{f}_o(\underline{x}_c,\ \underline{a},\ \underline{u}_o,\ \underline{c}) = \underline{0} \\ &\underline{a} = (j\omega I - F_{DF})^{-1}\ G_{DF}\ \underline{c} \end{aligned} \qquad (13)$$

These 2n nonlinear algebraic equations (n of which are complex-valued) can readily be solved using standard computer routines. In this case, one should be careful to ensure that F_{DF} does not have eigenvalues on or very close to the imaginary axis; otherwise limit cycles may exist, in contradiction to the assumption underlying (9). More examples and details of this approach to frequency response analysis are available in Taylor and Mohan (1980).

Statistical Performance Analysis of Stochastic Systems

The dynamics of a nonlinear continuous-time stochastic system can often be represented by a first-order vector differential equation in which $\underline{x}(t)$ is the system state vector and $\underline{w}(t)$ is a forcing function vector,

$$\dot{\underline{x}}(t) = \underline{f}(\underline{x},t) + G(t)\,\underline{w}(t) \tag{14}$$

The state vector is composed of any set of variables sufficient to describe the behavior of the system completely. The forcing function vector $\underline{w}(t)$ represents disturbances as well as control inputs that may act upon the system. In what follows, $\underline{w}(t)$ is assumed to be composed of a mean or deterministic value $\underline{b}(t)$ and a random input $\underline{u}(t)$, the latter being comprised of elements which are uncorrelated in time; that is, $\underline{u}(t)$ is a white noise process having the spectral density matrix $Q(t)$. Similarly, the state vector has a deterministic component $m(t)$ and a random part $\underline{r}(t)$; for simplicity, $\underline{m}(t)$ will usually be called the mean vector. Thus the state vector $\underline{x}(t)$ is described statistically by its mean vector and covariance matrix, $S(t)$.

The differential equations that govern the propagation of the mean vector and covariance matrix for the system described by (14) can be derived directly, as demonstrated in Jazwinski (1970), to be

$$\begin{aligned} \dot{\underline{m}} &= E\,[\underline{f}(\underline{x},t)] + \underline{G}(t)\underline{b} \\ &\triangleq \hat{\underline{f}} + G(t)\underline{b} \\ \dot{S} &= E\,[\underline{f}\,\underline{r}^T] + E\,[\underline{r}\,\underline{f}^T] + G(t)Q\,G^T(t) \triangleq NS + SN^T + GQG^T \end{aligned} \tag{15}$$

The equation for S has been put into a form analogous to the covariance equations corresponding to $\underline{f}$ being linear, by defining the auxiliary matrix N as above.

The quantities $\hat{\underline{f}}$ and N defined in (15) must be determined before one can proceed to solve for $\underline{m}$ and S. Evaluating the indicated expected values requires knowledge of the joint probability density function (joint pdf) of the state variables. While it is possible, in principle, to evolve the n-dimensional joint pdf $p(\underline{x},t)$ for a nonlinear system with random inputs by solving a set of partial differential equations known as the Fokker-Planck equation or the forward equation of Kolmogorov (Jazwinski, 1970), this procedure is generally not feasible from a practical point of view. The fact that the pdf is not available precludes the exact solution of (15).

One procedure for obtaining an approximate solution to (15) is to assume the form of the joint probability density function of the state variables in order to evaluate $\underline{f}$ and N. Although it is possible to use any joint pdf, all work to date has been based on the assumption that the state variables are jointly normal; the choice was made because it is both reasonable and convenient. The justifications of this approximation are given more fully in Taylor, Price, Siegel and Gelb (1980). The outcome is that $\hat{\underline{f}}$ and N are evaluated by using RIDF's, as in (2), and the numerical integration of (15) can be accomplished readily.

The following problem exemplifies the above RIDF/covariance analysis method of stochastic system performance assessment: Given

$$\ddot{x} + 2\dot{x} + x + x^3 = w, \quad w \sim N(0,q)$$

The zero-mean RIDF for x^3 is $3s_{11}$ where s_{11} is the variance of x (Atherton, 1975); defining the state vector to be $\underline{x}^T = [x \quad \dot{x}]$ gives us the following RIDF matrix for use in (15):

$$N = \begin{bmatrix} 0 & 1 \\ -(1 + 3s_{11}) & -2 \end{bmatrix}$$

Equation (15) then expands to yield the following differential equations for the elements of S:

$$\dot{s}_{11} = 2\, s_{12}$$

$$\dot{s}_{12} = s_{22} - (1 + 3\, s_{11})\, s_{11} - 2\, s_{12}$$

$$\dot{s}_{22} = -2\, (2\, s_{22} + (1 + 3\, s_{11})\, s_{12}) + q$$

Given suitable initial conditions, these differential equations can be integrated easily on a digital computer. The steady-state value of S can be obtained directly by setting $\dot{S} = 0$:

$$S_{ss} = \begin{bmatrix} \frac{\sqrt{1 + 3q} - 1}{6} & 0 \\ 0 & \frac{q}{4} \end{bmatrix}$$

More examples and detailed discussions of the theory and use of this technique are given in Taylor, Price, Siegel and Gelb (1980).

Summary and Conclusions

Before the foregoing recent advances in DF theory were made, it was generally conceded that direct simulation of highly nonlinear, high-order differential equations was the only available approach for studying their behavior. Unless a very large number of simulations are performed (with many initial conditions and input functions, for simulation time intervals long enough to allow transients to die away, etc.), the information gained by simulation is often spotty and of questionable value. The analysis tools outlined in this paper have proven to be of great value in supplementing and complementing direct simulation for understanding the behavior of nonlinear systems.

References

Atherton, D. P. (1975). Nonlinear Control Engineering. Van Nostrand Reinhold, London.

Gelb, A. and Vander Velde, W. E. (1968). Multiple-Input Describing Functions and Nonlinear System Design. McGraw-Hill, New York.

Gelb, A. and Warren, R. S. (1973). Direct Statistical Analysis of Nonlinear Systems: CADET. AIAA J., 11, 689-694.

Hannebrink, D. N., Lee, H. S. H., Weinstock, H. and Hedrick, J. K. (1977).

Influence of Axle Load, Track Gage, and Wheel Profile on Rail-Vehicle Hunting. Trans. ASME - J. Eng. Ind., Feb., 186-195.

Jazwinski, A. H. (1970). Stochastic Processes and Filtering Theory. Academic Press, New York.

Kazakov, I. E. (1965). Generalization of the Method of Statistic Linearization to Multidimensional Systems. Avtom. & Telemekh., 26, 1210-1215.

Taylor, J. H. (1975). An Algorithmic State-Space/Describing Function Technique for Limit Cycle Analysis. I.O.M., The Analytic Sciences Corp. (TASC), Reading, MA. (Also issued as TASC TIM-612-1 to the Office of Naval Research, Oct. 1975).

Taylor, J. H. (1980). Applications of a General Limit Cycle Analysis Method for Multi-variable Systems. Part 2, Ch. 3 of Nonlinear System Analysis and Synthesis, Vol. 2 Applications, Ed. by Ramnath, R. V., Hedrick, J. K. and Paynter, H. M., publ. by the ASME.

Taylor, J. H. and Mohan, S. R. (1980). Describing Function Analysis of the Frequency Response of Highly Nonlinear Systems. Submitted to the IEEE Trans. on Auto. Control.

Taylor, J. H., Price, C. F., Siegel, J. and Gelb, A. (1980). Covariance Analysis of Nonlinear Stochastic Systems via Statistical Linearization. Part 2, Ch. 7 of Nonlinear System Analysis and Synthesis, Vol. 2 Applications, Ed. by Ramnath, R. V., Hedrick, J. K. and Paynter, H. M., publ. by the ASME.

SOME RECENT QUALITATIVE RESULTS ON MODELS BASED ON DIFFERENCE EQUATIONS*

E. F. Infante
Lefschetz Center for Dynamical Systems
Division of Applied Mathematics
Brown University
Providence, R. I. 02912

ABSTRACT

The noncontinuous effect of changes in the delays on the asymptotic behavior of multidelay difference equations is discussed, together with its implications on the use of such equations as models of dynamical systems. Theoretical results are illustrated by simple examples.

KEYWORDS

Difference equations; asymptotic behavior; robustness to changes in delays; stability; modeling.

SUMMARY

The use of difference equations as mathematical models of large scale systems is well established not only in the physical and biological sciences but especially in the economic and social sciences. In many situations such difference equations result from the modeling of phenomena in which several delays occur; the value of these delays is, in general, approximated; the relative value of the delays then determines the dimension of the overall difference equation used as a mathematical model for analysis and computation. It has been recently noted that the asymptotic behavior and in particular stability properties of difference equations with several delays do not, in general depend continuously on the value of the delays; that is, small changes in the value of the delays will result in radical changes in the stability properties of the difference equations. This unfortunate behavior has serious implications on the use of difference equations as reasonable mathematical models of phenomena where asymptotic properties are critical as in multiplier computations in macroeconomic models and in dominance arguments in ecological systems.

A review of the very recent mathematical results obtained in this class of problems

---

* Invited lecture, Session on Modeling and Control of Large Scale and Distributed Parameter Systems, International Congress on Applied Systems Research and Cybernetics, Dec. 12-15, 1980, Acapulco, Mexico.

will be presented and illustrated through simple examples. A partial identification of the subclass of difference equations whose stability properties are robust to changes in the delays will also be given; any credible model of observable phenomena must necessarily be in this subclass.

ACKNOWLEDGEMENT

The research presented has been supported during the past years in part by the National Science Foundation (NSF-MCS 7905774), the Air Force Office of Scientific Research (AF-AFOSR-76-3092) and the U.S. Army Research Office (ARO-DAAG-29-79-C0161).

REFERENCES

Hale, J.K. (1977). Theory of Functional Differential Equations. Springer-Verlag, New York.
Infante, E.F. and J.A. Walker (1977). Proc. R. Soc. Edinburgh, A79, 307-316.
Infante, E.F. and W.B. Castelan (1978). J. Differential Equations, 29, 439-451.
Infante, E.F., L. Carvalho, and J.A. Walker (1980). Tohoku Math. J. 32, 283-297.
Silkowskii, R.A. (1979). Proc. R. Soc. Edinburgh, A81, 292-307

GUARANTEED ULTIMATE BOUNDEDNESS CONTROL FOR UNCERTAIN DYNAMIC SYSTEMS

G. Leitmann
University of California
Berkeley, California, USA

ABSTRACT

The problem of designing a state feedback control that guarantees the desired performance of a dynamic system containing uncertain elements is discussed in Gutman and Leitmann (1976) and Gutman (1979) among others. The desired performance is usually uniform asymptotic stability of an equilibrium state. Sometimes one is content with uniform ultimate boundedness in some set; in that case one can consider feedback based on uncertain state or output as in Leitmann (1980). The salient feature of the problem is the fact that it is a *deterministic* treatment of uncertainty in that one requires *certain* performance in the presence of *uncertain* information. The essential knowledge about the uncertain elements concerns only their *possible* sizes; that is, only the sets in which the values of the uncertain quantities can range are presumed to be known.

If some conditions are satisfied -- primary among which are the so-called "matching conditions" [for a relaxation of the matching conditions see Barmish and Leitmann (1980)] -- then all uncertain elements can be "lumped" and the system is described by

$$\dot{x}(t) = f(x(t),t) + B(x(t),t)\, u(t) + B(x(t),t)\, e(x(t),t) \ , \quad x(t_0) = x_0 \tag{1}$$

where $t \in \mathbb{R}$ is time, $x(t) \in \mathbb{R}^n$ is the state, $u(t) \in \mathbb{R}^m$ is the control, and $e(x(t),t)$ is the "lumped" uncertain element. The "nominal" system -- that is, the system without uncertainty -- is described by

$$\dot{x}(t) = f(x(t),t) + B(x(t),t)\, u(t) \ , \quad x(t_0) = x_0 \tag{2}$$

where $f(\cdot) : \mathbb{R}^n \times \mathbb{R} \to \mathbb{R}^n$ and $B(\cdot) : \mathbb{R}^n \times \mathbb{R} \to \mathbb{R}^{n \times m}$ are known.

Based solely on the knowledge of the maximum possible value (which may depend on x and t) of the norm $\| e(x,t) \|$, we seek a state feedback control $p(\cdot) : \mathbb{R}^n \times \mathbb{R} \to \mathbb{R}^m$ such that, given (x_0,t_0), a corresponding response $x(\cdot) : [t_0,\infty) \to \mathbb{R}^n$ exists, and every such response enters a neighborhood of $x = 0$ in finite time and remains within it thereafter, provided $u(t) = p(x(t),t)$ is employed; that is, we desire *uniform ultimate boundedness*.

In Gutman and Leitmann (1976) and Gutman (1979) it is required that $x = 0$ is asymptotically stable. The control that assures such behavior is in general a *discontinuous* function of the state, x, and hence is usually not realizable in practice. Furthermore, the analysis must be carried out for a *generalized dynamical system* since the usual concept of a solution is not adequate. The subsequent treatment is motivated, in part, by the fact that *generalized solutions* describe behavior in the limit; for instance, see Aizerman and Pyatnitskii (1978) and Hájek (1979). We propose a class of controls which depend *continuously* on the state, x, and which guarantee the desired behavior, ultimate boundedness; for details see Corless and Leitmann (1980).

KEYWORDS

Uncertain systems; guaranteed performance; ultimate boundedness; feedback control.

REFERENCES

Aizerman, M. A., and E. S. Pyatnitskii (1978), J. Franklin Inst., 306, 379-408.
Barmish, B. R., and G. Leitmann (1980), IEEE Trans. A.C., to appear.
Corless, M., and G. Leitmann (1980), IEEE Trans. A.C., to appear.
Gutman, S. (1979), IEEE Trans. A.C., 24, 437-443.
Gutman, S., and G. Leitmann (1976). Stabilizing Feedback Control for Dynamical Systems with Bounded Uncertainty. Proceedings IEEE Conf. Decision and Control, Clearwater Beach, Florida.
Hájek, O. (1979), J. Diff. Equations 32, 149-185.
Leitmann, G. (1980). On the efficacy of nonlinear control in uncertain linear systems, Proceedings ASME Winter Meeting, Chicago, Illinois.

ON THE NUMERICAL ANALYSIS OF DYNAMICAL SYSTEMS

by

Eusebius Doedel
Computer Science Department
Concordia University
1455 de Maisonneuve Blvd. W.
Montreal, Quebec H3G 1M8
Canada

ABSTRACT

Computational techniques for the bifurcation analysis of autonomous systems of ordinary differential equations are described. These techniques have been implemented in a computer program (AUTO). Results from the application of AUTO to a typical problem are given.

KEYWORDS

Ordinary differential equations, autonomous systems, bifurcation theory, numerical analysis, periodic solutions.

INTRODUCTION

The automatic bifurcation analysis of time dependent differential equations containing a free parameter is important in many areas of research. For example such problems arise in the study of chemical reactions (Poore,1976), in fluid flow problems (Sattinger,1973), in ecological models and in mathematical biology. (Specifically neurobiology. (Rinzel and Miller,1980).) Frequently one wants to determine both steady state and periodic solutions to the differential equations. Since the equations have a parameter, one generally has branches of solutions. A point of intersection of two branches is called a bifurcation point. The computational problem then consists of the numerical treatment of bifurcation points and the determination of entire solution branches. For tracing out any given solution branch the underlying differential equation typically must be solved hundreds of times.

Recently the author has developed a computer code (AUTO) for the automatic bifurcation analysis of autonomous systems of ordinary differential equations of the form

$$u'(t) = f(\, u(t)\, ,\, \lambda\,), \quad u,f \text{ in } R^n, \quad \lambda \text{ a parameter.} \qquad (1)$$

Briefly described, AUTO determines all reachable steady state and periodic solution branches of the differential equation. Switching branches at bifurcation points is done automatically, both at steady state bifurcation points and at Hopf bifurcation points. (The latter are points where periodic solutions branch off from a steady state.) Limit points on either type of branch do not present difficulties due to the special continuation procedure used. Both asymptotically stable and asymptotically unstable solutions can be computed. The differential equation is discretized using high order collocation methods with automatic adaptive mesh selection. In the following paragraphs we give a short description of the various capabilities of AUTO as well as an example of its application to a representative problem.

BRANCHES OF STEADY STATE SOLUTIONS

These are solutions $(u(s),\lambda(s))$ of the (generally) nonlinear system of equations

$$f(u, \lambda) = 0, \tag{2}$$

where s denotes some parametrization. AUTO assumes that a solution u_0 is known for some particular value λ_0. In most applications such a solution is easily found. The program automatically determines the direction $(\dot{u}_0,\dot{\lambda}_0)$ of the branch at the starting point, i.e., it computes a null vector of the n by n+1 matrix $(f_u(u_0,\lambda_0)|f_\lambda(u_0,\lambda_0))$. The nullspace is assumed to be one dimensional. This assumption excludes bifurcation points and turning points as starting value. The given branch is then traced out in a stepwise manner using the continuation technique of Keller (1977). Assuming (u_{j-1},λ_{j-1}) has been computed, this technique determines the next solution (u_j,λ_j) from the equation

$$f(u_j,\lambda_j) = 0, \tag{3}$$

$$||u_j - u_{j-1}||^2 + (\lambda_j - \lambda_{j-1})^2 = \Delta s^2,$$

where Δs is the stepsize along the branch and $||.||$ denotes the Euclidean distance. An initial approximation to (u_j,λ_j) is obtained by extrapolation from the two preceding solution points (u_{j-1},λ_{j-1}) and (u_{j-2},λ_{j-2}). The main advantage of using the inflated system (3) is its capability of computing past turning points on a branch, i.e., solution points where the direction of the branch does not have a component in λ-direction. The computation of any branch is discontinued if either λ or the norm of the solution vector reaches pre-set limits, if a user-specified maximum number of steps has been taken or if an error condition occurs.

STEADY STATE BIFURCATION POINTS

The program monitors the determinant of the Jacobian of (3) along the solution branch. An iteration procedure is used to accurately locate points (u_*,λ_*) on the solution branch where the determinant changes sign. Such points are potential steady state bifurcation points, and relevant data concerning the eigenspace of the Jacobian at this point are stored. After the computation of the given branch has been completed, the program returns to each bifurcation point in turn and repeats the computational procedure for the bifurcating branches.

HOPF BIFURCATION POINTS

During the computation of the steady states, the program also checks whether a complex conjugate pair of eigenvalues of the Jacobian of (2) crosses the imaginary axis. The values of u and λ for which this occurs are determined very accurately by means of an iteration procedure similar to the one used for locating steady state bifurcation points. A necessary condition for a branch of periodic solutions to bifurcate from a steady state branch is precisely the occurrence of the situation described above. (Crandall and Rabinowitz (1976).) Formulae for the asymptotic behaviour of the periodic solutions near a Hopf bifurcation point are well known

(Langford (1977).) and the relevant data obtained from these formulae are stored in a file. After completion of the computation of the steady states, the program automatically retrieves the information of each successively found Hopf bifurcation point and starts the determination of the corresponding branches of periodic solutions.

BRANCHES OF PERIODIC SOLUTIONS

Consider the problem of determining branches of time periodic solutions to the autonomous system (1). First note that not only the periodic solution u, but also its period ρ changes along such a branch. To fix the period, linearly map $[0,\rho]$ into $[0,2\pi]$. This transforms the differential equation into

$$u'(t) = \frac{\rho}{2\pi} f(u,\lambda), \tag{4}$$

where the unknown period ρ now appears explicitly and where 2π-periodic solutions are to be determined, that is, we impose the condition

$$u(0) = u(2\pi). \tag{5}$$

This effectively replaces the original problem by a boundary value problem with non-separated boundary conditions. In particular this makes the computation of asymptotically unstable solutions possible. Unstable and therefore unphysical solutions frequently must be computed in order to reach stable solutions elsewhere on the branch. Suppose $(\lambda_0,\rho_0,u_0(t))$ defines a known periodic solution of (4), (5). The objective is to set up the equations for finding a solution nearby on the branch. A remaining difficulty is the inherent non-uniqueness of u, due to the fact that a periodic solution can be translated freely in time. For numerical computation the new solution u must be "anchored". There are many possible choices for an additional equation to accomplish this. One is to simply fix one of the components of u at t=0. However the resulting set of equations has an isolated solution only under conditions that are not required for the underlying problem itself. For both theoretical and practical purposes a much better choice is the equation

$$\int_0^{2\pi} (u(t) - u_0(t))u_0'(t)\, dt = 0. \tag{6}$$

As was the case for branches of steady state solutions, we treat λ as one of the unknowns. Thus another equation is necessary in order to get a well posed system of equations. A simple choice for this equation is the one that requires the "distance" between the known solution (λ_0,ρ_0,u_0) and the solution (λ,ρ,u) to be determined to equal Δs, i.e.,

$$\begin{aligned} &c_1^2(\lambda-\lambda_0)^2/\Delta s + c_2^2(\rho-\rho_0)^2/\Delta s \\ &+ c_3^2\int_0^{2\pi} (u-u_0)^T (u-u_0)/\Delta s\, dt - \Delta s = 0. \end{aligned} \tag{7}$$

Here the constant weights c_i, i=1,2,3, are included for generality. Thus to proceed one step along a branch of periodic solutions requires the equations (4)-(7) to be solved. Of course, for numerical computation, the differential equation (4)-(5), as well as the integrals in (6) and (7) must be discretized.

DISCRETIZATION OF THE DIFFERENTIAL EQUATION

There are many possible choices for a discretization of the nonlinear boundary value problem (4)-(5). Since the solutions are required to be periodic, one might consider collocation methods with trigonometric polynomials. In the current implementation, however, the method of collocation with piecewise continuous polynomials is used. More precisely, an approximate solution is sought that is continuous on the entire interval $[0,2\pi]$ and that is a polynomial of degree at most m in each of the mesh-subintervals of $[0,2\pi]$. This piecewise polynomials function is then required to satisfy the differential equation at m points in each mesh interval. It is well known that the order of accuracy of such a collocation method increases significantly if the collocation points are chosen to coincide with the roots of the m-th degree Legendre polynomial relative to each mesh interval. (de Boor and Swartz (1973).) The approximate solution is also required to satisfy the periodic boundary conditions. There is a very general collocation code for mixed order systems of nonlinear boundary value problems in ordinary differential equations. (Ascher et al (1979).) In AUTO, however, we have implemented our own discretization in order to take full advantage of the specialized form of the problem to be solved. This specialization should naturally result in a somewhat more efficient code. It also keeps the size of the program within reasonable bounds. (The current version is about 3000 lines of Fortran coding, including comments.) The user specifies beforehand the number of mesh intervals as well as the number of collocation points per mesh interval. The distribution of the mesh point can be made to adapt itself automatically to the solution. This is necessary in many problems where solution components vary rapidly. Such behaviour is very common, for example, in the solutions to chemical reaction equations. The mesh selection strategy used is that of Russell and Christiansen (1978). Bifurcation computations with a fixed mesh are in general not advisable since the bifurcation diagram of the discretization may differ significantly from that of the continuous problem if the mesh size is insufficiently small. (Beyn and Doedel (1980.) Adapting the mesh then can be thought of as switching from one bifurcation diagram to another. If the two are not sufficiently close then the computations would encounter difficulties, thus warning the user that the number of mesh points or the order of the method may be too low. The finite set of nonlinear equations arising from the discretization of the differential equation and its side conditions are solved by Newton's method. For a more detailed description of the solution procedure see Doedel (1980). The stepsize along a solution branch can also be chosen to adapt ifself automatically to the number of Newton iterations needed for the preceding step. Maximum and minimum allowable stepsize must also be specified by the user in this case.

PLOTTING AND RESTART INFORMATION

AUTO produces two files containing plotting information. The names and values of user specified parameters as well as the complete bifurcation diagram are stored in one. The other contains the information that completely specifies the periodic solution at a user-selectable number of points along branches of periodic solutions. An interactive program has been written by a student, J. Chin, for the graphical investigation of such files. A third file is used to store information that makes the restarting of the computation of a branch of periodic solutions possible at certain points on a branch. This is especially useful for difficult problems that require much computer time. Another file is needed for storage and retrieval of the intermediate information regarding Hopf bifurcation points.

A NUMERICAL EXAMPLE

$$u_1' = (1-\lambda)\sin(u_1) - \sin(u_2) - u_3^3$$

$$u_2' = (2-\lambda)\sin(u_1) + u_2^2$$

$$u_3' = \sin(u_3) + u_2^2 - u_1^2 .$$

The bifurcation diagram generated by AUTO is given in the Fig. There is one steady state bifurcation point and there are three Hopf bifurcation points. The branch of periodic solutions (3) bifurcating from the zero steady state branch (1) at $\lambda=1$, reattaches itself to the nontrivial steady branch (2) at $\lambda=1.88$. The second periodic branch (4) in the diagram does not reattach itself to any steady state branch. This branch demonstrates the often occurring situation where the periodic solutions approach a limiting configuration with the period becoming infinite.

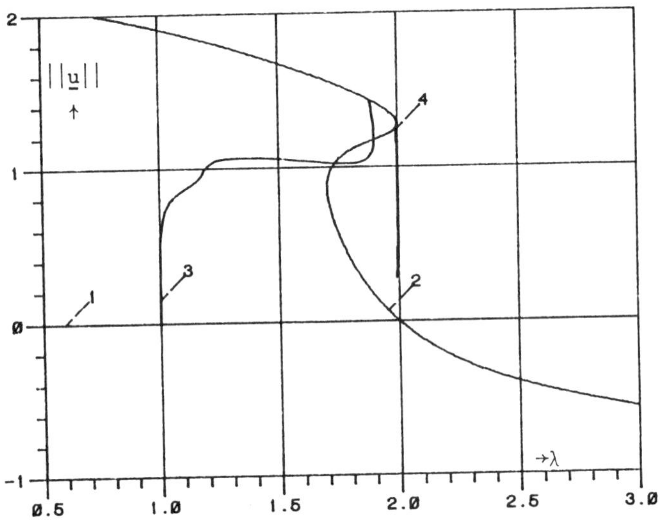

ACKNOWLEDGEMENT

This research was supported by NSERC (Canada) Grant A4274 and FCAC (Quebec) Grant EQ1438.

REFERENCES

Ascher, U., J. Christiansen and R. D. Russell (1979). Math. Of Comp. 33, 659-679.

Beyn, W.J., and E. J. Doedel (1980). Stability and multiplicity of solutions to discretizations of nonlinear ordinary differential equations, to appear in SIAM J. Sci. Stat. Comput.

de Boor, C., and B. Swartz (1973). SIAM J. Numer. Anal. 10, 582-606.

Crandall, M. G., and P. H. Rabinowitz (1976). TSR 1604, Mathematics Research Center, Madison, Wisconsin.

Doedel, E. J. (1980). The numerical computation of branches of periodic solutions, in: Mathematical methods for continuous systems simulation, T. D. Bui ed., The Soc. of Computer Simulation, to appear.

Keller, H. B. (1977). Numerical solution of bifurcation and nonlinear eigenvalue problems, in : Applications of bifurcation theory, P. H. Rabinowitz, ed., Academic Press, 359-384.

Langford, W. F. (1977). Numer. Math. 28, 171-190.

Poore, A. B. (1976). Arch. Rat. Mech. Anal. 60, 371-393.

Rinzel, J., and R. N. Miller (1980). Numerical calculation of stable and unstable periodic solutions to the Hodgkin-Huxley equations, preprint, Math. Research Branch, NIH, Bethesda, MD.

Russell, R. D., and J. Christiansen (1978). SIAM J. Numer. Anal. 15, 59-80.

Sattinger, D. H. (1973). Topics in stability and bifurcation theory, Lec. Notes in Math. 309, Springer Verlag, Berlin.

REAL-TIME SOLUTION OF STIFF LINEAR DIFFERENTIAL SYSTEMS

Karl H. Agar* and Wiley E. Thompson**

*Office of Missile Electronic Warfare
Electronic Warfare Laboratory (ERADCOM)
White Sands Missile Range, NM 88002
**Department of Electrical and Computer Engineering
New Mexico State University, Las Cruces, NM 88003

ABSTRACT

A new technique is developed for the real-time solution of stiff differential systems. The method is A-stable and is suitable for hardware-in-the-loop application. A computational architecture for implementation of the algorithm is presented, and properties of the techniques are examined.

KEYWORDS

Stiff linear differential systems; real-time solution; state-variable form; digital processing computational architecture; hardware-in-the-loop modeling.

INTRODUCTION

Many problems in systems and control engineering may be modeled by a set of ordinary differential equations expressed in state-variable form:

state equations: $\dot{\bar{x}}(t) = A\,\bar{x}(t) + B\,\bar{u}(t)$ (1)

output equations: $\bar{c}(t) = C\,\bar{x}(t) + D\,\bar{u}(t)$ (2)

initial conditions: $\bar{x}(0) = \bar{x}_0$

This paper presents the development of a real-time solution technique for stiff differential linear, time-invariant system models based on the above formulation. The technique is A-stable and capable of accepting input signals generated externally to the model. The implementation of the method is based on using a desktop-type host computer with the aid of a special purpose digital processor.

STIFF DIFFERENTIAL SYSTEMS

If the system model is linear and the eigenvalues λ of A are widely separated, that is, $\lambda_{max}/\lambda_{min} << 1$, the system is said to be "stiff." For the solution of stiff systems of ordinary differential equations, classical methods of solutions have limited applications. Recently, new implicit methods have been introduced that ensure stable as well as accurate solutions for stiff problems (Bickart, Burgess, Sloate, 1971; Liniger, 1975; Fowler and Warten, 1967; Liniger and Willoughby,

1967). These methods are all software based on very large general purpose computers, and typically involve limiting the values of $h\lambda$ (where h is the time duration between solution updates) to satisfy stability boundary requirements. In the case of hardware-in-the-loop applications, which require real-time solutions, a fixed h method is the most desirable approach to the solution method. For stiff differential equations, it is essential to use A-stable numerical methods to avoid stability boundary problems when $h\lambda$ is large.

SOLUTION METHOD

The problem associated with stiff systems narrows to two basic requirements: stability and accuracy. The use of A-stable methods avoids the stability problem, but for a reasonable step length h, the solution corresponding to the largest eigenvalue may be approximated inaccurately. The problems associated with real-time solutions to difference equations is the number of sequential operations that have to be performed to obtain a solution update during the time interval h. Considering these problems a review of solution methods judged applicable for the solution of stiff-systems indicates that the trapezoidal rule

$$\bar{x}(n+1) = \bar{x}(n) + h/2\,[\dot{\bar{x}}(n+1) + \dot{\bar{x}}(n)], \tag{3}$$

is a suitable method. This method is A-stable, but produces slowly decaying oscillatory errors of order h^3 for $\lambda h > 1.0$. A simple smoothing procedure which avoids the oscillatory error and reduces the total error significantly was proposed by Lindberg in 1971. It is based on the formula

$$\hat{x}(n) = {}^1/_4\,[x(n-1) + 2\,x(n) + x(n+1)] \tag{4}$$

where $\hat{x}(n)$ is used as the starting value for the forthcoming computation.

To demonstrate the sequential computational requirements, consider

$$\dot{x} = \lambda\, x(t) \tag{5}$$

with $x(0) = x_0$, λ = constant, and $t_0 \leq t \leq \beta$.

Using the trapezoidal rule with step size h, the solution is (Agar, 1980)

$$x(n) = \left[\frac{1 + h\lambda/2}{1 - h\lambda/2}\right]^n x_0. \tag{6}$$

When (5) is solved using the trapezoidal rule with smoothing for m < n steps

$$x(n) = \left[\frac{(1 - h\lambda/2)^{m-2}}{(1 - (h\lambda/2)^2)^m}\left(\frac{1 + h\lambda/2}{1 - h\lambda/2}\right)^n\right] x_0, \tag{7}$$

and using smoothing for every step of the integration process,

$$x(n) = \left[\frac{1}{1 - (h\lambda/2)^{2n}}\right] x_0. \tag{8}$$

The implementation of (6), (7), and (8) on a digital computing machine is straightforward and achievable with a minimum number of arithmetic steps.

To illustrate the quantization error incurred with the above algorithms, consider (5) with $x_0 = 1$. The total quantization errors for n steps of integration using (6), (7), and (8) are illustrated in Figs. 1, 2, and 3, respectively. The error performance observed is judged to be sufficient in most hardware-in-the-loop applications.

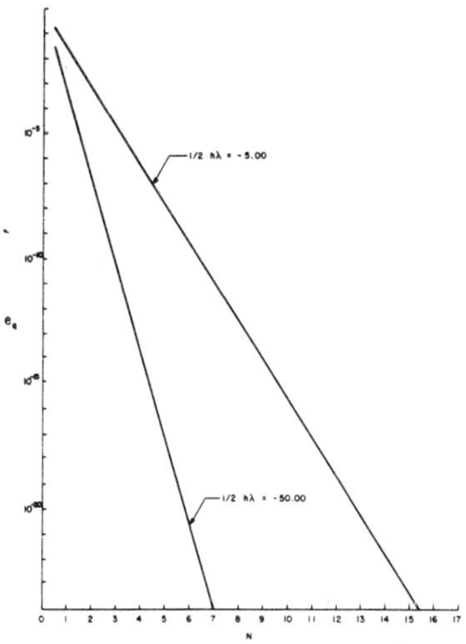

FIG. 2. QUANTIZATION ERROR FOR $|1/2\, h\lambda| > 1$, m = n.

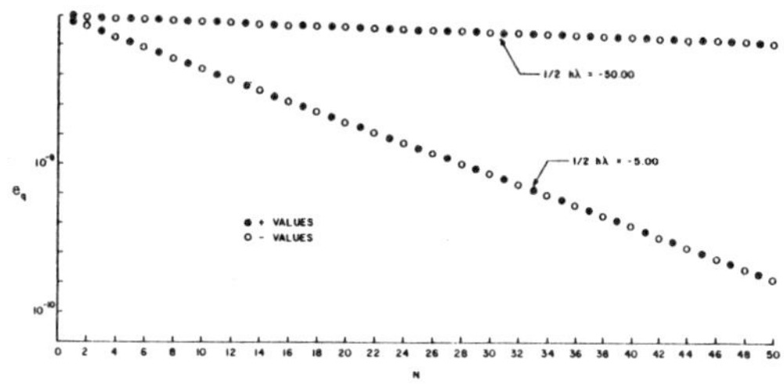

FIG. 1. QUANTIZATION ERROR FOR $|1/2\, h\lambda| > 1$, m = 0.

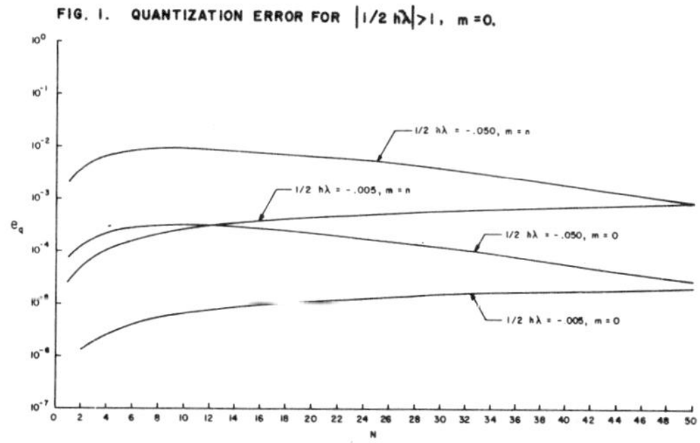

FIG. 3. QUANTIZATION ERROR FOR $|1/2\, h\lambda| < 1$.

SOLUTION TECHNIQUE

The state-variable formulation shown in (1), in general, because of the coupling nature introduced by the off-diagonal elements of the matrix A, does not lend itself to a direct decomposition that is practical to evaluate. By an appropriate change of variable

$$\bar{y} = T\,\bar{x}, \text{ and } \bar{A} = T\,\bar{A}\,T^{-1} \tag{9}$$

A is transformed into the Jordan canonical form. The Jordan canonical form does lend itself to the desired scalar decomposition, especially since most real-world problems encountered do not have multiple eigenvalues.

To illustrate the method, consider the fourth-order system already expressed in Jordan form with unique, real, negative eigenvalues.

$$\begin{bmatrix} \dot{y}_1 \\ \dot{y}_2 \\ \dot{y}_3 \\ \dot{y}_4 \end{bmatrix} = \begin{bmatrix} \lambda_1 & & & \\ & \lambda_2 & & \\ & & \lambda_3 & \\ & & & \lambda_4 \end{bmatrix} \begin{bmatrix} y_1 \\ y_2 \\ y_3 \\ y_4 \end{bmatrix} + \begin{bmatrix} \beta_1 \\ \beta_2 \\ \beta_3 \\ \beta_4 \end{bmatrix} u \tag{10}$$

$$C = [\gamma_1\ \gamma_2\ \gamma_3\ \gamma_4] \begin{bmatrix} y_1 \\ y_2 \\ y_3 \\ y_4 \end{bmatrix} + [d]\,u \tag{11}$$

with initial conditions $\bar{y}(t_0) = \bar{y}(0)$ and u a step function.

The solution to (10) is

$$y_1 = e^{\lambda_1 t}\, y_1(0) + \lambda_1^{-1}\beta_1(1-e^{\lambda_1 t})\ \ u \tag{12}$$

$$y_2 = e^{\lambda_2 t}\ y_2(0) + \lambda_2^{-1}\beta_2(1-e^{\lambda_2 t})\ \ u \tag{13}$$

$$y_3 = e^{\lambda_3 t}\ y_3(0) + \lambda_3^{-1}\beta_3(1-e^{\lambda_3 t})\ \ u \tag{14}$$

$$y_4 = e^{\lambda_4 t}\ y_4(0) + \lambda_4^{-1}\beta_4(1-e^{\lambda_4 t})\ \ u, \text{ and} \tag{15}$$

$$c = \gamma_1 y_1 + \gamma_2 y_2 + \gamma_3 y_3 + \gamma_4 y_4. \tag{16}$$

Applying the Trapezoidal-based algorithms (6), (7), and (8), the solutions for a general system are:

$$y_i(n + 1) = \Omega_i(n)\ (y_i(0) - \Phi_i(n)) + \Phi_i(n) \tag{17}$$

$$\bar{c}(n) = C\ \bar{y}(n) + D\ u(n) \tag{18}$$

where

$$\Omega_i(n) = \begin{cases} \left[\dfrac{1 + {}^1/_2 h\lambda_i}{1 - {}^1/_2 h\lambda_i}\right]^n, & \text{for } m = 0 \\ \left[\dfrac{1 + {}^1/_2 h\lambda_i}{1 - {}^1/_2 h\lambda_i}\right]^n \dfrac{(1 - {}^1/_2 h\ \lambda_i)^{m-2}}{(1 - ({}^1/_2 h\ \lambda_i)^2)^m}, & \text{for } m < n \\ \left[\dfrac{1}{(1 - {}^1/_2 h\lambda_i)^2}\right]^n, & \text{for } m = n \end{cases} \tag{19}$$

$$\Phi_i(n) = \lambda_i^{-1} \sum_{j=1}^{p} \beta_{ij} u_j(n), \quad i = 1, 2, \ldots, k \tag{20}$$

The following solution procedure applies to obtain a real-time solution to linear, time-invarient, stiff differential systems with unique eigenvalues:

Step 1: Reformat the state-variable formulation of the original problem into the Jordan canonical format by finding the eigenvalues and eigenvectors of A to determine the matrices T and T^{-1}. The elements of T are designated as τ_{ij}. Determine λ_i^{-1}.

Step 2: Based on the systems's operational requirements, i.e., basic data rate, select h and determine ${}^1/_2 h\lambda_i$. If $|{}^1/_2 h\lambda| = 1$, select a different h. Determine $\lambda_i^{-1}\ \beta_{ij}$.

Step 3: Calculate $\Omega_i(1)$ using (19).

Step 4: Sample and hold $\bar{u}(n)$, $n = 1, 2, \ldots, N$. Calculate Φ using (20) and $D\ \bar{u}(n)$. Determine $y_i(0) - \Phi_i$ and $\Omega_i(n)$.

Step 5: Calculate $y_i(n)$ using (17) and $\bar{c}(n)$ using (18).

Steps 1, 2, and 3 are preparatory type operations which may be considered as part of the problem setup prior to solution. The calculations in steps 4 and 5 are the ones that need to be performed as fast as possible. Therefore, if steps 1, 2, and 3 are performed on a general-purpose host computer prior to the start of the solution then steps 4 and 5 are performed on a specialized processor in real-time.

Agar (1980) has shown that for a linear time invariant k-dimensional state model in Jordan canonical form with unique real eigenvalues and input vector of size p is solvable by the trapezoidal rule algorithms (6), (7), and (8) using two input arithmetic computational devices in L arithmetic step, where

$$L = 6 + \log_2 s + \log_2 t \tag{22}$$

with $s = 2^n$ for $2^{n-1} < k \leq 2^n$, n a nonnegative integer and $t = 2^m$ for $2^{m-1} < p \leqslant 2^m$, m a nonnegative integer.

PROBLEM SOLVING CAPABILITIES

Processor chips having the capability of performing 32-bit floating-point addition and subtraction in 14μs and 32-bit floating-point multiplication in 37μs are commercially available from Advanced Micro Devices (1978). With the utilization of high-speed data transfer techniques it is possible, for example, to solve state-variable models of order 16 in less than 400μs (using two input logic) capable of handling eigenvalues limited only by the numerical capacity of the processor and the user's willingness to accept the quickly damping error incurred during the first few steps of the calculations.

CONCLUSIONS

Using desktop-type general purpose computers with floating-point number representation as a host computer for the real-time, hardware-in-the-loop solution technique makes it possible to solve problems in state-variable form without scaling or special preparations. The user has to specify only the size of h and ensure that the input and output signals to the special processor are compatible with the hardware. This procedure is a considerable improvement over existing methods using analog computers and digital analog replacements, all of which require scaling and special problem setups. The method developed is specifically designed for handling stiff problems in real-time.

ACKNOWLEDGEMENT

The authors express their gratitude to Ms. Angela Owen and Mr. Carl Hundley of the Office of Missile Electronic Warfare editorial staff. Without their invaluable assistance this paper could not have been prepared in the limited time available.

REFERENCES

Bickart, T., D. Burgess, and H. Sloate (1971). High order A-stable composite multistep methods for numerical integration of stiff differential equations. Procceedings 9th Annual Allentown Conference on Circuit and System Theory, pp. 465-473.

Liniger, W. (1975). High Order A-Stable Averaging Algorithms for Stiff Differential Systems. AD-A029729, IBM Thomas J. Watson Research Center, Yorktown Heights, New York.

Fowler, H. E., and R. M. Warten (1967). A Numerical Integration Technique for Ordinary Differential Equations with Widely Separated Eigenvalues, Vol. II. IBM J. Res. Dev., pp. 537-543.

Liniger, W., and R. A. Willoughby (1967). Efficient Numerical Integration of Stiff Systems of Ordinary Differential Equations. Report RC 1970, IBM Thomas J. Watson Research Center, Yorktown Heights, New York.

Lindberg, B. (1971). On smoothing and extrapolation for the trapezoidal rule. BIT 11, pp. 29-52.

Agar, K. H. (1980). A real-time solution technique for stiff linear differential systems. Doctorial Dissertation, New Mexico State University, Las Cruces, New Mexico.

AM 9511 Arithmetic Processor (Sales Brochure) (1978). Advanced Micro Devices, Incorporated.

FUNDAMENTAL STRUCTURES OF ALGORITHMIC SYSTEMS IN INDUSTRIAL ORGANIZATION

H.-G. Lauenroth

Dept. of Economic Cybernetics, Martin-Luther-University Halle-Wittenberg, GDR

ABSTRACT

Human task can be facilitated and improved qualitatively by rational organization with algorithmic methods.
Control algorithms are usable for modelling, optimization and automatization of control processes in large scale economic systems.
Classification of algorithms contains fundamental structures in the kind of

- elementary algorithms for detail processes,
- integrated systems of algorithms with horizontal and vertical structures,
- complex hierarchical systems of algorithms.

Content, elements, internal and external relations of these algorithms and their practical application in industrial organization are described.

KEYWORDS

Algorithmic systems; control algorithms; hierarchy; industrial organization; production control.

SCOPE OF ALGORITHMS FOR CONTROLLING PROCESSES OF A PLANT

The work of people in industrial organization includes the operations of preparation, observation, check and interpretation of controlling reproduction processes of a plant for fulfilling defined problems. The complexity and complicatedness of economic processes and their multilateral relations to the environment leads to a great number of internal and external facts of influence which doesn't meet an optimal decision-making only on the basis of empiricism and intuition. Moreover many situations of decision in the control of processes of a plant contains a lot of constantly repeating operations which raise the coefficient for the part in routine work within the whole process of controlling.

Both causes determine the necessity for growing utilization of algorithms to case and to improve the work of man qualitatively. The scope of control algorithms in industrial organization is therefore related to 3 fields:

- Modelling of control processes to discover the lawfulness of their way of function, structure and their input/output data bulk,
- optimization of control processes to maximize the results of an activity in an enterprise by means of minimizing the expenditure required of resources,
- automatization of control processes as information processes which are the "technical" basis of industrial management and organization.

The solution of the first task can be used to the qualitative further development and rationalizing of the runnings of process as well as - in form of a cybernetic system analysis - the basis for optimizing and automating of controlling. The second and third tasks require the use of computers with an active communication between man and machine. With view to the forms of model used the simulation methods and fuzzy-algorithms are becoming more and more important [1].

BASIC STRUCTURES OF INDUSTRIAL CONTROL ALGORITHMS

Economic processes are modelled as large scale systems. This requires to consider the activity of man in a controlled process and in one which is to control, the complexity and complicatedness, the stochastic process and the ability to adaption and development of system [2]. Corresponding to the extent and profoundness of research of controlling economic processes 4 categories of structures of industrial control algorithms are to derive:

1. <u>Elementary algorithms for controlling sub-processes</u>

The elementary control algorithm

$$\Pi_E^{(St)} = \left[O, R, E, Z \right] \tag{1}$$

includes the functional operators $O = \{O_1, O_2, O_3, O_4^*\}$ and the logical operators $E = \{E_1, E_2\}$, which form the basis of each process of controlling (Fig. 1). The sequence of its working out is determined by a set of relations

$$R = \{ R_1 (O_1, O_2^{\cdot} \vee O_1), R_2 (O_2, O_3 \vee O_1), R_3 (O_3, O_4^*)\} \tag{2}$$

and by a set of allocations $Z = \{Z_1, Z_2\}$

$$Z_1 = \{ e_1 (> O_1, < O_2), \bar{e}_1 (> O_1, < O_1)\} \tag{3}$$

$$Z_2 = \{ e_2 (> O_2, < O_3), \bar{e}_2 (> O_2, < O_1)\}. \tag{4}$$

The process variants resulting from that are systematically combined in form of a tabulation (TABLE 1).
Elementary algorithms can be elaborated for sub-processes of following process - classes:

- forecast and long-term planning
- constructive and technical preparation of production

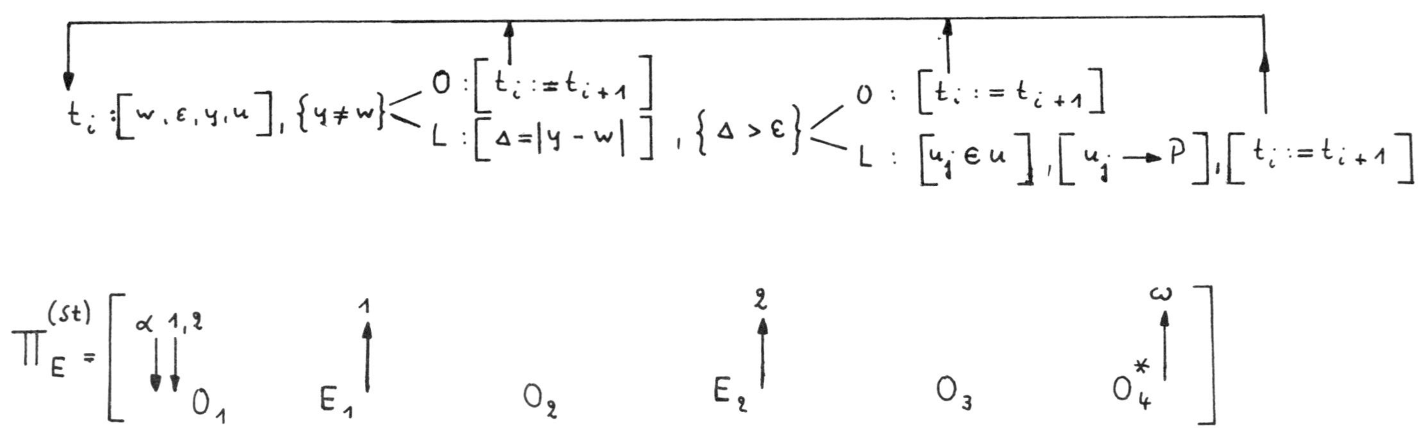

t_i	defined time interval of controlling
w	aims of process that are to achieve (reference values)
ε	tolerance of admissable deviations Δ between reference values and actual ones
u	quantity of possible control measures for new distribution of resources
y	really achieved process results (actual values)
P	controlled economic process
α	start of process
ω	end of process

Fig. 1. Algorithmic chain and operator algorithm of the elementary process of controlling

TABLE 1 Variants of Sequence of the Elementary Control Algorithm

No.	Combination E_i		Sequence O_j	Process Character	Control Measure
	E_1	E_2			
1	L	L	$O_1O_2O_3O_4^*$	substantially disturbed	necessary
2	I	O	$O_1O_2\overline{O_1O_2}\ldots$	unessentially disturbed	not necessary
3	O	L	$O_1\overline{O_1}\ldots$	undisturbed	not necessary
4	O	O	$O_1\overline{O_1}\ldots$	undisturbed	not necessary

- medium- and short-term planning
- preparation in organizing production
- operative control of production process
- marketing of products
- account and analyse of reproduction process.

2. Integrated systems of algorithms with horizontal structure

The condensation and linkage of sub-processes of a process class (e. g. of controlling production), of sub-processes of more process classes (e. g. of preparation and operative controlling production) or of process classes of one level of hierarchy (e. g. of preparation, implementation and check of production) leads to integrated horizontal systems of algorithms

$$\pi_{I'h}^{(St)} = \left[\left\{ \pi_{E_1}^{(St)}, \ldots, \pi_{E_n}^{(St)} \right\}, \left\{ R_{I'h} \right\} \right] . \qquad (5)$$

Such a system of algorithms was developed for automated controlling of production of a plant in the motor vehicle industry. It includes the controlling of premanufacturing departments (manufacturing of motor, machinery, carriage body) and installation departments [3].

3. Integrated systems of algorithms with vertical structure

The complexity of controlling in industrial organization is especially clear to see in its hierarchical character. The algorithms-hierarchy proves as a central category within the basic categories of hierarchy (aims-, decision-, algorithms-, information- and system-hierarchy).
A vertical integrated system of algorithms is to model as hierarchical algorithm $\pi_H^{(St)}$

$$\Pi^{(St)}_{I'v} = \Pi^{(St)}_{H} = \left[\left\{ \Pi^{(St)}_{o}, \Pi^{(St)}_{+m}, \Pi^{(St)}_{-n} \right\}, \left\{ R_H \right\} \right]. \quad (6)$$

It consists in the algorithm of the analysed level $\Pi^{(St)}_{o}$ - e. g. of controlling installation processes in a plant for motor vehicles - the algorithms of levels which are places over $\Pi^{(St)}_{+m}$ - e. g. of controlling production systems as a whole - and of algorithms of rearranged levels $\Pi^{(St)}_{-n}$, e.g. of controlling sub-installation processes for subassemblies, aggregates and finished products. The integration of these algorithms is defined by the hierarchical aggregate of relation R_H.

4. Complex hierarchical systems of algorithms

The control of reproduction processes of a plant at large unites integrated systems of algorithms of horizontal and vertical structure to complex hierarchical systems of algorithms

$$S = \left[M_{\Pi}, \ R_{\Pi} \right] \quad (7)$$

with the set of algorithms

$$M_{\Pi} = \left\{ \Pi^{(St)}_{I'h}, \Pi^{(St)}_{H} \right\} \quad (8)$$

and the set of relations R_{Π}, which determines the connection

- of elementary control algorithms $\Pi^{(St)}_{E_i}$ within horizontal systems of algorithms,
- of algorithms of different levels within the framework of vertical systems of algorithms,
- of systems of algorithms of horizontal and vertical structure.

As a result of a system analysis in the motor car production a complex hierarchical system of algorithms was elaborated. It consists of 10 subalgorithms Π_i

$$M_{\Pi} = \left\{ \Pi_1, \Pi_2, \Pi_3, \Pi_4, \Pi_5, \Pi_6, \Pi_7, \Pi_8, \Pi_9, \Pi_{10} \right\} \quad (9)$$

Π_1 Collection of relevant data from production, material economy and storage, transport, and sale in due time

Π_2 Preparation and provision of data for analytic and prognostic calculations

Π_3 Linking of data to the calculation of technological-economic parameters in defined check frequencies

Π_4 Situation analysis on the rate of deviations, tolerance limits exceeded and the necessity of control measures

Π_5 Analysis of equipment regarding groups of machines and techniques in reference to measures of maintenance

Π_6 Preparation of long-, medium- and short-term decisions by using optimization and simulation programmes for pre-manufacturing, assembly, transport, storage and maintenance

Π_7 Investigation of tendencies of changes in substantial technological-economic parameters affecting longer-term or operative control

Π_8 Calculation of anticipatory capacity developments in reference to investment projects and relations of cooperation

Π_9 Elaboration of control measures on the distribution of resources in equipment, material and working time as well as changes of command variables

Π_{10} Realization of decisions made about stabilizing perturbated processes and the further development of process parameters.

and of the relations between these algorithms

$$R_{\Pi} = \{ 1,(2),\ 2,(3,4,5),\ 3,(6,7),\ 4,(6),\ 5,(6,8),\ 6,(5,9),\ 7,(10),\ 8,(9),\ 9,(10) \} . \tag{10}$$

The connection of the level of hierarchy and of the subalgorithms is shown in TABLE 2.

TABLE 2 Hierarchical Allocation of Subalgorithms

No.	Level of Hierarchy	Allocated Algorithms
1	$\Pi_0^{(St)}$	Π_2, Π_4, Π_9
2	$\Pi_{+1}^{(St)}$	Π_3, Π_5, Π_6, Π_7, Π_8
3	$\Pi_{-1}^{(St)}$	Π_1, Π_{10}

In this way it will be possible

- to demonstrate the profile of division of labour, of scope of decision-making and communication among the levels exactly,
- to undertake the tuning of algorithms inside and among the levels unequivocally,
- to organize the provision of data aimed at proceeding individual algorithms over more levels.

This general structure is available as a principle for modelling reproduction processes in a branch, territory or planned national economy.

REFERENCES

[1] Stoica, M., and E. Scarlat (1975). Econ. Comput. and Econ. Cyb. Stud. and Res. (Bucharest) IX, 239-347.

[2] Golenko, D.I., H.-G. Lauenroth, H. Schultze, and G. Schulze (1977). Kybernetik zur Steuerung ökonomischer Prozesse - Grundlagen und Anwendungen. Akademie-Verlag, Berlin.

[3] Lauenroth, H.-G., and K. Böhm (1979). Kybernetik in der industriellen Organisation. Verlag Technik, Berlin.

ASYMPTOTIC STABILITY OF A CLASS OF STOCHASTIC INTERCONNECTED SYSTEMS

Dhemetrios Boussalis

Department of Electrical and Computer Engineering, Box 3-0
New Mexico State University
Las Cruces, New Mexico 88003

ABSTRACT

The stability of a class of stochastic interconnected systems is examined in the Lyapunov sense. Subsystem Lyapunov functions are used to construct Lyapunov functions for the composite system. The results presented are sufficient conditions for asymptotic stability with probability one.

INTRODUCTION

The systems involved in most engineering applications, and several classes of biological and socioeconomic systems are realized as interconnections of individual components or subsystems. A large-scale interconnected system can be completely defined in terms of the mathematical models of the unconstraint components and the constraints imposed by their interconnections. Due to the wide range of applications and the complexity involved, the stability analysis of such systems has received considerable attention during the last decade.

In earlier publications (Hahn 1963; Kalman 1960; Krasovskii 1963; Lyapunov 1947) the qualitative properties of nonlinear deterministic systems have been investigated by applying Lyapunov's Direct Method. When dealing with large-scale systems, however, one may encounter several difficulties in obtaining and applying stability criteria, since the construction of Lyapunov functions for higher-order nonlinear systems has been proven exceedingly difficult. Further, stability criteria, if obtained, are difficult to relate to the fundamental structural features of the system. To circumvent such difficulties several researchers (Michel 1974,; Ryaciotaki-Boussalis 1979; Siljak 1972; Thompson 1968) have shown that by viewing the large-scale system in terms of the individual components and the interconnection patterns, one may obtain stability results that are useful for actual design purposes, and for which computational difficulty has been considerably reduced.

Similar procedures have been established (Boussalis 1979a, 1979b, 1980; Michel 1975, 1976) for several classes of stochastic interconnected systems, where the individual components as well as the interconnections are influenced by white Gaussian noise. In the present paper some of these methods are employed to establish new stability results for a class of stochastic interconnected systems. Specifically, it is assumed that the isolated components are asymptotically stable in the large with probability one, and the associated Lyapunov functions are used to construct stochastic

Lyapunov functions for the interconnected system. The resulting stability criteria are sufficient conditions for asymptotic stability in the large with probability one.

NOTATION AND DEFINITIONS

Let $\mathcal{R}^n$ denote the n-dimensional Euclidean space of real n-vectors, $\mathcal{R}$ denote the real numbers, and $\mathcal{R}^+ \equiv [0,\infty)$. If $x \varepsilon \mathcal{R}^n$, let X^T denote its transpose $X^T \equiv [x_1,\ldots,x_n]$. The norm $|X|$ of a real n-vector is taken to be the Euclidean norm

$$|X| = (X^T X)^{1/2} = (\sum_{i=1}^{n} x_i^2)^{1/2}$$

Let $A = [a_{ij}]$ be an arbitrary $m \times n$ real matrix. Denote by A^T the transpose of A, and and let $\lambda_M(A)$ and $\lambda_m(A)$ be the largest and smallest eigenvalues of A, respectively. Then define the norms

$$|A| \equiv \{a : a|X| \geq |AX|, \text{ for all } X \varepsilon \mathcal{R}^n\} = [\lambda_m(A^T A)]^{1/2}$$

$$|A|_M = [tr(A^T A)]^{1/2}$$

and let M be the set of indices $M \equiv \{i : i = 1,2,\ldots,N\}$.

Definition 1: A continuous function $\phi : [0,r_1] \to \mathcal{R}^+$ (or $\phi : [0,\infty) \to \mathcal{R}^+$) is said to belong to class K, i.e., $\phi \varepsilon K$, if

- $\phi(0) = 0$, and
- $\phi(r)$ is strictly monotonically increasing with r on $[0,r_1]$ (or on $[0,\infty)$).

Definition 2: A continuous function $\phi : [0,r_1] \to \mathcal{R}^+$ (or $\phi : [0,\infty) \to \mathcal{R}^+$) is said to belong to class KR, i.e., $\phi \varepsilon KR$, if

- $\phi \varepsilon K$, and
- $\lim_{r\to\infty} \phi(r) = \infty$.

STOCHASTIC STABILITY

Consider the system of n stochastic differential equations

$$dX_t = f(X_t,t)dt + \sigma(X_t,t)dW_t, \quad f(0,t) = 0, \quad t_0 \leq t \leq T < \infty \tag{1}$$

where W_t is a $\mathcal{R}^m$-valued Wiener process, $f(X,t)$ is a $\mathcal{R}^n$-valued function, and $\sigma(X,t)$ is a $n \times m$ matrix-valued function. Further, assume that $f(.)$ and $\sigma(.)$ are continuously differentiable with respect to t, and that $\sigma(.)$ is twice continuously differentiable with respect to X, and define the differential operator of the process X_t with respect to system (1) as

$$L \equiv \frac{\partial}{\partial t} + \sum_{i=1}^{n} f_i(x_i,t)\frac{\partial}{\partial x_i} + \frac{1}{2}\sum_{i=1}^{n}\sum_{j=1}^{n} S_{ij}(X,t)\frac{\partial^2}{\partial x_i \partial x_j} . \tag{2}$$

where $S_{ij}(X,t) = \left[\sigma(X,t)\sigma(X,t)^T\right]_{ij}$.

Definition 3: The null solution of (1) is stable with probability one if and only if for any $\rho > 0$ and $\varepsilon > 0$, there exists a $\delta = \delta(\rho,\varepsilon)$, such that, if $|X_0| < \delta$, then

$$P_x\{\sup_{t\varepsilon R^+} |X_t| \geq \varepsilon \mid X_0 = X\} \leq \rho.$$

Definition 4: The null solution of (1) is asymptotically stable with probability one, if and only if it is stable with probability one and if $\lim X_t = 0$, $t \to \infty$, with probability one, for all X_0 in some neighbourhood R_0 of the origin. Further, if $R_0 = R^n$, then the origin is said to be asymptotically stable in the large with probability one.

Theorem 1: Consider the n-dimensional system of stochastic differential equations (1). If there exist a stochastic Lyapunov function $v(X_t,t)$, two functions $\phi_1, \phi_2 \varepsilon KR$, and a function $\phi_3 \varepsilon K$ such that

$$\begin{aligned} &\text{(a)}\ \phi_1(|X_t|) \leq v(X_t,t) \leq \phi_2(|X_t|) \\ &\text{(b)}\ Lv(X_t,t) \leq -\phi_3(|X_t|) \end{aligned} \tag{3}$$

for all $X_t \varepsilon Q_m \equiv \{(X,t) : v(X,t) < m \mid m > 0\}$, then the null solution of (1) is asymptotically stable in the large with probability one.

The proof of the above theorem can be found in most of the references (e.g. Boussalis 1979a), and is omitted.

SYSTEM DESCRIPTION

Of interest here are composite systems described by

$$dX_t = [f(X_t,t) + g(X_t,t)]dt + \sigma(X_t,t)dW_t \tag{4}$$

with

$$|g(X_t,t)| < e\ [\phi_3(|X_t|)]^{1/2} \tag{5}$$

where $X_t \varepsilon R^n$, $f: R^n \times R^+ \to R^n$, $g: R^n \times R^+ \to R^n$, $\sigma: R^n \times R^+ \to R^{n \times m}$, $W_t \varepsilon R^m$, $e > 0$, and $\phi_3 \varepsilon K$, with decomposition

$$dX_t^i = [f_i(X_t^i,t) + g_i(X_t,t)]dt + \sum_{j=1}^{N} \sigma_{ij}(X_t^j,t)dW_t^j \tag{6}$$

for all $i,j \varepsilon M$, where $X_t^i \varepsilon R^{n_i}$, $f_i: R^{n_i} \times R^+ \to R^{n_i}$, $g_i: R^n \times R^+ \to R^{n_i}$, $\sigma_{ij}: R^{n_j} \times R^+ \to R^{n_i \times m_j}$, $W_t^i \varepsilon R^{m_i}$,

$\sum_i n_i = n$, $\sum_i m_i = m$, and with isolated components given by

$$dX_t^i = f_i(X_t^i,t)dt + \sigma_{ii}(X_t^i,t)dW_t^i \tag{7}$$

for all $i,j \varepsilon M$.

ASYMPTOTIC STABILITY

The following theorems provide sufficient conditions for asymptotic stability of system (4), under the assumption that the isolated components (7) are asymptotically stable in the large with probability one. Also note that the matrix $A_{ii}(t)$ appearing in the sequel is defined by

$$A_{ii}(t) = \nabla^2 v_i(X_t^i,t) \tag{8}$$

Theorem 2: Consider interconnected systems (4). Let v_i, and $\phi_1^i, \phi_2^i \varepsilon KR$, $\phi_3^i \varepsilon K$, be stochastic lyapunov function and associated scalar functions satisfying (3) for the i-th component (7). Assume that in general $\sigma_{ij} \neq 0$, and let $b > 0$ be a constant such

$$|\nabla v(X_t,t)| \leq b[\phi_3(|X_t|)]^{1/2}. \tag{9}$$

Further, suppose there exist a positive parameter vector $K = [k_1,\ldots,k_N]^T$, and constants $d_{ij} > 0$ satisfying

$$|\sigma_{ij}(X_t^j,t)|_M^2 \leq d_{ij}\phi_3^i(|X_t|) \tag{10}$$

and

$$0 < e < \frac{1}{b} \min_i k_i(1 - \frac{1}{2} \sum_{j=1}^N \lambda_M^i d_{ij}), \qquad i \neq j, \tag{11}$$

for all $i,j \varepsilon M$, where e is given in (5), and λ_M^i is the largest eigenvalue of A_{ii}, then, the null solution of (4) is asymptotically stable in the large with probability one.

Proof: For interconnected system (4) assume Lyapunov functions of the form

$$v(X_t,t) = \sum_i k_i v_i(X_t^i,t), \quad i \varepsilon M. \tag{12}$$

Then one may easily show (Boussalis 1979a) with $\phi_1 = \sum_i \phi_1^i$ and $\phi_2 = \sum_i \phi_2^i$, (3a) of Theorem 1 is satisfied. Then applying the differential operator L to v, and using (3b), (5), (8), and (9) one has

$$Lv \leq \sum_{i=1}^N -k_i\phi_3^i + |\nabla v||g(X_t,t)| + \frac{1}{2} \sum_{i=1}^N \sum_{j=1}^N k_i\lambda_M^i d_{ij}\phi_3^j, \quad i \neq j \tag{13}$$

or

$$Lv \leq \sum_{i=1}^N -k_i\phi_3^i + be\phi_3^i + \frac{1}{2} \sum_{i=1}^N \sum_{j=1}^N k_i\lambda_M^i d_{ij}\phi_3^j \leq -c \sum_{i=1}^N \phi_3^i = -\phi_3(|X_t|) \tag{14}$$

where

$$c = \min_i k_i(1 - \frac{1}{2} \sum_i \lambda_M^i d_{ij}) - be \tag{15}$$

If c given by (15) is positive, it is implied that ϕ_3 in (14) satisfies (3b), and according to Theorem 1 the null solution of (4) is asymptotically stable in the large with probability one. But $c > 0$ if (11) is satisfied.

The foolowing theorem considers the special case in which the interconnecting structure is not corrupted by noise.

Theorem 3: Consider interconnected systems (4). Let $v_i, \phi_1^i, \phi_2^i, \varepsilon KR$, $\phi_3^i \varepsilon K$ be stochastic Lyapunov function and associated scalar functions satisfying (3a) for the i-th component (7). Further, assume that $\sigma_{ij}=0$, $i \neq j$, $i,j \varepsilon M$, and that there is a constant $b > 0$ satisfying (9). If there exists a positive parameter vector $K= [k_1, \ldots, k_N]^T$ such that

$$e < \frac{1}{b} \min_i k_i \tag{16}$$

where e is given by (5), then the null solution of (4) is asymptotically stable in the large with probability one.

Proof: The theorem is proven by assuming Lyapunov functions given by (12) and with the same reasoning as in the proof of Theorem 2.

CONCLUSIONS

The stability of a class of stochastic interconnected systems is examined. Sufficient conditions for asymptotic stability in the large with probability one for the interconnected system are obtained in terms of the stability properties of the isolated components. The results suggest the existence of a parameter vector K. In actual applications this vector may be obtained as the solution of a static parameter optimization problem.

REFERENCES

Boussalis, D. (1979a). Stochastic Stability of Interconnected Systems. Ph.D. dissertation, New Mexico State University, Las Cruces, New Mexico.

Boussalis, D., and W. E. Thompson (1979b). Stochastic Stability of Decentralized Systems. Proc. 18th IEEE CDC, pp. 672-673.

Boussalis, D., and W. E. Thompson (1980). Exponential Stability of Stochastic Decentralized Systems. Proc. 11th Annual Pitt. Conf. on Modeling and Simulation.

Hahn, W. (1963). Application of Lyapunovs Direct Method. Prentice-Hall, Englwood Cliffs.

Kalman, R. E., and J. E. Bertram (1960). Control System Analysis and Synthesis via the Second Method of Lyapunov. Trans. ASME Ser. J. Basic Engin., 82D, pp. 371-393.

Krasovskii, N. N. (1963). Stability of Motion. Stanford Univ. Press. Stanford, Ca.

Lyapunov, A. M. (1947). Probleme General de la Stabilite du Movement. Ann. Math. Studies, No 17. Princeton.

Michel, A. N. (1974). Stability Analysis of Interconnected Systems. SIAM J. Control, vol. 12, No. 3, pp. 554-579.

Ryaciotaki-Boussalis, H. A. (1979). Stability of Large-Scale Systems. Ph.D. Dissertation, New Mexico State University, Las Cruces, New Mexico.

Thompson, W. E. (1968). Stability of Interconnected Systems. Ph.D. Dissertation, Michigan State University, East Lansing, Michigan.

STOCHASTIC STABILITY OF A CLASS OF DECENTRALIZED SYSTEMS

Dhemetrios Boussalis and Wiley E. Thompson

Department of Electrical and Computer Engineering, Box 3-0
New Mexico State University
Las Cruces, New Mexico 88003

ABSTRACT

Stochastic stability of large-scale decentralized systems is examined in terms of their lower-order subsystems and interconnection patterns. The stability analysis is performed in the sense of Lyapunov, and the results presented are sufficient conditions for exponential stability with probability one.

INTRODUCTION

The study of decentralized control systems has received considerable attention in recent years because of the increasing number of applications in which such systems are involved. The feature of decentralization arizes naturally when dealing with large-scale systems such as power systems, urban traffic networks, and several classes of socioeconomic and biological systems. In a decentralized scheme, the interconnected system is usually characterized by the geographical separation of its components. With each component a local feedback controller and a local input are associated. Each controller observes the local outputs, controlls the local inputs, and contributes to the stabilization of the composite system. Recent papers have examined the stability, in the sense of Lyapunov, of deterministic (Thompson 1978) as well as of stochastic (Boussalis 1979a, 1979b) decentralized systems. In the present paper the same class of decentralized systems with the local inputs and the interconnecting structure involving stochastic disturbances (Kushner 1967; Michel 1977) is considered. The results presented, are sufficient conditions for exponential stability with probability one. The stability criteria for the decentralized system are obtained in terms of the fundamental structural features of the system (subsystems, and interconnecting structure), thus, relating the stability properties of the isolated components and local feedback controllers to those of the composite system.

NOTATION AND DEFINITIONS

Let $\mathcal{R}^n$ denote the n-dimentional Euclidean space of real n-vectors, $\mathcal{R}$ denote the real numbers, and $\mathcal{R}^+ = [0,\infty)$. If $X \in \mathcal{R}^n$, let X^T denote its transpose. The norm $|X|$ of a real n-vector is taken to be the Euclidean norm $|X| = (X^TX)^{1/2}$. Let $A = [a_{ij}]$ be an arbitrary $m \times n$ matrix. Denote by A^T the transpose of A, and let $\lambda_M(A)$ and $\lambda_m(A)$ be the largest and smallest eigenvalues of A, respectively. Then define the norms $|A| = \{a : a|X| \geq |AX|$, for all $X \in \mathcal{R}^n\} = [\lambda_M(A^TA)]^{1/2}$, and $|A|_M = [tr(A^TA)]^{1/2}$. Further, let M denote the set of indices $M = \{i : i=1,2,\ldots,N\}$.

STOCHASTIC STABILITY

Consider the system of n stochastic differential equations

$$dX_t = f(X_t,t)dt + \sigma(X_t,t)dW_t, \quad f(0,t) = 0, \quad t_0 \leq t \leq T < \infty, \tag{1}$$

where W_t is a $\mathcal{R}^m$-valued Wiener process, $f(X_t,t)$ is a $\mathcal{R}^n$-valued function, and $\sigma(X_t,t)$ is a $n \times m$ matrix-valued function. Further, assume that f(.), and σ(.) are continuously differentiable with respect to t, and that σ(.) is twice continuously differentiable with respect to X. Also define the differential operator of the process X_t with respect to system (1) as

$$L \equiv \frac{\partial}{\partial t} + \sum_{i=1}^{N} f_i(x_i,t)\frac{\partial}{\partial x_i} + \frac{1}{2}\sum_{j=1}^{N}\sum_{j=1}^{N} S_{ij}(X,t)\frac{\partial^2}{\partial x_i \partial x_j} \tag{2}$$

where $S_{ij}(X,t) = [\sigma(X,t)\sigma(X,t)^T]_{ij}$.

<u>Definition 1:</u> The null solution of (1) is stable with probability one, if and only if for any $\rho > 0$, and $\varepsilon > 0$, there exists a $\delta = \delta(\rho,\varepsilon)$, such that, if $|X_0| < \delta$, then

$$P_x\{\sup_{t\in\mathcal{R}^+} |X_t| \geq \varepsilon \mid X_0 = X\} \leq \rho.$$

<u>Definition 2:</u> The null solution of (1) is exponentially stable in the large with probability one if and only if it is stable with probability one, and for all $r > 0$, all $T \in \mathcal{R}^+$, and $\varepsilon > 0$, there exist positive constants $\beta = \beta(r)$, and α such that if $|X_0| < r$, then

$$P_x\{\sup_{T<t<\infty} |X_t| \geq \varepsilon \mid X_0 = X\} \leq \beta\exp(-\alpha T)$$

<u>Theorem 1:</u> Consider the n-dimensional system of stochastic differential equations (1). If there exist a stochastic Lyapunov function v and a set of positive constants c_1, c_2, c_3, and c_4 satisfying (a) and (b) of the relations

$$\begin{aligned} &(a) \quad c_1|X_t|^2 \leq v(X_t,t) \leq c_2|X_t|^2 \\ &(b) \quad Lv(X_t,t) \leq -c_3|X_t|^2 \\ &(c) \quad |\nabla v(X_t,t)| \leq c_4|X_t| \end{aligned} \tag{3}$$

for all $X_t \in \Omega_m \equiv \{(X,t) : v(X_t,t) < m \mid m > 0\}$, then the null solution of (1) is exponentially stable in the large with probability one.

<u>Theorem 2:</u> Consider the n-dimensional system of stochastic differential equations (1). Let v be a stochastic Lyapunov function, and let c_1, c_2, and α be positive constants such that

$$(a) \quad c_1|X_t|^2 \leq v(X_t,t) \leq c_2|X_t|^2$$

$$(b)\quad \mathcal{L}v(X_t,t) \le -\alpha v(X_t,t) \tag{4}$$

for all $X_t \in Q_m$. Then, the null solution of (1) is exponentially stable in the large with probability one.

STOCHASTIC DECENTRALIZED SYSTEMS

Consider the class of interconnected systems described by

$$\begin{aligned} \dot{X}_t &= f(X_t,t) + \sum_{j=1}^{N} B_i u_t^i \\ y_{ct}^i &= C_i(t)X_t + \zeta_t^i \end{aligned} \tag{5}$$

for all $i\varepsilon M$, where $X_t \varepsilon R^n$, $f{:}R^n \varepsilon R^+ \to R^n$, $B_i \varepsilon R^{n\times m_i}$, $u_t^i \varepsilon R^{m_i}$, $C_i \varepsilon R^{p_i \times n}$. To each of the N components of (1), a local feedback controller is associated, and described by

$$\begin{aligned} \dot{Z}_t^i &= h_i(Z_t^i,t) + R_i(t)y_{ct}^i \\ y_{zt}^i &= Q_i(t)Z_t^i + \nu_t^i \end{aligned} \tag{6}$$

where $Z_t^i \varepsilon R^{n_i}$, $h_i{:}R^{n_i}\times R^+ \to R^{n_i}$, $R_i \varepsilon R^{n_i \times p_i}$, $y_{zt}^i \varepsilon R^{m_i}$, $Q_i \varepsilon R^{m_i \times n_i}$. Further, let the input process u be described by

$$u_t^i = y_{zt}^i + K_i(t)y_{ct}^i + V_i(t) \tag{7}$$

where $K_i \varepsilon R^{m_i \times p_i}$, and $V_i \varepsilon R^{m_i}$ is a deterministic vector input to the i-th subsystem. The noise processes ζ_t and ν_t can be expressed (Arnold 1973, Boussalis 1979a) in terms of a Wiener process W_t, and (5),(6) and (7) after some algebraic manipulations yield

$$dX_t = [f(X_t,t) + BQ(t)Z_t + K(t)C(t)X_t]dt + \sigma_c(X_t,Z_t,t)dW_t \tag{8}$$

and

$$dZ_t = [h(Z_t,t) + R(t)C(t)X_t]dt + \sigma_z(X_t,t)dW_t \tag{9}$$

The i-th component and the i-th feedback controller are described, respectively, by

$$dX_t^i = f_i(X_t^i,t)dt + \sigma_c^{ii}(X_t^i,Z_t^i,t)dW_t^i \tag{10}$$

and

$$dZ_t^i = h_i(Z_t^i,t)dt + \sigma_z^{ii}(X_t^i,t)dW_t^i \tag{11}$$

or in matrix form, by combining (8)-(11), the system is described by

$$d\Psi_t = [F(\Psi_t,t) + M(t)\Psi_t]dt + \sigma(\Psi_t,t)dW_t \tag{12}$$

with component models

$$d\Psi_t^i = F_i(\Psi_t,t)dt + \sigma^{ii}(\Psi_t^i,t)dW_t^i \tag{13}$$

where $\Psi_t^T = [X_t^T \mid Z_t^T]$, $\Psi_t^{iT} = [X_t^{iT} \mid Z_t^{iT}]$, $F(\Psi_t,t)^T = [f(X_t,t)^T \mid h(Z_t.t)^T]$, $F_i(\Psi_t^i,t)^T = [f_i(X_t^i,t)^T \mid h_i(Z_t^i,t)^T]$,

$$M(t) = \left[\begin{array}{c|c} BK(t)C(t) & BQ(t) \\ \hline R(t)C(t) & 0 \end{array}\right]$$

and $\sigma(\Psi_t,t)^T = [\sigma_c^T \mid \sigma_z^T]$, $X_t^i \varepsilon R^{n_i}$, $\sum_i n_i = n$, and $f_i{:}R^{n_i}\times R^+ \to R^{n_i}$.

EXPONENTIAL STABILITY RESULT

The following theorem gives sufficient conditions for exponential stability with probability one of system (12). The result is obtained under the assumption that the isolated components (10) and the local feedback controllers (11) are exponentially stable in the large with probability one. Also note that the matrix A used in the sequel is defined by

$$A_{ii}(t) = \begin{bmatrix} \nabla_{x_i x_i} v_c^i & 0 \\ 0 & \nabla_{z_i z_i} v_z^i \end{bmatrix} \tag{14}$$

Theorem 3: Consider interconnected systems (12). Let v_c^i, α_1^i, α_2^i, α_3^i, α_4^i, and v_z^i, β_1^i, β_2^i, β_3^i, β_4^i be stochastic Lyapunov functions and associated positive constants satisfying (3) for the i-th component (10), and the i-th feedback controller (11), respectively. If there exist a positive parameter vector $K = |k_1, \ldots, k_N|^T$ and constants γ, $d_{ij} > 0$ satisfying

$$|M(t)\Psi_t| < \gamma |\Psi_t| \tag{15}$$

and

$$|\sigma_{ij}(\Psi_t^j, t)|_M^2 \leq d_{ij} |\Psi_t^j|^2 \tag{16}$$

for all $i,j \in M$, such that the relation

$$\gamma < \frac{\min\limits_i \left\{ \dfrac{\alpha_3^i + \beta_3^i}{\alpha_2^i + \beta_2^i} - \dfrac{1}{2} \sum\limits_{j=1}^{N} \dfrac{\lambda_M^i(A) d_{ij}}{\alpha_1^j + \beta_1^j} \right\}}{\max\limits_j \dfrac{\alpha_4^j + \beta_4^j}{\alpha_1^j + \beta_1^j}} \tag{17}$$

is satisfied, then the null solution of (12) is exponentially stable in the large with probability one.

Proof: For decentralized system (12) consider stochastic Lyapunov functions given by

$$v(\Psi_t, t) = \sum_{i=1}^{N} k_i v_i(\Psi_t^i, t) = \sum_{i=1}^{N} k_i \left[v_c^i(X_t^i, t) + v_z^i(Z_t^i, t) \right]. \tag{18}$$

Applying (3a), and (3c) to (18) one can show that

$$c_1 |\Psi_t|^2 \leq v(\Psi_t, t) \leq c_2 |\Psi_t|^2 \tag{19}$$

and

$$|\nabla v(\Psi_t, t)| \leq c_4 |\Psi_t| \tag{20}$$

where $c_1 = \min_i k_i(\alpha_1^i + \beta_1^i)$, $c_2 = \max_i k_i(\alpha_2^i + \beta_2^i)$, and $c_4 = \max_i k_i(\alpha_4^i + \beta_4^i)$. Then applying the differential operator L with respect to (12) to (18) one obtains (Boussalis 1979a):

$$Lv(X_t, t) \leq \sum_{i=1}^{N} -k_i(\alpha_3^i + \beta_3^i)|\Psi_t^i|^2 + |\nabla v| \, |M(t)\Psi_t| + \frac{1}{2} \sum_{i=1}^{N} \sum_{\substack{j=1 \\ j \neq i}}^{N} k_i \lambda_M^i(A) d_{ij} |\Psi_t^j|^2$$

where $_M^i(A)$ is the largest eigenvalue of A in (14), and d_{ij} are given by (16). Applying (15), (19), and (20) to the above gives:

$$Lv(\Psi_t,t) \le \sum_{i=1}^{N} -k_i \frac{\alpha_3^i + \beta_3^i}{\alpha_2^i + \beta_2^i} v_i + \frac{c_4}{c_1}\gamma v(\Psi_t,t) + \frac{1}{2}\sum_{i=1}^{N}\sum_{\substack{j=1 \\ j\neq i}}^{N} k_i \frac{\lambda_M^i(A)\ d_{ij}}{\alpha_1^j + \beta_1^j} v_j$$

$$\le -\alpha v(\Psi_t,t) \tag{21}$$

where

$$\alpha = \min_i\{\frac{\alpha_3^i + \beta_3^i}{\alpha_2^i + \beta_2^i} - \frac{1}{2}\sum_{\substack{j=1 \\ j\neq i}}^{N} \frac{\lambda_M^i(A)\ d_{ij}}{\alpha_1^j + \beta_1^j}\} - \gamma \max_j \frac{\alpha_4^j + \beta_4^j}{\alpha_1^j + \beta_1^j} \tag{22}$$

If α in (22) is positive, it is implied from (19), (21), and Theorem 2 that the null solution of (12) is exponentially stable in the large with probability one. But $\alpha > 0$ only if (17) is satisfied.

CONCLUSIONS

The stability of a class of stochastic decentralized systems is examined in the Lyapunov sense. Stability results are obtained in the form of sufficient conditions for exponential stability in the large with probability one of the decentralized system. The criteria given in Theorem 3 depend on the positive parameter vector K. In actual applications this vector may be determined as the solution of a static parameter optimization problem.

REFERENCES

Arnold L., (1973). Stochastic Differential Equations, Wiley, New York.

Boussalis D., (1979a). Stochastic Stability of Interconnected Systems, Ph.D. Dissertation, New Mexico State University, Las Cruces, New Mexico.

Boussalis D., and W. E. Thompson, (1979b). Stochastic Stability of Decentralized Systems. Proc. 18th IEEE CDC, pp. 672-673.

Kushner H. J., (1967). Stochastic Stability and Control, Academic Press, New York.

Michel A. N., and R. K. Miller, (1977). Qualitative Analysis of Large-Scale Systems, Academic Press, New York.

Thompson W. E., H. R. Boussalis, and D. Boussalis, (1978). Stability of a Class of Decentralized Systems. Proc. 22nd Midwest Symposium on CAS, pp 69-73.

ON THE CHOICE OF THE NUMBER OF INTEGRATIONS IN ARIMA MODELS

N.K. Sinha and O. Abul-Haggag Ibrahim
Department of Electrical and Computer Engineering
McMaster University
Hamilton, Canada

ABSTRACT

A simple procedure is proposed for selecting the number of integrations required in an autoregressive-integrated moving-average (ARIMA) model for nonstationary time series. Several examples have been included which provide a good insight into this important modeling problem.

KEY WORDS

Stochastic modeling; nonstationary time-series; ARIMA models; number of integrations.

INTRODUCTION

In a recent paper (Sinha and Prasad, 1979) several different types of stochastic models for estimating and forecasting were described. These models can be obtained from the given data using simple procedures which can be implemented even on a programmable pocket calculator. It was pointed out that for most cases a time-series model (Box and Jenkins, 1970) gives good results. However, there are two main questions which must be answered before attempting to obtain a model. First of all, we must determine whether the model should be stationary or nonstationary. A time-series is said to be stationary if it fluctuates randomly about a constant mean. A nonstationary time-series does not have a natural mean, but if it is differenced several times, eventually a stationary series is obtained. The number of times this differencing must be performed is called the number of integrations in the autoregressive moving-average (ARIMA) model. In other words, the given nonstationary time-series is obtained by integrating a stationary process the same number of times. To obtain a suitable model, therefore, one must determine this number of integrations. The second problem is the determination of the order of the model, i.e., the number of autoregressive and moving-average terms in a good parsimonious model. As far as the authors are aware, a systematic solution to these problems have not been presented in the literature on time series analysis.

In this paper, a simple procedure will be proposed for determining the number of integrations most suitable for a given time-series. Several examples will be utilized to provide a good insight into the problem.

STATIONARITY OF A TIME-SERIES

In conventional time-series analysis (Box and Jenkins, 1970), a process is considered stationary if its sample autocorrelation function tails off as the number of lags is increased. For a nonstationary time-series, on the other hand, the sample autocorrelation function will die down extremely slowly. Although this is a useful criterion, it is rather indirect and requires a large amount of computation. Furthermore, it does not tell us the number of times the data must be differenced to obtain a stationary series. A more direct approach is to utilize the basic definition of a stationary time-series, that it fluctuates randomly about a constant mean. To use this idea, we should examine the variation of the running average of the time series. For our purpose, the running average series may be obtained by first obtaining the average of a suitable fraction of the entire data, and then updating this average after each subsequent observation. Thus, we may define the running average as

$$A_k = \frac{1}{k} \sum_{n=1}^{k} R_n \tag{1}$$

where A_k = the kth value of the running average
R_n = the nth value of time-series to be modelled.

A recursive equation for updating the running average is easily obtained, and is given by

$$A_{k+1} = \frac{1}{k+1} (kA_k + R_{k+1}) \tag{2}$$

As an example, consider the time series F from Box and Jenkins (1970), which represents the yields from a chemical batch process. This series, and its running average have been plotted in Fig. 1, where the solid curve shows the time series and the dashed curve is a plot of the running average. It is seen that although the time series is varying randomly, its running average is nearly constant. Hence, it is evident that this is a stationary time series.

As another example, the population of the United States from annual data over the period 1948 to 1971 has been plotted in Fig. 2. The running average, shown by the dashed curve is far from constant. Hence, this cannot be regarded as a stationary time series. As a matter of fact, since the running average, in this example, appears to be increasing linearly, it will be reasonable to assume that the first difference of this series will be stationary, and that it may be modelled as a series with one integration.

ESTIMATION OF THE NUMBER OF INTEGRATIONS

From the above, it is evident that a plot of the running average of a time series can be used to determine not only whether it is non-stationary, but also to estimate the number of integrations required. The following simple rules are proposed.

(1) If the plot of the running average A_k against k is almost constant, i.e., a curve with zero slope, it represents a stationary series, and the number of integrations is equal to zero.

(2) If the slope of the plot of A_k against k is constant but nonzero, then the number of integrations is equal to one.

(3) If the slope of the plot of the running average A_k against k is changing, but the change in the slope is constant then the necessary number of integrations is equal to two.

We shall now apply these rules to the two practical examples.

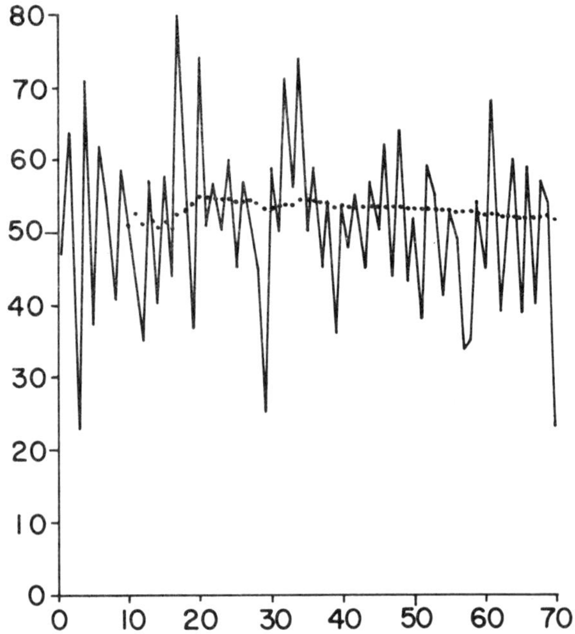

Fig. 1 Series F: (Yields from Batch Chemical Process) from Box and Jenkins.

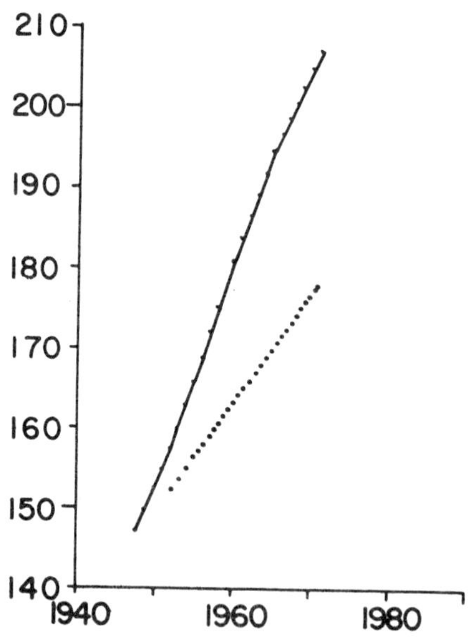

Fig. 2 Population of the United States (in millions)

EXAMPLE I: POPULATION OF THE UNITED STATES

As pointed out earlier, the plot of the running average shown in Fig. 2 suggests that this model is nonstationary, and one integration is required. Accordingly, the following ARIMA(1, 1, 0) model was obtained, utilizing the procedure described in an earlier paper (Sinha and Prasad, 1979)

$$\hat{x}_k = 1.05623x_{k-1} - 1.00785x_{k-2} - 0.048482\hat{x}_{k-1} \tag{3}$$

where

$\hat{x}_k$ = the estimated population in year k

x_{k-1} = the actual population in the previous year.

For the sake of comparison, two other models were also obtained, one using linear regression and the other was an ARMA(1, 1) model assuming that the model was stationary. These models are given below

Linear regression: $$\hat{x}_k = 144.813 + 2.687(k-1947) \tag{4}$$

ARMA(1, 1) $$\hat{x}_k = 1.0125x_{k-1} - 0.1072x_k + 0.0947\bar{x} \tag{5}$$

where $\bar{x}$ = the mean value of x calculated for the entire available data.

The actual population for the period, as well as the predicted values using these models are shown in Table I. For the two time-series models, the one-step ahead predicted values are shown for each case.

It is seen that the ARIMA(1, 1, 0) model gives the least mean-square error, while the ARMA(1, 1) model is rather poor. This emphasizes the need for using the correct number of integrations in an ARIMA model.

EXAMPLE 2: THE SLANT RANGE OF A SATELLITE

The data, consisting of 200 successive observations of the slant range of a communications satellite, was obtained from the Communications Research Centre in Ottawa, and it was desired to obtain a time-series model.

A plot of the running average is shown in Fig. 3. In this case, the running average was started with 90% of the data, and is plotted to a different scale. This change of scale does not alter the fact that the running average is increasing linearly. Hence, the parameters of an ARIMA(3, 1, 3) model were determined for this nonstationary series which gave a very good fit with mean-square error of the order of 0.14×10^{-5} (Sinha and Ibrahim, 1979).

CONCLUSIONS

A simple and straightforward method has been proposed for estimating the number of integrations in a nonstationary time-series model. The plot of the running averages is obtained in a fairly simple manner and because of its recursive nature the calculation can even be carried out on a programmable pocket calculator. From a number of test-cases carried out by the authors (only three have been reported in this paper), it was found that the best results were obtained when the running average was started with 90% of the data. The reason for this is currently being investigated.

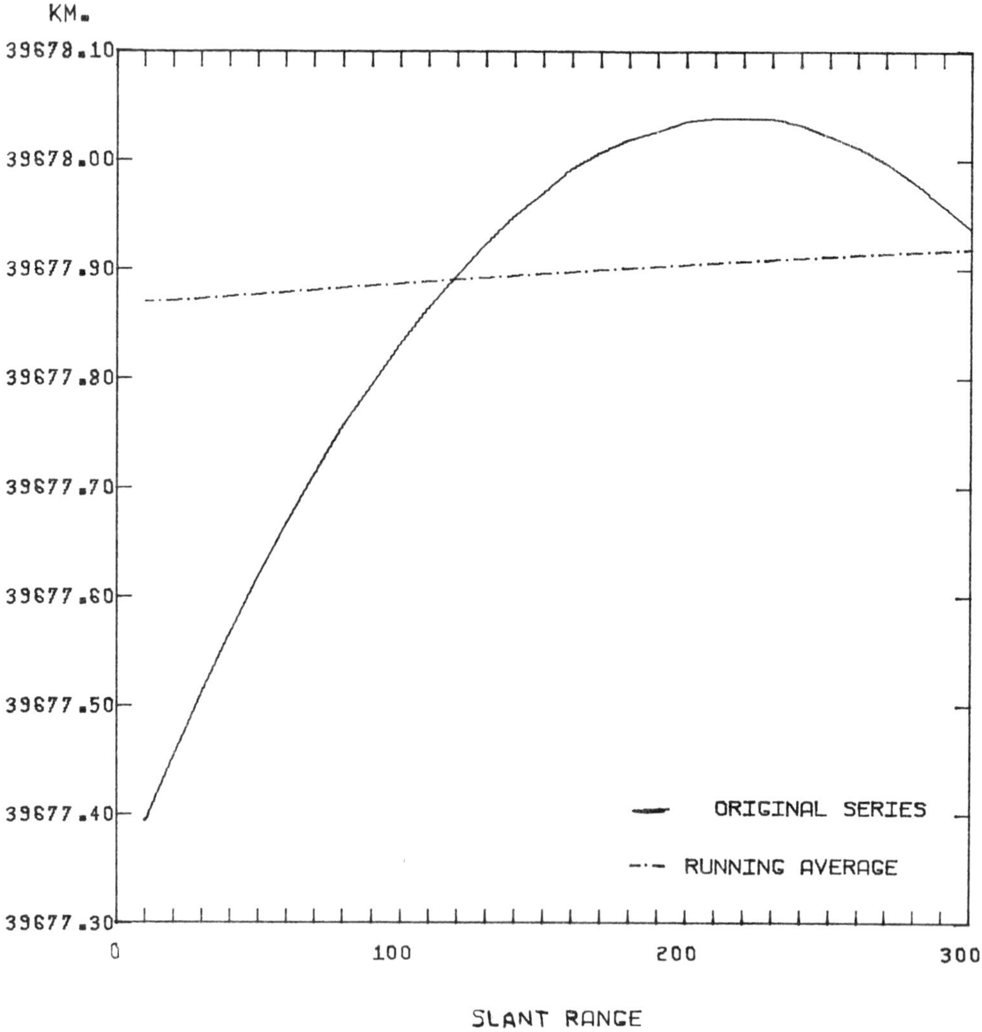

Fig. 3 Slant range of a satellite.

TABLE 1 Comparison of the actual and predicted population of the United States (in millions)

Year	Actual	Linear Regression	ARMA(1, 1)	ARIMA(1, 1, 0)
1948	147.2	147.5	-	-
1949	149.8	150.2	150.2	-
1950	152.3	152.9	152.5	152.4
1951	154.9	155.6	154.8	154.8
1952	157.6	158.2	157.1	157.5
1953	160.2	160.9	159.6	160.3
1954	163.0	163.6	162.0	162.8
1955	165.9	166.3	164.6	165.8
1956	168.9	169.0	167.2	168.8
1957	172.0	171.7	170.0	171.9
1958	174.9	174.4	172.8	175.1
1959	177.8	177.0	175.5	177.8
1960	180.7	179.7	178.1	180.9
1961	183.7	182.4	180.8	183.6
1962	186.5	185.1	183.5	186.7
1963	189.2	187.8	186.1	189.3
1964	191.9	190.5	188.5	191.9
1965	194.3	193.2	191.0	194.6
1966	196.6	195.9	193.1	196.7
1967	198.7	198.5	195.2	198.9
1968	200.7	201.2	197.1	200.8
1969	202.7	203.9	199.0	202.7
1970	204.9	206.6	200.8	204.7
1971	207.0	209.3	202.8	207.1
Mean Error	-	0	2.17	-0.038
Mean Square Error	-	0.985	6.55	0.020

ACKNOWLEDGEMENTS

Partial support of this work by the Communications Research Centre of Ottawa is gratefully acknowledged. The authors are indebted to S.P. Altman of the C.R.C. for many interesting discussions.

REFERENCES

Box, G.E.P., and Jenkins, G.M. (1970). Time series analysis-forecasting and control. Holden Day, San Francisco.

Sinha, N.K. and Ibrahim, O. Abul-Haggag (1979). Modelling and forecasting for slant range. Report prepared for the Communications Research Centre, Ottawa.

Sinha, N.K. and Prasad, T. (1979). Some stochastic modelling techniques and their applications. Applied Mathematical Modelling, 3, pp. 2-6.

STOCHASTIC MODEL RELATING HIGH SCHOOL PERFORMANCE TO THAT IN UNIVERSITY EDUCATION IN ENGINEERING

N.K. Sinha and O. Abul-Haggag Ibrahim

Department of Electrical and Computer Engineering
McMaster University, Hamilton, Canada

ABSTRACT

A stochastic model is presented relating the performance of students in a university program in engineering with their work in the high school. This model can provide useful information for admission, counselling and technical forecasting.

KEYWORDS

Mathematical modelling; stochastic models; linear regression; technical forecasting.

INTRODUCTION

In an earlier paper (Sinha and diCenzo, 1975a), a stochastic state space model was proposed for the progress of students in an educational system. The engineering school of McMaster University was used as an example of the model. Another approach to the same problem based on time-series analysis was presented in a different paper (Sinha and diCenzo, 1975b). These studies indicate that the success of a student in the engineering program is determined largely by his or her ability to handle the work in the first two years of the program. Moreover, the attrition rate was highest in the first year of the engineering program, which may, therefore, be considered as the dominant factor.

Since more than eighty-five percent of the students in the undergraduate engineering program at McMaster complete their admission requirements in the Ontario secondary school system, it is reasonable to assume that there should be some correlation between the performance of a student in high school and in the first year of the engineering program. However, since it is not necessary that all students doing very well in high school will also do very well in the engineering program at the university, any such relationship can only be of stochastic nature. Some other nondeterministic factors which affect the relationship are the nonuniformity of standards between different high schools in the area, the variation in the standards from year to year, and the change in motivation caused by the difference between the social and academic environments of a high school and a university.

In view of the above, a stochastic model relating secondary school and university performances was developed and presented at an earlier conference (Sinha and diCenzo, 1976). In that paper a linear model was obtained from data for six years (academic years 1969-70 to 1974-75) relating the average grades obtained by students in the first year of the engineering program to their average grade 13 mark in Mathematics, Physics, Chemistry and English. It may be pointed out that the average mark obtained by the student in these four subjects is used as the criterion for admission to the engineering school and a student must have an average of at least 64 percent to be eligible. Since then, the number of students in the engineering program has increased considerably. It was, therefore, considered desirable to re-examine the model utilizing the additional data obtained during the last four years. It is felt that such a model would be very useful and provide more information for guidance and counselling purposes. Furthermore, since most engineering schools in Ontario are planning to limit their enrolments, the model should make it possible to examine the validity of the admission requirements currently utilized and the modifications which might be considered.

FORMS OF THE MODEL

It is quite reasonable to assume that the performance of a student in the first year of the engineering program should be related in some way to the performance in the high school. Assuming a simple first-order linear autoregressive process, we get the following relationship

$$g_1 = \alpha\, g_0 + v \tag{1}$$

where

g_1 = average grade of the student at the end of the first year of the engineering program, expressed as a weighted percentage (a grade-point average may also be used).

g_0 = average mark obtained by the student in Physics, Chemistry, Mathematics and English in grade 13 of the Ontario high school system

α = model parameter to be determined

v = white Gaussian noise with zero-mean and constant variance.

The model described by equation (1) would be most useful if α is a constant, and v has zero mean as well as the smallest possible variance. It would be an acceptable model if the residuals, defined as the difference between the forecasts and the actual values, form a zero-mean white noise sequence (Sinha and Prasad, 1979). At the same time, it is reasonable to expect considerable variation in the grade g_1, dependent upon the effort put in by each individual student. Furthermore, there may be many other nondeterministic changes from year to year. Hence, at best, equation (1) can only provide an expected value of g_1 for a given g_0.

In the previous study (Sinha and diCenzo, 1976) it was observed that the value of the coefficient α was higher for students obtaining high school averages of over 75 percent. It may, therefore, be desirable to investigate a second-order nonlinear model of the form

$$g_1 = \alpha_1 g_0 + \alpha_2 g_0^2 + v \tag{2}$$

where the coefficients α_1 and α_2 have to be determined, with g_0, g_1 and v defined as before.

It may be noted that the model of equation (2) is more realistic in the sense that the students who obtain very high marks in the high school are more highly motivated, and are expected to maintain their high performance at the university.

DETERMINATION OF THE PARAMETERS OF THE MODELS

The value of the parameter α of the model, represented by equation (1), which will minimize the mean-square prediction error, is given by

$$\alpha = \frac{E[g_1 g_0]}{E[g_0^2]} \tag{3}$$

where E[ ] implies the expectation of the quantity within the brackets. The expectations indicated above can be approximated as

$$E[g_1\ g_0] = \frac{1}{N} \sum_{k=1}^{N} g_1(k)\ g_0(k) \tag{4}$$

and

$$E[g_0^2] = \frac{1}{N} \sum_{k=1}^{N} g_0^2(k) \tag{5}$$

where N is a large number.

The value of the parameters α_1 and α_2 of the second-order model, represented by equation (2), which will minimize the mean-square prediction error, are given by

$$\alpha_0 = \frac{E[g_1\ g_0] \cdot E[g_0^4] - E[g_1\ g_0^2] \cdot E[g_0^3]}{E[g_0^2] \cdot E[g_0^4] - (E[g_0^3])^2} \tag{6}$$

$$\alpha_1 = \frac{E[g_0^2] \cdot E[g_1\ g_0^2] - E[g_1\ g_0^2] \cdot E[g_0^3]}{E[g_0^2] \cdot E[g_0^4] - (E[g_0^3])^2} \tag{7}$$

The expectations can, again, be estimated in a manner similar to equations (4) and (5).

RESULTS FOR THE FIRST-ORDER MODEL

Since the value of α may vary from year to year as well as for different values of g_0, it was calculated in two different ways from data collected over a ten-year period. First, the value of α was determined using equation (3) for each integral value of the high school average starting from 64% and going up to 97%. These values are given in Table I, which also lists the average grade and the standard deviation for each case. It may be noted that in this case, since g_0 is constant, equation (3) is simplified to the following form

$$\alpha = \frac{E[g_1]}{g_0} \tag{8}$$

This implies that for each value of g_0, the constant α is simply the ratio of the mean value of g_1 and g_0. Moreover, for this case the standard deviation of g_1 is also a measure of the mean square error of the actual grades from the predicted value of αg_0. The large standard deviation emphasizes the stochastic nature of the model, and points out that the individual grades vary over a wide range.

TABLE I Separate Model for Each Percentage Grade

High School Grade g_0	Number of Students N	Average Year I Grade $\bar{g}_1$	Parameter α	Standard Deviation σ
64	48	53.74	0.8397	9.83
65	56	52.50	0.8077	9.31
66	97	50.97	0.7722	10.66
67	98	53.27	0.7951	11.39
68	97	53.73	0.7901	10.41
69	105	55.61	0.8059	10.25
70	84	55.29	0.7899	11.44
71	94	57.83	0.8145	8.68
72	94	55.46	0.7703	12.68
73	71	59.14	0.8102	11.75
74	81	60.08	0.8119	9.51
75	87	61.00	0.8134	9.82
76	79	62.86	0.8271	8.98
77	93	64.90	0.8428	8.62
78	63	61.32	0.7869	11.99
79	40	66.08	0.8365	8.94
80	100	66.27	0.8284	9.69
81	71	68.25	0.8425	9.12
82	75	68.73	0.8382	8.18
83	52	70.65	0.8512	6.30
84	42	69.76	0.8305	11.13
85	60	72.71	0.8554	9.38
86	55	72.87	0.8474	8.43
87	51	77.00	0.8851	6.17
88	52	75.58	0.8588	7.68
89	46	75.80	0.8517	8.08
90	47	80.11	0.8902	8.10
91	28	80.31	0.8826	7.53
92	28	83.88	0.9117	6.58
93	18	85.48	0.9118	6.64
94	13	85.54	0.9100	6.09
95	15	87.30	0.9189	6.25
96	4	89.95	0.9370	2.34
97	3	91.43	0.9426	3.60

It is interesting to note from the table that the value of the parameter α varies between 0.7720 and 0.851 for students with grade 13 average between 64% and 84%. For students with averages above 84%, the parameter α had a higher value, ranging from 0.847 to 0.943. For the entire population of 2,047 students, the value of α was found to be 0.8277.

Next, the value of α was determined separately for each of the last ten academic years. The results are shown in Table II on the following page.

In the earlier study (Sinha and diCenzo, 1976) the value for the parameter α was seen to be increasing steadily as the average high school grade of the students admitted to the engineering program was also increasing. This trend appears to have reversed after 1975. One explanation is the large jump in the first-year intake starting from the academic year 1975-1976, whcih might have affected the overall performance. This trend appears to have been arrested, so that by 1978,

TABLE II Separate Model for Each Year

Academic Year	Number of Students	Average High School Grade	Average Year I Grade	α
1969-1970	94	73.81	60.07	0.8138
1970-1971	121	74.44	61.38	0.8245
1971-1972	91	74.68	62.22	0.8332
1972-1973	117	75.52	63.72	0.8438
1973-1974	140	76.84	65.15	0.8479
1974-1975	199	78.52	67.49	0.8595
1975-1976	330	77.51	64.56	0.8329
1976-1977	288	78.19	64.54	0.8254
1977-1978	296	77.19	62.95	0.8155
1978-1979	371	75.36	61.65	0.8180

the university was prepared to handle larger numbers of students without an appreciable deterioration in the quality of education.

It would appear from these results that the simple stochastic model proposed in equation (1) is verified to be reasonably correct, with only small changes in the parameter α (a variation of less than ±4% from the overall mean of 0.8277) over a period of ten years and over wide variations in the high school average.

RESULTS FOR THE SECOND-ORDER MODEL

The second-order model described by equation (2) was estimated using equations (6) and (7). The resulting model, obtained from the entire data for the ten-year period is given below,

$$\hat{g}_1 = 0.5266g_0 + 0.003887g_0^2 \tag{9}$$

For the sake of comparison, it may be recalled that the linear model for the entire data is

$$\hat{g}_1 = 0.8277g_0 \tag{10}$$

A comparison of the two models is made in Table III on the following page, where for each year the predicted and the actual values of g_1 are shown.

A comparison of the mean square error between the actual and the predicted values shows that the second-order model is slightly better. However, the improvement is so small that one may wonder whether it is worthwhile to use this model instead of the simpler linear model.

TABLE III Comparison Between Two Models

Academic Year	Average High School Grade	Average Year I Grade	Predicted Year I Average From Linear Model	Predicted Year I Average from Second-order Model
1969-1970	73.81	60.04	61.09	60.07
1970-1971	74.44	61.38	61.61	60.74
1971-1972	74.68	62.22	61.81	61.00
1972-1973	75.52	63.72	62.51	61.94
1973-1974	76.84	65.15	63.60	63.41
1974-1975	78.52	67.49	64.99	65.37
1975-1976	77.51	64.56	64.16	64.17
1976-1977	78.19	64.54	64.72	64.94
1977-1978	77.19	62.95	63.89	63.81
1978-1979	75.36	61.75	62.38	61.76
Mean error			-0.307	-0.659
Mean Square Error			1.651	1.168

CONCLUSIONS

Two stochastic models have been considered which relate the performance of students in the first year of the engineering program to their performance in related subjects in the high school system. The parameters of both of these models were obtained using data collected for the past ten years at McMaster University. It is seen that both the models give good predictions, and there is a slight improvement when the second-order nonlinear model is used.

The results justify the present requirements for admission to the engineering program, since the failure rate is maximum near the cut-off point of 64% average in the high school. The fact that the standard deviation is of the order of 10% emphasizes the stochastic nature of the model.

From the point of view of counselling and admission requirements it is felt that although the present regulations are satisfactory, many students with high school averages near 64% do quite well, whereas some with marks as high as 80% have difficulties. One problem is that there is considerable variation in standards between different high schools. Perhaps, if the admissions to the engineering program were based on a well-designed aptitude test in addition to the requirement of at least 64% grade 13 marks in Physics, Chemistry, Mathemaitcs and English, the failure rate in the first year of the engineering program could be cut down considerably.

ACKNOWLEDGEMENTS

The authors express their gratitute to Dean M.B. Ives and former Dean C.D. diCenzo for providing data.

REFERENCES

Sinha, N.K. and diCenzo, C.D. (1975a). IEEE Trans. in Systems, Man and Cybernetics, SMC-3, 437-440.

Sinha, N.K. and diCenzo, C.D. (1975b). INt. J. Systems Science, 6, 1001-1007.

Sinha, N.K. and diCenzo, C.D. (1976). Proc. 7th Annual Pittsburgh Conf. on Modelling and Simulation, Pittsburgh, PA, 160-162.

Sinha, N.K. and Prasad, T. (1979). Applied Mathematical Modelling, 3, 2-6.

NEARLY-COMPLETELY DECOMPOSABLE STOCHASTIC PROCESSES

G. Louchard and G. Latouche

Université Libre de Bruxelles
Faculté des Sciences-CP 212
Laboratoire d'Informatique Théorique
boulevard du Triomphe
B-1050 Bruxelles-Belgium

ABSTRACT

Consider a Markov chain for which the states are grouped into disjoint aggregates, in such a way that the probabilities of transitions within aggregates are large compared to transitions between aggregates. Let Ω be a subset of one aggregate. Approximations are determined for quantities related to the time to reach Ω.

KEYWORDS

Markov chains; near-complete decomposability; stationary probability vector; random times; approximations; aggregation.

INTRODUCTION

We consider irreducible, nearly-completely decomposable Markov chains. Formally, let P be the transition probability matrix of the Markov chain. By a suitable permutation of rows and columns, we can get

$$P = P^* + \varepsilon C ,$$

where ε is a real number, small compared to the non-zero elements of P^*, P^* is stochastic and corresponds to a Markov chain decomposable into a number n (possibly infinite) of finite irreducible aperiodic subchains.

Let A,B,... be the sets of states in each of these subchains. A,B,... form a partition of the states space of the original Markov chain, and are called the aggregates. Let Ω be a proper subset of one of the aggregates, say A, and let T_i be the number of transitions, starting from state i, before reaching any state in Ω.

The object is to analyze the dependency of T_i on ε and the aggregates structure. This article only presents a brief summary. The detailed results are to be found in Courtois and Louchard (1976), and Latouche and Louchard (1978, 1980). The last reference cited may be obtained from the authors.

Notational Convention

Let M be an $n \times n$ matrix and A and B denote two subsets of $\{1,2,\ldots,n\}$ containing respectively $n(A)$ and $n(B)$ elements. We denote by ${}_A M_B$ the $n(A) \times n(B)$ matrix consisting of those elements of M at the intersection of rows A and columns B. Similarly, if $\underline{V}$ is a vector with n components, $\underline{V}_A$ is a vector with $n(A)$ components: those V_i's for which $i \in A$.

This convention will help us simplify notations in the sequel. Note, for instance, that if M and L are two $n \times n$ matrices, if we partition $\{1,2,\ldots,n\}$ into disjoined subsets (or aggregates) A,B,... and if R is the product of M and L,

$$ {}_A R_B = {}_A (M.L)_B = \sum_{X=A,B,\ldots} {}_A M_X \, {}_X L_B . $$

We denote by $\underline{1}$ or $\underline{0}$ vectors for which all components are equal to 1 or 0 respectively.

Finally,

$$ (M^+L)_{ij} = \sum_{k\in\Omega} M_{ik} L_{kj}, $$

$$ (M^-L)_{ij} = \sum_{k\notin\Omega} M_{ik} L_{kj}. $$

FINITE IRREDUCIBLE MARKOV CHAIN

In such a case, the Markov chain is clearly ergodic and we denote by $\underline{\pi}$ the stationary vector.

The Stationary Probability Vector π

It is proved that each entry of the vector $\underline{\pi}$ has a power series expansion in ε, for ε small enough. The subvector $\underline{\pi}_X$ corresponding to the states in the aggregate X is given by

$$ \underline{\pi}_X = S^*(X)\underline{\pi}^*_X + \varepsilon\underline{\pi}^{(1)}_X + O(\varepsilon^2), \quad \text{for all aggregate } X, $$

where $\underline{\pi}^*_A$ corresponds to the stationary probability vector of the Markov chain ${}_A P^*_A$.
The constants $S^*(A)$ and the vectors $\underline{\pi}^{(1)}_A$ are completely determined and easily computable.
This may be (loosely) interpreted as follows: in first approximation, the stationary probabilities for a given aggregate are determined by transitions within the aggregate, up to a multiplicative constant which depends on transitions between aggregates.

The First Two Moments of T_i

We define as $G_i(z)$ the generating function

$$ G_i(z) = (1-z) \sum_{n=o}^{\infty} z^n \, P[T_i > n]. $$

$G_i(z)$ is analytic in $0 \leqslant z \leqslant 1$, and if we write

$$G_i(z) = (1-z)h_i + (1-z)^2 j_i + O(1-z)^2,$$

we have that

$$E[T_i] = h_i, \quad E[T_i^2] = h_i - 2j_i \ (=k_i).$$

The vector $\underline{G}(z)$ is analyzed, and from the expressions

$$\underline{G}_A(z) = (I - z\ {}_A P_A^{-} I)^{-1}[(1-z)I + z \sum_{X \neq A} {}_A P_X\ \underline{G}_X(z)],$$

$$\underline{G}_B(z) = 1 + \varepsilon z(\sum_{n=o}^{\infty} z^n\ {}_B P_B^n)({}_B C_B\ \underline{1} + \sum_{Y \neq B}^{\infty} {}_B C_Y\ \underline{G}_Y(z) - {}_B C_A^+\ \underline{G}_A(z)),$$

for $B \neq A$.

One determines how the successive derivatives of $G_i(z)$, computed at z=1, depend on ε. From this, we prove that

$$\underline{h}_A = \sum_{i=o}^{\infty} \underline{h}_A^{(i)} \varepsilon^i,$$

and

$$\underline{h}_B = \sum_{i=-1}^{\infty} \underline{h}_B^{(i)} \varepsilon^i, \quad \text{for } B \neq A.$$

Similarly, we prove that

$$\underline{k}_A = \sum_{i=-1}^{\infty} \underline{k}_A^{(i)} \varepsilon^i,$$

$$\underline{k}_B = \sum_{i=-2}^{\infty} \underline{k}_B^{(i)} \varepsilon^i, \quad \text{for } B \neq A.$$

The leading terms in these power series expansions are determined in two ways: firstly, by systems of linear equations, which are well suited for numerical purposes; secondly by explicit expressions which give some insight in the behaviour of the system.

INFINITE TRANSIENT MARKOV CHAIN

In this case, there is a positive probability for (at least) some initial states of never reaching Ω. We examine the following quantities:

$$p_i = P[T_i = \infty],$$

$$\sum_{n=o}^{\infty} z^n\ P[n < T_i < \infty],$$

$$E[\tau_i],$$

and

$$\sum_{n=o}^{\infty} z^n\ P[n < U_i < \infty],$$

where τ_i is the total time spent in one of the states of Ω, and U_i is the last time the Markov chain is in one of the states of Ω, starting from i.

The **transient** Markov chains are much more difficult to analyze than the finite ergodic Markov chains. It is not possible, for transient chains, to prove that any of the quantities of interest has a power series expansion in ε.
However we determine their asymptotic behaviour.
As an example, we find conditions under which

$$\underline{p}_A = \varepsilon \underline{p}_A^{(1)} + o(\varepsilon),$$

$$\underline{p}_B = \underline{p}_B^{(0)} + \varepsilon \underline{p}_B^{(1)} + o(\varepsilon), \quad \text{for } B \neq A,$$

and we determine the terms $\underline{p}_A^{(1)}$, $\underline{p}_B^{(0)}$ and $\underline{p}_B^{(1)}$. In particular, we show that $\underline{p}_B^{(0)}$ does depends on A, but not on the precise subset Ω of A.

REFERENCES

Courtois, P.J., and Louchard, G. (1976). Approximations of eigencharacteristics in nearly-completely decomposable stochastic systems. *Stochastic Processes & Appl.*, *4*, 283-296.

Latouche, G., and Louchard, G. (1978). Return times in nearly-completely decomposable stochastic processes. *J. Appl. Probab.*, *15*, 251-267.

Latouche, G., and Louchard, G. (1980). Nearly-completely decomposable transient Markov chains. In preparation.

CLASSIFICATIONS ENDOWED BY TRANSFER OF STRUCTURES

J.P. AURAY Université LYON 1
G. DURU "
A. DUSSAUCHOY "
M. MOUGEOT Université de BESANCON

In some previous papers we described a methodology allowing to build a classification on a set E endowed with a pretopological structure or a poor structure. This method lies in building a partition of E , that is to say an equivalence relation on E associated to the less fine topology among the topologies associated to equivalence relation and finner than the given pretopological structure on E. It is shown that many equivalence relations may solve this problem.

As extension of this methodology we propose, in this paper, a methodology for building pretopological structures on E from a set E' whose elements are called descriptors and a correspondance X from E into E' ; for $x \in E$, X(x) is the set of all descriptors of x : then X(x) can be regarded as a complete characterisation of x. When known the pretopological structure on E built from X and E' , we can determine classifications on E by the above methodology described in [2] and [3].

We assume now that E' is endowed with a pretopological structure expressing, proximities, similarities or dissimilarities, beetwen descriptors. Let us denote by (E', a') or (E', i') this structure, where a' and i' are respectively the closure and the interior mappings. Then we want to endow E with a structure taking into account the structure on E' through the correspondance X . We study two methods : one lies in endowing E with the less fine pretopological structure making the correspondance X be lower-hemi-continuous (l.h.c) on E , the other lies in endowing E with the less fine pretopological structure making X be upper-hemi-continuous (u.h.c) on E . In the two cases we give an interpretation of the continuity.

1. STRUCTURE ENDOWED WITH l.h.c.

Let (E', a') or (E', i') the pretopological structure on E' . We know (see [1]) that the less fine pretopological structure making X l.h.c on E is the pretopological structure (E, a) or (E, i) such as :

$$\forall A \subset E \ , \ a(A) = \bigcup_{\substack{A \subset H \\ X(H) \subset a'(X(A))}} H$$

Moreover, if (E', i') is a $\mathcal{V}$-pretopological structure, the following properties are equivalent for $x \in E$:

i) X is l.h.c in x

ii) $\forall H' \subset E'$, $x \in X^W(i'(H')) \Rightarrow X^W(H') \in \mathcal{V}_x$.

Where X^W is the weak-inverse of X ($X^W(H') = \{x \in E \ / \ X(x) \cap H' \neq \emptyset\}$, and $\mathcal{V}_x$ is the set of preneigbourhood of x . $\mathcal{V}_x = \{V \subset E \ / \ x \in i(V)\}$).

We can prove the following :

Proposition :

If (E', a') or (E', i') is a $\mathcal{V}$-pretopological structure, the less fine pretopological structure making X l.h.c on E is also $\mathcal{V}$-pretopological, and it is such that for $x \in E$

$$\mathcal{V}_x = \{X^W(H')/x \in X^W(i'(H'))\}$$

is a base of preneighbourhoods of x.

If (E, a) or (E, i) denote this pretopological structure, then we have for $A \subset E$:

$$x \in i(A) \Leftrightarrow \exists \ x' \in X(x), \ \exists \ V' \in \mathcal{V}'_{x'} \ , \ X^W(V') \subset A$$

$$x \in a(A) \Leftrightarrow \forall x' \in X(x), \ \forall \ V' \in \mathcal{V}'_{x'} \ , \ X^W(V') \cap A \neq \emptyset \ .$$

Then $x \in i(A)$ means that there is x' a descriptor or a characteristic of x , and V' a preneigbourhood of x' (that is to say : a set of descriptors closed to x') such that each $y \in E$ having at least one of its descriptors in V' belongs to A ; $x \in a(A)$ means that for each x' descriptor of x , and each V' preneigbourhood of x' (that is to say for each set of descriptors closed to x') there is $y \in A$ whose one descriptor at least is in V'.

We have :

Proposition :

The reflexive relation $\mathcal{R}$ (see [2]) associated to the less fine pretopological structure making X l.h.c is defined by :

$$\forall x \in E, \ \forall y \in E, \ x \ \mathcal{R} \ y \Leftrightarrow X(x) \subset a'(X(y))$$

The reflexive and symmetric relation $\mathcal{S}$ associated (see [2]) is defined by :

$$\forall x \in E, \ \forall y \in E, \ x \ \mathcal{S} \ y \Leftrightarrow X(x) \subset a'(X(y)) \text{ and } X(y) \subset a'(X(x))$$

Finally if $\mathcal{E}$ is an equivalence relation associated to (E,a) (see [2]) the equivalence class of x is cl(x) such that

$$\forall y \in cl(x),\ \forall z \in cl(x),\ X(y) \subset a'(X(z)) \text{ and } X(z) \subset a'(X(y))$$

Some particular cases

If a' (or i') is idempotent clearly we have

$$\forall x \in E,\ \forall y \in E,\ y \in cl(x) \Leftrightarrow a'(X(x)) = a'(X(y))$$

Most particularly if (E', a') or (E', i') is associated to a partition $\{E'_j,\ j \in J\}$ of E' , we have :

$$\forall x \in E\ cl(x) = \{y \in E / \forall j \in J, X(y) \cap E_j = \emptyset \Leftrightarrow X(x) \cap E_j = \emptyset\}$$

In other words, the classification put together the elements of E whose characteristics are in the same elements of the partition.

If the partition of E' is discrete we obtain

$$\forall x \in E,\ cl(x) = \{y / X(x) = X(y)\}$$

and if it is anti-discrete then obviously.

$$\forall x \in E,\ cl(x) = E$$

Remarks :

We can also try to build classifications on E only taking into account the correspondance X , excluding the structuration of E' . For instance E can be endowed by one of the following pretopological structures defined either by closure mapping or by the bases of preneighbourhoods :

1 $\forall x \in E,\ \mathcal{B}_x = \{\{y \in E / X(x) \subset X(y)\}\}$
2 $\forall x \in E,\ \mathcal{B}_x = \{\{y \in E / X(y) \subset X(x)\}\} = \{X^s X(x)\}$
3 $\forall x \in E,\ \mathcal{B}_x = \{\{y \in E / X(x) \cap X(y) \neq \emptyset\}\} = \{X^W X(x)\}$
4 $\forall x \in E,\ \mathcal{B}_x = \{\{y \in E / X(x) \cap X(y) = \emptyset\} \cup \{x\}\}$
5 $\forall A \subset E,\ a(A) = \{x \in E / X(x) \supset \tilde{X}(A)\}$
6 $\forall A \subset E,\ a(A) = X^s X(A)$

Where $\tilde{X}(A) = \bigcap_{x \in A} X(x)$ and $X^s(A') = \{x \in E / X(x) \subset A'\}$ for $A \subset E$ and $A' \subset E'$.

It is easy to give an interpretation of this six pretopological structures. We give hereafter the endowed classifications on E , that is to say the equivalences relations & associated to each of this six pretopological structures.

For 1 and 2 we have

$$\forall x \in E,\ \forall y \in E,\ x \mathbin{\&} y \Leftrightarrow X(x) = X(y)$$

The closure mapping of the pretopological structure in 3 is defined by :

$$\forall A \subset E,\ a(A) = X^W X(A)$$

For an equivalence relation $\mathcal{E}$ associated to this pretopological structure we have for $x \in E$:

$$\forall y \in cl(x),\ \forall z \in cl(x),\ X(y) \cap X(z) \neq \emptyset$$

For an equivalence relation $\mathcal{E}$ associated to the pretopological structure defined in 4 , we have for $x \in E$

$$\forall y \in cl(x),\ \forall z \in cl(x),\ y \neq z,\ X(y) \cap X(z) = \emptyset$$

The structure defined in 5 can be considered as pretopological, provided that E' is such that

$$\forall x \in E\ ,\ X(x) \neq E'\ ;$$

the closure mapping for this structure is still defined by

$$\forall A \subset E.\ a(A) = \bigcap_{x' \in \tilde{X}(A)} X^{w}(x')$$

We can check that a is idempotent and that $\{a(A),\ A \in \mathcal{P}(E)\}$ is the set of the closed subsets of E in the Galois's lattice built from X on $\mathcal{P}(E) \times \mathcal{P}(E')$. Finally this structure is not very interesting, because the associated equivalence is $\mathcal{E}$ such that :

$$\forall x \in E,\ \forall y \in E,\ x\ \mathcal{E}\ y \Leftrightarrow X(x) = X(y).$$

Last, the structure defined in 6 is a pretopological structure provided that

$$\forall x \in E,\ X(x) \neq \emptyset$$

and the equivalence relation associated is the same that for 5 .

2 - STRUCTURE ENDOWED WITH u.h.c

In this part we assume that (E', a') or (E', i') is a $\mathcal{V}$-pretopological structure. We know (see [1]) that the less fine $\mathcal{V}$-pretopological structure making X u.h.c on E is defined by the bases of preneigbourhoods :

$$\forall x \in E\ ,\ \mathcal{B}_x = \{X^{s}(H')/X(x) \subset i'(H')\}$$

or

$$\forall x \in E\ ,\ \mathcal{B}_x = \{X^{s}(H')/H' \in \mathcal{V}'_{X(x)}\ \}$$

where $\mathcal{V}'_{X(x)} = \{V' \subset E'/X(x) \subset i'(H')\}$ is the set of preneigbourhoods of $X(x)$.

The equivalence relation $\mathcal{E}$ associated is such that for $x \in E$,

$$\forall y \in cl(x),\ \forall z \in cl(x),\ X(y) \subset \bigcap_{H' \in \mathcal{V}'_{X(z)}} H' \quad \text{and}$$

$$X(z) \subset \bigcap_{H' \in \mathcal{V}'_{X(y)}} H'$$

If the pretopological structure on E' is associated to an equivalence relation on E , then we have :

$$\forall y \in cl(x),\ \forall z \in cl(x),\ X(y) \subset X(\hat{z}) \text{ and } X(z) \subset X(\hat{y})$$

where $\hat{H}' = \bigcup_{x' \in H'} cl(x')$ for $H' \subset E'$. This classification has been obstained previously in the l.h.c case.

Instead of the less fine pretopological structure making X u.h.c , we can endow E with the $\mathcal{V}_{D}^{\star}$ -pretopological whose bases of preneighbourhoods are :

$$\forall x \in E\ ,\ \mathcal{B}_x^{\star} = \{x\} \cup \{X^s(i'(X(x))\}$$

By this pretopological structure, clearly X is u.h.c. , and for the associated equivalence relation & we have :

$$\forall y \in cl(x), \forall z \in cl(x),\ y \neq z,\ X(y) \subset i'(X(z)) \text{ and } X(z) \subset i'(X(y))$$

Again a particular case is obtained if the pretopological structure on E' is associated to an equivalence relation : then for the equivalence relation on E, we have

$$\forall x \in E,\ \forall y \in E,\quad x \neq y,\ x \,\&\, y \Leftrightarrow X(x) = X(y) = X(\hat{x}).$$

BIBLIOGRAPHY :

[1] AURAY J.P. : Jeux dynamiques sous forme de fonctions caractéristiques. Thèse en mathématique Appliquée, Université de LYON I.

[2] AURAY J.P., DURU G. , MOUGEOT M. : A problem of classification on a set endowed with a poor structure. V Symposium of Operation Research Cologne Août 1980 (a paraître).

[3] AURAY J.P., DURU G. , DUSSAUCHOY A. , MOUGEOT M. : Une méthode pour la recherche d'une partition associée à une structure. Actes du colloque "regards sur la théorie des graphes". Cerisy-la-Salle (Juin 1980).

[4] BARBUT M., MONTJARDET B. : Ordre et classification. Algèbre combinatoire, Hachette Université Paris 1970.

[5] DURU G. : Contribution à l'étude des structures des systèmes complexes dans les sciences humaines. Thèses de Doctorat ès Sciences (Mathématiques) Université de LYON I

[6] DUSSAUCHOY A. : Generalized information theory on some ordered set and structure of systems. 14 th. international conference on Information Sciences and System. Princeton University March 1980.

KEYWORDS

Pretopologcal structure, Classification, Partition, Relation, Galois 's lattice.

A NEW LATTICE THEORETICAL METHOD IN THE SYSTEM RESEARCH

K.Seitz
Technical University of Budapest

ABSTRACT

In this lecture we shall generalize the notion of structure-lattice which play an important role in the system theory of chemical engineering.

Certain modularity and distributivity conditions will be determined for the general structure-lattice by means of conditions concerning elements of the lattice in question.

For the study of the field treated in this paper, the reader is referred to the works [1], [2], [3], [4].

KEYWORDS

System, lattice, general structure-lattice, modularity, distrubutivity.

1.The general structure-lattice

Let T be a given non-empty finite set. Let denote $P(T)$ the power set of the set T.

Let $H \subseteq P(T)$. By the adjunction of the "ideal" elements O, I we obtain the set $\{H, O, I\}$, on which two operations will be defined as follows: if $a, b \in H$

$$a \sqcap b = \begin{cases} a \cap b & \text{if } a \cap b \in H, \\ O & \text{if } a \cap b \notin H, \end{cases} \qquad (1)$$

$$\underbrace{a \sqcap O}_{O \sqcap a} = O, \quad \underbrace{a \sqcap I}_{I \sqcap a} = a, \quad O \sqcap O = O, \quad \underbrace{O \sqcap I}_{I \sqcap O} = O, \quad I \sqcap I = I, \qquad (2)$$

$$a \sqcup b = \begin{cases} a \cup b, & \text{if } a \cup b \in H, \\ I, & \text{if } a \cup b \notin H \end{cases} \qquad (3)$$

$a \sqcup O = O \sqcup a = a, \quad a \sqcup I = I \sqcup a = I, \quad O \sqcup O = O,$
$O \sqcup I = I \sqcup O = I, \quad I \sqcup I = I. \qquad (4)$

In what follows we shall say that the set H is of property κ if and only if $a, b \in H$ and $O \subseteq c \subseteq b$ imply $c \in H$

It can be proved that if H is a subset of $P(T)$ having property κ then $\mathcal{L} = (\{H, O, I\}, \sqcap, \sqcup)$ is a lattice.

This lattice will be called κ- structure-lattice or shortly, structure-lattice.

It may be formulated a question: if for a set $H \subseteq P(T)$ the triplet $(\{H, O, I\}, \sqcap, \sqcup)$ is a lattice, what consequences are concerning the set H ? We shall see that the κ-property is not necessary for H in order that $\mathcal{L}$ is a lattice.

Theorem 1. The elements of the set $\{H, O, I\}$ under the operations $\sqcap, \sqcup$ form a lattice if and only if for any triplet $a, b, c \in H$ it holds the following two conditions:

1^*. $a \cap b \cap c \in H$ or if $a \cap b \cap c \notin H$ then either both $a \cap b$ and $b \cap c \in H$ or none of them belongs to H.

2^*. $a \cup b \cup c \in H$ or if $a \cup b \cup c \in H$ then either $a \cup b$ and $b \cup c \in H$ or none of them belongs to H.

In what follows we shall say that a non-empty subset H of the power set $P(T)$ has the property γ if for any triplet a, b, c of elements of H the two conditions 1^*. and 2^*. hold.

2. The study of modularity and distributivity conditions for the general structure-lattice

Theorem 2. The general structure-lattice $\mathcal{L}(\{H, O, I\}, \sqcap, \sqcup)$ is modular if and only if (H has the γ property) for $a, b, c \in H$ and $a \cap c = a$ it holds one of the following conditions:

1°. $a = c$, $b \cap c \in H$, $a \cup b \in H$.

2°. $b \cap c = a$, $b \cap c \notin H$, $a \cup b \in H$.

3°. $a \cup (b \cap c) = c$, $b \cap c \in H$, $a \cup b \notin H$.

4°. $b \cap c \in H$, $a \cup b \in H$.

Theorem 3. The general structure-lattice $\mathcal{L}(\{H, O, I\}, \sqcap, \sqcup)$ is distributive if and only if for $a, b, c \in H$ it holds $a \cap b \in H$ $a \cap c \notin H$, $a \cap (b \cup c) \notin H$ and $b \cup c \in H$ or one of the conditions $\bar{1}$ - $\bar{6}$ is satisfied,

$\bar{1}$ $b \cup c \notin H$, $a \cap b \notin H$, $a \cap c \in H$ and $a \cap c = a$,

$\bar{2}$ $a \cap (b \cup c) \in H$, $b \cup c \in H$, $a \cap b \notin H$, $a \cap c \in H$,

$\bar{3}$. $b\cup c\notin H$, $a\cap c\notin H$, $a\cap b\in H$ and $a\cap b=a$,

$\bar{4}$. $b\cup c\in H$, $a\cap(b\cup c)\in H$, $a\cap c\notin H$ $a\cap b\in H$ and $a\cap(b\cup c)=a\cap b$.

$\bar{5}$. $b\cup c\notin H$, $a\cap b\in H$, $a\cap c\in H$, $(a\cap b)\cup(a\cap c)\in H$ and $(a\cap b)\cup(a\cap c)=a$.

$\bar{6}$. $b\cup c\in H$, $a\cap b\in H$, $a\cap c\in H$, $a\cap(b\cup c)\in H$.

We remark that in the study of the distributivity of the general structure-lattice it can be used the well known criterion due to Birkhoff: A lattice is distributiv if and only if it has no sublattice isomorphic to one of the following to lattices:

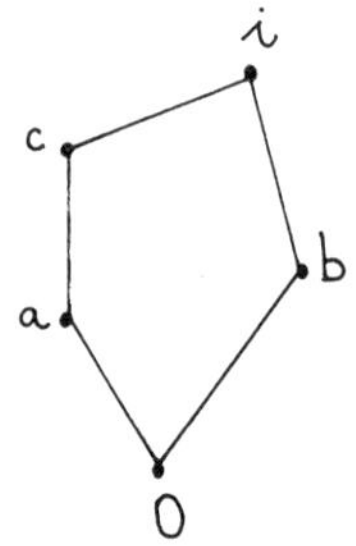

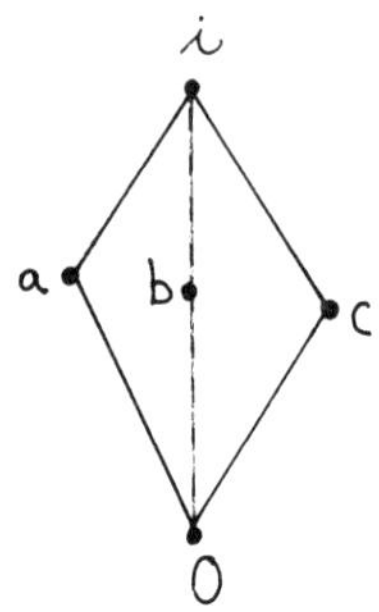

REFERENCES

[1]Birkhoff,G. 1948 .Lattice Theory. American Math.Soc.,Colloquium Publications, Vol.25, 2nd Edition, New York. 1948.

[2]Szász G. 1963 . Introduction to Lattice Theory, Academic Press, London, New-York, 1963.

[3]Grätzer,G. 1971 . Lattice Theory, W.H.Freeman and Co. San Francisco.

[4]Grätzer,G. General Lattice Theory, Birkhäuser,1978.

THE APPLICATION OF THE CONCEPT OF COMPLEX SPACE/TIME TO A MODEL OF ELECTRIC CHARGE

David W. Brisson

Rhode Island School of Design
Providence, Rhode Island

ABSTRACT

Based on concepts of the modulus of complex numbers differing from the classical definition, and of the definition of space/time as an identity with complex n-dimensional geometry, particles with positive and negative charge are identified with particles with real and imaginary radii.

KEYWORDS

The Minkowski plane; the complex plane; the modulus; Relativity; space/time; electric charge; imaginary and real radii.

INTRODUCTION

In my paper: "Complex Space/Time: An Alternative to the Special Theory of Relativity"[1], I presented a geometrical demonstration, based on plotting the equation for the unit circle in a four-dimensional complex graph, that the classical definition of the modulus of a complex number z = a + bi as :

$$\left| z \right| = +\sqrt{a^2 + b^2} \tag{1}$$

was a less mathematically satisfactory definition of the modulus than a new definition that I then proposed:

$$\left| z \right| = +\sqrt{a^2 + (bi)^2} \tag{2}$$

This new definition of the modulus may be generalized as:

---

[1] Brisson, David W., "Complex Space/Time: An Alternative to the Special Theory of Relativity", Proceedings of the 10th Annual Conference on Simulation and Modeling University of Pittsburgh, Instrument Society of America, Pittsburgh, 1980.

$$\left|z\right| = +\sqrt{\alpha^2 + b^2} \quad , \text{ where } \alpha = a \text{ or } ai \text{ , and } b = b \text{ or } bi \qquad (3)$$

which deals with the imaginary plane as well as the complex plane.

With the above definition of the modulus, it is seen that the Minkowski plane, although in classical terms is isomorphic with the complex plane, is actually based on a quite different system of measurement than the complex plane. The Minkowski plane, which is defined in terms of the equation of a rectangular hyperbola:

$$X^2 - t^2 = 1 \qquad (4)$$

at first glance seems to be isomorphic with the equation for the unit circle when plotted in the complex plane:

$$X^2 + t^2 = 1 \qquad (5)$$

However, since the equation (5) represents the locus a unit distance from the origin, it follows that the distance meaning of the hyperbola in the complex plane is quite different from that of the hyperbola in the Minkowski plane. This is the basis for the new definition of the modulus

When this new modulus is applied to Special Relativity it is found that the strange properties of distance and time dilatation disapear, and light thus is not necessarily the limit speed. Further, at the 11th Conference on Simulation and Modeling, at the University of Pittsburgh, I applied this same system to the problem of gravitation, by representing two bodies in complex six-dimensional space. Each body was represented as the equation of a sphere as plotted in complex six-space. The gravitational effect was shown to be the interactive result of the geometrical relation of such bodies in complex six-space. In this paper, I will show that if two bodies are defined as having respectively real and imaginary radii that their geometrical interaction is that of electric charge, as opposed to a simple gravitational relation.

THE MODULUS

Consider, for example, the equation for a unit circle with an imaginary radius:

$$X^2 + Y^2 = -1 \qquad (6)$$

If this equation is solve for various values of X and Y, the solutions form a partial table, such as Table 1, below:

TABLE 1 SAMPLE SOLUTIONS TO EQUATION (6)

X	$\pm i\sqrt{10}$	$\pm i\sqrt{5}$	$\pm i\sqrt{2}$	$\pm i$	0	± 1	± 2	± 3
Y	± 3	± 2	± 1	0	$\pm i$	$\pm i\sqrt{2}$	$\pm i\sqrt{5}$	$\pm i\sqrt{10}$

If the values in Table 1 are plotted in a four-dimensional graph, in which the co-ordinates represent respectively X, Xi, Y and Yi, then curves are plotted in the graph as in Fig. 1. Clearly the equation for a circle represents a <u>surface</u> in comlex four-space. Figure 2 represents the X,Yi complex co-ordinate plane of Fig. 1, rotated until it is parallel to the projection surface. Note that the hyperbola crosses the imaginary axis, and that the "circle" lies entirely in the imaginary plane.

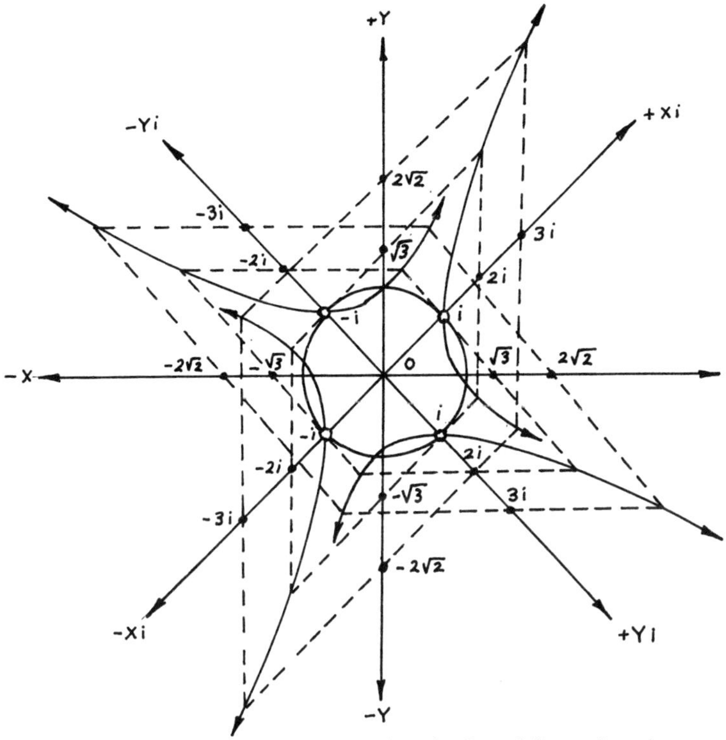

Figure 1. The graph of a unit circle with an imaginary radius.

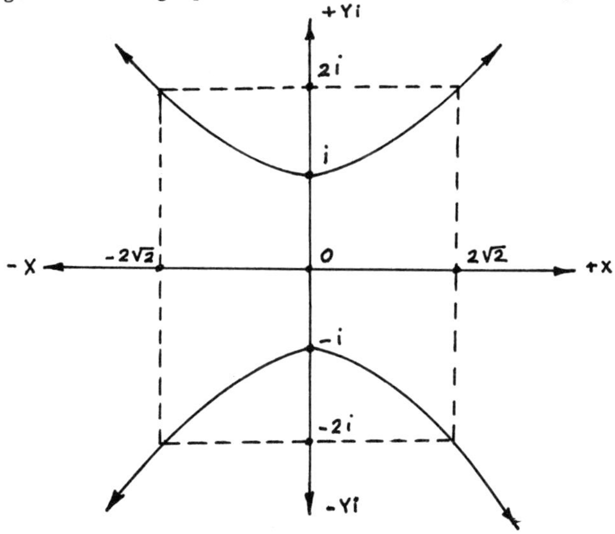

Figure 2. The X,Yi complex plane from Figure 1.

If equation (5) had been used, substituting Y for t, then an hyperbola would have appeared in Fig. 2, but crossing the real axis. If the equation were written:

$$X^2 + Y^2 = 0 \qquad (7)$$

and would have appeared as the asymptotes of both hyperbolas in Fig. 2. Similarly, any value on the right side of these equations results in an hyperbola in the X,Yi complex plane. Thus, we have a new method of measuring distances in the complex plane, since each one of these equations is the equation for a circle and thus a locus an equal distance from the origin: real, imaginary or multiples of the square root of zero. There is not space here, but it can further be shown that these distances can be moved about the complex plane, if they are maintained in their respective angle in relation to the co-ordinates, and that further, since rotation in the complex plane takes place in hyperbolas, that the Pythagorean Theorem holds for the complex plane. The result of this set of characteristics is the basis for the modulus in equations (2) and (3).

RELATIVITY

The application of this material to Special Relativity has already been described in the paper cited earlier. I will comment only that the basic rationale for using the complex plane in preference to the Minkowski plane, is that since space and time are incomensurable simply ordered number systems, that they are mathematically isomorphic to the co-ordinates of the complex plane and not to those of the two real co-ordinates of the Minkowski plane. With this comment, I will pass directly to the General Theory of Relativity and the concept of gravitation.

For a body with a real radius there must be an equivalent expression in imaginary terms. For example, for every circular section of a spherical body there must be an associated surface in complex space/time of which the circle is a cross-section. Figure 3 below, shows the complex section of two such bodies, some distance apart in space. Figure 3 also shows that the two bodies intersect at some future date by virtue of the necessary intersection of their hyperbolic branches. Finally, letting one of the bodies be a reference frame, as the second body moves along the time line, it will be noted that the distance between their respective centers becomes smaller and smaller as the angle approaches an isotropic line, i.e. parallel to the asymptotes, until at a given point of their intersection there is a real line connecting their centers the length of the sum of their radii.

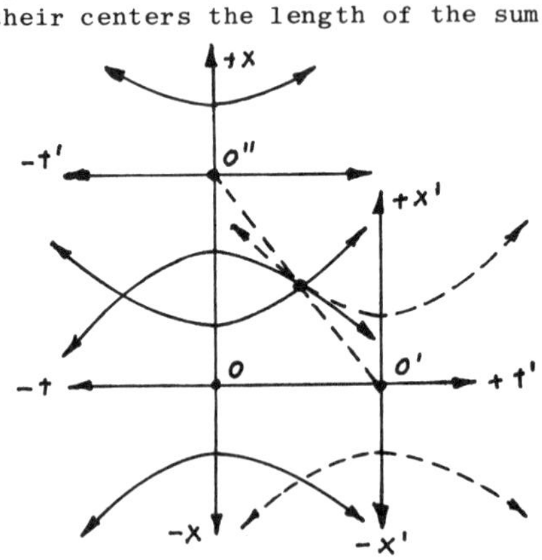

Figure 3. The graphic expression of the effect of gravitation.

CHARGE

On the basis of prior reasoning, positive and negative charge are isomorphic to complex numbers because of their incommensurability. In the above case of two bodies was given in respect to their relative diameters, i.e. their mass. Here we consider the relation between two "bodies" that are different with respect to being either "space-like" or "time-like" ,i.e. positively or negatively charged. In Fig. 4, below, we find the "space-like" bodies , i.e. positively charged bodies distributed over time, and the "time-like" bodies, i.e. negatively charged bodies distributed over space. The positively charged particle is thus described by equation (5), and the negatively charged particle by equation (6). Now the conception here, is that as the positively charged particle slides along the time axis, the negatively charged particle slides along the space axis, something like the trammel of Archimedes..

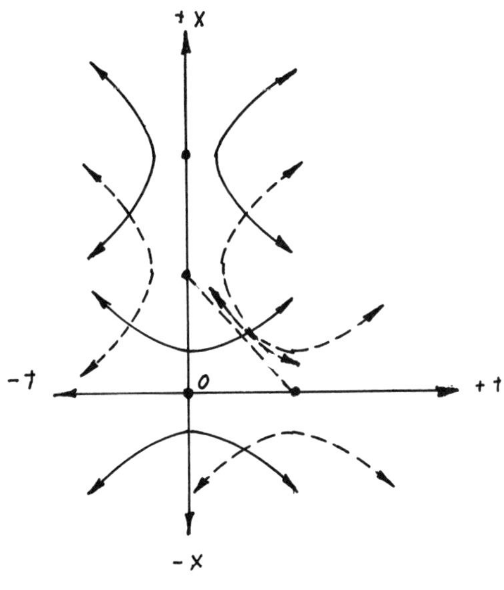

Figure 4. The geometrical representation of the relation of charge.

CONCLUSION

The difference between electromagnetic force and gravitational force is thus accounted for by the space/time difference in the two relations, and the difference in the meaning of a reference frame between an imaginary and a real particle, and that between two real and two imaginary particles. The gravitational relation is defined by the radii of the two particles, while the electromagnetic relation is dependant upon the relative distribution along the two co-ordinates.